# PER- AND POLYFLUOROALKYL SUBSTANCES (PFAS) AS ENVIRONMENTAL CONTAMINANTS

# PER- AND POLYFLUOROALKYL SUBSTANCES (PFAS) AS ENVIRONMENTAL CONTAMINANTS

## Materials and Technologies for Advancements in Water and Soil Remediation

*Edited by*

DIMITRIOS A. GIANNAKOUDAKIS

*Faculty of Chemistry, Institute of Chemical Sciences, Maria Curie-Sklodowska University, Lublin, Poland*

IOANNIS ANASTOPOULOS

*Department of Agriculture, University of Ioannina (UoI), Kostakii Campus, Arta, Greece*

ELSEVIER

Elsevier
Radarweg 29, PO Box 211, 1000 AE Amsterdam, Netherlands
125 London Wall, London EC2Y 5AS, United Kingdom
50 Hampshire Street, 5th Floor, Cambridge, MA 02139, United States



ISBN: 978-0-443-43856-1

For Information on all Elsevier publications
visit our website at https://www.elsevier.com/books-and-journals

*Publisher:* Peter Llewellyn
*Acquisitions Editor:* Pamela Sadhukhan
*Editorial Project Manager:* Waali Aafaq
*Production Project Manager:* Jenela Priscy
*Cover Designer:* Raman Kumar

Typeset by MPS Limited, Chennai, India

# Dedication

To my father, who taught me to observe, to question, and to seek what lies beneath the surface in science, a way of thinking that became my path into chemistry.

**—Dimitrios A. Giannakoudakis**

This book is dedicated to the memory of Prof. Anastasios Tsinas, who served as Dean and Head of the Department of Agriculture of the University of Ioannina, in recognition of his significant contribution, guidance, and the inspiration he offered to the academic community.

**—Ioannis Anastopoulos**

# Contents

# 2

# Remediation technologies

## 5. Adsorption-based per- and polyfluoroalkyl substances remediation technologies: principles and applications

Md Abdullah Al Masud, Mahlet M. Kebede, Sarah Ortbal, Md Refat Jahan Rakib, Hasara Samaraweera and Leigh G. Terry



## 6. Photocatalysis as an advanced per- and polyfluoroalkyl substances remediation technique

Stephen Sunday Emmanuel and Ademidun Adeola Adesibikan



## 7. Sonication-induced catalytic approaches for PFAS degradation in aqueous systems

Ahmed Malek Djaballah, Dimitrios A. Giannakoudakis, Ioannis Anastopoulos, Giacomo de Falco and Juan Carlos Colmenares



## 8. Electrochemical oxidation of per- and polyfluoroalkyl substances: treatment mechanisms, electrode materials, and future directions

Sajjad Hussain, Saima Gul, Hammad Khan, Arslan Maqbool, Nadeem Khan, Farrukh Faheem, Josiel Viana E.Viana and Sabir Khan

## 15. Metal–organic frameworks for the remediation of per- and polyfluoroalkyl substances

Stephen Sunday Emmanuel and Ademidun Adeola Adesibikan



## 16. Activated carbon for per- and polyfluoroalkyl substances remediation: properties, mechanisms, challenges, and future directions

Mohamed Zbair and Jocelyne Brendle



## 17. Biochar for the remediation of per- and polyfluoroalkyl substances

Navish Kataria, Sangita Yadav and Divya Dhillayan



## 18. Metal oxides: solid scaffolding for per- and polyfluoroalkyl substances removal

Samar Dhiman, Hemant Veer Jain and Nasreen Ghazi Ansari

## 19. Minerals/metal oxide-based photocatalysts for degradation/removal of per- and polyfluoroalkyl substances from water

Solomon Oluwaseun Akinnawo



## 20. Membrane processes and materials for remediation of per- and polyfluoroalkyl substances from water and wastewater

Ayah Jabr, Laith Jaber, Amani Al-Othman, Muhammad Tawalbeh and Mohammad Mahdi A. Shirazi



# 4

# Policy and regulation

## 21. Regulatory approaches to per- and polyfluoroalkyl substance contamination in the United States and Canada

Sai Sharanya Pulimamidi, Rushikesh Bandal, Srikanth Ponneganti, Saurabh Bodake, Dhanavath Dattu Naik and Roshan M. Borkar



## 22. Regulatory frameworks for per- and polyfluoroalkyl substances in Australia and the European Union

Sai Sharanya Pulimamidi, Soumi Saha, Srikanth Ponneganti, Dhanavath Dattu Naik and Roshan M. Borkar

## 23. Regulatory approaches to per- and polyfluoroalkyl substances in Asia: implications for remediation, enforcement, and public health

Sai Sharanya Pulimamidi, Srikanth Ponneganti, Saurabh Bodake, Suriyaprakash M., Prakash Kumar Panda and Roshan M. Borkar

# List of contributors

**Ademidun Adeola Adesibikan** Department of Industrial Chemistry, Faculty of Physical Sciences, University of Ilorin, Ilorin, Kwara, Nigeria

**Adewale Adewuyi** Department of Chemical Sciences, Faculty of Natural Sciences, Redeemer's University, Ede, Osun State, Nigeria

**Solomon Oluwaseun Akinnawo** Department of Chemical Sciences, Olusegun Agagu University of Science and Technology, Okitipupa, Ondo State, Nigeria

**Amani Al-Othman** Department of Chemical and Biological Engineering, American University of Sharjah, Sharjah, United Arab Emirates; Energy, Water and Sustainable Environment Research Center, American University of Sharjah, Sharjah, United Arab Emirates

**Ioannis Anastopoulos** Department of Agriculture, University of Ioannina (UoI), Kostakii Campus, Arta, Greece

**Amit Hasan Anik** Department of Environmental Science, Bangladesh University of Professionals (BUP), Dhaka, Bangladesh

**Nasreen Ghazi Ansari** Analytical Chemistry Laboratory, Analytical Sciences and Accredited Testing Services Group, Analytical Sciences & Services, Industrial Support through Technological Solutions (ASSIST) Division, CSIR-Indian Institute of Toxicology Research (CSIR-IITR), Lucknow, Uttar Pradesh, India; Academy of Scientific and Innovative Research (AcSIR), Ghaziabad, Uttar Pradesh, India

**Rushikesh Bandal** Department of Pharmaceutical Analysis, National Institute of Pharmaceutical Education and Research, Guwahati, Changsari, Assam, India

**Saurabh Bodake** Department of Pharmaceutical Analysis, National Institute of Pharmaceutical Education and Research, Guwahati, Changsari, Assam, India

**Roshan M. Borkar** Department of Pharmaceutical Analysis, National Institute of Pharmaceutical Education and Research, Guwahati, Changsari, Assam, India

**Jocelyne Brendle** Institut de Science des Mat´eriaux de Mulhouse CNRS UMR 7361, Université de Haute Alsace, Mulhouse, Grand Est, France; Université de Strasbourg, Strasbourg, Grand Est, France

**Juan Carlos Colmenares** Institute of Physical Chemistry, Polish Academy of Sciences, Warsaw, Poland

**Giacomo de Falco** New York City Department of Environmental Protection, New York, NY, United States

**Divya Dhillayan** Department of Environmental Science and Engineering, Guru Jambheshwar University of Science & Technology, Hisar, Haryana, India

**Samar Dhiman** Analytical Chemistry Laboratory, Analytical Sciences and Accredited Testing Services Group, Analytical Sciences & Services, Industrial Support through Technological Solutions (ASSIST) Division, CSIR-Indian Institute of Toxicology Research (CSIR-IITR), Lucknow, Uttar Pradesh, India; Academy of Scientific and Innovative Research (AcSIR), Ghaziabad, Uttar Pradesh, India

**Ahmed Malek Djaballah** Institute of Physical Chemistry, Polish Academy of Sciences, Warsaw, Poland; CNRS -UPR3079, Université d'Orléans, Conditions Extrêmes Matériaux Haute Température et Irradiation (CEMHTI), Orléans, France

**Josiel Viana E.Viana** Programa de Pós-Graduação em Química (PPGQ), Universidade Federal Do Amazonas (UFAM), Manaus, Amazonas (AM), Brazil

**Stephen Sunday Emmanuel** Department of Industrial Chemistry, Faculty of Physical Sciences, University of Ilorin, Ilorin, Kwara, Nigeria; New Life Research Foundation, Atlanta, GA, United States

**Farrukh Faheem** São Carlos Institute of Chemistry, University of São Paulo, São Carlos, SP, Brazil

**Stefano Freguia** Department of Chemical Engineering, The University of Melbourne, Parkville, VIC, Australia

**Sunantha Ganesan** Department of Biomedical Sciences, School of BioSciences and Technology, Vellore Institute of Technology, Vellore, Tamil Nadu, India

**Dimitrios A. Giannakoudakis** Faculty of Chemistry, Institute of Chemical Sciences, Maria Curie-Sklodowska University, Lublin, Poland

**Ganeshkumar Govindasamy** School of Chemical Engineering, Vellore Institute of Technology, Vellore, Tamil Nadu, India

**Saima Gul** Department of Chemistry, Islamia College Peshawar, Peshawar, Khyber Pakhtunkhwa, Pakistan

**Sajjad Hussain** Faculty of Materials and Chemical Engineering, GIK Institute of Engineering Sciences and Technology, Swabi, Khyber Pakhtunkhwa, Pakistan; Programa de Pós-Graduação em Química (PPGQ), Universidade Federal Do Amazonas (UFAM), Manaus, Amazonas (AM), Brazil

**Nandini I.V.** Department of Chemical Engineering, National Institute of Technology Karnataka (NITK), Mangalore, Karnataka, India

**Ioannis Ioannidis** Department of Chemistry, University of Cyprus, Nicosia, Cyprus

**Laith Jaber** Department of Chemical and Biological Engineering, American University of Sharjah, Sharjah, United Arab Emirates

**Ayah Jabr** Department of Chemical and Biological Engineering, American University of Sharjah, Sharjah, United Arab Emirates

**Hemant Veer Jain** Analytical Chemistry Laboratory, Analytical Sciences and Accredited Testing Services Group, Analytical Sciences & Services, Industrial Support through Technological Solutions (ASSIST) Division, CSIR-Indian Institute of Toxicology Research (CSIR-IITR), Lucknow, Uttar Pradesh, India; Academy of Scientific and Innovative Research (AcSIR), Ghaziabad, Uttar Pradesh, India

**Hrissi K. Karapanagioti** Department of Chemistry, University of Patras, Patras, Greece

**Navish Kataria** Department of Environmental Sciences, J.C. Bose University of Science and Technology, YMCA, Faridabad, Haryana, India

**Mahlet M. Kebede** Department of Civil, Construction, and Environmental Engineering, University of Alabama, Tuscaloosa, AL, United States

**Hammad Khan** Faculty of Materials and Chemical Engineering, GIK Institute of Engineering Sciences and Technology, Swabi, Khyber Pakhtunkhwa, Pakistan

**Nadeem Khan** São Carlos Institute of Chemistry, University of São Paulo, São Carlos, SP, Brazil

**Sabir Khan** Technological Development Center-CDTec, Postgraduate Program in Materials Science and Engineering, PPGCEM, Federal University of Pelotas- UFPel, Pelotas, RS, Brazil

**Vasiliki Kinigopoulou** Department of Environmental Engineering, School of Engineering, International Hellenic University, Sindos, Thessaloniki, Greece

**Sivagami Krishnasamy** School of Chemical Engineering, Vellore Institute of Technology, Vellore, Tamil Nadu, India

**Harshavardhan M.** Department of Chemical Engineering, National Institute of Technology Karnataka (NITK), Mangalore, Karnataka, India

**Monisha Manogharan** VIT School of Agricultural Innovations and Advanced Learning (VAIAL), Vellore Institute of Technology, Vellore, Tamil Nadu, India

**Arslan Maqbool** Faculty of Materials and Chemical Engineering, GIK Institute of Engineering Sciences and Technology, Swabi, Khyber Pakhtunkhwa, Pakistan

**Md Abdullah Al Masud** Department of Civil, Construction, and Environmental Engineering, University of Alabama, Tuscaloosa, AL, United States

**Paraskevi Mpeza** Department of Agriculture, University of Ioannina (UoI), Kostakii Campus, Arta, Greece

**Dhanavath Dattu Naik** Department of Pharmaceutical Analysis, National Institute of Pharmaceutical Education and Research, Guwahati, Changsari, Assam, India

**Vaishakh Nair** Department of Chemical Engineering, National Institute of Technology Karnataka (NITK), Mangalore, Karnataka, India

**Maria Nahin Oishy** Department of Environmental Science, Bangladesh University of Professionals (BUP), Dhaka, Bangladesh

**Sarah Ortbal** Department of Civil, Construction, and Environmental Engineering, University of Alabama, Tuscaloosa, AL, United States

**Prakash Kumar Panda** Department of Pharmaceutical Analysis, National Institute of Pharmaceutical Education and Research, Guwahati, Changsari, Assam, India

**Ioannis Pashalidis** Department of Chemistry, University of Cyprus, Nicosia, Cyprus

**Srikanth Ponneganti** Department of Food and Science and Technology, The Ohio State University, Columbus, OH, United States

**Gloria Post** New Jersey Department of Environmental Protection, Trenton, NJ, United States

**Sai Sharanya Pulimamidi** Department of Pharmaceutical Analysis, National Institute of Pharmaceutical Education and Research, Guwahati, Changsari, Assam, India

**Maimuna Binte Hasan Raisa** Department of Environmental Science, Bangladesh University of Professionals (BUP), Dhaka, Bangladesh

**Md Refat Jahan Rakib** School of Earth, Environmental and Marine Sciences, University of Texas - Rio Grande Valley, Brownsville, TX, United States

**Mohammed Rizwan** School of Chemical Engineering, Vellore Institute of Technology, Vellore, Tamil Nadu, India

**Soumi Saha** Department of Pharmaceutical Analysis, National Institute of Pharmaceutical Education and Research, Guwahati, Changsari, Assam, India

**Hasara Samaraweera** Department of Civil and Environmental Engineering, Western University, London, ON, Canada

**Erisa Saraçi** Institute of Catalysis Research and Technology, Karlsruhe Institute of Technology, Eggenstein-Leopoldshafen, Germany

**Mohammad Mahdi A. Shirazi** Department of Civil, Geological & Environmental Engineering, University of Saskatchewan, Saskatoon, SK, Canada

**Krishnasamy Sivagami** School of Chemical Engineering, Vellore Institute of Technology, Vellore, Tamil Nadu, India

**Sanne J. Smith** Department of Water Management, Delft University of Technology, Delft, The Netherlands

**Suriyaprakash M.** Department of Pharmaceutical Analysis, National Institute of Pharmaceutical Education and Research, Guwahati, Changsari, Assam, India

**Muhammad Tawalbeh** Sustainable and Renewable Energy Engineering Department, University of Sharjah, Sharjah, United Arab Emirates; Water and Environmental Conservation Center, RISE, University of Sharjah, Sharjah, United Arab Emirates

**Leigh G. Terry** Department of Civil, Construction, and Environmental Engineering, University of Alabama, Tuscaloosa, AL, United States

**Andreas Tzachristas** Department of Chemistry, University of Patras, Patras, Greece

**Angel Chyi En We** Department of Chemical Engineering, The University of Melbourne, Parkville, VIC, Australia

**Sangita Yadav** Department of Environmental Science and Engineering, Guru Jambheshwar University of Science & Technology, Hisar, Haryana, India

**Mohamed Zbair** Institut de Science des Mat´eriaux de Mulhouse CNRS UMR 7361, Université de Haute Alsace, Mulhouse, Grand Est, France; Université de Strasbourg, Strasbourg, Grand Est, France

**Foteini Zormpa** Institute of Catalysis Research and Technology, Karlsruhe Institute of Technology, Eggenstein-Leopoldshafen, Germany

# About the editors

**Dr. Dimitrios A. Giannakoudakis** obtained his PhD at the City University of New York, New York, United States (February 2017). Nowadays, he serves as an assistant professor at the University Marie Curie Sklodowska (UMCS), Lublin, Poland. His research is focused on the development of composite porous nanomaterials for (photo)(sono)(piezo) catalytic technologies for biomass valorization and (micro)plastics upcycling to green chemicals, as well as for air and water purification, linking physicochemical properties to performance and involved mechanisms. He coauthored more than 140 articles, 1 monograph, 7 edited books, and 17 book chapters with over 7000 citations. Since 2021, he has been included in the top 2% most influential scientists worldwide (Stanford list, Elsevier).

**Dr. Ioannis Anastopoulos** is an associate professor at the Department of Agriculture, University of Ioannina, Arta, Greece. He has conducted postdoctoral research at the Agricultural Research Institute (Cyprus), the University of Patras (Greece), the University of Cyprus (Cyprus), and the Hellenic Mediterranean University (Greece). His research focuses on the estimation of greenhouse gas emissions from agricultural soils following the application of organic and inorganic materials, the development of adsorbent materials for wastewater treatment, and the use of organic amendments for soil remediation. He has authored more than 130 articles in peer-reviewed journals with over 12,000 citations and has been included among the top 2% most influential scientists worldwide for the years 2019–24 (Stanford list, Elsevier).

# Foreword

**Prof. George Z. Kyzas,** *Hephaestus Laboratory, School of Chemistry, Faculty of Sciences, Democritus University of Thrace, Kavala, Greece*

Over the past few decades, a category of artificial chemicals named per- and polyfluoroalkyl substances (PFAS), or per- and polyfluoroalkyl compounds, has quite literally found its way into the fabric of our contemporary lives. These compounds are designed with a high degree of heat, water, and oil resistance and have found common use in cookware and waterproof fabrics as well as in firefighting foams and food wraps. PFAS are not easily broken down either in the environment or human body, which is why they have now become known as forever chemicals.

The properties of PFAS at the biological level include the capacity of the substances to bind proteins and build up with time. When compared to most other environmental toxins, which are either broken down or eliminated within relatively short periods of time, some PFAS compounds have a half-life in humans that is measured in years. This tenacity enables them to disrupt physiological systems in a rather subtle yet huge way. Studies have attributed the PFAS exposure to lipid metabolism and immunological and endocrine mechanisms. They have also been found to be associated with higher levels of cholesterol, slower vaccine reaction, developmental outcomes in infants, and high chances of some cancers; the strength and specificity of such relationships are still under study.

The fact that PFAS are diverse is what makes them complicated in their toxicological understanding. There are thousands of different compounds under this category whose chemical structures and biological behaviors are slightly different. Although certain legacy forms of PFAS have been phased out in certain areas of the world, newer forms have been used in their place whose long-term impacts are poorly known. This dynamic chemical environment makes it difficult to regulate, assess risk, and clean up.

PFAS have one of the characteristic toxicological properties, which is the ability to bind to proteins as opposed to fats. PFAS bind to the proteins in blood like albumin and circulate throughout the body after they enter the body, mainly by contaminated water, food items, or air. These build up in body organs such as liver and kidney and, in some instances, the developing child of the mother through placental transmission. Due to its slow rate of elimination, with repeated exposure, there is a progressive accumulation in the interior, although levels of exposure may seem to be low.

At the cellular level, PFAS have been known to engage receptors that control gene expression, especially those that control metabolism and inflammation. The most researched ones include the peroxisome proliferator-activated receptors, which are important in the homeostasis of lipids and glucose. PFAS are capable of causing an imbalance in regular metabolism by modulating these signaling pathways. This mechanistic observation is consistent with the epidemiological evidence indicating that exposure to high cholesterol status and potential increases in the risk of developing metabolic diseases.

PFAS seem highly sensitive to the immune system. Various studies have found that there were lower antibody responses after vaccination of exposed populations, indicating that the immune system is weakened. This not only brings to play the issue of vulnerability to infectious diseases but also the general regulation of immunity. Causal relationships between PFAS and autoimmune conditions are undergoing investigation, but the cause-and-effect relationships are hard to prove.

Another area of concern is the endocrine disruption. PFAS may disrupt hormone systems, such as thyroid hormones that are essential in growth, brain development, and metabolism. Even small changes in thyroid signaling during pregnancy or in babies can have long-term developmental effects. Certain PFAS have also been linked with disrupted reproductive hormones, which could have impacts on fertility and developmental outcomes.

One area that long-term research has focused on is carcinogenicity. Some of the most researched PFAS compounds were found to cause more risks of kidney and testicular cancers in populations most exposed to them. The mechanisms have not been completely clarified, but they could be oxidative stress, immune modulation, and cell growth regulation defects. However, it is important to note that the risk of cancer differs with regard to the particular compound, length of exposure, and individual susceptibility.

Another complication is that of mixture effects. Exposure to a single PFAS compound is uncommon in real-life situations. Rather, people are exposed to complex combinations, whose joint action can either be additive or even synergistic. The toxicological models continue to undergo changes whereby the conventional testing methods tend to measure chemicals individually.

New versions of PFAS raise more doubt. Although certain more recent compounds have been found to have longer biological half-lives, preliminary data indicate that they can still have a biological effect, especially at biological subsystems such as cellular signaling and tissue-specific effects. The lower persistence of the body may not necessarily mean low toxicity, particularly in continuous conditions of exposure.

Finally, PFAS toxicity has less of an immediate character than that of persistent and subtle disturbance and cumulative effects. It contests traditional beliefs regarding dose, timing, and risk and again reiterates the role of long-term monitoring and precautionary measures in the management of the chemicals. Due to the ongoing development of research, understanding of the interactions between these compounds and living systems—and how their risks can more appropriately be alleviated—will evolve as well.

The "story" of PFAS is more about responsibility than hazard and adaptation. It puts all of us (scientists, policymakers, industries, and communities) on the way to understanding why these chemicals are designed, developed, and handled.

This book will attempt to deliver a clear and solid comprehension of a complicated matter by following some steps such as the origin, use, environmental exposure, and health effects of PFAS. This book is divided into four sections (Section 1: Introduction and environmental occurrence and toxicity, Section 2: Remediation technologies, Section 3: Materials for PFAS remediation, and Section 4: Policy and regulation).

In Section 1 (Introduction and environmental occurrence and toxicity), four chapters are included regarding PFAS in drinking water (occurrence, health risks, and regulatory frameworks); in soil and plants; issues for the environmental mixtures, exposure pathways, and ecotoxicological risks; and microplastics as carriers of PFAS.

In Section 2 (Remediation technologies), eight chapters are included regarding major remediation techniques for PFAS, such as photocatalysis, ultrasonication (sonolysis), electrochemical oxidation, soil remediation strategies, soil oxidation techniques, phytoremediation of PFAS-contaminated soils, and foam fractionation for PFAS removal from wastewater.

In Section 3 (Materials for PFAS remediation), eight chapters are included regarding major materials for PFAS remediation, such as clay minerals, zeolites, MOFs/ZIFs, activated carbon, biochars, and metal oxides (mineral- and metal oxide-based photocatalysts). Additional contributions address membrane technologies.

The last section (Policy and regulation), three chapters are included regarding regulatory approaches to PFAS in North America (United States and Canada), Australia, European Union, and Asia.

I want also to congratulate all contributors of this book, as well as the distinguished editors (Prof. Ioannis Anastopoulos and Prof. Dimitrios A. Giannakoudakis) who made a great contribution to the PFAS theory, technology, knowledge, and understanding.

PART 1

# Introduction and environmental occurrence and toxicity

CHAPTER

# 1

# Per- and polyfluoroalkyl substances in drinking water: occurrence, health risks, and regulatory frameworks

*Giacomo de Falco*[1] *and Gloria Post*[2]

[1]New York City Department of Environmental Protection, New York, NY, United States
[2]New Jersey Department of Environmental Protection, Trenton, NJ, United States

## 1.1 Introduction

Per- and polyfluoroalkyl substances (PFAS) are a large and diverse class of synthetic chemicals first discovered in the 1930s (Lindstrom et al., 2011). These compounds can be divided into two main families: nonpolymer (perfluoroalkyl [totally fluorinated] and polyfluoroalkyl [partially fluorinated] substances) and polymer (including fluoropolymers, side-chain fluorinated polymers, and perfluoropolyethers), which can be further categorized into several classes based on their chemical structure (ITRC, 2023a or Buck et al., 2011). Although there is no universally accepted definition, PFAS are typically characterized by the presence of the $-C_nF_{2n+1}$ or $C_nF_{2n}-$ moiety (Adewuyi and Li, 2024). The Organization for Economic Cooperation and Development defines PFAS as substances containing at least one fully fluorinated methyl or methylene carbon atom (OECD, 2021), and many U.S. states have also adopted this definition (ECOS, 2025). In contrast, while the definitions used by different programs (Office of Pollution Prevention and Toxics, Office of Water) within the U.S. Environmental Protection Agency (USEPA) are not identical, they all exclude compounds with only one fully fluorinated carbon atom and are therefore more restrictive than (OECD, 2021; USEPA, 2021a, USEPA, 2022a, ITRC, 2023a).

Nonpolymer PFAS are of concern as drinking water contaminants, and perfluoroalkyl acids (PFAAs), with structures consisting of a totally fluorinated carbon chain and an anionic (carboxylate or sulfonate) functional group, are the most well-known group within this specific subclass. PFAAs can be grouped according to the number of carbons in the molecule: they are typically described as long-chain if the number of carbons is eight or more for perfluorocarboxylates (PFCAs) and six or more for perfluorosulfonates, short-chain if the number of carbon atoms is between four and seven for PFCAs and four or five for

*Per- and Polyfluoroalkyl Substances (PFAS) as Environmental Contaminants*
**DOI: https://doi.org/10.1016/B978-0-443-43856-1.00016-2**

perfluorosulfonates, or ultra-short-chain if there are three or fewer carbon atoms (USEPA, 2009a; ITRC, 2023a). While PFAAs were the primary focus of earlier PFAS studies, PFAS with other structures are now receiving increasing attention (Lu et al., 2019; Post, 2025).

Due to their exceptional chemical and thermal stability, PFAS have been widely used in a broad range of industrial and consumer applications. These include firefighting foams, mining operations, photolithography, hydraulic fluids, electronics manufacturing, floor polishes, photographic films, denture cleaners, soaps, shampoos, cosmetics, nonstick cookware coatings, and oil- and water-repellent treatments for paper, textiles, and leather (Glüge et al., 2020). Their extreme resistance to environmental degradation has earned them the name of "forever chemicals."

Long-chain PFAS are highly bioaccumulative in humans, and several, including perfluorooctanoic acid (PFOA), perfluorooctane sulfonic acid (PFOS), perfluorononanoic acid (PFNA), and perfluorohexane sulfonic acid (PFHxS), are detected in human blood serum of virtually all residents of the U.S. and many other nations (CDC, 2024; Kato et al., 2015). Toxicological effects of PFAS in laboratory animals include liver damage, immune system suppression, reproductive and developmental toxicity, and/or carcinogenicity (Fenton et al., 2021; ITRC, 2023b), and recent assessments conclude that long-chain PFAAs increases the risk of several health effects, including certain types of cancer for PFOA, even at exposure levels that occur within the general population (Fenton et al., 2021; Post, 2025).

In the early 2000s, 3M, one of the leading global PFAS manufacturers, voluntarily phased out production of PFOA, PFOS, and PFHxS due to concerns about their persistence and widespread human exposure (Butenhoff et al., 2009). In 2006, eight major PFAS manufacturers entered into a voluntary agreement with USEPA to phase out production of long-chain PFCAs (e.g., PFOA, PFNA) and their precursors by 95% by 2010 and completely by 2015 due to their "persistence, presence in the environment and in the blood of the general US population, long half-life in people, and developmental and other adverse effects in laboratory animals" (USEPA, 2024a).

Short-chain PFAS, including PFAAs (e.g., perfluorobutane sulfonate) and those with other structures (e.g., hexafluoropropylene oxide dimer acid [GenX], a six-carbon perfluoroether), have been introduced as replacements for the phased-out long-chain PFAAs (Post, 2025). The intended advantage of these short-chain PFAS is that they are less bioaccumulative. However, they are equally environmentally persistent as the phased-out long-chain PFAAs and cause similar toxicological effects, in some cases at similarly low doses (Post, 2025).

In 2009, the USEPA identified PFOA and PFOS as contaminants of potential concern in drinking water and added PFOA and PFOS to the contaminant candidate list. By that time, significant contamination had already been documented in several states, including West Virginia, Minnesota, and New Jersey (Post et al., 2009; Herrick et al., 2017; Goeden et al., 2019), and PFAS had also been detected in drinking water sources in Europe, Asia, and Canada (Rumsby et al., 2009).

This chapter aims to provide a comprehensive survey and critical analysis of the sources and occurrence of PFAS in drinking water. Given the increasing detection of PFAS in surface water and groundwater used as drinking water sources and in treated drinking water, it is essential to understand both the extent of contamination and the limitations of existing monitoring frameworks. Moreover, we will examine the development and application of

PFAS drinking water guidelines (i.e., standards and guidance values) worldwide. By comparing guidelines established by different regulatory authorities, the chapter underscores how scientific evidence and public health considerations are shaping global regulatory landscapes. Finally, this chapter also explores future regulatory directions and possible implications for water treatment.

## 1.2 Sources of per- and polyfluoroalkyl substances in drinking water

The release of PFAS into raw drinking water occurs in two ways: point sources and nonpoint sources (Kurwadkar et al., 2022). Industrial manufacturing facilities, firefighting training grounds, airports, municipal and industrial wastewater treatment plant (WWTP) effluents, and landfills' leachate are considered point sources, while atmospheric transport, surface runoff, and precipitation, and other diffuse sources of unknown locations are defined as nonpoint PFAS sources as shown in Fig. 1.1.

One of the main studied sources of PFAS in raw drinking water (surface water and/or groundwater) is WWTPs. PFAS are released from WWTPs through both effluent discharge and land application or landfill disposal of biosolids (Guo et al., 2010; Zhang et al., 2024). WWTPs can receive influent water from both sanitary and industrial sources, such as landfill leachate, which often contains a wide range of materials that can release PFAS into wastewater (Sabba et al., 2025). Conventional WWTPs typically do not fully remove PFAS from wastewater, and many studies reported an increase in concentrations of PFAAs after

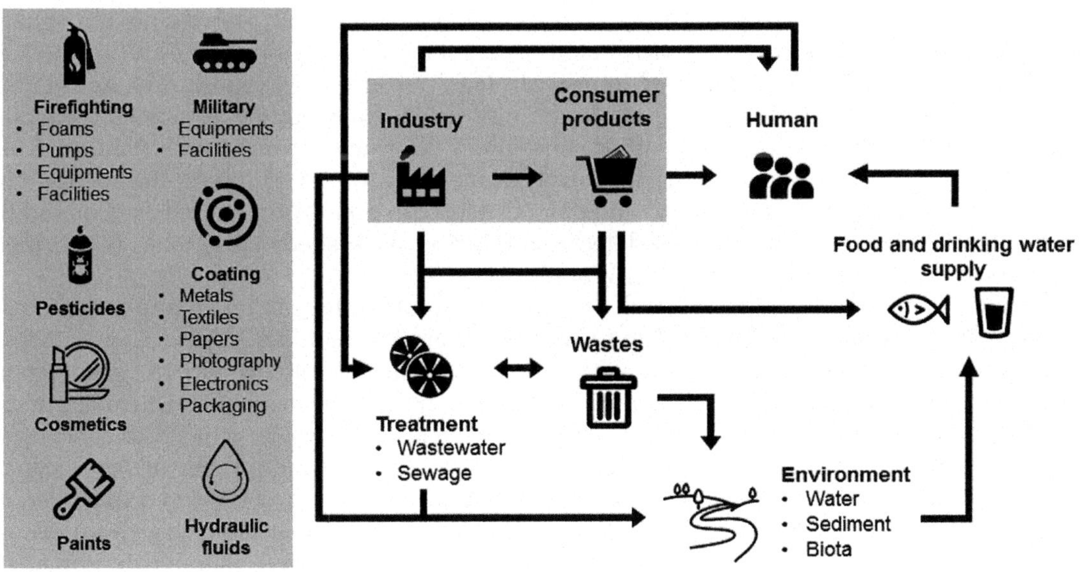


**FIGURE 1.1** Release and exposure pathways of per- and polyfluoroalkyl substances in the environment (Wee and Aris, 2023).

treatment (Lenka et al., 2021). This is due to the conversion of PFAS precursor, which are polyfluoroalkyl substances, to PFAAs (Yu et al., 2012). Tavasoli et al. (2021) studied the distribution and fate of 24 PFAS for six New Hampshire municipal WWTPs, which showed that the effluent concentration of PFAS was between 80 and 200 ng $L^{-1}$ for a sampling campaign during summer months (Tavasoli et al., 2021).

The presence of PFAS compounds in biosolids is a rapidly growing concern. For example, Maine has banned the land application of biosolids, and several other states limit the amounts that can be applied based on PFAS concentrations in the biosolids (ECOS, 2025). A recent study carried out by Peter and Lee investigated the PFAS contamination of surface water and groundwater of biosolid-applied agricultural areas with limited impact of other sources of PFAS contamination (e.g., industrial discharges, waste disposal, the use of aqueous film forming foams [AFFFs] and landfill leachates) (Peter and Lee, 2025). The authors showed that concentrations of quantifiable PFAS were generally higher in surface water compared to well water, and the site with the highest concentration was hydrologically connected to an active biosolids-applied field. The recent draft USEPA (2025a) risk assessment for PFOA and PFOS in sewage sludge (i.e., biosolids) concluded that health risks may result from groundwater contamination with PFAS from biosolids applied to agricultural land or stored in surface disposal sites. Additional investigations are needed to better understand the role of land-applied biosolid as a source of PFAS in raw drinking water.

As mentioned above, AFFF used in airports and/or in firefighting training exercises is an additional main source of PFAS in raw drinking water (Moody et al., 2002; Awad et al., 2011; Karrman et al., 2011). The US Department of Defense has identified 723 sites where PFAS may have been used and released in surface and or groundwater (USDoD, undated). Although the specific composition of AFFF can vary between formulations and production batches, studies indicate that PFAS concentrations, including both short- and long-chain compounds, may reach up to 5% (Król et al., 2012). Legacy AFFFs produced from the late 1960s until 2002 primarily contained long-chain (C8) perfluoroalkyl substances with PFOS being the predominant compound, while the new AFFF formulations exclusively contain short-chain (C6) fluorotelomers or other short-chain fluorinated compounds (Jahura et al., 2024). Hu et al. (2016) developed a statistical framework for investigating the increase of PFAS concentration in raw drinking water associated with point sources, and they observed a statistically significant increase of detectable PFAS in watersheds close to AFFF-certified airports and facilities (Hu et al., 2016; Wang et al., 2022).

Finally, as nonpoint sources, rainfall and surface runoff water can be an important cause of PFAS contamination in raw drinking water (Kim and Kannan, 2007; Podder et al., 2021). Murakami and Takada (2008) investigated the PFAS concentration in rainfall and stormwater runoff. A higher PFAS concentration was obtained in the case of stormwater runoff compared to rainfall, implying a contribution of vehicles and city dust to PFAS concentration. Interestingly, Cousins et al. (2022) investigated the concentrations of four legacy PFAS—PFOA, PFOS, PFHxS, and PFNA—in global precipitation and observed that even in remote regions such as the Tibetan Plateau, rainwater contains detectable concentrations of PFAS (e.g., median of ~55 pg $L^{-1}$). Recently, Kali et al. (2025) reported a significant impact of stormwater discharges on PFAS concentration in urban streams. Higher PFAS concentrations were observed for stormwater derived from airports, landfill sites, and WWTPs. These findings highlight that stormwater runoff is a significant source of PFAS contamination in raw drinking water sources. As stormwater can carry PFAS from various urban and industrial surfaces into lakes, rivers, and reservoirs, it represents an important and often underestimated pathway of environmental exposure. Therefore comprehensive strategies should be developed and implemented to mitigate PFAS inputs from stormwater. This may include source control measures, green infrastructure, and advanced stormwater treatment technologies aimed at reducing PFAS levels before they reach drinking water supplies.

## 1.3 Occurrence of per- and polyfluoroalkyl substances in raw and finished drinking water

PFAS have been consistently documented as ubiquitous contaminants in aquatic systems worldwide. Table 1.1 reports examples of PFAS contamination observed in raw and finished drinking water. Important to underline is that the sensitivity of PFAS monitoring varies considerably across studies, with notable differences in the specific compounds targeted and the corresponding method reporting limits. These discrepancies can significantly influence interstudy comparability and interpretation of occurrence data. In a comprehensive global synthesis, Ackerman Grunfeld et al., 2024 aggregated PFAS concentration data from a wide range of international monitoring studies and developed high-resolution geospatial maps to visualize contamination trends (Fig. 1.2). These maps not only delineate regional PFAS hotspots (i.e., areas with notably elevated concentrations) but also provide insights into the geographic density of sampling efforts, revealing significant gaps in global surveillance. This analysis demonstrated that PFAS contamination is not limited to sites of direct industrial discharge or manufacturing activity. Elevated concentrations were also observed in areas characterized by high PFAS usage, such as airports, firefighting training facilities, military installations, and urban zones with heavy runoff. As we pointed out in the previous paragraph, PFAS have even been detected in remote environments. For example, Wang et al. (2014) reported measurable concentrations in snow, glacial ice, and high-altitude lakes in remote regions of China, indicating atmospheric deposition as a major vector for long-range environmental transport.

Comparative studies have also shown a tendency for PFAS concentrations to be higher in groundwater than in surface waters. Sims et al. (2022) reviewed data on the occurrence of 24 PFAS from 371 monitoring studies worldwide that include 979 groundwater and 10,617 surface water data points. The number of data points for each of the 24 PFAS varied from 27 (perfluorononane sulfonate; PFNS) to 1724 for PFOA for groundwater and from 178 (PFNS) to 6813 (PFOA) for surface water. Detection frequencies varied from 1% for 2-(N-methylperfluorooctanesulfonamido) acetic acid (N-MeFOSAA) to 68% for pefluorohexanoic acid in groundwater and from 15% for N-MeFOSAA to 90% for PFOA in surface water. Sims et al. (2022) reported that the median concentration of individual PFAS in groundwater reached 6.71 ng $L^{-1}$, compared to 1.67 ng $L^{-1}$ in surface water. This disparity may reflect the persistence and leaching potential of PFAS compounds through soils and into aquifers. In the United States, groundwater near military bases with historical AFFF applications has shown extremely high PFAS levels, with the highest PFOA concentration exceeding

TABLE 1.1 Examples of per- and polyfluoroalkyl substances (PFAS) contamination observed in raw and finished drinking water.

| Geographic location | Main PFAS compound(s) | PFAS concentration | Water matrix | References |
|---|---|---|---|---|
| Asia (general) | PFOA | Surface water median: 11.0 ng $L^{-1}$; max: 2682 ng $L^{-1}$ | Surface water | Sims et al. (2022) |
| Oceania | PFOS, PFOA | Surface water median: 4.50 ng $L^{-1}$; max: 90 ng $L^{-1}$ | Surface water | Sims et al. (2022) |
| North America | PFOS, PFOA | Groundwater median: 601 ng $L^{-1}$; max: 7.4 mg $L^{-1}$ | Groundwater | Sims et al. (2022) |
| China (Yangtze River) | PFOA, PFOS | PFOA: up to 297.5 ng $L^{-1}$; PFOS: up to 37.8 ng $L^{-1}$ | Surface water | Wang et al. (2015) |
| India (Ganges River) | PFBS, PFHxA | PFBS: up to 10.2 ng $L^{-1}$<br>PFHxA: up to 4.7 ng $L^{-1}$ | Surface water | Sharma et al. (2016) |
| USA (drinking water) | PFOS, PFOA | PFOA–PFOS <1–1102 ng $L^{-1}$; median raw water: 21.4 ng $L^{-1}$ | Drinking water | Boone et al. (2019) |
| Netherlands (Dordrecht) | GenX (HFPO-DA) | River: 812 ng $L^{-1}$; drinking water: up to 11 ng $L^{-1}$ | Surface and drinking water | Gebbink et al. (2017) |
| Germany (Moehne River) | PFOA, PFOS | PFOA: 3640 ng $L^{-1}$; PFOS: 193 ng $L^{-1}$ | Surface water | Exner and Färber, (2006) |
| Sweden (nationwide) | PFOA, PFOS | PFOA: up to 130 ng $L^{-1}$, PFOS: up to 400 ng $L^{-1}$ | Drinking water | Banzhaf et al. (2016) |
| Maltese Islands | PFOA, PFOS | Surface water: PFOA up to 16 ng $L^{-1}$ | Surface water | Sammut et al. (2017) |
| Brazil and Colombia | PFOS, PFOA | PFOS: 0.35–2.82 ng $L^{-1}$, PFOA: 0.58–6.70 ng $L^{-1}$ | Drinking water | Schwanz et al. (2016) |
| Vietnam (surface water) | PFOA, PFOS | PFOA: 53.5 ng $L^{-1}$; PFOS: 40.2 ng $L^{-1}$ | Surface water | Lam et al. (2017) |
| Rhine River (Europe) | PFBA, PFHxA | PFBA: up to 335 ng $L^{-1}$; PFHxA: 49.6 ng $L^{-1}$ | Surface water | Li et al. (2020) |
| Plankenburg River, South Africa | PFBA, PFHpA, PFOA | Total PFAS: 62.3–186.4 ng $L^{-1}$ | Surface water | Fagbayigbo et al. (2018) |
| Cape Fear River, NC, USA | PFBA, PFBS, PFHxA, PFHpA | Total: 10–324 ng $L^{-1}$ in raw drinking water | Raw drinking water | Li et al. (2020) |
| Tangxun Lake, China | PFBA, PFBS | PFBA: 6280 ng $L^{-1}$; PFBS: 4520 ng $L^{-1}$ | Surface Water | Li et al. (2020) |
| Ghana | PFOA, PFOS | ΣPFAA: 200 ng $L^{-1}$ | Drinking water | Essumang et al. (2017) |

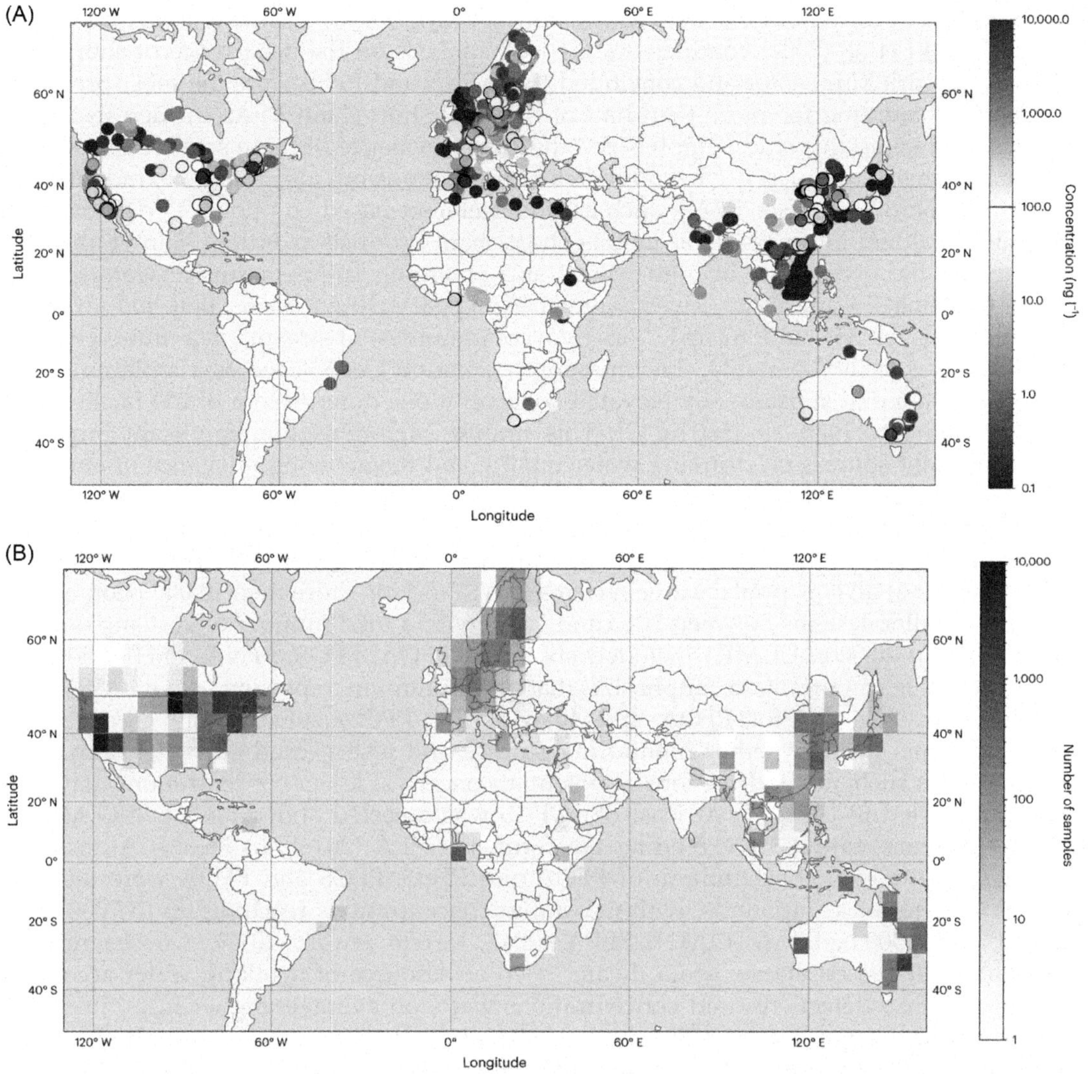


**FIGURE 1.2** Map of per- and polyfluoroalkyl substances concentration (A) and number of samples associated (B) in surface and groundwater (Ackerman Grunfeld et al., 2024). Map lines delineate study areas and do not necessarily depict accepted national boundaries.

6000 µg $L^{-1}$ (Moody and Field, 1999). Similar patterns were observed in surface waters impacted by direct PFAS discharges or AFFF usage, where concentrations were often an order of magnitude higher than those found in typical environmental waters. A striking example includes surface water in Osaka, Japan, where PFOA levels reached 1630 ng $L^{-1}$ (Jian et al., 2017).

Short-chain PFAAs have also been widely detected in raw drinking water. In North Carolina's Cape Fear River, short-chain PFAS, including PFBA, PFBS, PFHxA, and PFHpA, were detected at concentrations between 10 and 324 ng $L^{-1}$ in raw drinking water intakes (Li et al., 2020). Li et al. (2020) conducted a focused analysis on the occurrence of short-chain PFAS in raw drinking water and concluded that PFBA and PFBS were the most prevalent, frequently accounting for more than half of the total short-chain PFAS burden in aquatic systems. Moreover, this study emphasized that the prevalence of these smaller PFAS is not only due to their continued use in commercial formulations but also to the degradation and transformation of precursor PFAS, such as fluorotelomer-based compounds (e.g., 6:2 FTS), which break down into persistent, mobile short-chain products overtime. Sun et al. (2016) investigated the occurrence of both legacy and emerging PFAS in source waters across North Carolina's Cape Fear River watershed. Legacy PFAS were detected at mean concentrations of 355 and 62 ng $L^{-1}$ in two communities upstream of a fluorochemical manufacturing facility. Notably, the emerging compound GenX, used as a replacement for PFOA, was found at substantially elevated concentrations downstream of the facility, with levels reaching as high as 4560 ng $L^{-1}$. This finding highlights the significant impact of industrial point sources on drinking water quality and the growing relevance of emerging PFAS in environmental monitoring.

An important initial step in understanding PFAS occurrence in final drinking water in the U.S. was the USEPA third unregulated contaminant monitoring rule (UCMR3), which tested finished water at all U.S. public water systems (PWS) serving more than 10,000 people and a subset of smaller systems between 2013 and 2015 (with a total number of systems sampled equal to about 20,000). UCMR3 included six PFAS: PFOA, PFOS, PFNA, PFHxS, PFHpA, and PFBS. It is important to emphasize that the minimum reporting levels (MRLs) for UCMR3 (e.g., 20 ng $L^{-1}$ for PFOA and 40 ng $L^{-1}$ for PFOS) substantially exceeded the reporting limits in many other monitoring studies and also exceed current health-based guidelines. As such, many detections of potential concern at lower concentrations may have gone unreported in UCMR3. At least one PFAS was detected above the UCMR3 MRL in 4.0% of systems, with 1.3% exceeding the former USEPA (2016a, b) Health Advisory of 70 ng $L^{-1}$ for the total concentration of PFOA and PFOS. PFOS and PFOA were the most frequently detected compounds, with maximum concentrations reaching up to 7000 ng $L^{-1}$ (PFOS) and 1300 ng $L^{-1}$ (PFOA) (USEPA, 2017, Crone et al., 2019). Consistent with information discussed above, groundwater used as a source of drinking water accounted for 72% of PFAS detections, and concentrations were, on average, higher in groundwater than in surface water sources (210 ng $L^{-1}$ vs 90 ng $L^{-1}$). Interestingly, it was observed that PFAS cooccurrence was widespread; multiple PFAS were present in about 50% of samples with PFAS detections.

EPA has posted the results of UCMR5, which is being conducted in 2023–25, as they become available, with data reporting expected to be complete by 2026. UCMR5 is significantly expanded in scope compared to UCMR3 and will test all U.S PWS serving more than 3300 people and a subset of smaller systems for 29 PFAS (approximately 18,000 sampling locations from 6900 PWS) at much lower MRLs (e.g., 4 ng $L^{-1}$ for PFOA and PFOS). Based on the data available through April 2025 (USEPA, 2025b), representing about 75% of the complete study, about 8.5% of PWS (weighted nationally) are estimated to exceed one or more of the final USEPA (2024b) maximum contaminant levels (MCLs) (4 ng $L^{-1}$ for PFOA and PFOS; 10 ng $L^{-1}$ for PFHxS, PFNA, and GenX, and a hazard index of one for a mixture of PFHxS, PFNA, GenX, and PFBS). PFOA, PFOS, and PFHxS were detected above their MCLs in 8.9%, 7.4%, and 0.7%, respectively, of the ~6900 PWS that had complete UCMR5 results, while GenX and PFNA were detected above their MCLs in only 2% and 4% of the ~6900 systems, respectively. Of the 23 unregulated PFAS included in UCMR5, short-chain PFCAs were the most frequently detected. PFBA, PFPeA, and PFHxA were detected above their MRLs of 3–5 ng $L^{-1}$ in 18%, 20%, and 18% of the ~8700 PWS tested, respectively; none of these detections exceeded the UCMR5 health reference levels of 6000 ng $L^{-1}$ for PFBA and 3000 ng $L^{-1}$ for PFHxA, and there is no Health Reference Level for PFPeA. The other 20 unregulated PFAS included in UCMR5 were detected in zero to 0.05% of the ~8700 PWS. It is important to emphasize that the overall purpose of UCMR is to provide information that can be used by USEPA in determining future USEPA regulatory priorities. While UCMR5 data give critical insight into potential hotspots and nationwide trends in PFAS contamination, it is important to note that it is not used by USEPA to determine regulatory compliance.

Also, the European Union has significantly increased its efforts to monitor PFAS compounds in recent decades. According to a recent report of the European Environment Agency, more than 40,000 measurements were submitted to the European Union for 15 PFAS compounds in 2022 (EEA, 2024). By contrast, dedicated PFAS monitoring programs are limited in scope in developing countries of Africa, South America, and parts of Asia (Adewuyi and Li, 2024). Ssebugere et al. (2020) provided one of the first continent-wide syntheses of PFAS occurrences in African environmental waters. The authors highlighted the lack of analytical infrastructure, insufficient laboratory capacity, and constrained financial resources continue to impede the generation of reliable PFAS data across the continent. A similar pattern can be observed in Latin America and the Middle East as well. As an example, a detailed study of raw drinking water in the metropolitan area of Porto Alegre, Brazil, found 23 PFAS compounds, and the total PFAS concentration in groundwater ranged from 22 to 718 ng $L^{-1}$ (Stefano et al., 2023).

## 1.4 Development and application of per- and polyfluoroalkyl substances drinking water guidelines (USA and worldwide)

Drinking water guidelines for PFAS include enforceable regulatory standards (e.g., USEPA MCLs; USEPA, 2024c) and nonenforceable guidance values (e.g., USEPA Health Advisories; USEPA, 2025c). In general, health effects are the primary basis of drinking water guidelines for PFAS and other contaminants; additional considerations are discussed below.

Guidelines for PFAS in drinking water have been established by the USEPA, numerous U.S. states, and other nations, including Australia, several European countries, and Japan (OECD, undated; ITRC, 2025). Up-to-date information on guidelines for PFAS in drinking water and other environmental media established by the USEPA, U.S. states, and other nations is found in the Interstate Technology and Regulatory Council PFAS Environmental Media Values Table (ITRC, 2025).

PFAS drinking water guidelines were initially developed primarily for PFOA and PFOS, the two PFAS with the most extensive information on health effects and drinking water occurrence, starting in the early 2000s (Post, 2021). Guidelines have now been established for additional long-chain PFAAs (e.g., PFHxS, PFNA), as well as short-chain PFAAs (e.g.,

PFHxA, PFBS) and other types of PFAS (e.g., HFPO-DA [GenX]), with specific PFAS with guidelines varying among regulatory authorities (ITRC, 2025).

Variations in PFAS drinking water guidelines established by various authorities may be due to differences in the risk assessment approaches used, interpretations of scientific data, and/or information available when the guidelines were developed. The earliest U.S. values were 150,000 ng $L^{-1}$ for PFOA and 1000 ng $L^{-1}$ for PFOS, established in 2002 by the states of West Virginia and Minnesota, respectively (WVDEP, 2002; MDH, 2025). Federal (i.e., USEPA) and state guidelines for PFOA and PFOS in the U.S. have decreased by multiple orders of magnitude since that time (Post, 2021). For example, USEPA Health Advisories, which are based on noncarcinogenic effects, for PFOA and PFOS were initially set at 400 and 200 ng $L^{-1}$, respectively (USEPA, 2009b), were reduced to 70 ng $L^{-1}$ for the total of both compounds in 2016 (USEPA, 2016a; 2016b), and were later further reduced to 0.004 ng $L^{-1}$ for PFOA (USEPA, 2022b) and 0.02 ng $L^{-1}$ for PFOS (USEPA, 2022c). Similarly, Minnesota's initial guideline of 1000 ng $L^{-1}$ for PFOS was reduced to 600 ng $L^{-1}$ in 2006, and subsequently to 300 ng $L^{-1}$ in 2007, 27 ng $L^{-1}$ in 2017, 15 ng $L^{-1}$ in 2019, and 2.3 ng $L^{-1}$ in 2024 (MDH, 2025). Between 2016 and 2020, several states established PFOA and/or PFOS guidelines of 20 ng $L^{-1}$ or below (Post, 2021), lower than the USEPA (2016a, b) guideline at the time of 70 ng $L^{-1}$ for the total of both compounds, and USEPA adopted national MCLs of 4 ng $L^{-1}$, with health-based goals of zero, for both PFOA and PFOS in 2024 (USEPA, 2024b).

Health-based drinking water guidelines consider both the contaminant's toxicity and the assumed exposure from drinking water and other sources (Post, 2021). For guidelines based on noncarcinogenic effects, the toxicity factor (e.g., reference dose in the U.S.) is the daily dose of the contaminant (ng/kg body weight/day) at which no adverse effects are expected over a lifetime. For carcinogenic effects, some authorities, including USEPA, assume that any level of exposure poses a risk unless a threshold for carcinogenic effects is conclusively demonstrated. For numerical guidelines based on cancer risk, the daily dose (ng/kg/day) is the quotient of the cancer risk level (e.g., 1 in 1,000,000, 1 in 100,000; unitless), which is a policy decision, and the cancer slope factor ($[ng/kg/day]^{-1}$). Assumed exposure is based on the drinking water ingestion rate (L/kg body weight/day) for an adult or a higher ingestion rate for a specific subpopulation (e.g., infant, nursing woman) when relevant for the health effect of concern. For guidance based on noncarcinogenic effects, exposure from sources other than drinking water (e.g., relative source contribution) is also considered to ensure that total exposure from all sources does not exceed the acceptable value (e.g., reference dose).

Current drinking water guidelines consider toxicokinetic characteristics specific to PFAS. Guidelines for short-chain PFAS (e.g., PFHxA, PFBS) are generally substantially higher than for long-chain PFAS because long-chain PFAS are more bioaccumulative (i.e., excreted more slowly) than short-chain PFAS and therefore generally cause toxicity at lower administered doses (Post, 2021); an exception is the relatively low USEPA Health Advisory and MCL of 10 ng $L^{-1}$ for GenX, a short-chain PFAS which is not highly bioaccumulative and is used as a PFOA replacement (USEPA, 2022d; USEPA, 2024b). Additionally, guidelines based on effects in laboratory animals must account for the much higher internal dose (blood serum level, ng $L^{-1}$) in humans than in laboratory animals from the same administered dose; this results from the much greater bioaccumulation of PFAS in humans (Post, 2021). Finally, infants, particularly those who are breastfed, receive higher exposures to long-chain PFAS than older age groups. This occurs because long-chain PFAS concentrations are higher in breast milk than in the mother's drinking water, and infants also consume more fluid (breast milk or formula prepared with drinking water) on a body weight basis (L/kg/day) than older individuals (Post, 2022). These higher exposures to infants must be considered in drinking water guidelines based on developmental toxicity or other effects that may result from exposures at an early age.

Until 2020, PFAS toxicity factors and drinking water guidelines were based on noncancer effects such as hepatic toxicity, decreased immune system response, and delayed development in laboratory animal toxicology studies. More recently, toxicity factors and/or drinking water guidelines based on effects of long-chain PFAS (e.g., PFOA, PFOS, PFNA, PFHxS, PFDA) in human epidemiology studies have been developed by several authorities (EFSA, 2020; MDH, 2024a; MDH, 2024b; California EPA, 2024; USEPA, 2024b; USEPA, 2024e). Adverse health effects of long-chain PFAS in humans include increased blood serum levels of liver enzymes (indicative of liver damage) and cholesterol, decreased antibody response to vaccines, decreased birth weight, and, for PFOA, increased risk of certain types of cancer. These effects are observed in humans from much lower PFAS exposures than those that cause toxicological effects in laboratory animals, and drinking water guidelines based on human data are orders of magnitude lower than earlier guidelines based on animal data, well below 1 ng $L^{-1}$ (MDH, 2024a; MDH, 2024b; California EPA, 2024). Additionally, USEPA has concluded that PFOA and PFOS are likely human carcinogens (USEPA, 2024d; USEPA, 2024e), and USEPA policy is to set health-based drinking water levels (maximum contaminant level goals, MCLGs) for carcinogens at zero, an aspirational goal (USEPA, 2024f). The recently updated Australian drinking water guidelines of 200 ng $L^{-1}$ for PFOA, 8 ng $L^{-1}$ for PFOS, 30 ng $L^{-1}$ for PFHxS, and 1000 ng $L^{-1}$ for PFBS are an exception to this trend of reliance on human data (NHMRC, 2025). In developing these guidelines, it was concluded that all human epidemiology studies, and many of the animal toxicology studies used by other regulatory authorities, have limitations that preclude their consideration as the basis for guidelines.

In contrast to earlier health-based guidelines based on animal studies, the recent, more stringent health-based levels based on human studies, obviously including the USEPA MCLGs of zero for PFOA and PFOS, are not achievable with the analytical methods used for routine drinking water monitoring and/or available treatment removal technologies. In such cases, the enforceable regulatory standard must be set above the health-based guideline based on these practical considerations (Post, 2021). In addition, cost-benefit evaluations are considered by some authorities (e.g., USEPA). For example, the USEPA MCLs for PFOA and PFOS were set at the analytical practical quantitation level (the concentration at which they can be reliably measured by drinking water laboratories) of 4 ng $L^{-1}$; USEPA determined that this level can be achieved by available treatment removal technology and that the anticipated health benefits exceed the costs associated with meeting this standard (USEPA, 2024b).

While many drinking water guidelines address PFAS as individual contaminants, others apply to mixtures of specified PFAS. Approaches for risk assessment of PFAS mixtures are relevant to the development of drinking water guidelines and are reviewed in ITRC (2025). Several U.S. states (Maine, Massachusetts, Oregon, Rhode Island, Vermont) have established guidelines of 20 or 30 ng $L^{-1}$ for the total concentration of four to six long-chain PFAS (PFOA,

PFOS, PFNA, PFHxS, and, in some cases, PFHpA and/or PFDA) based on the assumptions that these PFAS are equally toxic and have additive toxicity (Maine legislature, 2021; MADEP, 2019; OHA, undated; RIDOH, 2023; VT DOH, 2018).

Other drinking water guidelines for PFAS mixtures also assume additive toxicity but consider differences in toxicological potency (dose at which toxicity occurs) among the included PFAS. For example, Minnesota's Health Risk Index, which groups cooccurring contaminants (including PFAS and others), is based on toxicity to the same target organ (MDH, 2020). Notably, USEPA (2024b) adopted an MCL for mixtures of four PFAS (PFHxS, PFNA, GenX, PFBS) based on the "general" Hazard Index approach that groups contaminants with reference doses based on differing target organs (USEPA, 2024g); individual MCLs were also established for PFHxS, PFNA, and GenX. It should be noted that USEPA (2025d) has announced its intent to rescind the individual and mixture-based MCLs for these four PFAS and reconsider the regulatory determinations for them.

Drinking water guidelines for the total concentration of a specified list of long- and short-chain PFAS have been established in other nations; these guidelines do not necessarily assume that the included PFAS are all equally toxic. These include the European Union guidelines of 100 ng $L^{-1}$ for the total concentration of 20 PFAAs with 4 to 13 carbon atoms and 500 ng $L^{-1}$ for total PFAS (see below) (EU, 2020), Denmark's guideline of 2 ng $L^{-1}$ for the total concentration of PFOA, PFOS, PFNA, and PFHxS and 100 ng $L^{-1}$ for the total concentration of 22 PFAS (Denmark Ministry of the Environment, 2023), Canada's guideline of 30 ng $L^{-1}$ for the total concentration 25 PFAS (Health Canada, 2024), and the United Kingdom recent guideline of 100 ng $L^{-1}$ for the total concentration of 48 PFAS (DWI, 2025).

## 1.5 Implications for water treatment and future regulatory directions

PFAS as drinking water contaminants are currently a focus of worldwide attention. This section briefly reviews some current issues relevant to PFAS in drinking water.

The World Health Organization (WHO) develops guidelines for drinking-water quality (GDWQ) as "guidance to countries and to others as to what constitutes safe drinking-water and safe water supply." The WHO GDWQs are based on scientific considerations and are intended to be protective of public health (WHO, 2023), and they "provide the scientific point of departure" for use in the development of nation-specific regulations and standards. WHO encourages countries to develop "locally relevant standards and regulations" and to "adapt their application of the GDWQ based on their own sociocultural, environmental and economic circumstances," including cost-benefit considerations.

WHO has undertaken an effort toward potentially developing GDWQs for PFAS (WHO, 2023). WHO's initial draft drinking water guidance values of 100 ng $L^{-1}$ for PFOA and PFOS, which were based on treatment removal considerations rather than health effects, received extensive public comment (WHO, 2025). WHO then decided to continue its review of PFAS, including consideration of more recent information (e.g., the International Agency for Research on Cancer's conclusions that PFOA is carcinogenic to humans and PFOS is possibly carcinogenic to humans (IARC, 2025)), consideration of other PFAS in addition to PFOS and

PFOA, and consideration of sources of exposure other than drinking water. WHO's PFAS review is being conducted in multiple phases and will likely continue for several years (WHO, 2025). The first phase is aimed at "identifying and prioritizing key ingested PFAS and key health effects of PFAS" and "development of risk assessment methodology for prioritized PFAS, including individual PFAS and as mixtures." Later phases will use these methodologies to evaluate selected PFAS and then develop "formal, normative health-based guidance values for key ingested PFAS."

Until the WHO completes its development of GDWQs for PFAS, its recommendations for addressing PFAS in drinking water are consistent with its general principles for addressing drinking water contaminants discussed above. Specifically, WHO recommends that efforts should be made to keep PFAS levels in drinking water as "low as reasonably practical," PFAS contamination should be addressed at the source and new sources of should be prevented, "non-essential uses of PFAS should be stopped," and risks from PFAS should be "balanced with other risks in the water supply including not having adequate supplies of drinking-water" (WHO, 2025).

The concept of addressing PFAS as an entire class (i.e., all chemicals defined as PFAS) is receiving increasing attention (ITRC, 2023b). The primary rationale for this approach is PFAS' common property of extreme environmental persistence, which "will result in increasing concentrations (in the environment) and increasing probabilities of the occurrence of known and unknown effects (on human and/or ecological health)" (Cousins et al., 2020), and it is not assumed that all PFAS are equally toxic. While most efforts related to PFAS as the entire class focus on products that contain intentionally added PFAS, some authorities have taken actions to address the entire class of PFAS in drinking water. For example, the European Union has adopted a drinking water guideline of 500 ng $L^{-1}$ for total PFAS (EU, 2020), and legislation in the state of California (CA Assembly, 2022) requires development of "a treatment-based regulation for the entire class of PFAS" in drinking water. Such efforts depend on the use of one or more analytical methods that can quantify or estimate total PFAS. The European Union has developed technical guidance for use of the total oxidizable precursor assay, extractable organic fluorine (EOF)–combustion ion chromatography, or after EOF, suspect and nontarget analysis using liquid chromatography–high-resolution mass spectrometry, along with an additional method for measurement of trifluoroacetic acid to estimate total PFAS (EU, 2024). Moreover, the state of California conducted a method comparison study to determine the "most appropriate broad-spectrum analytical method for characterizing the occurrence of 'total PFAS' in drinking water" (CA Water Board, 2024).

Developing nations can address the high costs and technical complexity of PFAS analysis by adopting emerging monitoring approaches that have been advanced in the United States and Europe. Treating PFAS as a chemical class through nontargeted techniques, such as total organic fluorine-based methods, offers a good screening tool. When complemented by new technologies, including sensor-based and field-deployable screening tools, these approaches can support rapid assessment, source identification, and prioritization of samples for further analysis, without requiring immediate access to high-resolution mass spectrometry (Huo et al., 2025). In parallel, the establishment of centralized regional hubs for PFAS analysis can significantly reduce overall costs while maintaining analytical rigor. Consolidating advanced instrumentation, specialized expertise, and quality assurance systems within shared centers of excellence allows multiple countries to benefit from economies of scale and harmonized methodologies.

Treatment of PFAS in drinking water is broadly categorized into nondestructive and destructive techniques (Tshangana et al., 2025). Adsorption, ion exchange, and membrane separation are widely used nondestructive treatment technologies for PFAS removal, and they constitute most full-scale implementations currently in operation. While effective at removing PFAS from drinking water, these technologies transfer the contaminants to secondary waste streams, such as spent adsorbers or concentrated brine, necessitating further treatments for these waste streams. A range of destructive technologies (complete mineralization of PFAS to carbon dioxide, fluoride ions, and water) has been under investigation (Hussain et al., 2025). These include electrochemical oxidation, supercritical water oxidation, and plasma-based treatments (Tshangana et al., 2025). Although many of these approaches have shown promising results at the laboratory and pilot scales, significant challenges remain. Key issues include high energy demands, limited scalability, varying efficacy across PFAS chemistries, and the need for robust control of transformation by-products. Moreover, the integration of these destructive methods into full treatment trains is still in development.

For the above reasons, prevention of PFAS contamination before it occurs is unquestionably preferable to the need for removal of PFAS from drinking water. While PFAS that have already entered the environment will persist indefinitely, restricting the manufacture and use of products containing PFAS and stopping the release of PFAS to drinking water sources can prevent further contamination of drinking water. Consistent with these goals, the WHO (WHO, 2023) recommends that "contamination of water sources should be minimized, including preventing new sources of contamination," and that "non-essential uses of PFAS should be stopped." Notably, the European Chemicals Agency has proposed to restrict the manufacture, sale, and use of all PFAS in the European Union, with exceptions for some uses for a limited time period while alternatives are developed (ECHA, 2020). Additionally, many U.S. states have proposed or adopted restrictions on the sale, use, and/or disposal of all or certain products that contain PFAS (ECOS, 2025).

A robust and enforceable pretreatment framework for wastewater discharges represents a critical component of the broader strategy to prevent PFAS contamination in drinking water sources. As highlighted earlier, WWTPs are largely ineffective at removing PFAS, and industrial and commercial dischargers remain significant contributors to PFAS loading in municipal systems (Lenka et al., 2021). To mitigate this upstream input, the implementation of source-specific pretreatment measures is essential. These may include targeted treatment technologies by dischargers such as activated carbon filtration, foam fractionation, or ion exchange, as well as pollutant load monitoring, discharge characterization, and revisions to industrial permitting (Kurwadkar et al., 2022). Such upstream controls will not only limit PFAS entry but will also substantially reduce the burden placed on downstream drinking water utilities, which would otherwise be responsible for costly and technically demanding removal processes at the point of use (Boone et al., 2019). In this context, pretreatment serves as both a preventative and cost-effective barrier, reinforcing the need for its integration into future PFAS management policies.

## Acknowledgments

This investigation does not reflect the views and policies of the current affiliations of the coauthor, GDF, and any mention of trade names or commercial products does not constitute the authors' endorsement or recommendation thereof.

## References

Ackerman Grunfeld, D., Gilbert, D., Hou, J., et al., 2024. Underestimated burden of per- and polyfluoroalkyl substances in global surface waters and groundwaters. Nature Geoscience 17, 340–346.

Adewuyi, A., Li, Q., 2024. Emergency of per- and polyfluoroalkyl substances in drinking water: Status, regulation, and mitigation strategies in developing countries. Eco-Environment & Health 3, 355–368.

Awad, E., Zhang, X.M., Bhavsar, S.P., Petro, S., Crozier, P.W., Reiner, E.J., et al., 2011. Long-term environmental fate of perfluorinated compounds after accidental release at Toronto Airport. Environmental Science & Technology 45, 8081–8089.

Banzhaf, S., Filipovic, M., Lewis, J., Sparrenbom, C.J., Barthel, R., 2016. A review of contamination of surface-, ground-, and drinking water in Sweden by perfluoroalkyl and polyfluoroalkyl substances (PFASs). Ambio 463, 335–346.

Boone, J.S., Vigo, C., Boone, T., Byrne, C., Ferrario, J., Benson, R., et al., 2019. Per- and polyfluoroalkyl substances in source and treated drinking waters of the United States. Science of the Total Environment 653, 359–369.

Buck, R., Franklin, J., Berger, U., Conder, J., Cousins, I., de Voogt, P., et al., 2011. Perfluoroalkyl and polyfluoroalkyl substances in the environment: terminology, classification, and origins. Integrated Environmental Assessment and Management 7, 513–541.

Butenhoff, J.L., Chang, S.C., Ehresman, D.J., York, R.G., 2009. Evaluation of potential reproductive and developmental toxicity of potassium perfluorohexanesulfonate in Sprague Dawley rats. Reproductive Toxicology 27 (3–4), 331–341.

CA Assembly, 2022. California Assembly Bill 178. Available at https://legiscan.com/CA/text/AB178/id/2600061.

CA Water Board, 2024. California State Water Resources Control Board. Broad-Spectrum Per- and Polyfluoroalkyl Substances (PFAS) Method Comparison Study Results and Selection of Broad-Spectrum Methods to Support Statewide Monitoring for the Class of PFAS. Available at https://www.waterboards.ca.gov/pfas/docs/broad-spectrum-pfas-method-comparison-study-summary-combined.pdf (accessed 24.04.24).

Calafat, A.M., Wong, L.Y., Kuklenyik, Z., Reidy, J.A., Needham, L.L., 2007. Polyfluoroalkyl chemicals in the U.S. population: data from the National Health and Nutrition Examination Survey (NHANES) 2003–2004 and comparisons with NHANES 1999–2000. Environmental Health Perspectives 115, 1596–1602.

CDC, 2024. Centers for Disease Control and Prevention. Biomonitoring data tables for environmental chemicals. Available at https://www.cdc.gov/exposurereport/data_tables.html.

California EPA, 2024. California Environmental Protection Agency. Perfluorooctanoic acid and perfluorooctane sulfonic acid in drinking water. Available at https://oehha.ca.gov/sites/default/files/media/downloads/air/public-health-goal-document/pfoapfosphgfinaldraft040524.pdf (accessed March 2024).

Cousins, I.T., DeWitt, J.C., Glüge, J., Goldenman, G., Herzke, D., Lohmann, R., et al., 2020. The high persistence of PFAS is sufficient for their management as a chemical class. Environmental Science. Processes & Impacts 22, 2307–2312. Available at https://doi.org/10.1039/d0em00355g.

Cousins, I.T., Johansson, J.H., Salter, M.E., Sha, B., Scheringer, M., 2022. Outside the safe operating space of a new planetary boundary for per- and polyfluoroalkyl substances (PFAS). Environmental Science & Technology 56 (16), 11172–11179. Available at https://doi.org/10.1021/acs.est.2c02765.

Crone, B.C., Speth, T.F., Wahman, D.G., Smith, S.J., Abulikemu, G., Kleiner, E.J., et al., 2019. Occurrence of per- and polyfluoroalkyl substances (PFAS) in source water and their treatment in drinking water. Critical Reviews in Environmental Science and Technology 49 (24), 2359–2396.

Denmark Ministry of the Environment, 2023. Decree on water quality and supervision of water supply facilities. Available at https://www.retsinformation.dk/eli/lta/2023/1023.

Drinking Water Inspectorate (DWI), 2025. PFAS and forever chemicals. Available at https://www.dwi.gov.uk/pfas-and-forever-chemicals/ (accessed 18.06.25).

European Chemicals Agency (ECHA), 2020. Consultation on a proposed restriction on the manufacture, placing on the market and use of per- and polyfluoroalkyl substances (PFAS). Available at https://echa.europa.eu/documents/10162/cad38c27-ede8-2268-00c6-939ea066743c.

Environmental Council of the States (ECOS), 2025. ECOS compendium of state PFAS Actions. Available at https://www.ecos.org/documents/ecos-compendium-of-state-pfas-actions/ (accessed 17.04.25).

European Environment Agency (EEA), 2024. PFAS pollution in European waters. Briefing No. 19/2024. Available at https://www.eea.europa.eu/en/analysis/publications/pfas-pollution-in-european-waters.

EFSA, 2020. European food safety authority panel on contaminants in the food chain. Scientific opinion on the risk to human health related to the presence of perfluoroalkyl substances in food. EFSA Journal 18 (9), 6223. Available at https://doi.org/10.2903/j.efsa.2020.6223.

Essumang, D.K., Eshun, A., Hogarh, J.N., Bentum, J.K., Adjei, J.K., Negishi, J., et al., 2017. Perfluoroalkyl acids (PFAAs) in the Pra and Kakum River basins and associated tap water in Ghana. Science of Total Environment 579, 729–735. Available at https://doi.org/10.1016/j.scitotenv.2016.11.035.

European Union (EU), 2020. Directive (EU) 2020/2184 of the European Parliament and of the Council of 16 December 2020 on the quality of water intended for human consumption (recast). Available at https://eur-lex.europa.eu/legal-content/EN/TXT/PDF/?uri=CELEX:32020L2184.

European Union (EU), 2024. Technical guidelines regarding methods of analysis for monitoring of per- and polyfluoroalkyl substances (PFAS) in water intended for human consumption (C/2024/4910). Official Journal of the European Union. Available at https://eur-lex.europa.eu/legal-content/EN/TXT/PDF/?uri=CELEX:52024XC04910 (accessed 08.07.24).

Exner, M., Färber, H., 2006. Perfluorinated surfactants in surface and drinking waters. Environmental Science & Pollution Research 135, 299–307.

Fagbayigbo, B., Opeolu, B., Fatoki, O., Olatunji, O., 2018. Validation and determination of nine PFCS in surface water and sediment samples using UPLC-QTOF-MS. Environmental Monitoring and Assessment 190, 346.

Fenton, S.E., Ducatman, A., Boobis, A., DeWitt, J.C., Lau, C., Ng, C., et al., 2021. Per- and polyfluoroalkyl substance toxicity and human health review: current state of knowledge and strategies for informing future research. Environmental toxicology and chemistry 40 (3), 606–630. Available at https://doi.org/10.1002/etc.4890.

Gebbink, W.A., van Asseldonk, L., van Leeuwen, S.P.J., 2017. Presence of emerging per- and polyfluoroalkyl substances (PFASs) in river and drinking water near a fluorochemical production plant in the Netherlands. Environmental Science & Technology 51, 11057–11065.

Glüge, J., Scheringer, M., Cousins, I.T., DeWitt, J.C., Goldenman, G., Herzke, D., et al., 2020. An overview of the uses of per- and polyfluoroalkyl substances (PFAS). Environmental Science: Processes & Impacts 22 (12), 2345–2373.

Goeden, H.M., Greene, C.W., Jacobus, J.A., 2019. A transgenerational toxicokinetic model and its use in derivation of Minnesota PFOA water guidance. Journal of Exposure Science & Environmental Epidemiology 29, 183–195.

Guo, R., Sim, W.-J., Lee, E.-S., Lee, J.-H., Oh, J.-E., 2010. Evaluation of the fate of perfluoroalkyl compounds in wastewater treatment plants. Water Research 44, 3476–3486.

Health Canada, 2024. Objective for Canadian drinking water quality: per- and polyfluoroalkyl substances. Health Canada. Available at https://www.canada.ca/content/dam/hc-sc/documents/services/publications/healthy-living/objective-drinking-water-quality-per-polyfluoroalkyl-substances/objective-for-canadian-drinking-water-quality-en-final.pdf. (accessed August 2024).

Herrick, R.L., Buckholz, J., Biro, F.M., Calafat, A.M., Ye, X., Xie, C., et al., 2017. Polyfluoroalkyl substance exposure in the Mid-Ohio River Valley, 1991–2012. Environmental Pollution 228, 50–60.

Hu, X.C., et al., 2016. Detection of poly- and perfluoroalkyl substances (PFASs) in U.S. drinking water linked to industrial sites, military fire training areas, and wastewater treatment plants. Environmental Science & Technology Letters 3, 344–350.

Huo, T., Sarkar, D., Andreescu, S., Du, H., 2025. Advances in nanosensor technologies for PFAS detection: a review. IEEE Sensors Reviews 2, 498–510. Available at https://doi.org/10.1109/SR.2025.3607886.

Hussain, H.N., Jilani, M.I., Imtiaz, F., Ahmed, T., Arshad, M.B., Mudassar, M., et al., 2025. Advances in the removal of polyfluoroalkyl substances (PFAS) from water using destructive and non-destructive methods. Green. Analytical Chemistry 12, 100123. Available at https://doi.org/10.1016/j.greeac.2025.100123.

IARC, 2025. International agency for research on cancer. Perfluorooctanoic acid (PFOA) and perfluorooctanesulfonic acid (PFOS). IARC Monographs on the Identification of Carcinogenic Hazards to Humans. 135, 1–754. Available at https://publications.iarc.who.int/636.

Interstate Technology and Regulatory Council (ITRC), 2023. PFAS technical and regulatory guidance document: Section 2.2 chemistry, terminology, and acronyms.

Interstate Technology and Regulatory Council (ITRC), 2023. PFAS technical and regulatory guidance document: Section 7.1 human health effects. Available at https://pfas-1.itrcweb.org/7-human-and-ecological-health-effects-of-select-pfas/#7_1.

Interstate Technology and Regulatory Council (ITRC), 2025. PFAS environmental media values table (Excel file, updated April 2025). *ITRC PFAS Fact Sheets*. Available at https://pfas-1.itrcweb.org/fact-sheets/ (accessed 19.06.25).

Jahura, F.T., Mazumder, N., Hossain, M.T., Kasebi, A., Girase, A., Ormond, R.B., 2024. Exploring the prospects and challenges of fluorine-free firefighting foams (F3) as alternatives to aqueous film-forming foams (AFFF): a review. ACS Omega 9, 37430–37444.

Jian, J.M., Guo, Y., Zeng, L., Liang-Ying, L., Lu, X., Wang, F., et al., 2017. Global distribution of perfluorochemicals (PFCs) in potential human exposure source – a review. Environment International 108, 51–62.

Kali, S., Osterlund, H., Viklander, M., Blecken, G., 2025. Stormwater discharges affect PFAS occurrence, concentrations, and spatial distribution in water and bottom sediment of urban streams. Water Research 271, 122973.

Karrman, A., Elgh-Dalgren, K., Lafossas, C., Moskeland, T., 2011. Environmental levels and distribution of structural isomers of perfluoroalkyl acids after aqueous fire-fighting foam (AFFF) contamination. Environmental Chemistry 8, 372–380.

Kato, K., Ye, X., Calafat, A.M., 2015. Chapter 3. PFASs in the general population. In: DeWitt, J., editor. Toxicological effects of perfluoroalkyl and polyfluoroalkyl substances. 10.1007/978-3-319-15518-0.

Kim, S.K., Kannan, K., 2007. Perfluorinated acids in air, rain, snow, surface runoff, and lakes: relative importance of pathways to contamination of urban lakes. Environmental Science & Technology 41, 8328–8334.

Król, B., Prochaska, K., Chrzanowski, L., 2012. Biodegradability of firefighting foams. Fire Technology 48, 173–181.

Kurwadkar, S., Dane, J., Kanel, S.R., Nadagouda, M.N., Cawdrey, R.W., Ambade, B., et al., 2022. Per- and polyfluoroalkyl substances in water and wastewater: a critical review of their global occurrence and distribution. Science of the Total Environment 809, 151003.

Lam, N.H., Cho, C.-R., Kannan, K., Cho, H.-S., 2017. A nationwide survey of perfluorinated alkyl substances in waters, sediment and biota collected from aquatic environment in Vietnam: Distributions and bioconcentration profiles. Journal of Hazardous Materials 323 (Part A), 116–127.

Lenka, S.P., Kah, M., Padhye, L.P., 2021. A review of the occurrence, transformation, and removal of poly- and perfluoroalkyl substances (PFAS) in wastewater treatment plants. Water Research 199, 117187. Available at https://doi.org/10.1016/j.watres.2021.117187.

Li, F., Duan, J., Tian, S., Ji, H., Zhu, Y., Wei, Z., et al., 2020. Short-chain per- and polyfluoroalkyl substances in aquatic systems: occurrence, impacts and treatment. Chemical Engineering Journal 380, 122506. Available at https://doi.org/10.1016/j.cej.2019.122506.

Lindstrom, A.B., Strynar, M.J., Libelo, E.L., 2011. Polyfluorinated compounds: past, present, and future. Environmental Science & Technology 45 (19), 7954–7961.

Lu, Y., Liang, Y., Zhou, Z., Wang, Y., Jiang, G., 2019. Possible fluorinated alternatives of PFOS and PFOA: ready to go? Environmental Science & Technology 53 (24), 14091–14092.

MADEP, 2019. Technical Support Document - Per- and Polyfluoroalkyl Substances (PFAS): An Updated Subgroup Approach to Groundwater and Drinking Water Values. Available at https://www.mass.gov/doc/per-and-polyfluoroalkyl-substances-pfas-an-updated-subgroup-approach-to-groundwater-and/download (accessed 26. 12.19).

Minnesota Department of Health (MDH), 2020. Evaluating concurrent exposures to multiple chemicals. Updated 3 October 2022. Available at https://www.health.state.mn.us/communities/environment/risk/guidance/gw/additivity.html (accessed 18.06.25).

Minnesota Department of Health (MDH), 2024a. Toxicological summary for perfluorooctanoate. Available at https://www.health.state.mn.us/communities/environment/risk/docs/guidance/gw/pfoa2024.pdf (accessed January 2024).

Minnesota Department of Health (MDH), 2024b. Toxicological summary for perfluorooctane sulfonate. Available at https://www.health.state.mn.us/communities/environment/risk/docs/guidance/gw/pfos.pdf (accessed February 2024).

Minnesota Department of Health (MDH), 2025. History of MDH activities – per- and polyfluoroalkyl substances (PFAS). Last updated 22 April 2025. Available at https://www.health.state.mn.us/communities/environment/hazardous/topics/history.html (accessed 19.06.25).

Maine legislature, 2021. S.P. 64 - L.D. 129. Resolve, to protect consumers of public drinking water by establishing maximum contaminant levels for certain substances and contaminants. Approved by Governor. Available at https://www.mainelegislature.org/legis/bills/getPDF.asp?paper=SP0064&item=3&snum=130 (accessed 21.06.21).

Moody, C.A., Field, J.A., 1999. Determination of perfluorocarboxylates in groundwater impacted by fire-fighting activity. Environmental Science & Technology 33 (16), 2800–2806.

Moody, C.A., Martin, J.W., Kwan, W.C., Muir, D.C. G., Mabury, S.C., 2002. Monitoring perfluorinated surfactants in biota and surface water samples following an accidental release of fire-fighting foam into Etobicoke Creek. Environmental Science & Technology 36, 545–551.

Murakami, M., Takada, H., 2008. Perfluorinated surfactants (PFSs) in size-fractionated street dust in Tokyo. Chemosphere 73, 1172–1177.

National Health and Medical Research Council (NHMRC), 2025. Australian drinking water guidelines: per- and polyfluoroalkyl substances (PFAS). Last updated 25 June 2025. Available at https://guidelines.nhmrc.gov.au/australian-drinking-water-guidelines/part-5/physical-chemical-characteristics/cas-numbers-1763-23-1-pfos-335-67-1-pfoa-355-46-4-pfhxs.

Organisation for Economic Co-operation and Development (OECD), n.d. Japan. Available at https://www.oecd.org/content/dam/oecd/en/topics/policy-sub-issues/risk-management-risk-reduction-and-sustainable-chemistry2/pfas-country-information/Japan.pdf (accessed 19.06.25).

Organisation for Economic Co-operation and Development (OECD), 2021. Reconciling terminology of the universe of per- and polyfluoroalkyl substances: recommendations and practical guidance. OECD Series on Risk Management of Chemicals. Available at https://www.oecd.org/content/dam/oecd/en/publications/reports/2021/07/reconciling-terminology-of-the-universe-of-per-and-polyfluoroalkyl-substances_a7fbcba8/e458e796-en.pdf.

Oregon Health Authority (OHA), n.d. Per- and polyfluoroalkyl substances (PFAS). Available at https://www.oregon.gov/oha/PH/HEALTHYENVIRONMENTS/DRINKINGWATER/OPERATIONS/Pages/PFAS.aspx (accessed 18.06.24).

Peter, L.G., Lee, L.S., 2025. Sources and pathways of PFAS occurrence in water sources: Relative contribution of land-applied biosolids in an agricultural dominated watershed. Environmental Science & Technology 59, 1344–1353.

Podder, A., Anwar Sadmani, A.H. M., Reinhart, D., Chang, N., Goel, R., 2021. Per and poly-fluoroalkyl substances (PFAS) as a contaminant of emerging concern in surface water: a transboundary review of their occurrences and toxicity effects. Journal of Hazardous Materials 419, 126361.

Post, G.B., 2021. Recent US state and federal drinking water guidelines for per- and polyfluoroalkyl substances. Environmental Toxicology and Chemistry 40 (3), 550–563. Available at https://doi.org/10.1002/etc.4863.

Post, G.B., 2022. Invited perspective: current breast milk PFAS levels in the United States and Canada indicate need for additional monitoring and actions to reduce maternal exposures. Environmental Health Perspectives 130, 21301. Available at https://doi.org/10.1289/EHP10730, https://ehp.niehs.nih.gov/doi/full/.

Post, G.B., 2025. Obtaining and evaluating information on the use, release, and health effects of two types of long-chain PFAS used as alternatives for legacy long-chain perfluoroalkyl acids: a case study. Environmental Health Perspectives 133 (5), 55002. Available at https://doi.org/10.1289/EHP15995.

Post, G.B., Louis, J.B., Cooper, K.R., Boros-Russo, B., Lippincott, R.L., 2009. Occurrence and potential significance of perfluorooctanoic acid (PFOA) detected in New Jersey public drinking water systems. Environmental Science & Technology 43, 4547–4554.

Rhode Island Department of Health (RIDOH), 2023. Advance notice of proposed rulemaking: Public drinking water (216-RICR-50-05-1). Available at https://risos-apa-production-public.s3.amazonaws.com/DOH/13079/PBN_13079_20231120151949736.pdf.

Rumsby, P.C., McLaughlin, C.L., Hall, T., 2009. Perfluorooctane sulphonate and perfluorooctanoic acid in drinking and environmental waters. Philosophical Transactions of the Royal Society A 3674119–3674413.

Sabba, F., Kassar, C., Zeng, T., Mallick, S.P., Downing, L., McNamara, P., 2025. PFAS in landfill leachate: Practical considerations for treatment and characterization. Journal of Hazardous Materials 481, 136685.

Sammut, G., Sinagra, E., Helmus, R., de Voogt, P., 2017. Perfluoroalkyl substances in the Maltese environment – (I) surface water and rain water. Science of the Total Environment 589, 182–190.

Schwanz, T.G., Llorca, M., Farré, M., Barceló, D., 2016. Perfluoroalkyl substances assessment in drinking waters from Brazil, France, and Spain. Science of the Total Environment 539, 143–152.

Sharma, B.M., Bharat, G.K., Tayal, S., Larssen, T., Bečanová, J., Karásková, P., et al., 2016. Perfluoroalkyl substances (PFAS) in river and ground/drinking water of the Ganges River basin: Emissions and implications for human exposure. Environmental Pollution 208, 704–713.

Sims, J.L., Stroski, K.M., Kim, S., et al., 2022. Global occurrence and probabilistic environmental health hazard assessment of per- and polyfluoroalkyl substances (PFASs) in groundwater and surface waters. The Science of the Total Environment 816, 151535. Available at https://doi.org/10.1016/j.scitotenv.2021.151535.

Ssebugere, P., Sillanpää, M., Matovu, H., Wang, Z., Schramm, K.-W., Omwoma, S., et al., 2020. Environmental levels and human body burdens of per- and poly-fluoroalkyl substances in Africa: A critical review. Science of Total Environment 258, 127340.

Stefano, P.H.P., Roisenberg, A., D'Anna Acayaba, R., Roque, A.P., Bandoria, D.R., Soares, A., et al., 2023. Occurrence and distribution of per-and polyfluoroalkyl substances (PFAS) in surface and groundwaters in an urbanized and agricultural area, Southern Brazil. Environmental Science and Pollution Research International 30, 6159–6169. Available at https://doi.org/10.1007/s11356-022-22603-x.

Sun, M., Arevalo, E., Strynar, M., Lindstrom, A., Richardson, M., Kearns, B., et al., 2016. Legacy and emerging perfluoroalkyl substances are important drinking water contaminants in the Cape Fear River Watershed of North Carolina. Environmental Science & Technology Letters 3 (12), 415–419. Available at https://doi.org/10.1021/acs.estlett.6b00398.

Tavasoli, E., Luek, J.L., Malley, J.P., Mouser, P.J., 2021. Distribution and fate of per- and polyfluoroalkyl substances (PFAS) in wastewater treatment facilities. Environmental Science: Processes & Impacts 23, 903–913.

Tshangana, C.S., Nhlengethwa, S.T., Glass, S., Denison, S., Kuvarega, A.T., Nkambule, T.T.I., et al., 2025. Technology status to treat PFAS-contaminated water and limiting factors for their effective full-scale application. npj Clean Water 8, 45. Available at https://doi.org/10.1038/s41545-025-00457-3.

U.S. Department of Defense (USDoD), n.d. Cleanup data for PFAS. Available at https://www.acq.osd.mil/eie/eer/ecc/pfas/data/cleanup-pfas.html (accessed June 2025).

U.S. Environmental Protection Agency (USEPA), 2009a. Long-chain perfluorinated chemicals (PFCs) action plan. Available at https://www.epa.gov/sites/default/files/2016-01/documents/pfcs_action_plan1230_09.pdf (accessed 30.12.09).

U.S. Environmental Protection Agency (USEPA), 2009b. Provisional health advisories for perfluorooctanoic acid (PFOA) and perfluorooctane sulfonate (PFOS). Office of Water. Available at http://www.epa.gov/waterscience/criteria/drinking/pha-PFOA_PFOS.pdf (accessed 08.01.09).

United States Environmental Protection Agency (USEPA), 2016a. Drinking water health advisory for perfluorooctanoic acid (PFOA). Office of Water. Available at https://www.epa.gov/sites/default/files/2016-05/documents/pfoa_health_advisory_final_508.pdf (accessed May 2016).

United States Environmental Protection Agency (USEPA), 2016b. Drinking Water Health Advisory for Perfluorooctane Sulfonate (PFOS). Office of Water. EPA 822-R-16-005. Available at https://www.epa.gov/sites/default/files/2016-05/documents/pfos_health_advisory_final_508.pdf (accessed May 2016).

U.S. Environmental Protection Agency (USEPA), 2017. The third unregulated contaminant monitoring rule (UCMR 3): data summary. Available at https://www.epa.gov/sites/production/files/2017-02/documents/ucmr3-data-summary-january-2017.pdf (accessed January 2017).

U.S. Environmental Protection Agency (USEPA), 2021a. Toxic Substances Control Act reporting and recordkeeping requirements for perfluoroalkyl and polyfluoroalkyl substances. Available at https://www.epa.gov/assessing-and-managing-chemicals-under-tsca/tsca-section-8a7-reporting-and-recordkeeping.

U.S. Environmental Protection Agency (USEPA), 2022a. Drinking water contaminant candidate list 5—final. *Federal Register* 87(218), 68060–68085. Washington (DC).

U.S. Environmental Protection Agency (USEPA), 2022b. Interim drinking water health advisory: perfluorooctanoic acid (PFOA) (CASRN 335-67-1). EPA/822/R-22/003. Available at https://www.epa.gov/system/files/documents/2022-06/interim-pfoa-2022.pdf (accessed June 2022).

U.S. Environmental Protection Agency (USEPA), 2022c. Interim drinking water health advisory: perfluorooctane sulfonic acid (PFOS) (CASRN 1763-23-1). EPA/822/R-22/004. Available at https://www.epa.gov/system/files/documents/2022-06/interim-pfos2022.pdf (accessed June 2022).

U.S. Environmental Protection Agency (USEPA), 2022d. Drinking water health advisory: hexafluoropropylene oxide (HFPO) dimer acid (CASRN 13252-13-6) and HFPO dimer acid ammonium salt (CASRN 62037-80-3), also known as "GenX chemicals". Available at https://www.epa.gov/system/files/documents/2022-06/drinking-water-genx-2022.pdf (accessed June 2022).

U.S. Environmental Protection Agency (USEPA), 2024a. Fact sheet: 2010/2015 PFOA Stewardship Program. Available at https://www.epa.gov/assessing-and-managing-chemicals-under-tsca/fact-sheet-20102015-pfoa-stewardship-program#launch.

U.S. Environmental Protection Agency (USEPA), 2024b. PFAS national primary drinking water regulation. Federal Register 89(82), 32532–32757. Available at https://www.govinfo.gov/content/pkg/FR-2024-04-26/pdf/2024-07773.pdf (accessed 26.04.24).

U.S. Environmental Protection Agency (USEPA), 2024c. National primary drinking water regulations. Last updated 12 December 2024. Available at https://www.epa.gov/ground-water-and-drinking-water/national-primary-drinking-water-regulations (accessed 19.06.25).

U.S. Environmental Protection Agency (USEPA), 2024e. Final human health toxicity assessment for perfluorooctane sulfonic acid (PFOS) and related salts. EPA Document No. 815R24007. Available at https://www.epa.gov/system/files/documents/2024-04/main_final-toxicity-assessment-for-pfos_2024-04-09-refs-formatted_508c.pdf (accessed April 2024).

U.S. Environmental Protection Agency (USEPA), 2024f. How EPA regulates drinking water contaminants. Last updated 21 October 2024. Available at https://www.epa.gov/sdwa/how-epa-regulates-drinking-water-contaminants (accessed 18.06.25).

U.S. Environmental Protection Agency (USEPA), 2025a. Draft sewage sludge risk assessment for perfluorooctanoic acid (PFOA) and perfluorooctane sulfonic acid (PFOS) (CASRN 1763-23-1). Available at https://www.epa.gov/system/files/documents/2025-01/draft-sewage-sludge-risk-assessment-pfoa-pfos.pdf (accessed January 2025).

U.S. Environmental Protection Agency (USEPA), 2024g. Maximum contaminant level goals (MCLGs) for three individual per- and polyfluoroalkyl substances (PFAS) and a mixture of four PFAS. Available at https://www.epa.gov/system/files/documents/2024-04/pfas-hi-mclg_final508.pdf (accessed April 2024).

U.S. Environmental Protection Agency (USEPA), 2025b. The fifth unregulated contaminant monitoring rule (UCMR 5) data summary. Available at https://www.epa.gov/system/files/documents/2023-08/ucmr5-data-summary_0.pdf (accessed April 2025).

U.S. Environmental Protection Agency (USEPA), 2025c. Drinking water health advisories (HAs). Last updated 8 April 2025. Available at https://www.epa.gov/sdwa/drinking-water-health-advisories-has (accessed 19.06.25).

U.S. Environmental Protection Agency (USEPA), 2025d. Per- and polyfluoroalkyl substances (PFAS): final PFAS national primary drinking water regulation. Last updated 21 May 2025. Available at https://www.epa.gov/sdwa/and-polyfluoroalkyl-substances-pfas (accessed 26.06.25).

Vermont Department of Health (VT DOH), 2018. Drinking water health advisory for five PFAS (per- and polyfluorinated alkyl substances). Available at https://www.healthvermont.gov/sites/default/files/documents/pdf/ENV_DW_PFAS_HealthAdvisory.pdf (accessed 10.07.18).

Wee, S.Y., Aris, A.Z., 2023. Revisiting the "forever chemicals", PFOA and PFOS exposure in drinking water. npj Clean Water 6, 57. Available at https://doi.org/10.1038/s41545-023-00274-6.

World Health Organization (WHO), 2023. Guidelines for drinking-water quality – frequently asked questions. 19 January 2023. Available at https://www.who.int/news-room/questions-and-answers/item/guidelines-for-drinking-water-quality—frequently-asked-questions (accessed 18.06.25).

World Health Organization (WHO), 2025. PFOS and PFOA in drinking-water: background document for development of WHO Guidelines for Drinking-water Quality. 15 September 2025. Available at https://www.who.int/teams/environment-climate-change-and-health/water-sanitation-and-health/chemical-hazards-in-drinking-water/per-and-polyfluoroalkyl-substances (accessed 07.12.25).

West Virginia Department of Environmental Protection (WVDEP), 2024. Final ammonium perfluorooctanoate (C8) assessment of toxicity team (CATT) report. Available at https://dep.wv.gov/WWE/watershed/wqmonitoring/Documents/C-8/C-8_FINAL_CATT_REPORT_8-02.pdf (accessed August 2002).

Wang, T., Wang, P., Meng, J., Liu, S., Lu, Y., Khim, J.S., et al., 2015. A review of sources, multimedia distribution and health risks of perfluoroalkyl acids (PFAAs) in China. Chemosphere 129, 87–99.
Wang, X., Halsall, C., Codling, G., Xie, Z., Xu, B., Zhao, Z., et al., 2014. Accumulation of perfluoroalkyl compounds in tibetan mountain snow: temporal patterns from 1980 to 2010. Environmental Science & Technology 48 (1), 173–181. Available at https://doi.org/10.1021/es4044775.
Wang, Y., Kim, J., Huang, C., Hawkins, G., Li, K., Chen, Y., et al., 2022. Occurrence of per- and polyfluoroalkyl substances in water: a review. Environmental Science: Water Research & Technology 8, 1136–1151.
Yu, Y., Che, S., Ren, C., Jin, B., Tian, Z., Liu, J., et al., 2012. Microbial defluorination of unsaturated per- and polyfluorinated carboxylic acids under anaerobic and aerobic conditions: a structure specificity study. Environmental Science & Technology 56, 4894–4904.
Zhang, Y., Zhou, Y., Dong, R., Song, N., Hong, M., Li, J., et al., 2024. Emerging and legacy per- and polyfluoroalkyl substances (PFAS) in fluorochemical wastewater along full-scale treatment processes: source, fate, and ecological risk. Journal of Hazardous Materials 465.

CHAPTER

# 2

# Per- and polyfluoroalkyl substances in soil and plants

*Adewale Adewuyi*

**Department of Chemical Sciences, Faculty of Natural Sciences, Redeemer's University, Ede, Osun State, Nigeria**

## 2.1 Introduction

Per- and polyfluoroalkyl substances (PFAS) are used in several industrial and consumer products owing to their fire resistance, waterproofing, heat resistance, and nonstick properties. Due to its several applications in product formulation, it has found its way into the environment and has been detected in soil and plants (Evich et al., 2022). The continuous use of PFAS coupled with its persistent nature in the environment has aided in its widespread spread in the environment and, as a result, getting into the food chain. The widespread use of PFAS is a serious concern that requires urgent attention to develop mitigation and monitoring strategies. The persistence of PFAS in the environment is linked to its strong chemical and thermal stability due to its carbon–fluorine (C–F) bond (Evich et al., 2022; Lyu et al., 2022). Furthermore, its amphiphilic property gives it an edge over other chemical molecules in industrial applications. PFAS may be classified into polymer and nonpolymer groups, as shown in Fig. 2.1 (Buck et al., 2011).

Despite its novel applications, PFAS can bioaccumulate to pose a threat to humans and ecological health, which has resulted in listing perfluorooctanoic acid (PFOA) and perfluorooctane sulfonic acid (PFOS), their associated molecules, and salts on the Stockholm Convention's list of persistent organic pollutants (Cai et al., 2022). Studies have revealed the detection of PFAS in soil from different countries, as presented in Table 2.1. Due to the potential to pose risks, it is necessary to understand the ecological risk assessment with in-depth knowledge of the fate of PFAS in soil and plants, which are the primary exposure routes for humans. Therefore this study focuses on understanding the presence of PFAS in soil and plants.

The behavior and fate of PFAS in the environment differ from that of most organic pollutants. There are many PFAS with different properties, making it problematic to deal with. The fully fluorinated forms persist in the environment and can be metabolized in

*Per- and Polyfluoroalkyl Substances (PFAS) as Environmental Contaminants*
DOI: https://doi.org/10.1016/B978-0-443-43856-1.00007-1

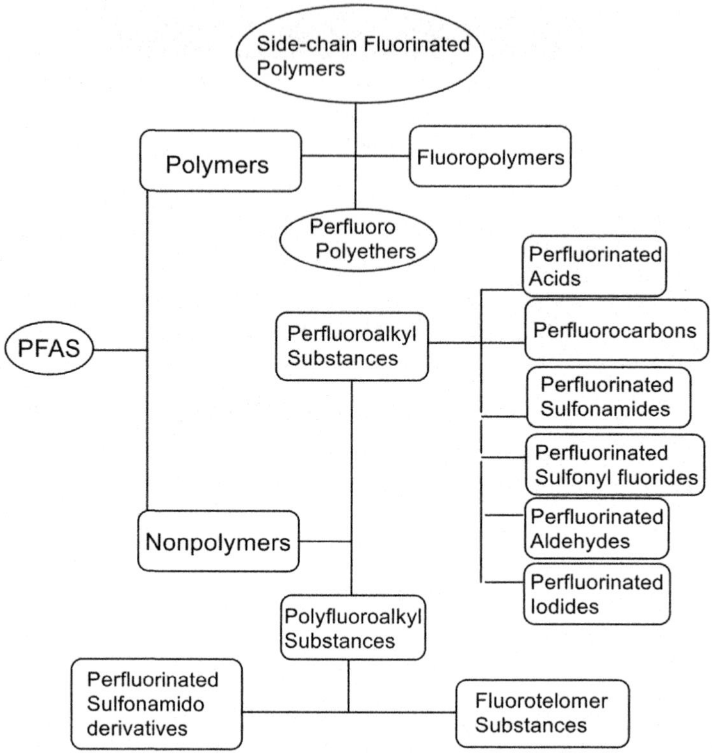


**FIGURE 2.1** Classification of PFAS.

humans, animals, and plants. They are referred to as "forever chemicals." Initially, PFAS was considered safe; however, after some time, researchers discovered that it was not as secure as it seemed, as it is persistent and difficult to degrade under environmental conditions. It can bioaccumulate in the food chain and has been found in the top predators, which include human blood (Panieri et al., 2022; Habib et al., 2024). Even when its concentration is low in humans during exposure, it can bioaccumulate with time to cause serious problems (Lewis et al., 2022; Khan et al., 2023). PFAS gets into humans mainly through food consumption and inhalation of dust and air. The presence of PFAS in the human body is dangerous and undesired. It was revealed that the presence of PFAS in human blood at elevated concentrations can reduce the vaccine's efficacy in children (Grandjean et al., 2012). A study in Germany revealed PFAS in blood samples, detecting the presence of PFOS and PFOA (Göckener et al., 2020). Due to the danger that PFAS may cause in humans and animals, many countries have set thresholds or guidelines for drinking water to help safeguard against the risks that this may bring (Post, 2021; Biegel-Engler and Frauenstein, 2024). Furthermore, for the most studied PFAS group—PFOA and PFOS, the European Food Safety Agency has developed a tolerable weekly intake rate, which was updated in 2018 (Authority, 2008; Chain et al., 2018; Chain et al., 2020) to control and manage the contamination and public effect that this may cause. This has been the driving force for

**TABLE 2.1** Record of per- and polyfluoroalkyl substances (PFAS) detection in soil and plants from different countries.

| Potential sources | Media | Potential source | PFAS | Conc (ng $g^{-1}$) | References |
|---|---|---|---|---|---|
| China | Soil | POSF manufacturing facilities | $\Sigma_{18}$PFAS | 2,6300 | Jia et al. (2023) |
| Australia | Soil | Biosolids | $\Sigma_{38}$PFAS | 10.4–104 | Sivaram et al. (2023) |
| South Korea | Soil | Industrial complexes | $\Sigma_{15}$PFAS | 0.346–11.7 | Sim et al. (2021) |
| USA | Soil | Irrigation water | $\Sigma_{7}$PFAS | 0.71–29.64 | Scher et al. (2018) |
| Norway | Soil | AFFF | $\Sigma_{12}$PFAS | 0.3–6500 | Høisæter et al. (2019) |
| Finland | Soil | AFFF | $\Sigma_{23}$PFAS | 2800 | Reinikainen et al. (2022) |
| China | Soil | AFFF | $\Sigma_{11}$PFAS | 3.4–531.7 | Cao et al. (2022) |
| Australia | Soil | AFFF | PFOS | 0.2–1400 | Tefera et al. (2022) |
| China | Soil | Fluoropolymer production facility | $\Sigma_{18}$PFAS | 4.40–189 | Xie et al. (2021) |
| USA | Soil | Waste incineration | Σ8PFAS | 0.05–8.3 | Martin et al. (2023) |
| Ireland | Soil | Landfill | $\Sigma_{5}$PFAS | 0.15–6 | Harrad et al. (2020) |
| South Korea | Soil | Landfill | $\Sigma_{15}$PFAS | 0.5–10.4 | Sim et al. (2021) |
| Netherlands | Soil | Atmospheric deposition | PFOS and PFOA | 3.3 | Gerardu et al. (2023) |
| France | Soil | Atmospheric deposition | $\Sigma_{11}$PFAS | 42–347 | Dauchy (2023) |

*Conc*, concentration, *AFFF*, aqueous film-forming foam.

developing European Union (EU) drinking water directives for PFAS, which has come up with a limit of 0.5 $\mu g\, L^{-1}$ (Coperchini et al. 2024) for total PFAS (Biegel-Engler and Frauenstein, 2024). This is a pointer to the need to monitor PFAS in the environment to control and manage the threat it may pose to the health of plants, humans, and animals.

As PFAS enters the environment, it remains resident in the soil for a long time. When in soil, it is picked up by plants for nutrition and may get into the ecological food chain. It is essential to understand PFAS in soil and plants. Therefore this chapter focuses on the sources of PFAS in soil and plants and the mitigation strategies to manage and remove PFAS from soil, one of the routes by which it gets into plants.

## 2.2 Sources of per- and polyfluoroalkyl substances in soil and plants

Poor and unregulated disposal of PFAS-containing waste is one of the significant sources of PFAS in soil and plants. This situation is common in poor developing countries with poorly managed solid waste disposal. Improper disposal of solid domestic waste continues to release PFAS into the environment. Most homes and government waste handling agencies in developing countries practice landfills for solid waste disposal. This practice allows the release of PFAS from PFAS-containing waste when such wastes are exposed to certain environmental conditions, like temperature, pressure, etc. PFAS can leach from such waste, getting into the soil and leading to contamination. If plants are grown on such contaminated soil, PFAS is picked up and may be translocated to different parts of the plant. Apart from this, the PFAS can travel through the aquifer, where it may become deposited in agricultural land, causing plant contamination.

Wastewater treatment plant (WWTP) is also known as a source of PFAS (Nguyen et al., 2022; Gobelius et al., 2023). Many known WWTPs cannot completely remove PFAS during wastewater purification (Coggan et al., 2019; Lenka et al., 2021; Gobelius et al., 2023). The sewage sludge generated in the WWTP contains PFAS, which contaminates the soil when disposed of in the environment. The sewage sludge, usually called biosolids, is used in agricultural practice to improve soil properties (Kumar et al., 2017; Boudjabi and Chenchouni, 2021). Unfortunately, the biosolids from WWTPs may contain PFAS (Pepper et al., 2021), which contaminates the soil and any plant grown on such soil. A recent study investigating 10 farms in the United States confirmed the mobility of PFAS and change in soil physicochemical properties due to contamination of soil by PFAS-containing biosolid, revealing a vertical migration of PFAS in surface and subsurface layers (Oviedo-Vargas et al., 2025). Other studies have also shown related results of PFAS from biosolids at multiple soil depths, indicating the capacity of PFAS to move through the soil or aquifer (Pepper et al., 2021; Johnson, 2022; Alvarez-Ruiz et al., 2024). This suggests the need for continuous monitoring of WWTPs' sewage sludge to control the release of PFAS into the environment.

Aqueous film-forming foam (AFFF) has been identified as a source of PFAS in the environment (McGarr et al., 2023; Yan et al., 2024). AFFF is a fire suppressant used in firefighting to extinguish flames. It is preferred in AFFF because of its resilience to pressure, heat, and oxidation. PFAS remains one of the direct or indirect components of AFFF. During firefighting activities, PFAS may be released into the environment. Studies have shown that most fire training sites are usually contaminated with AFFF, which contains PFAS (Place and Field, 2012). PFOS and PFOA are long-chain PFAS, which in some AFFF products are not directly used but are found as breakdown products emanating from precursor molecules. Other unintended PFAS molecules are obtainable from such breakdown. The class B AFFF, which is divided into PFAS-containing and fluorine-free (F3) foams, is designed to combat flammable fuel fires. When the PFAS-containing groups are used, the environment may become at risk of being contaminated with PFAS. Firefighting activities with the PFAS-containing groups of the class B AFFF introduce PFAS to the environment and eventually get into the soil. Over time, AFFF has evolved into different formulations that rely almost entirely on PFAS for adequate foam performance. A study on the Norwegian firefighting training facility revealed that the soil was contaminated with PFAS, with PFOS being the most contaminated group, with concentrations in the range <0.3 to 6500 $\mu g\ kg^{-1}$ (Høisæter et al., 2019). Although the use of PFOS in AFFF is banned in the United Kingdom, Canada, and the European Union (Seow, 2013; Sontake and Wagh, 2014), the consequences of its historical contamination remain a concern. Most firefighting training sites are contaminated with PFAS, which is a concern because of the threat this may pose to environmental health.

Atmospheric deposition of PFAS and PFAS by-products from solid waste incineration or burning is identified as a source of PFAS in the environment. Since PFAS is thermally stable, it may escape into the atmosphere during solid waste incineration and later become deposited on soil and plant surfaces. PFAS may escape intact without being broken down by fire or may break down to their precursor molecules, which are also toxic to humans, plants, and animals in the environment. When deposited on soil, PFAS tend to seep into the underground water system and may get into the surface water system as well (Wang et al., 2023b). When this seeps into the underground water system occurs, PFAS undergoes adsorption–desorption on soil, which is influenced by the physicochemical properties of PFAS and the soil's composition. The influence of some of these properties on the adsorption–desorption of PFAS in soil is presented in Table 2.2.

**TABLE 2.2** Factors that influence the sorption of per- and polyfluoroalkyl substances (PFAS) on soil.

| Property | Selected property | Effect on adsorption of PFAS | References |
|---|---|---|---|
| **Soil** | | | |
| | Organic carbon | High adsorption of PFAS at high organic carbon content | Ahrens et al. (2011), Chen et al. (2012), Milinovic et al. (2015), Luft et al. (2022), Dauchy (2023) |
| | Cation exchange capacity | High adsorption of PFAS with an increase in cation exchange capacity | Barzen-Hanson et al. (2017), Xiao et al. (2019), Mejia-Avendaño et al. (2020), Nguyen et al. (2020) |
| | Anion exchange capacity | High adsorption of PFAS with an increase in anion exchange capacity | Barzen-Hanson et al. (2017), Li et al. (2019a) |
| | Humic acid | High adsorption of PFAS with low humic acid in high soil content. However, low adsorption of PFAS with high humic acid in low soil content | Higgins and Luthy (2006), Zhang et al. (2014) |
| | Minerals | High adsorption of PFAS with cationic mineral-soil content | Hellsing et al. (2016) |
| **Water quality** | | | |
| | pH | Low adsorption capacity at high pH value | Kwadijk et al. (2010), Martz et al. (2019), Oliver et al. (2019) |
| | Cation | High adsorption capacity in the presence of lower mono and divalent cations | Kwadijk et al. (2010), You et al. (2010), Chen et al. (2013), Cai et al. (2022) |
| **PFAS** | | | |
| | Functional groups | Perfluoroalkane sulfonic acid is better adsorbed than Perfluoroalkyl carboxylic acid | Gellrich et al. (2012), Zhao et al. (2014), Loganathan and Wilson (2022), Luft et al. (2022), Wang et al. (2022a,c), Wang et al. (2023a) |

The atmospheric deposition of PFAS is mainly through dry and wet deposition processes. In the dry deposition, the ionic PFAS are attached to particulate matter on settling and may be adsorbed by soil and taken up by plants (Lin et al., 2020a; Lin et al., 2020b; Shimizu et al., 2021). The wet precipitation involves precipitation of PFAS in rain and snow, which is the major route of PFAS movement from the atmosphere to the soil and later picked up by plants (Chen et al., 2019; Gewurtz et al., 2019). A study revealed that the annual flux deposition of PFAS by the wet deposition process is 20 times greater than that of the dry deposition process (Shimizu et al., 2021). However, deposition by wet deposition is affected by local climate and the physicochemical properties of the PFAS being deposited. Perfluoroalkane sulfonic acid (PFSA) and perfluoroalkyl carboxylic acid (PFCA) are the most studied PFAS groups in this respect. Interestingly, PFSA is less soluble in water than PFCA due to the modest hydrophobicity of its sulfate group (Zhao et al., 2014). However, the PFSA binds readily with the surface of particulate matter (Rayne and Forest, 2009; Lin et al., 2020a), and when this happens, it migrates to the soil where plants may pick it up. The long-chain PFAS are preferably deposited, whereas the short-chain's precursors degrade during atmospheric transportation (Wang et al., 2023b).

The toxicity of PFAS is a serious issue globally. The toxicity of PFOA and PFOS has been studied and has been shown to induce diseases in humans relating to neurotoxicity, renal toxicity, genotoxicity, endocrine disruption toxicity and immunotoxicity (Li et al., 2017; Zeng et al., 2019; Li et al., 2022). The tendency of PFAS to cause havoc suggests the need to understand its sorption in plants and soil, which will help develop solutions for its mitigation and management.

## 2.3 Sorption of per- and polyfluoroalkyl substances in soil and plant

### 2.3.1 Sorption in soil

The sorption of PFAS in soil and its translocation in plants is determined by its functional group and chain length (Campos-Pereira et al., 2022). Its distribution between the soil/liquid and solid phase via adsorption and desorption goes a long way in understanding its environmental fate. Many studies have tried to understand the sorption and desorption of PFAS on soil and sediment (Knight et al., 2019; Nguyen et al., 2020; Oliver et al., 2020). Besides, the organic carbon content, cationic and anionic exchange capacity, cationic valence, soil texture, and soil ionic strength play a vital role in determining the sorption of PFAS on soils (Li et al., 2018; Liu et al., 2020; Nguyen et al., 2020; Campos-Pereira et al., 2022; Uwayezu et al., 2022; Wang et al., 2023b). A study revealed that long-chain perfluoroalkyl acid (PFAA) adsorbs firmly to the hydroxyl group on the surface of kaolinite clay, forming large aggregate clusters, which may be attributed to the interaction between the kaolinite hydroxyl group and the PFAA functional group (Loganathan and Wilson, 2022). However, although the short-chain also interacts with the hydroxyl group on the surface of soil kaolinite, they do not form clusters. PFSA is tightly held to the surface of soil or kaolinite than PFCA because the sulfate group ($SO_3^-$) in PFSA forms a stronger bond with the surface than the carboxyl group ($COO^-$) of the PFCA (Loganathan and Wilson, 2022).

Similarly, pyrophyllite, hematite, and montmorillonite exhibited strong adsorption of PFAA forming aggregates in the case of less ionic and large PFAA molecules (Zhao et al., 2014; Luft et al., 2022). This tendency shows the ability of soil to hold and transport PFAS within its structure.

The adsorption capacity of soil towards PFAS is determined by the PFAS solution pH, soil organic carbon content, cation, and anion exchange capacity. A study has shown a statistical correlation between organic carbon content and perfluorooctane sulfonic acid (PFSA) for sediment samples collected from rivers (Ahrens et al., 2011), oceans (Chen et al., 2012), and lakes (You et al., 2010). This included soils obtained from farmland (Milinovic et al., 2015) and wetland (Chen et al., 2013; Milinovic et al., 2015). Organic carbon in soil enhances PFAS sorption, which may be linked to the strong hydrophobic interaction between the soil's organic carbon and the PFAS's perfluoroalkyl tails. However, the adsorption may be affected by the arrangement of the organic carbon and the surface accessibility of the soil (Mahinroosta and Senevirathna, 2020). A study has further shown that soil's organic carbon may boost soil's cationic exchange capacity because most organic carbon present in the soil exist in their charged forms such as phenolic and carboxylic form (Schwarzenbach et al., 2003; Higgins and Luthy, 2006; Wang et al., 2023b); therefore any factor that affects the ionization of soil's organic carbon will affect the sorption of PFAS on soil. It is important to note that the sorption of PFAS on soil molecules is not only affected by organic carbon content but also by other interactions within the soil-PFAS environment, such as electrostatic interaction and ionic strength (Becker et al., 2008; Pan and You, 2010).

The charges on the soil's surface may be attributed to the type of its organic carbon and mineral content, mainly at its surface. Generally, variable and permanent charges are the two known types of charges that exist at the soil's surface (Wen et al., 2020). Usually, the soil's surface is negatively charged. However, some minerals at their surface may carry a positive charge, which could change the soil's surface charge and become positively charged if the mineral has a substantial presence. This soil's surface charges drive the sorption of PFAS via hydrophobic and electrostatic interactions as previously described (Zhang et al., 2015; Li et al., 2018). The soil's mineral component, predominantly cations such as $Ca^{2+}$, can facilitate the sorption of PFAS via a charge-based mechanism involving a cation-bridge system (You et al., 2010). Therefore the type of mineral available at the soil's surface determines the magnitude and nature of interaction that results in the sorption of PFAS (Hellsing et al., 2016; Mejia-Avendaño et al., 2020). For example, cations like $Al^{3+}$, $Na^{+}$, or $Ca^{2+}$ at the surface of clay soils may facilitate the formation of a bridge via electrostatic interaction with anionic PFAS in its surrounding (Hellsing et al., 2016).

Studies have shown an increase in PFAS sorption with an increase in the soil's cation concentration. For example, previous studies have shown an increase in sorption of PFAS with increasing $Ca^{2+}$ on sediments from marine and freshwater systems (Higgins and Luthy, 2006; Chen et al., 2012). Furthermore, PFOS's sorption on lake sediment has increased with increasing $Ca^{2+}$ concentration from 5–50 mM at a pH of 8 (You et al., 2010). Besides $Ca^{2+}$, other divalent cations like $Mg^{2+}$ have shown similar results; for instance, when the concentration of $Mg^{2+}$ was increased from 0.5–50 mM in a study, the sorption of PFOS on ocean sediment was significantly increased (Chen et al., 2012). Increasing the concentration of divalent cations and ionic strength favors the sorption of long-chain PFAS on soil (Cai et al., 2022). The exhibited capacity to adsorb long-chain PFAS was attributed to hydrophobic interaction, which may be explained as the screening for negative charges on the PFAS functional group and the reorientation of molecules existing at the interface between the organic carbon surface and the soil solution, which results in charge neutralization at the soil surface. Increased sorption of PFAS was recorded on wastewater sludge and organic carbon-enriched soil in the presence of cations (Wang et al., 2015; Pereira et al., 2018). While divalent cations have shown consistent performance of promoting the sorption of PFAS on increased concentrations, results obtained from monovalent cations have been inconsistent (Chen et al., 2013; Hong et al., 2013; Cai et al., 2022; Wang et al., 2023b), which may require more investigation for a better understanding. Like cation-rich soils, sorbs anion PFAS, smectite mineral-soil system with negatively charged surface will sorb cationic PFAS via electrostatic interaction (Mejia-Avendaño et al., 2020). Additionally, the interlayer structure of clay soil may be involved in the sorption process; a study revealed that the sorption of PFOS by montmorillonite involved both its surface and interlayer system (Zhang et al., 2014).

PFAS has been reported in landfill soil due to leaching of PFAS from PFAS-enriched wastes disposed at such sites (Masoner et al., 2020; Yu et al., 2025b). Efforts are in place to control or minimize the migration of PFAS as leachate to the underground water system (Shankar and Muthukumar, 2017). Still, this may sometimes not work efficiently (Xiang et al., 2019). This may involve a layer of clay as a hydraulic barrier. Some studies have shown that pH may change the dynamics of the landfill composition, which, on the contrary, facilitates the migration of PFAS even into the underground water system. An example is a study that showed that increasing the pH increased the soil's hydraulic conductivity, facilitating the movement of organic molecules (Eltarabily et al., 2015). This is true for bentonite and kaolinite clays (Faraghati et al., 2019). The increased hydraulic conductivity of the soil led to increased pore size in an alkaline environment (high pH), which eventually allowed easy flow of organic molecules. Since, the pH at landfill site may be high, this situation may lead to soil structure deterioration favoring the retainment of cationic PFAS and less sorption of anionic PFAS and a few cationic groups because they become more ionized and less hydrophobic for interaction with soil's surface (Kah and Brown, 2006; Wang et al., 2023b), furthermore, these PFAS migrate easily to underground water system which becomes a serious challenge. Since PFAS may have charges, they are known to also contribute to increasing hydraulic conductivity of landfill soil; their low dielectric properties increase soil's permeability (Eltarabily et al., 2015), which contaminates the underground water system; therefore landfill quality control management is a must.

Unfortunately, many landfill sites fail over time (Pecci, 2018). Such failures are common among poor developing countries, although several countries globally have challenges of PFAS in their landfill leachates (Stoiber et al., 2020). Studies reporting PFAS from landfill sites from China, the United States, and Australia (Yan et al., 2015; Gallen et al., 2017; Lang et al., 2017; Wang et al., 2020a; Helmer et al., 2022) revealed 20 PFAS in a range of $10^{-2}$–$10^{5}$ ng $L^{-1}$ from 58 studied sites, among which PFCA and PFSA are the most frequently detected. Similar results have been reported from other studied landfill sites (Fuertes et al., 2017; Wei et al., 2019). Developing a strategy to adequately handle or phase out PFAS in landfills is essential, as it will go a long way in reducing the presence of PFAS in the environment. A study approaches this by recommending the removal of PFAS from some domestic products. The study revealed that this approach will reduce landfill PFAS accumulation by 4% (Chen et al., 2020). The concentration and type of PFAS in landfill soil will depend on the type of landfill and the nature of waste disposed of at such a landfill site. For example, a study reported higher concentrations of PFAS at a construction site landfill than at other studied sites, such as landfills for ash obtained from the incineration of solid municipal wastes, which was attributed to the use of PFAS in water repellent in building construction materials and sealants (Solo-Gabriele et al., 2020). It is appropriate to conclude that landfills do not remove PFAS but delay their manifestation in the real-time environment. Therefore landfills are a long-term source or reservoir of PFAS in soil.

### 2.3.2 Per- and polyfluoroalkyl substances in plant

Mobility of PFAS in plants is through translocation, which takes place when plants pick up PFAS from the soil or when direct atmospheric deposition of PFAS takes place. Such uptake of PFAS by plants is one of the significant exposure routes of humans and animals to PFAS (Jiao et al., 2021). PFAS molecules adsorbed on soil are available to plants for uptake (Mei et al., 2021; Qi et al., 2022); the hydrophilic PFAS are readily available for uptake than the hydrophobic group (Gellrich et al., 2012; Krippner et al., 2014; Lesmeister et al., 2021). It has been reported that the long-chain PFAS are difficult to translocate through the plant structure because of their high hydrophobicity, which leaves them strongly bonded to the root (Schnoor et al., 1995). In this regard, several studies have revealed uneven distribution of PFAS in different plant parts, which may be associated with the hydrophilicity, hydrophobicity, and chain length of the PFAS groups (Gredelj et al., 2020b; Lin et al., 2020a,b; McDonough et al., 2021). PFAS concentration is always high in the root, which may be due to the high lipid content of the root (Zhao et al., 2018a,b; Lesmeister et al., 2021). Roots are the primary interface between the soil and plants, therefore the compositions (such as the lipids) of the root, including the root-derived dissolved organic matter, play an essential role in the binding of PFAS. The lipophilic end of PFAS exhibits enhanced binding with the lipid composition of the root, promoting the bioavailability of PFAS in the root; furthermore, the root-derived dissolved organic matter suppresses the mobility of PFAS through complexation or macromolecular blockage of membrane transport, thus making PFAS more available in the root compared to other parts of plants (Qi et al., 2022; Liu et al., 2023; Cao et al., 2025). Therefore a significant amount is retained in the root compared to the amount that travels through the xylem to other parts of the plant. In addition, a study described the translocation of PFAS through plants to occur through active transport, which includes single carriers and passive diffusion (Zhang et al., 2020b; Fawad et al., 2026). Several factors may affect the movement of PFAS from the soil to the plant and its translocation in the plant. High concentrations of PFAS have been reported in plants grown on PFAS-contaminated sites or sites close to PFAS-contaminated sites. For example, PFAS was detected in edible parts of crops grown on a fluorochemical industrial park's soil (Liu et al., 2019); similarly, PFAS was detected in wheat-grass grown on AFFF-contaminated soil (Bräunig et al., 2019). A study on the distribution of PFAS in aquatic plant-based systems revealed a concentration of 68.9 ng $g^{-1}$ for PFOS, 18.5 ng $g^{-1}$ for PFOA, and 13.4 ng $g^{-1}$ for PFHxS in the root of *Typha angustifolia* (Zhang et al., 2020a). Bioaccumulation of PFAS in plants from PFAS-contaminated sites has been studied; the study revealed ΣPFASs to range from ng $g^{-1}$ to μg $g^{-1}$ with a maximum concentration of 8085.18 ng $g^{-1}$ dw reported in soybean, while perfluoroheptanoic acid (PFHpA), PFOA, perfluoropentanoic acid (PFPeA), perfluorobutanoic acid (PFBA), and perfluorohexanoic acid (PFHxA) ranged from 211.80 to 3966.62 ng $g^{-1}$ dw (Liu et al., 2019).

PFAS in plants may lead to the generation of reactive oxygen species (ROS) within the plants' system; this has been reported in plants such as red chicory (*Cichorium intybus*), thale cress (*Arabidopsis thaliana*), wheat (*Triticum aestivum*), soft rush (*Juncus effusus*), algae (*Scenedesmus obliquus*) and lettuce (*Lactuca sativa*) when they were exposed to PFAS (Yang et al., 2015; Liu et al., 2018; Zhang et al., 2019; Gredelj et al., 2020a; Lin et al., 2020b; Li and Li, 2021). The presence of ROS in plants disrupts metabolic activities related to carbon and nitrogen metabolism, seed germination, gene expression, and photosynthesis (Choudhary et al., 2020), leading to hypersensitive expressions from damaged cell structure, metabolic disorder, biomass inhibition, and lipid peroxidation (González-Naranjo and Boltes, 2014; Zhou et al., 2016; Sharma et al., 2020; Li et al., 2021a,b). ROS in plants due to PFAS has caused serious concerns, the consequences of which have been shown to depend on plant exposure time to PFAS, concentration of PFAS exposed to, and plant species (Qian et al., 2019; Sharma et al., 2020). Unfortunately, no formidable solution is currently known to substantially detoxify plants from ROS generated from PFAS bioaccumulation.

Root uptake of PFAS from the nutrition matrix and leaf absorption from atmospheric deposition of PFAS are the main routes of plant contamination. The uptake involves physiological mechanisms, transmembrane transport, and symplastic and apoplastic pathways (Karamat et al., 2024; Yu et al., 2025a). The contamination from PFAS-polluted soil was reported to range from ng $g^{-1}$ to μg $g^{-1}$ (Liu et al., 2019; Niu et al., 2019; Li and Li, 2021). A study on PFOA and PFOS has revealed that these PFAS are migrated to the root cortex from the nutrient solution via apoplastic and symplastic routes (Wang et al., 2020b). Afterwards, they accumulate in root organelles, the intercellular space in the cortex, and the cell wall. The absorbed PFAS by leaves is metabolized into PFCAs and perfluoroalkyl sulfonic acids (PFSAs) (Ghisi et al., 2019). The bioaccumulation of some selected PFAS in plants is shown in Table 2.3. This includes edible and nonedible plants, which corroborates that PFAS already exists in the food chain and the urgent need to take action to prevent the public health problems this may cause. The bioaccumulation of PFAS in plants is influenced by factors including physiological characteristics of the plant, physicochemical properties of PFAS, and certain environmental factors like soil pH, organic matter, temperature, and salinity (Li et al., 2022). The bioaccumulation factor (BAF), which is the ratio of PFAS concentration in the plant to growth matrix, is an essential index used in characterizing the uptake and translocation efficiency of PFAS in plants (Li et al., 2022). For example, a study has shown that BAF increased with increasing PFAS chain length (Lesmeister et al., 2021). However, the BAF in the reproductive organs and the vegetative parts exhibits an inversely proportional relationship with an increase in PFAS chain length, which may be described as higher retainment of long-chain PFAS in biological macromolecules and cell membranes (Lan et al., 2018b; Zhang et al., 2019). This revealed that BAF is a vital tool in recommending plant materials for animal and human consumption. PFAS movement with the plant can be described by the translocation factor (TF). For example, the movement of PFAS from the shoot to the root may be determined from its bioaccumulation ratio in the shoot and root as $TF_{\frac{Shoot}{Root}}$ (Lechner and Knapp, 2011; Blaine et al., 2014; Lesmeister et al., 2021). The short-chain

**TABLE 2.3** Bioaccumulation of some selected per- and polyfluoroalkyl substances (PFAS) in plants.

| Plant | Source | PFBA (ng $g^{-1}$ dw) | PFPeA (ng $g^{-1}$ dw) | PFHxA (ng $g^{-1}$ dw) | PFOA (ng $g^{-1}$ dw) | PFOS (ng $g^{-1}$ dw) | References |
|---|---|---|---|---|---|---|---|
| Soyabean | FMP | 2378.31 | 992.62 | 211.80 | 3966.62 | 2.35 | Liu et al. (2019) |
| Balsam pear | FMP | 3.54 | 0.25 | - | 0.16 | - | Zhang et al. (2020b) |
| Cereals corn | FMP | 1448.59 | 387.68 | 116.06 | 2478.44 | 1.07 | Liu et al. (2019) |
| Rice | WWTP | - | - | - | 1.73 | - | Kim et al. (2019) |
| Cabbage | FMP | 17.85 | 1.79 | 0.56 | 1.94 | 0.43 | Zhang et al. (2020b) |
| Carrot | FMP | 2552.74 | 852.31 | 196.85 | 1468.08 | 1.31 | Liu et al. (2019) |
| Cucumber | FMP | 63 | 0.85 | 0.32 | 2.60 | 0.12 | Bao et al. (2020) |
| Tomato | FMP | 87 | 1.30 | 0.56 | 1.70 | 0.15 | Bao et al. (2020) |
| White melon | FMP | 8.50 | - | - | 1.20 | - | Li et al. (2019b) |
| Fruits grape | FMP | 9.80 | - | 1 | 1.60 | - | Li et al. (2019b) |
| Muskmelon | FMP | 2.90 | - | - | 1 | - | Li et al. (2019b) |
| Watermelon | FMP | 3.60 | - | - | 7.90 | - | Li et al. (2019b) |
| Grass | FMP | - | - | - | 520 | - | Zhu and Kannan (2019) |
| Tree | FMP | - | - | - | 700 | - | Zhu and Kannan (2019) |
| Zucchini | FMP | 1.54 | 0.21 | 0.11 | 0.64 | - | Zhang et al. (2020b) |

-, Not available, *FMP*, fluorochemical manufacturing park, *WWTP*, wastewater treatment plant.

PFAS are preferably accumulated in the leaf and fruit, while the long-chain PFAS are better accumulated in the root (Christou et al., 2019; Verley et al., 2025). This helps in understanding the accumulation behaviors of the PFAS and their preferential plant parts for bioaccumulation. It was reported that the uptake and translocation of PFAS is species-specific and it is influenced by the PFAS chain length (Fawad et al., 2026).

Atmospheric PFAS includes both ionic and neutral groups and are mainly released from landfill sites, wastewater treatment sites, and fluorochemical industries (Wang et al., 2021b; Li et al., 2022); previous studies have reported a maximum of 1297 pg $m^{-3}$ (Li et al., 2011) and 1440 pg $m^{-3}$ (Guo et al., 2018; Guo et al., 2020) in the atmosphere. The accumulation of PFAS in plants is known. For instance, soybeans obtained from a site 0.3 km from a fluorochemical park accumulated PFAS at a maximum of 8085.18 ng $g^{-1}$ dw (Liu et al., 2019). The bioaccumulation factor characterizes the uptake and translocation of PFAS in plants (Li et al., 2022). The bioaccumulation is influenced by environmental factors, physicochemical properties of PFAS, and the plant's physiological properties (Lesmeister et al., 2021; Li et al., 2022). The bioaccumulation factor decreases with an increase in PFAS chain length, which may be attributed to high retention of PFAS molecules by biomacromolecules and cell membranes as hydrophobicity increases due to chain length (Felizeter et al., 2012; Lan et al., 2018a; Zhang et al., 2019). A study has shown that the root may be the main plant organ for bioaccumulating PFAS, with a TF less than 1 (Costello and Lee, 2020).

High exposure of plants to PFAS may damage cell morphology. A study revealed a distortion in root cell shape and rupture of the cell wall of *Cyperus alternifolius* and *Eichhornia crassipes* when exposed to 10 mg $L^{-1}$ of PFOS. Another study observed a distortion in the leaf cell wall of water spinach (*Ipomoea aquatica* Forsk) when exposed to 1 mg $kg^{-1}$ F-53B, which further caused an enlarged intercellular space (Tang et al., 2020). Elsewhere, higher plants' cell membrane permeability was damaged on exposure to a 1.49 to 250 mg $L^{-1}$ concentration range of PFAS (Qu et al., 2010; Xu et al., 2017; Liu et al., 2018; Lin et al., 2020b). PFAS in plants has significantly negatively impacted photosynthetic content and gas exchange (Li et al., 2022). The thylakoid membrane in lettuce (*Lactuca sativa*) leaves was damaged when exposed to PFOA and PFOS at a concentration of 500–5000 ng $L^{-1}$ (Li et al., 2020a). ROS in plants may cause several oxidative damages. A study reported an increase in the contents of 8-hydroxy-deoxyguanosine (an indicator for DNA oxidative injury) in lettuce (*Lactuca sativa*) leaves when exposed to PAFS (Li et al., 2020b). Some other studies have also observed an increase in malondialdehyde contents in some plants exposed to PFAS, which is an indication of lipid peroxidation within the plant cell structure (Yang et al., 2015; Zhao et al., 2017; Liu et al., 2018; Lin et al., 2020b). It is essential to develop strategies to remove PFAS from soil before they get into plants to become translocated and bioaccumulated, which may threaten not just the plant but also other animals and humans that feed on them.

## 2.4 Remediation strategy for per- and polyfluoroalkyl substances in soil and plants

The risk associated with PFAS in soil and plants suggests the need to develop an efficient strategy to get rid of it before it causes harm. Different remediation strategies have

**TABLE 2.4** Comparison of the remediation strategies for removing per- and polyfluoroalkyl substances (PFAS) from soil.

| Remediation strategy | | Mechanism | Merit | Demerit | References |
|---|---|---|---|---|---|
| Type | Approach | | | | |
| Destruction | Thermal treatment | Combustion and pyrolysis | High degradation | High cost, disruption to soil | Duchesne et al. (2020), Alinezhad et al. (2022) |
| | Ultrasound | Oxidation by hydroxyl radicals | High efficiency | High cost, requires addition of chemical, depends on PFAS property | Lei et al. (2020), Kewalramani et al. (2022) |
| | Chemical oxidation | Chemical oxidation in water | High degradation for some PFAS, mild operational conditions | Soil and agent property-dependence, partial degradation, requires addition of chemical, requires consistent monitoring | Dombrowski et al. (2018) |
| | Ball milling | Chemical reaction in close contact | High degradation, low cost | Slow rate and small treatment volume | Baláž et al. (2013), Zhang et al. (2013,2016), Cagnetta et al. (2017) |
| | Enzymatic humification | Enzyme-catalyzed oxidative humification reactions | Low disruption, low cost and large treatment volume | Soil property-dependent, partial degradation, slow rate | Luo et al. (2018) |
| Immobilization | Activated carbon | Electrostatic interaction, Ion exchange, and hydrophobic interaction | High capacity, easy operation, low cost | Low efficiency in the presence of other compounds | Hale et al. (2017), Sorengard et al. (2019), Sörengård et al. (2021) |
| | Mineral | Electrostatic interaction and hydrophobic interaction | Sustainability, low cost | Low capacity | Das et al. (2013), Hale et al. (2017), ITRC (2020) |
| | Biochar | Electrostatic interaction, Ion exchange, and hydrophobic interaction | Sustainability, easy operation, low cost | Low capacity, biodegradation | Kupryianchyk et al. (2016), Silvani et al. (2019), Askeland et al. (2020), Sørmo et al. (2021) |

*(Continued)*

TABLE 2.4 (Continued)

| Remediation strategy | | Mechanism | Merit | Demerit | References |
|---|---|---|---|---|---|
| Separation | Soil washing | High PFAS solubility in water or organic solvents | High removal efficiency | High cost, requires secondary treatment | Høisæter et al. (2021), Quinnan et al. (2022) |
| | Soil liquefraction | Soil liquefaction and foam fraction | High removal efficiency | High cost, necessary secondary treatment | Bolan et al. (2021), Pang et al. (2022) |
| | Phytoremediation | Bioaccumulation | Low disruption, low cost, easy monitoring | Requires secondary treatment, root depth-dependent, depends on plant species, slow rate | Blaine et al. (2013), Wang et al. (2019), Gredelj et al. (2020b) |

been developed for removing organic molecules from soil; these strategies are also applicable for removing PFAS from soil (Kumar et al., 2021). These strategies can be categorized into destruction, immobilization, and separation as compared in Table 2.4. With the destruction-based strategy, the PFAS is lost and not recoverable at the end of the remediation process, making it an irreversible degradation approach, whereas for the containment-based strategy (immobilization and separation), the PFAS is recoverable and can be reused at the end of the process.

### 2.4.1 Immobilization

The carbon chain length contributes to the sorption of PFAS on surfaces, with the long-chain length being preferably adsorbed and may sometimes displace the shorter chain length (Higgins and Luthy, 2006; Maimaiti et al., 2018; Fabregat-Palau et al., 2021; Abraham et al., 2022). The remediation of PFAS-contaminated soil sites is challenging due to the high stability of the C–F bond of PFAS and heterogeneous sorption capacity (Bierbaum et al., 2023). However, immobilization of PFAS is a cost-effective way of mitigating the leaching of PFAS from soil and its uptake by plants (Hale et al., 2017; Bolan et al., 2021; Juhasz et al., 2022). The approach includes mixing immobilization material with PFAS-contaminated soil to increase sorption capacity or reduce the permeability for leaching (Sleep and Juhasz, 2021; Wang et al., 2023c). Immobilization is an in situ or ex situ process where an adsorbent is included in soil to stabilize PFAS to prevent PFAS movement and bioavailability in the underground water system (Sleep and Juhasz, 2021). Activated carbon, mineral, or biochar-based materials may be used as a sorbent for soil stabilization, promoting PFAS via electrostatic and hydrophobic interactions. A study has shown that PFAS-contaminated soil treatment by activated carbon increased the soil/water partitioning coefficient (Hale et al., 2017). Some studies have also used the combination of adsorbent and binder (bentonite, lime, etc.) to achieve this purpose (Sörengård et al., 2019; Barth et al., 2021).

The immobilization efficiency and the long-term stability are essential criteria that must be considered when selecting a material for the immobilization process. Several studies have investigated the performance of the immobilization method to understand the leachability of PFAS from contaminated soil (Hale et al., 2017; Sörengård et al., 2019; Bräunig et al., 2021; Kabiri and McLaughlin, 2021; Juhasz et al., 2022). A study reported a high performance of the immobilization method using activated carbon-based additives in combination with aluminum hydroxide and clay mineral (Kabiri and McLaughlin, 2021). The approach achieved a 99.9% performance and concluded that a long-term stability of PFAS in the soil system is possible using the new strategy. A two-phase study was conducted using a chemical stabilization agent combined with cement solidification to treat PFAS-contaminated soil (Barth et al., 2021). The authors used iron (Fe)-amended biochar, granular activated carbon, biochar, activated carbon-clay blend, and modified clay as adsorbents. The first phase involved PFAS adsorption using these adsorbents, while the second phase entails chemical stabilization treatment; the physical solidification was achieved using cement binder. Among the adsorbents used, the granular activated carbon gave the best results, substantially reducing PFAS leaching from the soil. The outstanding performance of activated carbon-based materials, like the granular activated carbon for this process, may be linked to their porous structure and high surface area. They pick up PFAS via hydrophobic interaction, ionic exchange, and electrostatic interaction (Wang et al., 2023b). The surface of the carbon-based adsorbents may be modified to improve the performance for the immobilization process. Biochar has shown high cation exchange capacity and large surface area for the immobilization process; however, the physicochemical properties exhibited by any biochar will depend on its source (Mahinroosta and Senevirathna, 2020; Sleep and Juhasz, 2021). Since biochar can biodegrade, using it for long-term purposes is vital. A study has revealed that biochar is less effective for PFAS when compared with activated carbon (Wang et al., 2023b).

Some authors have reported the use of noncarbon-based clay minerals, including hydrotalcite, zeolite, kaolinite, bentonite, and montmorillonite, as sorbents for PFAS due to their high cation exchange capacity, chemical stability, and high surface area (Sarkar et al., 2019). Clay minerals have low efficiency in removing bulk contaminants from a polluted system, suggesting the need for modification to improve their capacity for removing bulk pollutants. The mineral clays usually combine with other adsorbents and/or minerals to form a composite. Examples are RemBind®, which contains activated carbon, aluminum hydroxide, and kaolinite (Sorengard et al., 2019), Fluorolock®, containing sepiolite coated with the cationic polymer polydiallyldimethyl-ammonium chloride (Gunarathne et al., 2025a), MatCARE™, containing oleylamine-modified clay (Das et al., 2013), etc. To corroborate this, a study reported the modification of bentonite clay for treating PFAS-contaminated soil with efficiency reaching 99% (Wang et al., 2021a). A recent study reportedly used clay mineral sepiolite coated with the cationic polymer polydiallyldimethylammionium chloride for in situ remediation of PFAS-contaminated soil (Gunarathne et al., 2025a). The results obtained for the clay mineral sepiolite coated with the cationic polymer polydiallyldimethylammionium chloride were compared with colloidal activated carbon, which revealed a superiority of the activated carbon-based treatment, which was attributed to higher carbon content and surface area of the activated carbon. Studies have shown the use of composite and polymer materials as sorbents in the immobilization process (Aly et al., 2019; Anderson et al., 2019; Sleep and Juhasz, 2021). Polydiallyldimethylammonium chloride and polyquaternium polymers are materials that can effectively mitigate PFAS in soil (Anderson et al., 2019).

Reusability of a PFAS immobilization agent is vital. Efficient remobilization is vital in determining the economic viability of the process. Process cost is reduced when the immobilization agent can be reused several times. There are several factors that may affect the effectiveness of the remobilization process; they include soil chemistry (pH, organic matter content, and ionic strength), aging of the adsorbent, and environmental fate of the PFAS at the soil–water interface (Gunarathne et al., 2025b). It is essential to take these factors into consideration when designing an immobilization process for PFAS remediation. The more the mobilization process is repeated, the better and cost-effective it is for large-scale application.

### 2.4.2 Separation

Soil washing can remove PFAS from soil without creating a side product (Ross et al., 2018a,b). The approach removes PFAS from soil into a liquid phase and can be achieved by washing or liquefaction (Wang et al., 2023b). The PFAS are detached from the soil, and based on solubility and partition coefficient, they migrate to the liquid phase. While the washing excavates the PFAS from soil, the liquefaction combines foam fractionation and soil liquefaction. The method is suitable for treating a bulk contaminated system (Liu et al., 2021). The approach is appropriate for large-scale treatment of PFAS-contaminated soil (Senevirathna et al., 2021; Quinnan et al., 2022; Grimison et al., 2023). The efficiency of soil washing may be improved by adding additives like surfactants, oxalate, and alkaline solution (Tang et al., 2017; Liu et al., 2020). However, the overall performance depends on the soil's properties (Li et al., 2019a; Bolan et al., 2021; Quinnan et al., 2022). One of the challenges of this method is the requirement for a large volume of water and the need for secondary treatment of the PFAS-contaminated leachate (Uwayezu et al., 2024). A recent study combining washing, foam fractionation, and electrochemical oxidation achieved a 95% removal of PFOS from contaminated soil via washing before applying the other treatments (Uwayezu et al., 2024). Likewise, a study has treated samples collected from an Air Force base using this approach (Quinnan et al., 2022). The samples collected were contaminated with PFOA, PFOS, and perfluorobutane sulfonate. Elsewhere, a study discovered the partition coefficient for some PFAS to follow the order PFCAs < PFSAs < FTS < FOSA (Hubert et al., 2023). Depending on the more suitable need for the process, the liquefraction method can adopt both ex situ and in situ approaches (Pang et al., 2022). Liquefraction requires more energy than washing, making it more expensive to adopt.

### 2.4.3 Destruction

This method uses biological (bioremediation), advanced chemical, and physical (thermal and ball milling) methods to convert or breakdown PFAS into less toxic and safer molecules. Among these methods, the advanced oxidation process (AOP) of the advanced chemical method is the most reported. The AOP relies on the formation of highly reactive species such as hydroxyl radical ($^{\cdot}OH$) that breaks down the PFAS molecules. The approach effectively removes stubborn and difficult PFAS (Barisci and Suri, 2021). Some known AOPs for removing PFAS include chemical oxidation, electrochemical oxidation, and plasma oxidation/reduction (Barisci and Suri, 2021). Most of the reported chemical oxidation processes were photo-induced to generate the reactive species needed for the degradation of PFAS. The reactive species collide with the PFAS molecules to perturb the PFAS molecule, which knocks off an electron from the PFAS molecule to initiate the degradation. Cooxidants like hydrogen peroxide and persulphate are added to enhance the oxidation process (Uwayezu et al., 2023). On the other hand, the plasma oxidation generates reactive species such as radicals, ions, and electrons by applying an electric field to the gas to ionize it. The plasma process occurs at ambient pressure and temperature (Nzeribe et al., 2019; Khan et al., 2022). The thermal treatment involves PFAS vaporization at relatively low temperature (<600°C) and desorption from the soil. The desorbed PFAS molecules are degraded at elevated temperature (>1000°C) to safe, low molecular weight molecules. It is an ex situ method with little success at large-scale application due to some challenges such as requiring high energy, high cost to set up, death of soil microorganisms, and destruction of soil structure (Mahinroosta and Senevirathna, 2020; Wang et al., 2022a,b).

The chemical oxidation method may be applied in situ or ex situ for PFAS degradation in soil. The process is efficient in porous soil with low organic and high-water content at high temperatures (Dombrowski et al., 2018; Mahinroosta and Senevirathna, 2020). The performance of chemical oxidation depends on the effective contact between the PFAS in soil and the chemical oxidants used during the process. In chemical oxidation, ·OH is required for the oxidation process; it attacks PFAS to produce less toxic molecules. During this process, ·OH is generated from a catalyst under UV light irradiation, ozone ($O_3$), or hydrogen peroxide. Although the ·OH is short-lived but less likely to be contaminated or poisoned, making it a rugged oxidant for organic molecule degradation, apart from ·OH, persulphate with a redox potential of 2.01 V has been reported as a strong oxidant for degrading PFAS through electron transfer (Nzeribe et al., 2019; Wei et al., 2019). A study has used thermally activated persulphate to degrade PFAS (Yin et al., 2016), a process achieved by generating sulfate radicals during the thermal activation of persulphate. Some studies have reported partial degradation efficiency of PFAS using UV and Fenton's reagent because of the strong C–F bond in an advanced oxidation process (Schröder and Meesters, 2005; Liu et al., 2012). However, there have been several efforts to improve the performance of the chemical oxidation process (Dombrowski et al., 2018; Nzeribe et al., 2019).

The electrochemical oxidation uses electrical current in an electrolytic cell containing a cathode and an anode (Sivagami et al., 2023), unlike AOP, which generates ·OH from the catalyst under UV light irradiation, ozone ($O_3$), or hydrogen peroxide. Electrochemical oxidation generates ·OH from electrochemistry, which makes it environmentally benign. The processes may occur via direct and indirect electrochemical oxidation as shown in Fig. 2.2. In the direct electrochemical oxidation process, PFAS may become oxidized by direct electron transfer from the anode's surface to the adsorbed PFAS molecules without using any chemical agents, whereas in the indirect electrochemical oxidation, chemical reactive agents like ·OH, hydrogen peroxide, chlorine reactive species, persulphate ions, etc., are used. These chemical agents are produced at the anode's surface and get involved in the degradation of the PFAS in the bulk contaminated solution (Mojiri et al., 2023; Iovino et al.,

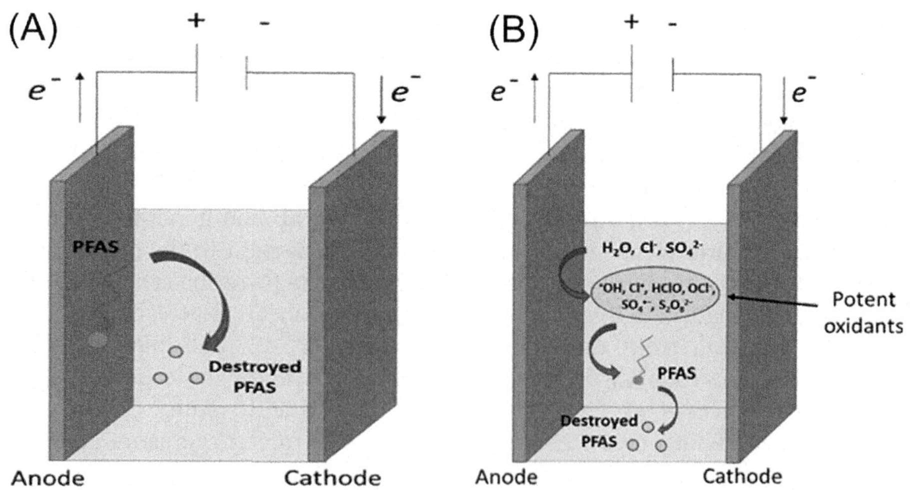


FIGURE 2.2 Schematic representation of the electrochemical oxidation mechanisms for PFAS: (a) direct oxidation and (b) Indirect oxidation (Sanzana et al., 2025).

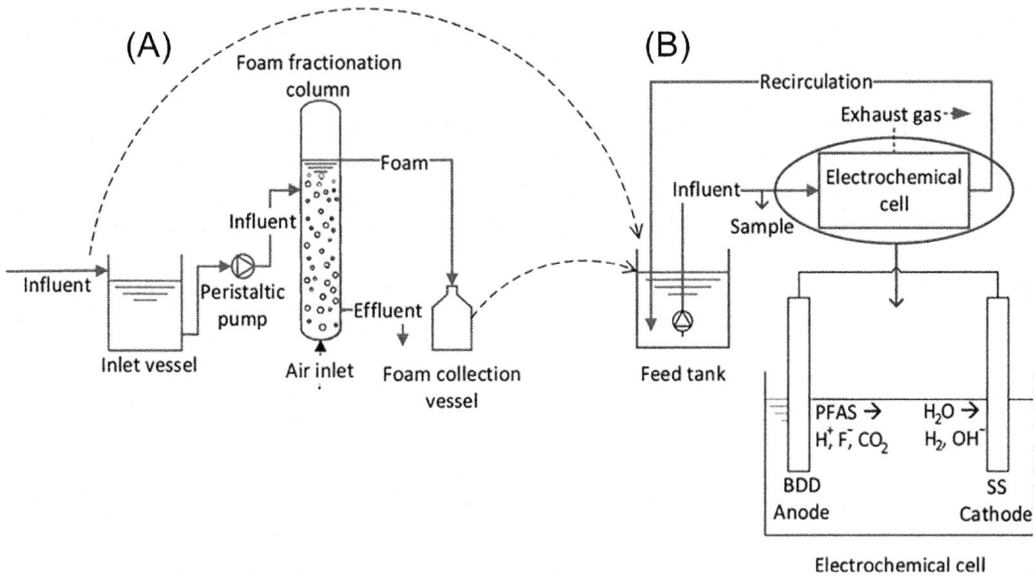


**FIGURE 2.3** Diagrammatic overview of (a) the foam fractionation FF and (b) the electrochemical oxidation process (Smith et al., 2023).

2024; Rekik et al., 2024). Studies have shown promising degradation of PFAS by electrochemical oxidation (Yang et al., 2022; Londhe and Venkatesan, 2024; Samuel et al., 2024). A study reported the degradation of PFAS from groundwater and landfill leachate by combining an upscaled experimental approach with characterization strategies as described in Fig. 2.3. The authors reported a poor removal of the short-chain PFAS group in the foam fractionation step, whereas the degradation of the long-chain group was better.

The choice of electrode is crucial in the design of an electrochemical oxidation process. The electrode is a conductive metal coated with a reactive catalyst material. Materials of different reactivity have been studied as possible electrodes for electrochemical oxidation; they include precious metals (platinum, silver, iridium, gold, etc.), titanium oxide, lead, tin, boron-doped diamond, and mixed metal oxides (Moreira et al., 2017; Barisci and Suri, 2021; Sukeesan et al., 2021; Veciana et al., 2022; Iovino et al., 2023; Sivagami et al., 2023; Sanzana et al., 2025). Although the reactivity of the electrode is essential during selection, the service life and replacement interval must be seriously considered. Poor service life and frequent replacement make the process expensive and unaffordable, even with high reactivity. Apart from this, the electrical performance should be considered as well. Before choosing an electrode, there should be an adequate balance of service life, replacement, and electrical costs. The mechanism of operation of the electrochemical oxidation process involves a complex chemical reaction pathway, which is influenced by electrolyte composition, temperature, electrode spacing, pH, and electrical potential. An illustration of the electrochemical oxidation of perfluorinated compounds is given in Fig. 2.4, which proposes four distinct degradation pathways (circles A–D). A study reported using boron-doped diamond

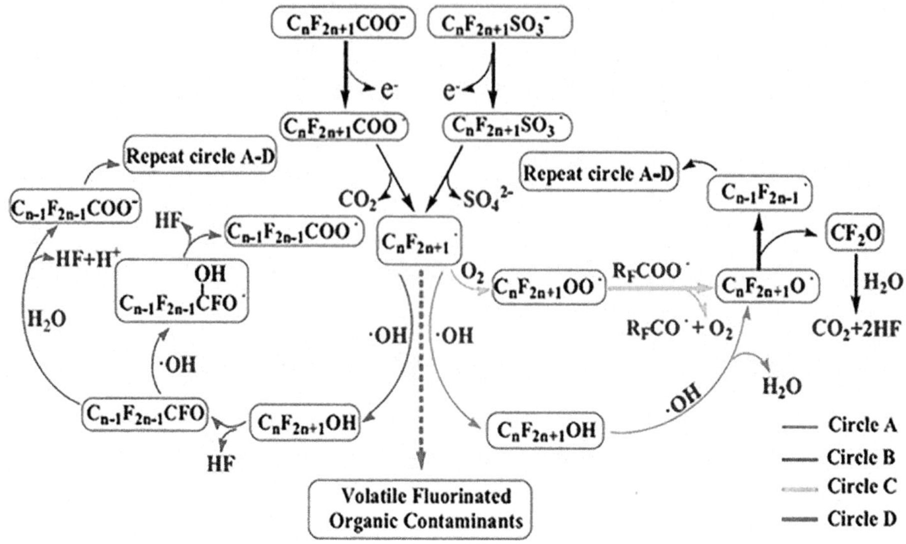


**FIGURE 2.4** Degradation pathways for perfluorinated compounds via electrochemical oxidation (Niu et al., 2016).

(BDD) and iron electrodes in a small-scale study (Öhberg, 2022). Four electrochemical cells containing PFAS-contaminated soils were used. The study further revealed the formation of short-chain PFAS in the BDD electrode, a phenomenon which did not occur in the iron electrode-containing cell. This further affirms the importance of selecting a highly efficient electrode for electrochemical oxidation.

The biodegradation process involves the use of microorganisms to break down the PFAS in soil through enzymatic activity. Many microorganisms have shown capacity for biodefluorination and biotransformation (Zhang et al., 2022a; Berhanu et al., 2023). Although the action can be achieved under aerobic and anaerobic conditions, the aerobic condition is most preferred (Zhang et al., 2022b). The activity is affected by factors such as other metabolites, the type of microbial species involved, and environmental conditions. Despite the success, the process faces the challenge of producing short-chain PFAS during biodegradation. Furthermore, it is time-consuming, and treatment conditions may be challenging. However, the biodegradation process has merits, including environmental friendliness and availability of the microorganisms. The biodegradation process primarily uses bacteria, which break down PFAS under specific environmental conditions. Interestingly, a study revealed that high concentrations of PFAS are not toxic to the microorganism during the biodegradation process (Grgas et al., 2023), which suggests the safety of the microorganism at high concentration treatment. The mechanism of action of the microorganism depends on the vulnerability of the PFAS's functional group to biotransformation and defluorination (Sanzana et al., 2025). Therefore some of these microorganisms apply their metabolic pathway to attack weak points of the PFAS molecule to biotransform the PFAS into less fluorinated products and short-chain molecules (Sanzana et al., 2025); examples of such microorganisms are *Rhodococcus, Acidimicrobium* sp. *A6* and *Pseudomonas*. The organisms release fluorine ions ($F^-$) when they break the C–F bond via reductive defluorination.

The biotransformation and reductive fluorination processes are catalyzed by enzymes, oxidoreductases, and desulfonases, respectively, in a slow process, producing intermediates that may still be recalcitrant if allowed in the environment (Berhanu et al., 2023). *Pseudomonas aeruginosa* and *Pseudomonas putida* have demonstrated the capacity to transform PFOA and PFOS (Chiriac et al., 2023). During this study, *P. aeruginosa* transformed 27.9% of PFOA and 47.3% of PFOS in 96 h, while *P. putida* transformed 19.0% of PFOA and 46.9% of PFOS. When considering *P. aeruginosa*, PFHxA was the principal biotransformation product for PFOS, while PFPeA, PFPxA, and PFHpA were the principal products for *P. putida*. *Synechocystis* sp. PCC6803 degraded 88% PFOA and 37% PFOS at a 0.15 and 0.36 mg $L^{-1}$ $day^{-1}$ removal rate, respectively (Marchetto et al., 2021). In a study by Harris et al. (Harris et al., 2022), soil samples from PFAS-contaminated sites were cultured for microbial isolation to identify strains with potential for PFAS degradation. Interestingly, *Delftia acidovorans* was isolated at the site with dehalogenase enzyme activities, suggesting the possibility of *D. acidovorans* to degrade PFOA or other PFAS via bioremediation strategies (Harris et al., 2022).

Several organisms have been identified with capacity to degrade PFAS; some of these organisms further include *P. aeruginosa*, *Desulfosporosinus*, *Sulfurospirillum*, *Aspergillus niger*, *Phanerochaete chrysosporium*, *P. chrysosporium*, *Pleurotus ostreatus*, *Trametopsis cervine*, *Pseudeurotium* sp. and *Geomyces* sp, *Phanerochaete chrysosporium*, *P. chrysosporium*, *T. versicolor* and *Scenedesmus quadricauda* (Tang and Kristanti, 2022; Chiriac et al., 2023; Douna and Yousefi, 2023; Grgas et al., 2023; Merino et al., 2023; Zhou et al., 2023; Grimberg et al., 2024; Lorah et al., 2024; Tang et al., 2024; Thapa et al., 2024; Ha et al., 2025). The performance of some selected microorganisms for the degradation of PFAS is presented in Table 2.5. Microalgae have shown capacity for the degradation of PFAS (Ha et al., 2025; Liao et al., 2025); this capacity is based on cell surface adsorption with electrostatic interaction between polar regions of cell membranes and polar functional groups of PFAS (Amen et al., 2023; Sanzana et al., 2025). The function depends on the chemical structure of PFAS, medium pH, membrane polarity, and medium temperature (Amen et al., 2023). Another study has attributed PFAS transformation by microalgae to the actions of phase I and II enzymes (Douna and Yousefi, 2023). Enzymatic degradation converts PFAS into simple components that are easy to manage (Thapa et al., 2024). *S. quadricauda* has demonstrated partial capacity to degrade PFOA via a chain-reduction defluorination pathway to produce transformation products such as PFHxA, PFBA, PFHpA, and PFPeA (Ha et al., 2025).

**TABLE 2.5** Performance of some selected microorganisms towards per- and polyfluoroalkyl substances (PFAS) degradation.

| Organism | Target PFAS | Performance (%) | References |
|---|---|---|---|
| *Pseudomonas aeruginosa*, *Pseudomonas putida* | PFOA, PFOS | 27.9–47.3 | Chiriac et al. (2023) |
| Dehalogenating culture WBC-2 | PFOS, PFOA, 6:2 FTS | 46.4 | Lorah et al. (2024) |
| *Phanerochaete chrysosporium* | PFOA | 69.23 | Zhou et al. (2023) |
| Native microbial community | TFA, PFOA, HFPO-DA | 8.03–13.52 | Tang et al. (2024) |
| PFOS-reducing microbial consortium | PFOS | 52–80 | Sorn et al. (2023) |
| Iron-reducing bacteria (FeRB) | PFOA, PFOS | 33.0–40.8 | Kang et al. (2023) |

Biodegradation of PFAS offers an affordable and sustainable, scalable effort. However, the biodegradation process is limited by a low degradation rate and undesired side-product production. The fate of some of the side-products during biodegradation of PFAS is unknown; many are fluorinated, easily mobilized, and potentially toxic (Sanzana et al., 2025). These limitations continue to suggest the need for more investigations geared towards upgrading the efficiency and performance of the process. In addition, it is vital to include studies that try to understand the ecological implications of the biodegradation process.

## 2.5 Future perspectives

PFAS in plants and soil continues to be a challenge that requires continuous attention in monitoring and mitigation. The contamination of soil by PFAS is monitored less in many countries. There is a need to focus on understanding the PFAS pollution index globally to develop a sustainable means to tackle the problem. The challenge is more concerning in poor developing countries because many of these countries lack the facilities to conduct cutting-edge research to monitor the presence and impact that PFAS may have. Governments in poor developing countries must fund research activities to mitigate the current environmental and health problems posed by PFAS. More studies are still required to better understand the phytotoxicity of PFAS at physiological, biochemical, and molecular levels. There are conflicting results regarding the effect of PFAS concentration on plant metabolic activities (Li et al., 2022). Some studies believed that low concentration supports plant growth, while others indicated danger even at high concentration; however, it is certain that when PFAS becomes bioaccumulated, it inflicts some hazard on the plant's wellness. Although some plant defense mechanisms, such as activation of antioxidant systems, alteration of plasma membrane, repair of DNA injury, and increase of tolerance have been proposed against PFAS (Li et al., 2020b; Li et al., 2021b), there is no outstanding strategy to detoxify PFAS-contaminated plants or a strategy to handle oxidative stress in plants. This calls for more research studies to help prevent PFAS in the food chain and manage its negative impact on agricultural activities.

Most studies on removing PFAS from soil were conducted at the laboratory scale. There is a need to conduct a large-scale application of the proposed strategies in real life. During immobilization method as a strategy for removing PFAS from soil, transformation of PFAS or its precursor may occur posttreatment through biotic or abiotic oxidation (Ross et al., 2018a). Although little research has been conducted to understand the postimmobilization occurrences, there is a need for further detailed studies to be undertaken. Implementing hybrid technology by combining two or more remediation strategies may be necessary to obtain outstanding performance in removing PFAS from the environment.

## 2.6 Conclusion

PFAS are chemical molecules of global concern that pose health concerns. PFAS have been detected in the environment; they are persistent, difficult to degrade, and environmentally unfriendly. Therefore this chapter evaluates the presence of PFAS in soil and plants and the possible remediation strategies. The study identified poor and unregulated disposal of PFAS-containing waste, WWTPs, AFFF, and atmospheric deposition of PFAS and PFAS by-products from solid waste incineration or burning as sources of PFAS in soil and plants. Furthermore, destruction, immobilization, and separation are possible remediation strategies to help manage PFAS in plants and soil. Currently, most of the remediation studies are laboratory-based, which indicates the need for large-scale studies. Furthermore, despite the proposed plant defense mechanisms for managing PFAS contamination, there is still no outstanding strategy to detoxify PFAS-contaminated plants or to handle oxidative stress in plants. This calls for more research studies to help prevent PFAS in the food chain and manage its negative impact on agricultural activities.

## Acknowledgment

The author is grateful to the support received from the Tekena Tamuno Library, Redeemer's University.

## References

Abraham, J.E., Mumford, K.G., Patch, D.J., Weber, K.P., 2022. Retention of PFOS and PFOA mixtures by trapped gas bubbles in porous media. Environmental Science & Technology 56, 15489–15498.

Ahrens, L., Yeung, L.W., Taniyasu, S., Lam, P.K., Yamashita, N., 2011. Partitioning of perfluorooctanoate (PFOA), perfluorooctane sulfonate (PFOS) and perfluorooctane sulfonamide (PFOSA) between water and sediment. Chemosphere 85, 731–737.

Alinezhad, A., Challa Sasi, P., Zhang, P., Yao, B., Kubátová, A., Golovko, S.A., et al., 2022. An investigation of thermal air degradation and pyrolysis of per-and polyfluoroalkyl substances and aqueous film-forming foams in soil. ACS ES&T Engineering 2, 198–209.

Alvarez-Ruiz, R., Lee, L.S., Choi, Y., 2024. Fate of per-and polyfluoroalkyl substances at a 40-year dedicated municipal biosolids land disposal site. Science of the Total Environment 954, 176540.

Aly, Y.H., McInnis, D.P., Lombardo, S.M., Arnold, W.A., Pennell, K.D., Hatton, J., et al., 2019. Enhanced adsorption of perfluoro alkyl substances for in situ remediation. Environmental Science: Water Research & Technology 5, 1867–1875.

Amen, R., Ibrahim, A., Shafqat, W., Hassan, E.B., 2023. A critical review on PFAS removal from water: removal mechanism and future challenges. Sustainability 15, 16173.

Anderson, E.L., Mousavi, M.P., Aly, Y.H., Chen, X.V., Simcik, M.F., Bühlmann, P., 2019. Remediation of perfluorooctylsulfonate contamination by in situ sequestration: direct monitoring of PFOS binding to polyquaternium polymers. ACS omega 4, 1068–1076.

Askeland, M., Clarke, B.O., Cheema, S.A., Mendez, A., Gasco, G., Paz-Ferreiro, J., 2020. Biochar sorption of PFOS, PFOA, PFHxS and PFHxA in two soils with contrasting texture. Chemosphere 249, 126072.

Authority, E.F.S., 2008. Perfluorooctane sulfonate (PFOS), perfluorooctanoic acid (PFOA) and their salts scientific opinion of the panel on contaminants in the food chain. Efsa Journal 6, 653.

Baláž, P., Achimovičová, M., Baláž, M., Billik, P., Cherkezova-Zheleva, Z., Criado, J.M., et al., 2013. Hallmarks of mechanochemistry: from nanoparticles to technology. Chemical Society Reviews 42, 7571–7637.

Bao, J., Li, C.-L., Liu, Y., Wang, X., Yu, W.-J., Liu, Z.-Q., et al., 2020. Bioaccumulation of perfluoroalkyl substances in greenhouse vegetables with long-term groundwater irrigation near fluorochemical plants in Fuxin, China. Environmental Research 188, 109751.

Barisci, S., Suri, R., 2021. Electrooxidation of short-and long-chain perfluoroalkyl substances (PFASs) under different process conditions. Journal of Environmental Chemical Engineering 9, 105323.

Barth, E., McKernan, J., Bless, D., Dasu, K., 2021. Investigation of an immobilization process for PFAS contaminated soils. Journal of Environmental Management 296, 113069.

Barzen-Hanson, K.A., Davis, S.E., Kleber, M., Field, J.A., 2017. Sorption of fluorotelomer sulfonates, fluorotelomer sulfonamido betaines, and a fluorotelomer sulfonamido amine in national foam aqueous film-forming foam to soil. Environmental Science & Technology 51, 12394–12404.

Becker, A.M., Gerstmann, S., Frank, H., 2008. Perfluorooctanoic acid and perfluorooctane sulfonate in the sediment of the Roter Main river, Bayreuth, Germany. Environmental Pollution 156, 818–820.

Berhanu, A., Mutanda, I., Taolin, J., Qaria, M.A., Yang, B., Zhu, D., 2023. A review of microbial degradation of per- and polyfluoroalkyl substances (PFAS): biotransformation routes and enzymes. Science of the Total Environment 859, 160010.

Biegel-Engler, A., Frauenstein, J., 2024. PFAS in soil and groundwater: comprehensive challenges and progress in regulation and management in Germany. International Yearbook of Soil Law and Policy 2022 285–304.

Bierbaum, T., Klaas, N., Braun, J., Nürenberg, G., Lange, F.T., Haslauer, C., 2023. Immobilization of per-and polyfluoroalkyl substances (PFAS): comparison of leaching behavior by three different leaching tests. Science of the Total Environment 876, 162588.

Blaine, A.C., Rich, C.D., Hundal, L.S., Lau, C., Mills, M.A., Harris, K.M., et al., 2013. Uptake of perfluoroalkyl acids into edible crops via land applied biosolids: field and greenhouse studies. Environmental Science & Technology 47, 14062–14069.

Blaine, A.C., Rich, C.D., Sedlacko, E.M., Hundal, L.S., Kumar, K., Lau, C., et al., 2014. Perfluoroalkyl acid distribution in various plant compartments of edible crops grown in biosolids-amended soils. Environmental Science & Technology 48, 7858–7865.

Bolan, N., Sarkar, B., Yan, Y., Li, Q., Wijesekara, H., Kannan, K., et al., 2021. Remediation of poly-and perfluoroalkyl substances (PFAS) contaminated soils–to mobilize or to immobilize or to degrade? Journal of Hazardous Materials 401, 123892.

Boudjabi, S., Chenchouni, H., 2021. On the sustainability of land applications of sewage sludge: how to apply the sewage biosolid in order to improve soil fertility and increase crop yield? Chemosphere 282, 131122.

Bräunig, J., Baduel, C., Barnes, C.M., Mueller, J.F., 2019. Leaching and bioavailability of selected perfluoroalkyl acids (PFAAs) from soil contaminated by firefighting activities. Science of the Total Environment 646, 471–479.

Bräunig, J., Baduel, C., Barnes, C.M., Mueller, J.F., 2021. Sorbent assisted immobilisation of perfluoroalkyl acids in soils–effect on leaching and bioavailability. Journal of Hazardous Materials 412, 125171.

Buck, R.C., Franklin, J., Berger, U., Conder, J.M., Cousins, I.T., De Voogt, P., et al., 2011. Perfluoroalkyl and polyfluoroalkyl substances in the environment: terminology, classification, and origins. Integrated Environmental Assessment and Management 7, 513–541.

Cagnetta, G., Zhang, Q., Huang, J., Lu, M., Wang, B., Wang, Y., et al., 2017. Mechanochemical destruction of perfluorinated pollutants and mechanosynthesis of lanthanum oxyfluoride: a Waste-to-Materials process. Chemical Engineering Journal 316, 1078–1090.

Cai, W., Navarro, D.A., Du, J., Ying, G., Yang, B., McLaughlin, M.J., et al., 2022. Increasing ionic strength and valency of cations enhance sorption through hydrophobic interactions of PFAS with soil surfaces. Science of the Total Environment 817, 152975.

Campos-Pereira, H., Makselon, J., Kleja, D.B., Prater, I., Kögel-Knabner, I., Ahrens, L., et al., 2022. Binding of per- and polyfluoroalkyl substances (PFASs) by organic soil materials with different structural composition–charge- and concentration-dependent sorption behavior. Chemosphere 297, 134167.

Cao, C., Huo, Q., Tang, Q., Guo, Y., Zeng, L., Cheng, Y., et al., 2025. Linking vegetable per-and polyfluoroalkyl substance accumulation with root chemical traits. ACS Environmental Au 6, 105–115.

Cao, L., Xu, W., Wan, Z., Li, G., Zhang, F., 2022. Occurrence of PFASs and its effect on soil bacteria at a fire-training area using PFOS-restricted aqueous film-forming foams. Iscience 25.

Chain, E.Po.CitF., Knutsen, H.K., Alexander, J., Barregård, L., Bignami, M., Brüschweiler, B., et al., 2018. Risk to human health related to the presence of perfluorooctane sulfonic acid and perfluorooctanoic acid in food. EFSA Journal 16, e05194.

Chain, E.Po.CitF., Schrenk, D., Bignami, M., Bodin, L., Chipman, J.K., del Mazo, J., et al., 2020. Risk to human health related to the presence of perfluoroalkyl substances in food. EFSA Journal 18, e06223.

Chen, H., Zhang, C., Yu, Y., Han, J., 2012. Sorption of perfluorooctane sulfonate (PFOS) on marine sediments. Marine Pollution Bulletin 64, 902–906.

Chen, Y.-C., Lo, S.-L., Li, N.-H., Lee, Y.-C., Kuo, J., 2013. Sorption of perfluoroalkyl substances (PFASs) onto wetland soils. Desalination and Water Treatment 51, 7469–7475.

Chen, H., Zhang, L., Li, M., Yao, Y., Zhao, Z., Munoz, G., et al., 2019. Per-and polyfluoroalkyl substances (PFASs) in precipitation from mainland China: contributions of unknown precursors and short-chain (C2C3) perfluoroalkyl carboxylic acids. Water Research 153, 169–177.

Chen, J., Tang, L., Chen, W.-Q., Peaslee, G.F., Jiang, D., 2020. Flows, stock, and emissions of poly-and perfluoroalkyl substances in California carpet in 2000–2030 under different scenarios. Environmental Science & Technology 54, 6908–6918.

Chiriac, F.L., Stoica, C., Iftode, C., Pirvu, F., Petre, V.A., Paun, I., et al., 2023. Bacterial biodegradation of perfluorooctanoic acid (PFOA) and perfluorosulfonic acid (PFOS) using pure Pseudomonas strains. Sustainability 15, 14000.

Choudhary, A., Kumar, A., Kaur, N., 2020. ROS and oxidative burst: roots in plant development. Plant Diversity 42, 33–43.

Christou, A., Papadavid, G., Dalias, P., Fotopoulos, V., Michael, C., Bayona, J.M., et al., 2019. Ranking of crop plants according to their potential to uptake and accumulate contaminants of emerging concern. Environmental Research 170, 422–432.

Coggan, T.L., Moodie, D., Kolobaric, A., Szabo, D., Shimeta, J., Crosbie, N.D., et al., 2019. An investigation into per- and polyfluoroalkyl substances (PFAS) in nineteen Australian wastewater treatment plants (WWTPs). Heliyon 5.

Coperchini, F., Greco, A., Rotondi, M., 2024. Changing the structure of PFOA and PFOS: a chemical industry strategy or a solution to avoid thyroid-disrupting effects? Journal of Endocrinological Investigation 47, 1863–1879.

Costello, M.C.S., Lee, L.S., 2020. Sources, fate, and plant uptake in agricultural systems of per-and polyfluoroalkyl substances. Current Pollution Reports 1–21.

Das, P., Arias, E., V.A., Kambala, V., Mallavarapu, M., Naidu, R., 2013. Remediation of perfluorooctane sulfonate in contaminated soils by modified clay adsorbent—a risk-based approach. Water, Air, & Soil Pollution 224, 1–14.

Dauchy, X., 2023. Evidence of large-scale deposition of airborne emissions of per-and polyfluoroalkyl substances (PFASs) near a fluoropolymer production plant in an urban area. Chemosphere 337, 139407.

Dombrowski, P.M., Kakarla, P., Caldicott, W., Chin, Y., Sadeghi, V., Bogdan, D., et al., 2018. Technology review and evaluation of different chemical oxidation conditions on treatability of PFAS. Remediation Journal 28, 135–150.

Douna, B.K., Yousefi, H., 2023. Removal of PFAS by biological methods. Asian Pacific Journal of Environment and Cancer 6, 53–68.

Duchesne, A.L., Brown, J.K., Patch, D.J., Major, D., Weber, K.P., Gerhard, J.I., 2020. Remediation of PFAS-contaminated soil and granular activated carbon by smoldering combustion. Environmental Science & Technology 54, 12631–12640.

Eltarabily, M.G.A., Negm, A.M., Saavedra Valeriano, O.C., Gafar, K.E., 2015. Effects of di-ammonium phosphate on hydraulic, compaction, and shear strength characteristic of sand and clay soils. Arabian Journal of Geosciences 8, 10419–10432.

Evich, M.G., Davis, M.J., McCord, J.P., Acrey, B., Awkerman, J.A., Knappe, D.R., et al., 2022. Per-and polyfluoroalkyl substances in the environment. Science 375, eabg9065.

Fabregat-Palau, J., Vidal, M., Rigol, A., 2021. Modelling the sorption behaviour of perfluoroalkyl carboxylates and perfluoroalkane sulfonates in soils. Science of the Total Environment 801, 149343.

Faraghati, S.N., Sadeghpour Haji, M., Ghadimi Hamzeh Kolaei, A., 2019. Effect of pH changes on the geotechnical properties of clay liners in landfill. Advance Researches in Civil Engineering 1, 49–57.

Fawad, M., Wu, T., Xu, X., Wang, J., Zhou, X., Rajput, V.D., et al., 2026. Mechanisms of PFAS uptake and bioaccumulation in plants. Ecotoxicology and Environmental Safety 311, 119888.

Felizeter, S., McLachlan, M.S., De Voogt, P., 2012. Uptake of perfluorinated alkyl acids by hydroponically grown lettuce (Lactuca sativa). Environmental Science & Technology 46, 11735–11743.

Fuertes, I., Gómez-Lavín, S., Elizalde, M.P., Urtiaga, A., 2017. Perfluorinated alkyl substances (PFASs) in northern Spain municipal solid waste landfill leachates. Chemosphere 168, 399–407.

Gallen, C., Drage, D., Eaglesham, G., Grant, S., Bowman, M., Mueller, J., 2017. Australia-wide assessment of perfluoroalkyl substances (PFASs) in landfill leachates. Journal of Hazardous Materials 331, 132–141.

Gellrich, V., Stahl, T., Knepper, T., 2012. Behavior of perfluorinated compounds in soils during leaching experiments. Chemosphere 87, 1052–1056.

Gerardu, T., Dijkstra, J., Beeltje, H., van Duivenbode, Av.R., Griffioen, J., 2023. Accumulation and transport of atmospherically deposited PFOA and PFOS in undisturbed soils downwind from a fluoropolymers factory. Environmental Advances 11, 100332.

Gewurtz, S.B., Bradley, L.E., Backus, S., Dove, A., McGoldrick, D., Hung, H., et al., 2019. Perfluoroalkyl acids in Great Lakes precipitation and surface water (2006–2018) indicate response to phase-outs, regulatory action, and variability in fate and transport processes. Environmental Science & Technology 53, 8543–8552.

Ghisi, R., Vamerali, T., Manzetti, S., 2019. Accumulation of perfluorinated alkyl substances (PFAS) in agricultural plants: a review. Environmental Research 169, 326–341.

Gobelius, L., Glimstedt, L., Olsson, J., Wiberg, K., Ahrens, L., 2023. Mass flow of per-and polyfluoroalkyl substances (PFAS) in a Swedish municipal wastewater network and wastewater treatment plant. Chemosphere 336, 139182.

Göckener, B., Weber, T., Rüdel, H., Bücking, M., Kolossa-Gehring, M., 2020. Human biomonitoring of per-and polyfluoroalkyl substances in German blood plasma samples from 1982 to 2019. Environment International 145, 106123.

González-Naranjo, V., Boltes, K., 2014. Toxicity of ibuprofen and perfluorooctanoic acid for risk assessment of mixtures in aquatic and terrestrial environments. International Journal of Environmental Science and Technology 11, 1743–1750.

Grandjean, P., Andersen, E.W., Budtz-Jørgensen, E., Nielsen, F., Mølbak, K., Weihe, P., et al., 2012. Serum vaccine antibody concentrations in children exposed to perfluorinated compounds. JAMA. 307, 391–397.

Gredelj, A., Nicoletto, C., Polesello, S., Ferrario, C., Valsecchi, S., Lava, R., et al., 2020a. Uptake and translocation of perfluoroalkyl acids (PFAAs) in hydroponically grown red chicory (*Cichorium intybus* L.): growth and developmental toxicity, comparison with growth in soil and bioavailability implications. Science of the Total Environment 720, 137333.

Gredelj, A., Nicoletto, C., Valsecchi, S., Ferrario, C., Polesello, S., Lava, R., et al., 2020b. Uptake and translocation of perfluoroalkyl acids (PFAA) in red chicory (*Cichorium intybus* L.) under various treatments with pre-contaminated soil and irrigation water. Science of the Total Environment 708, 134766.

Grgas, D., Petrina, A., Štefanac, T., Bešlo, D., Landeka Dragičević, T., 2023. A review: per-and polyfluoroalkyl substances—Biological degradation. Toxics 11, 446.

Grimberg, F., Holsen, T.M., Fernando, S., Wang, S., 2024. Biotransformation of 6: 2 fluorotelomer sulfonate (6: 2 FTS) in sulfur-rich media by Trametopsis cervina. Frontiers of Environmental Science & Engineering 18, 107.

Grimison, C., Knight, E.R., Nguyen, T.M.H., Nagle, N., Kabiri, S., Bräunig, J., et al., 2023. The efficacy of soil washing for the remediation of per-and poly-fluoroalkyl substances (PFASs) in the field. Journal of Hazardous Materials 445, 130441.

Gunarathne, V., Melo, T.M., Schauerte, M., Groth, F., Slaný, M., Rinklebe, J., 2025a. Immobilization of per-and polyfluorinated alkyl substances (PFAS) from field contaminated groundwater by a novel organo-clay vs. colloidal activated carbon under flow conditions. Journal of Hazardous Materials 137273.

Gunarathne, V., Schauerte, M., Boie, F., Yang, X., Melo, T.M., Rinklebe, Jr, 2025b. Poly-/perfluoroalkyl substance (PFAS) immobilization using a novel adsorptive organoclay compared with colloidal-activated carbon in soil under dynamic redox conditions and the response of microbial communities. ACS Omega 10, 54441–54454.

Guo, M., Lyu, Y., Xu, T., Yao, B., Song, W., Li, M., et al., 2018. Particle size distribution and respiratory deposition estimates of airborne perfluoroalkyl acids during the haze period in the megacity of Shanghai. Environmental Pollution 234, 9–19.

Guo, Q., He, Z., Liu, X., Liu, B., Zhang, Y., 2020. High-throughput non-targeted metabolomics study of the effects of perfluorooctane sulfonate (PFOS) on the metabolic characteristics of A. thaliana leaves. Science of the Total Environment 710, 135542.

Ha, G.-S., Sim, M.-G., Jeon, B.-H., Baek, G., 2025. Bioremediation of perfluorooctanoic acid using microalgae with a transcriptomic approach. Journal of Hazardous Materials 137508.

Habib, Z., Song, M., Ikram, S., Zahra, Z., 2024. Overview of per-and polyfluoroalkyl substances (PFAS), their applications, sources, and potential impacts on human health. Pollutants 4, 136–152.

Hale, S.E., Arp, H.P.H., Slinde, G.A., Wade, E.J., Bjørseth, K., Breedveld, G.D., et al., 2017. Sorbent amendment as a remediation strategy to reduce PFAS mobility and leaching in a contaminated sandy soil from a Norwegian firefighting training facility. Chemosphere 171, 9–18.

Harrad, S., Drage, D.S., Sharkey, M., Berresheim, H., 2020. Perfluoroalkyl substances and brominated flame retardants in landfill-related air, soil, and groundwater from Ireland. Science of the Total Environment 705, 135834.

Harris, J.D., Coon, C.M., Doherty, M.E., McHugh, E.A., Warner, M.C., Walters, C.L., et al., 2022. Engineering and characterization of dehalogenase enzymes from Delftia acidovorans in bioremediation of perfluorinated compounds. Synthetic and Systems Biotechnology 7, 671–676.

Hellsing, M.S., Josefsson, S., Hughes, A.V., Ahrens, L., 2016. Sorption of perfluoroalkyl substances to two types of minerals. Chemosphere 159, 385–391.

Helmer, R.W., Reeves, D.M., Cassidy, D.P., 2022. Per-and polyfluorinated alkyl substances (PFAS) cycling within Michigan: contaminated sites, landfills and wastewater treatment plants. Water Research 210, 117983.

Higgins, C.P., Luthy, R.G., 2006. Sorption of perfluorinated surfactants on sediments. Environmental Science & Technology 40, 7251–7256.

Høisæter, Å., Pfaff, A., Breedveld, G.D., 2019. Leaching and transport of PFAS from aqueous film-forming foam (AFFF) in the unsaturated soil at a firefighting training facility under cold climatic conditions. Journal of Contaminant Hydrology 222, 112–122.

Høisæter, Å., Arp, H.P.H., Slinde, G., Knutsen, H., Hale, S.E., Breedveld, G.D., et al., 2021. Excavated vs novel in situ soil washing as a remediation strategy for sandy soils impacted with per-and polyfluoroalkyl substances from aqueous film forming foams. Science of the Total Environment 794, 148763.

Hong, S., Khim, J.S., Park, J., Kim, M., Kim, W.-K., Jung, J., et al., 2013. In situ fate and partitioning of waterborne perfluoroalkyl acids (PFAAs) in the Youngsan and Nakdong River Estuaries of South Korea. Science of the Total Environment 445, 136–145.

Hubert, M., Arp, H.P.H., Hansen, M.C., Castro, G., Meyn, T., Asimakopoulos, A.G., et al., 2023. Influence of grain size, organic carbon and organic matter residue content on the sorption of per-and polyfluoroalkyl substances in aqueous film forming foam contaminated soils-Implications for remediation using soil washing. Science of the Total Environment 875, 162668.

Iovino, P., Chianese, S., Fenti, A., Blotevogel, J., Musmarra, D., 2023. An innovative approach for atrazine electrochemical oxidation modelling: process parameter effect, intermediate formation and kinetic constant assessment. Chemical Engineering Journal 474, 146022.

Iovino, P., Lavorgna, M., Orlo, E., Russo, C., De Felice, B., Campolattano, N., et al., 2024. An integrated approach for the assessment of the electrochemical oxidation of diclofenac: by-product identification, microbiological and eco-genotoxicological evaluation. Science of the Total Environment 909, 168511.

ITRC, W., 2020. Per-and Polyfluoroalkyl Substances (PFAS). TIT Council and Regulatory, Washington, DC, USA.

Jia, X., Li, X., Zhou, L., Hui, Y., Li, W., Cai, Y., et al., 2023. Variations of the level, profile, and distribution of PFAS around POSF manufacturing facilities in China: an overlooked source of PFCA. Environmental Science & Technology 57, 5264–5274.

Jiao, X., Shi, Q., Gan, J., 2021. Uptake, accumulation and metabolism of PFASs in plants and health perspectives: a critical review. Critical Reviews in Environmental Science and Technology 51, 2745–2776.

Johnson, G.R., 2022. PFAS in soil and groundwater following historical land application of biosolids. Water Research 211, 118035.

Juhasz, A.L., Kastury, F., Herde, C., Tang, W., 2022. Application of soil amendments for reducing PFAS leachability and bioavailability. Environmental Pollution 307, 119498.

Kabiri, S., McLaughlin, M.J., 2021. Durability of sorption of per-and polyfluorinated alkyl substances in soils immobilized using common adsorbents: 2. Effects of repeated leaching, temperature extremes, ionic strength and competing ions. Science of the Total Environment 766, 144718.

Kah, M., Brown, C., 2006. Adsorption of ionisable pesticides in soils. Reviews of Environmental Contamination and Toxicology: Continuation of Residue Reviews 149–217.

Kang, Y., Guo, Z., Ma, H., Wu, H., Zhang, J., 2023. Enhanced removal of perfluorooctanoic acid (PFOA) and perfluorooctane sulfonate (PFOS) in constructed wetlands: iron cycling and microbial mechanisms. ACS ES&T Water 3, 287–297.

Karamat, A., Tehrani, R., Foster, G.D., Van Aken, B., 2024. Plant responses to per-and polyfluoroalkyl substances (PFAS): a molecular perspective. International Journal of Phytoremediation 26, 219–227.

Kewalramani, J.A., Wang, B., Marsh, R.W., Meegoda, J.N., Freire, L.R., 2022. Coupled high and low-frequency ultrasound remediation of PFAS-contaminated soils. Ultrasonics Sonochemistry 88, 106063.

Khan, M.J., Jovicic, V., Zbogar-Rasic, A., Poser, A., Freichels, K., Delgado, A., 2022. Effectiveness of non-thermal plasma induced degradation of per-and polyfluoroalkyl substances from water. Water 14, 1408.

Khan, B., Burgess, R.M., Cantwell, M.G., 2023. Occurrence and bioaccumulation patterns of per-and polyfluoroalkyl substances (PFAS) in the marine environment. ACS ES&T Water 3, 1243–1259.

Kim, H., Ekpe, O.D., Lee, J.-H., Kim, D.-H., Oh, J.-E., 2019. Field-scale evaluation of the uptake of Perfluoroalkyl substances from soil by rice in paddy fields in South Korea. Science of the Total Environment 671, 714–721.

Knight, E.R., Janik, L.J., Navarro, D.A., Kookana, R.S., McLaughlin, M.J., 2019. Predicting partitioning of radiolabelled 14C-PFOA in a range of soils using diffuse reflectance infrared spectroscopy. Science of the Total Environment 686, 505–513.

Krippner, J., Brunn, H., Falk, S., Georgii, S., Schubert, S., Stahl, T., 2014. Effects of chain length and pH on the uptake and distribution of perfluoroalkyl substances in maize (Zea mays). Chemosphere 94, 85–90.

Kumar, V., Chopra, A., Kumar, A., 2017. A review on sewage sludge (Biosolids) a resource for sustainable agriculture. Archives of Agriculture and Environmental Science 2, 340–347.

Kumar, M., Bolan, N.S., Hoang, S.A., Sawarkar, A.D., Jasemizad, T., Gao, B., et al., 2021. Remediation of soils and sediments polluted with polycyclic aromatic hydrocarbons: to immobilize, mobilize, or degrade? Journal of Hazardous Materials 420, 126534.

Kupryianchyk, D., Hale, S.E., Breedveld, G.D., Cornelissen, G., 2016. Treatment of sites contaminated with perfluorinated compounds using biochar amendment. Chemosphere 142, 35–40.

Kwadijk, C., Korytar, P., Koelmans, A., 2010. Distribution of perfluorinated compounds in aquatic systems in the Netherlands. Environmental Science & Technology 44, 3746–3751.

Lan, Z., Zhou, M., Yao, Y., Sun, H., 2018a. Plant uptake and translocation of perfluoroalkyl acids in a wheat–soil system. Environmental Science and Pollution Research 25, 30907–30916.

Lan, Z., Zhou, M., Yao, Y., Sun, H., 2018b. Plant uptake and translocation of perfluoroalkyl acids in a wheat–soil system. Environmental Science and Pollution Research 25, 30907–30916.

Lang, J.R., Allred, B.M., Field, J.A., Levis, J.W., Barlaz, M.A., 2017. National estimate of per-and polyfluoroalkyl substance (PFAS) release to US municipal landfill leachate. Environmental Science & Technology 51, 2197–2205.

Lechner, M., Knapp, H., 2011. Carryover of perfluorooctanoic acid (PFOA) and perfluorooctane sulfonate (PFOS) from soil to plant and distribution to the different plant compartments studied in cultures of carrots (Daucus carota ssp. Sativus), potatoes (Solanum tuberosum), and cucumbers (Cucumis Sativus). Journal of Agricultural and Food Chemistry 59, 11011–11018.

Lei, Y.-J., Tian, Y., Sobhani, Z., Naidu, R., Fang, C., 2020. Synergistic degradation of PFAS in water and soil by dual-frequency ultrasonic activated persulfate. Chemical Engineering Journal 388, 124215.

Lenka, S.P., Kah, M., Padhye, L.P., 2021. A review of the occurrence, transformation, and removal of poly-and perfluoroalkyl substances (PFAS) in wastewater treatment plants. Water Research 199, 117187.

Lesmeister, L., Lange, F.T., Breuer, J., Biegel-Engler, A., Giese, E., Scheurer, M., 2021. Extending the knowledge about PFAS bioaccumulation factors for agricultural plants–a review. Science of the Total Environment 766, 142640.

Lewis, A.J., Yun, X., Spooner, D.E., Kurz, M.J., McKenzie, E.R., Sales, C.M., 2022. Exposure pathways and bioaccumulation of per-and polyfluoroalkyl substances in freshwater aquatic ecosystems: key considerations. Science of the Total Environment 822, 153561.

Li, P., Li, J., 2021. Perfluorooctanoic acid (PFOA) caused oxidative stress and metabolic disorders in lettuce (*Lactuca sativa*) root. Science of the Total Environment 770, 144726.

Li, J., Del Vento, S., Schuster, J., Zhang, G., Chakraborty, P., Kobara, Y., et al., 2011. Perfluorinated compounds in the Asian atmosphere. Environmental Science & Technology 45, 7241–7248.

Li, K., Gao, P., Xiang, P., Zhang, X., Cui, X., Ma, L.Q., 2017. Molecular mechanisms of PFOA-induced toxicity in animals and humans: implications for health risks. Environment International 99, 43–54.

Li, Y., Oliver, D.P., Kookana, R.S., 2018. A critical analysis of published data to discern the role of soil and sediment properties in determining sorption of per and polyfluoroalkyl substances (PFASs). Science of the Total Environment 628, 110–120.

Li, F., Fang, X., Zhou, Z., Liao, X., Zou, J., Yuan, B., et al., 2019a. Adsorption of perfluorinated acids onto soils: kinetics, isotherms, and influences of soil properties. Science of the Total Environment 649, 504–514.

Li, P., Oyang, X., Zhao, Y., Tu, T., Tian, X., Li, L., et al., 2019b. Occurrence of perfluorinated compounds in agricultural environment, vegetables, and fruits in regions influenced by a fluorine-chemical industrial park in China. Chemosphere 225, 659–667.

Li, P., Oyang, X., Xie, X., Guo, Y., Li, Z., Xi, J., et al., 2020a. Perfluorooctanoic acid and perfluorooctane sulfonate co-exposure induced changes of metabolites and defense pathways in lettuce leaves. Environmental Pollution 256, 113512.

Li, P., Oyang, X., Xie, X., Li, Z., Yang, H., Xi, J., et al., 2020b. Phytotoxicity induced by perfluorooctanoic acid and perfluorooctane sulfonate via metabolomics. Journal of Hazardous Materials 389, 121852.

Li, P., Sun, J., Xie, X., Li, Z., Huang, B., Zhang, G., et al., 2021a. Stress response and tolerance to perfluorooctane sulfonate (PFOS) in lettuce (*Lactuca sativa*). Journal of Hazardous Materials 404, 124213.

Li, P., Xiao, Z., Xie, X., Li, Z., Yang, H., Ma, X., et al., 2021b. Perfluorooctanoic acid (PFOA) changes nutritional compositions in lettuce (*Lactuca sativa*) leaves by activating oxidative stress. Environmental Pollution 285, 117246.

Li, J., Sun, J., Li, P., 2022. Exposure routes, bioaccumulation and toxic effects of per-and polyfluoroalkyl substances (PFASs) on plants: a critical review. Environment International 158, 106891.

Liao, J., Lu, Y., Liu, Y., Sun, B., Zhang, K., Wang, C., et al., 2025. How heatwaves impact microalgae in the presence of environmentally relevant PFAS concentration: metabolic shifts and challenges posed. Journal of Hazardous Materials 484, 136640.

Lin, H., Taniyasu, S., Yamazaki, E., Wei, S., Wang, X., Gai, N., et al., 2020a. Per-and polyfluoroalkyl substances in the air particles of Asia: levels, seasonality, and size-dependent distribution. Environmental Science & Technology 54, 14182–14191.

Lin, Q., Zhou, C., Chen, L., Li, Y., Huang, X., Wang, S., et al., 2020b. Accumulation and associated phytotoxicity of novel chlorinated polyfluorinated ether sulfonate in wheat seedlings. Chemosphere 249, 126447.

Liu, C., Shih, K., Wang, F., 2012. Oxidative decomposition of perfluorooctanesulfonate in water by permanganate. Separation and Purification Technology 87, 95–100.

Liu, W., Li, J., Gao, L., Zhang, Z., Zhao, J., He, X., et al., 2018. Bioaccumulation and effects of novel chlorinated polyfluorinated ether sulfonate in freshwater alga Scenedesmus obliquus. Environmental Pollution 233, 8–15.

Liu, Z., Lu, Y., Song, X., Jones, K., Sweetman, A.J., Johnson, A.C., et al., 2019. Multiple crop bioaccumulation and human exposure of perfluoroalkyl substances around a mega fluorochemical industrial park, China: implication for planting optimization and food safety. Environment International 127, 671–684.

Liu, Y., Qi, F., Fang, C., Naidu, R., Duan, L., Dharmarajan, R., et al., 2020. The effects of soil properties and co-contaminants on sorption of perfluorooctane sulfonate (PFOS) in contrasting soils. Environmental Technology & Innovation 19, 100965.

Liu, J., Zhao, L., Liu, Q., Li, J., Qiao, Z., Sun, P., et al., 2021. A critical review on soil washing during soil remediation for heavy metals and organic pollutants. International Journal of Environmental Science and Technology 1–24.

Liu, S., Zhou, J., Guo, J., Xue, M., Shen, L., Bai, S., et al., 2023. Impact mechanisms of humic acid on the transmembrane transport of per-and polyfluoroalkyl substances in wheat at the subcellular level: the important role of slow-type anion channels. Environmental Science & Technology 57, 8739–8749.

Loganathan, N., Wilson, A.K., 2022. Adsorption, structure, and dynamics of short-and long-chain PFAS molecules in kaolinite: molecular-level insights. Environmental Science & Technology 56, 8043–8052.

Londhe, K., Venkatesan, A.K., 2024. Effect of chain length, electrolyte composition and aerosolization on the removal of per-and polyfluoroalkyl substances during electrochemical oxidation. Environmental Science: Water Research & Technology 10, 1377–1388.

Lorah, M.M., He, K., Blaney, L., Akob, D.M., Harris, C., Tokranov, A., et al., 2024. Anaerobic biodegradation of perfluorooctane sulfonate (PFOS) and microbial community composition in soil amended with a dechlorinating culture and chlorinated solvents. Science of the Total Environment 932, 172996.

Luft, C.M., Schutt, T.C., Shukla, M.K., 2022. Properties and mechanisms for PFAS adsorption to aqueous clay and humic soil components. Environmental Science & Technology 56, 10053–10061.

Luo, Q., Liang, S., Huang, Q., 2018. Laccase induced degradation of perfluorooctanoic acid in a soil slurry. Journal of Hazardous Materials 359, 241–247.

Lyu, X., Xiao, F., Shen, C., Chen, J., Park, C.M., Sun, Y., et al., 2022. Per-and polyfluoroalkyl substances (PFAS) in subsurface environments: occurrence, fate, transport, and research prospect. Reviews of Geophysics 60 e2021RG000765.

Mahinroosta, R., Senevirathna, L., 2020. A review of the emerging treatment technologies for PFAS contaminated soils. Journal of Environmental Management 255, 109896.

Maimaiti, A., Deng, S., Meng, P., Wang, W., Wang, B., Huang, J., et al., 2018. Competitive adsorption of perfluoroalkyl substances on anion exchange resins in simulated AFFF-impacted groundwater. Chemical Engineering Journal 348, 494–502.

Marchetto, F., Roverso, M., Righetti, D., Bogialli, S., Filippini, F., Bergantino, E., et al., 2021. Bioremediation of per- and poly-fluoroalkyl substances (PFAS) by Synechocystis sp. PCC 6803: a chassis for a synthetic biology approach. Life 11, 1300.

Martin, K.V., Hilbert, T.J., Reilly, M., Christian, W.J., Hoover, A., Pennell, K.G., et al., 2023. PFAS soil concentrations surrounding a hazardous waste incinerator in East Liverpool, Ohio, an environmental justice community. Environmental Science and Pollution Research 30, 80643–80654.

Martz, M., Heil, J., Marschner, B., Stumpe, B., 2019. Effects of soil organic carbon (SOC) content and accessibility in subsoils on the sorption processes of the model pollutants nonylphenol (4-n-NP) and perfluorooctanoic acid (PFOA). Science of the Total Environment 672, 162–173.

Masoner, J.R., Kolpin, D.W., Cozzarelli, I.M., Smalling, K.L., Bolyard, S.C., Field, J.A., et al., 2020. Landfill leachate contributes per-/poly-fluoroalkyl substances (PFAS) and pharmaceuticals to municipal wastewater. Environmental Science: Water Research & Technology 6, 1300–1311.

McDonough, A.M., Bird, A.W., Freeman, L.M., Luciani, M.A., Todd, A.K., 2021. Fate and budget of poly-and perfluoroalkyl substances in three common garden plants after experimental additions with contaminated river water. Environmental Pollution 285, 117115.

McGarr, J.T., Mbonimpa, E.G., McAvoy, D.C., Soltanian, M.R., 2023. Fate and transport of per-and polyfluoroalkyl substances (PFAS) at aqueous film forming foam (AFFF) discharge sites: a review. Soil Systems 7, 53.

Mei, W., Sun, H., Song, M., Jiang, L., Li, Y., Lu, W., et al., 2021. Per-and polyfluoroalkyl substances (PFASs) in the soil–plant system: sorption, root uptake, and translocation. Environment International 156, 106642.

Mejia-Avendaño, S., Zhi, Y., Yan, B., Liu, J., 2020. Sorption of polyfluoroalkyl surfactants on surface soils: effect of molecular structures, soil properties, and solution chemistry. Environmental Science & Technology 54, 1513–1521.

Merino, N., Wang, N., Gao, Y., Wang, M., Mahendra, S., 2023. Roles of various enzymes in the biotransformation of 6: 2 fluorotelomer alcohol (6: 2 FTOH) by a white-rot fungus. Journal of Hazardous Materials 450, 131007.

Milinovic, J., Lacorte, S., Vidal, M., Rigol, A., 2015. Sorption behaviour of perfluoroalkyl substances in soils. Science of the Total Environment 511, 63–71.

Mojiri, A., Zhou, J.L., Ozaki, N., KarimiDermani, B., Razmi, E., Kasmuri, N., 2023. Occurrence of per-and polyfluoroalkyl substances in aquatic environments and their removal by advanced oxidation processes. Chemosphere 330, 138666.

Moreira, F.C., Boaventura, R.A., Brillas, E., Vilar, V.J., 2017. Electrochemical advanced oxidation processes: a review on their application to synthetic and real wastewaters. Applied Catalysis B: Environmental 202, 217–261.

Nguyen, H.T., McLachlan, M.S., Tscharke, B., Thai, P., Braeunig, J., Kaserzon, S., et al., 2022. Background release and potential point sources of per-and polyfluoroalkyl substances to municipal wastewater treatment plants across Australia. Chemosphere 293, 133657.

Nguyen, T.M.H., Bräunig, J., Thompson, K., Thompson, J., Kabiri, S., Navarro, D.A., et al., 2020. Influences of chemical properties, soil properties, and solution pH on soil–water partitioning coefficients of per-and polyfluoroalkyl substances (PFASs). Environmental Science & Technology 54, 15883–15892.

Niu, Z., Na, J., Xu, Wa, Wu, N., Zhang, Y., 2019. The effect of environmentally relevant emerging per-and polyfluoroalkyl substances on the growth and antioxidant response in marine *Chlorella* sp. Environmental Pollution 252, 103–109.

Nzeribe, B.N., Crimi, M., Mededovic Thagard, S., Holsen, T.M., 2019. Physico-chemical processes for the treatment of per-and polyfluoroalkyl substances (PFAS): a review. Critical Reviews in Environmental Science and Technology 49, 866–915.

Öhberg, A., 2022. Electrochemical oxidation of PFAS in soil conditions: using Boron-doped diamond electrodes and iron electrodes.

Oliver, D.P., Li, Y., Orr, R., Nelson, P., Barnes, M., McLaughlin, M.J., et al., 2019. The role of surface charge and pH changes in tropical soils on sorption behaviour of per-and polyfluoroalkyl substances (PFASs). Science of the Total Environment 673, 197–206.

Oliver, D.P., Li, Y., Orr, R., Nelson, P., Barnes, M., McLaughlin, M.J., et al., 2020. Sorption behaviour of per-and polyfluoroalkyl substances (PFASs) in tropical soils. Environmental Pollution 258, 113726.

Oviedo-Vargas, D., Anton, J., Coleman-Kammula, S., Qin, X., 2025. Quantification of PFAS in soils treated with biosolids in ten northeastern US farms. Scientific Reports 15, 5582.

Pan, G., You, C., 2010. Sediment–water distribution of perfluorooctane sulfonate (PFOS) in Yangtze River Estuary. Environmental pollution 158, 1363–1367.

Pang, H., Dorian, B., Gao, L., Xie, Z., Cran, M., Muthukumaran, S., et al., 2022. Remediation of poly-and perfluoroalkyl substances (PFAS) contaminated soil using gas fractionation enhanced technology. Science of the Total Environment 827, 154310.

Panieri, E., Baralic, K., Djukic-Cosic, D., Buha Djordjevic, A., Saso, L., 2022. PFAS molecules: a major concern for the human health and the environment. Toxics 10, 44.

Pecci, K., 2018. All landfills leak, and our health and environment pay the toxic price. Conservation Law Foundation 23.

Pepper, I.L., Brusseau, M.L., Prevatt, F.J., Escobar, B.A., 2021. Incidence of Pfas in soil following long-term application of class B biosolids. Science of the Total Environment 793, 148449.

Pereira, H.C., Ullberg, M., Kleja, D.B., Gustafsson, J.P., Ahrens, L., 2018. Sorption of perfluoroalkyl substances (PFASs) to an organic soil horizon–effect of cation composition and pH. Chemosphere 207, 183–191.

Place, B.J., Field, J.A., 2012. Identification of novel fluorochemicals in aqueous film-forming foams used by the US military. Environmental Science & Technology 46, 7120–7127.

Post, G.B., 2021. Recent US state and federal drinking water guidelines for per-and polyfluoroalkyl substances. Environmental Toxicology and Chemistry 40, 550–563.

Qi, Y., Cao, H., Pan, W., Wang, C., Liang, Y., 2022. The role of dissolved organic matter during Per-and Polyfluorinated Substance (PFAS) adsorption, degradation, and plant uptake: a review. Journal of Hazardous Materials 436, 129139.

Qian, J., Lu, B., Chen, H., Wang, P., Wang, C., Li, K., et al., 2019. Phytotoxicity and oxidative stress of perfluorooctanesulfonate to two riparian plants: Acorus calamus and Phragmites communis. Ecotoxicology and Environmental Safety 180, 215–226.

Qu, B., Zhao, H., Zhou, J., 2010. Toxic effects of perfluorooctane sulfonate (PFOS) on wheat (*Triticum aestivum* L.) plant. Chemosphere 79, 555–560.

Quinnan, J., Morrell, C., Nagle, N., Maynard, K.G., 2022. Ex situ soil washing to remove PFAS adsorbed to soils from source zones. Remediation Journal 32, 151–166.

Rayne, S., Forest, K., 2009. Perfluoroalkyl sulfonic and carboxylic acids: A critical review of physicochemical properties, levels and patterns in waters and wastewaters, and treatment methods. Journal of Environmental Science and Health Part A 44, 1145–1199.

Reinikainen, J., Perkola, N., Äystö, L., Sorvari, J., 2022. The occurrence, distribution, and risks of PFAS at AFFF-impacted sites in Finland. Science of the Total Environment 829, 154237.

Rekik, H., Arab, H., Pichon, L., El Khakani, M.A., Drogui, P., 2024. Per-and polyfluoroalkyl (PFAS) Eternal pollutants: sources, environmental impacts and treatment processes. Chemosphere 142044.

Ross, I., Houtz, E., Kalve, E., McDonough, J., Hurst, J., Miles, J., 2018a. PFASs and the TOP assay. Industrial Fire Journal 1, 26–27.

Ross, I., McDonough, J., Miles, J., Storch, P., Thelakkat Kochunarayanan, P., Kalve, E., et al., 2018b. A review of emerging technologies for remediation of PFASs. Remediation Journal 28, 101–126.

Samuel, M.S., Kadarkarai, G., Ryan, D.R., McBeath, S.T., Mayer, B.K., McNamara, P.J., 2024. Enhanced perfluorooctanoic acid (PFOA) degradation by electrochemical activation of peroxydisulfate (PDS) during electrooxidation for water treatment. Science of the Total Environment 942, 173736.

Sanzana, S., Fenti, A., Iovino, P., Panico, A., 2025. A review of PFAS remediation: separation and degradation technologies for water and wastewater treatment. Journal of Water Process Engineering 74, 107793.

Sarkar, B., Rusmin, R., Ugochukwu, U.C., Mukhopadhyay, R., Manjaiah, K.M., 2019. Modified clay minerals for environmental applications. Modified Clay and Zeolite Nanocomposite Materials. Elsevier, pp. 113–127.

Scher, D.P., Kelly, J.E., Huset, C.A., Barry, K.M., Hoffbeck, R.W., Yingling, V.L., et al., 2018. Occurrence of perfluoroalkyl substances (PFAS) in garden produce at homes with a history of PFAS-contaminated drinking water. Chemosphere 196, 548–555.

Schnoor, J.L., Light, L.A., McCutcheon, S.C., Wolfe, N.L., Carreia, L.H., 1995. Phytoremediation of organic and nutrient contaminants. Environmental Science & Technology 29, 318A–323A.

Schröder, H.F., Meesters, R.J., 2005. Stability of fluorinated surfactants in advanced oxidation processes—a follow up of degradation products using flow injection–mass spectrometry, liquid chromatography–mass spectrometry and liquid chromatography–multiple stage mass spectrometry. Journal of Chromatography A 1082, 110–119.

Schwarzenbach, R.P., Gschwend, P.M., Imboden, D.M., 2003. Organic liquid–water partitioning. Environmental Organic Chemistry. Wiley, Hoboken, New Jersey, pp. 213–244.

Senevirathna, S., Mahinroosta, R., Li, M., KrishnaPillai, K., 2021. In situ soil flushing to remediate confined soil contaminated with PFOS-an innovative solution for emerging environmental issue. Chemosphere 262, 127606.

Seow, J., 2013. Fire Fighting Foams with Perfluorochemicals-environmental Review. Hemming Information Services, London, UK.

Shankar, M.U., Muthukumar, M., 2017. Comprehensive review of geosynthetic clay liner and compacted clay liner. IOP Conference Series: Materials Science and Engineering. IOP Publishing, pp. 032026.

Sharma, N., Barion, G., Shrestha, I., Ebinezer, L.B., Trentin, A.R., Vamerali, T., et al., 2020. Accumulation and effects of perfluoroalkyl substances in three hydroponically grown Salix L. species. Ecotoxicology and Environmental Safety 191, 110150.

Shimizu, M.S., Mott, R., Potter, A., Zhou, J., Baumann, K., Surratt, J.D., et al., 2021. Atmospheric deposition and annual flux of legacy perfluoroalkyl substances and replacement perfluoroalkyl ether carboxylic acids in Wilmington, NC, USA. Environmental Science & Technology Letters 8, 366–372.

Silvani, L., Cornelissen, G., Smebye, A.B., Zhang, Y., Okkenhaug, G., Zimmerman, A.R., et al., 2019. Can biochar and designer biochar be used to remediate per-and polyfluorinated alkyl substances (PFAS) and lead and antimony contaminated soils? Science of the Total Environment 694, 133693.

Sim, W., Park, H., Yoon, J.-K., Kim, J.-I., Oh, J.-E., 2021. Characteristic distribution patterns of perfluoroalkyl substances in soils according to land-use types. Chemosphere 276, 130167.

Sivagami, K., Sharma, P., Karim, A.V., Mohanakrishna, G., Karthika, S., Divyapriya, G., et al., 2023. Electrochemical-based approaches for the treatment of forever chemicals: removal of perfluoroalkyl and polyfluoroalkyl substances (PFAS) from wastewater. Science of the Total Environment 861, 160440.

Sivaram, A.K., Lee, E., Curnow, A., Surapaneni, A., Kannan, K., Megharaj, M., 2023. Uptake, accumulation, and toxicity of per-and polyfluoroalkyl substances in *Allium cepa* grown in soils amended with biosolids. Environmental Challenges 10, 100670.

Sleep, J.A., Juhasz, A.L., 2021. A review of immobilisation-based remediation of per-and poly-fluoroalkyl substances (PFAS) in soils. Current Pollution Reports 1–16.

Solo-Gabriele, H.M., Jones, A.S., Lindstrom, A.B., Lang, J.R., 2020. Waste type, incineration, and aeration are associated with per-and polyfluoroalkyl levels in landfill leachates. Waste Management 107, 191–200.

Sontake, A.R., Wagh, S.M., 2014. The phase-out of perfluorooctane sulfonate (PFOS) and the global future of aqueous film forming foam (AFFF), innovations in fire fighting foam. Fire Engineer 39, 19–23.

Sörengård, M., Gago-Ferrero, P., Kleja, D.B., Ahrens, L., 2021. Laboratory-scale and pilot-scale stabilization and solidification (S/S) remediation of soil contaminated with per-and polyfluoroalkyl substances (PFASs). Journal of Hazardous Materials 402, 123453.

Sorengard, M., Kleja, D.B., Ahrens, L., 2019. Stabilization of per-and polyfluoroalkyl substances (PFASs) with colloidal activated carbon (PlumeStop®) as a function of soil clay and organic matter content. Journal of Environmental Management 249, 109345.

Sörengård, M., Kleja, D.B., Ahrens, L., 2019. Stabilization and solidification remediation of soil contaminated with poly-and perfluoroalkyl substances (PFASs). Journal of Hazardous Materials 367, 639–646.

Sørmo, E., Silvani, L., Bjerkli, N., Hagemann, N., Zimmerman, A.R., Hale, S.E., et al., 2021. Stabilization of PFAS-contaminated soil with activated biochar. Science of the Total Environment 763, 144034.

Sorn, S., Hara-Yamamura, H., Vet, S., Xiao, M., Hoek, E.M., Honda, R., 2023. Biological treatment of perfluorooctanesulfonic acid (PFOS) using microbial capsules of a polysulfone membrane. Chemosphere 329, 138585.

Stoiber, T., Evans, S., Naidenko, O.V., 2020. Disposal of products and materials containing per-and polyfluoroalkyl substances (PFAS): A cyclical problem. Chemosphere 260, 127659.

Sukeesan, S., Boontanon, N., Boontanon, S.K., 2021. Improved electrical driving current of electrochemical treatment of per-and polyfluoroalkyl substances (PFAS) in water using boron-doped diamond anode. Environmental Technology & Innovation 23, 101655.

Tang, J., Zhang, Y., Zha, Y., Li, X., Fan, S., 2017. Oxalate enhances desorption of perfluorooctane sulfonate from soils and sediments. Water, Air, & Soil Pollution 228, 1–11.

Tang, K.H.D., Kristanti, R.A., 2022. Bioremediation of perfluorochemicals: current state and the way forward. Bioprocess and Biosystems Engineering 45, 1093–1109.

Tang, T., Liu, X., Wang, L., Zuh, A.A., Qiao, W., Huang, J., 2020. Uptake, translocation and toxicity of chlorinated polyfluoroalkyl ether potassium sulfonate (F53B) and chromium co-contamination in water spinach (*Ipomoea aquatica* Forsk). Environmental Pollution 266, 115385.

Tang, Z., Vogel, T.M., Wang, Q., Wei, C., Ali, M., Song, X., 2024. Microbial defluorination of TFA, PFOA, and HFPO-DA by a native microbial consortium under anoxic conditions. Journal of Hazardous Materials 465, 133217.

Tefera, Y.M., Gaskin, S., Mitchell, K., Springer, D., Mills, S., Pisaniello, D., 2022. Food grown on fire stations as a potential pathway for firefighters' exposure to per-and poly-fluoroalkyl substances (PFAS). Environment International 168, 107455.

Thapa, B.S., Pandit, S., Mishra, R.K., Joshi, S., Idris, A.M., Tusher, T.R., 2024. Emergence of per-and poly-fluoroalkyl substances (PFAS) and advances in the remediation strategies. Science of the Total Environment 916, 170142.

Uwayezu, J.-N., Yeung, L.W., Bäckström, M., 2022. Sorption of perfluorooctane sulfonate (PFOS) including its isomers on hydrargillite as a function of pH, humic substances and Na2SO4. Journal of Environmental Sciences 111, 263–272.

Uwayezu, J.N., Carabante, I., van Hees, P., Karlsson, P., Kumpiene, J., 2023. Validation of UV/persulfate as a PFAS treatment of industrial wastewater and environmental samples. Journal of Water Process Engineering 53, 103614.

Uwayezu, J.N., Ren, Z., Sonnenschein, S., Leiviskä, T., Lejon, T., van Hees, P., et al., 2024. Combination of separation and degradation methods after PFAS soil washing. Science of the Total Environment 907, 168137.

Veciana, M., Bräunig, J., Farhat, A., Pype, M.-L., Freguia, S., Carvalho, G., et al., 2022. Electrochemical oxidation processes for PFAS removal from contaminated water and wastewater: fundamentals, gaps and opportunities towards practical implementation. Journal of Hazardous Materials 434, 128886.

Verley, J.C., McLennon, E., Rein, K.S., Dikgang, J., Kankarla, V., 2025. Current trends and patterns of PFAS in agroecosystems and environment: a review. Journal of environmental quality 54, 80–107.

Wang, F., Shih, K., Ma, R., Li, X.-y, 2015. Influence of cations on the partition behavior of perfluoroheptanoate (PFHpA) and perfluorohexanesulfonate (PFHxS) on wastewater sludge. Chemosphere 131, 178–183.

Wang, P., Zhang, M., Lu, Y., Meng, J., Li, Q., Lu, X., 2019. Removal of perfluoalkyl acids (PFAAs) through fluorochemical industrial and domestic wastewater treatment plants and bioaccumulation in aquatic plants in river and artificial wetland. Environment International 129, 76–85.

Wang, B., Yao, Y., Chen, H., Chang, S., Tian, Y., Sun, H., 2020a. Per-and polyfluoroalkyl substances and the contribution of unknown precursors and short-chain (C2–C3) perfluoroalkyl carboxylic acids at solid waste disposal facilities. Science of the Total Environment 705, 135832.

Wang, T.-T., Ying, G.-G., Shi, W.-J., Zhao, J.-L., Liu, Y.-S., Chen, J., et al., 2020b. Uptake and translocation of perfluorooctanoic acid (PFOA) and perfluorooctanesulfonic acid (PFOS) by wetland plants: tissue-and cell-level distribution visualization with desorption electrospray ionization mass spectrometry (DESI-MS) and transmission electron microscopy equipped with energy-dispersive spectroscopy (TEM-EDS). Environmental Science & Technology 54, 6009–6020.

Wang, C., Yan, B., Munoz, G., Sauvé, S., Liu, J., 2021a. Modified clays reduce leaching of per-and polyfluoroalkyl substances from AFFF-contaminated soils. AWWA Water Science 3, e1241.

Wang, P., Zhang, M., Li, Q., Lu, Y., 2021b. Atmospheric diffusion of perfluoroalkyl acids emitted from fluorochemical industry and its associated health risks. Environment International 146, 106247.

Wang, J., Lin, Z., He, X., Song, M., Westerhoff, P., Doudrick, K., et al., 2022a. Critical review of thermal decomposition of per-and polyfluoroalkyl substances: mechanisms and implications for thermal treatment processes. Environmental Science & Technology 56, 5355–5370.

Wang, Y., Kim, J., Huang, C.-H., Hawkins, G.L., Li, K., Chen, Y., et al., 2022b. Occurrence of per-and polyfluoroalkyl substances in water: a review. Environmental Science: Water Research & Technology 8, 1136–1151.

Wang, Y., Warner, M., Li, K., Hawkins, G.L., Huang, Q., 2022c. Assessing explicit models of per-and polyfluoroalkyl substances adsorption on anion exchange resins by rapid small-scale column tests. Chemosphere 300, 134547.

Wang, Y., Ji, Y., Tishchenko, V., Huang, Q., 2023a. Removing per-and polyfluoroalkyl substances (PFAS) in water by foam fractionation. Chemosphere 311, 137004.
Wang, Y., Munir, U., Huang, Q., 2023b. Occurrence of per-and polyfluoroalkyl substances (PFAS) in soil: sources, fate, and remediation. Soil & Environmental Health 1, 100004.
Wang, Y., Munir, U., Huang, Q., 2023c. Occurrence of per-and polyfluoroalkyl substances (PFAS) in soil: sources, fate, and remediation. Soil & Environmental Health 1 (1), 100004.
Wei, Z., Xu, T., Zhao, D., 2019. Treatment of per-and polyfluoroalkyl substances in landfill leachate: status, chemistry and prospects. Environmental Science: Water Research & Technology 5, 1814–1835.
Wen, X., Wang, Y., Cheng, P., Liu, D., Ma, S., Zhang, C., et al., 2020. Surface charge properties of variable charge soils influenced by environmental factors. Applied Clay Science 189, 105522.
Xiang, R., Xu, Y., Liu, Y.-Q., Lei, G.-Y., Liu, J.-C., Huang, Q.-F., 2019. Isolation distance between municipal solid waste landfills and drinking water wells for bacteria attenuation and safe drinking. Scientific Reports 9, 17881.
Xiao, F., Jin, B., Golovko, S.A., Golovko, M.Y., Xing, B., 2019. Sorption and desorption mechanisms of cationic and zwitterionic per-and polyfluoroalkyl substances in natural soils: thermodynamics and hysteresis. Environmental Science & Technology 53, 11818–11827.
Xie, L.-N., Wang, X.-C., Dong, X.-J., Su, L.-Q., Zhu, H.-J., Wang, C., et al., 2021. Concentration, spatial distribution, and health risk assessment of PFASs in serum of teenagers, tap water and soil near a Chinese fluorochemical industrial plant. Environment International 146, 106166.
Xu, D., Chen, X., Shao, B., 2017. Oxidative damage and cytotoxicity of perfluorooctane sulfonate on *Chlorella vulgaris*. Bulletin of environmental contamination and toxicology 98, 127–132.
Yan, H., Cousins, I.T., Zhang, C., Zhou, Q., 2015. Perfluoroalkyl acids in municipal landfill leachates from China: occurrence, fate during leachate treatment and potential impact on groundwater. Science of the Total Environment 524, 23–31.
Yan, P.-F., Dong, S., Pennell, K.D., Cápiro, N.L., 2024. A review of the occurrence and microbial transformation of per-and polyfluoroalkyl substances (PFAS) in aqueous film-forming foams (AFFFs)-impacted environments. Science of the Total Environment 171883.
Yang, X., Ye, C., Liu, Y., Zhao, F.-J., 2015. Accumulation and phytotoxicity of perfluorooctanoic acid in the model plant species *Arabidopsis thaliana*. Environmental Pollution 206, 560–566.
Yang, L.-H., Yang, W.-J., Lv, S.-H., Zhu, T.-T., Sharif, H.M.A., Yang, C., et al., 2022. Is HFPO-DA (GenX) a suitable substitute for PFOA? A comprehensive degradation comparison of PFOA and GenX via electrooxidation. Environmental Research 204, 111995.
Yin, P., Hu, Z., Song, X., Liu, J., Lin, N., 2016. Activated persulfate oxidation of perfluorooctanoic acid (PFOA) in groundwater under acidic conditions. International Journal of Environmental Research and Public Health 13, 602.
You, C., Jia, C., Pan, G., 2010. Effect of salinity and sediment characteristics on the sorption and desorption of perfluorooctane sulfonate at sediment-water interface. Environmental Pollution 158, 1343–1347.
Yu, P.-F., Liu, B.-L., Zhao, H.-M., Liu, X., Li, Y.-W., Xiang, L., et al., 2025a. Novel insights into the mechanisms of perfluorooctanesulfonate (PFOS) accumulation in lettuce (*Lactuca sativa* L.): importance of detoxification efflux carriers. Environmental Science & Technology 59, 17286–17297.
Yu, X., Tang, L., Yan, R., Wang, J., Zhao, W., Lyu, S., et al., 2025b. Quantifying PFAS contamination and environmental risk in municipal solid waste landfill refuse: implications for landfill reuse. Water Research 123881.
Zeng, Z., Song, B., Xiao, R., Zeng, G., Gong, J., Chen, M., et al., 2019. Assessing the human health risks of perfluorooctane sulfonate by in vivo and in vitro studies. Environment international 126, 598–610.
Zhang, K., Huang, J., Yu, G., Zhang, Q., Deng, S., Wang, B., 2013. Destruction of perfluorooctane sulfonate (PFOS) and perfluorooctanoic acid (PFOA) by ball milling. Environmental Science & Technology 47, 6471–6477.
Zhang, R., Yan, W., Jing, C., 2014. Mechanistic study of PFOS adsorption on kaolinite and montmorillonite. Colloids and Surfaces A: Physicochemical and Engineering Aspects 462, 252–258.
Zhang, R., Yan, W., Jing, C., 2015. Experimental and molecular dynamic simulation study of perfluorooctane sulfonate adsorption on soil and sediment components. Journal of Environmental Sciences 29, 131–138.
Zhang, B., Song, X., Zhang, Y., Ma, Y., Tang, C., Yang, L., et al., 2016. The interaction between surface water and groundwater and its effect on water quality in the Second Songhua River basin, northeast China. Journal of Earth System Science 125, 1495–1507.

Zhang, L., Sun, H., Wang, Q., Chen, H., Yao, Y., Zhao, Z., et al., 2019. Uptake mechanisms of perfluoroalkyl acids with different carbon chain lengths (C2-C8) by wheat (*Triticum acstivnm* L). Science of the Total Environment 654, 19–27.

Zhang, D.Q., Wang, M., He, Q., Niu, X., Liang, Y., 2020a. Distribution of perfluoroalkyl substances (PFASs) in aquatic plant-based systems: From soil adsorption and plant uptake to effects on microbial community. Environmental Pollution 257, 113575.

Zhang, M., Wang, P., Lu, Y., Lu, X., Zhang, A., Liu, Z., et al., 2020b. Bioaccumulation and human exposure of perfluoroalkyl acids (PFAAs) in vegetables from the largest vegetable production base of China. Environment International 135, 105347.

Zhang, Y., Qv, Z., Wang, J., Yang, Y., Chen, X., Wang, J., et al., 2022a. Natural biofilm as a potential integrative sample for evaluating the contamination and impacts of PFAS on aquatic ecosystems. Water Research 215, 118233.

Zhang, Z., Sarkar, D., Biswas, J.K., Datta, R., 2022b. Biodegradation of per-and polyfluoroalkyl substances (PFAS): a review. Bioresource Technology 344, 126223.

Zhao, L., Bian, J., Zhang, Y., Zhu, L., Liu, Z., 2014. Comparison of the sorption behaviors and mechanisms of perfluorosulfonates and perfluorocarboxylic acids on three kinds of clay minerals. Chemosphere 114, 51–58.

Zhao, S., Fan, Z., Sun, L., Zhou, T., Xing, Y., Liu, L., 2017. Interaction effects on uptake and toxicity of perfluoroalkyl substances and cadmium in wheat (*Triticum aestivum* L.) and rapeseed (*Brassica campestris* L.) from co-contaminated soil. Ecotoxicology and Environmental Safety 137, 194–201.

Zhao, H., Guan, Y., Qu, B., 2018a. PFCA uptake and translocation in dominant wheat species (*Triticum aestivum* L.). International Journal of Phytoremediation 20, 68–74.

Zhao, S., Zhou, T., Zhu, L., Wang, B., Li, Z., Yang, L., et al., 2018b. Uptake, translocation and biotransformation of N-ethyl perfluorooctanesulfonamide (N-EtFOSA) by hydroponically grown plants. Environmental Pollution 235, 404–410.

Zhou, L., Xia, M., Wang, L., Mao, H., 2016. Toxic effect of perfluorooctanoic acid (PFOA) on germination and seedling growth of wheat (*Triticum aestivum* L.). Chemosphere 159, 420–425.

Zhou, L., Li, W., Zhang, J., Mao, H., 2023. Removal of perfluorooctanoic acid (PFOA) in the liquid culture of *Phanerochaete chrysosporium*. Chemosphere 345, 140427.

Zhu, H., Kannan, K., 2019. Distribution and partitioning of perfluoroalkyl carboxylic acids in surface soil, plants, and earthworms at a contaminated site. Science of the Total Environment 647, 954–961.

CHAPTER

# 3

# Environmental mixtures, exposure pathways, and ecotoxicological risks of per- and polyfluoroalkyl substances: a special focus on the aquatic environment

*Maria Nahin Oishy, Maimuna Binte Hasan Raisa, and Amit Hasan Anik*

**Department of Environmental Science, Bangladesh University of Professionals (BUP), Dhaka, Bangladesh**

## 3.1 Introduction

Per- and polyfluoroalkyl substances (PFAS) are a class of man-made chemicals that are widely used in commercial and industrial processes. Because of their molecular structure, which is composed of carbon–fluorine bonds, they are incredibly resistant to deterioration from heat, water, and oils. Although they have many uses, their durability adds to their environmental persistence and potential to endanger ecosystems and human health (Banyoi et al., 2022; Wang et al, 2022a,b). Firefighting foams, nonstick cookware, stain-resistant textiles, food packaging, and personal care products are just a few of the thousands of products that contain PFAS since they were first introduced in the 1940s. PFAS can move through the air, water, and soil after being released into the environment. They can then buildup in aquatic systems, where they are taken up by living things and passed on through the food chain (Nayak et al., 2023; Zarębska et al., 2024).

PFAS have been widely found in aquatic ecosystems and other environmental compartments in recent decades (Alazaiza et al., 2025). Aquatic environments, encompassing freshwater rivers, lakes, estuaries, and marine systems, are vital components of the

*Per- and Polyfluoroalkyl Substances (PFAS) as Environmental Contaminants*
DOI: https://doi.org/10.1016/B978-0-443-43856-1.00014-9

Earth's biosphere and perform various essential functions. However, per- and polyfluoroalkyl substances (PFAS) from various sources, including landfill leachates, firefighting foam applications, and industrial outfalls, are posing an increasing threat to these crucial ecosystems (Anik et al., 2025a,b; Kurwadkar et al., 2021). PFAS are also introduced into aquatic systems by nonpoint sources, such as atmospheric deposition and urban or agricultural runoff (Ahrens et al., 2011; Taniyasu et al., 2013). These contaminants have demonstrated the capacity for long-range environmental transport via air and water pathways, resulting in detectable concentrations even in remote regions far from direct sources of pollution (Young and Mabury, 2010; Ahrens and Bundschuh, 2014). The environmental behavior and fate of PFAS in aquatic systems are intricate and affected by numerous physicochemical parameters, such as chain length, functional groups, and solubility (Ahrens and Bundschuh, 2014; Mussabek et al., 2020). Their hydrophobic and lipophobic characteristics influence their sorption behavior, hence determining their bioavailability and potential uptake by aquatic species (Higgins and Luthy, 2006). The potential for PFAS to bioaccumulate and biomagnify within aquatic food webs is particularly concerning. Many studies have shown that PFAS, particularly long-chain perfluoroalkyl acids (PFAAs), such as perfluorooctane sulfonate (PFOS) and perfluorooctanoic acid (PFOA), can accumulate in a range of aquatic organisms, including top predators and invertebrates (Kurwadkar et al., 2021; Wang et al., 2020). This bioaccumulation not only threatens the health and survival of individual species but also disrupts the structure and functioning of entire aquatic ecosystems. Furthermore, the ingestion of contaminated aquatic organisms is a major human exposure pathway, which raises serious concerns about the risks to public health and food safety (Thompson et al., 2022).

Recent reviews on PFAS have highlighted critical challenges in assessing their environmental risks, including their persistence and mobility, complex behavior in aquatic systems, detection difficulties at low concentrations across diverse matrices, uncertainties in bioaccumulation and trophic transfer, and limitations in current analytical and risk assessment frameworks (Anik et al., 2025a; Kurwadkar et al., 2021; Lewis et al., 2022). The pathways through which PFAS enter aquatic environments, ranging from industrial discharge to atmospheric deposition, are diverse and not yet fully quantified. While existing research has largely focused on the toxicity of PFAS to individual species, a comprehensive understanding of their effects at the community and ecosystem levels remains limited. In particular, the impact of PFAS on microbial interactions, despite their essential role in nutrient cycling and organic matter decomposition, has received insufficient attention. The primary objective of this chapter is to provide a comprehensive overview of PFAS contamination in aquatic environments, with emphasis on environmental mixtures, exposure pathways, and ecotoxicological risks. By synthesizing current research, the chapter examines how PFAS are introduced into waterbodies, interact with biotic and abiotic components, and affect aquatic organisms and ecosystem stability, including emerging insights into their disruption of microbial communities. This chapter also explores these underexamined dynamics, emphasizing how PFAS-induced changes to microbial processes and foundational trophic interactions can threaten overall aquatic biodiversity and ecological balance.

## 3.2 Environmental mixtures of per- and polyfluoroalkyl substances in aquatic systems

### 3.2.1 Sources and distribution of per- and polyfluoroalkyl substances in waterbodies

PFAS are globally ubiquitous in aquatic systems due to their widespread industrial and consumer use, extreme environmental persistence, and mobility. Their presence in waterbodies is attributed to multiple point and nonpoint sources, as well as long-range atmospheric transport. Understanding the sources and patterns of PFAS distribution is critical for assessing exposure risk and informing mitigation strategies.

In environmental samples, the most common PFAS chemicals include PFOS, PFOA, perfluorohexane sulfonic acid (PFHxs), perfluoroundecanoic acid (PFUnA), perfluorodecanoic acid (PFDA), perfluoroheptanoic acid (PFHpA), perfluorononanoic acid (PFNA), perfluoropentanoic acid (PFPeA), perfluorohexanoic acid (PFHxA), and perfluorotridecanoic acid (PFTrDA). Four of these are classified as short-chain PFAS, and six are classified as long-chain PFAS (Dimitrakopoulou et al., 2024). PFAS contamination originates from various industrial processes and product life cycles, including manufacture, use, and disposal. The two main production processes, electrochemical fluorination and fluorotelomerization, release precursor compounds that can degrade into persistent PFAS in the environment. One of the most well-documented sources is the use of aqueous film-forming foams (AFFFs) in firefighting (Anik et al., 2025a). PFAS contamination in surface and groundwater largely stems from AFFFs used in firefighting. By forming a vapor-sealing layer that resists oxidation, heat, and pressure and prevents re-ignition, fluorinated surfactants in AFFFs make PFAS extremely persistent in the environment (Leeson et al., 2020). A significant consumer of AFFF, the US military has discovered almost 200 polluted locations, including airports and military posts (Kurwadkar et al., 2021). Wastewater treatment plants (WWTPs) and biosolids are significant sources of PFAS contamination in the environment, after AFFFs. PFAS have been found in sludge, influent, and effluent in WWTPs around the world, which has led to contamination of groundwater and soil (see Table 3.1). Although the reuse of treated wastewater for drinking and irrigation is widely encouraged, it's crucial to acknowledge that such water can still contain trace levels of micro-pollutants, such as PFAS, that pose potential risks to human health and may degrade the quality of both surface and groundwater (Xiao, 2017; Szabo et al., 2018; Thompson et al., 2022). In a study from Werribee South, Australia, PFASs were detected in every groundwater sample collected from fields irrigated with recycled water, underscoring the persistence and mobility of these compounds in aquatic ecosystems (Szabo et al., 2018). Similarly, a nationwide analysis in the United States showed that even after treatment, certain PFAS, such as PFOA, were present at higher concentrations in effluent than in influent, indicating that treatment processes may be ineffective at fully removing or may even transform precursors into more stable PFAS compounds. Despite overall declining trends over the past two decades, measurable concentrations of PFAS continue to be discharged into the environment via WWTP effluent, reinforcing the role of treated wastewater as an ongoing source of contamination in aquatic systems (Thompson et al., 2022). Beyond the commonly studied linear PFAS compounds, certain specialized applications

**TABLE 3.1** Sources and abundances of per- and polyfluoroalkyl substances compounds in various aquatic ecosystems with different affected organisms.

| Location | Dominant sources | Common PFAS compound | Detected concentration | Affected aquatic organisms | References |
|---|---|---|---|---|---|
| Red Sea, Saudi Arabia | Wastewater treatment plant (WWTP) effluents | PFOS (most abundant), PFBA, 6:2 FTS, PFBS, PFOA, PFNA, PFHxS | – Sediment: 0.57 to 2.6 ng $g^{-1}$ dw<br>– Fish muscle: 3.89–7.63 ng $g^{-1}$ dw<br>– Fish liver: 17.9–58.5 ng $g^{-1}$ dw | Fish | Ali et al. (2021) |
| Bahia estuary, Brazil | Multiple sources (inferred from biodilution and diverse exposure routes) | PFOS, PFTrDA, FOSA, EtFOSA, PFNA | – Biota (whole-body): 0.01–1.97 ng $g^{-1}$ ww (individual compounds) | Fish, crustaceans, and bivalves | Miranda et al. (2020) |
| Caohai Wetland, China | Nonpoint sources | PFBA, PFOA, PFBS, F-53B | – Surface water: 1.78–112.21 ng $L^{-1}$ (ΣPFAS) | Aquatic organisms | Peng et al. (2022) |
| Lake Tana, Ethiopia | Not specified | PFOS (risk assessed), total PFASs | – Water: 0.073 to 5.6 ng $L^{-1}$<br>– Sediment: 0.22 to 0.55 ng $g^{-1}$ dw<br>– Fish: 5.8 ng $g^{-1}$ ww | Fish | Ahrens et al. (2016) |
| Meiliang Bay, Taihu Lake, China | Not specified (environmental load, dietary exposure route implied) | PFOS, PFOSA, 6:2 diPAP | HR of ΣPFASs in muscle: 1.04 | Edible fish species (e.g., crucian, bitterling) | Chen et al. (2020) |
| Lake Erie, USA | Environmental contamination (surface water runoff, food web exposure) | PFOS (dominant), C9–C11 PFCAs, PFBA | – PFOS in fish: up to 79 ± 4.8 ng $g^{-1}$ ww<br>– ΣPFAS in fish: 88 ± 5.2 ng $g^{-1}$ ww | Fish species (e.g., yellow perch), benthos, zooplankton | Ren et al. (2022) |
| Orge River, Paris, France | Not specified | PFOS, PFCAs, PFSAs | 2.0–147 ng $g^{-1}$ ww | Biofilm, invertebrates, fish | Simonnet-Laprade et al. (2019) |
| Tampa Bay, Florida, USA | Airports, industry, WWTPs, fire fighting training areas, military installations | PFOS (dominant) | – Sediment: 0.0368–2.99 ng $g^{-1}$ dw<br>– Fish: 0.307–33.6 ng $g^{-1}$ ww | Fish (24 species) | Pulster et al. (2022) |

| | | | | | |
|---|---|---|---|---|---|
| Seine River Basin, France | Urban/industrial discharges (higher contamination downstream) | PFCAs, PFOS, FOSA, fluorotelomer compounds | – Fish 0.22–140 ng $g^{-1}$ wet weight | Fish (European chub) | Macorps et al. (2022) |
| Vaal River, South Africa | Industrial and mining pollution | PFOS (dominant in biota), PFPeA (higher in water) | – Water: maximum 38.5 ng $L^{-1}$;<br>– Sediment: generally <LOQ except PFOS (2.36 ng $g^{-1}$ dw);<br>– Biota: highest concentration 289 ng $g^{-1}$ ww (liver) | Fish, invertebrates | |
| Zhangjiang River estuary, China | Not specified | In the ZRE, the ΣPFAS concentrations in water samples varied from 6.51 to 12.54 ng $L^{-1}$ | – Water: ΣPFAS concentrations ranged from $6.51 \times 10^3$ to $12.54 \times 10^3$ ng $g^{-1}$<br>– Fish: PFAS levels ranged from 0.45 to 17.67 ng $g^{-1}$ ww | Fish (various tissues, highest in viscera and carcass) | Xie et al. (2024) |
| Llobregat River, Spain | Industrial discharge | PFBA, PFOA, L-PFOS | – Water: 0.01–233 ng $L^{-1}$<br>– Sediment: 0.01–3.67 ng $g^{-1}$ | Algae, *Daphnia* sp., fish (no risk indicated) | Campo et al. (2014) |
| Ganges River, India | Industrial and urban wastewater discharge | PFHxA, PFBS, PFBA, PFOA, PFOS | – Surface water: 10.2 ng $g^{-1}$<br>– Groundwater: 9.2 ng $g^{-1}$ | Not specified | Sharma et al. (2015) |
| Xiaoqing River estuary, China | Industrial discharge of PFAS alternatives | PFMOAA, HFPO-TrA, HFPO-TeA, PFO5DoA, PFOA | – PFMOAA: up to 16,700 ng $g^{-1}$<br>– HFPO-TrA: <1 to 188 ng $g^{-1}$<br>– PFOA: 23.5 to 682 ng $g^{-1}$ | Conch, various aquatic species | Li et al. (2021) |

have introduced more structurally complex variants into the environment. One such example is perfluoroethylcyclohexanesulfonate (PFECHS), a cyclic PFA historically used as an erosion inhibitor in aircraft hydraulic fluids. Despite its niche use, PFECHS has been shown to persist in the environment with similar resistance to degradation as traditional aliphatic PFAS (Kurwadkar et al., 2021). For instance, PFECHS was first detected in 2011 in both surface waters (ranging from 0.16 to 5.7 ng $L^{-1}$) and whole-body samples of predatory fish (up to 3.7 ng $g^{-1}$ wet weight) in the Great Lakes (De Silva et al., 2011). Its detection in both surface waters and aquatic biota in the Great Lakes region highlights its potential as an emerging contaminant and source of PFAS pollution in freshwater ecosystems.

Atmospheric deposition, both wet and dry, serves as a significant nonpoint source of PFASs in the environment. In contrast, primary point sources include industrial manufacturing facilities, WWTPs, landfills, and domestic activities (Ahrens et al., 2011). Volatile PFASs, such as fluorotelomer alcohols (FTOHs), perfluoroalkyl sulfonamides, and their esters, have been detected in ambient air samples, including in isolated and remote locations (Jahnke et al., 2006). Because of their volatility, these substances easily enter the atmosphere and travel great distances (Young and Mabury, 2010). Apart from atmospheric inputs, a significant nonpoint source of PFASs is runoff from contaminated soils or urban areas (Taniyasu et al., 2013). This runoff, which eventually washes pollutants into rivers, lakes, and coastal waters, may come from surfaces treated with biosolids, firefighting foams, or other materials containing PFAS. The diffuse and persistent presence of PFASs in aquatic environments is largely caused by the combined effects of surface runoff and atmospheric deposition.

With recent progress in analytical chemistry, scientists are now able to identify and measure PFAS at remarkably low concentrations, even in the picogram per liter (pg $L^{-1}$) range. These improvements in detection capabilities have spurred worldwide efforts to monitor PFAS presence across various environmental compartments. PFAS exhibit widespread and uneven distribution in aquatic systems ranging from surface and groundwater to even remote locations, reflecting both the intensity and proximity of source activities (see Table 3.1).

High concentrations are often detected in surface and groundwater bodies adjacent to industrial facilities, airports, and urban areas, while rural and remote regions may show lower levels (Alazaiza et al., 2025). For instance, the most elevated PFAS levels have been observed in groundwater from eastern Canada and eastern China, while countries such as India, the Maltese Islands, and South Korea tend to exhibit relatively low contamination levels, suggesting intense industrial input or inadequate containment measures (Benaafi and Bafaqeer, 2024). A study conducted at Canadian airports identified over 90 PFAS classes in soil and groundwater, with the highest diversity and concentrations observed at active firefighter training areas (Liu et al., 2021). Even remote surface waterbodies are not exempt from PFAS contamination, as demonstrated by a study analyzing 13 samples from isolated regions across China. Trace concentrations of PFOA and PFOS were detected in nearly all locations, with levels reaching up to 0.4 and 2.4 ng $L^{-1}$, respectively (Jin et al., 2009). The presence of these substances in remote lakes and rivers underscores the potential for long-range environmental transport of PFAS, likely through atmospheric deposition or riverine pathways (Kurwadkar et al., 2021). Another survey conducted across Wisconsin found PFAS contamination in 71% of shallow private wells, including those in agricultural and rural areas. The presence of PFAS in these regions is likely due to factors such as contaminated precipitation, septic systems, and the application of biosolids as fertilizer (Survey finds PFAS in 71% of shallow private wells across Wisconsin | AP News, 2023). The widespread distribution of PFAS even in remote areas underscores the persistence, mobility, and global reach of these compounds. This pervasive presence in aquatic environments not only complicates remediation efforts but also elevates concerns over long-term ecological and human health risks, necessitating continued monitoring and regulatory intervention.

### 3.2.2 Interaction with sediments and organic matter

Once PFAS enter aquatic systems, their environmental behavior is strongly influenced by interactions with sediments and organic matter. These interactions dictate the degree to which PFAS persist in the aqueous phase versus being sequestered in benthic substrates, thereby influencing their long-term environmental fate, bioavailability, and potential for remobilization (Ahrens and Bundschuh, 2014). PFAS, particularly PFAAs such as PFOA and PFOS, exhibit varying degrees of sorption to sediments depending on their molecular structure. Long-chain PFAS tend to bind more readily to sediments than short-chain analogs, largely due to increased hydrophobicity and van der Waals interactions (Higgins and Luthy, 2006). Researchers found that long-chain PFAS tend to correlate with certain elements in the sediment, such as lead, arsenic, iron, titanium, and sulfur. Mineral content in the sediments (such as iron and sulfur) likely carries charges that attract or repel PFAS, depending on the PFAS's properties (Mussabek et al., 2020). The partitioning behavior is often quantified using distribution coefficients (Kd) and organic carbon-normalized partition coefficients (Koc), both of which are significantly higher for long-chain PFAS. Recent investigations of a PFAS-impacted lake affected by AFFF reported ΣPFAS concentrations ranging from 95–100 ng $L^{-1}$ in water and 3.0–61 μg $kg^{-1}$ in sediments, with PFHxS and PFOS as the dominant compounds. Partitioning coefficients (log Kd) ranged from 0.6–2.3 for perfluorocarboxylic acids (PFCAs) and 0.9–5.6 for perfluorosulfonic acids (PFSAs), indicating a stronger affinity of sulfonates for sediment (Mussabek et al., 2020). Beyond hydrophobic interactions, electrostatic forces also play a critical role in PFAS binding to solid phases (Jeon et al., 2011). These intermolecular interactions depend heavily on the surface properties of sediments and suspended particles, such as charge distribution, mineral composition, and organic content (Hellsing et al., 2016).

In natural systems, the heterogeneity of sediments and organic matter introduces further complexity, as both ionic interactions and the zeta potential of particles can affect the strength and nature of PFAS sorption. The sorption of PFAS is also influenced by both the chemical properties of the solution and the sediment characteristics. Studies show that increasing pH leads to a decrease in the sorption coefficient (Kd). This is because higher pH alters the zeta potential of sediment surfaces, making them more negatively charged (Mussabek et al., 2020). Since PFOS are anionic, it experiences electrostatic repulsion from similarly charged surfaces, resulting in reduced sorption at higher pH levels. Additionally, sorption tends to be higher in sediments with more organic carbon, as PFAS are hydrophobic and prefer binding to organic matter over staying dissolved in water (Chen et al., 2020). The ionic composition of the solution, especially the presence of divalent cations such as calcium ($Ca^{2+}$), can significantly enhance PFAS sorption. Calcium ions can bridge the negative charges of PFOS and the sediment surface, reducing repulsion and promoting stronger binding

(You et al., 2010). So, sediments and organic matter significantly influence the environmental dynamics of PFAS in aquatic ecosystems. These interactions affect not only the immediate partitioning of PFAS but also their potential for long-term environmental persistence and remobilization. Understanding these processes is essential for accurate risk assessment and the design of effective remediation strategies.

## 3.3 Routes and fate of per- and polyfluoroalkyl substances in aquatic environments

### 3.3.1 Direct discharges from industrial and municipal sources

Direct industrial waste discharge, municipal sources, and weak sewage treatment activities have historically been major contributors of PFAS contamination in surrounding waterbodies. An analysis of 171 PFAS-contaminated sites in Michigan found that 75% of the pollution originated from municipal (Helmer et al., 2021). Waste from these activities is typically low and is commonly managed through strategies that include landfilling, wastewater treatment, and cremation. However, these approaches do not eliminate PFAS; rather, they simply shift them from one area to another (Stoiber et al., 2020). A study in the United States analyzed influent and effluent samples from three landfill-WWTP systems and compared them to two WWTPs not receiving landfill leachate. PFAS were detected more frequently in leachate (92%) compared to influent samples (55%), indicating higher PFAS presence in landfill-associated waste streams (Masoner et al., 2020). Perfluorinated PFAS, which have all of their carbon atoms linked to fluorine, are exceedingly durable and practically hard to degrade. In contrast, polyfluorinated PFAS have only some of their carbon atoms fluorinated, making them less stable and more easily degraded. However, they eventually degrade into the more stable and long-lasting perfluorinated compounds (PFC) (Kurwadkar et al., 2021).

Industrial facilities contribute significantly to environmental contamination with PFAS, per- and polyfluoroalkyl compounds. These facilities can be characterized as either primary or secondary manufacturers. Primary manufacturing facilities create or employ PFAS as processing aids, particularly in the creation of fluoropolymers. In some circumstances, PFAS may be employed during the production process rather than as part of the end product. Secondary manufacturing facilities, on the other hand, use PFAS-containing materials or chemical feedstocks to produce finished products, such as coated textiles (Anik et al., 2025b). Furthermore, PFAS may be employed in some industrial contexts to improve worker safety, such as employing PFOS based products to suppress hazardous chemical mist. Manufacturing facilities are significantly responsible for PFAS contamination in the aquatic environment due to facilities such as leather and textile coating, electroplating, polymer industry (Möller et al., 2010), metal finishing, wire manufacturing, and plating operations. Additionally, manufacturers that use substances such as surfactants, resin, molds, plastics, photolithography processes, or produce semiconductors (Liu et al., 2022a,b). In 2010 EU-wide survey of 90 WWTP effluents found that among 156 polar organic contaminants, PFAS were prevalent, with PFOA detected in 99% of samples, followed by PFHpA (94%), PFOS (93%), and PFNA, PFDA, PFHxA, and PFHxS detected in 70%–89% of cases (Moneta et al., 2023). Industrial operations can emit PFAS into the environment through a variety of means, including wastewater discharges, on-site and off-site waste disposal, unintentional leaks or spills, and emissions from factory stacks (see Fig. 3.1). In eastern China, 17 wastewater and corresponding river water samples from industries showed that PFAS levels in WWTP influents (0.172–20.6 $\mu g\ L^{-1}$) and effluents (0.167–93.6 $\mu g\ L^{-1}$) were significantly elevated. These concentrations were much higher than those found in upstream (0.0158–7.05 $\mu g\ L^{-1}$) and downstream (0.0237–1.82 $\mu g\ L^{-1}$) receiving water (Kurwadkar et al., 2021).

Stack emissions can cause PFAS to settle onto neighboring soil and surface water through air deposition, traveling both near and long-range transit; those attached to particles or aerosols tend to settle close to the source, whereas PFAS in vapor form are more likely to travel longer distances, potentially infiltrating groundwater. This process occurs in the facility's surrounding air zone or airshed. Some additional sources can be areas of firefighting training, storage areas for AFFF, or buildings equipped with AFFF-based fire

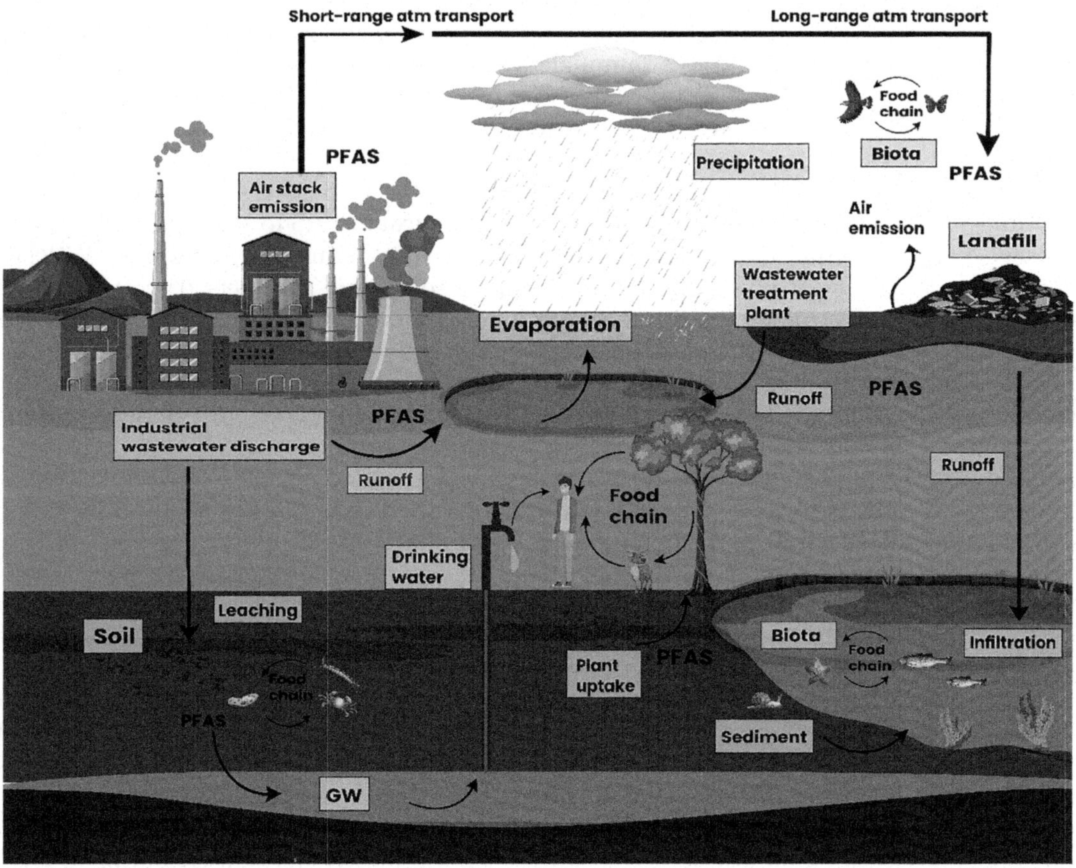


**FIGURE 3.1** Different sources, fate, and pathways of per- and polyfluoroalkyl substances (PFAS) in the aquatic environment.

suppression systems. The specific types of PFAS released from a facility depend on the kinds of PFAS being produced or used (Anik et al., 2025b). For instance, textile coating processes may involve water-based emulsions or powdered materials that contain higher levels of PFCAs compared to PFSAs. The highest recorded PFAS concentration, at 6690 ng $L^{-1}$, was found in the effluent of a Chinese textile industry WWTP involved in yarn manufacturing, bleaching, and dyeing processes (Heydebreck et al., 2016). Unlike dedicated AFFF sites, industrial facilities generally release fewer cocontaminants that could alter the chemical behavior of PFAS in the environment, unless those sites have also experienced AFFF-related releases from past firefighting activities. All of these industrial contaminated wastes, WWTP, treated municipal effluent, and sludge can contain PFAS, as it is not equipped to remove PFAS from household, institutional, and industrial sewage, resulting in entering surrounding waterbodies or seeping into groundwater (Fig. 3.1). Additionally, landfill leachate and stormwater runoff from urban areas can migrate PFAS contaminants to both terrestrial and aquatic systems. PFAS can leak through soils, especially short-chain kinds, and eventually reach groundwater. Wang et al. (2022a,b) observed an average concentration of ΣPFAS in surface and groundwater downstream of industrial plants, and WWTPs at 102 ng $L^{-1}$. They may also be absorbed by plants or accumulate throughout the food chain. On the other hand, PFAS of waterbodies also tend to accumulate in aquatic organisms, eventually entering the food chain through trophic level (see Fig. 3.1) (Ehsan et al., 2024; Barber et al., 2025).

### 3.3.2 Atmospheric deposition and runoff

Atmospheric deposition is a major contributor of toxicant PFAS to soil, lakes, and remote oceans. PFAS can become airborne through industrial emissions, landfills, WWTPs, firefighting foams, PFAS-treated household and personal care products. The atmosphere, especially near industrial areas, contains both ionic (i-PFAS) and neutral (n-PFAS) forms of PFAS, with neutral PFAS typically present at significantly higher concentrations than their ionic counterparts (Bossi, 2023). n-PFAS (such as FTOHs, FOSA, and FOSE) and i-PFAS (such as PFCAs, such as PFOA, PFSAs such as PFOS, and newer types such as GenX) behave differently in the air. A study in China, Japan, and Malaysia found atmospheric n-PFASs ranging from 4.8 to 1400 pg $m^{-3}$ (average 170 pg $m^{-3}$), dominated by 8:2 FTOH, while i-PFASs ranged from 3.7 to 330 pg $m^{-3}$ (average 49 pg $m^{-3}$) (Wang et al, 2022a,b). Neutral PFAS can exist as gases or attach to particles in the air, depending on their vapor pressure and the temperature. Warmer temperatures make them stay more in the gas form, while cooler temperatures help them stick to particles, especially those with longer carbon chains. On the other hand, i-PFAS are mostly found attached to particles in the air because they exist as nonvolatile (nonevaporating) anions, or negatively charged forms (Bossi, 2023).

With the onset of winter, atmospheric conditions enhance pollution levels and promote the movement of pollutants, such as smog and associated contaminants, from the air to soil, waterbodies, and even into the environment and human body. However, studies have shown that neutral and i-PFAS tend to display noticeable seasonal patterns, with their concentrations typically higher during the summer months (Manzetti, 2013). PFAS are capable of far-reaching atmospheric transport, enabling these persistent chemicals to reach regions far from their sources. The distance they travel through the atmosphere largely depends on their physical form: gas-phase PFAS, such as volatile precursors, can remain airborne for extended periods and be transported over long distances, whereas particle-bound PFAS tend to settle more rapidly due to gravitational forces, limiting their transport range (Yao et al., 2025). PFAS are removed from the atmosphere through both short and long-range transport with wet and dry deposition, which occur when particle-bound PFAS are scavenged or when gaseous PFAS partition into water droplets. These mechanisms typically operate over a timescale of several days and result in the presence of PFAS concentration in rain and snow, resulting in exposure to surrounding surface water and groundwater, contaminating drinking water, causing death of aquatic organisms, and disrupting the overall ecosystem (Environmental Fate and Transport for Per- and Polyfluoroalkyl Substances, 2018).

### 3.3.3 Leaching from contaminated soils and sediments

PFAS present in unsaturated permeable soils and low vegetative cover can move downward through leaching, especially during high rainfall or irrigation events that dissolve PFAS bound to soil particles (Wang et al., 2023). Sediments serve as both storage and release sources for PFAS, with leaching facilitating their transport from surface-origin sources into groundwater and surface waters (Ehsan et al., 2024b) Leaching may also contribute to PFAS uptake by plants and their movement from landfill sites lacking proper leachate containment (Gassmann et al., 2020).

Research in the United States discovered that in matched leachate and WWTP influent samples, 32 of 36 identified PFAS had considerably higher amounts in leachate, with an average of 23 times more than in influent. The maximum values of PFHxA (8300 ng $L^{-1}$), MeFPeSAA (7600 ng $L^{-1}$), PFHpA (6500 ng $L^{-1}$), and PFOA (4800 ng $L^{-1}$) in leachate were significantly greater than those in influent (470, 48, 350, and 1400 ng $L^{-1}$, respectively). Additionally, 29 of the 36 PFAS detected in leachate accounted for more than 90% of the total PFAS content in the matched samples (Masoner et al., 2020). The extent of PFAS leaching depends on characteristics of both the surrounding environment, such as pH, redox conditions, and the presence of organic-rich soils, and the chemical structure of the PFAS itself, including its charge and chain length. While several studies have documented PFAS movement through leaching, other research has found that longer-chain PFAS tend to remain in shallow soil layers even after prolonged water infiltration (Sharma et al., 2024). Although this soil persistence may restrict some exposure pathways, such as contaminating drinking water, it also implies that PFAS might linger in the environment for lengthy stretches of time, possibly serving as a long-term source of pollution (Environmental Fate and Transport for Per- and Polyfluoroalkyl Substances, 2018).

Still, its mobility nature can leach contaminants into water systems, enter the food chain via plants and animals, and spread widely even at low concentrations, posing threats to ecosystems. Human health is also concerned due to its exposure at any stage of its life cycle through the consumption of fruits and vegetables, fish, crops, etc. (Zhou et al., 2019).

### 3.3.4 Bioaccumulation and trophic transfer in aquatic food webs

PFAS are found in a wide range of biota, including plants, fish, and humans, through the bioaccumulation process (see Fig. 3.2). Among these, PFAAs, especially PFOA and PFOS,

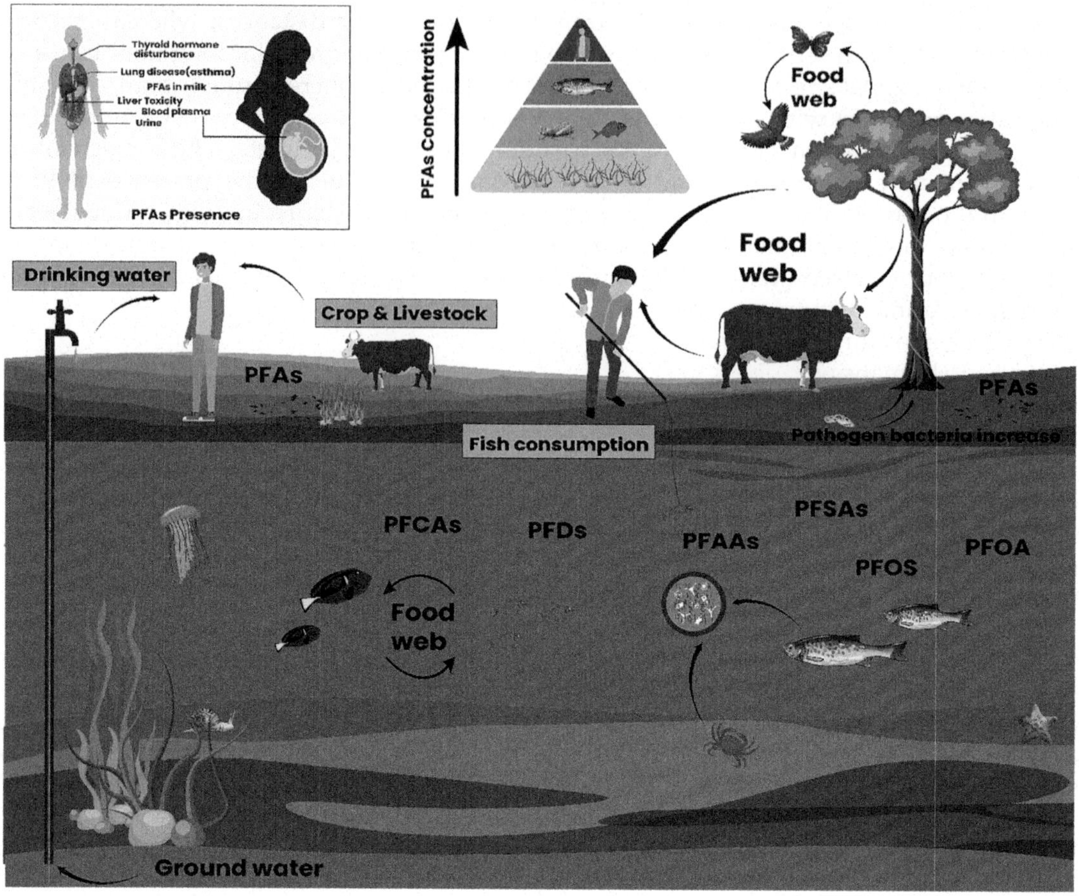


**FIGURE 3.2** Bioaccumulation and trophic transfer of per- and polyfluoroalkyl substances in different aquatic food chains and potential health concerns.

are the most commonly found compounds in biota, which tend to collect in the tissues of fish, invertebrates, and other aquatic organisms that are difficult to degrade (Hossain et al., 2024). Trophic transmission is the spread of PFAS across various levels of the food chain as predators consume contaminated animals, and PFAS can biomagnify, indicating that their concentrations grow at higher trophic levels (Zhang et al., 2024). For example, top predators such as large fish, birds, or marine mammals may contain significantly higher PFAS levels than the organisms lower in the food chain (see Fig. 3.2). Plants can absorb and accumulate PFAS through soil, raising concern about entering persistent PFAs contaminants in the food chain through crops and livestock (Biswas et al., 2025). Again, invertebrates that inhabit both the water column and sediment layers form the foundation of aquatic food webs and serve as key carriers of PFAS, playing a central role in the process of biomagnification (Wu et al., 2024).

Fish accumulate PFAS, notably PFOS, longer-chain PFCAs (with eight or more carbons), and perfluorodecane sulfonate (PFDS) (see Fig. 3.2). PFOS is the most commonly found

PFAS in fish due to its widespread use and bioaccumulation potential. PFDS, long-chain PFCAs, and other PFAS have been detected in fish . PFAS buildup in fish is linked to their diet and shows that fish tend to accumulate more of the longer-chain PFAS, which have a higher molecular weight (Giari et al., 2023). Shorter-chain PFCAs and PFSAs (less than eight and six carbons, respectively) are not easily bioconcentrated or accumulated. However, as perfluoroalkyl chain length increases, PFSAs become more bioaccumulative than PFCAs with the same number of carbons (Giari et al., 2023). The presence of toxic PFAS negatively impacts the microbiome by reducing populations of beneficial bacteria while promoting the growth of pathogenic bacteria, leading to disruptions in carbohydrate metabolism, lipid imbalances, and disturbances in hormonal regulation. So it is evident that PFAs' existence and movement through aquatic food webs pose risks not only to aquatic ecosystem health but also to humans who consume contaminated fish and seafood, raising significant concerns for both environmental and public health management. (Hong et al., 2025) PFAS are commonly found in fish, with PFOS being the dominant compound, making up 62% of total PFAS in fish from the lower Po River, consistent with results from other studies across Europe (Giari et al., 2023). PFAS is also exposed to humans at any stage of life through the food chain, resulting in its presence in blood, milk, urine, tissue, and organs. In 2006, Holzer et al. found high PFC levels in the blood plasma of children and adults in North Rhine-Westphalia exposed to polluted drinking water. To summarize, the widespread detection of PFAS in air, groundwater, freshwater, marine water, drinking water, and soil highlights their potential to cause ecotoxic effects across aquatic and terrestrial ecosystems, with bioaccumulation and trophic transfer impacting organisms at all levels of the food web, from primary producers to secondary consumers (Panieri et al., 2022).

## 3.4 Ecotoxicological risks of per- and polyfluoroalkyl substances in aquatic ecosystems

### 3.4.1 Persistency of per- and polyfluoroalkyl substances in the aquatic environment

PFAS are a class of synthetic chemicals characterized by fluorinated carbon chains, in which carbon atoms are bonded to fluorine atoms. Because of their remarkable chemical and thermal stability, which is one of the strongest bonds in organic chemistry, PFAS are extremely resistant to breaking down in natural settings. Because of their extremely persistent nature in the environment, PFASs are frequently referred to as "forever chemicals." PFAS can be classified based on their carbon chain length. Long-chain PFAS are defined as those containing ≥6 carbon atoms for sulfonic acids and ≥8 for carboxylic acids, while short-chain PFAS have fewer than these thresholds (Fang et al., 2023). This classification is not only structural but also affects their environmental behavior and persistence. Long-chain PFAS tend to be more bioaccumulative and have lower solubility, making them more likely to persist in sediments and biota, while short-chain PFAS are more mobile and soluble but still highly resistant to degradation (Fang et al., 2023).

The persistence of PFAS is primarily due to the strong carbon–fluorine (C–F) bonds present in their structures. PFAS possesses a carbon chain characterized by fluorine atoms and fluorocarbon bonds. The fluorocarbon bond, with a bond energy of 488 kJ $mol^{-1}$, is among the strongest covalent bonds found in organic compounds. The relative attraction of fluorine to carbon and its electronegativity are the causes of the carbon–fluorine (C–F) bond's strength. This attraction produces a partial positive charge on carbon and a partial negative charge on fluorine, and because of this strong bond, PFAS are highly stable and resistant to breaking down. This inherent strength contributes to the exceptional thermodynamic and chemical stability of PFAS, rendering them highly resistant to degradation (Fang et al., 2023). The hydrophobic and oleophobic properties of PFAS greatly contribute to their persistence by preventing interactions with substances that normally promote degradation. The low polarization of fluorine atoms contributes to the hydrophobic and oleophobic nature of PFAS, and these properties are influenced by the length of the carbon chain. Long-chain PFAS exhibit greater chemical stability, enhanced hydrophobicity, and increased lipophilicity (Meegoda et al., 2020; Leung et al., 2023). In contrast, short-chain PFAS demonstrate higher water solubility and a lower tendency to adsorb onto surfaces (Gagliano et al., 2019). Additionally, the chemical characteristics and environmental behavior of PFAS are influenced by functional groups, such as hydroxyls, sulfonates, and carboxyls. PFAS with carboxyl groups (PFCAs) breakdown more readily than those with sulfonate groups (PFSAs) of the same carbon length. These functional groups have the ability to dissociate into ionic forms in aqueous solutions, which alters how they interact with water and soil (Meegoda et al., 2020). Due to these properties, PFAS are globally distributed and have been detected in remote environments such as the Arctic and deep ocean sediments (Ahrens and Bundschuh, 2014).

### 3.4.2 Toxicity and ecological disruptions in aquatic organisms

PFAS are a major concern to aquatic life because they are resistant to degradation and can accumulate having a detrimental effect on various processes. Aquatic organisms, ranging from algae and invertebrates to fish and amphibians, are particularly vulnerable to PFAS toxicity through direct uptake from water, ingestion of contaminated food or sediment, and, in some cases, maternal transfer (see Fig. 3.3). Invertebrates, which play crucial roles in aquatic food webs and nutrient cycling, show consistent sensitivity to PFAS exposure (Byns et al., 2024). *Daphnia magna*, a key freshwater invertebrate, exhibits dose- and time-dependent adverse effects such as reduced survival, growth, reproduction, and altered enzyme activity when exposed to PFOS and PFOA (Ji et al., 2008). PFOS exposure in the freshwater mussel *Hyriopsis cumingii* leads to oxidative stress, tissue damage, and programmed cell death, primarily in the gill and hepatopancreas, reflecting significant biochemical and structural toxicity. Similarly, PFOS exposure reduced survival and increased oxidative stress in crustaceans such as *Macrophthalmus japonicus* (Park et al., 2015) and caused dose- and time-dependent structural damage to the hepatopancreas and gills of *Eriocheir sinensis*. Further, PFAS exposure triggered oxidative stress in *Neocaridina denticulata* and *Gammarus insensibilis*, with *N. denticulata* showing a dose-dependent antioxidant enzyme response and *G. insensibilis* exhibiting cellular but not neurotoxic effects (Touaylia et al., 2019). Several echinoderm species have also been investigated to assess the toxicological impacts of PFAS at physiological and genetic levels. In *Glyptocidaris crenularis*, PFOS caused tissue-level morphological and behavioral impairments, dose-dependent antioxidant enzyme

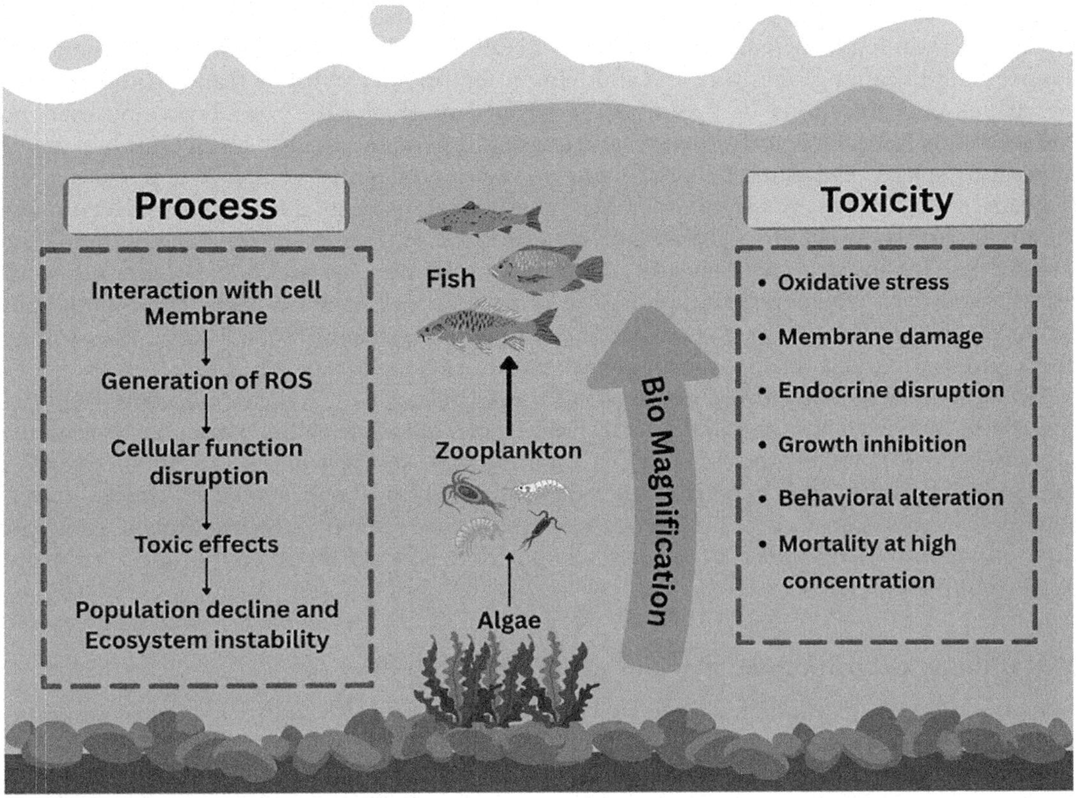


**FIGURE 3.3** per- and polyfluoroalkyl substances-induced biomagnification and toxicity pathways in aquatic organisms.

changes (notably in SOD and ALP), and increased DNA methylation, indicating potential genotoxic effects and long-term biological alterations (Ma et al., 2023).

Among vertebrates, zebrafish (*Danio rerio*) have been extensively studied as a model organism for PFAS toxicity. PFOS exposure in zebrafish induces liver damage, lipid membrane disruption, embryonic apoptosis, and altered muscle-related functions (Ma et al., 2023). One study revealed the misregulation of over 1200 genes and significant changes in miRNA profiles, suggesting widespread developmental, metabolic, and cytotoxic effects, including potential oncogenicity (Zhang et al., 2010). Developmental exposure to 58 individual PFAS (from 14 functional classes) and 2 mixtures at a single low dose in zebrafish revealed that 21 individual PFAS and 1 mixture induced abnormal larval behavior, and further testing of 10 PFCAs at 0–100 μM showed abnormal LPR in all cases except PFDoA (Rericha et al., 2021).

Amphibians are particularly susceptible to PFAS pollution due to their permeable skin and reliance on aquatic habitats during critical life stages (Hong et al., 2025). In a study on *Rana catesbeiana* tadpoles, exposure to PFOS and PFOA led to a notable reduction in snout length, indicating developmental toxicity (Flynn et al., 2019). Amphibians are particularly susceptible to PFAS-induced immunotoxicity due to their environmental exposure and sensitive physiology. In *Rana nigromaculata*, PFOS and PFOA altered immune function by suppressing immunoglobulin A and inducing inflammatory cytokines and oxidative stress at various concentrations (0, 1, and 10 μ g $L^{-1}$) (Liu et al., 2022a,b). Emerging evidence suggests that PFAS can also adversely impact the nervous system in amphibians. In *Lithobates pipiens*, exposure to PFOS and PFOA significantly reduced dopamine levels (Foguth et al., 2019), while another study reported notable declines in serotonin and glutamate levels, indicating PFAS-induced neurotoxicity (Foguth et al., 2020). Research has shown that freshwater shellfish are also highly vulnerable to PFOS exposure, especially during their early developmental phases. Exposure to even low levels of PFOS significantly reduced survival during the glochidia stage, while a concentration of 4.5 μg $L^{-1}$ over 36 days also caused a marked decline in marsupial survival (Hazelton et al., 2012).

Collectively, these findings demonstrate that PFAS cause diverse and often species-specific toxic effects across aquatic taxa. From molecular and cellular damage to impaired growth, behavior, and reproduction, PFAS interfere with key physiological systems, including the endocrine, immune, hepatic, and nervous systems. These sublethal effects, though sometimes subtle, can accumulate across generations and scale up to cause broader ecological disruptions, including shifts in species composition, loss of biodiversity, and diminished ecosystem resilience.

### 3.4.3 Impacts on biodiversity and ecosystem stability

Beyond individual-level toxicity, PFAS contamination can significantly disrupt fundamental ecological processes, posing long-term threats to ecosystem functionality and stability. These disruptions are primarily driven by alterations in microbial activity, species composition, nutrient cycling, and interspecies interactions.

PFAS can exert both direct toxic effects on aquatic life and indirect ecological consequences by altering the quality of primary food sources. In aquatic environments, algae serve as the foundational producers, forming the base of the food web by generating biomass through photosynthesis (see Fig. 3.3). However, the presence of PFAS can disrupt this role by interfering with algal cell membrane integrity (Latała et al., 2009). This disruption tends to be more severe with PFAS compounds that have longer carbon chains and in algae with less structural protection, such as those with lower cellulose content in their cell walls (Ahrens and Bundschuh, 2014). Such physiological impacts on algae don't just affect the organisms directly; they can ripple through the ecosystem, especially impacting animals that feed on algae, such as filter-feeding or grazing invertebrates. This can result in reduced nutritional quality of algae, which may in turn affect the reproductive success and development of key aquatic species, such as the commonly studied water flea *D. magna* (Colombo et al., 2008), as well as broader zooplankton populations (Sanderson et al., 2002). Similarly, in heterotrophic food webs, where energy flows through the decomposition of organic matter, such as leaf litter, PFAS can disrupt microbial-mediated processes essential for nutrient cycling (Ahrens and Bundschuh, 2014). Leaf litter decomposition begins with a conditioning phase, where fungi and bacteria breakdown complex plant materials and enhance their nutritional value for detritivorous invertebrates (Schneider et al., 2010b). PFAS contamination may interfere with this microbial activity by compromising cell membrane integrity and inducing oxidative stress in microbes, particularly bacteria. Research indicates that PFAS, particularly PFCAs, have the potential to significantly impair microbial function, especially by compromising cell membrane integrity and inducing oxidative stress (Reistad et al., 2013). Additionally, such interference may alter the synergistic (Schneider et al., 2010b) or antagonistic (Gulis and Suberkropp, 2003) interactions between fungi and bacteria, further affecting the conditioning efficiency. As a result, leaf-shredding invertebrates may experience reduced food quality, leading to physiological stress and slower breakdown of organic matter, ultimately hindering ecosystem processes such as detritus turnover and secondary production (Fenoy et al., 2021). So, all of these impacts demonstrate the pervasive threat PFAS pose to aquatic biodiversity and ecological balance. By weakening foundational trophic interactions and disrupting microbial ecosystem services, PFAS contamination can destabilize entire aquatic food webs.

## 3.5 Challenges and future directions

Understanding the ecotoxicological hazards of PFAS in aquatic ecosystems poses a complex variety of challenges. One main issue is the complexity and diversity of PFAS combinations in the environment. Aquatic samples often contain a wide range of PFAS, including legacy chemicals, short-chain analogs, and precursors, many of which remain analytically unresolved due to limitations in existing detection techniques (Kurwadkar et al., 2021). This hinders precise exposure characterization and risk evaluations, especially in intricate environmental matrices including sediments, suspended particles, and biota. A further significant challenge is the insufficiency of mixture-based ecotoxicological research. The majority of existing research focuses on specific PFAS compounds (e.g., PFOS, PFOA), but natural environments exhibit intricate combinations with uncertain additive or synergistic effects (Kurwadkar et al., 2021). Moreover, while current studies have predominantly focused on the toxicity of PFAS to individual organisms, a thorough comprehension of their impacts at the community and ecological levels is still lacking. The influence of PFAS on microbial populations and their interactions, which is crucial for nutrient cycling, organic matter decomposition, and ecosystem stability, has been inadequately addressed. Alterations in microbial dynamics may lead to cascade effects on biogeochemical processes; however, these impacts are inadequately described in the majority of aquatic systems. For example, while leaf-associated bacteria have been investigated in PFAS exposure studies, aquatic fungi, particularly hyphomycetes involved in litter decomposition, are significantly unexplored (Shittu et al., 2023; Ahrens and Bundschuh, 2014). These fungi play a crucial role in the decomposition of organic matter in aquatic systems (Tant et al., 2015), yet the possible ecotoxicological effects of PFAS on their population, enzymatic function, and ecological roles remain largely unknown. The lack of data hinders our capacity to analyze microbial-mediated ecosystem processes under PFAS stress and underscores the necessity for increased fungal incorporation in future ecotoxicological evaluations. In summary, the major gaps are:

- Inadequate analytical capability to identify the full range of PFAS chemicals, particularly ultra-short-chain and precursor types

- Inadequate investigation of PFAS mixtures, as most investigations emphasize individual compounds despite environmental exposures being complex and combined
- Lack of understanding of PFAS impacts at the community and ecosystem levels, beyond individual species toxicity.
- Significant lack of information regarding aquatic fungi, including hyphomycetes, despite their essential function in organic matter decomposition

So, by addressing these knowledge gaps, future research can contribute to a more integrated understanding of PFAS as emerging contaminants, their ecological consequences, potential risks to the aquatic environment and human health, and support the development of more effective mitigation and management strategies.

## 3.6 Concluding remarks

The ubiquitous presence and persistence of PFAS in aquatic ecosystems constitute a significant and urgent environmental issue. Their remarkable chemical stability, along with various sources including industrial discharges, wastewater effluents, atmospheric deposition, and leaching from contaminated soils, facilitates their widespread distribution in both local and distant waterbodies. PFAS exhibit significant mobility in the environment, facilitating long-range transport and accumulation in areas that are distant from direct sources. Moreover, their interaction with sediments and organic matter affects their fate and mobility, while partitioning behavior determines their bioavailability and long-term environmental persistence. One of the most concerning aspects of PFAS contamination is their ability to bioaccumulate in aquatic organisms and biomagnify through food webs, and ultimately posing risks to ecosystem health and human safety through trophic transfer. The ecological consequences include adverse physiological, biochemical, and reproductive effects on aquatic species, with potential disruptions to ecosystem stability, nutrient cycling, and biodiversity. These results highlight the critical need for more studies on the fate, behavior, and toxicological effects of PFAS in the environment, as well as the development of efficient mitigation and remediation plans. In summary, an integrated strategy involving sophisticated monitoring methods, thorough risk assessments, and strong regulatory frameworks is needed to address the complex hazards connected to PFAS in aquatic environments. In order to protect aquatic ecosystems and the general public from the long-term effects of PFAS contamination, such initiatives are crucial.

## References

Ahrens, L., Bundschuh, M., 2014. Fate and effects of poly- and perfluoroalkyl substances in the aquatic environment: a review. Environmental Toxicology and Chemistry 33 (9), 1921–1929. Available at https://doi.org/10.1002/etc.2663.

Ahrens, L., Gashaw, H., Sjöholm, M., Gebrehiwot, S.G., Getahun, A., Derbe, E., et al., 2016. Poly- and perfluoroalkylated substances (PFASs) in water, sediment and fish muscle tissue from Lake Tana, Ethiopia and implications for human exposure. Chemosphere 165, 352–357. Available at https://doi.org/10.1016/j.chemosphere.2016.09.007.

Ahrens, L., Shoeib, M., Harner, T., Lee, S.C., Guo, R., Reiner, E.J., 2011. Wastewater treatment plant and landfills as sources of polyfluoroalkyl compounds to the atmosphere. Environmental Science & Technology 45 (19), 8098–8105. Available at https://doi.org/10.1021/es1036173.

Alazaiza, M.Y., Alzghoul, T.M., Ramu, M.B., Amr, S.S.A., Abushammala, M.F., 2025. PFAS contamination and mitigation: a comprehensive analysis of research trends and global contributions. Case Studies in Chemical and Environmental Engineering 11, 101127. Available at https://doi.org/10.1016/j.cscee.2025.101127.

Ali, A.M., Sanden, M., Higgins, C.P., Hale, S.E., Alarif, W.M., Al-Lihaibi, S.S., Ræder, E.M., Langberg, H.A., Kallenborn, R., 2021. Legacy and emerging per- and polyfluorinated alkyl substances (PFASs) in sediment and edible fish from the Eastern Red Sea. Environmental Pollution 280, 116935. Available at https://doi.org/10.1016/j.envpol.2021.116935.

Anik, A.H., Basir, M.S., Sultan, M.B., Alam, M., Rahman, M.M., Tareq, S.M., 2025a. Unveiling the emerging concern of per- and polyfluoroalkyl substances (PFAS) and their potential impacts on estuarine ecosystems. Marine Pollution Bulletin 212, 117554. Available at https://doi.org/10.1016/j.marpolbul.2025.117554.

Anik, A.H., Aurnab, I.T., Sultan, M.B., Basir, M.S., Murshed, M.F., Jarin, J.M., Sorker, R., Alam, M., Tareq, S.M., 2025b. A State-of-the-Art review of the environmental mixtures, exposure, and health risks of PFAS: a special focus on developing countries. Water Air & Soil Pollution 236 (9). Available at https://doi.org/10.1007/s11270-025-08253-y.

Banyoi, S., Porseryd, T., Larsson, J., Grahn, M., Dinnétz, P., 2022. The effects of exposure to environmentally relevant PFAS concentrations for aquatic organisms at different consumer trophic levels: systematic review and meta-analyses. Environmental Pollution 315, 120422. Available at https://doi.org/10.1016/j.envpol.2022.120422.

Barber, L.B., Miller, S.A., Blaney, L., Bradley, P.M., Faunce, K.E., Fleck, J.A., Frick, M., He, K., Hollins, R.D., Lewellyn, C.J., Majcher, E.H., McAdoo, M.A., Smalling, K.L., 2025. Municipal and industrial wastewater treatment plant effluent contributions to per- and polyfluoroalkyl substances in the potomac river: a basin-scale measuring and modeling approach. Environmental Science & Technology. Available at https://doi.org/10.1021/acs.est.4c12167.

Benaafi, M., Bafaqeer, A., 2024. Comprehensive review of global perspectives on per- and Polyfluoroalkyl compounds: occurrence, fate, and remediation in groundwater systems. Water 16 (11), 1583. Available at https://doi.org/10.3390/w16111583.

Biswas, B., Joseph, A., Parveen, N., Ranjan, V.P., Goel, S., Mandal, J., Srivastava, P., 2025. Contamination of per- and poly-fluoroalkyl substances in agricultural soils: a review. Journal of Environmental Management 380, 124993. Available at https://doi.org/10.1016/j.jenvman.2025.124993.

Bossi, R., 2023. Per- and polyfluoroalkyl Substances (PFAS) in atmosphere and deposition. Available at https://dce.au.dk/fileadmin/dce.au.dk/Udgivelser/Notater_2023/N2023_20.pdf.

Byns, C., Groffen, T., Bervoets, L., 2024. Aquatic macroinvertebrate community responses to pollution of perfluoroalkyl substances (PFAS): can we define threshold body burdens? The Science of the Total Environment 917, 170611. Available at https://doi.org/10.1016/j.scitotenv.2024.170611.

Campo, J., Pérez, F., Masiá, A., Picó, Y., Farré, M., Barceló, D., 2014. Perfluoroalkyl substance contamination of the Llobregat River ecosystem (Mediterranean area, NE Spain). The Science of the Total Environment 503–504, 48–57. Available at https://doi.org/10.1016/j.scitotenv.2014.05.094.

Chen, M., Zhu, L., Wang, Q., Shan, G., 2020. Tissue distribution and bioaccumulation of legacy and emerging per- and polyfluoroalkyl substances (PFASs) in edible fishes from Taihu Lake, China. Environmental Pollution 268, 115887. Available at https://doi.org/10.1016/j.envpol.2020.115887.

Colombo, I., De Wolf, W., Thompson, R.S., Farrar, D.G., Hoke, R.A., L'Haridon, J., 2008. Acute and chronic aquatic toxicity of ammonium perfluorooctanoate (APFO) to freshwater organisms. Ecotoxicology and Environmental Safety 71 (3), 749–756. Available at https://doi.org/10.1016/j.ecoenv.2008.04.002.

De Silva, A.O., Spencer, C., Scott, B.F., Backus, S., Muir, D.C.G., 2011. Detection of a cyclic perfluorinated acid, perfluoroethylcyclohexane sulfonate, in the Great Lakes of North America. Environmental Science & Technology 45 (19), 8060–8066. Available at https://doi.org/10.1021/es200135c.

Dimitrakopoulou, M., Karvounis, M., Marinos, G., Theodorakopoulou, Z., Aloizou, E., Petsangourakis, G., et al., 2024. Comprehensive analysis of PFAS presence from environment to plate. Npj Science of Food 8 (1). Available at https://doi.org/10.1038/s41538-024-00319-1.

Ehsan, M.N., Riza, M., Pervez, M.N., Li, C., Zorpas, A.A., Naddeo, V., 2024. PFAS contamination in soil and sediment: contribution of sources and environmental impacts on soil biota. Case Studies in Chemical and Environmental Engineering 9, 100643. Available at https://doi.org/10.1016/j.cscee.2024.100643.

Environmental Fate and Transport for Per- and Polyfluoroalkyl Substances, 2018, March. Available at https://www.nccoast.org/wp-content/uploads/2018/10/ITRC-PFAS-Transport-Fact-Sheet.pdf.

Fang, J., Li, S., Gu, T., Liu, A., Qiu, R., Zhang, W., 2023. Treatment of per- and polyfluoroalkyl substances (PFAS): a review of transformation technologies and mechanisms. Journal of Environmental Chemical Engineering 12 (1), 111833. Available at https://doi.org/10.1016/j.jece.2023.111833.

Fenoy, E., Rubio-Ríos, J., González, J.M., Salinas, M.J., Moyano, F.J., Casas, J.J., 2021. Strategies of shredders when feeding on low-quality leaf-litter: Local population adaptations or fixed species traits? Limnology and Oceanography 66(5). Available at https://doi.org/10.1002/lno.11745.

Flynn, R.W., Chislock, M.F., Gannon, M.E., Bauer, S.J., Tornabene, B.J., Hoverman, J.T., et al., 2019. Acute and chronic effects of perfluoroalkyl substance mixtures on larval American bullfrogs (*Rana catesbeiana*). Chemosphere 236, 124350. Available at https://doi.org/10.1016/j.chemosphere.2019.124350.

Foguth, R.M., Flynn, R.W., De Perre, C., Iacchetta, M., Lee, L.S., Sepúlveda, M.S., et al., 2019. Developmental exposure to perfluorooctane sulfonate (PFOS) and perfluorooctanoic acid (PFOA) selectively decreases brain dopamine levels in Northern leopard frogs. Toxicology and Applied Pharmacology 377, 114623. Available at https://doi.org/10.1016/j.taap.2019.114623.

Foguth, R., Hoskins, T., Clark, G., Nelson, M., Flynn, R., De Perre, C., et al., 2020. Single and mixture per- and polyfluoroalkyl substances accumulate in developing Northern leopard frog brains and produce complex neurotransmission alterations. Neurotoxicology and Teratology 81, 106907. Available at https://doi.org/10.1016/j.ntt.2020.106907.

Gagliano, E., Sgroi, M., Falciglia, P.P., Vagliasindi, F.G., Roccaro, P., 2019. Removal of poly- and perfluoroalkyl substances (PFAS) from water by adsorption: role of PFAS chain length, effect of organic matter and challenges in adsorbent regeneration. Water Research 171, 115381. Available at https://doi.org/10.1016/j.watres.2019.115381.

Gassmann, M., Weidemann, E., Stahl, T., 2020. Combined leaching and plant uptake simulations of PFOA and PFOS under field conditions. Environmental Science and Pollution Research 28 (2), 2097–2107. Available at https://doi.org/10.1007/s11356-020-10594-6.

Giari, L., Guerranti, C., Perra, G., Cincinelli, A., Gavioli, A., Lanzoni, M., et al., 2023. PFAS levels in fish species in the Po River (Italy): new generation PFAS, fish ecological traits and parasitism in the foreground. The Science of the Total Environment 876, 162828. Available at https://doi.org/10.1016/j.scitotenv.2023.162828.

Gulis, V., Suberkropp, K., 2003. Interactions between stream fungi and bacteria associated with decomposing leaf litter at different levels of nutrient availability. Aquatic Microbial Ecology 30, 149–157. Available at https://doi.org/10.3354/ame030149.

Hazelton, P.D., Cope, W.G., Pandolfo, T.J., Mosher, S., Strynar, M.J., Barnhart, M.C., et al., 2012. Partial life-cycle and acute toxicity of perfluoroalkyl acids to freshwater mussels. Environmental Toxicology and Chemistry 31 (7), 1611–1620. Available at https://doi.org/10.1002/etc.1866.

Hellsing, M.S., Josefsson, S., Hughes, A.V., Ahrens, L., 2016. Sorption of perfluoroalkyl substances to two types of minerals. Chemosphere 159, 385–391. Available at https://doi.org/10.1016/j.chemosphere.2016.06.016.

Helmer, R.W., Reeves, D.M., Cassidy, D.P., 2021. Per- and Polyfluorinated Alkyl Substances (PFAS) cycling within Michigan: contaminated sites, landfills and wastewater treatment plants. Water Research 210, 117983. Available at https://doi.org/10.1016/j.watres.2021.117983.

Heydebreck, F., Tang, J., Xie, Z., Ebinghaus, R., 2016. Emissions of per- and polyfluoroalkyl substances in a textile manufacturing plant in China and their relevance for workers' exposure. Environmental Science & Technology 50 (19), 10386–10396. Available at https://doi.org/10.1021/acs.est.6b03213.

Higgins, C.P., Luthy, R.G., 2006. Sorption of perfluorinated surfactants on sediments. Environmental Science & Technology 40 (23), 7251–7256. Available at https://doi.org/10.1021/es061000n.

Hong, M., Lee, J., Lee, M., Lee, J., 2025. Ecotoxicological effects of per- and polyfluoroalkyl substances in aquatic organisms: a review. Marine Pollution Bulletin 214, 117678. Available at https://doi.org/10.1016/j.marpolbul.2025.117678.

Hossain, M.M., Nawaz, M.Z., Dar, M.A., Geng, A., Alghamdi, H.A., Han, S., et al., 2024. Per- and polyfluoroalkyl substances (PFAS) exposure in biota and remediation strategies: toxicological and biochemical perspectives. Journal of Hazardous Materials Advances 17, 100579. Available at https://doi.org/10.1016/j.hazadv.2024.100579.

Jahnke, A., Ahrens, L., Ebinghaus, R., Temme, C., 2006. Urban versus remote air concentrations of fluorotelomer alcohols and other polyfluorinated alkyl substances in Germany. Environmental Science & Technology 41 (3), 745–752. Available at https://doi.org/10.1021/es0619861.

Jeon, J., Kannan, K., Lim, B.J., An, K.G., Kim, S.D., 2011. Effects of salinity and organic matter on the partitioning of perfluoroalkyl acid (PFAs) to clay particles. Journal of Environmental Monitoring 13 (6), 1803. Available at https://doi.org/10.1039/c0em00791a.

Ji, K., Kim, Y., Oh, S., Ahn, B., Jo, H., Choi, K., 2008. Toxicity of perfluorooctane sulfonic acid and perfluorooctanoic acid on freshwater macroinvertebrates (*Daphnia magna* and *Moina macrocopa*) and fish (*Oryzias latipes*). Environmental Toxicology and Chemistry 27 (10), 2159–2168. Available at https://doi.org/10.1897/07-523.1.

Jin, Y.H., Liu, W., Sato, I., Nakayama, S.F., Sasaki, K., Saito, N., et al., 2009. PFOS and PFOA in environmental and tap water in China. Chemosphere 77 (5), 605–611. Available at https://doi.org/10.1016/j.chemosphere.2009.08.058.

Kurwadkar, S., Dane, J., Kanel, S.R., Nadagouda, M.N., Cawdrey, R.W., Ambade, B., et al., 2021. Per- and polyfluoroalkyl substances in water and wastewater: a critical review of their global occurrence and distribution. The Science of the Total Environment 809, 151003. Available at https://doi.org/10.1016/j.scitotenv.2021.151003.

Latała, A., Nędzi, M., Stepnowski, P., 2009. Acute toxicity assessment of perfluorinated carboxylic acids towards the Baltic microalgae. Environmental Toxicology and Pharmacology 28 (2), 167–171. Available at https://doi.org/10.1016/j.etap.2009.03.010.

Leeson, A., Thompson, T., Stroo, H.F., Anderson, R.H., Speicher, J., Mills, M.A., et al., 2020. Identifying and managing aqueous film-forming foam-derived per- and polyfluoroalkyl substances in the environment. Environmental Toxicology and Chemistry 40 (1), 24–36. Available at https://doi.org/10.1002/etc.4894.

Leung, S.C.E., Wanninayake, D., Chen, D., Nguyen, N., Li, Q., 2023. Physicochemical properties and interactions of perfluoroalkyl substances (PFAS) - challenges and opportunities in sensing and remediation. The Science of the Total Environment 905, 166764. Available at https://doi.org/10.1016/j.scitotenv.2023.166764.

Lewis, A.J., Yun, X., Spooner, D.E., Kurz, M.J., McKenzie, E.R., Sales, C.M., 2022. Exposure pathways and bioaccumulation of per- and polyfluoroalkyl substances in freshwater aquatic ecosystems: key considerations. The Science of the Total Environment 822, 153561. Available at https://doi.org/10.1016/j.scitotenv.2022.153561.

Li, Y., Yao, J., Zhang, J., Pan, Y., Dai, J., Ji, C., et al., 2021. First report on the bioaccumulation and trophic transfer of perfluoroalkyl ether carboxylic acids in estuarine food web. Environmental Science & Technology 56 (10), 6046–6055. Available at https://doi.org/10.1021/acs.est.1c00965.

Liu, M., Munoz, G., Duy, S.V., Sauvé, S., Liu, J., 2021. Per- and polyfluoroalkyl substances in contaminated soil and groundwater at airports: a Canadian case study. Environmental Science & Technology 56 (2), 885–895. Available at https://doi.org/10.1021/acs.est.1c04798.

Liu, S., Jin, B., Arp, H.P.H., Chen, W., Liu, Y., Zhang, G., 2022a. The fate and transport of chlorinated polyfluorinated ether sulfonates and other PFAS through industrial wastewater treatment facilities in China. Environmental Science & Technology 56 (5), 3002–3010. Available at https://doi.org/10.1021/acs.est.1c04276.

Liu, Z., Lin, H., Zheng, Y., Feng, Y., Shi, C., Zhu, R., et al., 2022b. Perfluorooctanoic acid and perfluorooctanesulfonic acid induce immunotoxicity through the NF-κB pathway in black-spotted frog (*Rana nigromaculata*). Chemosphere 313, 137622. Available at https://doi.org/10.1016/j.chemosphere.2022.137622.

Ma, T., Wu, P., Wang, L., Li, Q., Li, X., Luo, Y., 2023. Toxicity of per- and polyfluoroalkyl substances to aquatic vertebrates. Frontiers in Environmental Science 11. Available at https://doi.org/10.3389/fenvs.2023.1101100.

Macorps, N., Menach, K.L., Pardon, P., Guérin-Rechdaoui, S., Rocher, V., Budzinski, H., et al., 2022. Bioaccumulation of per- and polyfluoroalkyl substance in fish from an urban river: occurrence, patterns and investigation of potential ecological drivers. Environmental Pollution 303, 119165. Available at https://doi.org/10.1016/j.envpol.2022.119165.

Manzetti, S., 2013. Polycyclic aromatic hydrocarbons in the environment: environmental fate and transformation. Polycyclic Aromatic Compounds 33 (4), 311–330. Available at https://doi.org/10.1080/10406638.2013.781042.

Masoner, J.R., Kolpin, D.W., Cozzarelli, I.M., Smalling, K.L., Bolyard, S.C., Field, J.A., et al., 2020. Landfill leachate contributes per-/poly-fluoroalkyl substances (PFAS) and pharmaceuticals to municipal wastewater. Environmental Science Water Research & Technology 6 (5), 1300–1311. Available at https://doi.org/10.1039/d0ew00045k.

Meegoda, J.N., Kewalramani, J.A., Li, B., Marsh, R.W., 2020. A review of the applications, environmental release, and remediation technologies of per- and polyfluoroalkyl substances. International Journal of Environmental Research and Public Health 17 (21), 8117. Available at https://doi.org/10.3390/ijerph17218117.

Miranda, D.A., Benskin, J.P., Awad, R., Lepoint, G., Leonel, J., Hatje, V., 2020. Bioaccumulation of Per- and polyfluoroalkyl substances (PFASs) in a tropical estuarine food web. The Science of the Total Environment 754, 142146. Available at https://doi.org/10.1016/j.scitotenv.2020.142146.

Möller, A., Ahrens, L., Surm, R., Westerveld, J., Van Der Wielen, F., Ebinghaus, R., et al., 2010. Distribution and sources of polyfluoroalkyl substances (PFAS) in the River Rhine watershed. Environmental Pollution 158 (10), 3243–3250. Available at https://doi.org/10.1016/j.envpol.2010.07.019.

Moneta, B.G., Feo, M.L., Torre, M., Tratzi, P., Aita, S.E., Montone, C.M., et al., 2023. Occurrence of per- and polyfluorinated alkyl substances in wastewater treatment plants in Northern Italy. The Science of the Total Environment 894, 165089. Available at https://doi.org/10.1016/j.scitotenv.2023.165089.

Mussabek, D., Persson, K.M., Berndtsson, R., Ahrens, L., Nakagawa, K., Imura, T., 2020. Impact of the sediment organic vs. mineral content on distribution of the per- and polyfluoroalkyl substances (PFAS) in lake sediment. International Journal of Environmental Research and Public Health 17 (16), 5642. Available at https://doi.org/10.3390/ijerph17165642.

Nayak, S., Sahoo, G., Das, I.I., Mohanty, A.K., Kumar, R., Sahoo, L., et al., 2023. Poly- and perfluoroalkyl substances (PFAS): do they matter to aquatic ecosystems? Toxics 11 (6), 543. Available at https://doi.org/10.3390/toxics11060543.

Panieri, E., Baralic, K., Djukic-Cosic, D., Djordjevic, A.B., Saso, L., 2022. PFAS molecules: a major concern for the human health and the environment. Toxics 10 (2), 44. Available at https://doi.org/10.3390/toxics10020044.

Park, K., Nikapitiya, C., Kwak, T., Kwak, I., 2015. Antioxidative-related genes expression following perfluorooctane sulfonate (PFOS) exposure in the intertidal mud crab, *Macrophthalmus japonicus*. Ocean Science Journal 50 (3), 547–556. Available at https://doi.org/10.1007/s12601-015-0050-0.

Peng, J., Huang, H., Xu, C., Li, J., Lin, S., Zhao, H., et al., 2022. Occurrence, distribution, and ecological risks of per- and polyfluoroalkyl substances (PFASs) in the surface water of Caohai wetland, Guizhou Province, China. Research Square (Research Square). Available at https://doi.org/10.21203/rs.3.rs-2358484/v1.

Pulster, E.L., Rullo, K., Gilbert, S., Ash, T.M., Goetting, B., Campbell, K., et al., 2022. Assessing per- and polyfluoroalkyl substances (PFAS) in sediments and fishes in a large, urbanized estuary and the potential human health implications. Frontiers in Marine Science 9. Available at https://doi.org/10.3389/fmars.2022.1046667.

Reistad, T., Fonnum, F., Mariussen, E., 2013. Perfluoroalkylated compounds induce cell death and formation of reactive oxygen species in cultured cerebellar granule cells. Toxicology Letters 218 (1), 56–60. Available at https://doi.org/10.1016/j.toxlet.2013.01.006.

Ren, J., Point, A., Baygi, S.F., Fernando, S., Hopke, P.K., Holsen, T., et al., 2022. Bioaccumulation of perfluoroalkyl substances in the Lake Erie Food web. SSRN Electronic Journal. Available at https://doi.org/10.2139/ssrn.4224037.

Rericha, Y., Cao, D., Truong, L., Simonich, M., Field, J.A., Tanguay, R.L., 2021. Behavior effects of structurally diverse per- and polyfluoroalkyl substances in zebrafish. Chemical Research in Toxicology 34 (6), 1409–1416. Available at https://doi.org/10.1021/acs.chemrestox.1c00101.

Sanderson, H., Boudreau, T.M., Mabury, S.A., Cheong, W., Solomon, K.R., 2002. Ecological impact and environmental fate of perfluorooctane sulfonate on the zooplankton community in indoor microcosms. Environmental Toxicology and Chemistry 21 (7), 1490–1496. Available at https://doi.org/10.1002/etc.5620210722.

Schneider, T., Gerrits, B., Gassmann, R., Schmid, E., Gessner, M.O., Richter, A., et al., 2010b. Proteome analysis of fungal and bacterial involvement in leaf litter decomposition. Proteomics 10 (9), 1819–1830. Available at https://doi.org/10.1002/pmic.200900691.

Sharma, B.M., Bharat, G.K., Tayal, S., Larssen, T., Bečanová, J., Karásková, P., et al., 2015. Perfluoroalkyl substances (PFAS) in river and ground/drinking water of the Ganges River basin: emissions and implications for human exposure. Environmental Pollution 208, 704–713. Available at https://doi.org/10.1016/j.envpol.2015.10.050.

Sharma, N., Kumar, V., Sugumar, V., Umesh, M., Sondhi, S., Chakraborty, P., et al., 2024. A comprehensive review on the need for integrated strategies and process modifications for per- and polyfluoroalkyl substances (PFAS) removal: current insights and future prospects. Case Studies in Chemical and Environmental Engineering 9, 100623. Available at https://doi.org/10.1016/j.cscee.2024.100623.

Shittu, A.R., Iwaloye, O.F., Ojewole, A.E., Rabiu, A.G., Amechi, M.O., Herve, O.F., 2023. The effects of per- and polyfluoroalkyl substances on environmental and human microorganisms and their potential for bioremediation. Archives of Industrial Hygiene and Toxicology 74 (3), 167–178. Available at https://doi.org/10.2478/aiht-2023-74-3708.

Simonnet-Laprade, C., Budzinski, H., Maciejewski, K., Menach, K.L., Santos, R., Alliot, F., et al., 2019. Biomagnification of perfluoroalkyl acids (PFAAs) in the food web of an urban river: assessment of the trophic transfer of targeted and unknown precursors and implications. Environmental Science Processes & Impacts 21 (11), 1864–1874. Available at https://doi.org/10.1039/c9em00322c.

Stoiber, T., Evans, S., Naidenko, O.V., 2020. Disposal of products and materials containing per- and polyfluoroalkyl substances (PFAS): a cyclical problem. Chemosphere 260, 127659. Available at https://doi.org/10.1016/j.chemosphere.2020.127659.

Survey finds PFAS in 71% of shallow private wells across Wisconsin | AP News, 2023, November 3. AP News. Available at https://apnews.com/article/pfas-pollution-wells-wisconsin-shallow-groundwater-4e7e7c901ace10abe8050c51530aa504.

Szabo, D., Coggan, T.L., Robson, T.C., Currell, M., Clarke, B.O., 2018. Investigating recycled water use as a diffuse source of per- and polyfluoroalkyl substances (PFASs) to groundwater in Melbourne, Australia. The Science of the Total Environment 644, 1409–1417. Available at https://doi.org/10.1016/j.scitotenv.2018.07.048.

Taniyasu, S., Yamashita, N., Moon, H., Kwok, K.Y., Lam, P.K., Horii, Y., et al., 2013. Does wet precipitation represent local and regional atmospheric transportation by perfluorinated alkyl substances? Environment International 55, 25–32. Available at https://doi.org/10.1016/j.envint.2013.02.005.

Tant, C.J., Rosemond, A.D., Mehring, A.S., Kuehn, K.A., Davis, J.M., 2015. The role of aquatic fungi in transformations of organic matter mediated by nutrients. Freshwater Biology 60 (7), 1354–1363. Available at https://doi.org/10.1111/fwb.12573.

Thompson, K.A., Mortazavian, S., Gonzalez, D.J., Bott, C., Hooper, J., Schaefer, C.E., et al., 2022. Poly- and perfluoroalkyl substances in municipal wastewater treatment plants in the United States: seasonal patterns and meta-analysis of long-term trends and average concentrations. ACS ES&T Water 2 (5), 690–700. Available at https://doi.org/10.1021/acsestwater.1c00377.

Touaylia, S., Khazri, A., Mezni, A., Bejaoui, M., 2019. Effects of emerging persistent organic pollutant perfluorooctane sulfonate (PFOS) on the *CrustaceanGammarus insensibilis*. Human and Ecological Risk Assessment an International Journal 25 (8), 2133–2141. Available at https://doi.org/10.1080/10807039.2018.1489717.

Wang, P., Lu, Y., Su, H., Su, C., Johnson, A.C., Yu, L., et al., 2020. Managing health risks of perfluoroalkyl acids in aquatic food from a river-estuary-sea environment affected by fluorochemical industry. Environment International 138, 105621. Available at https://doi.org/10.1016/j.envint.2020.105621.

Wang, Q., Ruan, Y., Jin, L., Lu, G., Ma, L., Yeung, L.W., et al., 2022a. Oysters for legacy and emerging per- and polyfluoroalkyl substances (PFASs) monitoring in estuarine and coastal waters: phase distribution and bioconcentration profile. The Science of the Total Environment 846, 157453. Available at https://doi.org/10.1016/j.scitotenv.2022.157453.

Wang, S., Lin, X., Li, Q., Liu, C., Li, Y., Wang, X., 2022b. Neutral and ionizable per-and polyfluoroalkyl substances in the urban atmosphere: occurrence, sources and transport. The Science of the Total Environment 823, 153794. Available at https://doi.org/10.1016/j.scitotenv.2022.153794.

Wang, Y., Munir, U., Huang, Q., 2023. Occurrence of per- and polyfluoroalkyl substances (PFAS) in soil: sources, fate, and remediation. Soil & Environmental Health 1 (1), 100004. Available at https://doi.org/10.1016/j.seh.2023.100004.

Wu, X., Yu, S., Zeng, J., Zheng, X., Ren, Z., Shu, Y., et al., 2024. Biomagnification of persistent organic pollutants (POPs) in detritivorous, phytophagous, and predatory invertebrates: how POPs enter terrestrial food web? The Science of the Total Environment 924, 171677. Available at https://doi.org/10.1016/j.scitotenv.2024.171677.

Xiao, F., 2017. Emerging poly- and perfluoroalkyl substances in the aquatic environment: a review of current literature. Water Research 124, 482–495. Available at https://doi.org/10.1016/j.watres.2017.07.024.

Xie, X., Lu, Y., Lei, H., Cheng, J., An, X., Wang, W., et al., 2024. Bioaccumulation and trophic transfer of per- and polyfluoroalkyl substances in a subtropical mangrove estuary food web. The Science of the Total Environment 927, 172094. Available at https://doi.org/10.1016/j.scitotenv.2024.172094.

Yao, Y., Wang, X., Liu, F., Zhang, W., Artigas, F.J., Gao, Y., 2025. Distributions and partitioning of airborne Per- and polyfluoroalkyl substances (PFAS) in urban atmosphere of Northern New Jersey. The Science of the Total Environment 970, 179037. Available at https://doi.org/10.1016/j.scitotenv.2025.179037.

You, C., Jia, C., Pan, G., 2010. Effect of salinity and sediment characteristics on the sorption and desorption of perfluorooctane sulfonate at sediment-water interface. Environmental Pollution 158 (5), 1343–1347. Available at https://doi.org/10.1016/j.envpol.2010.01.009.

Young, C.J., Mabury, S.A., 2010. Atmospheric perfluorinated acid precursors: chemistry, occurrence, and impacts. Reviews of Environmental Contamination and Toxicology 1–109. Available at https://doi.org/10.1007/978-1-4419-6880-7_1.

Zarębska, M., Bajkacz, S., Hordyjewicz-Baran, Z., 2024. Assessment of legacy and emerging PFAS in the Oder River: occurrence, distribution, and sources. Environmental Research 251, 118608. Available at https://doi.org/10.1016/j.envres.2024.118608.

Zhang, D., Liu, W., Xin, Y., Liu, X., Zhang, Z., Liu, Y., 2024. Trophic transfer of PFAS potentially threatens vulnerable Saunders's gull (*Larus saundersi*) via the food chain in the coastal wetlands of the Yellow Sea, China. Frontiers in Marine Science 11. Available at https://doi.org/10.3389/fmars.2024.1467022.

Zhang, L., Li, Y., Zeng, H., Wei, J., Wan, Y., Chen, J., et al., 2010. MicroRNA expression changes during zebrafish development induced by perfluorooctane sulfonate. Journal of Applied Toxicology 31 (3), 210–222. Available at https://doi.org/10.1002/jat.1583.

Zhou, J., Li, Z., Guo, X., Li, Y., Wu, Z., Zhu, L., 2019. Evidences for replacing legacy per- and polyfluoroalkyl substances with emerging ones in Fen and Wei River basins in central and western China. Journal of Hazardous Materials 377, 78–87. Available at https://doi.org/10.1016/j.jhazmat.2019.05.050.

CHAPTER

# 4

# Microplastics as carriers of per- and polyfluoroalkyl compounds

*Andreas Tzachristas*[1], *Ioannis Ioannidis*[2], *Vasiliki Kinigopoulou*[3], *Paraskevi Mpeza*[4], *Dimitrios A. Giannakoudakis*[5], *Ioannis Anastopoulos*[4], *Ioannis Pashalidis*[2], *and Hrissi K. Karapanagioti*[1]

[1]Department of Chemistry, University of Patras, Patras, Greece [2]Department of Chemistry, University of Cyprus, Nicosia, Cyprus [3]Department of Environmental Engineering, School of Engineering, International Hellenic University, Sindos, Thessaloniki, Greece [4]Department of Agriculture, University of Ioannina (UoI), Kostakii Campus, Arta, Greece [5]Faculty of Chemistry, Institute of Chemical Sciences, Maria Curie-Sklodowska University, Lublin, Poland

## 4.1 Introduction

Microplastics (MPs) can act as per- and polyfluoroalkyl compounds (PFAS) carriers. This can be true either due to the presence of PFAS in the initial plastic item that degraded to produce MPs or due to the sorption of PFAS onto the MPs during their transport through polluted areas where both pollutants coexist. To better understand this topic, the definitions of both MPs and PFAS are provided below.

### 4.1.1 Microplastics

MPs are small plastic particles with all their dimensions smaller than 5 mm (GESAMP, 2015). There are primary MPs that were produced in this small size, and there are secondary MPs that result from the degradation of bigger plastic items. The MP polymer types that can be found predominantly in the environment are polyethylene (PE), polypropylene (PP), polyethylene terephthalate (PET), and polystyrene (PS). Based on their appearance, they can be characterized as fragments, foams, fibers, films, microbeads, and plastic pellets. This last characterization can potentially help determine the sources of MPs in the environment.

*Per- and Polyfluoroalkyl Substances (PFAS) as Environmental Contaminants*
DOI: https://doi.org/10.1016/B978-0-443-43856-1.00009-5

MPs have been detected in an increasing number of environmental systems, including wastewater, freshwater bodies such as rivers, lakes, sediments (Karapanagioti and Kalavrouziotis, 2020), terrestrial ecosystems (Isari et al., 2021), marine ecosystems (Tziourrou et al., 2019), the atmosphere (O'Brien et al., 2023), as well as within organisms (Simantiris et al., 2024). More recently, their presence has been confirmed in human food (Mamun et al., 2023) and tissues (Yang et al., 2023). MPs, similar to bigger plastics, are vectors of hazardous chemicals (Takada and Karapanagioti, 2019) such as persistent organic pollutants (POPs) and metals. MPs carry chemical additives that were added to them during their production. These additives can be colors, plasticizers, fillings, flame retardants, etc. At the same time, since MPs are organic phases and usually hydrophobic, they can sorb organic pollutants when in contact with polluted water. These chemicals tend to be desorbed when in contact with other lipophilic media, such as fish oil or organic solvents, either in the environment or inside organisms' bodies.

### 4.1.2 Per- and polyfluoroalkyl compounds

PFAS are a category of exceedingly fluorinated chemical species that have been widely produced since 1940 and have found application in industrial processes and everyday products such as flame retardants, surfactants, polymer additives, lubricants, etc. Moreover, PFAS are generated as by-products and residues in different processes (Buck et al., 2011). Perfluorooctanoic acid (PFOA) and perfluorooctane sulfonate (PFOS), the skeletal formulas of which are shown in Fig. 4.1, are two of the most used PFAS compounds and the stable end products of other PFAS forms (Dickman et al., 2022).

Despite their global phase out, PFAS are still widely spread in various environmental compartments (e.g., atmosphere, waters, and soils) (Wee et al., 2023) because they are very stable and resistant to degradation (POPs), and are therefore commonly referred to as "forever chemicals" and are listed in the Stockholm Convention (Cousins et al., 2020; Secretariat of the Stockholm Convention, 2019). In addition, PFAS exhibit a broad range of toxic effects, such as carcinogenicity, hormonal and developmental toxicity, etc. (Schiavone et al., 2023).

PFAS are both hydrophilic and lipophilic compounds, resulting in complex behavior and transport in the environment and particularly in aquatic and soil environments. Depending on their chain length and polarity, PFAS can be found dissolved in aqueous media and adsorbed on solid matrices (e.g., mineral surfaces) and colloidal particles (inorganic, organic, and biogenic particles) and are capable of bioaccumulation and biomagnification in living organisms, thereby posing severe toxicity to affected organisms, including humans.

Due to their bioaccumulation and biomagnification properties, and the discovery of a wide range of health problems related to PFAS, monitoring and understanding their

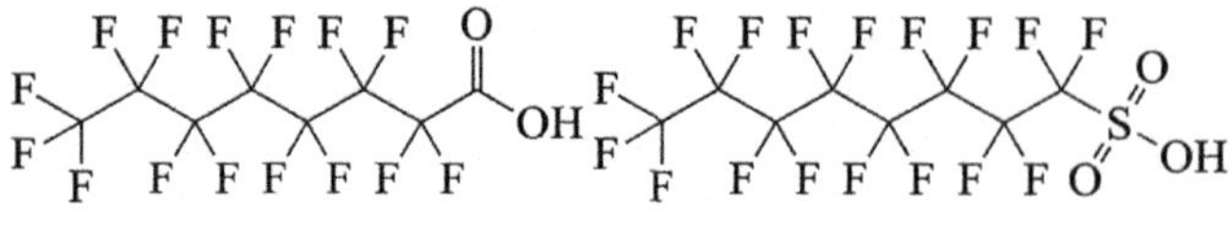

FIGURE 4.1 Skeletal formula of perfluorooctanoic acid and perfluorooctane sulfonate.

chemical behavior and migration in PFAS-contaminated environments is obligatory and of fundamental importance regarding their environmental fate and the development of efficient remediation technologies. In this context, the dispersion of PFAS contained in MPs, as well as after interaction in the environmental compartments to form secondary pollutants, is of particular interest.

## 4.2 Relevant literature

Bibliometric evidence (Fig. 4.2) demonstrates the rapid rise and diversification of research on MPs over the past decade. From fewer than 100 publications in 2015, the annual output surged to nearly 6800 papers in 2024, establishing MPs as one of the most intensively studied emerging contaminants worldwide. Within this expanding field, more than 1000 papers in 2024 specifically investigated MPs as adsorptive materials or carriers, highlighting increasing recognition of their ability to accumulate, transport, and redistribute pollutants across ecosystems. In parallel, PFAS research has evolved into a well-established domain of environmental science, with more than 1600 publications reported in 2024 alone.

Despite these robust individual trajectories, the overlap between MPs and PFAS research remains strikingly limited. Fewer than 50 papers per year directly address PFAS–MP interactions, compared to hundreds of studies that pair MPs with more conventional pollutants such as dyes, metals, and pesticides. This discrepancy underscores a profound imbalance in the literature. Considering the persistence, mobility, and toxicity of PFAS, their interactions with MPs establish a critical research frontier with direct implications for contaminant fate, ecological exposure, and human health. The current rarity of studies at this interface highlights both the novelty of the topic and the urgency of advancing systematic investigations in this area.

As summarized in the heatmap (Fig. 4.3), the following trends are visually evident: dyes and metals dominate the publication landscape, peaking with more than 180 annual studies, whereas PFAS, pesticides, and radionuclides consistently register below 40 studies per year. This bibliometric evidence reinforces the conclusion that while traditional pollutant classes are relatively well understood, interactions of MPs with emerging contaminants remain severely underexplored, providing a clear rationale for prioritizing them in future research agendas.

The heatmap visualizes the dominance of dyes and metals (dark cells; peaks >180 papers per year) and the persistent scarcity of studies on PFAS, pesticides, and radionuclides (<40 papers per year), supporting the gap analysis discussed in the text.

Beyond their scientific implications, these bibliometric patterns also reflect the trajectory of funding priorities and regulatory concerns. The heavy concentration of work on dyes and metals suggests that research has historically aligned with pollutants that are relatively easy to detect, quantify, and model. In contrast, the limited attention to PFAS, pesticides, and radionuclides points to analytical and methodological barriers, such as the need for advanced instrumentation, complex standards, and specialized facilities. Overcoming these barriers will be essential if future studies are to capture the full extent of risks posed by MPs in real-world contaminant mixtures.

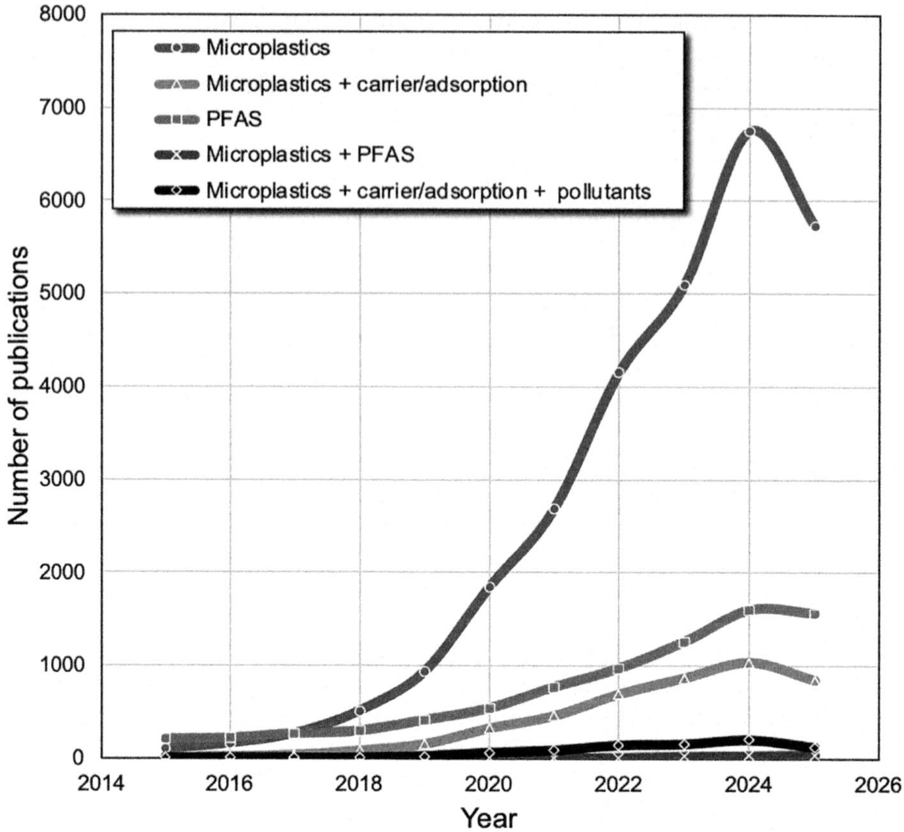


**FIGURE 4.2** Evolution of publications (2015–25) retrieved from Scopus for: Microplastics (microplastic*); Microplastics + carrier/adsorption (microplastic* AND (adsorb* OR transport OR vector* OR "sorption" OR carrier)); PFAS (PFAS OR "per- and polyfluoroalkyl" OR "perfluoroalkyl"); Microplastics + PFAS (microplastic* AND (PFAS OR "perfluoroalkyl" OR "polyfluoroalkyl" OR "per- and polyfluoroalkyl")); Microplastics + carrier/adsorption + pollutants (microplastic* AND (adsorb* OR transport OR vector* OR "sorption" OR carrier) AND (PFAS OR "perfluoroalkyl" OR "polyfluoroalkyl" OR "per- and polyfluoroalkyl" OR dye* OR colorant* OR pollutant* OR radionuclide* OR uranium OR thorium OR plutonium OR cesium OR strontium OR metal* OR heavy metal* OR pesticide*)). PFAS, Per- and polyfluoroalkyl compounds.

The abovementioned imbalance highlights an important opportunity for innovation. Integrating advanced characterization techniques, environmental modeling, and mechanistic studies will allow the field to move beyond descriptive observations and toward a predictive understanding of how MPs mediate contaminant fate. Such progress would not only fill a fundamental gap in environmental chemistry but also inform risk assessments and regulatory strategies targeting the combined effects of plastics and hazardous pollutants. Ultimately, expanding research beyond conventional model systems toward high-priority contaminants like PFAS and radionuclides can generate impactful knowledge with direct societal and policy relevance.

| | 2025 | 2024 | 2023 | 2022 | 2021 | 2020 | 2019 | 2018 | 2017 | 2016 | 2015 |
|---|---|---|---|---|---|---|---|---|---|---|---|
| **PFAS** | 10 | 16 | 10 | 5 | 3 | 1 | 1 | 1 | 0 | 0 | 0 |
| **dyes** | 34 | 30 | 19 | 13 | 11 | 5 | 1 | 0 | 0 | 0 | 0 |
| **radionuclides** | 10 | 9 | 9 | 9 | 2 | 5 | 3 | 1 | 0 | 0 | 0 |
| **heavy metals** | 112 | 183 | 147 | 135 | 78 | 53 | 19 | 5 | 3 | 5 | 2 |
| **pesticides** | 18 | 32 | 21 | 13 | 14 | 10 | 6 | 1 | 1 | 1 | 0 |

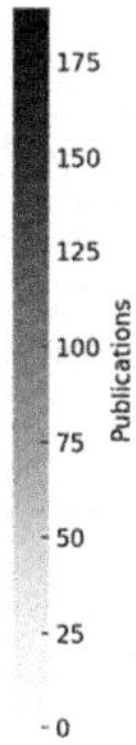


**FIGURE 4.3** Heatmap of the number of Scopus-indexed publications on microplastics as pollutant carriers for per- and polyfluoroalkyl compounds (PFAS), dyes, radionuclides, heavy metals, pesticides from 2015 to August 2025. Cell values are annual counts; color from light yellow to dark blue reflects increasing publication numbers (see scale bar, right). Search syntax included: microplastic* AND (adsorb* OR transport OR vector* OR sorption OR carrier) combined with each pollutant term set (e.g., PFAS OR "perfluoroalkyl" OR "polyfluoroalkyl" OR "per- and polyfluoroalkyl"; dye* OR colorant*; "heavy metal*" OR metal*; pesticide*; radionuclide* OR uranium OR thorium OR plutonium OR cesium OR strontium).

## 4.3 Factors influencing the sorption of per- and polyfluoroalkyl compounds by microplastics

MPs and PFAS coexist in the environment (in freshwater, seawater, groundwater, wastewater, soils, and even in the air) and interact with each other, with MPs in some cases acting as carriers, as PFAS are sorbed onto them. Several factors influence the sorption of PFAS by MPs, including the properties of the MPs, the characteristics of the PFAS, as well as environmental factors.

The physicochemical characteristics of MPs in the environment are highly variable and diverse. Factors such as polymer type, particle size, shape, polarity, crystallinity, and functional groups play a crucial role in determining their sorption behavior.

Commonly used nondegradable polymers such as PP, PS, PE, and polytetrafluoroethylene originate from a wide variety of sources, and plastic items made of them can be released into the environment through both direct and indirect pathways. These plastics undergo processes such as transport, transformation, and degradation in natural settings, eventually forming MPs that may pose significant risks to surrounding ecosystems (Ding et al., 2024). Nonpolar polymers are characterized by covalent bonds between carbon and hydrogen atoms, giving them high structural stability and strong chemical resistance. Polarity can be introduced into these polymers by incorporating specific functional groups. Among the most common MP types, their relative polarity follows the order: PP < PE < PS < polyvinyl chloride (PVC) < PET < polyamide (PA).

Considering PFAS sorption by MPs, it is not only influenced by the polarity of the polymers but also by other factors (Fig. 4.4). For example, nonpolar perfluorooctanesulfonamide shows a strong affinity for PE due to its high partitioning capability, while polar PFOS also displays greater sorption to PE and PVC than to PS. This is attributed to the

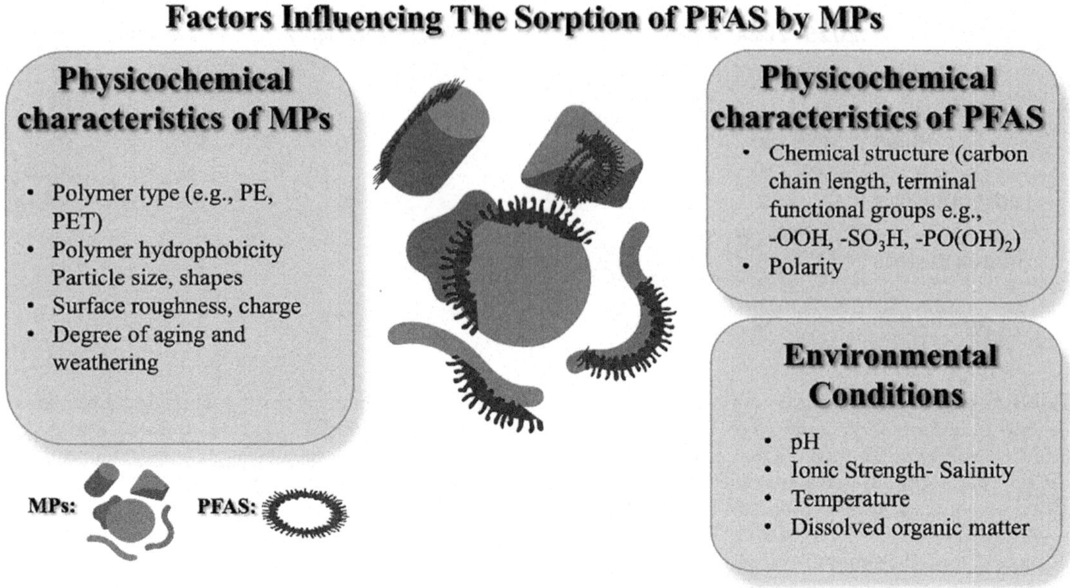


FIGURE 4.4 Various factors affecting per- and polyfluoroalkyl compounds (PFAS) sorption by microplastics (MPs) (Fu et al., 2021; Gagliano et al., 2020; Ateia et al., 2020). PE, Polyethylene; PET, polyethylene terephthalate.

structural characteristics of PS, which provide limited space for molecular interaction due to its compact nature. Similarly, PFOS and PFOA are more easily sorbed by PP, while PE exhibits relatively lower sorption efficiency. Additionally, PS and PS-COOH have greater distribution coefficients for PFAS compared to HDPE (Islam et al., 2025). The crystallinity of a polymer, referring to the degree of structural order within its molecular arrangement, can significantly influence the sorption behavior of MPs. Higher crystallinity typically results in a reduced capacity and slower rate of contaminant uptake into the polymer matrix. This is likely because sorption tends to occur more readily in the amorphous (less ordered) regions of the polymer, where there is greater molecular mobility and more available spaces, as opposed to the tightly packed crystalline regions (Yu et al., 2024).

Except for the type of MPs, particle size and specific surface area are also important factors that affect their sorption ability. The size of MP particles plays a significant role in their ability to sorb PFAS. Studies have shown that smaller MPs tend to exhibit higher rates of pollutant exchange compared to larger MPs (Bere et al., 2024). This enhanced sorption capability is largely attributed to their greater specific surface area and closer interaction with the polymer matrix, allowing for more efficient interaction (Tanzil et al., 2025).

Furthermore, the degree of weathering and aging of MPs is another important factor that influences their ability to sorb various pollutants, including PFAS, and to act as secondary pollutants in the environment. The surface characteristics of MPs undergo alterations after exposure to the natural environment, which subsequently influences the sorption behavior of pollutants on MPs. Environmental exposure leads to the weakening and roughening of MPs' surfaces, along with an increase in the presence of surface fragments and oxygen-containing functional groups, because of degradation processes (Ding et al., 2024; Barhoumi et al., 2024).

## 4.4 Mechanisms of per- and polyfluoroalkyl compound sorption by microplastics

The sorption of PFAS by MPs is governed by a complex interplay of physicochemical mechanisms, the most prominent being hydrophobic interactions between the fluorinated carbon chains of PFAS and the hydrophobic polymer matrix of the MPs. In parallel, electrostatic interactions occur between the negatively charged head groups of PFAS molecules (such as carboxylates or sulfonates) and the functionalized or oxidized surface groups of MPs, especially those that have undergone environmental aging (Islam et al., 2025; Sarkar et al., 2021). Weathering introduces polar oxygen-containing moieties on the MP surface, which significantly enhances sorption potential through increased polarity and available reactive sites (Sarkar et al., 2021).

In addition to these primary mechanisms, hydrogen bonding, dipole–dipole interactions, and π–π stacking may also contribute to the overall sorption, particularly for aged MPs with chemically heterogeneous surfaces (Sarkar et al., 2021; Joo et al., 2021). The importance of PFAS molecular structure is also well established, and long-chain PFAS (e.g., PFOA and PFOS) typically exhibit stronger affinity for MPs than their short-chain analogs due to higher hydrophobicity and lower aqueous solubility (Sarkar et al., 2021; Snyder, 2021). The aliphatic (nonpolar) portions of organic molecules, including hydrocarbons, play a key role in their partitioning behavior. In general, PFASs with longer carbon chains exhibit stronger sorption onto MPs (Yu et al., 2024). For instance, one study observed that the sorption of perfluorocarboxylic acids (PFCAs) on PA MPs increased with longer chain lengths. Additionally, different functional groups—such as carboxyl and sulfonyl groups—also influence PFAS sorption. The presence and nature of these groups affect the polarity and hydrophobicity of the molecules. Specifically, the hydrophobic character of PFASs tends to rise with the extension of their fluorinated carbon chains, thereby enhancing their interaction with MPs (Ding et al., 2024).

Sorption of PFAS compounds follows a consistent trend from less polar to more polar species, with hydrophobic interactions being the dominant removal mechanism. Among the different types of polymers studied, PA66 exhibited the highest sorption efficiency. This is attributed to the hydrogen donor/acceptor capacity of the –NH groups in the polymer structure. The observed sorption trend (PFOS > PFOA > hexafluoropropylene oxide dimer acid, also known as GenX) further supports the role of hydrophobic interactions as the primary mechanism of uptake (Ateia et al., 2020). However, aquatic chemistry has a significant impact on PFAS sorption in seawater that has an overwhelming impact on hydrophobic interactions, resulting in no increasing trend in sorption coefficients with carbon-chain length, which is opposed to freshwater (Parashar et al., 2023). A study was conducted to examine the sorption of six perfluorinated compounds on PA MPs, and it revealed that PFOS adsorbed to PA MPs to the largest degree (0.873 mg $g^{-1}$), followed by PFOA (0.235 mg $g^{-1}$), perfluoroheptanoic acid (0.231 mg $g^{-1}$), perfluorohexanoic acid (0.201 mg $g^{-1}$), perfluoropentanoic acid (PFPeA) (0.192 mg $g^{-1}$), and perfluorobutanoic acid (0.188 mg $g^{-1}$) isothermally at pH 5.9. This also supports the fact that PFASs with longer chains and increased hydrophobic characteristics sorb more effectively. Generally, PFOS portrayed a greater inclination to sorb onto PA MPs in comparison to perfluoroalkyl carboxylic acids (PFCAs) (Mejías et al., 2023). However, for very long-chain PFCAs such as perfluorohexadecanoic acid and perfluorooctadecanoic acid, sorption is not favorable due to the formation of semimicelles or higher affinity to dissolved organic matter (DOM) in water (Ma et al., 2025).

Environmental conditions such as pH, salinity, ionic strength, and temperature play a crucial role in modifying the sorption behavior. Lower pH levels promote protonation of MPs' surface groups, which reduces electrostatic repulsion between anionic PFAS and the polymer surface, thereby enhancing sorption (Sarkar et al., 2021). For example, at acidic pH, amino groups present in certain polymers can become protonated, forming positively charged species such as quaternary amines or amides (Freilinger et al., 2025). This protonation increases the electrostatic attraction between the positively charged adsorbent surface and negatively charged PFAS molecules. Different studies have shown that the sorption of PFOS onto PE MPs and PS MPs increases significantly as the pH drops from 7 to 3, rising from 0.6 to 1.8 $\mu g\, g^{-1}$ for PE and from 0 to 0.3 $\mu g\, g^{-1}$ for PS (Fu et al., 2021). Conversely, at higher pH levels, such as pH 10, the surface of MPs tends to carry a net negative charge, leading to electrostatic repulsion with anionic PFASs and thus a decrease in sorption efficiency (Dogra et al., 2025).

Furthermore, elevated salinity compresses the electrical double layer surrounding MPs, allowing closer approach of PFAS molecules and increasing sorption affinity (Sarkar et al., 2021; Ma et al., 2025; Scott et al., 2021). This indicates that increased salinity can enhance the sorption of PFASs on MPs, primarily due to strengthened hydrophobic interactions and the salting-out effect (Fu et al., 2021; Joo et al., 2021). This phenomenon reduces the solubility of hydrophobic compounds in water, thereby promoting their partitioning onto plastic surfaces. For instance, higher salinity levels significantly increased the sorption of PFOS onto both PE MPs and PS MPs (Tanzil et al., 2025).

Additionally, temperature can influence both the kinetics and thermodynamics of PFAS sorption: higher temperatures may accelerate PFAS diffusion to the MP surface but may also reduce sorption capacity if the process is exothermic (Al Harraq and Bharti, 2021). However, excessively high temperatures may also alter the physicochemical properties of both the pollutants and the MPs, potentially impacting the overall sorption process. For example, Salawu et al. (2024) reported a notable increase in PFAS sorption by MPs as the temperature rose from 25°C to 50°C, indicating a temperature-dependent enhancement in sorption capacity.

Experimental studies support that PFAS sorption by MPs generally follows pseudo-second-order kinetics, implying that chemisorption, rather than simple physical adsorption, may be the rate-controlling mechanism (Sarkar et al., 2021). Sorption equilibrium is typically achieved after several hours to days, depending on the type of PFAS, MPs properties, and environmental conditions. Sorption isotherms most often fit the Freundlich model, indicative of multilayer sorption on heterogeneous surfaces, rather than monolayer, homogeneous adsorption as described by the Langmuir model. These findings align with the chemically complex and variable nature of environmental MPs.

Importantly, the intrinsic chemical stability and environmental persistence of PFAS, rooted in their strong carbon–fluorine bonds, make them resistant to degradation processes such as photolysis, hydrolysis, and biodegradation (Al Harraq and Bharti, 2021; Lionetto and Esposito Corcione, 2021). Once sorbed onto MPs, PFAS may be further shielded from environmental breakdown, contributing to their long-term persistence and mobility in ecosystems. As a result, PFAS–MPs complexes may act as mobile vectors for contaminant transport, increasing the risk of bioaccumulation and trophic transfer in aquatic and terrestrial environments (Lionetto and Esposito Corcione, 2021).

## 4.5 Analytical techniques used to investigate per- and polyfluoroalkyl compound–microplastic interactions

A wide array of analytical techniques has been employed to investigate the interactions between PFAS and MPs in both laboratory and environmental contexts. Time-of-flight secondary ion mass spectrometry is frequently used for qualitative and semiquantitative surface analysis, enabling detection of PFAS-specific fragments directly on MP surfaces. For example, ions such as $C_7F_{15}^-$ (from PFOA) and $C_8F_{17}O_3S^-$ (from PFOS) have been detected on exposed MPs, confirming surface sorption with high spatial and molecular resolution (Sarkar et al., 2021). Complementary to this, X-ray photoelectron spectroscopy allows for the determination of surface elemental composition (e.g., atomic %F, %C, and %O) and enables researchers to estimate surface coverage, monitor oxidation states, and evaluate changes resulting from PFAS uptake.

To characterize the type and surface condition of the polymer, Fourier transform infrared spectroscopy (FTIR) is widely used. FTIR spectra help identify polymer composition and detect signs of aging or oxidation, which are known to influence PFAS sorption by introducing polar or reactive groups on MP surfaces (Sarkar et al., 2021). For the quantification of PFAS sorbed onto MPs, liquid chromatography coupled with tandem mass spectrometry (LC-MS/MS) is considered the gold standard. Typically, PFAS are extracted from the plastic matrix using solvent-based or alkaline digestion methods prior to LC-MS/MS analysis (Sarkar et al., 2021; Richardson and Ternes, 2021).

In environmental monitoring studies, MPs are first isolated from water or sediment using density separation techniques, often followed by filtration, microscopy, and FTIR analysis for visual and spectral identification of polymer type (Scott et al., 2021; Kuhn and Sokefun, 2021). Subsequent extraction and quantification of PFAS allow for assessment of real-world PFAS–MP associations. The integration of these analytical approaches ensures robust and comprehensive evaluation of sorption behavior, facilitates comparison across different polymer types, and supports the differentiation between field-aged and laboratory-treated MPs.

## 4.6 Microplastics and per- and polyfluoroalkyl compound properties and environmental conditions associated with higher $K_d$ values

Higher distribution coefficients ($K_d$), that reflect the partitioning of PFAS from the aqueous phase onto MPs, are typically observed under specific physicochemical conditions and MP characteristics. Weathered or aged MPs, which undergo environmental exposure such as UV radiation, mechanical abrasion, and oxidative processes, tend to exhibit oxidized, roughened, and more polar surfaces. These surface modifications increase the available binding sites and enhance both hydrophobic and electrostatic interactions with PFAS, leading to significantly higher $K_d$ values compared to pristine polymers (Sarkar et al., 2021; Scott et al., 2021). Moreover, polymer type plays a critical role: MPs with higher crystallinity and polarity, such as PET and PS, tend to retain PFAS more effectively than low-polarity, smooth-surfaced MPs like pristine PE (Sarkar et al., 2021).

In terms of PFAS characteristics, chain length is a major determinant of sorption behavior. Long-chain PFAS (e.g., PFOS and PFOA) consistently show higher $K_d$ values than short-chain analogs due to their stronger hydrophobic interactions and lower aqueous solubility (Sarkar et al., 2021; Snyder, 2021). This effect is amplified in microenvironments with increased ionic strength or high salinity, which compresses the electrical double layer surrounding the MPs and facilitates closer approach and interaction between PFAS and the MP surface (Sarkar et al., 2021; Scott et al., 2021). Similarly, low pH enhances protonation of surface groups on MPs, thereby reducing electrostatic repulsion and promoting sorption of anionic PFAS species (Sarkar et al., 2021).

DOM can strongly influence PFAS sorption on MPs. DOM may compete for sorption sites, block pores, or form PFAS–DOM complexes, while adsorbed DOM layers can also provide new hydrophobic binding domains for perfluoroalkyl tails (Ning et al., 2024). Since DOM is largely anionic, its sorption can also generate electrostatic repulsion (Gagliano et al., 2020). DOM content decreases PFAS sorption on MPs through competition for binding sites and complexation with hydrophobic moieties of humic acid (HA) and fulvic acid (FA), thereby modifying partitioning between water and solid phases. For instance, Parashar et al. (2023) reported that HA ($25\ mg\ L^{-1}$) reduced PFOS sorption on PA MPs from 92% to 26% and PFOA from 20% to 3.2%, while short-chain PFAS (e.g., PFBuA, PFPeA) were less affected. Overall, the combined effects of DOM and ionic strength are complex and highly context dependent.

Laboratory studies typically use HA or FA as DOM models, but natural organic matter (NOM) is more complex, comprising proteins, amino acids, and carboxylic acids in addition to HA and FA. NOM may compete for sorption sites on the MP surface, leading to reduced PFAS uptake, or may alter surface characteristics through coating or conditioning, thereby facilitating sorption indirectly (Sarkar et al., 2021). Experimental studies also highlight the importance of contact time, as longer exposure durations allow PFAS molecules to diffuse into the polymer matrix and establish stronger, more stable binding, particularly in the case of high molecular weight or long-chain compounds.

In addition to laboratory conditions, field studies corroborate these trends. Environmentally exposed MPs often show enhanced sorption behavior due to cumulative effects of weathering, biofouling, and exposure to cocontaminants (Sarkar et al., 2021; Scott et al., 2021; Kuhn and Sokefun, 2021). Moreover, once PFAS are adsorbed onto MPs, they may be shielded from degradation processes such as photolysis, hydrolysis, or biodegradation (Al Harraq and Bharti, 2021). This persistence is of particular concern given that fluorinated compounds are inherently resistant to breakdown due to the strength of the carbon–fluorine bond (Al Harraq and Bharti, 2021; Lionetto and Esposito Corcione, 2021). As a result, PFAS–MP complexes may act as mobile, long-lived contaminant carriers, potentially enabling the long-range transport of PFAS in the environment and their bioavailability to aquatic and terrestrial organisms through ingestion of contaminated plastics (Richardson and Ternes, 2021; Lionetto and Esposito Corcione, 2021).

## 4.7 Conclusions

The coexistence of PFAS and MPs in the environment poses a significant and growing ecological concern. MPs can act as carriers for PFAS, facilitating their persistence, transport, and potential bioaccumulation. PFAS sorption onto MPs is influenced by polymer characteristics (e.g., type, polarity, crystallinity, and aging), PFAS molecular properties (e.g., chain length and functional groups), and environmental factors such as pH, salinity, temperature, and DOM. Hydrophobic interactions, electrostatic attraction, and hydrogen bonding are the dominant sorption mechanisms, with long-chain PFAS and aged, oxidized MPs generally showing higher sorption capacities.

Once adsorbed, PFAS are shielded from degradation, allowing PFAS–MP complexes to persist and migrate across ecosystems. This interaction amplifies the environmental risks associated with both pollutants, emphasizing the need for a deeper mechanistic understanding and the development of effective monitoring and remediation strategies to mitigate their combined impact on aquatic and terrestrial systems.

## References

Al Harraq, A., Bharti, B., 2021. Microplastics through the lens of colloid science. ACS Environmental Au 2, 3–10.

Ateia, M., Zheng, T., Calace, S., et al., 2020. Sorption behavior of real microplastics (MPs): insights for organic micropollutants adsorption on a large set of well-characterized MPs. The Science of the Total Environment 720, 137634.

Barhoumi, B., Metian, M., Alonso-Hernández, C.M., et al., 2024. Insight into the effect of natural aging of polystyrene microplastics on the sorption of legacy and emerging per-and polyfluorinated alkyl substances in seawater. Heliyon 10.

Bere, K., Bakk, B., Illés, E., et al., 2024. Role of fluorocarbon chain length in the adsorption of perfluoroalkyl substances on nanoplastic particles. ACS ES&T Water 4, 5114–5121.

Buck, R.C., et al., 2011. Perfluoroalkyl and poly fluoroalkyl substances in the environment: terminology, classi cation, and origins. Integrated Environmental Assessment and Management 7, 513–541.

Cousins, I.T., et al., 2020. The high persistence of PFAS is suf cient for their management as a chemical class. Environmental Science: Processes & Impacts 22, 2307–2312.

Dickman, R.A., Aga, D.S., et al., 2022. A review of recent studies on toxicity, sequestration, and degradation of per- and polyfluoroalkyl substances (PFAS). Journal of Hazardous Materials 436, 129120. Available at https://doi.org/10.1016/j.jhazmat.2022.129120.

Ding, X., Liu, S., Tang, S., et al., 2024. The role of microplastics as carriers of per-and polyfluoroalkyl substances and their fate in the environment. Trends in Analytical Chemistry 181, 118003.

Dogra, K., Kumar, M., Singh, S., Bahukhandi, K.D., 2025. Decoding the interactions between microplastics, polyfluoroalkyl substances, and endocrine disruptors: sorption kinetics and toxicity. Current Opinion in Chemical Engineering 48, 101126.

Freilinger, J., Kappacher, C., Huter, K., et al., 2025. Interactions between perfluorinated alkyl substances (PFAS) and microplastics (MPs): findings from an extensive investigation. Journal of Hazardous Materials Advances 18, 100740.

Fu, L., Li, J., Wang, G., et al., 2021. Adsorption behavior of organic pollutants on microplastics. Ecotoxicology and Environmental Safety 217, 112207. Available at https://doi.org/10.1016/J.ECOENV.2021.112207.

Gagliano, E., Sgroi, M., Falciglia, P.P., et al., 2020. Removal of poly-and perfluoroalkyl substances (PFAS) from water by adsorption: role of PFAS chain length, effect of organic matter and challenges in adsorbent regeneration. Water Research 171, 115381.

GESAMP, 2015. Chapter 3.1.2 Defining 'microplastics'. Sources, fate and effects of microplastics in the marine environment: a global assessment. Chapter 3.1.2 Defining 'microplastics. In: Kershaw, P.J. (Ed.), (IMO/FAO/

UNESCO IOC/UNIDO/WMO/IAEA/UN/UNEP/UNDP Joint Group of Experts on the Scientific Aspects of Marine Environmental Protection (GESAMP)). Rep. Stud. GESAMP, London. No. 90, 96p, London.

Isari, E.A., Papaioannou, D., Kalavrouziotis, I.K., Karapanagioti, H.K., 2021. Microplastics in agricultural soils: a case study in cultivation of watermelons and canning tomatoes. Water V. 13, 2168.

Islam, M.S., Kekre, K.M., Shah, T.A., et al., 2025. Unraveling the complexities of microplastics and PFAS synergy to foster sustainable environmental remediation and ecosystem protection: a critical review with novel insights. Journal of Hazardous Materials Advances, 100621.

Joo, S.H., Liang, Y., Kim, M., et al., 2021. Microplastics with adsorbed contaminants: mechanisms and treatment. Environmental Challenges 3, 100042.

Karapanagioti, HK, Kalavrouziotis, IK, 2020. Microplastics in water and wastewater. Revised with a Comprehensive Study Guide, 2nd Edition. IWA Publishing, London, pp. 267. ISBN: 9781789061680.

Kuhn, J.N., Sokefun, Y.O., 2021. Aqueous-phase photocatalytic degradation of emerging forever chemical contaminants. ChemistrySelect 6, 5225–5240.

Lionetto, F., Esposito Corcione, C., 2021. An overview of the sorption studies of contaminants on poly (Ethylene Terephthalate) microplastics in the marine environment. Journal of Marine Science and Engineering 9, 445.

Mamun, A.A., Prasetya, T.A.E., Dewi, I.R., Ahmad, M., 2023. Microplastics in human food chains: food becoming a threat to health safety. Science of the Total Environment 858 (Part 1), 159834.

Ma, M., Coulon, F., Tang, Z., et al., 2025. Unveiling the truth of interactions between microplastics and per-and polyfluoroalkyl substances (PFASs) in wastewater treatment plants: microplastics as a carrier of PFASs and beyond. Environmental Science & Technology 59, 2211–2221.

Mejías, C., Martín, J., Santos, J.L., et al., 2023. Adsorption of perfluoroalkyl substances on polyamide microplastics: effect of sorbent and influence of environmental factors. Environmental Research 216, 114834.

Ning, Z., Zhou, S., Yang, Y., et al., 2024. Adsorption behaviors of perfluorooctanoic acid on aged microplastics. Water Environment Research: A Research Publication of the Water Environment Federation 96, e11080.

O'Brien, S., Cassandra Rauert, C., et al., 2023. There's something in the air: a review of sources, prevalence and behaviour of microplastics in the atmosphere. Science of the Total Environment 874, 162193.

Parashar, N., Mahanty, B., Hait, S., 2023. Microplastics as carriers of per-and polyfluoroalkyl substances (PFAS) in aquatic environment: interactions and ecotoxicological effects. Water Emerging Contaminants & Nanoplastics 2. N-A.

Richardson, S.D., Ternes, T.A., 2021. Water analysis: emerging contaminants and current issues. Analytical Chemistry 94, 382–416.

Salawu, O.A., Olivares, C.I., Adeleye, A.S., 2024. Adsorption of PFAS onto secondary microplastics: a mechanistic study. Journal of Hazardous Materials 470, 134185.

Sarkar, B., Mukhopadhyay, R., Ramanayaka, S., et al., 2021. The role of soils in the disposition, sequestration and decontamination of environmental contaminants. Philosophical Transactions of the Royal Society B 376, 20200177.

Schiavone, C., Portesi, C., et al., 2023. PFAS: a review of the state of the art, from legislation to analytical approaches and toxicological aspects for assessing contamination in food and environment and related risks. Applied Sciences 13, 6696.

Scott, J.W., Gunderson, K.G., Green, L.A., et al., 2021. Perfluoroalkylated substances (PFAS) associated with microplastics in a lake environment. Toxics 9, 106.

Secretariat of the Stockholm Convention, 2019. All POPs listed in the Stockholm Convention. Available from: http://chm.pops.int/TheConvention/ThePOPs/AllPOPs/tabid/2509/Default.aspx.

Simantiris, N., Cladas, Y., et al., 2024. First evaluation of microplastics in juveniles of the invasive blue crab Callinectes sapidus from a Mediterranean coastal lagoon. Estuarine, Coastal and Shelf Science 302, 108775.

Snyder A., 2021. Surface Chemical Studies of Environmental Microplastics from the Laurentian Great Lakes: Field and Laboratory Samples.

Takada, H., Karapanagioti, H., 2019. Hazardous Chemicals Associated with Plastics in the Marine Environment. The Handbook of Environmental Chemistry. Springer, Cham. 78.

Tanzil, K., Ealias, A.M., George, G., Panigrahi, S., 2025. Microplastics: a hidden carrier of per-and polyfluoroalkyl substances and their effect on soil properties. Journal of Environmental Chemical Engineering 13, 114974.

Tziourrou, P., Megalovasilis, P., Tsounia, M., Karapanagioti, H.K., 2019. Characteristics of microplastics on two beaches affected by different land used in Salamina Island in Saronikos Gulf, east Mediterranean. Marine Pollution Bulletin 149. UNSP 110531.
Wee, S.Y., Aris, A.Z., et al., 2023. Revisiting the "forever chemicals", PFOA and PFOS exposure in drinking water. npj Clean Water 6, 57.
Yang, Y., Xie, E., et al., 2023. Environmental Science & Technology 57 (30), 10911–10918.
Yu, F., Wu, J., Wang, H., et al., 2024. Interaction of microplastics with perfluoroalkyl and polyfluoroalkyl substances in water: a review of the fate, mechanisms and toxicity. The Science of the Total Environment 948, 175000.

PART 2

# Remediation technologies

CHAPTER

# 5

# Adsorption-based per- and polyfluoroalkyl substances remediation technologies: principles and applications

*Md Abdullah Al Masud*[1], *Mahlet M. Kebede*[1], *Sarah Ortbal*[1], *Md Refat Jahan Rakib*[2], *Hasara Samaraweera*[3], and *Leigh G. Terry*[1]

**[1]Department of Civil, Construction, and Environmental Engineering, University of Alabama, Tuscaloosa, AL, United States [2]School of Earth, Environmental and Marine Sciences, University of Texas - Rio Grande Valley, Brownsville, TX, United States [3]Department of Civil and Environmental Engineering, Western University, London, ON, Canada**

## 5.1 Introduction

Per- and polyfluoroalkyl substances (PFAS) comprise over 14,000 synthetic organofluorine compounds whose strong carbon–fluorine (C–F) bonds confer exceptional thermal, chemical, and biological stability, resulting in their environmental persistence and bioaccumulation (Panieri et al., 2022). These compounds have been widely used in a range of industrial products, commercial goods, and consumer products, including firefighting foams, nonstick cookware, waterproof clothing, and food packaging, from the 1940s onwards (Boone et al., 2019; Palma et al., 2022). PFAS are persistent compounds that are resistant to degradation, which can lead to significant levels of contamination in the environment, especially aquatic ecosystems (Groffen et al., 2021; Koch et al., 2020; Ma et al., 2023). PFAS are recognized as widespread environmental contaminants and found in surface water, groundwater, soil, sediments, and biological tissues, including human blood and breast milk (Alazaiza et al., 2025; Boone et al., 2019).

Short-chain and long-chain PFAS differ primarily in the length of their carbon–fluorine backbone, which influences their environmental behavior and toxicity. Long-chain PFAS,

*Per- and Polyfluoroalkyl Substances (PFAS) as Environmental Contaminants*
DOI: https://doi.org/10.1016/B978-0-443-43856-1.00020-4

such as perfluorooctanoic acid (PFOA) and perfluorooctane sulfonate (PFOS), have eight or more carbon atoms and are known for their persistence, bioaccumulation, longer half-lives in humans, and potential adverse health effects including liver toxicity, immune disruption developmental issues, metabolic disorders, endocrine disruption, and cancer (Feng et al., 2024; Fenton et al., 2021; Wee and Aris, 2023). In contrast, short-chain PFAS (fewer than six carbon atoms) are less bioaccumulative and have a shorter half-life in humans, but they are still highly persistent in the environment and can be more mobile in water systems (Lukić Bilela et al., 2023). Although short-chain PFAS are considered less bioaccumulative and toxic than their long-chain counterparts, potential health risks have still been associated with their exposure, and their removal using conventional treatment technologies has been found to be more challenging (Brendel et al., 2018). Due to human and environmental concerns, many nations and international organizations have developed stringent PFAS regulations, including drinking water regulations for specific PFAS chemicals, that is, PFOA and PFOS, at 4 parts per trillion (US EPA, 2024a).

Despite growing awareness and regulatory efforts, mitigating PFAS contamination is a formidable challenge. Traditional water treatment processes such as coagulation, filtration, and biological treatment are largely ineffective in removing PFAS due to their chemical inertness and water solubility (Leung et al., 2022). Advanced treatment technologies such as membrane filtration, ion exchange, sonolysis, electrochemical oxidation, and advanced oxidation processes (AOPs) have demonstrated some efficacy but are often limited by high operational costs, energy demands, and difficulties in treating short-chain PFAS (Hussain et al., 2025; Verma et al., 2021). Among available treatment options, adsorption has appeared as a particularly promising and scalable approach for PFAS remediation (Abeysinghe et al., 2025; Liang et al., 2024). The advantages include operational simplicity, cost-effectiveness, adaptability to existing treatment systems, and high removal efficiency for a large number of PFAS compounds (Wang et al., 2023). The adsorption of PFAS is primarily governed by electrostatic interactions, hydrophobic forces, anion exchange, and hydrogen bonding, depending on the physicochemical properties of both the PFAS molecules and the adsorbent materials (Leung et al., 2023). A wide variety of adsorbents have been explored, ranging from conventional materials such as activated carbon and anion-exchange resins to advanced engineered sorbents such as metal–organic frameworks (MOFs), mesoporous carbons, and layered double hydroxides (LDHs) (Abeysinghe et al., 2025; Cantoni et al., 2021; Chavan et al., 2024; Ilango et al., 2025; Lei et al., 2023). Recently, biochar has attracted considerable attention as a green and sustainable alternative (Chavan et al., 2024). Produced through the pyrolysis of agricultural and forestry biomass, biochar offers several environmental and economic benefits, including carbon sequestration potential, low production cost, high surface area, and tunable surface chemistry (Behnami et al., 2024; Chavan et al., 2024).

Moreover, recent studies have shown that biochar's PFAS adsorption performance can be significantly enhanced through surface modifications such as acid or alkaline activation, metal impregnation, ball-milling, and thermal treatments (Behnami et al., 2024; Chavan et al., 2024). These techniques improve pore structure, surface charge distribution, and functional group availability, which are crucial factors for enhancing interactions with both long- and short-chain PFAS (Leung et al., 2023). In addition, the integration of machine learning and computational models has begun to facilitate the prediction of adsorption efficiency and optimization of biochar properties for targeted PFAS removal (Kwon et al., 2023). Despite these advancements, several challenges persist in real-world applications. The presence of natural organic matter (NOM), pH variability, and competitive ions (e.g., sulfate, chloride, bicarbonate) can interfere with PFAS sorption (Leung et al., 2023; Mukherjee et al., 2023). Furthermore, issues related to regeneration, reuse, and safe disposal of spent adsorbents are still significant bottlenecks in the development of sustainable remediation solutions (Baskar et al., 2022; Fouda-Mbanga et al., 2024).

Several recent reviews have addressed PFAS remediation using adsorption technologies. Vu and Wu (2022) provided a broad overview of adsorptive removal mechanisms and materials, focusing on the physicochemical interactions between PFAS and adsorbents. Lei et al. (2023) emphasized engineering applications and challenges in scaling up adsorption systems, while Abbasian Chaleshtari and Foudazi (2022) explored molecular-level interactions and separation mechanisms. Liang et al. (2024) and Chavan et al. (2024) focused on biochar-based adsorbents, discussing their preparation methods and environmental sustainability.

While these studies offer valuable insights, our chapter distinguishes itself by integrating both experimental and computational perspectives, presenting a comprehensive synthesis of traditional and emerging adsorbents, including MOFs, covalent organic frameworks (COFs), and nanocomposites, and evaluating their performance under diverse environmental conditions such as pH, ionic strength, and NOM. Furthermore, we emphasize recent advancements in adsorbent modification strategies, regeneration techniques, and the application of machine learning for performance prediction and optimization. This integrative and forward-looking approach provides a more holistic understanding of adsorption-based PFAS remediation and identifies critical knowledge gaps for future research.

This chapter attempts to give a thorough review of PFAS, emphasizing their widespread presence, environmental persistence, and related health hazards. Particular focus is given to remediation solutions based on adsorption, namely the application of new sorbent materials including biochar, MOFs, and engineered carbons. Key scientific ideas supporting PFAS adsorption are described in this chapter, together with an assessment of the effectiveness and uses of different adsorbents and a discussion of the practical and technical difficulties of full-scale operation. To improve the effectiveness, sustainability, and scalability of PFAS treatment systems, this chapter also looks at potential future research pathways. By synthesizing current knowledge and finding critical gaps, this work facilitates the development of novel and efficient approaches to mitigating PFAS contamination in aquatic environments.

## 5.2 Key concepts: explanation of the main ideas for the remediation technologies and class of materials

PFAS constitute a structurally diverse class of synthetic chemicals characterized by varying chain lengths and functional groups (Rekik et al., 2024). Typical examples include PFOA, perfluorotetradecanoic acid (PFTeDA), perfluorodecanoic acid (PFDA), PFOS, and perfluorohexanoic acid (PFHxA). Based on their functional groups and environmental conditions, PFAS can exist in different ionic forms such as anionic, zwitterionic, cationic, and neutral species, influencing their sorption mechanisms (Leung et al., 2023). The strength of the carbon–fluorine (C–F) bond (energy 110 kcal $mol^{-1}$) and the exceptional thermal stability make them highly resistant to chemical or thermal degradation, environmentally persistent, and commonly referred to as "forever chemicals" (Vu and Wu, 2022). PFAS exist as long (C = >8), short (C = 4–7), and ultra-short (C = 2–3) chain lengths (Ateia et al., 2019). Shorter-chain PFAS (e.g., PFHxA) are more water-soluble and mobile, making them harder to capture with conventional adsorbents, while longer-chain PFAS tend to bind more readily to organic matter and surfaces. Structure analysis of PFAS is crucial prior to the selection of ideal remediation technology.

Physical separation of PFAS from contaminated media includes adsorption, filtration, or ion exchange (removal processes). Adsorption is a cost-effective and robust way to remove PFAS from different environments. Adsorption onto granular activated carbon (GAC) is a common strategy for PFAS removal in water treatment. However, GAC does not selectively remove PFAS and will adsorb multiple contaminants in solution. Thus the adsorption capacity of GAC can become quickly exhausted in complex sample matrices, which can lead to frequent GAC replacement for a high level of PFAS removal efficiency. Many low-cost adsorbents containing both polar and nonpolar groups have been designed to avoid these complications and to remove amphiphilic PFAS molecules, achieving above 95% removal (Zhang et al., 2024b). Examples included carbon-based materials such as biochar, ion-exchange resins (IXs)/polymeric materials, MOFs, inorganic and mineral-based materials, cyclodextrin-based materials, nanomaterials, and nanocomposites, etc. The effectiveness of PFAS removal is largely dictated by the adsorbent materials, and a site-specific characterization is often required to determine the best adsorptive material for treatment.

### 5.2.1 Carbon-based materials

Carbon-based materials, including activated carbon, biochar, and carbon nanotubes (CNTs) (Table 5.1), are typically used to remove PFAS, in which the removal of longer-chain compounds (e.g., PFOS and PFOA) is governed through hydrophobic interactions and $\pi$–$\pi$ stacking. To further enhance adsorption capacity and selectivity, adsorbent surfaces are modified. Siriwardena et al. (2019) employed hydrochloric acid (HCl), sodium hydroxide (NaOH), heat activated persulfate (PS), and hydrogen peroxide catalyzed with iron ($H_2O_2$/Fe) for treating two commercial GAC types and found that increased adsorption of both PFHxS and PFOA (7%–8% in charcoal-based Filtrasorb-F400 and 6%–9% in coconut-based carbon) with the treatments of HCl, no significant changes with NaOH, and decreased with PS (22%–25% in F400 and 27%–35% in CBC) and $H_2O_2$/Fe (4%–8% in F400 and 12%–13% in CBC) (Siriwardena et al., 2019). Furthermore, treated GAC had lower Brunauer–Emmett–Teller (BET) surface area compared to untreated GAC, impairing the adsorption. In addition, decrease in the GAC hydrophobicity resulted from treatments, which also caused lower PFAS adsorption.

The removal of long-chain PFAS is typically governed by anion exchange, which is less effective for short-chain PFAS, causing them to remain in the environment (Behnami et al., 2024). Short-chain PFAS are challenging to treat through these conventional methods and new research must be conducted to optimize removal. Biochars' surface area and hydrophobicity are key properties governing PFAS adsorption. Commercial biochar showed more PFAS adsorption than biosolids biochar and removed 88.06%–100% (for single PFAS adsorption)

**TABLE 5.1** Papers reviewed in this study, along with adsorbent types, the experimental conditions, the adsorption capacity, and the regeneration method used (if any).

| Adsorbent type | PFAS targeted | Water matrix | Setup type | Experimental conditions | Removal % | Kinetic model | Adsorption capacity ($mg\ g^{-1}$) | Isotherm model | Mechanisms | Regeneration | References |
|---|---|---|---|---|---|---|---|---|---|---|---|
| Activated carbon (AC) | PFSA and PFCA | Drinking water | Full scale | 0–8 $ng\ L^{-1}$ | >99% | N/A | N/A | N/A | N/A | N/A | Belkouteb et al. (2020) |
| AC (coconut shell based) | PFOA | DIW | Batch | [PFAS] o = 25–360 $mg\ L^{-1}$; pH = 7 | | PFO and PSO | 55 (GAC); 331 (PAC) | Freundlich and Langmuir | Electrostatic and hydrophobic interactions | N/A | Kanchanapiya et al. (2022) |
| GAC with chitosan | PFOA | DIW | Batch | [PFAS] o = 10–200 $mg\ L^{-1}$ | N/A | PSO | 149–298 | Freundlich and Langmuir | Hydrophobic interaction | Ethanol | Liu et al. (2022) |
| AC | PFOA and PFOS | DIW | Batch | [PFAS] o = 100 $mg\ L^{-1}$ | N/A | PSO | 178, 202 (PFOA); 390, 535 (PFOA) | Freundlich and Langmuir | Hydrophobic interaction | N/A | Chen et al., 2017) |
| Powder-activated carbon (PAC) | 9 PFAS (PFCAs and PFSAs, short- and long-chain) | Surface water | Batch | pH = 5.5–10; [PFAS] o = 100 $ng\ L^{-1}$ | ~90% | PFO and PSO | N/A | N/A | Acidic protonation | Thermal | Son et al. (2020) |
| AC–ECTFE–PEI (polymer coated on activated carbon) | PFOA | DIW | Column | pH = 6.5; [PFAS] o = 2.4 $mg\ L^{-1}$ | N/A | N/A | 380 | N/A | Electrostatic and hydrophobic interactions | 90% MeCN + 10% 0.1 M MSA | Xie et al. 2024) |
| PEI–PVC–NF | PFCA and PFSA | DIW | N/A | Co = 81–120 $mg\ L^{-1}$ | 30%–60% | PSO | 98–320 | Langmuir | Electrostatic intraparticle diffusion | N/A | Kang et al. (2023) |
| β-zeolites (coated) | PFCA and PFSA | DIW | Batch | [PFAS] o = 5–500 ppm | | PSO | 27–181 | Freundlich and Langmuir | Electrostatic interactions | N/A | Hossain et al. (2023) |
| MIEX® GOLD resin | PFCA and PFSA | Surface water and WW | Batch | pH = 3–11; [PFAS]o = 1 ppm | >99% | N/A | 1010–1050 | N/A | N/A | 70% MeOH + NaCl | Tamanna et al. (2023) |
| Polyethyleneimine-chitosan beads | PFCA and PFSA | DIW | Batch | pH = 3–11; [PFAS] o = 100 ppb | >90% | PFO and PSO | 229–305 | Freundlich and Langmuir | Electrostatic and hydrogen bonding | Ethanol | Shahrokhi and Park (2024) |

*(Continued)*

TABLE 5.1 (Continued)

| Adsorbent type | PFAS targeted | Water matrix | Setup type | Experimental conditions | Removal % | Kinetic model | Adsorption capacity ($mg\ g^{-1}$) | Isotherm model | Mechanisms | Regeneration | References |
|---|---|---|---|---|---|---|---|---|---|---|---|
| Anion exchange resin (IX) | PFCA | Surface and groundwater | RSSCT | N/A | N/A | Intraparticle diffusion | N/A | Freundlich | N/A | N/A | Cheng and Knappe (2024) |
| A532E, A600 (gel-type), and A860 (macroporous) | PFPrA, PFHxA, PFOA, PFOS, and GenX | DIW | Batch | pH = 3–5.5; [PFAS] o = 100 ppb | 70%–100% | N/A | N/A | N/A | Electrostatic (short-chain) and hydrophobic (long-chain) | 5% NaCl and 5% NH | Ren et al. (2024) |
| AC andion exchange resin | PFCA and PFSA | Tap and groundwater | Batch, column | [PFAS] o = 100 ppb | N/A | N/A | N/A | Freundlich and Langmuir | N/A | N/A | Riegel et al. (2023) |
| AER1 and AER2 | Emerging, legacy PFAS | Drinking and wwt | RSSCT | [PFAS]o = 30–300 $ng\ L^{-1}$ | N/A | N/A | 340 | N/A | Ion exchange (electrostatic) | 10% NaCl | Liu et al. (2024) |
| Macroporous and gel-type AER | PFCA and polyfluorinated substances | AFFF | Batch | pH = 8; [PFAS] o = 0.7–97 $\mu g\ L^{-1}$ | N/A | N/A | 20–280 | Langmuir | Hydrophobic effects | N/A | Fang et al. (2021) |
| GAC, IRA910, and DOWEX MB-50 (mixed resin) | PFOS, PFOA, and PFHxS | DIW | Batch | pH = neutral; [PFAS] o = 10–1000 $\mu g\ L^{-1}$ | N/A | PSO | N/A | Freundlich | Hydrophobic interactions and ion exchange | N/A | Sukeesan et al. (2023) |
| MOF | PFOS, PFHxS, and PFBS | DIW | Batch | pH = 3 l; [PFAS] o = 100 ppm | >99% | PSO | 395–963 | Sips model | Hydrophobic ligands | 0.1 M HCL/MeOH | Zhang et al. (2024 |
| PAC | GenX | DIW | Batch | pH = 3–9; [PFAS] o = 270 ppm | N/A | PFO and PSO | 300 | Freundlich and Langmuir | N/A | Ethanol | Wang et al. (2019) |
| Polyaniline-derived polymers | PFOA | DIW | Batch | pH = 3.5–10; [PFAS]o = 60 ppt | N/A | N/A | 0.5 | Freundlich and Langmuir | Electrostatic and hydrophobic interactions | N/A | Yang et al. (2020) |

Note: DIW = deionized water; N/A = not applicable.

versus 59.09%–100% (multiple PFAS adsorption) (Nguyen et al., 2023). Zhang et al. (2021) discovered lower PFOS and PFOA adsorption on biochar than alum sludge. The removal is highest at pH 3 and occurred via electrostatic attraction and hydrophobic interaction (Table 5.1).

### 5.2.2 Ion-exchange resins

IXs, typically functionalized with quaternary ammonium groups, are effective for capturing anionic PFAS such as PFOS and PFOA. In a recent study (Murray et al., 2023), IX exhibited the best performance over GAC and clay in treating PFAS-impaired waters, including groundwater, landfill leachate, membrane concentrate, and wastewater effluent.

### 5.2.3 Metal–organic frameworks and covalent organic frameworks

MOFs and COFs are crystalline, porous materials that offer precise control over pore size and surface chemistry, allowing for customized interactions with different PFAS compounds, including those with sulfonate and carboxylate groups. Moreover, MOFs and COFs contain high surface area and high PFAS selectivity, which are found in most industrially favored PFAS adsorbents. In a comparative study, Liu et al. (2015) found that MIL-101 MOFs based on postsynthetic modification (PSM) are better than preassembled modification (PAM), and PSM offers more PFOA removal than PAM ($1.9\ mg\ g^{-1}$ vs $1.2\ mg^{-1}g$). PAM places functional groups onto the aperture of the nanopore, hindering the entrance of organic contaminants, thus saw less removal. Moreover, the adsorption equilibrium was achieved within 60 min using both methods. The major PFAS removal mechanisms included anion exchange, Lewis's acid/base complexation, and adsorption. Zeolites, LDHs, and clays provide ion-exchange and surface sorption mechanisms for PFAS remediation. Surface functionalization improves their performance toward smaller, more mobile compounds such as PFHxA.

### 5.2.4 Cyclodextrin-based materials

Cyclodextrins form host-guest complexes with a variety of PFAS, including short-chain compounds like PFHxA and long-chain ones like PFOS, depending on polymer structure and functionalization. β-cyclodextrin polymer has a toroid shape cyclic oligosaccharide consisting of seven units of 1,4-linked glucopyranosides (Fig. 5.1A), which can form host-guest complexes with organic molecules (e.g., PFAS) due to its internal hydrophobic cavity. Cross-functionalized β-cyclodextrin polymers (β-CDPs) are widely used to adsorb PFAS molecules (Ching et al., 2022; Shapiro et al., 2025). Shapiro et al. (2025) discovered that a styrene-functionalized β-CD adsorbent (StyDex) has rapid PFAS sorption kinetics that are maintained in wastewater. Further, StyDex preferentially adsorbed longer-chain PFAS compounds, PFOA, PFHxA, and PFHxS, over shorter-chain PFAS compounds, perfluorobutanoic acid (PFBA) and GenX, and pharmaceuticals, BEZ and DCF. However, some CDPs have poorly defined structures, especially with respect to the random substitution patterns of the cyclodextrins as well as side reactions that modify the structures of some cross-linkers (Wang et al., 2022). A design made by Wang et al. (2022) overcame this issue by copolymerizing a StyDex derivative with various styrenic or methacrylic comonomers (Fig. 5.1B–D). The new design is porous and

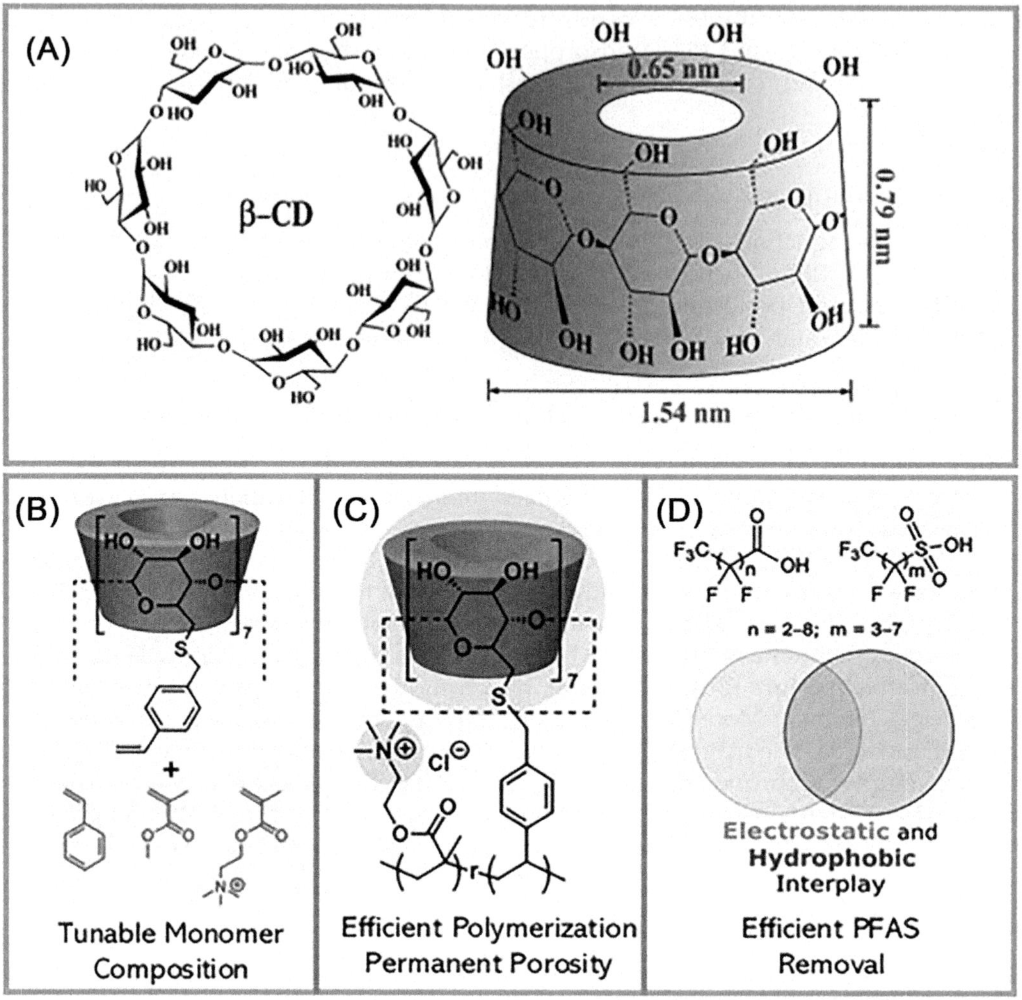


FIGURE 5.1 The structure of β cyclodextrin polymer (β-CD) (A), copolymerizing a styrene-functionalized β-CD derivative with various styrenic or methacrylic comonomers to give permanently porous, cross-linked molecules with a more uniform β-CD binding environment (B) and (C), and per- and polyfluoroalkyl substances (PFASs) removal from copolymerized β-CD (D). *Adapted from Noreña-Caro and Alvarez-Láinez (2016) and Wang et al. (2022).*

forms a more uniform β-CD binding environment, which is easily tunable and contains both hydrophobic and charged comonomers.

## 5.2.5 Nanomaterials and hybrid composites

Nanostructured materials such as metal oxides, graphene derivatives, and hybrid composites offer both high reactivity and customizable surface chemistry. These materials are increasingly being used for both adsorption and catalytic degradation of structurally diverse PFAS. In a recent study, polyethyleneimine modified graphene oxide exhibited high adsorption capacity ($q_e$ = 368.2 mg $g^{-1}$) with a maximum monolayer amount of 231.2 mg $g^{-1}$ (Lei et al., 2022).

Fujioka et al. (2024) showed that nanomembranes rejected PFAS with a carboxylic acid (–$CO_2H$) to a greater extent than PFAS with a sulfate group (–$SO_3H$). PFAS with fully dissociated carboxylate ions can cause high electrostatic repulsion from the carboxylate ions of the nanofiltration (NF) membranes. In this particular case, the other two rejection mechanisms, size exclusion and adsorption, have a lower priority when determining functional group-based remediation.

### 5.2.6 Material-enabled destruction technologies

Destructive technologies are often targeted to permanently eliminate PFAS. Up to 99% destruction can be achieved using advanced oxidation and reduction, electrochemical degradation, and thermal treatment. In AOPs/ARPs, techniques such as UV/sulfite systems, photocatalysis using $TiO_2$ or doped semiconductors, and plasma treatment are designed to generate reactive species capable of cleaving the strong C–F bonds in compounds like PFOS, PFDA, and PFTeDA.

Electrochemical degradation utilizes materials such as boron-doped diamond electrodes and metal-oxide catalysts for electrochemical oxidation and reduction of PFAS, with variable success depending on the chain length and functional groups of the PFAS. Moreover, high-temperature methods such as incineration and pyrolysis are used to destroy PFAS in solid matrices. These are particularly effective for long-chain compounds like PFTeDA and PFDA but may require stringent conditions to avoid the formation of short-chain PFAS as byproducts.

As summarized in Table 5.1, the most extensively investigated PFAS in recent adsorption studies are the legacy long-chain compounds PFOS and PFOA, followed by PFHxS, PFHxA, and the emerging alternatives such as GenX. These species dominate the literature due to their high environmental occurrence and persistence. Among the materials evaluated, GAC/PAC remain the benchmark adsorbents, achieving >90%–99% removal efficiencies for long-chain PFAS in both deionized and surface water systems, primarily via hydrophobic and π–π interactions. However, their performance decreases markedly for short-chain PFAS, which are more hydrophilic and poorly retained. IXs show comparable or superior performance for both long- and medium-chain PFAS, reaching adsorption capacities up to 1010–1050 mg $g^{-1}$, driven mainly by electrostatic attraction between quaternary ammonium groups and anionic PFAS species. MOFs display the highest reported capacities (395–963 mg $g^{-1}$), attributable to their large surface area and tunable pore chemistry, although their large-scale application remains limited by cost and stability. Carbon-based biochars and polymer composites (e.g., chitosan–GAC and β-CDs) provide promising low-cost and regenerable options with high selectivity for longer-chain PFAS. Collectively, the current evidence highlights that adsorption efficiency strongly correlates with PFAS chain length, functional group polarity, and adsorbent surface chemistry. Long-chain PFAS are best treated by hydrophobic carbonaceous adsorbents, whereas short-chain analogs and emerging PFAS benefit from ion-exchange or functionalized hybrid materials.

## 5.3 Methods and applications of high-performing adsorbent materials

Among all developed PFAS remediation technologies, including AOPs (Gar Alalm and Boffito, 2022), bioremediation (LaFond et al., 2023), and photocatalytic reduction (Xu et al., 2017), adsorption has been identified as the most viable and widely adopted approach (Vu and Wu, 2022). This preference can be attributed to the operational simplicity and the relatively mature development of the technology (Lei et al., 2023). The effectiveness of PFAS removal via adsorption has been evaluated at both pilot and full-scale treatment facilities (Belkouteb et al., 2020; Chow et al., 2022; Liu et al., 2019). Moreover, adsorption-based systems have been widely implemented in point-of-use and point-of-entry treatment applications (Patterson et al., 2019)

A wide range of adsorbent materials has been examined for PFAS removal, including GAC, PAC, colloidal activated carbon (CAC), biochar, MOFs, IXs, natural and modified clays, and CNT-based composites (Cantoni et al., 2021; Gunarathne et al., 2025; Ilango et al., 2025). The adsorption behavior of each material has been found to depend on specific characteristics such as surface area, pore structure, surface functionality, and hydrophobicity.

### 5.3.1 Experimental setup for performance evaluation of adsorbents

The performance of adsorbent materials for PFAS removal is typically evaluated through a combination of laboratory-scale experimental setups. These tests are essential to determine adsorption capacity, removal efficiency, selectivity, equilibrium time, and material regeneration potential. The effectiveness of an adsorbent often depends on its physicochemical characteristics, such as surface area, porosity, surface chemistry, particle size, and hydrophobicity. Importantly, some adsorbents can be reversibly regenerated using appropriate desorption procedures, enhancing their reusability and operational cost-effectiveness. To comprehensively assess adsorbent performance, three common experimental approaches are widely employed: batch experiments, column experiments, and rapid small-scale column tests (RSSCTs). These setups differ in complexity, scale, and representativeness of field conditions, providing distinct insights into PFAS adsorption behavior.

#### *5.3.1.1 Batch adsorption experiments*

Batch adsorption tests are the simplest and most widely used laboratory technique to evaluate the performance of adsorbents preliminarily. In this setup, a known mass of adsorbent is added to a fixed volume of PFAS-containing solution, and the mixture is agitated for a predetermined period. The remaining PFAS concentration in the solution is then analyzed over time to assess removal efficiency (kinetic analysis), equilibrium time (isotherm analysis), and adsorption capacity under controlled conditions.

Adsorption kinetics describe the rate at which PFAS molecules are adsorbed onto adsorbent surfaces over time, providing insight into the adsorption mechanism and potential rate-controlling steps. Several kinetic models are commonly applied to experimental data:

- **Pseudo-first-order model (PFO)**: assumes the rate of occupancy of adsorption sites is proportional to the number of unoccupied sites (Lagergren, 1898).

$$q_t = q_e(1 - e^{-k_1 t}) \tag{5.1}$$

Where $q_e$ and $q$ are the adsorption capacity at equilibrium and at any time of the adsorption (mg g$^{-1}$) respectively, $k_1$ is the PFO model rate constant (min$^{-1}$), and $t$ is time (min).

- **Pseudo-second-order model (PSO)**: assumes adsorption follows second-order chemisorption kinetics, with the rate depending on the availability of both adsorbent sites and adsorbate concentration (Ho and McKay, 1999).

$$q_t = \frac{q_e^2 k_2 t}{1 + q_e k_2 t} \quad (5.2)$$

Where $k_2$ is the PSOM rate constant (g mg$^{-1}$ min$^{-1}$).

- **Intraparticle diffusion model**: accounts for the possibility that the rate-limiting step could be the diffusion of PFAS molecules within the pores of the adsorbent (Morris and Weber, 1964).

$$q_t = k_d t^{0.5} + C \quad (5.3)$$

Where $k_d$ is the IPDM rate constant (mg g$^{-1}$ min$^{-1}$), and $C$ is a constant of the experiment (mg g$^{-1}$).

- **Elovich model**: describes chemisorption processes on heterogeneous surfaces (Peers, 1965).

$$q_t = \frac{1}{\beta}(\alpha\beta) + \frac{1}{\beta} lnt \quad (5.4)$$

Where $\alpha$ initial adsorption rate; $\beta$ = desorption constant relating to surface coverage and activation energy for chemosorption.

Adsorption isotherms describe the equilibrium relationship between the concentration of PFAS in solution and the amount adsorbed onto a material at constant temperature. Several models are widely employed to fit experimental isotherm data.

- **Langmuir isotherm**: assumes monolayer adsorption onto a homogeneous surface with a finite number of adsorption sites, each possessing equal affinity for the adsorbate (Langmuir, 1918).

$$q = \frac{q_{mL} b_L C}{1 + b_L C} \quad (5.5)$$

Where $q_{mL}$ (μmol g$^{-1}$) and $b_L$ (L μmol$^{-1}$) are the Langmuir model parameters that represent maximum adsorption capacity and the site energy factor, respectively.

- **Freundlich isotherm:** describes adsorption on a heterogeneous surface, where adsorption sites vary in affinity, with the most energetically favorable sites being occupied preferentially (Freundlich, 1922).

$$q = K_F C^{1/n} \quad (5.6)$$

Where $K_F$ is the Freundlich adsorption coefficient [(μmol g$^{-1}$) (μmol L$^{-1}$)$^{-(1/n)}$] and 1/n (–) is the Freundlich exponent.

- **Temkin isotherm**: considers adsorbate-adsorbent interactions and a linear decrease in heat of adsorption with surface coverage (Temkin and Pyzhev, 1940).

$$q = \frac{RT}{b} \ln(K_m C) \quad (5.7)$$

- **Sips isotherm**: a hybrid of Langmuir and Freundlich models, used for heterogeneous systems. This model is used for monolayer adsorption of PFAS, which is valid for localized sorption without interaction of the adsorbent and adsorbate (Saadi et al., 2015).

$$q = \frac{k_s C^{\beta_s}}{1 + \alpha_s C^{\beta_s}} \tag{5.8}$$

Where $\beta_s$ is a model exponent; $\alpha_s$ and $k_s$ are model constants ($L\,mg^{-1}$) and $L\,g^{-1}$, respectively.

- **Redlich–Peterson isotherm**: another hybrid model for both homogenous and heterogeneous systems. Due to this, it does not assume ideal monolayer adsorption behaviors (Saadi et al., 2015).

$$q = \frac{K_f C}{1 + \alpha C^{\beta}} \tag{5.9}$$

Where $K_f$ is the Redlich–Peterson isotherm constant in $L\,g^{-1}$, $\beta$ is an exponent that range between 0 (Langmuir isotherm) and 1 (Freundlich isotherm), and $\alpha$ is a constant in $L\,mg^{-1}$.

#### 5.3.1.2 *Column and rapid small-scale column tests adsorption experiments*

While batch experiments provide fundamental data, column studies simulate continuous-flow conditions encountered in full-scale water treatment systems, offering more realistic performance assessments.

Column experiments involve packing columns with adsorbent materials and operating them under controlled flow rates. Effluent samples are collected at predetermined time intervals to monitor the breakthrough of PFAS, the point at which PFAS begins to appear in the effluent, and to construct breakthrough curves. These curves graphically depict the concentration of PFAS in the effluent as a function of time or bed volumes treated, enabling the determination of key parameters such as breakthrough time, exhaustion time, and dynamic adsorption capacity ($mg\,g^{-1}$). These metrics are critical for evaluating adsorbent performance in continuous-flow systems and informing the design and scaling of full-scale treatment applications. Previous studies that compared the results of batch and column experiments reported similar results between the two approaches (Brusseau et al., 2019; Van Glubt et al., 2021).

RSSCTs were developed in the 1980s as a cost-effective, accelerated alternative to full-scale column testing, particularly for GAC (Crittenden et al., 1991). RSSCTs use much smaller columns and finer adsorbent particles to accelerate testing while preserving the hydrodynamic similarity and adsorption behavior observed in full-scale systems. The RSSCT relies on maintaining similarity between the small test column and the full-scale system through dimensionless parameters: dimensionless time ($t/\tau$) and Reynolds number (Re) (Crittenden et al., 1986, 1987).

### 5.3.2 Performance assessment of adsorbents under different conditions

The adsorption performance of materials used for PFAS removal varies significantly depending on the physicochemical properties of both the adsorbent (Fig. 5.2) and the aqueous environment (Table 5.1). The key metrics commonly used to evaluate adsorbent

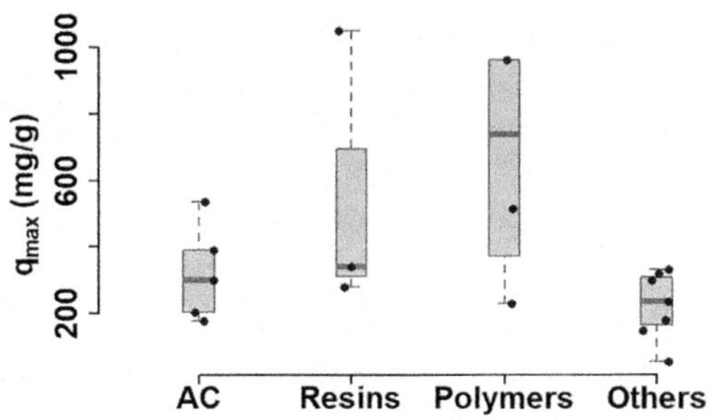


FIGURE 5.2 Box-and-whiskers plot of adsorption capacity of long-chain per- and polyfluoroalkyl substances on different adsorbent materials AC: activated carbon; resin; polymers and others (including hybrid biomaterials, and nano filter membranes). Previous studies' data used to generate this figure are all included in Table 5.1.

performance include removal efficiency, adsorption capacity, selectivity, and regenerability. These metrics are strongly influenced by factors such as PFAS chain length, solution pH, ionic strength, NOM, temperature, and contact time. Most studies are conducted at ambient temperature (20°C–25°C), which reflects typical environmental and drinking water treatment conditions. A few studies also reported that increasing temperature slightly enhanced adsorption rates, likely due to improved intraparticle diffusion (Shan et al., 2024), although adsorption capacity was generally less sensitive to temperature variations compared to other parameters such as pH and initial PFAS concentration. Whereas contact time plays a critical role, with most adsorption systems reaching equilibrium within 6–48 h in batch setups. Full-scale GAC systems demonstrate sustained removal over weeks to months, depending on bed life and influent concentration (Belkouteb et al., 2020; Rodowa et al., 2020).

Across the reviewed studies, PSO and PFO models were most commonly used to describe PFAS adsorption, with PSO often providing the best fit, indicating chemisorption or strong electrostatic interaction. Freundlich and Langmuir models were predominant, with Freundlich often better representing heterogeneous surface adsorption. Adsorbent surface chemistry (hydrophobicity and surface charge), pH, and water chemistry are identified as key parameters controlling removal efficiency and are further discussed in the following subsections.

#### 5.3.2.1 *Effect of per- and polyfluoroalkyl substancescharacteristics*

PFAS exhibit a wide range of physicochemical properties due to variations in their molecular structure, functional groups, and carbon chain lengths, all of which significantly influence their adsorption behavior. In general, long-chain PFAS (e.g., PFOA, PFOS) demonstrate stronger adsorption affinities compared to their short-chain analogs (e.g., PFBA, PFBS), primarily due to enhanced hydrophobic interactions and van der Waals forces (Table 5.1). The nature of the functional group also plays a critical role in adsorption performance, with sulfonic acid groups typically imparting greater hydrophobicity and stronger adsorption than carboxylic acid groups. Additionally, the initial PFAS concentration affects its distribution between aqueous and solid phases, subsequently influencing the adsorption capacity of the material. Higher adsorbate-to-adsorbent ratios can lead to increased competition for available adsorption sites, potential desorption, and eventual site saturation, all contributing to a reduction in overall adsorption capacity (Abeysinghe et al., 2025). A summary of key physicochemical properties of selected PFAS compounds commonly studied in adsorption processes is presented in Table 5.2.

#### 5.3.2.2 *Effect of pH*

Solution pH plays a critical role in influencing both the ionization state of PFAS compounds and the surface charge characteristics of adsorbents. Most PFAS remain anionic under environmentally relevant pH conditions ($pK_a < 1$), while adsorbent surfaces such as activated carbon become increasingly negatively charged as pH rises. Consequently, adsorption typically decreases at higher pH values due to intensified electrostatic repulsion between the negatively charged PFAS molecules and the adsorbent surface. Optimal adsorption performance is often observed under acidic to near-neutral pH conditions. While some previous studies have accounted for the variability of environmental conditions, many investigations that characterize novel adsorbents and evaluate adsorption kinetics and isotherms have not systematically examined the influence of varying pH levels (Table 5.1).

#### 5.3.2.3 *Effect of ionic strength and competing ions*

The presence of coexisting ions in aqueous systems plays a significant role in determining PFAS adsorption performance. Increased ionic strength, the presence of $Ca^{2+}$, $Cu^{2+}$, and $Pb^{2+}$, can compress the electrical double layer surrounding adsorbent surfaces, reduce electrostatic repulsion between negatively charged PFAS molecules and adsorbent sites, and consequently enhance adsorption via electrostatic attraction and cation bridging effects (Abeysinghe et al., 2025; Mohd Jais et al., 2024). Several studies demonstrated that increasing salt concentrations can either suppress or enhance PFAS adsorption depending on the adsorbent type and system chemistry. For example, Ali et al. (2024) reported that increasing ionic strength (0–100 mM NaCl) reduced PFOS and PFOA removal efficiencies on rGO–nZVI–NH, primarily due to competition for active sites and compression of the electrical double layer. Interestingly, while most adsorbents exhibited decreased PFAS adsorption at higher ionic strength, certain materials showed the opposite trend. Molé et al. (2024) observed CAC demonstrated a substantial increase in PFAS adsorption capacity at elevated NaCl concentrations due to the salting-out effect and decreased electrostatic repulsion

The valency of cations also influences the extent of adsorption enhancement, with divalent cations like $Ca^{2+}$ showing a stronger effect than monovalent cations such as $Na^+$ (Sleep et al., 2023). On the other hand, the presence of competing anions such as $Cl^-$, $SO_4^{2-}$, and $HCO_3^-$ can inhibit PFAS adsorption by occupying active adsorption sites (Lei et al., 2023).

#### 5.3.2.4 *Effect of natural organic matter*

NOM is ubiquitous in most water bodies and exerts complex, often competing, effects on PFAS adsorption. The functional groups in NOM can directly compete with PFAS for adsorption sites, resulting in decreased removal efficiencies (Mukherjee et al., 2023). Furthermore, NOM can absorb onto adsorbent surfaces, causing pore blockage and altering surface charge characteristics, which diminishes PFAS uptake, particularly for activated carbons.

**TABLE 5.2** Selected physicochemical properties of representative per- and polyfluoroalkyl substances (PFAS) compounds.

| PFAS | Formula | Functional group | Molecular weight ($g\ mol^{-1}$) | log $K_{ow}$ | Chain length | Molecular length (nm) | $pK_a$ | Water solubility ($g\ L^{-1}$) |
|---|---|---|---|---|---|---|---|---|
| Perfluorooctanic acid | $C_8HF_{15}O_2$ | Carboxylic acid | 414.07 | 4.59 | 8 | 1.11 | 2.8 | 9.5 |
| Perfluorooctane sulfonic acid | $C_8HF_{17}SO_3$ | Sulfonic acid | 538.22 | 5.26 | 8 | 1.33 | -3.27 | 0.6 |
| Perfluorobutanic acid | $C_4HF_7O_2$ | Carboxylic acid | 214.04 | 2.32 | 4 | 0.6 | 0.4 | 0.4 |
| Perfluorobutane sulfonic acid | $C_4HF_9SO_3$ | Sulfonic acid | 338.19 | 2.73 | 4 | 0.81 | 0.14 | 0.5 |

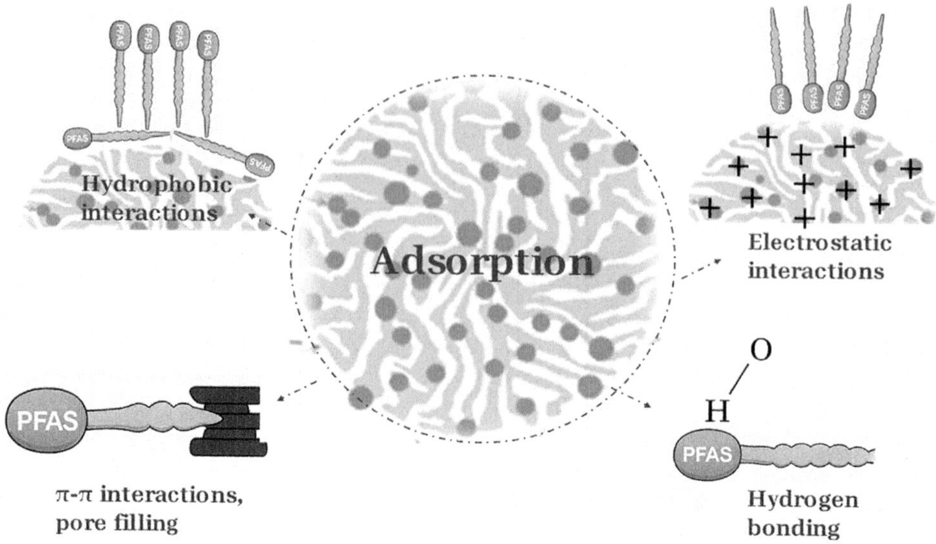


FIGURE 5.3 Adsorption mechanisms of per- and polyfluoroalkyl substances (PFAS) onto adsorbents.

The impact of NOM also varies with adsorbent type and pore structure. For instance, mesoporous activated carbon exhibited greater resilience to NOM-induced fouling compared to microporous carbon, attributed to reduced pore-blocking effects (Cantoni et al., 2021).

#### *5.3.2.5 Regeneration and reusability*

The practical application of adsorbents also depends on their regenerability. Thermal regeneration is often effective for GAC but may degrade the material over multiple cycles. For other materials (e.g., resins), solvent-based or alkaline regeneration is used (Table 5.1), though it may not fully restore performance. Materials with high regeneration efficiency and stability offer better life-cycle cost-effectiveness. Alcohol-based solvents (e.g., ethanol, methanol), saline solutions (e.g., NaCl), and mixed solvents (e.g., MeOH + acid) were the most effective regenerants for AC and ion- exchange resins (Table 5.1)

### 5.3.3 Mechanism of adsorption

Understanding the mechanisms involved in PFAS adsorption is critical for material design and process optimization. PFAS adsorption is a multifaceted process that often involves a combination of physical and chemical interactions (Fig. 5.3) depending on the structure of the PFAS and the properties of the adsorbent. The most suggested mechanism for adsorption is through hydrophobic interactions, particularly for long-chain PFAS, and when there is minimal competition from other organics. Electrostatic interactions, pore filling and size exclusion, π–π interactions and hydrogen bonding, and van der Waals and dipole–dipole interactions can also play central roles depending on the specific PFAS present, the water matrix, other cooccurring compounds, and the adsorption technique.

#### 5.3.3.1 *Hydrophobic interactions*

Hydrophobic interactions play a central role, particularly for long-chain PFAS, due to their long-fluorinated backbone. Materials with nonpolar, carbonaceous surfaces (e.g., activated carbon, biochar) facilitate the partitioning of hydrophobic PFAS tails into the adsorbent matrix (Fig. 5.3). These interactions are especially dominant in environments with low minimal competition from other organic compounds (Mohd Jais et al., 2024).

#### 5.3.3.2 *Electrostatic interactions*

PFAS are negatively charged in environmental pH ranges; thus electrostatic attraction is a key mechanism for materials bearing positively charged surface functional groups (e.g., IXs, metal centers in MOFs). Surface charge modification can significantly improve performance by enhancing these interactions (Zhang et al., 2024a). This is particularly important for short-chain PFAS, which lack the strong hydrophobic interactions of their longer-chain counterparts; electrostatic interactions are more critical for the adsorption of short-chain PFAS. Conversely, in unmodified adsorbents like GAC, electrostatic repulsion reduces adsorption at neutral to alkaline pH.

#### 5.3.3.3 *Pore filling and size exclusion*

Microporous and mesoporous materials may adsorb PFAS through pore filling, especially when pore diameters are similar to the molecular dimensions of the PFAS (Abeysinghe et al., 2025). In these cases, physical confinement within the pore structure enhances adsorption. However, pore size must be optimized as overly small pores hinder diffusion and overly large pores reduce contact efficacy.

#### 5.3.3.4 *π–π interactions and hydrogen bonding*

In addition to hydrophobic and electrostatic interactions, π–π interactions and hydrogen bonding have been recognized as secondary but contributory mechanisms in PFAS adsorption onto carbonaceous materials. π–π interactions occur between the aromatic rings present in certain PFAS structures, particularly those containing phenyl or other conjugated moieties—and the graphitic domains of activated carbon surfaces. These interactions are facilitated by the delocalized π-electron systems in both the adsorbate and adsorbent, enabling stacking or alignment that enhances adsorption affinity (Menezes et al., 2025).

Hydrogen bonding, on the other hand, involves the interaction between polar functional groups of PFAS molecules (e.g., carboxylate or sulfonate groups) and surface hydroxyl, carboxyl, or amine groups present on the activated carbon. These bonds are typically weaker than covalent or ionic interactions but can contribute to adsorption stability, especially under acidic to neutral pH conditions where proton donors and acceptors are more active. The presence of oxygen-containing functional groups on the carbon surface—often introduced or enhanced through oxidation or ball milling can increase the likelihood of hydrogen bonding (Masud et al., 2022, 2023). Although these mechanisms are generally considered secondary compared to hydrophobic partitioning and electrostatic attraction, their role becomes more pronounced in systems where PFAS molecules possess aromatic or highly polar structures, or where the carbon surface has been chemically modified to enhance polarity and functional group density.

#### 5.3.3.5 *Van der Waals and dipole–dipole interactions*

Van der Waals and dipole–dipole interactions have been recognized as contributing mechanisms in PFAS adsorption, particularly when high surface area adsorbents are employed. These interactions arise from transient or permanent dipoles within PFAS molecules and the adsorbent surface, facilitating molecular attraction and retention. In practical applications, PFAS adsorption is governed by a combination of mechanisms, including hydrophobic partitioning, electrostatic attraction, pore filling, hydrogen bonding, and $\pi$–$\pi$ interactions. For example, GAC primarily relies on hydrophobic interactions and pore-filling effects, whereas IXs utilize electrostatic attraction and hydrogen bonding to achieve PFAS removal. Mechanistic insights into these interactions are typically inferred through surface characterization techniques. Zeta potential measurements provide information on surface charge behavior, while Fourier transform infrared spectroscopy and X-ray photoelectron spectroscopy reveal changes in functional groups and bonding environments. BET surface area analysis and scanning electron microscopy are also employed to assess surface morphology and porosity before and after adsorption. Collectively, these analytical methods validate the presence and contribution of multiple adsorption mechanisms in PFAS remediation systems.

## 5.4 Challenges in per- and polyfluoroalkyl substances adsorption: scientific and practical constraints

The adsorption of PFAS from aqueous media has emerged as one of the most promising remediation approaches due to its cost-effectiveness and relative simplicity. However, despite the advancement in adsorbent design and performance evaluation, several scientific and practical challenges persist, hindering its field-scale implementation and long-term sustainability. These challenges range from molecular-level interactions, for example, chemical affinity and surface charge, to broader operational and environmental factors, including scalability, cost, and regulatory limitations. This section elaborates on the major hurdles currently associated with PFAS adsorption methods, including limitations in adsorbent selectivity, reduced efficiency in complex water matrices, and the persistent challenge of regenerating spent materials. It also addresses difficulties in scaling up laboratory successes to real-world applications, variability in environmental conditions that affect adsorption performance, and the lack of reliable predictive models. By examining these interconnected challenges, the discussion highlights critical gaps that must be addressed to advance the practical implementation of PFAS adsorption technologies (Lei et al., 2023).

### 5.4.1 Chemical diversity and complexity of per- and polyfluoroalkyl substances compounds

PFAS comprises a broad range of synthetic chemicals, encompassing over 4700 alternatives that include perfluoroalkyl acids, fluoropolymers, and fluorotelomer-based substances. These compounds vary in carbon chain length, terminal functional groups (e.g., carboxylate, sulfonate), and degrees of branching and fluorination, resulting in significantly different physical and chemical behaviors. This molecular diversity means that adsorbents which are effective for long-chain compounds such as PFOA or PFOS may be significantly less effective against short-chain variants such as PFBA or GenX (Abbasian Chaleshtari and Foudazi, 2022). Designing a universal adsorbent that can selectively and efficiently remove a wide range of PFAS is still an ongoing scientific challenge (Wanninayake, 2021).

Additionally, the ionic nature of PFAS affects how they interact with adsorbents. Most PFAS in the environment are anionic and adsorb better onto positively charged surfaces through electrostatic attraction. In contrast, cationic and zwitterionic PFAS exhibit more complex or variable adsorption behavior depending on the pH and surface charge of the adsorbent (Abbasian Chaleshtari and Foudazi, 2022; Wanninayake, 2021).

### 5.4.2 Factors limiting per- and polyfluoroalkyl substances adsorption efficiency

Short-chain PFAS have gained prominence as replacements for long-chain compounds due to regulatory restrictions (Chavan et al., 2024). The higher water solubility, lower molecular weight, and reduced hydrophobicity of short-chain PFAS result in weaker interactions with many commonly used adsorbents such as GAC and PAC (Chavan et al., 2024). Short-chain PFAS often exhibit poor adsorption affinity and tend to break through adsorbent beds rapidly. Many adsorbents lack the ability to effectively adsorb short-chain PFAS, significantly compromising treatment efficacy, particularly in systems dealing with PFAS cooccurrence in a single system.

In environmental matrices, PFAS coexist with a wide variety of organic and inorganic substances, including NOM, humic acids, bicarbonates, phosphates, and sulfates. These coexisting compounds can compete with PFAS for active adsorption sites or cause pore blockage (fouling), significantly reducing the performance of the adsorbents. NOM has been shown to mask adsorption sites through fouling or binding to positively charged sites that would otherwise interact with anionic PFAS (Wanninayake, 2021). Lei et al. (2023) highlight that high NOM in waters reduces PFAS sorption capacity.

The adsorption of PFAS is strongly influenced by environmental conditions such as pH, temperature, ionic strength, and the presence of competing ions. Most PFAS exist as negatively charged species in environmental waters (pH levels 6–8) due to their low $pK_a$ values (e.g., $pK_a = -3.27$ for PFOS and -0.2 for PFOA), making them susceptible to electrostatic interactions with positively charged adsorbent surfaces (Abbasian Chaleshtari and Foudazi, 2022). However, the effectiveness of various adsorbents for PFAS removal is significantly influenced by pH levels, and low pH levels may lead to electrostatic repulsion and reduced PFAS uptake. Short-chain PFAS (e.g., PFBA, PFPeA) are more affected by high ionic strength due to weaker hydrophobic interactions, while long-chain PFAS (e.g., PFOA, PFOS) are less impacted and retain better adsorption performance (Lei et al., 2023). Many high surface area adsorbents, especially those with microporous structures, show limited adsorption kinetics due to internal diffusion constraints. This issue is particularly significant for long-chain PFAS, which must migrate through narrow pores to reach active sites. The adsorption process can take hours or even days to reach equilibrium, making it unsuitable for high-throughput systems (Chavan et al., 2024). In continuous-flow systems, such as fixed-bed columns used in water treatment plants, short contact times can result in incomplete adsorption, highlighting the need for optimized reactor design and adsorbent particle size distribution.

### 5.4.3 Regeneration and desorption constraints

Effective PFAS treatment not only involves capturing these contaminants but also managing the spent adsorbents. Many adsorbents bind PFAS through strong hydrophobic or electrostatic interactions, making desorption and regeneration challenging. Frequently used techniques include thermal treatment, solvent washing, and alkaline regeneration. These are either energy intensive, environmentally hazardous, or result in partial regeneration with a significant loss in adsorption capacity, making them unsustainable in the long-term (Lei et al., 2023; Wanninayake, 2021). The regeneration process can lead to the release of concentrated PFAS waste streams, which pose further disposal challenges. These issues underscore the need for adsorbents that combine high performance with easy regenerability and environmentally safe disposal pathways. Since adsorption does not destroy PFAS but merely concentrates them out of solution, it is crucial to combine this method with destruction technologies such as AOPs, high-temperature incineration, or electrochemical oxidation. This integrated strategy ensures the efficient removal of PFAS from aqueous systems while enabling their breakdown into less harmful compounds, thereby preventing their re-entry into the environment (Abbasian Chaleshtari and Foudazi, 2022).

### 5.4.4 Economic and scalability barriers

GAC is the most used adsorbent for PFAS removal due to its availability, affordability, and high surface area. However, GAC often shows reduced efficiency for short-chain PFAS and requires frequent replacement, generating secondary waste that needs proper disposal. Ion-exchange resins are also widely applied and particularly effective for certain PFAS types, though they similarly require regeneration and produce concentrated waste streams. To overcome these limitations, novel adsorbents such as MOFs, COFs, CNTs, and functionalized polymers have illustrated exceptional adsorption capacities and selectivities under laboratory conditions. However, their high production costs, stability issues, and difficulties in large-scale manufacturing limit their practical applications (Chavan et al., 2024). Even more established materials like activated carbon require frequent replacement and generate secondary waste that must be managed appropriately. Additionally, most advanced adsorbents have not been extensively field-tested, raising questions about their reliability and robustness under operational stress and real water chemistry conditions.

### 5.4.5 Inadequacy of isotherm and kinetic models for field prediction

Laboratory studies typically use isotherm models such as Langmuir and Freundlich to describe PFAS adsorption behavior (Lei et al., 2023). These models assume idealized conditions (e.g., monolayer coverage, uniform surface energy) for both adsorbents and aqueous environment conditions that rarely reflect the complexity of real-world water systems or adsorbent surfaces.

As a result, they often fail to accurately predict adsorbent performance in dynamic conditions such as varying influent concentrations, flow rates, or adsorbent aging. Most models in literature are simplistic and do not account for competitive adsorption, surface fouling, or temperature effects, which limits their applicability in real-world design

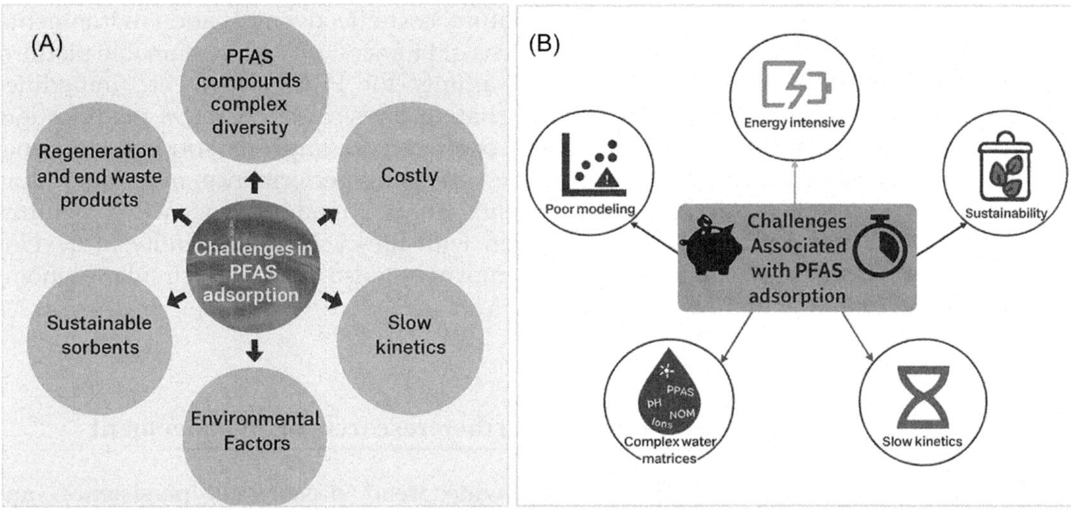


**FIGURE 5.4** (A) Major challenges in per- and polyfluoroalkyl substances (PFAS) adsorption and (B) key challenges associated with PFAS adsorption technologies.

(Abbasian Chaleshtari and Foudazi, 2022). The lack of predictive models capable of integrating these complexities requires extensive bench and pilot-scale testing to optimize site-specific adsorption needs, increasing costs and time.

### 5.4.6 Regulatory and analytical challenges

The increased awareness and regulation of PFAS in drinking waters across the globe have led to the development of several analytical methods capable of detecting PFAS at low levels in water (i.e., US EPA methods 533, 537.1, and 1633). However, these methods involve extensive sample preparation and expensive analytical equipment, such as a liquid chromatograph–tandem mass spectrometer (LC–MS/MS). These methods fail to incorporate certain short-chain and emerging PFAS, while also being costly and time-intensive. This limits the ability of smaller utilities or researchers to perform routine monitoring and implement pilot testing. Furthermore, regulatory limits differ across regions and are rapidly evolving, which adds uncertainty to the design approach and evaluation of treatment technologies (Lei et al., 2023; Wanninayake, 2021).

### 5.4.7 Environmental sustainability of adsorbent materials

The environmental sustainability of adsorbent materials is a critical consideration in PFAS removal by adsorption technologies (Fig. 5.4A). While advanced materials such as MOFs, COFs, and engineered nanomaterials have demonstrated superior adsorption capacities, their long-term environmental impacts remain questionable due to high energy demands for synthesis, limited biodegradability, and potential toxicity risks (Chavan et al., 2024; Lei et al., 2023). In contrast, bio-based adsorbents like biochar have emerged as promising alternatives due to their renewable nature, cost-effectiveness, and environmental compatibility. Biochar, produced through pyrolysis of biomass, offers a sustainable platform with tunable surface functionalities and high affinity for PFAS. However, unmodified biochars may show limited performance, necessitating physical or chemical modifications such as amination, magnetization, or surface oxidation to improve sorption efficiency (Chavan et al., 2024). These modifications, while enhancing performance, may reintroduce concerns about energy input and the environmental burden of added reagents. Thus future research must balance performance optimization with life-cycle sustainability to develop adsorbents that are not only effective in PFAS removal but also align with circular economy principles and environmental safety standards.

## 5.5 Future directions: areas for further research or improvement

PFAS pose a global issue due to their widespread occurrence, persistence, and complexity. Thousands of PFAS exist, but the vast majority of these are unable to be analyzed due to a lack of standards and methods for PFAS. Different PFAS compounds have distinct structures, physical properties, and chemical properties, making a one-size-fits-all approach to PFAS adsorption nearly impossible. Additionally, cooccurring compounds can increase the difficulty of selective PFAS removal. The use of decision-making tools, machine learning, and life-cycle assessment could be beneficial for addressing the large-scale applicability of novel adsorbents before full-scale implementation. Future directions and areas for improvement include scalability assessments, utilizing life-cycle analysis and AI tools such as machine learning, adopting the use of sustainable adsorbents, and including comprehensive testing when analyzing PFAS removals through adsorption.

### 5.5.1 Scalability

Several PFAS adsorbents are still in the bench-scale testing phase. While RSSCTs can provide quick insight into adsorption characteristics and performance (Mulugeta et al., 2025), practical constraints, such as cost, adsorbent synthesis, and regeneration may be excluded during this analysis. Thus utilizing pilot-scale operations before full-scale implementation is imperative to assess both larger-scale adsorption performance, the ability to synthesize the adsorbent at a quick and cost-effective rate, and the potential regeneration and reuse of the adsorbent. Further, promising adsorbents should be able to effectively remove a variety of PFAS in various water matrices, which can be challenging with the vast array of PFAS contaminants and complex water matrices.

### 5.5.2 Life cycle assessments and decision analysis

Adsorbents, such as various types of biochar, are promising sustainable alternatives for PFAS adsorption (Chavan et al., 2024). Using life-cycle assessments and cost analysis for different adsorbents can help assess the overall environmental footprint and cradle-to-grave cost, in addition to removal performance, as each aspect plays an important role when assessing an adsorbent (Ellis et al., 2023). Machine learning can be useful in predicting PFAS removal using certain adsorbents and inputting specific water quality parameters (Rahman et al., 2025). Multicriteria decision analysis is a decision-making tool that considers both monetary and nonmonetary factors, including nonquantifiable aspects such as social impacts, that can aid in making complex decisions by weighting criteria and ranking options, and increasing decision efficiency (Lei et al., 2023).

### 5.5.3 Sustainability

Some novel adsorbents, such as MIPs, nanoparticles, and functionalized membranes, have been demonstrated to show selective PFAS removal; however, their proper disposal is challenging, and scale-up is difficult (Fig. 5.4B) (Abbasian Chaleshtari and Foudazi, 2022). Membranes such as reverse osmosis and NF can be energy intensive (costly and unsustainable), as well as produce a concentrated PFAS-laden waste stream, which needs to be properly treated and disposed (Abbasian Chaleshtari and Foudazi, 2022). Adsorbents only serve to remove the PFAS, leaving a concentrated solid or liquid waste stream, which needs to be further treated or disposed. Destruction technology options include chemical oxidation and thermal degradation, but incomplete defluorination and harmful byproducts need to be studied in more detail before their large-scale implementation (Fang et al., 2024).

### 5.5.4 Comprehensive per- and polyfluoroalkyl substances testing for wholistic evaluation of removal

In the early 2000s, the EPA and PFAS manufacturers worked together to phase out two legacy PFAS, PFOA and PFOS (US EPA, 2024b). Due to this phaseout, the creation of novel PFAS and PFAS precursors to replace conventional ones has led to a rise in the number of PFAS that exist. Novel PFAS typically have shorter-chain lengths, making them more difficult to remove through adsorption. Tens of thousands of different PFAS exist, yet only up to ~50 can be tested for in a single method, leading to a frequent underestimation of total PFAS in various environmental water sources (McDonough et al., 2019). This underestimation has led to an increase in the use of and need for total PFAS screening methods, such as the total oxidizable precursor (TOP) assay and organic fluorine measurements (Smith et al., 2024). The TOP assay was developed to measure PFAS precursors and can assist in giving a more comprehensive PFAS analysis by accounting for PFCA precursors (Houtz and Sedlak, 2012). Precursors analyzed using the TOP assay were determined to be the reason for more frequent adsorbent changeout and higher operating costs in a full-scale system, which could have been avoided if precursors were taken into account at the bench scale (Ateia et al., 2023; Rodowa et al., 2020). Organic fluorine measurements, such as total organic fluorine (TOF), adsorbable organic fluorine, and extractable organic fluorine, can also be used as a nontarget analysis to analyze the total amount of carbon–fluorine bonds in a sample. These methods are less specific than targeted analysis and the TOP assay, but they can give a broad picture of contamination and can be useful when evaluating PFAS removal in environmental water samples. Typically, targeted PFAS analytes only account for a small fraction of the organic fluorine in a sample, so utilizing both measurements can provide both specific analyte information as well as a broader overview. In a study assessing adsorbents, nontargeted PFAS (using TOP and TOF) were less effectively removed than targeted analytes, suggesting the presence of novel PFAS and the importance of using nontargeted methods (O. Menezes et al., 2024).

## 5.6 Conclusion

PFAS have emerged as persistent and pervasive environmental contaminants due to their strong C–F bonds, chemical inertness, and extensive historical use across industrial and consumer applications. Addressing the widespread occurrence and toxicological risks posed by both long- and short-chain PFAS demands the development and implementation of effective, scalable, and sustainable remediation strategies.

Among the various technologies investigated, adsorption remains one of the most practical and cost-effective methods for PFAS remediation in both water and soil systems. A wide range of adsorbent materials—including GAC and PAC, biochar, IXs, MOFs, and cyclodextrin-based materials—have demonstrated varying degrees of success. Each material class offers unique advantages depending on the PFAS chain length, water chemistry, and operational conditions. However, no single solution has proven universally effective, particularly for short-chain and emerging PFAS.

This chapter has comprehensively reviewed the mechanistic insights, material modifications, adsorption models, and performance evaluation strategies relevant to PFAS adsorption. While laboratory studies show promise, real-world applications face multiple challenges, including competitive sorption with cocontaminants, pH and ionic strength effects, adsorbent fouling by NOM, and difficulties in regeneration and reuse of spent adsorbents. These challenges, alongside economic and regulatory constraints, limit the transition of high-performing materials from the bench to full-scale deployment.

Future research should prioritize:

- The development of multifunctional, regenerable, and environmentally benign adsorbents.
- Integration of adsorption with destructive technologies to achieve complete PFAS removal.
- Application of machine learning and decision analysis for adsorbent optimization.
- Broader implementation of nontargeted analytical methods (e.g., TOP assay, organofluorine) to better capture total PFAS loads.

A holistic and multidisciplinary approach—encompassing materials science, environmental chemistry, process engineering, and regulatory frameworks—is essential to overcome current barriers and to design next-generation solutions for PFAS remediation that are not only effective but also scalable and sustainable.

## References

Abbasian Chaleshtari, Z., Foudazi, R., 2022. A Review on Per- and Polyfluoroalkyl Substances (PFAS) Remediation: Separation Mechanisms and Molecular Interactions. ACS ES and T Water 2. American Chemical Society, pp. 2258–2272. Available at https://doi.org/10.1021/acsestwater.2c00271 Issue 12.

Abeysinghe, H., Ma, X., Tsige, M., 2025. PFAS removal via adsorption: A synergistic review on advances of experimental and computational approaches. Chemosphere 377, 144323. Available at https://doi.org/10.1016/J.CHEMOSPHERE.2025.144323.

Alazaiza, M.Y.D., Alzghoul, T.M., Ramu, M.B., Abu Amr, S.S., Abushammala, M.F.M., 2025. PFAS contamination and mitigation: A comprehensive analysis of research trends and global contributions. Case Studies in Chemical and Environmental Engineering 11, 101127. Available at https://doi.org/10.1016/J.CSCEE.2025.101127.

Ali, M.A., Thapa, U., Antle, J., Tanim, E.U.H., Aguilar, J.M., Bradley, I.M., et al., 2024. Influence of water chemistry and operating parameters on PFOS/PFOA removal using rGO-nZVI nanohybrid. Journal of Hazardous Materials 469, 133912. Available at https://doi.org/10.1016/J.JHAZMAT.2024.133912.

Ateia, M., Maroli, A., Tharayil, N., Karanfil, T., 2019. The overlooked short- and ultrashort-chain poly- and perfluorinated substances: A review. Chemosphere 220, 866–882. Available at https://doi.org/10.1016/J.CHEMOSPHERE.2018.12.186.

Ateia, M., Chiang, D., Cashman, M., Acheson, C., 2023. Total Oxidizable Precursor (TOP) Assay—Best Practices, Capabilities and Limitations for PFAS Site Investigation and Remediation. In Environmental Science and Technology Letters 10 (4). Available at https://doi.org/10.1021/acs.estlett.3c00061.

Baskar, A.V., Bolan, N., Hoang, S.A., Sooriyakumar, P., Kumar, M., Singh, L., et al., 2022. Recovery, regeneration and sustainable management of spent adsorbents from wastewater treatment streams: A review. Science of The Total Environment 822, 153555. Available at https://doi.org/10.1016/J.SCITOTENV.2022.153555.

Behnami, A., Pourakbar, M., Ayyar, A.S.R., Lee, J.W., Gagnon, G., Zoroufchi Benis, K., 2024. Treatment of aqueous per- and poly-fluoroalkyl substances: A review of biochar adsorbent preparation methods. Chemosphere 357, 142088. Available at https://doi.org/10.1016/J.CHEMOSPHERE.2024.142088.

Belkouteb, N., Franke, V., McCleaf, P., Köhler, S., Ahrens, L., 2020. Removal of per- and polyfluoroalkyl substances (PFASs) in a full-scale drinking water treatment plant: Long-term performance of granular activated carbon (GAC) and influence of flow-rate. Water Research 182, 115913. Available at https://doi.org/10.1016/J.WATRES.2020.115913.

Boone, J.S., Vigo, C., Boone, T., Byrne, C., Ferrario, J., Benson, R., et al., 2019. Per- and polyfluoroalkyl substances in source and treated drinking waters of the United States. Science of The Total Environment 653, 359–369. Available at https://doi.org/10.1016/J.SCITOTENV.2018.10.245.

Brendel, S., Fetter, É., Staude, C., Vierke, L., Biegel-Engler, A., 2018. Short-chain perfluoroalkyl acids: environmental concerns and a regulatory strategy under REACH. Environmental Sciences Europe 30 (1). Available at https://doi.org/10.1186/s12302-018-0134-4.

Brusseau, M.L., Yan, N., Van Glubt, S., Wang, Y., Chen, W., Lyu, Y., et al., 2019. Comprehensive retention model for PFAS transport in subsurface systems. Water Research 148, 41–50. Available at https://doi.org/10.1016/J.WATRES.2018.10.035.

Cantoni, B., Turolla, A., Wellmitz, J., Ruhl, A.S., Antonelli, M., 2021. Perfluoroalkyl substances (PFAS) adsorption in drinking water by granular activated carbon: Influence of activated carbon and PFAS characteristics. Science of The Total Environment 795, 148821. Available at https://doi.org/10.1016/J.SCITOTENV.2021.148821.

Chavan, D., Mayilswamy, N., Chame, S., Kandasubramanian, B., 2024. Biochar Adsorption: A Green Approach to PFAS Contaminant Removal. CleanMat 1 (1), 52–77. Available at https://doi.org/10.1002/clem.16.

Chen, W., Zhang, X., Mamadiev, M., Wang, Z., 2017. Sorption of perfluorooctane sulfonate and perfluorooctanoate on polyacrylonitrile fiber-derived activated carbon fibers: In comparison with activated carbon. RSC Advances 7 (2), 927–938. Available at https://doi.org/10.1039/c6ra25230c.

Cheng, L., Knappe, D.R.U., 2024. Removal of Per- and Polyfluoroalkyl substances by anion exchange resins: Scale-up of rapid small-scale column test data. Water Research 249, 120956. Available at https://doi.org/10.1016/J.WATRES.2023.120956.

Ching, C., Ling, Y., Trang, B., Klemes, M., Xiao, L., Yang, A., et al., 2022. Identifying the physicochemical properties of β-cyclodextrin polymers that determine the adsorption of perfluoroalkyl acids. Water Research 209, 117938. Available at https://doi.org/10.1016/J.WATRES.2021.117938.

Chow, S.J., Croll, H.C., Ojeda, N., Klamerus, J., Capelle, R., Oppenheimer, J., et al., 2022. Comparative investigation of PFAS adsorption onto activated carbon and anion exchange resins during long-term operation of a pilot treatment plant. Water Research 226, 119198. Available at https://doi.org/10.1016/J.WATRES.2022.119198.

Crittenden, J.C., Berrigan, J.K., Hand, D.W., 1986. Design of rapid small-scale adsorption tests for a constant diffusivity. Journal of the Water Pollution Control Federation 58 (4).

Crittenden, J.C., Berrigan, J.K., Hand, D.W., Lykins, B., 1987. Design of Rapid Fixed-Bed Adsorption Tests for Nonconstant Diffusivities. Journal of Environmental Engineering 113 (2). Available at https://doi.org/10.1061/(asce)0733-9372(1987)113:2(243).

Crittenden, J.C., Reddy, P.S., Arora, H., Trynoski, J., Hand, D.W., Perram, D.L., et al., 1991. Predicting GAC Performance With Rapid Small-Scale Column Tests. Journal AWWA 83 (1). Available at https://doi.org/10.1002/j.1551-8833.1991.tb07088.x.

Ellis, A.C., Boyer, T.H., Fang, Y., Liu, C.J., Strathmann, T.J., 2023. Life cycle assessment and life cycle cost analysis of anion exchange and granular activated carbon systems for remediation of groundwater contaminated by per- and polyfluoroalkyl substances (PFASs). Water Research 243, 120324. Available at https://doi.org/10.1016/J.WATRES.2023.120324.

Fang, Y., Ellis, A., Choi, Y.J., Boyer, T.H., Higgins, C.P., Schaefer, C.E., et al., 2021. Removal of Per- And Polyfluoroalkyl Substances (PFASs) in Aqueous Film-Forming Foam (AFFF) Using Ion-Exchange and Nonionic Resins. Environmental Science and Technology 55 (8). Available at https://doi.org/10.1021/acs.est.1c00769.

Fang, J., Li, S., Gu, T., Liu, A., Qiu, R., Zhang, W.X., 2024. Treatment of per- and polyfluoroalkyl substances (PFAS): A review of transformation technologies and mechanisms. Journal of Environmental Chemical Engineering 12 (1), 111833. Available at https://doi.org/10.1016/J.JECE.2023.111833.

Feng, S., Lu, X., Ouyang, K., Su, G., Li, Q., Shi, B., et al., 2024. Environmental occurrence, bioaccumulation and human risks of emerging fluoroalkylether substances: Insight into security of alternatives. In Science of the Total Environment 922. Available at https://doi.org/10.1016/j.scitotenv.2024.171151.

Fenton, S.E., Ducatman, A., Boobis, A., DeWitt, J.C., Lau, C., Ng, C., et al., 2021. Per- and Polyfluoroalkyl Substance Toxicity and Human Health Review: Current State of Knowledge and Strategies for Informing Future Research. In Environmental Toxicology and Chemistry 40 (3). Available at https://doi.org/10.1002/etc.4890.

Fouda-Mbanga, B.G., Onotu, O.P., Tywabi-Ngeva, Z., 2024. Advantages of the reuse of spent adsorbents and potential applications in environmental remediation: A review. Green Analytical Chemistry 11, 100156. Available at https://doi.org/10.1016/J.GREEAC.2024.100156.

Freundlich, H., 1922. Colloid & capillary chemistry. E.P. Dutton, New York.

Fujioka, T., Takeuchi, H., Tahara, H., Murakami, H., Boivin, S., 2024. Effects of functional groups of polyfluoroalkyl substances on their removal by nanofiltration. Water Research X 24, 100233. Available at https://doi.org/10.1016/J.WROA.2024.100233.

Gar Alalm, M., Boffito, D.C., 2022. Mechanisms and pathways of PFAS degradation by advanced oxidation and reduction processes: A critical review. Chemical Engineering Journal 450, 138352. Available at https://doi.org/10.1016/J.CEJ.2022.138352.

Groffen, T., Nkuba, B., Wepener, V., Bervoets, L., 2021. Risks posed by per- and polyfluoroalkyl substances (PFAS) on the African continent, emphasizing aquatic ecosystems. Integrated Environmental Assessment and Management 17 (4). Available at https://doi.org/10.1002/ieam.4404.

Gunarathne, V., Melo, T.M., Schauerte, M., Groth, F., Slaný, M., Rinklebe, J., 2025. Immobilization of per- and polyfluorinated alkyl substances (PFAS) from field contaminated groundwater by a novel organo-clay vs. colloidal activated carbon under flow conditions. Journal of Hazardous Materials 488, 137273. Available at https://doi.org/10.1016/J.JHAZMAT.2025.137273.

Ho, Y.S., McKay, G., 1999. Pseudo-second order model for sorption processes. Process Biochemistry 34 (5). Available at https://doi.org/10.1016/S0032-9592(98)00112-5.

Hossain, S., Stuart, T.D., Ramarao, B.V., VanLeuven, C.C., Wriedt, M., Kiemle, D., et al., 2023. Investigation into Cationic Surfactants and Polyelectrolyte-Coated β-Zeolites for Rapid and High-Capacity Adsorption of Short- and Long-Chain PFAS. Industrial and Engineering Chemistry Research 62 (21). Available at https://doi.org/10.1021/acs.iecr.3c00468.

Houtz, E.F., Sedlak, D.L., 2012. Oxidative conversion as a means of detecting precursors to perfluoroalkyl acids in urban runoff. Environmental Science and Technology 46 (17). Available at https://doi.org/10.1021/es302274g.

Hussain, H.N., Jilani, M.I., Imtiaz, F., Ahmed, T., Arshad, M.B., Mudassar, M., et al., 2025. Advances in the removal of Polyfluoroalkyl Substances (PFAS) from water using destructive and non-destructive methods. Green Analytical Chemistry 12, 100225. Available at https://doi.org/10.1016/J.GREEAC.2025.100225.

Ilango, A.K., Mekkat, R., Jeyalakshmi, V., Pervez, M.N., Jiang, T., Chand, P., et al., 2025. Enhanced removal of PFAS in water using activated ZIF-8 carbons: High adsorption efficiency, repeatable regenerability and reusability. Chemical Engineering Journal 507, 160192. Available at https://doi.org/10.1016/J.CEJ.2025.160192.

Kanchanapiya, P., Songngam, S., Tantisattayakul, T., 2022. The Adsorption of Perfluorooctanoic Acid on Coconut Shell Activated Carbons. AIMS Environmental Science 9 (2), 128–139. Available at https://doi.org/10.3934/environsci.2022010.

Kang, S.B., Wang, Z., Zhang, W., Kim, K.Y., Won, S.W., 2023. Removal of short- and long-chain PFAS from aquatic systems using electrostatic attraction of polyethylenimine-polyvinyl chloride electrospun nanofiber adsorbent. Separation and Purification Technology 326, 124853. Available at https://doi.org/10.1016/J.SEPPUR.2023.124853.

Koch, A., Jonsson, M., Yeung, L.W.Y., Kärrman, A., Ahrens, L., Ekblad, A., et al., 2020. Per- and Polyfluoroalkyl-Contaminated Freshwater Impacts Adjacent Riparian Food Webs. Environmental Science and Technology 54 (19). Available at https://doi.org/10.1021/acs.est.0c01640.

Kwon, H., Ali, Z.A., Wong, B.M., 2023. Harnessing Semi-Supervised Machine Learning to Automatically Predict Bioactivities of Per- and Polyfluoroalkyl Substances (PFASs). Environmental Science and Technology Letters 10 (11). Available at https://doi.org/10.1021/acs.estlett.2c00530.

LaFond, J.A., Hatzinger, P.B., Guelfo, J.L., Millerick, K., Jackson, W.A., 2023. Bacterial transformation of per- and poly-fluoroalkyl substances: a review for the field of bioremediation. Environmental Science: Advances 2. Royal Society of Chemistry, pp. 1019–1041. Available at https://doi.org/10.1039/d3va00031a 8.

Lagergren, S., 1898. Zur theorie der sogenannten adsorption gelöster Stoffe. Stockholm Kongl. svenska vetenskaps-akad. Handlinger 24 (4).

Langmuir, I., 1918. The adsorption of gases on plane surfaces of glass, mica and platinum. Journal of the American Chemical Society 40 (9). Available at https://doi.org/10.1021/ja02242a004.

Lei, X., Lian, Q., Zhang, X., Wang, T., Gee, M., Holmes, W., et al., 2022. Removal of perfluorooctanoic acid via polyethyleneimine modified graphene oxide: Effects of water matrices and understanding mechanisms. Chemosphere 308, 136379. Available at https://doi.org/10.1016/J.CHEMOSPHERE.2022.136379.

Lei, X., Lian, Q., Zhang, X., Karsili, T.K., Holmes, W., Chen, Y., et al., 2023. A review of PFAS adsorption from aqueous solutions: Current approaches, engineering applications, challenges, and opportunities. In Environmental Pollution 321. Available at https://doi.org/10.1016/j.envpol.2023.121138.

Leung, S.C.E., Shukla, P., Chen, D., Eftekhari, E., An, H., Zare, F., et al., 2022. Emerging technologies for PFOS/PFOA degradation and removal: A review. Science of The Total Environment 827, 153669. Available at https://doi.org/10.1016/J.SCITOTENV.2022.153669.

Leung, S.C.E., Wanninayake, D., Chen, D., Nguyen, N.T., Li, Q., 2023. Physicochemical properties and interactions of perfluoroalkyl substances (PFAS) - Challenges and opportunities in sensing and remediation. Science of The Total Environment 905, 166764. Available at https://doi.org/10.1016/J.SCITOTENV.2023.166764.

Liang, D., Li, C., Chen, H., Sørmo, E., Cornelissen, G., Gao, Y., et al., 2024. A critical review of biochar for the remediation of PFAS-contaminated soil and water. Science of The Total Environment 951, 174962. Available at https://doi.org/10.1016/J.SCITOTENV.2024.174962.

Liu, K., Zhang, S., Hu, X., Zhang, K., Roy, A., Yu, G., 2015. Understanding the Adsorption of PFOA on MIL-101(Cr)-Based Anionic-Exchange Metal-Organic Frameworks: Comparing DFT Calculations with Aqueous Sorption Experiments. Environmental Science and Technology 49 (14). Available at https://doi.org/10.1021/acs.est.5b00802.

Liu, C.J., Werner, D., Bellona, C., 2019. Removal of per- And polyfluoroalkyl substances (PFASs) from contaminated groundwater using granular activated carbon: A pilot-scale study with breakthrough modeling. Environmental Science: Water Research and Technology 5 (11), 1844–1853. Available at https://doi.org/10.1039/c9ew00349e.

Liu, W., Lin, T., Zhang, X., Jiang, F., Yan, X., Chen, H., 2022. Adsorption of perfluoroalkyl acids on granular activated carbon supported chitosan: Role of nanobubbles. Chemosphere 309, 136733. Available at https://doi.org/10.1016/J.CHEMOSPHERE.2022.136733.

Liu, B., Liu, Y.L., Sun, M., 2024. Remove legacy perfluoroalkyl acids and emerging per- and polyfluoroalkyl ether acids by single-use and regenerable anion exchange resins: Rapid small-scale column tests and model fits. Water Research 257, 121661. Available at https://doi.org/10.1016/J.WATRES.2024.121661.

Lukić Bilela, L., Matijošytė, I., Krutkevičius, J., Alexandrino, D.A.M., Safarik, I., Burlakovs, J., et al., 2023. Impact of per- and polyfluorinated alkyl substances (PFAS) on the marine environment: Raising awareness, challenges, legislation, and mitigation approaches under the One Health concept. In Marine Pollution Bulletin 194. Available at https://doi.org/10.1016/j.marpolbul.2023.115309.

Ma, T., Wu, P., Wang, L., Li, Q., Li, X., Luo, Y., 2023. Toxicity of per- and polyfluoroalkyl substances to aquatic vertebrates. Frontiers in Environmental Science 11 Frontiers Media S.A. Available at https://doi.org/10.3389/fenvs.2023.1101100.

Masud, M.A.A., Kim, D.G., Shin, W.S., 2022. Degradation of phenol using Fe(II)-activated CaO2: effect of ball-milled activated carbon (ACBM) addition. Environmental Research 214, 113882. Available at https://doi.org/10.1016/J.ENVRES.2022.113882.

Masud, M.A.A., Annamalai, S., Shin, W.S., 2023. Remediation of ciprofloxacin in soil using peroxymonosulfate activated by ball-milled seaweed kelp biochar: Performance, mechanism, and phytotoxicity. Chemical Engineering Journal 465, 142908. Available at https://doi.org/10.1016/J.CEJ.2023.142908.

McDonough, C.A., Guelfo, J.L., Higgins, C.P., 2019. Measuring total PFASs in water: The tradeoff between selectivity and inclusivity. Current Opinion in Environmental Science & Health 7, 13–18. Available at https://doi.org/10.1016/J.COESH.2018.08.005.

Menezes, O., Srivastava, K., Ferreira, B., Field, J.A., Root, R.A., Chorover, J., et al., 2024. Assessing strategies to measure hidden per- and polyfluoroalkyl substances (PFAS) in groundwater and to evaluate adsorption remediation efficiencies. Chemosphere 369, 143887. Available at https://doi.org/10.1016/J.CHEMOSPHERE.2024.143887.

Menezes, A., Goveas, L.C., Vinayagam, R., Selvaraj, R., 2025. Harnessing carbon-based adsorbents for poly- and perfluorinated substance removal: A comprehensive review. Journal of Water Process Engineering 69, 106621. Available at https://doi.org/10.1016/J.JWPE.2024.106621.

Mohd Jais, F., Ibrahim, M.S.I., El-Shafie, A., Choong, C.E., Kim, M., Yoon, Y., et al., 2024. Updated review on current approaches and challenges for poly- and perfluoroalkyl substances removal using activated carbon-based adsorbents. Journal of Water Process Engineering 64, 105625. Available at https://doi.org/10.1016/J.JWPE.2024.105625.

Molé, R.A., Velosa, A.C., Carey, G.R., Liu, X., Li, G., Fan, D., et al., 2024. Groundwater solutes influence the adsorption of short-chain perfluoroalkyl acids (PFAA) to colloidal activated carbon and impact performance for in situ groundwater remediation. Journal of Hazardous Materials 474, 134746. Available at https://doi.org/10.1016/J.JHAZMAT.2024.134746.

Morris, J.C., Weber, W.J., 1964. Removal of Biologically-Resistant Pollutants from Waste Waters by Adsorption. In Advances in Water Pollution Research. Available at https://doi.org/10.1016/b978-1-4832-8391-3.50032-4.

Mukherjee, P., Sathiyan, K., Zidki, T., Nadagouda, M.N., Sharma, V.K., 2023. Electrochemical degradation of per- and poly-fluoroalkyl substances in the presence of natural organic matter. Separation and Purification Technology 325, 124639. Available at https://doi.org/10.1016/J.SEPPUR.2023.124639.

Mulugeta, T.G., Ersan, M.S., Garcia-Segura, S., Ersan, G., 2025. Predicting full-scale performance of adsorbents for per- and polyfluoroalkyl substances adsorption: The role of rapid small-scale column tests. Science of The Total Environment 974, 178944. Available at https://doi.org/10.1016/J.SCITOTENV.2025.178944.

Murray, C.C., Safulko, A., Vatankhah, H., Liu, C.J., Tajdini, B., Marshall, R.E., et al., 2023. PFAS adsorbent selection: The role of adsorbent use rate, water quality, and cost. Journal of Hazardous Materials 454, 131481. Available at https://doi.org/10.1016/J.JHAZMAT.2023.131481.

Nguyen, M.D., Sivaram, A.K., Megharaj, M., Webb, L., Adhikari, S., Thomas, M., et al., 2023. Investigation on removal of perfluorooctanoic acid (PFOA), perfluorooctane sulfonate (PFOS), perfluorohexane sulfonate (PFHxS) using water treatment sludge and biochar. Chemosphere 338, 139412. Available at https://doi.org/10.1016/J.CHEMOSPHERE.2023.139412.

Noreña-Caro, D., Alvarez-Láinez, M., 2016. Experimental design as a tool for the manufacturing of filtering media based on electrospun polyacrylonitrile/ β -cyclodextrin fibers. International Journal on Interactive Design and Manufacturing 10 (2). Available at https://doi.org/10.1007/s12008-014-0241-4.

Palma, D., Richard, C., Minella, M., 2022. State of the art and perspectives about non-thermal plasma applications for the removal of PFAS in water. Chemical Engineering Journal Advances 10, 100253. Available at https://doi.org/10.1016/J.CEJA.2022.100253.

Panieri, E., Baralic, K., Djukic-Cosic, D., Djordjevic, A.B., Saso, L., 2022. PFAS Molecules: A Major Concern for the Human Health and the Environment. In Toxics 10 (2). Available at https://doi.org/10.3390/toxics10020044.

Patterson, C., Burkhardt, J., Schupp, D., Krishnan, E.R., Dyment, S., Merritt, S., et al., 2019. Effectiveness of point-of-use/point-of-entry systems to remove per- and polyfluoroalkyl substances from drinking water. AWWA Water Science 1 (2). Available at https://doi.org/10.1002/aws2.1131.

Peers, A.M., 1965. Elovich adsorption kinetics and the heterogeneous surface. Journal of Catalysis 4 (4). Available at https://doi.org/10.1016/0021-9517(65)90054-0.

Rahman, M.H.U., Sikder, R., Tonmoy, T.A., Hossain, M.M., Ye, T., Aich, N., et al., 2025. Transforming PFAS management: A critical review of machine learning applications for enhanced monitoring and treatment. Journal of Water Process Engineering 70, 106941. Available at https://doi.org/10.1016/J.JWPE.2025.106941.

Rekik, H., Arab, H., Pichon, L., El Khakani, M.A., Drogui, P., 2024. Per-and polyfluoroalkyl (PFAS) eternal pollutants: Sources, environmental impacts and treatment processes. Chemosphere 358, 142044. Available at https://doi.org/10.1016/J.CHEMOSPHERE.2024.142044.

Ren, Z., Zhang, R., Xu, X., Li, Y., Wang, N., Leiviskä, T., 2024. Sorption/desorption and degradation of long- and short-chain PFAS by anion exchange resin and UV/sulfite system. Environmental Pollution 361, 124847. Available at https://doi.org/10.1016/J.ENVPOL.2024.124847.

Riegel, M., Haist-Gulde, B., Sacher, F., 2023. Sorptive removal of short-chain perfluoroalkyl substances (PFAS) during drinking water treatment using activated carbon and anion exchanger. Environmental Sciences Europe 35 (1). Available at https://doi.org/10.1186/s12302-023-00716-5.

Rodowa, A.E., Knappe, D.R.U., Chiang, S.Y.D., Pohlmann, D., Varley, C., Bodour, A., et al., 2020. Pilot scale removal of per- and polyfluoroalkyl substances and precursors from AFFF-impacted groundwater by granular activated carbon. Environmental Science: Water Research and Technology 6 (4), 1083–1094. Available at https://doi.org/10.1039/c9ew00936a.

Saadi, R., Saadi, Z., Fazaeli, R., Fard, N.E., 2015. Monolayer and multilayer adsorption isotherm models for sorption from aqueous media. In Korean Journal of Chemical Engineering 32 (5). Available at https://doi.org/10.1007/s11814-015-0053-7.

Shahrokhi, R., Park, J., 2024. Enhanced removal of short- and long-chain per- and poly-fluoroalkyl substances from aqueous phase using crushed grafted chitosan beads: Performance and mechanisms. Environmental Pollution 340, 122836. Available at https://doi.org/10.1016/J.ENVPOL.2023.122836.

Shan, Y., Hao, H., Yin, Y., Hu, N., Zhan, M., Ma, D., et al., 2024. Effects of Temperature and DC Electric Fields on Perfluorooctanoic Acid Sorption Kinetics to Activated Carbon. Environmental Science and Technology 58 (13). Available at https://doi.org/10.1021/acs.est.3c10590.

Shapiro, E.F., Lin, Z.W., Cifuentes, E.S., Barajas-Rodriguez, F.J., Gwinn, R., Dichtel, W.R., et al., 2025. Removal of PFAS and pharmaceuticals from municipal wastewater using a novel β-cyclodextrin adsorbent over distinct contact times. Water Research 282, 123631. Available at https://doi.org/10.1016/J.WATRES.2025.123631.

Siriwardena, D.P., Crimi, M., Holsen, T.M., Bellona, C., Divine, C., Dickenson, E., 2019. Changes in adsorption behavior of perfluorooctanoic acid and perfluorohexanesulfonic acid through chemically-facilitated surface modification of granular activated carbon. Environmental Engineering Science 36 (4). Available at https://doi.org/10.1089/ees.2018.0319.

Sleep, J.A., Miklavcic, S.J., Juhasz, A.L., 2023. Modelling of PFAS-surface interactions: Effect of surface charge and solution ions. Chemosphere 319, 137910. Available at https://doi.org/10.1016/J.CHEMOSPHERE.2023.137910.

Smith, S.J., Lauria, M., Higgins, C.P., Pennell, K.D., Blotevogel, J., Arp, H.P.H., 2024. The Need to Include a Fluorine Mass Balance in the Development of Effective Technologies for PFAS Destruction. Environmental Science and Technology 58. American Chemical Society, pp. 2587–2590. Available at https://doi.org/10.1021/acs.est.3c10617 Issue 6.

Son, H., Kim, T., Yoom, H.S., Zhao, D., An, B., 2020. The adsorption selectivity of short and long per-and polyfluoroalkyl substances (Pfass) from surface water using powder-activated carbon. Water (Switzerland) 12 (11). Available at https://doi.org/10.3390/w12113287.

Sukeesan, S., Boontanon, N., Fujii, S., Boontanon, S.K., 2023. Evaluation of the adsorption behavior of mixed perfluoroalkyl and polyfluoroalkyl substances onto granular activated carbon and styrene-divinylbenzene resins. Remediation 33 (4), 297–308. Available at https://doi.org/10.1002/rem.21766.

Tamanna, T., Mahon, P.J., Hockings, R.K., Alam, H., Raymond, M., Smith, C., et al., 2023. Ion Exchange MIEX® GOLD Resin as a Promising Sorbent for the Removal of PFAS Compounds. Applied Sciences (Switzerland) 13 (10). Available at https://doi.org/10.3390/app13106263.

Temkin, M.J., Pyzhev, V., 1940. Recent modifications to Langmuir isotherms. Acta Physiochim URSS 12, 217–225.

US EPA. (2024a). EPA Finalizes PFAS Drinking Water Regulation.

US EPA. (2024b). Risk Management for Per- and Polyfluoroalkyl Substances (PFAS) under TSCA.

Van Glubt, S., Brusseau, M.L., Yan, N., Huang, D., Khan, N., Carroll, K.C., 2021. Column versus batch methods for measuring PFOS and PFOA sorption to geomedia. Environmental Pollution 268, 115917. Available at https://doi.org/10.1016/J.ENVPOL.2020.115917.

Verma, S., Varma, R.S., Nadagouda, M.N., 2021. Remediation and mineralization processes for per- and polyfluoroalkyl substances (PFAS) in water: A review. Science of The Total Environment 794, 148987. Available at https://doi.org/10.1016/J.SCITOTENV.2021.148987.

Vu, C.T., Wu, T., 2022. Recent progress in adsorptive removal of per- and poly-fluoroalkyl substances (PFAS) from water/wastewater. Critical Reviews in Environmental Science and Technology 52 (1). Available at https://doi.org/10.1080/10643389.2020.1816125.

Wang, W., Maimaiti, A., Shi, H., Wu, R., Wang, R., Li, Z., et al., 2019. Adsorption behavior and mechanism of emerging perfluoro-2-propoxypropanoic acid (GenX) on activated carbons and resins. Chemical Engineering Journal 364, 132–138. Available at https://doi.org/10.1016/J.CEJ.2019.01.153.

Wang, R., Lin, Z.W., Klemes, M.J., Ateia, M., Trang, B., Wang, J., et al., 2022. A Tunable Porous β-Cyclodextrin Polymer Platform to Understand and Improve Anionic PFAS Removal. ACS Central Science 8 (5), 663–669. Available at https://doi.org/10.1021/acscentsci.2c00478.

Wang, M., Yan, J., Diao, Y., Zhou, X., Luo, T., Wang, H., et al., 2023. Ball milled Mg/Al hydroxides modified nitrogen-rich biochar for arsenic removal: performance and governing mechanism. Carbon Research 2 (1). Available at https://doi.org/10.1007/s44246-023-00063-3.

Wanninayake, D.M., 2021. Comparison of currently available PFAS remediation technologies in water: A review. Journal of Environmental Management 283, 111977. Available at https://doi.org/10.1016/J.JENVMAN.2021.111977.

Wee, S.Y., Aris, A.Z., 2023. Revisiting the "forever chemicals", PFOA and PFOS exposure in drinking water. In npj Clean Water 6 (1). Available at https://doi.org/10.1038/s41545-023-00274-6.

Xie, R., Beckman, M.T., Almquist, C.B., Berberich, J.A., Danielson, N.D., 2024. Fixed-bed adsorption of perfluorooctanoic acid from water by a polyamine-functionalized polychlorotrifluoroethylene-ethylene polymer coated on activated carbon. Journal of Environmental Chemical Engineering 12 (4), 113001. Available at https://doi.org/10.1016/J.JECE.2024.113001.

Xu, B., Ahmed, M.B., Zhou, J.L., Altaee, A., Wu, M., Xu, G., 2017. Photocatalytic removal of perfluoroalkyl substances from water and wastewater: Mechanism, kinetics and controlling factors. Chemosphere 189, 717–729. Available at https://doi.org/10.1016/J.CHEMOSPHERE.2017.09.110.

Yang, A., Ching, C., Easler, M., Helbling, D.E., Dichtel, W.R., 2020. Cyclodextrin Polymers with Nitrogen-Containing Tripodal Crosslinkers for Efficient PFAS Adsorption. ACS Materials Letters 2 (9). Available at https://doi.org/10.1021/acsmaterialslett.0c00240.

Zhang, D., He, Q., Wang, M., Zhang, W., Liang, Y., 2021. Sorption of perfluoroalkylated substances (PFASs) onto granular activated carbon and biochar. Environmental Technology (United Kingdom) 42 (12). Available at https://doi.org/10.1080/09593330.2019.1680744.

Zhang, Y., Kong, K., Wu, Q., Ma, T., Liang, J., Wang, R., 2024a. A Porphyrinic Metal-Organic Framework with Cooperative Adsorption Domains for PFAS Removal from Water. ChemSusChem 17 (9). Available at https://doi.org/10.1002/cssc.202400069.

Zhang, Y., Wang, B., Ma, S., Zhang, Q., 2024b. Adsorption of per- and polyfluoroalkyl substances (PFAS) from water with porous organic polymers. Chemosphere 346, 140600. Available at https://doi.org/10.1016/J.CHEMOSPHERE.2023.140600.

CHAPTER

# 6

# Photocatalysis as an advanced per- and polyfluoroalkyl substances remediation technique

*Stephen Sunday Emmanuel*[1,2], *and Ademidun Adeola Adesibikan*[1]

[1]Department of Industrial Chemistry, Faculty of Physical Sciences, University of Ilorin, Ilorin, Kwara, Nigeria [2]New Life Research Foundation, Atlanta, GA, United States

## 6.1 Introduction

Photocatalysis has emerged as a versatile and sustainable advanced oxidation technology for environmental remediation, particularly the elimination of pollutants in water and wastewater systems, among other techniques, like adsorption, presented in Fig. 6.1a (Rahul et al., 2026; Yao et al., 2026; Bayode et al., 2025a). This process is driven by light-induced activation of semiconductor materials, which generate highly reactive species capable of mineralizing contaminants into benign end-products such as $CO_2$, $H_2O$, and inorganic ions (Badamasi et al., 2024; Dhanasekaran et al., 2026; Esfanddarani and Panigrahi, 2026). Notably, since the pioneering of this technique, it has been extensively investigated for various applications spanning hydrogen production (Zang et al., 2026; Wang et al., 2026; Dankawu et al., 2026), degradation of organic pollutants(Li et al., 2026; Najafidoust et al., 2026; Sun et al., 2026; Kourmousi et al., 2026), due to its various advantages, as summarized in Fig. 6.1b.

Notably, among these organic pollutants of emerging concern, per- and polyfluoroalkyl substances (PFAS) represent one of the most challenging classes due to their exceptional chemical stability and persistence (Kumar et al., 2026; Tong et al., 2026). PFAS comprise a large family of synthetic fluorinated compounds characterized by fully or partially fluorinated carbon chains, typically terminated with functional groups such as carboxylates or sulfonates. The extremely strong C–F bond (485–546 kJ $mol^{-1}$) and the shielding effect of fluorine atoms confer high thermal, chemical, and biological resistance, earning PFAS the designation of "forever chemicals" (Kumar et al., 2026; Hanphaiboon et al., 2026; Zhao et al.,

*Per- and Polyfluoroalkyl Substances (PFAS) as Environmental Contaminants*
DOI: https://doi.org/10.1016/B978-0-443-43856-1.00022-8

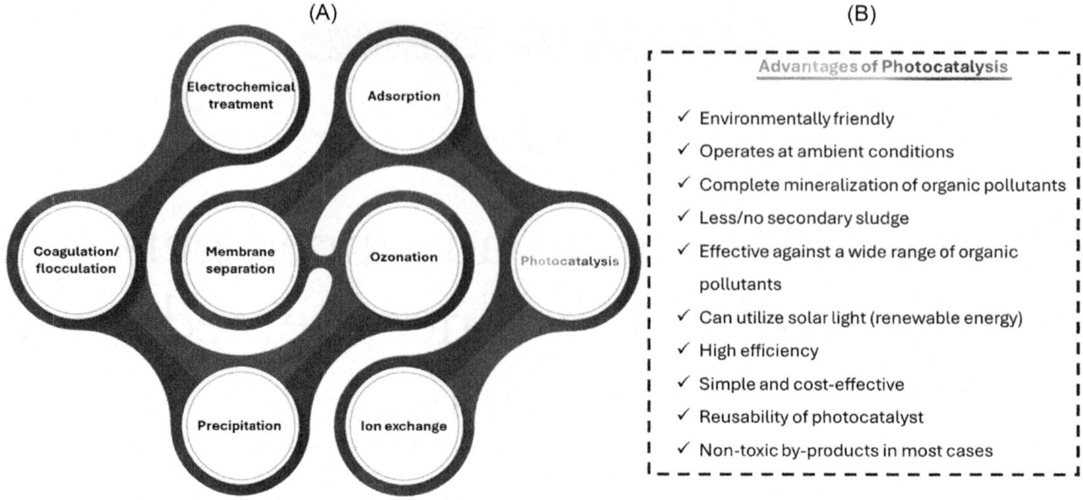


**FIGURE 6.1** (A) Treatment methods employed for the removal of per- and polyfluoroalkyl substances (PFAS). (B) Key advantages of photocatalysis for PFAS remediation.

2026; Vietri et al., 2026). These chemicals are widely used in industrial processes and consumer products, including firefighting foams, nonstick cookware, water-repellent textiles, and food packaging, and thus have become globally distributed in aquatic environments, soils, wildlife, and human populations (Kumar et al., 2026; Zhao et al., 2026). These compounds can bioaccumulate and thus pose risks to the ecosystem and human health, especially when exposed to polluted water sources for a long time (Zhao et al., 2026; Van Tri et al., 2026; Ke et al., 2026).

Conventional water treatment technologies, including adsorption, coagulation, and membrane filtration, primarily transfer PFAS between phases rather than destroying them, and thus often show poor effectiveness (Vietri et al., 2026). Notably, given the increasing regulatory pressure and environmental risks associated with PFAS contamination, photocatalysis is gaining recognition as a potentially effective destructive treatment technology capable of complementing existing separation-based approaches (Huang et al., 2026; Shashikumar et al., 2025). This chapter provides an overview of photocatalytic degradation-oriented techniques for PFAS remediation, including fundamental principles/degradation mechanisms, performance of various photocatalyst materials, and reusability efficiency in order to provide insight into potential scalability, economic, and environmental benefits. Emphasis is placed on recent material innovations, ranging from binary to ternary systems. At the end, prospects and challenges that can serve as future research directions for practical implementation in water treatment applications were presented. This chapter will be of interest to environmentalists, water experts, and various industries where PFAS are used.

Holistically, this chapter adds to the basic understanding of photocatalysis for PFAS remediation. In addition, because this chapter is relevant to the various sustainable development goals (SDGs) presented in Fig. 6.2, it is a useful resource for SDG stakeholders, researchers, engineers, decision-makers, and industries seeking to develop scalable,

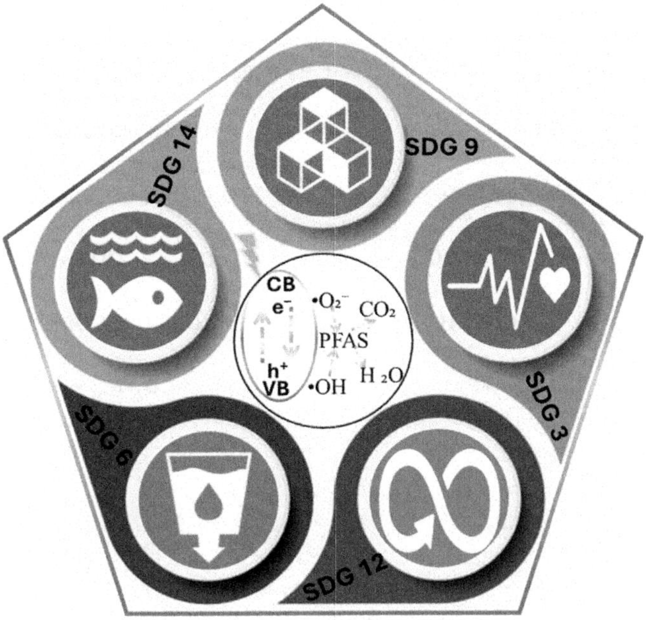


**FIGURE 6.2** Contribution to sustainable development goals (SDGs).

sustainable, and effective water treatment systems to reduce water insecurity and enhance environmental sustainability.

## 6.2 Fundamental mechanism of per- and polyfluoroalkyl substances photocatalytic degradation

Photocatalytic degradation is a light-driven process that employs semiconductor nanocatalysts to decompose persistent organic contaminants such as PFAS into less harmful products, including $F^-$, $CO_2$, $H_2O$, and shorter-chain intermediates, as illustrated in Fig. 6.3 (Cao et al., 2024). When a nanophotocatalyst is irradiated with photons possessing energy equal to or greater than its band gap, electrons ($e^-$) in the valence band (VB) are excited into the conduction band (CB), leaving behind positively charged holes ($h^+$). These photogenerated charge carriers migrate to the catalyst surface and initiate redox reactions with adsorbed species. Conventionally, CB electrons reduce dissolved $O_2$ to superoxide radicals ($\bullet O_2^-$), while VB holes oxidize $H_2O$ or $OH^-$ to hydroxyl radicals ($\bullet OH$) (Cao et al., 2024; Gomez-Ruiz et al., 2018; Gu et al., 2025).

However, one dilemma is that, unlike many organic pollutants, PFAS molecules are highly resistant to electrophilic $\bullet OH$ attack due to the strong C–F bond and electron-withdrawing fluorine sheath. Consequently, PFAS photocatalysis proceeds predominantly through reductive pathways initiated by photogenerated electrons or hydrated electrons. These electrons attack the electron-deficient carbon atoms adjacent to functional groups (e.g., $-COO^-$ or $-SO_3^-$) in PFAS such as perfluorooctanoic acid (PFOA) or perfluorooctane

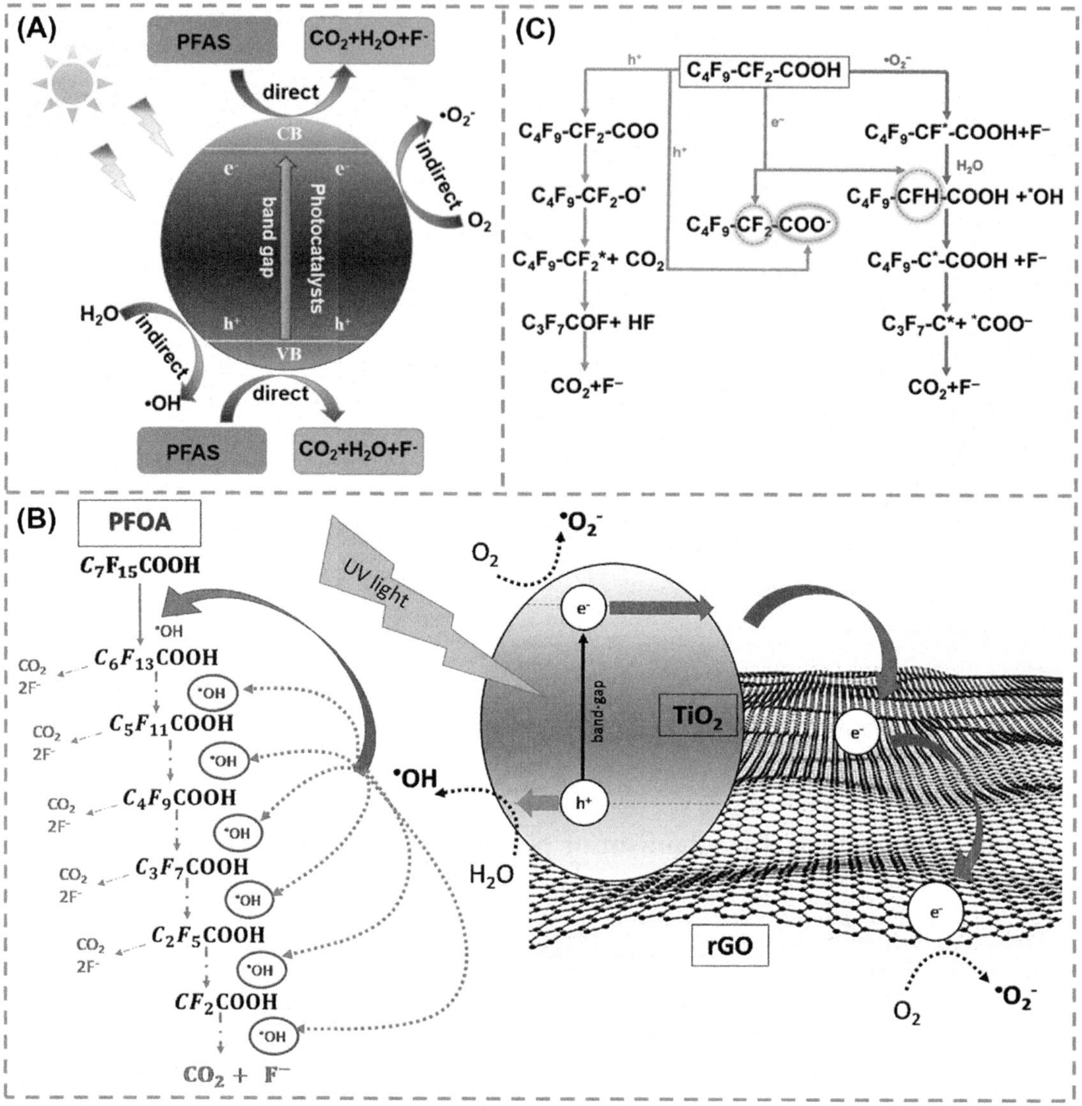


**FIGURE 6.3** (A) Overview of the degradation mechanism of per- and polyfluoroalkyl substances (PFAS) by the photocatalyst (Cao et al., 2024). (B) Photocatalytic degradation mechanism and pathways of perfluorooctanoic acid (PFOA) using the $TiO_2$@rGO (Gomez-Ruiz et al., 2018). *(A) Reused with permission from Elsevier (Authorization code: 6240970548960). (B) Reused with permission from Elsevier (Authorization code: 6240970834811). (C) PFH$_x$A photocatalytic degradation pathways by $Bi_2WO_6$@Zn (Hojamberdiev et al., 2023) (Open access).*

sulfonate, triggering stepwise C–F bond cleavage and release of fluoride ions as shown in Fig. 6.3B-C (Gomez-Ruiz et al., 2018; Gu et al., 2025; Hojamberdiev et al., 2023; Dehghani et al., 2022).

Mechanistically, the initial reductive defluorination generates unstable carbon-centered radicals ($C_nF_{2n+1}\bullet$), which undergo sequential reactions including decarboxylation or desulfonation, chain shortening, and reaction with dissolved oxygen to form perfluoroalkyl peroxyl radicals. These intermediates further transform into shorter-chain perfluorocarboxylic acids, carbonyl fluorides, and eventually $CO_2$ and $F^-$ upon continued irradiation (Huang et al., 2026; Gomez-Ruiz et al., 2018; Wen et al., 2025). In chromatographic and mass-spectrometric analyses reported across PFAS photocatalytic studies, the parent PFAS peak typically diminishes progressively after irradiation onset, accompanied by the emergence of new peaks corresponding to shorter-chain homologs and defluorinated fragments. With prolonged reaction time, these intermediate peaks also decline as mineralization advances, indicating ongoing defluorination and carbon backbone fragmentation as illustrated in Fig. 6.4 (Wen et al., 2025). Generally, PFAS photocatalytic degradation differs fundamentally from conventional organic pollutant oxidation in that reductive electron-driven defluorination dominates the early stages, while oxidative species ($\bullet OH$, $\bullet O_2^-$, $h^+$) mainly participate in subsequent transformation and mineralization of intermediate species. This dual reductive-oxidative pathway underpins current catalyst design strategies aimed at maximizing electron availability and suppressing charge recombination for efficient PFAS destruction.

## 6.3 Photocatalytic degradation efficiency

In this section, the degradation performance of various photocatalysts for PFAS remediation under light irradiation is discussed and summarized in Table 6.1 alongside optimal conditions. In a particular study, Xia et al. (2022) investigated the photocatalytic degradation of perfluorinated compounds using ZVI under UVC irradiation. Under acidic conditions (pH 3) and exposure to 112 W UVC light, ZVI achieved approximately 90% degradation of PFNA and 88% of PFOS within 120 min. This material was also tested for PFAS degradation in wastewater effluent obtained from two local WWTPs in Southern Illinois, and it performed excellently. These results demonstrated the strong reductive and photocatalytic capability of ZVI toward persistent perfluoroalkyl substances under optimized conditions. The high degradation efficiency was attributed to the enhanced generation of reactive species and accelerated electron transfer processes in the acidic medium, which promoted defluorination and breakdown of the PFAS molecules. The findings highlight the potential of ZVI-based systems as effective materials for the rapid treatment of recalcitrant fluorinated contaminants under ultraviolet irradiation (Xia et al., 2022).

Recently, the use of binary composite materials has been given attention for PFAS degradation. For example, Ong et al. (2017) compared the solar-driven degradation of PFOA using ZnO/rGO nanocomposite and pristine ZnO under identical operating conditions (initial concentration $=100\,mg\,L^{-1}$, catalyst dose $=1000\,mg\,L^{-1}$, pH 7). The ZnO/rGO composite exhibited markedly higher activity, achieving approximately 90.91% PFOA degradation within 100 min, whereas bare ZnO resulted in only 56.49% removal over the same period (Ong et al., 2017). Similar PFOA photocatalytic degradation efficiency of over

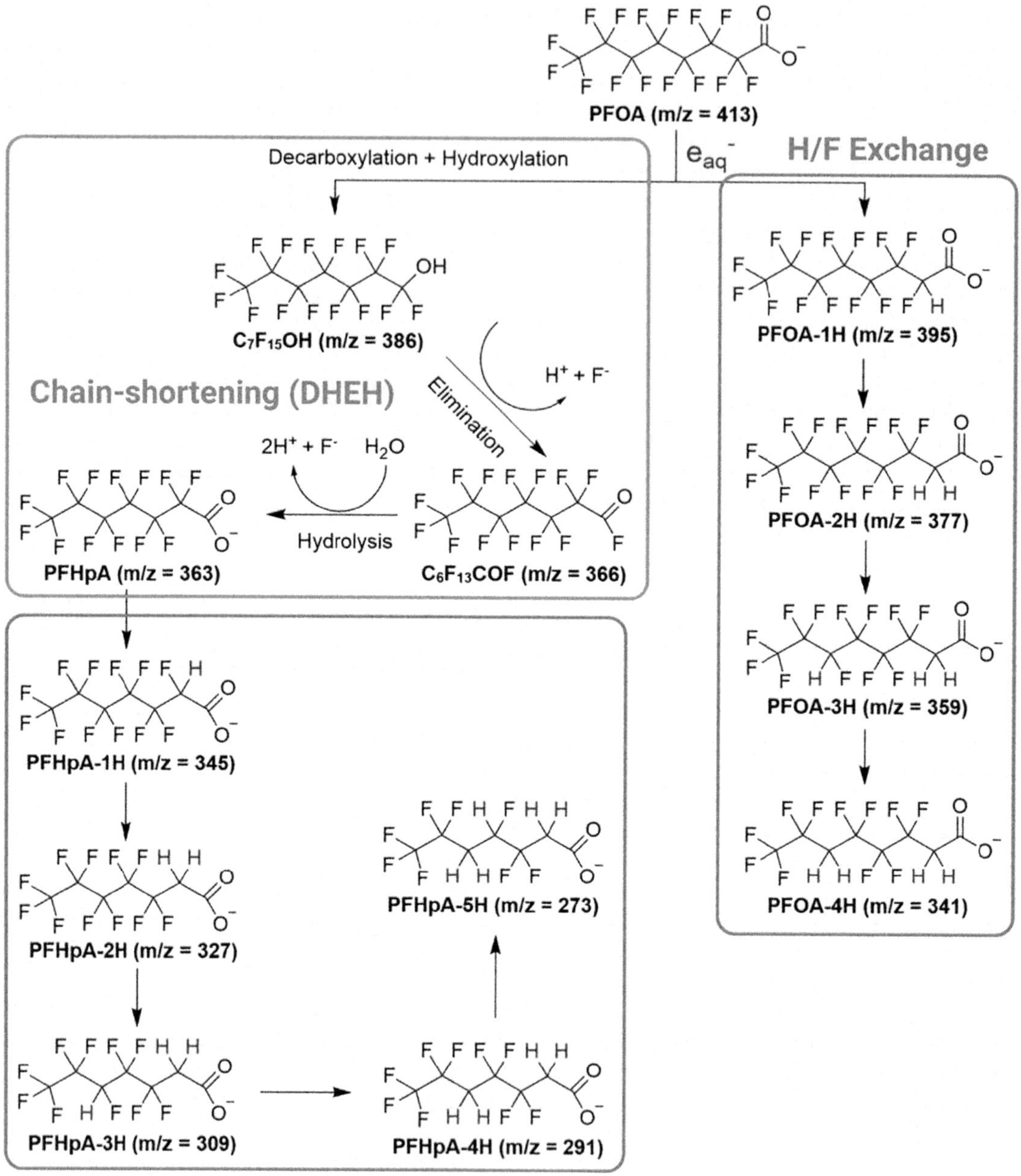


**FIGURE 6.4** Photocatalytic degradation pathways of perfluorooctanoic acid (PFOA) by MIL-177-HT (Wen et al., 2025). *Reused with permission from Elsevier (Authorization code: 6240971273224).*

**TABLE 6.1** Summary of photocatalytic performance of various nanomaterials for per- and polyfluoroalkyl substances (PFAS) degradation.

| Photocatalyst | PFAS | Light source | Optimal condition | Degradation efficiency (%) | Time (mins) | References |
|---|---|---|---|---|---|---|
| ZVI | PFNA | 112 W UV-C light bulbs | pH = 3, initial conc. =0.5 $\mu g\ L^{-1}$, dose =100 $mg\ L^{-1}$ | 90 | 120 | Xia et al. (2022) |
| ZVI | PFOS | 112 W UV-C light bulbs | pH = 3, initial conc. =0.5 $\mu g\ L^{-1}$, dose =100 $mg\ L^{-1}$ | 88 | 120 | Xia et al. (2022) |
| ZVI | PFOA | 112 W UV-C light bulbs | pH = 3, initial conc. =0.5 $\mu g\ L^{-1}$, dose =100 $mg\ L^{-1}$ | 46 | 120 | Xia et al. (2022) |
| Ga/TNTs@AC | PFOS | UV light | pH = 3.5 | 88.6 | 1440 | Zhu et al. (2026) |
| Ga/TNTs@AC | PFNS | UV light | pH = 3.5 | 99 | 1440 | Zhu et al. (2026) |
| Ga/TNTs@AC | PFPeS | UV light | pH = 3.5 | 73.4 | 1440 | Zhu et al. (2026) |
| $Bi_2WO_6$@Zn | PFHxA | 150 W mercury lamp | Dose =0.4 $mg\ L^{-1}$, initial conc. =5 $mg\ L^{-1}$ | 57 | 45 | Hojamberdiev et al. (2023) |
| Fe/$TiO_2$@AC | PFOA | UV light | pH = 7, dose =1000 $mg\ L^{-1}$, initial conc. =100 $\mu g\ L^{-1}$ | >90 | 240 | Li et al. (2020) |
| In/TNT@AC | PFOA | UV light | pH = 7, initial conc. =0.1 $mg\ L^{-1}$, dose =1000 $mg\ L^{-1}$ | >99 | 240 | Arana Juve et al. (2022) |
| $In_2O_3$@graphene | PFOA | 15 W mercury lamp | pH = 7, initial conc. =30 $mg\ L^{-1}$, dose =500 $mg\ L^{-1}$ | 100 | 180 | Li et al. (2013) |
| $In_2O_3$ | PFOA | 15 W mercury lamp | pH = 7, initial conc. =30 $mg\ L^{-1}$, dose =500 $mg\ L^{-1}$ | 87.6 | 180 | Li et al. (2013) |
| SiC/GQDs | PFOS | 5 W UV lamp | Dose =500 $mg\ L^{-1}$, pH = 7 | 93.9 | 1200 | Huang et al. (2017) |
| $TiO_2$@graphene | PFOA | 8 W UV-C Phillip lamp | pH = 4.64–4.95, dose =200 $mg\ L^{-1}$, initial conc. =12 μM | 99.2 | 480 | Panchangam et al. (2018) |
| GO@$TiO_2$ | PFOA | 8 W UV-C Phillip lamp | pH = 4.64–4.95, dose =200 $mg\ L^{-1}$, initial conc. =12 μM | 93.6 | 480 | Panchangam et al. (2018) |

(*Continued*)

TABLE 6.1 (Continued)

| Photocatalyst | PFAS | Light source | Optimal condition | Degradation efficiency (%) | Time (mins) | References |
|---|---|---|---|---|---|---|
| GO@P25 | PFOA | 8 W UV-C Phillip lamp | pH = 4.64–4.95, dose =200 mg $L^{-1}$, initial conc. =12 μM | 96 | 480 | Panchangam et al. (2018) |
| ZnO/rGO | PFOA | Solar irradiation | Initial conc. =100 mg $L^{-1}$, dose =1000 mg $L^{-1}$, pH = 7 | 90.91 | 100 | Ong et al. (2017) |
| ZnO | PFOA | Solar irradiation | Initial conc. =100 mg $L^{-1}$, dose =1000 mg $L^{-1}$, pH = 7 | 56.49 | 100 | Ong et al. (2017) |
| GO/$TiO_2$ | PFOA | 8 W UV lamp | pH = 3, initial conc. =0.121 μM $L^{-1}$ | 83 | 150 | Park et al. (2018) |
| $TiO_2$@rGO | PFOA | 150 W mercury lamp | pH = 3.8, initial conc. =0.24 mmol $L^{-1}$, dose =100 mg $L^{-1}$ | 93 | 720 | Gomez-Ruiz et al. (2018) |
| Pb–$BiFeO_3$/rGO | PFOA | 5 W UV light | pH = 2, initial conc. =50 mg $L^{-1}$, dose =100 mg $L^{-1}$ | 69.6 | 480 | Shang et al. (2018) |
| $TiO_2$@rGO | PFOA | 150 W mercury lamp | pH = 3.8, initial conc. =0.24 mmol $L^{-1}$, dose =100 mg $L^{-1}$ | 86 | 720 | Rivero et al. (2020) |
| $TiO_2$@MWCNTs | PFOA | 300 W mercury lamp | pH = 2, initial conc. =30 mg $L^{-1}$, dose =1.6 g $L^{-1}$ | 100 | 720 | Song et al. (2012) |
| $CdIn_2S_4$ | PFOS | 300 W Xe light | Dose =5 mg, initial conc. =~40 ppm | 99 | 1440 | Hamza et al. (2025) |
| FeOCl/(Bi-MOF–BiOCl)$_{100}$ | PFOA | UV light | Dose =75 mg $L^{-1}$, pH = 5.85, initial conc. =50 mg $L^{-1}$ | ~100 | 30 | Gu et al. (2025) |
| $TiO_2$–Pb/rGO | PFOA | UV lamp | Dose =330 mg $L^{-1}$ | 98 | 1440 | Chowdhury and Choi (2023) |
| $TiO_2$–Pb | PFOA | UV lamp | Dose =330 mg $L^{-1}$ | 80 | 1440 | Chowdhury and Choi (2023) |

| | | | | | | |
|---|---|---|---|---|---|---|
| $TiO_2$/rGO | PFOA | UV lamp | Dose =330 mg $L^{-1}$ | 70 | 1440 | Chowdhury and Choi (2023) |
| $TiO_2$ | PFOA | UV lamp | Dose =330 mg $L^{-1}$ | <10 | 1440 | Chowdhury and Choi (2023) |
| $TiO_2$@boron nitride | PFOA | Natural sunlight | Dose =500 mg $L^{-1}$, initial conc. =50 ppm | ~100 | 240 | Duan et al. (2022) |
| BiOHP@carbon sphere | PFOA | 18 W Hg lamp | pH = 7, dose =1000 mg $L^{-1}$, initial conc. =200 μg $L^{-1}$ | ~90 | 60 | Xu et al. (2020) |
| BiOCl | PFOA | 500 W Hg lamp | pH = 4, dose =500 mg $L^{-1}$, initial conc. =0.12 mM | ~100 | 40 | Wu et al. (2021) |
| rGO@$WO_3$ | PFNA | Diode-pumped solid-state laser | NA | 100 | 1320 | Gates et al. (2025) |
| MIL-177-HT | PFOA | 450 W mercury lamp | Initial conc. =100 μg $L^{-1}$, dose =2.5 g $L^{-1}$ | 83 | 1440 | Wen et al. (2025) |
| $In_2O_3$/$Bi_2O_3$ | PFOA | Mercury lamp | NA | >90 | 120 | Huang et al. (2026) |
| Bi/Ti coordination polymer | PFOA | 14 W UV lamps | pH = 3, initial conc. =10 mg $L^{-1}$, dose =5 mg $L^{-1}$ | ~100 | 300 | Bai et al. (2026) |
| $Fe_3O_4$/ZnO@Ag | PFAS | Visible light | pH = 3 | 87.37 | NA | Pouresmaeil et al. (2025) |

NA = not available.

80% was also recorded using $GO/TiO_2$ (Park et al., 2018) and $rGO/TiO_2$ (Gomez-Ruiz et al., 2018) at acidic pH. The superior performance of the hybrid catalyst was attributed to the incorporation of rGO, which enhanced charge separation and electron transport while also improving the adsorption of PFOA molecules on the catalyst surface. This synergistic interaction between ZnO and rGO promoted more efficient utilization of solar irradiation and facilitated the generation of reactive species capable of attacking the highly stable C–F bonds. Consequently, the ZnO/rGO system demonstrated substantially improved photocatalytic degradation efficiency compared with pristine ZnO under neutral conditions.

In another investigation, Zhu et al. (2026) reported the photocatalytic degradation of perfluoroalkyl sulfonates using a Ga/TNTs@AC composite under UV irradiation. At pH 3.5 and after 1440 min of exposure, the material achieved degradation efficiencies of approximately 88.6% for PFOS, 99% for PFNS, and 73.4% for PFPeS. These results demonstrated the broad applicability of the Ga-modified titanate nanotube/activated carbon system for the removal of structurally diverse PFAS compounds. The high degradation performance was attributed to the synergistic effects of Ga incorporation, titanate nanotubes, and activated carbon, which together enhanced light absorption, adsorption affinity, and charge transfer efficiency. This cooperative interaction facilitated the generation of reactive species and promoted the breakdown of the highly stable C–F bonds in the target pollutants, leading to effective degradation under acidic UV-assisted conditions (Zhu et al., 2026).

In addition, from a structure-function relationship perspective, the chemical structures of PFAS may greatly influence their remediation output, and this includes chain length, terminal or other functional groups (Cao et al., 2024). For example, Panchangam et al. (2009) observed that long-chain PFCAs, specifically PFOA and PFNA, degraded faster than PFHpA using $TiO_2$ under a 16 W mercury lamp. This behavior was attributed to their higher hydrophobicity, which enhances their interaction with the photocatalyst and thus promotes more efficient degradation (Panchangam et al., 2009). Tian and Gu (2018) reported comparable findings, showing that longer-chain PFAS exhibit higher degradation rates using organoclay under a 36 W mercury lamp (Tian and Gu, 2018). Similarly, another study showed that under a 500 W mercury lamp, In-MOF/BiOF completely degraded PFOS, PFOA, and 6:2 Cl-PFESA (15 $mg\ L^{-1}$) within 90, 150, and 150 min of irradiation, respectively, while HFPO-TA (15 $mg\ L^{-1}$) reached ~92% degradation in 180 min. The observed rate constants followed the trend: PFOS > 6:2 Cl-PFESA > PFOA > HFPO-TA, which further corroborates the difference in the degradation output (Wang et al., 2022).

Summarily, the findings here show that while some single-component nanomaterials deliver good PFAS degradation efficiency, their binary and ternary composites exhibited superior efficiency. However, more work still needs to be done to strongly validate and establish these recent advances, including the use of natural sunlight. Moreover, currently, the relationship between the chemical structure of PFAS and their degradation efficiency is still not well understood. Since multiple PFAS compounds often coexist in real natural water bodies, more research is needed to explore how their structural features influence degradation performance when photocatalytic methods are used. In the future, theoretical calculations like density functional theory simulation and in-depth analyses will be essential to reveal the underlying mechanisms behind these structure-function relationships (Cao et al., 2024).

## 6.4 Recyclability and reusability

From industrial, economic, environmental, and operational perspectives, photocatalyst reusability is a critical requirement for developing sustainable pollutant (PFAS in this case) remediation technologies. Reusing spent photocatalysts minimizes secondary solid waste generation, lowers the cost associated with synthesizing fresh catalytic materials, and reduces the environmental risks linked to the disposal of intermediate or fluorinated contaminant-laden catalysts (Emmanuel and Adesibikan, 2025; Emmanuel et al., 2026; Bayode et al., 2025b; Wahyuni et al., 2026). Consequently, recyclability and structural stability are key indicators of photocatalyst robustness and suitability for long-term PFAS remediation in realistic water treatment systems. The recyclability process typically involves recovery of the photocatalyst from the treated solution (e.g., by centrifugation or filtration), followed by washing and drying, and subsequent reuse in successive PFAS degradation cycles (Emmanuel and Adesibikan, 2025; Emmanuel et al., 2026; Bayode et al., 2025b; Zaaboul et al., 2026).

Herein, the reusability performance of various PFAS photocatalysts, summarized in Table 6.2, demonstrates encouraging stability across multiple cycles. For instance, FeOCl/(Bi-MOF–BiOCl)$_{100}$ maintained over 98% of its initial PFOA degradation efficiency after four consecutive reuse cycles, confirming excellent structural integrity and resistance to deactivation (Gu et al., 2025). Similarly, ZnO/cellulose nanofiber used for PFOS degradation exhibited only approximately 10% performance decline after three cycles, indicating good recyclability with minimal active-site loss (Dehghani et al., 2022). Comparable stability was reported for $TiO_2$@boron nitride, which sustained photocatalytic activity for at least four reuse rounds while maintaining approximately 70% total defluorination efficiency toward PFOA (Duan et al., 2022). In another case, as shown in Fig. 6.5c, $In_2O_3/Bi_2O_3$ preserved more than 80% of its original PFOA degradation performance after five cycles. As displayed in Fig. 6.5, the SEM and $N_2$ adsorption-desorption isotherms analysis further established that the structure and microscopic morphology of the spent photocatalyst remained unaltered after the reusability study, demonstrating good structural stability and durable catalytic functionality under repeated operation (Huang et al., 2026).

The findings here indicate that PFAS photocatalysts can typically be reused for 3–5 cycles while retaining >70% of their initial degradation efficiency. The slight decline in activity observed after successive cycles is generally attributed to unavoidable catalyst loss during recovery, surface fouling by fluorinated intermediates or fluoride ions, and partial aggregation or structural modification. Nevertheless, postreuse characterizations reported show negligible changes in crystal structure and morphology, confirming that these materials remain largely intact after repeated PFAS degradation. These highlight the promising economic and eco-friendliness potential of nanophotocatalysts for PFAS remediation. However, to further advance stability and recyclability, future studies should include quantitative assessment of fluoride accumulation and PFAS intermediate adsorption on catalyst surfaces, long-duration cycling tests under real water matrices, and development of regeneration strategies that restore active sites without generating secondary fluorinated waste.

**TABLE 6.2** Overview of reusability and recyclability tests.

| Photocatalyst | PFAS | No. of cycles | Remarks | References |
|---|---|---|---|---|
| $FeOCl/(Bi\text{-}MOF\text{–}BiOCl)_{100}$ | PFOA | 4 | Achieved over 98% of initial efficiency after the 4th cycle | Gu et al. (2025) |
| ZnO/cellulose nanofiber | PFOS | 3 | The spent photocatalyst showed approximately 10% decrease after the 3rd cycle | Dehghani et al. (2022) |
| $TiO_2$@boron nitride | PFOA | 4 | The photocatalyst sustained activity for at least 4 rounds with 70% total defluorination | Duan et al. (2022) |
| $In_2O_3/Bi_2O_3$ | PFOA | 5 | The photocatalyst sustained activity for five cycles with >80% of the initial performance | Huang et al. (2026) |

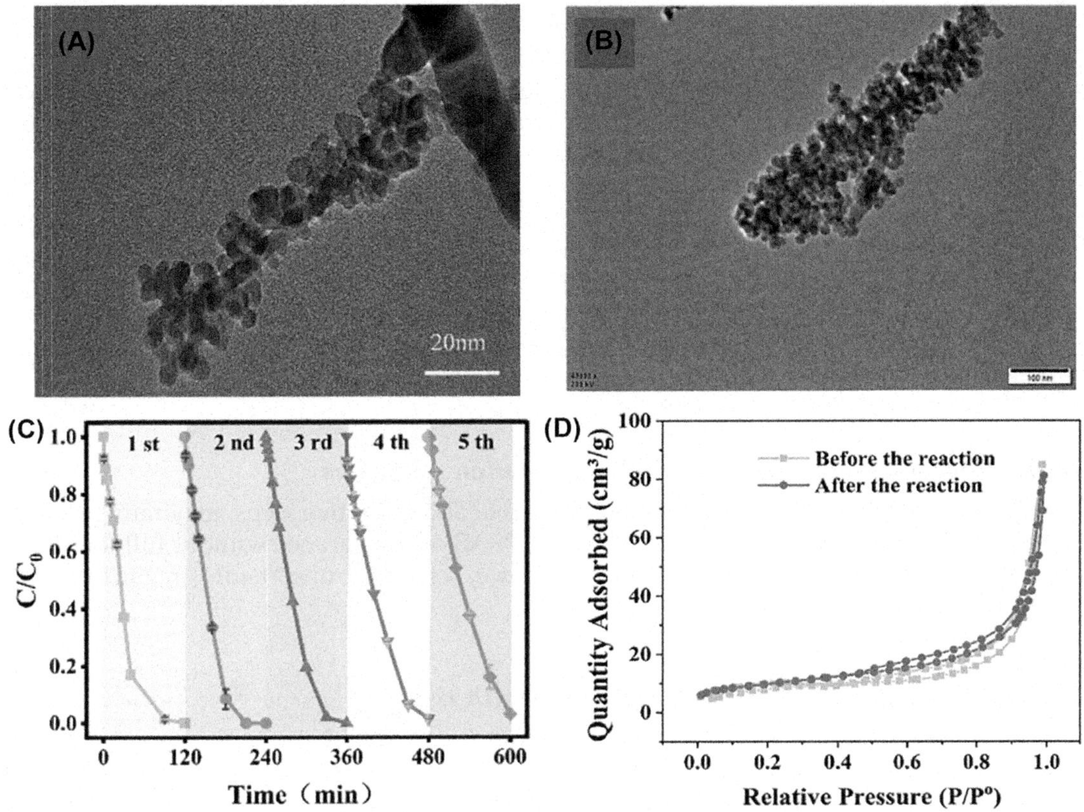


FIGURE 6.5 SEM/TEM picture of $In_2O_3/Bi_2O_3$ (A) before and (B) after perfluorooctanoic acid (PFOA) photocatalytic degradation. (C) Photocatalytic degradation reusability performance of $In_2O_3/Bi_2O_3$ for PFOA. (D) $N_2$ adsorption-desorption isotherms (Huang et al., 2026). *Reused with permission from Elsevier (Authorization code: 6240980157305).*

## 6.5 Challenges and future research directions

### 6.5.1 Challenges

Despite the promising potential of various photocatalysts for the degradation and defluorination of PFAS, several critical challenges remain that must be addressed to enable large-scale and practical deployment:

#### *6.5.1.1 Limited visible-light activity and reductive power*

Many conventional photocatalysts (e.g., $TiO_2$, ZnO) are primarily UV-responsive and also favor oxidative pathways, whereas effective PFAS degradation requires strong reductive species (e.g., hydrated electrons). Since UV constitutes only ~5% of solar energy, reliance on UV limits solar applicability and energy efficiency in real systems.

#### 6.5.1.2 Catalyst stability and fluorine poisoning

Fluoride ions and strongly adsorbing PFAS intermediates can accumulate on photocatalyst surfaces, blocking active sites and altering electronic properties. Photocorrosion, agglomeration, and leaching of dopants or cocatalysts during repeated use further reduce long-term stability and recovery efficiency.

#### 6.5.1.3 Real-water matrix interference

Natural organic matter, bicarbonate, chloride, sulfate, and coexisting contaminants compete for reactive species or occupy adsorption sites, suppressing PFAS degradation compared with ultrapure laboratory conditions. This discrepancy limits translation of lab efficiencies to field performance.

#### 6.5.1.4 Incomplete understanding of degradation pathways

PFAS photocatalysis involves complex reductive and oxidative steps, generating diverse fluorinated intermediates. Many studies report PFAS disappearance without full identification of intermediates or fluorine mass balance, creating uncertainty regarding true mineralization and toxicity evolution.

#### 6.5.1.5 Nanomaterial environmental and health risks

Some high-performance photocatalysts contain heavy metals or engineered nanostructures whose fate, transport, and ecotoxicity in aquatic environments are insufficiently understood. Potential nanoparticles released during operation or disposal raise safety and regulatory concerns.

#### 6.5.1.6 Lack of techno-economic and energy assessments

Most PFAS photocatalysis studies remain laboratory-scale and performance-focused, with limited evaluation of energy demand (e.g., UV sources), catalyst cost, recovery strategies, and reactor design. The absence of techno-economic analysis hampers scale-up and stakeholder adoption.

### 6.5.2 Way forward and future research directions

To overcome the above challenges and accelerate progress toward practical PFAS photocatalytic treatment, future research should prioritize the following:

#### 6.5.2.1 Development of visible-light-driven and reductive photocatalysts

Design Z-scheme and S-scheme heterojunctions, plasmon-enhanced systems, and electron-rich photocatalysts capable of generating hydrated electrons under solar irradiation, enabling efficient reductive C–F bond activation without energy-intensive UV sources.

#### 6.5.2.2 Comprehensive elucidation of degradation pathways and fluorine balance

Employ advanced analytics (LC–MS/MS, HRMS, ion chromatography, TOF–SIMS, and total organic fluorine analysis) to track intermediates, quantify released fluoride, and establish a complete fluorine mass balance and toxicity evolution during photocatalysis.

#### 6.5.2.3 Integration with hybrid and synergistic treatment systems

Couple photocatalysis with electrochemical reduction, persulfate activation, membrane concentration, or biodegradation to enhance PFAS capture, activation, and mineralization in sequential or tandem processes.

#### 6.5.2.4 Scalable reactor and natural sunlight-driven system development

Advance immobilized photocatalyst configurations, flow-through photoreactors, optical fiber or thin-film reactors, and solar-concentrator systems suitable for pilot- and field-scale PFAS treatment with improved photon utilization and catalyst recovery.

#### 6.5.2.5 Antifouling and fluoride-resistant photocatalyst design

Engineer surface architectures and cocatalysts that resist fluoride poisoning and organic fouling, including self-cleaning surfaces, conductive supports, and regenerable active sites to maintain long-term activity in real water matrices.

#### 6.5.2.6 Green synthesis and eco-safety evaluation

Develop bio-derived or low-toxicity photocatalysts using sustainable precursors like plant or biomass waste and green synthesis routes, alongside life cycle assessment and ecotoxicological testing to ensure environmental compatibility and regulatory acceptance.

#### 6.5.2.7 Focus on mixed per- and polyfluoroalkyl substances and cocontaminant systems

Future studies should target realistic PFAS mixtures (short/long-chain, sulfonates, and carboxylates) and cooccurring pollutants (e.g., PFAS + pharmaceuticals or dyes) commonly present in industrial and municipal waters.

#### 6.5.2.8 Integration of artificial intelligence and computational modeling

Apply machine learning to correlate photocatalyst physicochemical properties with PFAS adsorption and degradation performance, predict optimal band structures, and guide material discovery. Use DFT/MD simulations to map C–F activation pathways and reactive-site interactions in complex aqueous environments.

#### 6.5.2.9 Techno-economic and life cycle assessment frameworks

Incorporate standardized techno-economic analysis, energy metrics ($kWh\ m^{-3}$), photocatalyst lifetime, and cost-benefit evaluation alongside performance data. Establish comparative benchmarks for photocatalyst materials and reactor designs to support commercialization and policy decisions.

## 6.6 Conclusion

This chapter reviewed diverse research efforts on the photocatalytic degradation of PFAS using various photocatalysts. Notably, the findings demonstrate that photocatalysis-driven technologies represent a transformative approach for the destructive removal of PFAS from contaminated water systems. Specifically, it was observed that in most cases, different photocatalysts can achieve approximately >50–>90% PFAS degradation within reaction times ranging from 30 to 1440 min under optimized conditions. Moreover, nanocomposites (binary and ternary photocatalysts) consistently exhibited superior performance compared with single-component materials. This is due to enhanced light absorption, improved photocurrent response, suppressed electron-hole recombination, and more efficient charge carrier separation and migration associated with narrowed band gaps and interfacial electric fields. The study further shows that photocatalysts can be reused 3–5 cycles, underscoring their potential cost-effectiveness, environmental compatibility, and scalability potential. To accelerate field translation, future research should prioritize the rational design of Z-scheme and S-scheme photocatalysts tailored for reductive PFAS defluorination, research on regenerability and reusability, pilot-scale and natural sunlight-driven reactor studies, hybrid photocatalysis (e.g., photo-electrocatalytic processes), development of bio-derived photocatalysts, and comprehensive techno-economic and life cycle assessments supported by data-driven and artificial intelligence tools. Conclusively, strengthened regulatory control and phase-out strategies for high-risk PFAS, particularly legacy compounds such as PFOA and perfluorooctane sulfonate, are recommended, especially in developing regions where monitoring and treatment infrastructure remain limited.

## References

Arana Juve, J.-M., Li, F., Zhu, Y., Liu, W., Ottosen, L.D.M., Zhao, D., et al., 2022. Concentrate and degrade PFOA with a photo-regenerable composite of In-doped TNTs@AC. Chemosphere 300, 134495. Available at https://doi.org/10.1016/j.chemosphere.2022.134495.

Badamasi, H., Sanni, S.O., Ore, O.T., Bayode, A.A., Koko, D.T., Akeremale, O.K., et al., 2024. Eggshell waste materials-supported metal oxide nanocomposites for the efficient photocatalytic degradation of organic dyes in water and wastewater: a review. Bioresour Technol Reports 26, 101865. Available at https://doi.org/10.1016/j.biteb.2024.101865.

Bai, L., Onwumere, J., Ezugwu, C.I., Saad, A., Huang, Z., Xie, Z., et al., 2026. Activating PFAS for efficient photocatalytic defluorination using ultralow-dose Ti-doped bismuth coordination catalysts with unsaturated metal sites. Applied Catalysis B: Environmental and Energy 384. Available at https://doi.org/10.1016/j.apcatb.2025.126230 126230.

Bayode, A.A., Emmanuel, S.S., Terlanga, K.D., 2025b. Chapter 19 - Nanomaterials and their derivative biocomposites for dye adsorption. In: Jagaba, A.H., Mohamed Kutty, S.R., Isa, M.H., Birniwa AHBT-EB for, D.A. (Eds.), Eng. Biocomposites Dye Adsorpt. Elsevier, pp. 313–337. Available at https://doi.org/10.1016/B978-0-443-29877-6.00019-6.

Bayode, A.A., Sunday Emmanuel, S., Sanni, S.O., Lakhdar, F., Fu, L., Shang, J., et al., 2025a. Biogenic fabrication of spinel nickel ferrite imprinted on *Bifurcaria bifurcata* Macro-Alga activated carbon for the adsorption of ciprofloxacin and metronidazole. Chemical Engineering Science 302, 120843. Available at https://doi.org/10.1016/j.ces.2024.120843.

Cao, C.-S., Wang, J., Yang, L., Wang, J., Zhang, Y., Zhu, L., 2024. A review on the advancement in photocatalytic degradation of poly/perfluoroalkyl substances in water: insights into the mechanisms and structure-function relationship. The Science of the Total Environment 946, 174137. Available at https://doi.org/10.1016/j.scitotenv.2024.174137.

Chowdhury, N., Choi, H., 2023. Photocatalytic degradation of perfluorooctanoic acid on Pb-doped $TiO_2$ coated with reduced graphene oxide. Water Environment Research: a Research Publication of the Water Environment Federation 95, e10871. Available at https://doi.org/10.1002/wer.10871.

Dankawu, U.M., Hafeez, H.Y., Tayeb, R.A., Alabbad, E.A., Almohamadi, H., Alrashood, J.S., et al., 2026. Synergistic effect of $Ti_3C_2T_x$ MXene on g-$C_3N_4$/$ZnFe_2O_4$ nano-composite for photocatalytic solar fuel (H2) production via water-splitting. Materials Chemistry and Physics, 132054.

Dehghani, M., Naseri, M., Nadeem, H., Banaszak Holl, M.M., Batchelor, W., 2022. Sunlight-driven photocatalytic per- and polyfluoroalkyl substances degradation over zinc oxide/cellulose nanofiber catalyst using a continuous flow reactor. Journals of Environmental Chemistry Engineering 10, 108686. Available at https://doi.org/10.1016/j.jece.2022.108686.

Dhanasekaran, E., Sriharan, N., Senthil, T.S., Sivakumar, N., 2026. Efficient photocatalytic degradation of Zn and Gd-doped BiOBr 3D microspheres using reactive dyes under visible light. Materials Sciences Engineering B 323, 118779. Available at https://doi.org/10.1016/j.mseb.2025.118779.

Duan, L., Wang, B., Heck, K.N., Clark, C.A., Wei, J., Wang, M., et al., 2022. Titanium oxide improves boron nitride photocatalytic degradation of perfluorooctanoic acid. Chemical Engineering Journals 448, 137735. Available at https://doi.org/10.1016/j.cej.2022.137735.

Emmanuel, S.S., Adesibikan, A.A., 2025. Nanoremediation of chemical pollutants using green carbon nanomaterials. Sustain Carbon Nanomaterial Their Applied 5–29. Available at https://doi.org/10.1088/978-0-7503-6325-9ch5.

Emmanuel, S.S., Suleiman, A.I., Itopa, A.A., Adesibikan, A.A., Idris, M.O., 2026. Zero-valent iron and its derivative composites for heavy metal adsorption. Compos. Biocomposites Heavy Met. Adsorpt. Elsevier, pp. 73–88.

Esfanddarani, H.M., Panigrahi, M., 2026. Photocatalytic degradation of azo dyes assisted by biosynthesized AgNPs loaded in physically cross-linked PVA/CS hydrogel nanocomposites. Material Chemistry Physics 353, 132129. Available at https://doi.org/10.1016/j.matchemphys.2026.132129.

Gates, K., Rai, S., Pramanik, A., Kolawole, O.P., Kundu, S., Ucak-Astarlioglu, M., et al., 2025. Insight into the photocatalytic degradation mechanism for "Forever Chemicals" PFNA by reduced graphene oxide/$WO_3$ nanoflower heterostructures. ACS Omega 10, 10675–10684. Available at https://doi.org/10.1021/acsomega.5c00054.

Gomez-Ruiz, B., Ribao, P., Diban, N., Rivero, M.J., Ortiz, I., Urtiaga, A., 2018. Photocatalytic degradation and mineralization of perfluorooctanoic acid (PFOA) using a composite $TiO_2$ –rGO catalyst. Journal of Hazardous Materials 344, 950–957. Available at https://doi.org/10.1016/j.jhazmat.2017.11.048.

Gu, M., Duan, L., Zhang, Z., Jiang, Y., Zhang, Q., Zhou, L., et al., 2025. Multi-charge bridge transfer guiding design of photocatalyst with oxygen defects for effective degradation of PFAS in fluoropolymer production wastewater. Chemical Engineering Journal 505, 159124. Available at https://doi.org/10.1016/j.cej.2024.159124.

Hamza, M.A., Keltie, A.J., Matthews, R.K., Day, M.L., Shearer, C.J., 2025. CdIn2S4 Micro-pyramids for reductive photocatalytic degradation of perfluorooctanesulfonic acid. Small (Weinheim an der Bergstrasse, Germany) 21, e04601. Available at https://doi.org/10.1002/smll.202504601.

Hanphaiboon, P., Liu, X., Ahmad, R., Zhang, J., Pan, B., Wang, Y., 2026. Sustainable nitrogen-self-doped coffee ground-derived activated carbon for efficient adsorption of short- and long-chain PFAS: mechanistic insights and practical applications. Sustainable Materials and Technologies 47, e01857. Available at https://doi.org/10.1016/j.susmat.2026.e01857.

Hojamberdiev, M., Larralde, A.L., Vargas, R., Madriz, L., Yubuta, K., Sannegowda, L.K., et al., 2023. Unlocking the effect of $Zn^{2+}$ on crystal structure, optical properties, and photocatalytic degradation of perfluoroalkyl substances (PFAS) of Bi2WO6. Environmental Science: Water Research & Technology 9, 2866–2879. Available at https://doi.org/10.1039/D3EW00430A.

Huang, D., Yin, L., Lu, X., Lin, S., Niu, Z., Niu, J., 2017. Directional electron transfer mechanisms with graphene quantum dots as the electron donor for photodecomposition of perfluorooctane sulfonate. Chemistry Engineering Journals 323, 406–414. Available at https://doi.org/10.1016/j.cej.2017.04.124.

Huang, D., Wu, J., Feng, S., Zhu, X., Chen, W., Wu, J., et al., 2026. High-efficiency $In_2O_3$/$Bi_2O_3$ heterojunction photocatalyst for PFAS removal: experiments, mechanism and application in membrane distillation. Separation and Purification Technology 393, 137270. Available at https://doi.org/10.1016/j.seppur.2026.137270.

Ke, Z.-W., Liu, Y.-X., Wei, S.-J., Shen, P., Yang, K., Chen, Y.-M., et al., 2026. Hexadecyltrimethylammonium-modified bentonite for enhanced adsorption of per- and polyfluoroalkyl substances: experiment and thermodynamic mechanism. Applied Clay Sciences 279, 108022. Available at https://doi.org/10.1016/j.clay.2025.108022.

Kourmousi, M., Tsitsou, E., Tzevelekidis, P., Kostakis, M.G., Linganiso-Dziike, C.E., Efthimiadou, E.K., et al., 2026. Visible-light-driven photocatalytic degradation of dyes in wastewater using l-lysine-based carbon Dot/$TiO_2$ nanocomposites. Carbon Trends, 100620.

Kumar, A., Georgi, A., Huaccallo-Aguilar, Y., Meier, M., Kryk, H., Reinecke, S.F., et al., 2026. Degradation and defluorination of perfluorooctane sulfonate (PFOS) forever chemical in water using hydrodynamic cavitation treatment. Chemistry Engineering Journals Advances 25, 101046. Available at https://doi.org/10.1016/j.ceja.2026.101046.

Li, Z., Zhang, P., Li, J., Shao, T., Jin, L., 2013. Synthesis of $In2O_3$-graphene composites and their photocatalytic performance towards perfluorooctanoic acid decomposition. Journals of Photochemistry and Photobiologyl A: Chemistry 271, 111–116. Available at https://doi.org/10.1016/j.jphotochem.2013.08.012.

Li, F., Wei, Z., He, K., Blaney, L., Cheng, X., Xu, T., et al., 2020. A concentrate-and-destroy technique for degradation of perfluorooctanoic acid in water using a new adsorptive photocatalyst. Water Research 185, 116219. Available at https://doi.org/10.1016/j.watres.2020.116219.

Li, X., Wang, Y., Ma, Y., Ding, L., Chu, M., He, D., et al., 2026. $TiO_2/BiVO_4$ photocatalyst based on S-scheme charge transfer and hierarchical porous structure: efficient ibuprofen degradation and pathway analysis. Journals of Alloys Compound 1051, 186043. Available at https://doi.org/10.1016/j.jallcom.2026.186043.

Najafidoust, A., Ghanbari, R., Farahbakhsh, J., Permala, R., Iglauer, S., Zargar, M., 2026. Emerging trends in photocatalyst-based metal-organic frameworks for PFAS degradation. Coordination Chemistry Reviews 549, 217264.

Ong, C.B., Mohammad, A.W., Ng, L.Y., Mahmoudi, E., Azizkhani, S., Hayati Hairom, N.H., 2017. Solar photocatalytic and surface enhancement of ZnO/rGO nanocomposite: degradation of perfluorooctanoic acid and dye. Process Safety and Environmental Protection 112, 298–307. Available at https://doi.org/10.1016/j.psep.2017.04.031.

Panchangam, S.C., Lin, A.Y.-C., Shaik, K.L., Lin, C.-F., 2009. Decomposition of perfluorocarboxylic acids (PFCAs) by heterogeneous photocatalysis in acidic aqueous medium. Chemosphere 77, 242–248.

Panchangam, S.C., Yellatur, C.S., Yang, J.-S., Loka, S.S., Lin, A.Y.C., Vemula, V., 2018. Facile fabrication of $TiO_2$-graphene nanocomposites (TGNCs) for the efficient photocatalytic oxidation of perfluorooctanoic acid (PFOA). Journals of Environmental Chemistry Engineering 6, 6359–6369. Available at https://doi.org/10.1016/j.jece.2018.10.003.

Park, K., Ali, I., Kim, J.-O., 2018. Photodegradation of perfluorooctanoic acid by graphene oxide-deposited $TiO_2$ nanotube arrays in aqueous phase. Journal of Environmental Management 218, 333–339. Available at https://doi.org/10.1016/j.jenvman.2018.04.016.

Pouresmaeil, H., Nabi Bidhendi, G.R., Baghdadi, M., 2025. The removal of PFAS by $Fe_3O_4$/ZnO@Ag nano-photocatalyst under visible light irradiation. International Journal of Environmental Science and Technology 22, 10295–10304. Available at https://doi.org/10.1007/s13762-025-06520-3.

Rahul, Shukla, A., Vamshi, C., Chauhan, V.S., Vaish, R., 2026. Solar-light-active photocatalytic $SrBi_2Nb_2O_9$-$Li_2O$-$B_2O_3$ glass-ceramic plates. Solar Energy 307, 114370. Available at https://doi.org/10.1016/j.solener.2026.114370.

Rivero, M.J., Ribao, P., Gomez-Ruiz, B., Urtiaga, A., Ortiz, I., 2020. Comparative performance of $TiO_2$-rGO photocatalyst in the degradation of dichloroacetic and perfluorooctanoic acids. Separation and Purification Technology 240, 116637. Available at https://doi.org/10.1016/j.seppur.2020.116637.

Shang, E., Li, Y., Niu, J., Li, S., Zhang, G., Wang, X., 2018. Photocatalytic degradation of perfluorooctanoic acid over Pb-$BiFeO_3$/rGO catalyst: kinetics and mechanism. Chemosphere 211, 34–43. Available at https://doi.org/10.1016/j.chemosphere.2018.07.130.

Shashikumar, U., Goel, H., Rana, I., Tsai, P.-C., Manjappa, K.B., Pham, P.V., et al., 2025. Breaking the 'forever' bond: photocatalytic degradation of per- and polyfluoroalkyl substances using ternary photocatalysts. Reviews in Environmental Science and Bio/Technology 24, 607–659. Available at https://doi.org/10.1007/s11157-025-09734-6.

Song, C., Chen, P., Wang, C., Zhu, L., 2012. Photodegradation of perfluorooctanoic acid by synthesized $TiO_2$–MWCNT composites under 365nm UV irradiation. Chemosphere 86, 853–859. Available at https://doi.org/10.1016/j.chemosphere.2011.11.034.

Sun, Q., Gao, Y., Wang, W., Song, Y., Lv, J., Zhang, G., et al., 2026. Nanoscale structural heterogeneity tuning of sputtered CuZr metallic glass for superior azo dye photocatalytic degradation. Applied Surface Science 729, 166254. Available at https://doi.org/10.1016/j.apsusc.2026.166254.

Tian, H., Gu, C., 2018. Effects of different factors on photodefluorination of perfluorinated compounds by hydrated electrons in organo-montmorillonite system. Chemosphere 191, 280–287.

Tong, F., Zhang, R., Li, D., Qu, H., Li, X., Gao, Q., et al., 2026. Fluorinated amino covalent organic frameworks for efficient adsorption of per- and polyfluoroalkyl substances in water. Environmental Research 295, 123931. Available at https://doi.org/10.1016/j.envres.2026.123931.

Van Tri, D., Nhu, H.D.H., Anh, N.T., Nhi, D.T.K., Le Luu, T., 2026. Evaluation of per-/polyfluoroalkyl substances (PFAS) contamination in Binh Duong surface water and correlation with physico-chemical indicators. The Science of the Total Environment 1014, 181360. Available at https://doi.org/10.1016/j.scitotenv.2026.181360.

Vietri, V., Vaiano, V., Sacco, O., Mancuso, A., 2026. Photocatalysis and electro-oxidation for PFAS degradation: mechanisms, performance, and energy efficiency. Catalysts 16, 145. Available at https://doi.org/10.3390/catal16020145.

Wahyuni, S., Kadarwati, S., Alauhdin, M., Priatmoko, S., Sani, N.H., Prasetyo, R., et al., 2026. Enhanced visible-light photocatalytic degradation of methylene blue using $TiO_2$/g-$C_3N_4$ composites synthesized via co-calcination. Results Chemistry 24, 103230. Available at https://doi.org/10.1016/j.rechem.2026.103230.

Wang, J., Cao, C.-S., Wang, J., Zhang, Y., Zhu, L., 2022. Insights into highly efficient photodegradation of poly/perfluoroalkyl substances by In-MOF/BiOF heterojunctions: Built-in electric field and strong surface adsorption. Applied Catalysis B: Environmental 304, 121013.

Wang, Q., Jin, Z., Dai, S., Yang, Y., Liu, L., Li, Y., et al., 2026. Janus SiS2/SeWS heterojunction as a high-performance direct Z-scheme photocatalyst for spontaneous overall water splitting. Fuel 414, 138307.

Wen, Y., Kirchon, A., Day, G.S., Lin, H., Smith, M.F., Boehme, A., et al., 2025. Photocatalytic degradation of perfluorooctanoic acid (PFOA) by metal organic framework MIL-177-HT: New insights into the role of specific surface area, charge separation and dimensionality. Separation and Purification Technology 354, 128877.

Wu, Y., Hu, Y., Han, M., Ouyang, Y., Xia, L., Huang, X., et al., 2021. Mechanism insights into the facet-dependent photocatalytic degradation of perfluorooctanoic acid on BiOCl nanosheets. Chemistry Engineering Journals 425, 130672. Available at https://doi.org/10.1016/j.cej.2021.130672.

Xia, C., Lim, X., Yang, H., Goodson, B.M., Liu, J., 2022. Degradation of per- and polyfluoroalkyl substances (PFAS) in wastewater effluents by photocatalysis for water reuse. Journals of Water Process Engineering 46, 102556. Available at https://doi.org/10.1016/j.jwpe.2021.102556.

Xu, T., Zhu, Y., Duan, J., Xia, Y., Tong, T., Zhang, L., et al., 2020. Enhanced photocatalytic degradation of perfluorooctanoic acid using carbon-modified bismuth phosphate composite: effectiveness, material synergy and roles of carbon. Chemistry Engineering Journals 395, 124991. Available at https://doi.org/10.1016/j.cej.2020.124991.

Yao, M., Liu, Z., Guo, W., Wu, G., Gao, D., Jia, G., et al., 2026. Sodium periodate-driven one pot iodine doping $TiO_2$ for superior visible-light photocatalytic degradation. Materials & Design 263, 115585. Available at https://doi.org/10.1016/j.matdes.2026.115585.

Zaaboul, F., Canle, M., Haoufazane, C., Santaballa, J.A., Azzaoui, K., Hammouti, B., et al., 2026. Study of the photocatalytic degradation of the RB49 dye in the presence of ZnO electrodeposited on Carbon felt (CF) and Titanium (Ti). Materials Chemistry Physics 356, 132343. Available at https://doi.org/10.1016/j.matchemphys.2026.132343.

Zang, S., Qian, Z., Zhu, Y., Gu, H., Chai, M., Zhang, B., 2026. Knowledge mapping-based trend analysis of research on photocatalytic hydrogen generation from wastewater. Fuel 415, 138448.

Zhao, X., Bi, H., Zhang, S., Wang, T., An, L., Zhu, B., et al., 2026. Dual-driven magnetic nanocomposites enabling universal adsorption of perfluorocarboxylic acids (PFCAs) across ultrashort to long chains. Journal of Colloid and Interface Science 712, 140139. Available at https://doi.org/10.1016/j.jcis.2026.140139.

Zhu, Y., Leary, R.N., Xu, T., He, K., Blaney, L., Hao, X., et al., 2026. Photocatalytic degradation of PFAS under field water matrix conditions using an adsorptive photocatalyst. Water Research X 30, 100485. Available at https://doi.org/10.1016/j.wroa.2026.100485.

CHAPTER

# 7

# Sonication-induced catalytic approaches for PFAS degradation in aqueous systems

*Ahmed Malek Djaballah*[1,2], *Dimitrios A. Giannakoudakis*[3], *Ioannis Anastopoulos*[4], *Giacomo de Falco*[5], and *Juan Carlos Colmenares*[1]

[1]Institute of Physical Chemistry, Polish Academy of Sciences, Warsaw, Poland [2]CNRS -UPR3079, Université d'Orléans, Conditions Extrêmes Matériaux Haute Température et Irradiation (CEMHTI), Orléans, France [3]Faculty of Chemistry, Institute of Chemical Sciences, Maria Curie-Sklodowska University, Lublin, Poland [4]Department of Agriculture, University of Ioannina (UoI), Kostakii Campus, Arta, Greece [5]New York City Department of Environmental Protection, New York, NY, United States

## 7.1 Introduction

Per- and polyfluoroalkyl substances (PFAS) are a large and diverse class of synthetic fluorinated compounds that have been used extensively in industrial and consumer products for over six decades (Paul et al., 2009). Their unique physicochemical properties—including extreme thermal resistance, hydrophobicity, and oleophobicity—are primarily due to the high strength and stability of the carbon–fluorine (C–F) bond (Prevedouros et al., 2006; Xu et al., 2021). These same properties, however, render PFAS highly persistent in the environment, earning them the nickname "forever chemicals." They resist natural degradation, bioaccumulate in living organisms, and have been detected in remote environments, far from their sources of manufacture or use (Wen et al., 2020; Boiteux et al., 2017; Zhao et al., 2020). PFAS are commonly classified into subgroups such as perfluoroalkyl carboxylic acids (PFCAs), perfluoroalkyl sulfonic acids (PFSAs), fluorotelomer alcohols, and fluoropolymers (Stahl et al., 2011). Among them, perfluorooctanoic acid (PFOA) and perfluorooctanesulfonic

*Per- and Polyfluoroalkyl Substances (PFAS) as Environmental Contaminants*
DOI: https://doi.org/10.1016/B978-0-443-43856-1.00013-7

acid (PFOS) are two of the most studied due to their environmental ubiquity and health implications. These compounds exhibit high water solubility and mobility, allowing them to infiltrate drinking water systems, surface water bodies, and soil (US EPA, 2017). PFAS have been identified in food products, drinking water, breast milk, wildlife, and human serum, indicating widespread exposure and raising significant public health concerns (Stahl et al., 2011; Trudel et al., 2008; Weinfurtner et al., 2008).

Human exposure to PFAS has been linked to adverse health outcomes, including immunotoxicity, endocrine disruption, developmental toxicity, and certain cancers (Crone et al., 2019; Cordner et al., 2020; US EPA, 2016, 2018; Bartlett and Davis, 2018). In response, regulatory agencies have established increasingly stringent guidelines. The U.S. Environmental Protection Agency released drinking water health advisories as early as 2009 and, as of 2024, finalized maximum contaminant levels of $4\,ng\,L^{-1}$ for PFOA and PFOS. Year after year, regulations are being introduced across countries, as discussed in greater detail in subsequent chapters. Despite regulatory actions, many conventional water treatment technologies, including granular activated carbon (GAC), ion exchange resins, and membrane processes, are insufficient for complete PFAS removal, particularly for short-chain variants (Nzeribe et al., 2019; Kucharzyk et al., 2017; Zheng et al., 2019; Woodard et al., 2017).

Recent research has explored advanced treatment technologies, including photocatalysis, plasma treatment, and electrochemical oxidation (EO) (Ahmed et al., 2020; Vo et al., 2020; Rahman et al., 2014). Among these, ultrasonication (or sonolysis) has emerged as a promising destructive technique. Ultrasonication involves the application of high-frequency sound waves to aqueous systems, generating acoustic cavitation; rapid formation, growth, and implosive collapse of microbubbles. This collapse creates extreme localized conditions (temperatures >5000 K and pressures >1000 atm), forming transient "hot spots" capable of initiating chemical transformations through pyrolysis and radical generation.

The amphiphilic structure of PFAS, comprising a hydrophobic fluorinated tail and a hydrophilic functional head, enables their preferential localization at the gas–liquid interface of cavitation bubbles. Here, volatile PFAS species may enter the gas phase and undergo pyrolytic cleavage, while others may interact with hydroxyl and other reactive radicals formed through water sonolysis. This dual pathway enhances the degradation of persistent PFAS molecules, with several studies reporting efficient defluorination and breakdown of PFOA and PFOS (Vecitis et al., 2010a). Moreover, ultrasonication does not require added chemicals or catalysts, making it a comparatively simple and sustainable technology.

This chapter examines ultrasonication (sonolysis) for PFAS degradation, outlining mechanisms, experimental evidence, and degradation pathways, and evaluating the efficiency of different ultrasonic configurations. Additionally, hybrid systems combining ultrasonication with other advanced oxidation processes (AOPs) are discussed to explore their synergistic potential.

In the context of mounting environmental and regulatory pressures, ultrasonication represents a viable and innovative tool for the remediation of PFAS-contaminated waters. Its integration into future water treatment systems may offer a scalable, energy-efficient solution to address one of the most challenging groups of environmental contaminants.

## 7.2 Fundamental theory of ultrasonic cavitation

The discovery that ultrasonic waves can accelerate chemical reactions dates back to the 1940s when researchers at Princeton University first observed this phenomenon (Kumar et al., 2024). However, this finding remained relatively underexplored until the mid-1980s, when the development of high-powered ultrasound equipment renewed scientific interest in the chemical engineering applications of ultrasound. Traditional wastewater treatment methods, particularly for persistent organic pollutants, often proved ineffective, while recently cavitation effects, most notably those induced by ultrasound, have shown significant potential to enhance treatment efficiency (Kumar et al., 2024; Wang and Cheng, 2023; Crispim et al., 2022). As a result, ultrasonic cavitation degradation has emerged as a powerful technique for tackling recalcitrant organic contaminants (Djellabi et al., 2022; Hu et al., 2023; Shen and Ma, 2022). This approach is valued for its ability to detoxify pollutants, minimize secondary contamination, and efficiently degrade highly toxic, resistant compounds with relatively simple equipment (Low et al., 2022). Depending on the generation mechanism, cavitation can be categorized into ultrasonic, hydrodynamic, optical, and particle cavitation (Movahed et al., 2023; Kalinina et al., 2019). Among these, ultrasonic and hydrodynamic cavitation have been predominantly studied for water treatment applications. Hydrodynamic cavitation itself can be further divided into jet and vortex cavitation types (Ye et al., 2021; Gholami et al., 2021). This review primarily concentrates on ultrasonic cavitation, detailing its fundamental principles, the underlying mechanisms driving cavitation effects, and key factors influencing ultrasonic degradation performance. Furthermore, it surveys the integration of sonochemical degradation with other advanced treatment technologies targeting diverse pollutants. In essence, sonochemical degradation, driven by cavitation generated through ultrasonic waves typically within the frequency range of 15 kHz to 1 MHz, represents an effective physico-chemical pathway for breaking down organic contaminants in aqueous environments.

## 7.3 Mechanisms underpinning the ultrasonic cavitation effect

Ultrasonic cavitation is a fundamental physical process that occurs in liquids upon exposure to ultrasound waves, marked by the swift generation and collapse of microscopic bubbles due to rapid fluctuations in pressure. This phenomenon, illustrated in Fig. 7.1 (Pirsaheb and Moradi, 2020), encompasses several dynamic phases: the initial oscillation of bubble nuclei, their expansion and contraction, followed by their ultimate collapse (Iida et al., 2004). This entire sequence is essentially an adiabatic process, where acoustic energy is intensely focused and then rapidly discharged. To fully comprehend the mechanisms underlying sonochemical degradation of organic pollutants, two principal theoretical frameworks play a vital role, as detailed in the following subsections (Fatyukhin et al., 2022).

### 7.3.1 The hotspot theory

In ultrasonic-driven chemical reactions, the cavitation phenomenon concentrates acoustic energy into extremely small regions, creating localized "hotspots" characterized by very

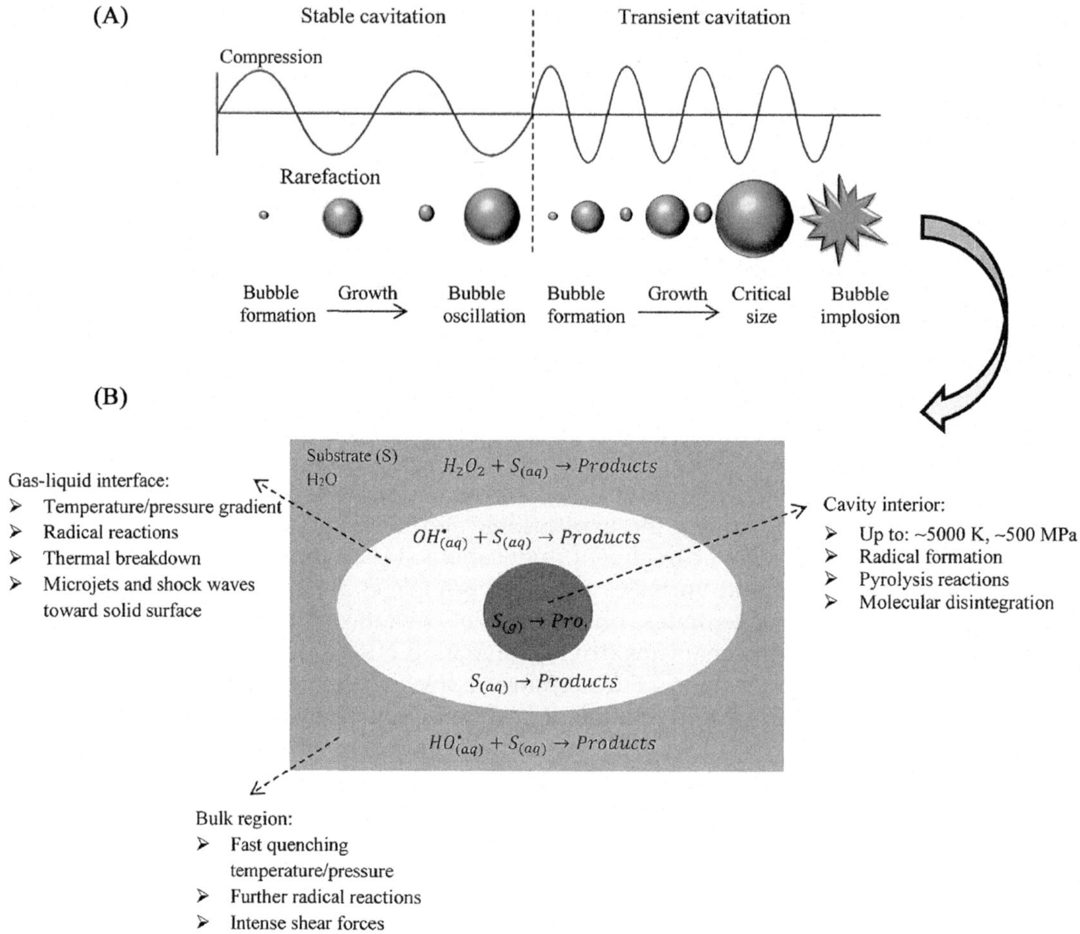


**FIGURE 7.1** Schematic representation of (A) cavitation bubbles displaying stable and transient cavitation and (B) reaction zones in the cavitation process. *Adapted from Pirsaheb, M., & Moradi, N. (2020). Sonochemical degradation of pesticides in aqueous solution: investigation on the influence of operating parameters and degradation pathway-a systematic review. RSC Advances, 10(13), 7396–7423. Licensed under CC BY 3.0. © The Royal Society of Chemistry.*

high temperatures and pressures. These intense conditions significantly enhance catalytic chemical reactions (Suslick et al., 1990). Organic compounds trapped within these cavitation bubbles are subjected to severe thermal decomposition due to the extreme internal environment, offering a novel pathway for specific chemical transformations (Kerboua and Hamdaoui, 2020). The hotspot model explains that ultrasound induces cavitation in liquids, triggering intricate physicochemical processes (L´eveque et al., 2018). Ultrasonic waves cause liquid molecules to oscillate at the molecular level. During the compression (positive) phase, molecules densify, whereas in the rarefaction (negative) phase, they spread apart, reducing density (Mason, 2015). When the ultrasonic amplitude surpasses a critical threshold, tiny gas nuclei grow into cavitation bubbles. However, not all bubbles collapse violently to produce heat; some dissolve back into the liquid, while others ascend and evaporate at the surface. A portion of bubbles, especially those not in resonance with the ultrasonic field, collapse abruptly, generating extreme conditions with temperatures exceeding 5000 K, pressures around 500 MPa, and high-velocity shock waves and jets (Bhatnagar and Cheung, 1994). These harsh localized environments enable both direct and indirect degradation of organic contaminants in water, forming the basis for ultrasonic degradation of aquatic pollutants.

### 7.3.2 The free radical oxidation theory

The sonochemical theory of free radical generation posits that when ultrasonic cavitation bubbles collapse rapidly, typically within 100 ns, the resulting extremely high temperatures and pressures cause the dissociation of water molecules ($H_2O$) in aqueous media, leading to the formation of highly reactive hydrogen ($H^\bullet$) and hydroxyl ($^\bullet OH$) radicals. These radicals may recombine to form molecular hydrogen ($H_2$) and water ($H_2O$), or they can interact with dissolved oxygen ($O_2$), resulting in the production of hydrogen peroxide ($H_2O_2$) or hydroperoxyl radicals ($HOO^\bullet$). The radicals generated during this cavitation process exhibit strong oxidative properties, initiating redox reactions with surrounding organic compounds, thereby contributing to their degradation (Dehghani et al., 2023).

The primary degradation mechanism of organic pollutants during ultrasonication involves oxidation by these reactive species, with cavitation being the key driver. Nam et al. (2003) provided a simplified reaction pathway illustrating the sequence of radical formation and their interactions in the sonolytic degradation of nonvolatile organics.

$$H_2O \rightarrow \cdot OH + H\cdot$$

$$H\cdot + O_2 \rightarrow HOO\cdot$$

$$\cdot OH + \cdot OH \rightarrow H_2O_2$$

$$H\cdot + H\cdot \rightarrow H_2$$

$$H\cdot + \cdot OH \rightarrow H_2O$$

$$HOO\cdot + HOO\cdot \rightarrow H_2O_2 + O_2$$

These radicals further participate in secondary reactions with hydrogen peroxide, producing additional radicals and molecules such as:

$$H\cdot + H_2O_2 \rightarrow H_2O + \cdot OH$$

$$H\cdot + H_2O_2 \rightarrow H_2 + HOO\cdot$$

$$\cdot OH + H_2O \rightarrow HOO\cdot + H_2O_2$$

$$\cdot OH + H_2 \rightarrow H_2O + H\cdot$$

When oxygen is dissolved in the solution, it can thermally decompose to yield atomic oxygen [O], which can subsequently participate in radical-generating reactions, for instance:

$$O_2 \rightarrow 2[O]$$

$$H\cdot + O_2 \rightarrow \cdot OH + [O]$$

$$[O] + H_2 \rightarrow \cdot OH + H\cdot$$

$$[O] + HOO\cdot \rightarrow \cdot OH + O_2$$

$$[O] + H_2O_2 \rightarrow \cdot OH + HOO\cdot$$

The generated •OH and H• radicals, due to their unpaired electrons, are highly reactive and can engage in multiple reactions, resulting in more stable compounds and significantly altering the surrounding chemical environment. These processes may accelerate preexisting reactions or lead to the formation of unexpected products.

According to the sonochemical free radical and hotspot theories (Yusof et al., 2022; Nanzai et al., 2023), radical generation and related reactions predominantly occur within cavitation bubbles, the surrounding interfacial liquid, and other regions of the solution. Volatile organic compounds (VOCs) primarily undergo thermal degradation within the high-temperature interiors of these collapsing bubbles, analogous to combustion, leading to their complete decomposition. In contrast, nonvolatile or low-volatility substances mainly interact with free radicals either at the bubble interface or within the bulk liquid. Following cavitation events, the recombination of radicals can release photons, resulting in a phenomenon known as sonoluminescence, a weak emission of light from the liquid medium, extensively characterized by Suslick and Flannigan (2008).

Further empirical support is provided by Dewulf et al. (2001), who examined the ultrasonic degradation of chlorobenzene and trichloroethylene. Their findings revealed that at low (micromolar) concentrations, free radical activity accounted for nearly 48.5% of chlorobenzene degradation. However, at higher (millimolar) concentrations, degradation was dominated by thermal mechanisms, reaching efficiencies over 99.97%, thereby diminishing the relative role of radicals. The kinetics of this process followed a first-order model.

In another important study, Kruus et al. (1997) investigated the ultrasonic degradation of phenol, confirming that its primary degradation pathway involves oxidation by •OH radicals. This interaction leads to the formation of dihydroxycyclohexadiene radicals. In the presence of oxygen, these radicals further react to form peroxyl radicals, which undergo complex transformations, including aromatic rearrangement, yielding intermediates such as hydroquinone and catechol. These intermediates can then degrade further into short-chain organic acids, such as carboxylic acids, and eventually mineralize into water and carbon dioxide, contributing to environmental detoxification. Additional validation of these oxidation pathways has been provided by Serpone and colleagues (Hua, 1995; Hua, Hoechemer, et al., 1995).

## 7.4 Degradation treatment technologies for per- and polyfluoroalkyl substances

The degradation of PFAS is only desirable if the resulting by-products pose minimal environmental and health risks (Cheremisinoff, 2017). PFAS compounds are notably resistant to conventional methods due to their exceptionally strong carbon–fluorine bonds, which have a bond energy of approximately 110 kcal $mol^{-1}$. Moreover, the oxidation potential of hydroxyl radicals (66 kcal $mol^{-1}$) is significantly lower than the average energy of carbon–carbon bonds (82.9 kcal $mol^{-1}$), rendering PFAS highly recalcitrant to degradation by hydroxyl radicals (Ross et al., 2018).

Thermal degradation has successfully achieved mineralization of PFAS at temperatures exceeding 900°C, primarily during waste treatment processes (Ross et al., 2018). However, such thermal methods are mainly suitable for solid wastes or liquids of low volume, which limits their practical application for widespread PFAS remediation (Wang et al., 2015). In addition, microbial degradation approaches, including fungal degradation and anaerobic biodegradation, have been investigated to cleave carbon–fluorine bonds in PFAS molecules, but these methods remain under evaluation for efficacy and scalability (Merino et al., 2016).

This section reviews mineralization processes for PFAS degradation, covering sonochemical, photochemical, electrochemical, chemical oxidation and reduction, and plasma-based treatments. The characteristics, advantages, and limitations of these technologies are summarized in Table 7.1.

### 7.4.1 Sonolysis or sonochemical mineralization

Sonochemical degradation of organic pollutants primarily occurs through the sonolysis of water molecules, generating reactive species such as hydroxyl radicals, hydrogen, and oxygen atoms, as introduced above (Kotronaru et al., 1991; Pang et al., 2011). Sonochemistry describes the influence of an acoustic field on chemical reactions within a solution (Zupanc et al., 2019). When ultrasonic waves propagate through water, they induce the formation of vapor bubbles that, upon collapse, produce extremely high localized temperatures, up to approximately 4000 K, facilitating the mineralization of PFAS at the bubble water interface, where temperatures range between 600 K and 1000 K (Hao et al., 2014). The ultrasound frequencies used typically vary from 20 to 1000 kHz. They were among the first to report sonochemical degradation of PFOA and PFOS as surfactants at initial concentrations of 20 μM under an argon atmosphere, observing pseudo-first-order degradation rate constants of 0.32 $min^{-1}$ for PFOA and 0.16 $min^{-1}$ for PFOS (Lin et al., 2012). They proposed that sonolysis involves cavitation phenomena where hydroxyl radicals dissociate surfactants in the aqueous phase through pyrolytic reactions. However, subsequent studies indicated that hydroxyl radicals alone do not effectively mineralize PFOA and PFOS (Benotti et al., 2009; Schröder and Meesters, 2005; Moriwaki et al., 2006; Rayne and Forest, 2009). Instead, Moriwaki et al. determined that the primary pyrolytic degradation occurs at the interfacial region between the cavitation bubbles and the bulk solution (Cheng et al., 2008a).

Further evidence of the effectiveness of sonolysis was provided by treating groundwater contaminated with PFOA and PFOS from a landfill site using high-frequency ultrasound at 354 kHz (Vecitis et al., 2008a). Their studies showed mineralization efficiencies of 44% for PFOA and 39% for PFOS across a wide initial concentration range (0.0002 to 0.2 M).

**TABLE 7.1** Characteristics of selected degradation treatment technologies for the removal of per- and polyfluoroalkyl substances (PFAS).

| Entry | Technology/ composite | Scale | Protocol | Justification | PFAS type and concentration | Yield | Source |
|---|---|---|---|---|---|---|---|
| 1 | Ultrasound | Bench | Sonolysis | Acoustic cavitation bubble interfaces | Perfluorooctanoic acid (PFOA) (30 ppb), perfluorooctanesulfonic acid (PFOS) (60 ppb) | 44% PFOA, 39% PFOS | Ahmad (2012) |
| 2 | Ultrasound | Bench | Sonolysis | PFAS dissociation by acoustic field | PFOS (65 ppb to 13,100 ppb) | 73% | Rodriguez-Freire et al. (2015) |
| 3 | ECOHR | Bench | Biological remediation | Microbial mineralization | PFOA (0.01 ppm to 10 ppm) | 30% | Vecitis et al. (2008)b |
| 4 | Microbial | Bench | Biological remediation | Microbial mineralization | PFOA, PFOS, 6:2 fluorotelomer alcohols (30 μM) | 95% | Vecitis et al. (2010)a |
| 5 | *P. chrysosporium* and vaults | Bench | Biological remediation | Microbial mineralization | PFOA | 45% | Qingguo (2013) |
| 6 | Heat | Bench | Thermal degradation | PFAS bond cleavage via an endothermic process | Fluorotelomer-based acrylic polymers (1600 mg) | 99%, 950°C | Key et al. (1998) |
| 7 | $O_3$ and $O_3/H_2O_2$ | Bench | Chemical oxidation | Hydroxyl radical-driven degradation | PFOA, PFOS (50 ppb) | 85%–100% degradation | Ochoa-Herrera et al. (2008) |
| 8 | UV/$TiO_2$ photocatalysis | Pilot | Chemical oxidation | OH radicals attack C–F bonds | PFOS (0.330 ± 0.092 ppb) | PFOS removal constant 0.044 ± 0.0668 | Tseng (2012) |
| 9 | $O_3$, $O_3$/UV, $O_3/H_2O_2$ | Bench | Chemical oxidation | OH radicals + ozone degradation | Mixed PFAS (10,000 ppm) | PFOS unchanged | Yamada et al. (2005) |

The degradation primarily resulted from pyrolytic reactions occurring at the vapor phase and surface of transiently collapsing cavitation bubbles. A detailed mechanism was proposed where PFAS molecules first lose their anionic functional groups at the bubble interface, followed by pyrolytic breakdown of the residual fragments into carbon monoxide, fluoride ions ($F^-$), sulfate ions ($SO_4^{2-}$), and carbon dioxide (Cheng et al., 2008a). Half-lives of approximately 30 min for both PFOA and PFOS were reported in these studies. Under ultrasonic irradiation at 354 kHz, in an argon atmosphere and at 10°C, the degradation rate constants for PFOA and PFOS were 0.41 and 0.27 $min^{-1}$, respectively, with observed formation of volatile fluorinated species during sonolysis (Ahmad, 2012).

The investigation of the sonolysis of aqueous film-forming foams (AFFFs) using a dual-transducer system combining ultrasonic and megasonic frequencies, demonstrated its potential for bulk treatment of these complex mixtures (Bruton and Sedlak, 2017; Keswani, 2015; Vecitis et al., 2008a). Other studies treating groundwater matrices containing PFAS by sonolysis reported that the presence of additional organic contaminants did not significantly affect PFAS degradation rates (Vecitis et al., 2008a). Inorganic constituents also influence sonolytic degradation; for example, bicarbonate and sulfate ions tend to inhibit PFAS breakdown, whereas nitrate and perchlorate ions enhance degradation rates (Vecitis et al., 2008b). Inorganic cations show negligible effects, but acidic conditions ($pH < 4$) increase degradation rates, likely due to proton exchange interactions at the water-bubble interface (Keswani, 2015). The role of surfactants on sonolysis has also been examined by PFOA degradation in the presence of different surfactants: the anionic sodium dodecyl sulfate (SDS), the cationic hexadecyltrimethylammonium bromide (CTAB), and the nonionic Triton X-100. Since PFOA itself acts as an anionic surfactant, the addition of CTAB significantly enhanced degradation, achieving 79% mineralization after 2 h at 25°C and under acidic conditions. Conversely, SDS and Triton X-100 inhibited sonolytic mineralization. These results suggest that the charge properties of surfactants strongly influence sonolysis, outweighing effects related to critical micelle concentration. Thus cationic surfactants such as CTAB may be employed to improve PFAS degradation efficiency during ultrasonic treatment.

Sonochemical mineralization of PFAS in AFFFs has also been investigated, demonstrating the method's potential for treating these complex and challenging mixtures (Rodriguez-Freire et al., 2015). The preparation FC-600 contains hydrocarbons (HC), anionic hydrocarbon surfactants, fluorochemical (FC) constituents, anionic fluorinated surfactants, fluorinated amphiphilic surfactants, and starch as a thickening agent. Sonochemical degradation of PFOS using aqueous dilutions of FC-600 at 65 ppb demonstrated that the carbon–sulfur (C–S) bond undergoes initial pyrolytic cleavage at the bubble–water interface, consistent with earlier studies conducted in Milli-Q water.

In 2011 an investigation of the sorption of PFAS onto GAC under 20 kHz ultrasonic irradiation. They reported that the combination of GAC with ultrasound enhanced the sorption kinetics of 50 ppm perfluorinated compounds (PFCs) by 2.5 to 9 times, demonstrating a clear synergistic effect between ultrasonic cavitation and adsorbent-mediated removal (Zhao et al., 2020). Sonication-mediated oxidative photocatalytic degradation of PFOA in a basic medium has been demonstrated using sol–gel-derived titanium oxide (Lin et al., 2012). While this approach enabled partial mineralization, the reaction kinetics were slow, achieving only ~70% mineralization of 50 ppm PFOA after 8 h of ultrasonic treatment, highlighting the limitations of sonophotocatalytic methods for rapid PFAS remediation (Vecitis et al., 2010a). More recently, Yang et al. examined the combined effect of vacuum UV irradiation at 185 nm with ultrasonic defluorination of aqueous PFOS (10 ppm). The study demonstrated that this integrated approach significantly outperformed sonolysis alone, as the synergy between high-energy UV photons and ultrasonic cavitation accelerated reactive species formation and promoted more complete PFOS degradation (Panchangam et al., 2009). Collectively, these studies emphasize that coupling ultrasound with complementary treatment strategies such as adsorption, photocatalysis, or UV irradiation can overcome kinetic limitations and improve PFAS removal efficiency.

Overall, numerous studies confirm that sonolysis at higher ultrasound frequencies enhances PFAS mineralization. The energy consumption for sonolysis is moderate, typically requiring between 100 and 300 Wh per liter of water treated. However, high capital and operational costs currently limit the widespread applicability of sonolysis (Fernandez et al., 2016). To date, sonolysis has primarily been demonstrated at the laboratory scale and has not yet been scaled up for commercial applications. Effective large-scale deployment requires overcoming challenges related to transducer-generated cavitation bubble distribution, which depends heavily on reactor geometry, power dissipation, transducer placement, and ultrasound frequency. Addressing these factors is essential for improving the scalability and efficiency of sonochemical PFAS treatment technologies and maximizing the synergistic benefits observed in combined or hybrid approaches (Yang et al., 2013) (Fig. 7.2).

### 7.4.2 Per- and polyfluoroalkyl substances degradation mechanisms at mid-high ultrasonic frequencies (100–1000 kHz)

The initial report of PFAS degradation by sonolysis occurred at a frequency of 200 kHz and a power density of 3333 W $L^{-1}$ (Pflieger et al., 2019). Ultrasonic irradiation of water induces the formation and growth of cavitation bubbles (Pflieger et al., 2019; Mason and Lorimer, 2002). These bubbles expand until reaching a critical size, then migrate to antinode regions where they collapse violently, causing the dissociation and ionization of water, gases, and other species inside the bubble, along with the generation of highly localized high temperatures. Since PFAS molecules exhibit surfactant-like behavior in aqueous environments (Wood et al., 2017), sonolytic degradation is believed to predominantly occur at the air–water interface of these bubbles. Prior to bubble collapse, the hydrophilic head groups of PFAS orient toward the bulk liquid, while their hydrophobic perfluorinated tails extend into the gas phase within the bubble (Bell et al., 2019) Fig. 7.2. PFAS molecules are comparatively small (approximately 1 nm in length (Moriwaki et al., 2005), with molar volumes between 75 and 477 $cm^3 mol^{-1}$ (Chen et al., 2017) relative to ultrasonic cavitation bubbles, which have ambient radii of roughly 3.2–3.5 μm and molar volumes around $10^{14} cm^3$ at 355 kHz under pressures of 2.5 to 4.6 bar (Brusseau, 2019; Yasui et al., 2008). Given this size disparity, PFAS molecules are likely "adsorbed" onto the bubble surface rather than penetrating or evaporating into the bubble core prior to collapse. Consequently, adsorption at the bubble interface is considered the initiating step in PFAS sonolysis, with diffusion of PFAS molecules to the bubble surface often identified as the rate-limiting process (Dehane et al., 2021; Lin et al., 2016a). Some studies, however, question the diffusion limitation hypothesis because rectified bubble growth occurs on time scales much faster than

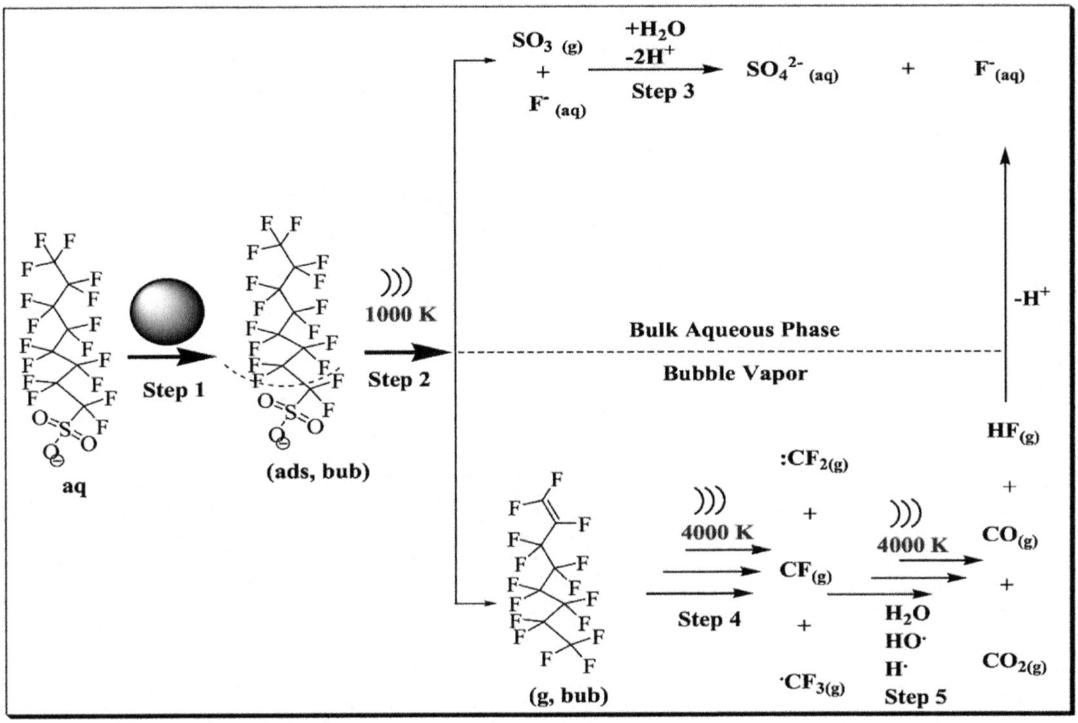


FIGURE 7.2 Sonolysis of perfluorooctanesulfonic acid into inorganic compounds using high-frequency ultrasound (Verma et al., 2021). *Adapted with permission from Verma, S., Varma, R.S., Nadagouda, M.N., 2021. Remediation and mineralization processes for per-and polyfluoroalkyl substances (PFAS) in water: a review. Science of the Total Environment. 794, 148987. Copyright (2021).*

molecular diffusion would allow, suggesting that the bubble interface moves toward the PFAS molecules rather than the reverse. Increasing PFAS concentration generally enhances degradation rates until a plateau is reached at sufficiently high concentrations, indicating saturation of the bubble interfacial area (Sekiguchi et al., 2017; Campbell et al., 2009; Campbell and Hoffmann, 2015). Conversely, when starting with high initial PFAS concentrations, the reaction kinetics shift from zero-order to pseudo-first-order as concentrations decrease during sonolysis (Fernandez et al., 2016). This change in reaction order supports the hypothesis that PFAS adsorption at the bubble interface is the rate-limiting and reaction-initiating step. Additional evidence for interfacial adsorption arises from experiments showing that adding another anionic surfactant decreases the degradation rate of PFOA at 40 kHz and 500 W $L^{-1}$, suggesting competitive adsorption at the bubble surface.

Moreover, sonolytic degradation rates of various PFAS, including perfluorooctane, PFOA, and perfluoropropionic acid, correlate with their hydrophobicity, which is attributed to their molecular orientation within the bulk liquid, bubble interface, and bubble gas core, respectively (Vecitis et al., 2008a). These microenvironments experience increasing temperatures and pressures closer to the bubble core during collapse (Fig. 7.3) (Shende et al., 2019). Table 7.2 collects the log ($K_{OW}$) values for different PFAS and related carboxylic

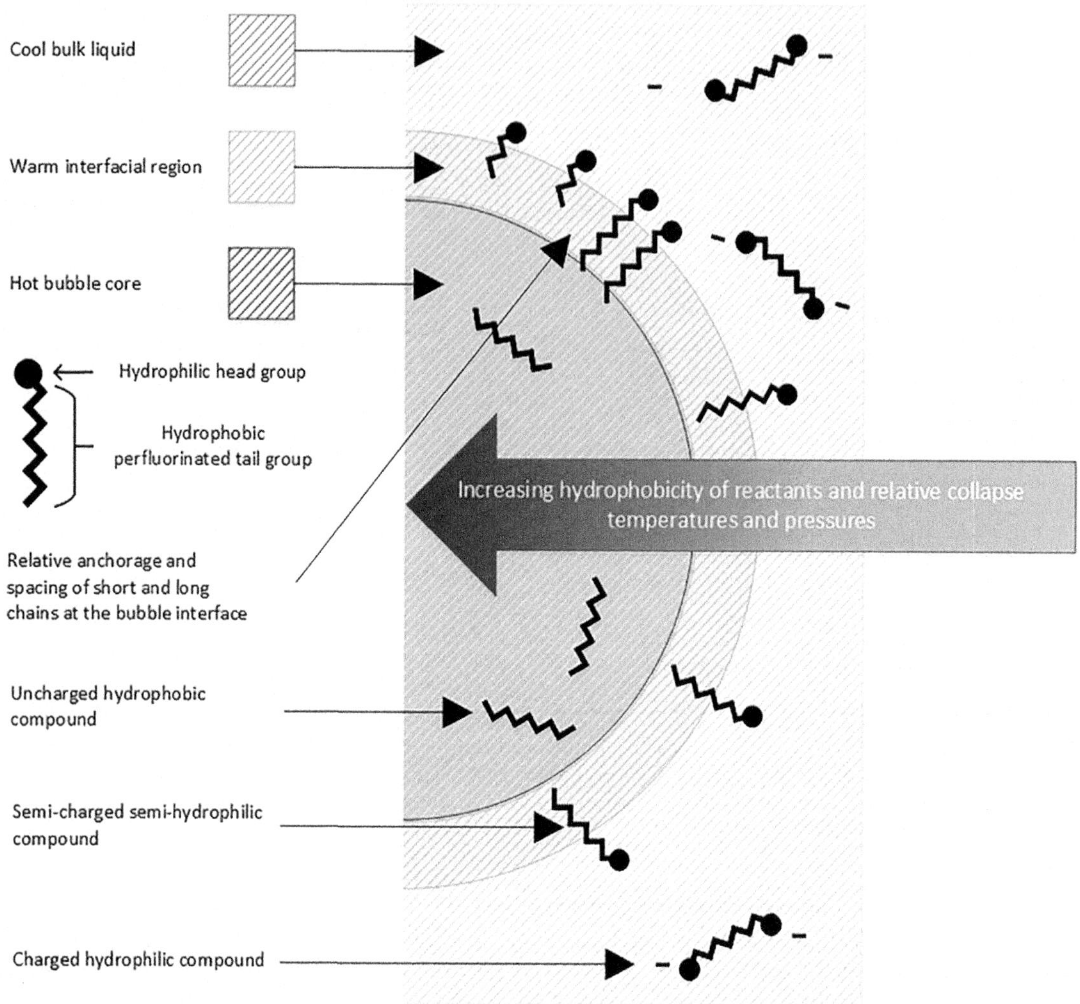


**FIGURE 7.3** Position and relative collapse temperatures and pressures experienced by hydrophilic, semihydrophilic, and hydrophobic surfactants, as well as relative anchoring and spacing of short and long chains at ultrasonic cavities. *Reproduced from Sidnell, T., et al., 2022. Sonolysis of per- and polyfluoroalkyl substances (PFAS): A review of reaction mechanisms, kinetics, and applications. Science of the Total Environment. 806, 150597.*

acids from prior studies (Cheng et al., 2010a; Hu et al., 2018), highlighting trends in hydrophobicity.

Consistent with earlier sonolysis findings, hydrophobicity increases with PFAS chain length (Vecitis et al., 2010b) and substitution of fluorine atoms with hydrogen (Kim et al., 2009). It is important to note, however, that traditional measurements of ($K_{OW}$) are challenging for surfactant PFAS due to their interfacial affinity, necessitating alternative measurement approaches (Love et al., 2005). Semihydrophobic PFAS such as PFOA and PFOS tend to accumulate at the bubble interface and are subjected to localized temperatures

**TABLE 7.2** $K_{OW}$ (octanol–water partition coefficient) of PFCAs and equivalent carboxylic acids.

| Carbon chain length | Per- and polyfluoroalkyl substances | Log ($K_{OW}$) | Carboxylic acid | Log ($K_{OW}$) |
|---|---|---|---|---|
| 1 | | | Formic acid | -0.54 |
| 2 | | | Acetic acid | -0.17 |
| 3 | | | Propanoic acid | 0.33 |
| 4 | PFBA | -0.62 | Butanoic acid | 0.79 |
| 5 | PFPeA | -0.02 | Pentatonic acid | 1.4 |
| 6 | PFHxA | 0.59 | Hexanoic acid | 1.9 |
| 7 | PFHpa | 1.19 | Heptanoic acid | 2.4 |
| 8 | PFOA | 1.79 | Octanoic acid | 3.1 |
| 9 | PFNA | 2.40 | Nonanoic acid | 3.4 |
| 10 | PFDA | 3.00 | | |
| 11 | PFUna | 3.51 | | |
| 12 | PFDoA | 4.21 | | |

around 1000 K and pressures near 100 bar during bubble collapse (Kotronarou et al., 1991). Previous research on surfactant sonolysis indicates that all parts of a surfactant molecule experience roughly the same collapse temperature (Sostaric and Riesz, 2001), aligning with the relative size scales of bubbles and PFAS molecules described here.

## 7.5 Key factors influencing sonochemical degradation of per- and polyfluoroalkyl substances

### 7.5.1 Effects of frequency

Acoustic power is a critical parameter in sonochemistry, directly influencing cavitation phenomena such as bubble nucleation, oscillation, collapse intensity, and subsequent turbulence within the reaction medium (). These phenomena play a key role in the formation of reactive species necessary for the defluorination of persistent pollutants such as PFOA. In a detailed study (Khoshyan, 2024), the effect of various parameters, such as frequency, power, initial PFOA concentration, and volume of the reactor, on the defluorination by ultrasonication was explored, and the results are shown in Fig. 7.4. The defluorination efficiency of PFOA was evaluated in both polypropylene (PP) and glass reactors under varying ultrasonic power input settings. Calorimetric measurements were used to estimate the actual energy delivered to the medium. At low power levels (below 20%), defluorination was negligible, particularly in the PP reactor. However, when the power was increased to 50% and 80% in the glass reactor, defluorination efficiency increased significantly to 70% ± 1.1% and 90% ± 0.9%, respectively. In contrast, under identical conditions, the PP

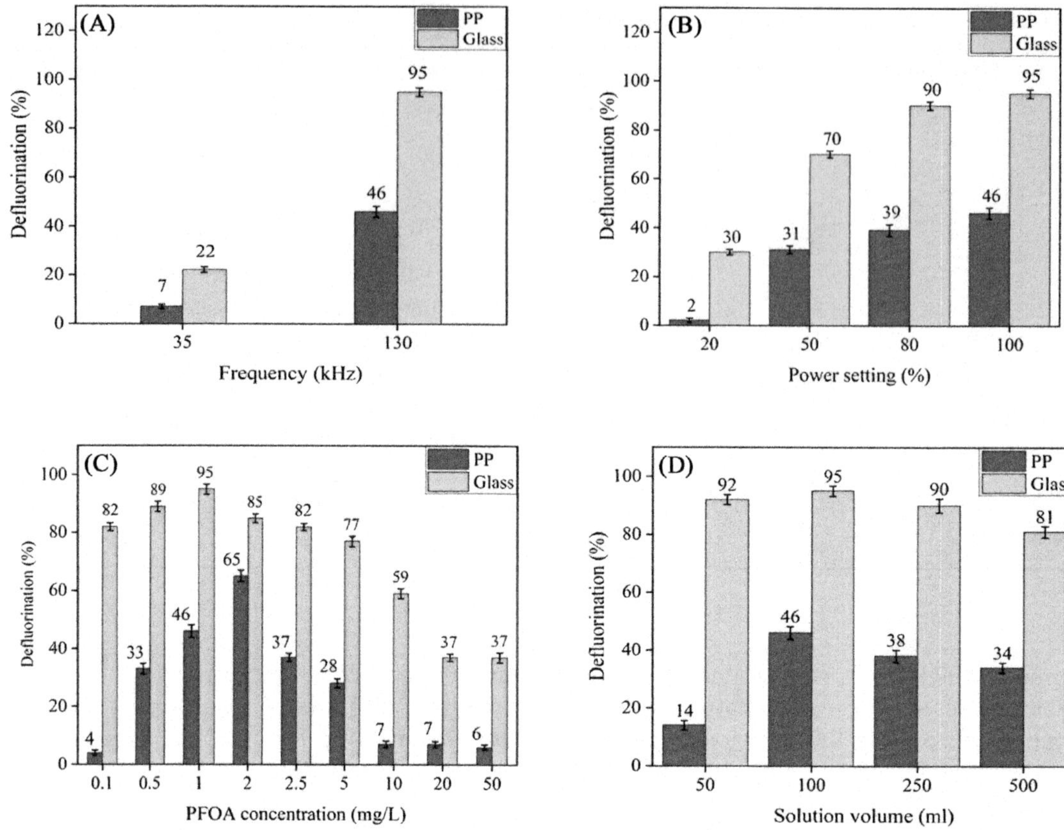


**FIGURE 7.4** Defluorination of perfluorooctanoic acid (PFOA) by ultrasonication under different operational conditions. Frequency (A), power input (B), concentration (C), and solution volume (D). (Regardless of the investigated parameters, all others were set as the default ones $[PFOA]_0 = 1\ mg\ L^{-1}$, (2.4 μM), frequency = 130 kHz, power setting = 100%, Milli-Q water = 100 mL, reaction for 3 h without controlling temperature). Degradation of per- and poly-fluoroalkyl substances using ultrasonication: effect of reactor materials. *Reproduced from Khoshyan, A., et al., 2024. Operating conditions are optimised: frequency (35 kHz vs. 130 kHz), power input, PFOA initial concentration, and solution volume. Journal of Water Process Engineering. 63 (2024), 105511.*

reactor achieved only about 15% degradation. Notably, further increasing the power input to 100% yielded minimal additional benefit, with a modest improvement of only 5% ± 1.4% in the glass reactor and 7% ± 1.7% in the PP reactor.

## 7.5.2 Effects of acoustic power

These results imply that increasing the power input improves cavitation only up to a certain point; beyond this optimal level, further power input becomes counterproductive and decreases efficiency. Excessive bubble formation can cause wave scattering, coalescence due to Bjerknes forces, and heat dissipation, thereby limiting the sonochemical effect (Kewalramani et al., 2023). Moreover, increased dissolved gas content from intense agitation

**TABLE 7.3** Calculated calorimetric powers in glass and polypropylene (PP) beakers.

| **Operational condition** | **Frequency (kHz)** | | | |
|---|---|---|---|---|
| **Applied power (%)** | **35** | | **130** | |
| | PP | Glass | PP | Glass |
| 20 | 13 ± 0.4 | 9.3 ± 0.7 | 13 ± 0.9 | 9.5 ± 0.4 |
| 50 | 39.53 ± 0.8 | 32.6 ± 0.3 | 23.2 ± 0.3 | 20.0 ± 0.7 |
| 80 | 49.4 ± 0.8 | 46.1 ± 0.6 | 47.2 ± 1 | 41.1 ± 0.9 |
| 100 | 59.1 ± 1.2 | 56.7 ± 1.3 | 55.1 ± 1 | 51.1 ± 0.7 |

can further suppress cavitation efficiency. Interestingly, although the applied power density in this study (51.1 ± 0.7 W $L^{-1}$) was lower than in previous works (e.g., 101–202 W $L^{-1}$), defluorination efficiency was significantly higher (Table 7.3). For instance, previous studies reported only 150 μM of fluoride ion release at 101 W $L^{-1}$ with 10 mg $L^{-1}$ initial PFOA, whereas the current study achieved 213.4 ± 2.26 μM fluoride release under lower power density and the same initial PFOA concentration Table 7.3.

## 7.5.3 Effects initial concentration

The influence of the initial PFOA concentration was also studied over a range from 0.1 to 50 mg $L^{-1}$ (Hu et al., 2018). In the glass reactor, over 75% defluorination was achieved for initial concentrations between 0.1 and 5 mg $L^{-1}$ under optimized conditions. For example, at 1 mg $L^{-1}$, the absolute degradation rate was approximately 0.98 mg $L^{-1}$ per hour, with a total of 0.095 mg of defluorinated PFOA. The PP reactor performed optimally at 2 mg $L^{-1}$, achieving 65% ± 3.1% defluorination and an absolute removal of ~0.13 mg. Beyond this concentration, especially above 2.5 mg $L^{-1}$, the PP reactor exhibited a sharp decline in defluorination efficiency, indicating saturation of active interfacial sites on cavitating bubbles and reduced adsorption capacity. However, at optimized concentrations, adsorbed PFOA molecules appeared to stabilize cavitation bubbles via electrostatic repulsion, preventing bubble coalescence and enhancing sonochemical activity.

## 7.5.4 Solution volume

The effect of solution volume was also assessed, as it influences power distribution and cavitation intensity. Experiments were conducted by Hu et al. (2018) in both reactors using volumes from 50 to 500 mL with 1 mg $L^{-1}$ of PFOA. In the glass reactor, defluorination efficiency remained high (≥90%) up to 250 mL and only slightly decreased at 500 mL. At this larger volume, the released fluoride concentration was 29.32 ± 1.43 μM (~0.56 mg $L^{-1}$), yielding an absolute degradation of 0.0405 mg, which demonstrates the potential of the glass reactor for scale-up. In contrast, the PP reactor showed increased efficiency when the volume rose from 50 mL (5.2 ± 0.74 μM fluoride) to 100 mL (16.81 ± 1.23 μM), but performance declined at higher volumes. This behavior is attributed to increased acoustic attenuation in PP materials and the reduced intensity of cavitation per unit volume as liquid volume increases (Cheng et al., 2008b; Kalra et al., 2021). Overall, the glass reactor demonstrated superior performance at all volumes, confirming its suitability for larger-scale sonochemical applications due to its favorable acoustic transmission properties and minimal energy loss.

### 7.5.5 Effects of organic matter

Research on PFAS sonolysis in groundwater samples collected from landfill sites and AFFFs has highlighted the significant influence of organic constituents on degradation and defluorination efficiencies in various reported works (Vecitis et al., 2010c; Lin et al., 2016b). Specifically, the degradation rates of PFOA and PFOS in groundwater were found to be approximately 2.3 and 2.6 times slower, respectively, compared to rates observed in pure Milli-Q water, with this decrease primarily attributed to the presence of organic matter in groundwater (Vecitis et al., 2010c). Among dissolved organic matter components, humic acid and fulvic acid exhibited minimal effects on sonolysis efficiency, likely due to their moderate surface activity and low volatility. Conversely, VOCs played a major role in inhibiting PFAS degradation. The decrement in degradation rates is explained by competitive adsorption of VOCs at the bubble–water interface and their evaporation into the bubble–vapor phase, which notably lowers the temperature at this interface during cavitation (Vecitis et al., 2010c).

Similarly, studies using diluted AFFFs containing HC and various FC surfactants (1%–5%) revealed a negligible impact of organic matter on the sonolysis rate of PFOS. However, defluorination efficiency was considerably lower in these complex matrices compared to pure water, suggesting that the fluorocarbon tail of PFOS remains partially resistant to degradation in the presence of organic compounds (Lin et al., 2016b). Further investigations into surfactant effects demonstrated that the addition of cationic (CTAB) and nonionic (Triton X-100) surfactants enhanced PFOA degradation, with the cationic surfactant showing a more pronounced effect (Cheng et al., 2010b). This improvement is attributed to the increased electrostatic attraction and accumulation of PFOA molecules at the bubble–water interface induced by the positively charged CTAB, thereby facilitating degradation. In contrast, the introduction of an anionic surfactant, SDS, inhibited PFOA sonolysis due to competitive adsorption at the bubble interface, reducing the availability of sites for PFAS molecules (Cheng et al., 2010b).

### 7.5.6 Effects of inorganic matter

The influence of inorganic anions on PFAS sonolysis can be either promotive or inhibitory, depending largely on the specific anion type and its concentration. Key factors governing these effects include anion partitioning at the bubble–water interface, interactions with the interface, and the potential formation of reactive species. Lin et al. (2015) reported that the sonolysis rate constants for PFOA and PFOS, such as (+ add which compounds were tested), followed the order: perchlorate ($ClO4^-$) > nitrate ($NO_3^-$) ≈ chloride ($Cl^-$) > Milli-Q water > bicarbonate ($HCO_3^-$) ≈ sulfate ($SO_4^{2-}$). Perchlorate ions, enriched at the bubble–-water interface, increase the negative surface charge, thereby reducing bubble coalescence and generating a greater number of smaller bubbles with a larger reactive surface area for PFAS pyrolysis (Fig. 7.5, **C-1**). Similarly, the addition of sodium chloride (NaCl) enhances sonolysis by increasing the hydrophobicity of PFAS molecules and their accumulation at the bubble–water interface (Fig. 7.5. **C-2**). In contrast, bicarbonate ions adversely affect interfacial pyrolysis conditions, slowing PFAS degradation (Fig. 7.5. **C-3**). However, Lin et al. (2016b) observed a contrary effect where bicarbonate addition promoted PFOA sonolysis via the formation of carbonate radicals ($CO_3^{\bullet-}$) generated from the reaction of bicarbonate with hydroxyl radicals (Fig. 7.5. **C-4**).

The role of sulfate in PFAS sonolysis is concentration-dependent, as shown by Cheng et al. (2010b). Low sulfate concentrations (5–10 mM) increase surface tension, limiting surfactant migration to the interface and slowing initial PFOA degradation. However, higher sulfate levels (above 10 mM) accelerate decomposition by promoting sulfate radical generation, which counteracts the surface tension effects.

Cation presence generally reduces PFAS sonolysis rates, but this effect is frequency-dependent. At low ultrasonic frequencies (40 kHz), metal ions such as Cu(II), Fe(II), and Fe (III) significantly inhibit PFOA degradation by forming metal-PFOA complexes in a permanganate–ultrasonic system (Thi et al., 2014). Conversely, at higher frequencies (612 kHz), the addition of $Na^+$, $NH_4^+$, $Ca^{2+}$, and $Mg^{2+}$ ions has minimal impact on

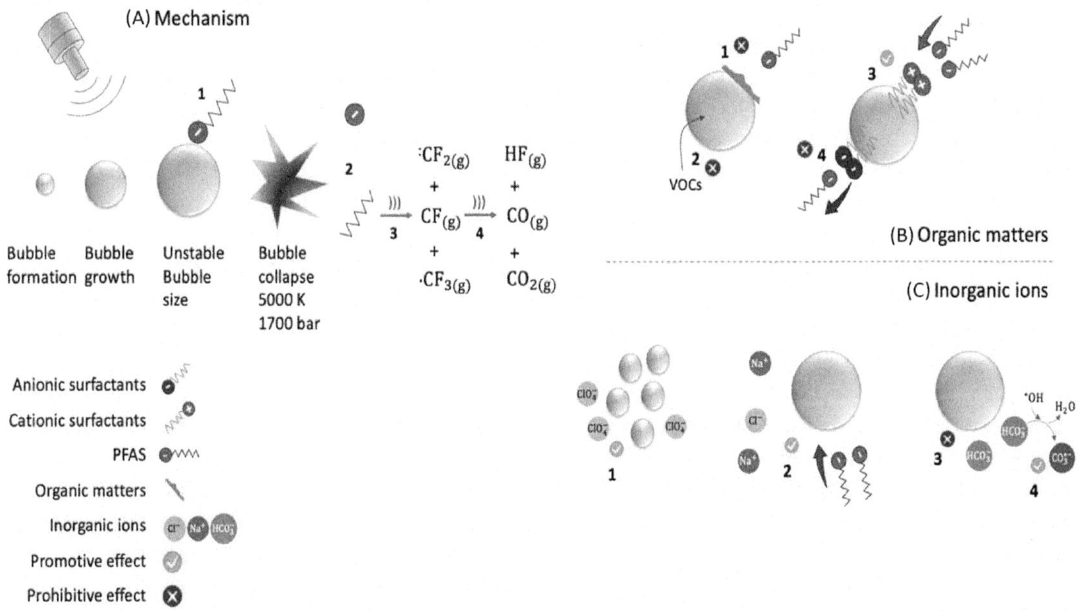


**FIGURE 7.5** Per- and polyfluoroalkyl substances (PFAS) degradation using sonolysis: A. Mechanism (1. PFAS adsorption to the water–bubble interface, 2. pyrolytic decomposition of PFAS through functional group cleavage, 3. pyrolysis of fluoro–carbon chain into fluoro-radicals, 4. transformation of fluoro-radicals within the bubble vapor to CO, $CO_2$, and HF). B. Effects of organic matters (1. competitive adsorption onto bubble–water interface, 2. volatile organic compounds evaporation to bubble phase, 3. attractive effect of cationic surfactant, 4. competitive adsorption of anionic surfactant). C. Effects of inorganic ions (1. increasing available surface area by $ClO_4^-$, 2. increasing hydrophobicity by NaCl, 3. adverse effect of $HCO_3$ on interfacial pyrolysis conditions, 4. formation of $CO_3^{\bullet-}$). *Adapted with permission from Zeidabadi, F.A., Esfahani, E.B., Mohseni, M., 2023. Effects of water matrix on per-and polyfluoroalkyl substances (PFAS) treatment: Physical-separation and degradation processes–a review. Journal of Hazardous Materials Advances. 10, 100322. Copyright (2023).*

degradation rates (Lin et al., 2015). This variation is attributed to the dominant mechanisms at each frequency: radical-mediated degradation at low frequency versus bubble dynamics and PFAS adsorption at high frequency (Sidnell et al., 2022).

Supporting this, Kalra et al. (2021) demonstrated that PFAS sonolysis rates in groundwater are influenced by total dissolved solids (TDS) concentration. Degradation of sulfonates and short-chain PFAS ($C < 8$) was 30%–60% higher in low TDS groundwater ($0.4\ g\ L^{-1}$; $k_{avg} = 0.029\ min^{-1}$) compared to deionized water ($k_{avg} = 0.021\ min^{-1}$), whereas it decreased in high TDS groundwater ($10.2\ g\ L^{-1}$; $k_{avg} = 0.012\ min^{-1}$). Low ion concentrations may facilitate PFAS transfer to the bubble interface by reducing surface tension, while high salt concentrations can neutralize the charge of cavitation bubbles, causing collapse or aggregation. Additionally, elevated salt levels and organic matter compete with PFAS for reactive radicals and sonic energy, diminishing sonolysis efficiency (Kalra et al., 2021).

## 7.6 Comparative assessment of per- and polyfluoroalkyl substances remediation technologies

To date, over 20 different processes have been explored for the remediation of PFAS, each employing distinct removal mechanisms. A range of both experimental and commercially available technologies exists to extract, but not necessarily destroy, PFAS from contaminated soil and water, as extensively reviewed in prior literature (Merino et al., 2016; McDonough et al., 2018; Horst et al., 2018; Arias Espana et al., 2015; Banks et al., 2019). These include adsorption-based methods such as GAC (Hansen et al., 2010; Carbon, n.d.); membrane-based approaches, such as reverse osmosis and nanofiltration (Dickenson and Higgins, 2016); chemical treatments, including ion exchange resins (Woodard et al., 2017; Rahman et al., 2014; Rusinek and Michigan, 2019; Dudley, 2012), and physical removal techniques, such as thermal desorption for soils (Norris et al., 1999) and ozofractionation for aqueous media (McDonough et al., 2018).

However, these separative technologies do not degrade PFAS, necessitating integration with destructive processes (McDonough et al., 2018). Notably, many destructive technologies are only effective at PFAS concentrations well above environmental levels, such as those found in concentrated AFFFs, which require a prior separation or concentration step (McDonough et al., 2018; Ross et al., 2016). Additionally, the presence of cocontaminants can hinder degradation efficiency, and the effluents generated from these destructive treatments may themselves demand further posttreatment before safe environmental discharge.

Given these constraints, large-scale PFAS remediation is increasingly expected to rely on an integrated "treatment train" approach (McDonough et al., 2018). Accordingly, this section evaluates the effectiveness and practical viability of sonolysis compared to other destructive technologies, excluding separative methods.

### 7.6.1 Biological degradation

Biological remediation, widely utilized for various aqueous pollutants, has shown limited efficacy in PFAS treatment (McDonough et al., 2018). Despite the high-energy density, PFAS could theoretically offer microbial systems (Ahrens, 2011), their synthetic nature and strong carbon–fluorine bonds hinder microbial adaptation and enzymatic degradation (McDonough et al., 2018; Parsons et al., 2002; Harper et al., 2003). The exceptionally high C–F bond dissociation energy and protective C–C bond environments present formidable barriers to microbial attack.

Although full biodegradation or mineralization of PFAS has not yet been achieved, partial biotransformation has been observed. Certain polyfluorinated compounds can be converted into shorter-chain perfluoroalkyl acids (PFAAs) through microbial attack on alpha-carbon or nonfluorinated regions (Key et al., 1998), a process comparable to radical-based degradation pathways. However, such biotransformations are typically slow and incomplete. For example, PFOA biotransformation has yielded only 28% fluoride release over 157 days, with a significant conversion to PFPeA (Luo et al., 2015). Although shorter-chain PFAAs tend to be less bioaccumulative (Conder et al., 2008), they are also more stable in aquatic environments and more resistant to removal by conventional treatment methods (Bruton and Sedlak, 2017; Eschauzier et al., 2012; Lee et al., 2012).

## 7.6.2 Chemical degradation

PFAS compounds, though chemically inert under most conditions, can be degraded via specific chemical pathways. AOPs, which are commonly applied to degrade organic contaminants, have demonstrated limited success with PFAS, particularly perfluorosulfonic acids (PFSAs) (Bruton and Sedlak, 2017). Several AOPs tested for PFAS remediation exhibit low reactivity and incomplete degradation, similar to the performance of low-frequency ultrasound (Arias Espana et al., 2015). For instance, oxidation using heated persulfate produces degradation rates near 1.12 $\mu M\ h^{-1}$ and generates shorter PFAAs. However, its efficacy diminishes significantly in complex matrices containing organic matter or sediments, and it is particularly ineffective against PFSAs. The process's high-temperature requirements (85°C) and low pH operation increase risks such as HF and perchlorate formation, thereby limiting practical applicability (Bruton and Sedlak, 2017; Lee et al., 2012).

An alternative approach using activated carbon in conjunction with persulfate at lower temperatures (≤45°C) has been shown to reduce the activation energy barrier, although fluorine release remains low and degradation appears incomplete, possibly yielding fluorinated by-products (Lee et al., 2013; Hori et al., 2006).

EO, which utilizes high-voltage electrodes to generate hydroxyl radicals in the presence of chemical additives, offers improved degradation rates and greater resistance to organic interferences (Uwayezu et al., 2021). For example, EO with sodium sulfate ($Na_2SO_4$) has demonstrated degradation of approximately 7 $\mu M\ h^{-1}$ for perfluoroalkyl oxoacids (PFOX). The introduction of boron-doped diamond electrodes has further enhanced EO efficacy, making it a promising technology for PFAS mineralization (Uwayezu et al., 2021; Liao and Farrell, 2009; Sukeesan et al., 2021; Nienhauser et al., 2022). Nevertheless, the persistence of partially defluorinated by-products remains problematic, with around 50% of fluorine retained in organic form (Uwayezu et al., 2021). Complete mineralization may be achievable through a multistep process, though it often favors the formation of short-chain intermediates. The short-lived nature of hydroxyl radicals, which exist for only milliseconds and react near the electrode surface, also limits scalability for large-volume treatment.

Photochemical oxidation, which involves UV-enhanced oxidation, has shown varied degradation times ranging from hours to days and generally requires 2–30 times more energy per degraded PFAS molecule than ultrasound-based treatment (Hori et al., 2004) (Table 7.4). This method also leads to the accumulation of short-chain PFAAs (Hori et al., 2004; Wang et al., 2008; Song et al., 2013; Singh et al., 2019a).

A novel physicochemical approach involves aqueous plasma treatment, which, similar to EO, generates reactive radicals (e.g., O• and OH•) and solvated electrons using high voltage (Singh et al., 2020a, 2019b, 2020b). Plasma degradation has achieved rapid PFOA degradation rates (up to 36 μM $h^{-1}$) with high-energy efficiency even in the presence of cocontaminants, outperforming sonolysis in some respects (Table 7.4) (Stratton et al., 2017). Additionally, plasma processes are not constrained by electrode surface area, as EO is. However, plasma treatment tends to produce high concentrations of short-chain PFAS and other small organic acids, limiting its effectiveness for complete mineralization. In several studies, over 97.5% of fluorine was retained in the aqueous phase as organics, with the remainder forming gaseous by-products (Aleksandrov et al., 2019). Increasing plasma pulse energy may reduce short-chain formation (Taylor et al., 2014), potentially enhancing the relative competitiveness of sonolysis. Plasma-based PFAS treatment has been successfully tested at volumes up to 4 L (Meng et al., 2017; Winchell et al., 2021; Turner et al., 2019; Yu et al., 2017; Zhang et al., 2016, 2013), indicating promising potential for future scale-up.

### 7.6.3 Physical/thermal degradation

Historically, incineration has been the primary method for the destruction of PFAS following their separation from contaminated media, as high-temperature combustion theoretically enables complete molecular breakdown (Zhang et al., 2013). However, real-world applications have shown that this ideal is not always met. Many studies fail to provide a comprehensive analysis of flue gas emissions or evaluate the overall energy demands associated with the process. In one reported case, 300 g per hour of polytetrafluoroethylene were treated using a system fueled by 100 kg per hour of wood pellets. Analysis revealed that between 56% and 78% of the initial fluorine content was emitted in gaseous and solid phases, with 22% to 44% remaining unaccounted for, suggesting the formation of unidentified or unmonitored by-products (Yamamoto et al., 2007). Another study involving PFAS incineration at 1000°C showed that only 21.6% to 45.1% of the fluorine was converted into hydrogen fluoride (HF), a highly toxic compound (Jin et al., 2014). Incomplete combustion may also result in the atmospheric release of intact PFAS compounds, particularly if incinerator temperatures or residence times are insufficient (Yasuoka et al., 2011).

Some controlled incineration processes, such as those conducted by PFAS manufacturer 3M, have reported effluents containing only low-molecular-weight fluorinated by-products (C1–C2), indicating that more complete degradation may be achievable under optimized conditions (Park et al., 2011). Nevertheless, these by-products are often more thermally stable than the original PFAS, demanding even higher incineration temperatures for their destruction (Park et al., 2011). Recent advancements have been made in using cement kilns to thermally treat chlorofluorocarbons, hydrofluorocarbons, and PFAS, offering a potentially more effective pathway for thermal destruction (Thangavadivel et al., 2013). However, incineration remains impractical for most PFAS contamination scenarios, which

**TABLE 7.4** Comparison of G-values for various perfluorooctanesulfonic acid degradation works and conditions.

| Technology | Treatment time (h) | Initial concentration (mg $L^{-1}$) | Efficiency (g $kW^{-1}$ $h^{-1}$) | Short-chain formation | References |
|---|---|---|---|---|---|
| **Electrochemical** | 4 | 0.0152 | 0.00033 | Up to 50% | Uwayezu et al. (2021) |
| **Photochemical** | 240 | 20.0 | 0.0133 | Observed a significant quantity indicated (71% $F^-$ release) | Yamamoto et al. (2007) |
| **Photochemical, ferric ion** | 60 | 10.0 | 0.00290 | –14% of initial mass | Jin et al. (2014) |
| **Electrochemical** | 2 | 8.00 | 0.00566 | Not discussed | Sukeesan et al. (2021) |
| **Sonication, 618 kHz** | 3 | 5.00 | 0.00801 | Almost none implied (–100% $F^-$ release) | Vecitis et al. (2008)b |
| **Photochemical, persulfate** | 2 | 10.0 | 0.00900 | Observed a significant quantity indicated (76% $F^-$ release) | Park et al. (2011) |
| **Photochemical, propanol** | 24 | 20.0 | 0.01520 | Not discussed | Yamamoto et al. (2007) |
| **Sonication, 400 kHz** | 4 | 9.42 | 0.01550 | 1% of initial mass | Wood et al. (2017) |
| **Plasma** | 4 | 50.0 | 0.02600 | Not discussed, none implied | Yasuoka et al. (2011) |
| **Sonication, 400 kHz** | 2 | 9.42 | 0.02610 | 13% of initial mass | Wood et al. (2017) |
| **Sonication, 358 kHz** | 3 | 59.5 | 0.04790 | Not discussed | Vecitis et al. (2008)a |
| **Plasma** | 0.5 | 0.0001 | 0.06900 | Observed, 5.65% of initial mass after 40 min | Stratton et al. (2017) |
| **Plasma** | 1 | 100 | 0.62100 | Observed a significant quantity indicated (≈30% $F^-$ release) | Lewis et al. (2020) |

are typically characterized by low PFAS concentrations in large volumes of aqueous media. The centralized nature of incineration infrastructure also poses logistical challenges. Effective use of this method for widespread environmental PFAS contamination would require on-site preconcentration of the pollutants, followed by transportation to incineration facilities, increasing operational complexity and costs (Wood et al., 2017).

An emerging alternative physical degradation method under investigation is mechanochemical destruction via ball milling, particularly for solid-phase PFAS. This process involves the mechanical breakdown of PFAS-containing materials in a rotating drum filled with hardened balls, where degradation occurs through repeated impact and attrition forces (Wood et al., 2019). Experimental studies have demonstrated complete degradation (up to 100%) of pure PFOX powders, with reportedly no formation of harmful by-products (Stratton et al., 2017; Khan et al., 2020). While these results are promising, this technique remains at an early stage of development. Key information regarding its energy efficiency, applicability to complex environmental samples such as AFFFs, and large-scale feasibility is still lacking.

### 7.6.4 Sonolysis

Sonolysis represents a hybrid degradation approach that integrates both physical and chemical mechanisms, and has been extensively studied under various operational parameters. Among existing technologies, it demonstrates notably high degradation rates, reaching up to 61.3 $\mu M\ h^{-1}$, which positions it among the fastest PFAS treatment methods currently available. When evaluated based on G-value (energy yield per amount of PFAS degraded), sonolysis ranks as the second most energy-efficient process, surpassed only by plasma-based treatments (Wood et al., 2017) (Table 7.4). However, when energy efficiency is recalculated in terms of fluoride mass released per kilowatt-hour, the efficiency gap between ultrasound and plasma narrows significantly (Wood et al., 2017; Campbell et al., 2009; Campbell and Hoffmann, 2015; Stratton et al., 2017; Hori et al., 2005). It is important to note that reported energy values in many comparative studies may only reflect the consumption of the core experimental device (e.g., ultrasonic amplifier, plasma reactor, electrochemical cell), potentially excluding auxiliary systems such as heaters, chillers, or fluid pumps, which contribute to overall energy demand (Wood et al., 2017; Campbell and Hoffmann, 2015). A further advantage of sonolysis lies in its nonselective chain-length degradation. Unlike some technologies, such as EO, incineration, or plasma systems, which may preferentially degrade certain PFAS species or chain lengths, sonolysis proceeds sequentially from longer to shorter chains and tends to degrade perfluorocarboxylic acids (PFCAs) before PFSAs (). Given its favorable combination of degradation efficiency, broad-spectrum applicability, and scalability, sonolysis emerges as a highly pragmatic solution for large-scale PFAS remediation. Nevertheless, plasma-based technologies continue to show strong competitive potential, particularly in terms of energy efficiency and rapid reaction rates (Lewis et al., 2020).

## 7.7 Conclusion

PFAS represent some of the most persistent and challenging environmental contaminants due to their chemical stability, bioaccumulative potential, and resistance to conventional remediation methods. Among advanced treatment strategies, ultrasonication, or sonolysis, has emerged as a particularly promising approach. By coupling physical disruption with chemical degradation via acoustic cavitation, sonolysis enables the breakdown of even the most recalcitrant PFAS compounds. The extreme localized temperatures and pressures within cavitation bubbles facilitate sequential defluorination, allowing effective degradation without the need for harsh reagents or extreme operating conditions. Compared to other destructive technologies such as plasma treatment, EO, or thermal incineration, sonolysis offers a favorable balance between degradation efficiency, moderate energy consumption, and nonselectivity toward PFAS chain length, while avoiding the limitations of biological or conventional chemical treatments, such as slow reaction rates or incomplete mineralization.

Although plasma processes may theoretically achieve slightly higher energy efficiency, sonolysis is advantageous in terms of operational simplicity, scalability, and adaptability for integration into modular treatment systems. Its nonselective degradation pattern from long- to short-chain PFAS further underscores its versatility across a wide spectrum of PFAS chemistries. Nevertheless, practical challenges remain. Future research should aim to optimize reactor designs, refine energy input relative to cavitation dynamics, and account comprehensively for auxiliary energy demands in lifecycle assessments. Moreover, the influence of water matrix constituents, including organic matter, surfactants, and inorganic ions, requires careful evaluation to ensure reliable performance under field conditions.

In conclusion, sonolysis stands out as a pragmatic, robust, and environmentally sustainable technology for PFAS remediation. As regulatory limits tighten and contamination spreads globally, ultrasound-assisted degradation offers a scalable and effective solution. Importantly, ultrasonication can be combined with complementary catalytic or "green" processes, such as photocatalysis and electrocatalysis, to create hybrid treatment systems. These integrated approaches often yield pronounced synergistic effects, enhancing reaction kinetics, improving energy efficiency, and expanding the overall applicability of PFAS remediation strategies. By accelerating degradation, maximizing resource utilization, and aligning with principles of sustainable chemistry, hybrid sonolytic processes represent a promising pathway for the next generation of water treatment technologies.

## Data availability statement

Data are contained within the article and supplementary materials.

## Acknowledgments

We would like to express our sincere appreciation to the National Science Center, Poland, for their generous support under the research project OPUS 20 No. 2020/39/B/ST5/ 00076.

## References

Ahmad, M., 2012. Innovative Oxidation Pathways for the Treatment of Traditional and Emerging Contaminants. Washington State University.

Ahmed, M.B., Alam, M.M., Zhou, J.L., Xu, B., Johir, M.A.H., Karmakar, A.K., et al., 2020. Advanced treatment technologies efficacies and mechanism of per-and poly-fluoroalkyl substances removal from water. Process Safety and Environmental Protection 136, 1–14.

Ahrens, L., 2011. Polyfluoroalkyl compounds in the aquatic environment: a review of their occurrence and fate. Journal of Environmental Monitoring: JEM 13 (1), 20–31. Available at https://doi.org/10.1039/c0em00373e.

Aleksandrov, K., Gehrmann, H.J., Hauser, M., et al., 2019. Waste incineration of polytetrafluoroethylene (PTFE) to evaluate potential formation of per- and poly- fluorinated alkyl substances (PFAS) in flue gas. Chemosphere 226, 898–906. Available at https://doi.org/10.1016/j.chemosphere.2019.03.191.

Arias Espana, V.A., Mallavarapu, M., Naidu, R., 2015. Treatment technologies for aqueous perfluorooctanesulfonate (PFOS) and perfluorooctanoate (PFOA): a critical review with an emphasis on field testing. Environmental Technology & Innovation 4, 168–181. Available at https://doi.org/10.1016/j.eti.2015.06.001.

Banks, D., Jun, B., Heo, J., Her, N., Min, C., 2019. Selected advanced water treatment technologies for perfluoroalkyl and polyfluoroalkyl substances: a review. Separation and Purification Technology 2020 (231). Available at https://doi.org/10.1016/j.seppur.2019.115929.

Bartlett, S.A., Davis, K.L., 2018. Evaluating PFAS cross contamination issues. Remediation 28, 53–57.

Bell, C., Gentile, M., Kalve, E., et al., 2019. In: Boca Raton, F.L. (Ed.), Emerging Contaminants Handbook. Taylor abd Francis;, Boca Ranton.

Benotti, M.J., Stanford, B.D., Wert, E.C., Snyder, S.A., 2009. Evaluation of a photocatalytic reactor membrane pilot system for the removal of pharmaceuticals and endocrine disrupting compounds from water. Water Research 43, 1513–1522.

Bhatnagar, A., Cheung, H.M., 1994. Sonochemical destruction of chlorinated C1 and C2 volatile organic compounds in dilute aqueous solution. Environmental Science & Technology 28, 1481–1486.

Boiteux, V., Dauchy, X., Bach, C., Colin, A., Hemard, J., Sagres, V., et al., 2017. Concentrations and patterns of perfluoroalkyl and polyfluoroalkyl substances in a river andthreedrinkingwatertreatment plants nearandfar from amajorproduction source. The Science of the Total Environment 583, 393–400.

Brusseau, M.L., 2019. The influence of molecular structure on the adsorption of PFAS to f luid-fluid interfaces: Using QSPR to predict interfacial adsorption coefficients. Water Research 152, 148–158. Available at https://doi.org/10.1016/j.watres.2018.12.057.

Bruton, T.A., Sedlak, D.L., 2017. Treatment of Aqueous Film-Forming Foam by Heat- Activated Persulfate under Conditions Representative of in Situ Chemical Oxidation. Environmental Science & Technology 51 (23), 13878–13885. Available at https://doi.org/10.1021/acs.est.7b03969.

Campbell, T., Hoffmann, M.R., 2015. Sonochemical degradation of perfluorinated surfactants: Power and multiple frequency effects. Separation and Purification Technology 156 (April), 1019–1027. Available at https://doi.org/10.1016/j.seppur.2015.09.053.

Campbell, T.Y., Vecitis, C.D., Mader, B.T., Hoffmann, M.R., 2009. Perfluorinated surfactant chain-length effects on sonochemical kinetics. The Journal of Physical Chemistry. A 113 (36), 9834–9842. Available at https://doi.org/10.1021/jp903003w L. Rodriguez-Freire, R. Balachandran, R. Sierra-Alvarez, M. Keswani, Effect of sound frequency and initial concentration on the sonochemical degradation of perfluorooctane sulfonate (PFOS), J Hazard Mater. 300 (2015) 662–669, https:// doi.org/10.1016/j.jhazmat.2015.07.077.

C. Carbon. Emerging contaminants: PFOS AND PFOA. Available at https://www.calgoncarbon.com/app/uploads/ccc_pfoa_pfos_pfc_specsheet2018.pdf.

Cheng, J., Vecitis, C.D., Park, H., Mader, B.T., Hoffmann, M.R., 2008a. Sonochemical degrada tion of perfluorooctane sulfonate (PFOS) and perfluorooctanoate (PFOA) in landfill groundwater: environmental matrix effects. Environmental Science & Technology 42, 8057–8063.

Cheng, J., et al., 2008b. Sonochemical degradation of perfluorooctane sulfonate and perfluorooctanoate in groundwater beneath a landfill. Environmental Science & Technology 42 (24), 9338–9343.

Cheng, J., Vecitis, C.D., Park, H., Mader MRH, B.T., Cheng, J.I.E., Vecitis, C.D., et al., 2010a. Sonochemical degradation of perfluorooctane sulfonate (PFOS) and perfluorooctanoate (PFOA) in groundwater: kinetic effects of matrix inorganics. Environmental Science & Technology 44 (1), 445–450. Available at https://doi.org/10.1021/es902651g.

Cheng, J., Vecitis, C.D., Park, H., Mader, B.T., Hoffmann, M.R., 2010b. Sonochemical degradation of perfluorooctane sulfonate (PFOS) and perfluorooctanoate (PFOA) in groundwater: kinetic effects of matrix inorganics. Environmental Science & Technology 44 (1), 445–450.

Chen, W., Zhang, X., Mamadiev, M., Wang, Z., 2017. Sorption of perfluorooctane sulfonate and perfluorooctanoate on polyacrylonitrile fiber-derived activated carbon fibers: In comparison with activated carbon. RSC Advances 7 (2), 927–938. Available at https://doi.org/10.1039/c6ra25230c.

Cheremisinoff, N.P., 2017. Groundwater Remediation: A Practical Guide for Environmen tal Engineers and Scientists, Chapter 5, Perfluorinated Compounds (PFCS). Wiley On line Books.

Conder, J.M., Hoke, R.A., Wolf, W.D., Russell, M.H., Buck, R.C., 2008. Are PFCAs bioaccumulative? A critical review and comparison with regulatory criteria and persistent lipophilic compounds. Environmental Science & Technology 42 (4), 995–1003. Available at https://doi.org/10.1021/es070895g.

Cordner, A., Vanessa, Y., Schaider, L.A., Rudel, R.A., Richter, L., Brown, P., 2020. Correction: guideline levels for PFOA and PFOS in drinking water: the role of scientific uncer tainty, risk assessment decisions, and social factors. Journal of Exposure Science & Environmental Epidemiology 30, 585–586.

Crispim, A.C., de Paiva, S.D.S.M., de Araújo, D.M., Souza, F.L., Dos Santos, E.V., 2022. Ultrasound and UV technologies for wastewater treatment using boron-doped diamond anodes. Current Opinion in Electrochemistry 33, 100935.

Crone, B.C., Speth, T.F., Wahman, D.G., Smith, S.J., Abulikemu, G., Kleiner, E.J., et al., 2019. Occurrence of per-and polyfluoroalkyl substances (PFAS) in source water andtheir treatment in drinking water. Critical Reviews in Environmental Science and Technology 49, 2359–2396.

Dehane, A., Merouani, S., Hamdaoui, O., Alghyamah, A., 2021. A comprehensive numerical analysis of heat and mass transfer phenomenons during cavitation sono-process. Ultrasonics Sonochemistry 73. Available at https://doi.org/10.1016/j.ultsonch.2021.105498.

Dehghani, M.H., Karri, R.R., Koduru, J.R., Manickam, S., Tyagi, I., Mubarak, N.M., 2023. Recent trends in the applications of sonochemical reactors as an advanced oxidation process for the remediation of microbial hazards associated with water and wastewater: A critical review. Ultrasonics Sonochemistry, 106302.

Dickenson, E., Higgins, C., 2016. Treatment mitigation strategies for poly- and perfluorinated chemicals. Available at http://www.waterrf.org/PublicReportLibrary/4322.pdf

Djellabi, R., Aboagye, D., Galloni, M.G., Andhalkar, V.V., Nouacer, S., Nabgan, W., et al., 2022. Combined conversion of lignocellulosic biomass into high-value products with ultrasonic cavitation and photocatalytic produced reactive oxygen species-a review. Bioresource Technology, 128333.

Dudley, L.-A., 2012. Removal of perfluorinated compounds by powdered activated carbon, superfine powdered activated carbon, and anion exchange resins. 185. Available at. http://www.lib.ncsu.edu/resolver/1840.16/7654.

Eschauzier, C., Beerendonk, E., Scholte-Veenendaal, P., Voogt, P.D., 2012. Impact of treatment processes on the removal of perfluoroalkyl acids from the drinking water production chain. Environmental Science & Technology 46 (3), 1708–1715. Available at https://doi.org/10.1021/es201662b.

Fatyukhin, D.S., Nigmetzyanov, R.I., Prikhodko, V.M., Sukhov, A.V., Sundukov, S.K., 2022. A comparison of the effects of ultrasonic cavitation on the surfaces of 45 and 40Kh steels. Metals 12, 138.

Fernandez, N.A., Rodriguez-Freire, L., Keswani, M., Sierra-Alvarez, R., 2016. Effect of chemical structure on the sonochemical degradation of perfluoroalkyl and polyfluoroalkyl substances (PFASs). Environmental Science: Water Research & Technology 2 (6), 975–983. Available at https://doi.org/10.1039/c6ew00150e.

Gholami, A., Pourfayaz, F., Maleki, A., 2021. Techno-economic assessment of biodiesel production from canola oil through ultrasonic cavitation. Energy Reports 7, 266–277.

Hansen, M.C., Børresen, M.H., Schlabach, M., Cornelissen, G., 2010. Sorption of perfluorinated compounds from contaminated water to activated carbon. Journal of Soils Sediments 10 (2), 179–185. Available at https://doi.org/10.1007/s11368-009-0172-z.

Hao, F., Guo, W., Wang, A., Leng, Y., Li, H., 2014. Intensification of sonochemical degrada tion of ammonium perfluorooctanoate by persulfate oxidant. Ultrasonics Sonochemistry 21, 554–558.

Harper, D.B., O'Hagan, D., Murphy, C.D., 2003. Fluorinated natural products: occurrence and biosynthesis. In: Gribble, G. (Ed.), Natural Production of Organohalogen Compounds. Springer Berlin Heidelberg, Berlin, Heidelberg, Heidelberg, pp. 141–169. Available at 10.1007/b10454.

Hori, H., Hayakawa, E., Einaga, H., et al., 2004. Decomposition of environmentally persistent perfluorooctanoic acid in water by photochemical approaches. Environmental Science & Technology 38 (22), 6118–6124. Available at https://doi.org/10.1021/es049719n.

Hori, H., Yamamoto, A., Hayakawa, E., et al., 2005. Efficient decomposition of environmentally persistent perfluorocarboxylic acids by use of persulfate as a photochemical oxidant. Environmental Science & Technology 39 (7), 2383–2388. Available at https://doi.org/10.1021/es0484754.

Hori, H., Nagaoka, Y., Yamamoto, A., et al., 2006. Efficient decomposition of environmentally persistent perfluorooctanesulfonate and related fluorochemicals using zerovalent iron in subcritical water. Environmental Science & Technology 40 (3), 1049–1054. Available at https://doi.org/10.1021/es0517419.

Horst, J., McDonough, J., Ross, I., et al., 2018. Water treatment technologies for PFAS: the next generation. Groundwater Monitoring & Remediation 38 (2), 13–23. Available at https://doi.org/10.1111/gwmr.12281.

Hua, I., 1995. Sonolytic hydrolysis of p-Nitrophenol in a parallel-plate near-field acoustical processor. Environmental Science & Technology 29, 2790–2796.

Hua, I., Hoechemer, R.H., Hoffmann, M.R., 1995. Sonolytic hydrolysis of p-nitrophenyl acetate: the role of supercritical water. The Journal of Physical Chemistry 99, 2335–2342.

Hu, Y.B., Lo, S.L., Li, Y.F., Lee, Y.C., Chen, M.J., Lin, J.C., 2018. Autocatalytic degradation of perfluorooctanoic acid in a permanganate-ultrasonic system. Water Research 140, 148–157. Available at https://doi.org/10.1016/j.watres.2018.04.044.

Hu, Y., Wei, J., Shen, Y., Chen, S., Chen, X., 2023. Barrier-breaking effects of ultrasonic cavitation for drug delivery and biomarker release. Ultrasonics Sonochemistry 94, 106346.

Iida, Y., Yasui, K., Tuziuti, T., Sivakumar, M., Endo, Y., 2004. Ultrasonic cavitation in microspace. Chemical Communications 2280–2281.

Jin, L., Zhang, P., Shao, T., Zhao, S., 2014. Ferric ion mediated photodecomposition of aqueous perfluorooctane sulfonate (PFOS) under UV irradiation and its mechanism. Journal of Hazardous Materials 271, 9–15. Available at https://doi.org/10.1016/j.jhazmat.2014.01.061.

Kalinina, I., Naumenko, N., Fatkullin, R., 2019. Perspectives of using of ultrasonic cavitation in water treatment technology for the food productions. IOP Conf. Ser.: Earth Environ. Sci. IOP Publishing, pp. 032077.

Kalra, S.S., et al., 2021. Sonolytic destruction of per- and polyfluoroalkyl substances in groundwater, aqueous film-forming foams, and investigation-derived waste. Chemical Engineering Journal 406, 127142.

Kerboua, K., Hamdaoui, O., 2020. Sonochemistry in green processes: modeling, experiments, and technology. Sustainable Green Chemical Processes and Their Allied Applications. Springer, pp. 409–460.

Keswani, M., 2013-2015. Complete Mineralization of Fluorochemicals in Aqueous Fire f ightin Foams Using a Novel Dual-frequency Based Sonochemical Process. Air Force Civil Engineering Center (AFCEC).

Kewalramani, J.A., et al., 2023. Kinetics effects of the power density and initial concentration on the sonochemical degradation of PFOS and PFOA in concentrated waste. Journal of Water Process Engineering 53, 103752.

Key, B.D., Howell, R.D., Criddle, C.S., 1998. Defluorination of organofluorine sulfur compounds by *Pseudomonas* sp. strain D2. Environmental Science & Technology 32 (15), 2283–2287. Available at https://doi.org/10.1021/es9800129.

Khan, M.Y., So, S., da Silva, G., 2020. Decomposition kinetics of perfluorinated sulfonic acids. Chemosphere 238. Available at https://doi.org/10.1016/j.chemosphere.2019.124615.

Khoshyan, A., et al., 2024. Operating conditions are optimised: frequency (35 kHz vs. 130 kHz), power input, PFOA initial concentration, and solution volume. Journal of Water Process Engineering 63 (2024), 105511.

Kim, M., Li, L.Y., Grace, J.R., Yue, C., 2009. Selecting reliable physicochemical properties of perfluoroalkyl and polyfluoroalkyl substances (PFASs) based on molecular descriptors. Environmental Pollution (Barking, Essex: 1987) 2015 (196), 462–472. Available at https://doi.org/10.1016/j.envpol.2014.11.008.

Kotronarou, A., Mills, G., Hoffmann, M.R., 1991. Ultrasonic irradiation of p-nitrophenol in aqueous solution. The Journal of Physical Chemistry 95 (9), 3630–3638. Available at https://doi.org/10.1021/j100162a037.

Kotronaru, A., Mills, G., Hoffmann, M.R., 1991. Ultrasonic irradiation of Para-nitrophenol in aqueous solution. The Journal of Physical Chemistry 95, 3630–3638.

Kruus, P., Burk, R.C., Entezari, M.H., Otson, R., 1997. Sonication of aqueous solutions of chlorobenzene. Ultrasonics Sonochemistry 4, 229–233.

Kucharzyk, K.H., Darlington, R., Benotti, M., Deeb, R., Hawley, E., 2017. Novel treatment technologies for PFAS compounds: a critical review. Journal of Environmental Management 15, 757–764.

Kumar, S., Prajapat, A., Panja, S.K., Shukla, M., 2024. Ultrasound irradiation: fundamental theory, electromagnetic spectrum, important properties, and physical principles. Green Chemical Synthesis with Microwaves and Ultrasound 1–19.

Lee, Y.C., Lo, S.L., Kuo, J., Lin, Y.L., 2012. Persulfate oxidation of perfluorooctanoic acid under the temperatures of 20–40 ∘ C. Chemical Engineering Journal 198–199, 27–32. Available at https://doi.org/10.1016/j.cej.2012.05.073.

Lee, Y.C., Lo, S.L., Kuo, J., Huang, C.P., 2013. Promoted degradation of perfluorooctanic acid by persulfate when adding activated carbon. Journal of Hazardous Materials 261, 463–469. Available at https://doi.org/10.1016/j.jhazmat.2013.07.054.

Lewis, A.J., Joyce, T., Hadaya, M., et al., 2020. Rapid degradation of PFAS in aqueous solutions by reverse vortex flow gliding arc plasma. Environmental Science: Water Research & Technology 6 (4), 1044–1057. Available at https://doi.org/10.1039/c9ew01050e.

Liao, Z., Farrell, J., 2009. Electrochemical oxidation of perfluorobutane sulfonate using boron-doped diamond film electrodes. Journal of Applied Electrochemistry 39 (10), 1993–1999. Available at https://doi.org/10.1007/s10800-009-9909-z.

Lin, A.Y., Panchangam, S.C., Chang, C.Y., Hong, P.K., Hsueh, H.F., 2012. Removal of perfluorooctanoic acid and perfluorooctane sulfonate via ozonation under alkaline condition. Journal of Hazardous Materials 243, 272–277.

Lin, J.C., et al., 2015. Enhanced sonochemical degradation of perfluorooctanoic acid by sulfate ions. Environmental Science & Technology 49 (5), 2870–2877.

Lin, J.C., Hu, C.Y., Lo, S.L., 2016a. Effect of surfactants on the degradation of perfluorooctanoic acid (PFOA) by ultrasonic (US) treatment. Ultrasonics Sonochemistry 28, 130–135. Available at https://doi.org/10.1016/j.ultsonch.2015.07.007.

Lin, J.C., et al., 2016b. Effect of surfactants on the degradation of perfluorooctanoic acid by ultrasonic treatment. Environmental Technology 37 (6), 742–749.

Love, A.H., Hanna, M.L., Reynolds, J.G., 2005. Engineering surface functional groups on silica aerogel for enhanced cleanup of organics from produced water. Separation Science and Technology 40 (1–3), 311–320. Available at https://doi.org/10.1081/SS-200042242.

Low, Z.L., Low, D.Y.S., Tang, S.Y., Manickam, S., Tan, K.W., Ban, Z.H., 2022. Ultrasonic cavitation: An effective cleaner and greener intensification technology in the extraction and surface modification of nanocellulose. Ultrasonics Sonochemistry, 106176.

Luo, Q., Lu, J., Zhang, H., et al., 2015. Laccase-catalyzed degradation of perfluorooctanoic acid. Environmental Science & Technology Letters 2 (7), 198–203. Available at https://doi.org/10.1021/acs.estlett.5b00119.

L´eveque, J.-M., Cravotto, G., Delattre, F., Cintas, P., 2018. Organic Sonochemistry: Challenges and Perspectives for the 21st Century. Springer.

Mason, T.J., 2015. Some neglected or rejected paths in sonochemistry–a very personal view. Ultrasonics Sonochemistry 25, 89–93.

Mason, T.J., Lorimer, J.P., 2002. Applied Sonochemistry 2–3. Available at https://doi.org/10.1002/352760054x.

McDonough, J., Ross, Miles, J., et al., 2018. A review of emerging technologies for remediation of PFASs. Remediation 28 (2), 101–126. Available at https://doi.org/10.1002/rem.21553.

Meng, J., Lu, Y., Wang, T., et al., 2017. Life cycle analysis of perfluorooctanoic acid (PFOA) and its salts in China. Environmental Science and Pollution Research 24 (12), 11254–11264. Available at https://doi.org/10.1007/s11356-017-8678-1.

Merino, N., Qu, Y., Deeb, R.A., Hawley, E.L., Hoffmann, M.R., Mahendra, S., 2016. Degradation and removal methods for perfluoroalkyl and polyfluoroalkyl substances in water. Environmental Engineering Science 33 (9), 615–649. Available at https://doi.org/10.1089/ees.2016.0233.

Moriwaki, H., Takagi, Y., Tanaka, M., Tsuruho, K., Okitsu, K., Maeda, Y., 2005. Sonochemical decomposition of perfluorooctane sulfonate and perfluorooctanoic acid. Environmental Science & Technology 39 (9), 3388–3392. Available at https://doi.org/10.1021/es040342v.

Moriwaki, M., Takagi, Y., Tanaka, M., Tsuruho, K., Okitsu, K., Maeda, Y., 2006. Sonochemical decomposition of perfluorooctano sulfonate and perfluorooctanoic acid. Environmental Science & Technology 39, 3388–3392.

Movahed, S.M.A., Calgaro, L., Marcomini, A., 2023. Trends and characteristics of employing cavitation technology for water and wastewater treatment with a focus on hydrodynamic and ultrasonic cavitation over the past two decades: a Scientometric analysis. The Science of the Total Environment 858, 159802.

Nam, S.-N., Han, S.-K., Kang, J.-W., Choi, H., 2003. Kinetics and mechanisms of the sonolytic destruction of non-volatile organic compounds: investigation of the sonochemical reaction zone using several OH monitoring techniques. Ultrasonics Sonochemistry 10, 139–147.

Nanzai, B., Mochizuki, A., Wakikawa, Y., Masuda, Y., Oshio, T., Yagishita, K., 2023. Sonoluminescence intensity and ultrasonic cavitation temperature in organic solvents: effects of generated radicals. Ultrasonics Sonochemistry 95, 106357.

Nienhauser, A.B., Ersan, M.S., Lin, Z., Perreault, F., Westerhoff, P., Garcia-Segura, S., 2022. Boron-doped diamond electrodes degrade short- and long-chain per- and polyfluorinated alkyl substances in real industrial wastewaters. Journal of Environmental Chemical Engineering 10 (2). Available at https://doi.org/10.1016/j.jece.2022.107192.

Norris, G., Al-Dhahir, Z., Birnstingl, J., Plant, S.J., Cui, S., Mayell, P., 1999. A case study of the management and remediation of soil contaminated with polychlorinated biphenyls. Engineering Geology 53 (2), 177–185. Available at https://doi.org/10.1016/S0013-7952(99)00031-9.

Nzeribe, B.N., Crimi, M., Mededovic Thagard, S., Holsen, T.M., 2019. Physico-chemical pro cesses for the treatment of per-and polyfluoroalkyl substances (PFAS): a review. Critical Reviews in Environmental Science and Technology 49, 866–915.

Ochoa-Herrera, V., Sierra-Alvarez, R., Somogyi, A., Jacobsen, N.E., Wysocki, V.H., Field, J.A., 2008. Reductive defluorination of perfluorooctane sulfonate. Environmental Science & Technology 42, 3260–3264.

Panchangam, S.C., Lin, A.Y., Tsai, J., Lin, C., 2009. Sonication-assisted photocatalytic decom position of perfluorooctanoic acid. Chemosphere 75, 654–660.

Pang, Y.L., Abdullah, A.Z., Bhatia, S., 2011. Review on sonochemical methods in the pres ence of catalysts and chemical additives for treatment of organic pollutants in waste water. Desalination 277, 1–14.

Park, H., Vecitis, C.D., Cheng, J., Dalleska, N.F., Mader, B.T., Hoffmann, M.R., 2011. Reductive degradation of perfluoroalkyl compounds with aquated electrons generated from iodide photolysis at 254 nm. Photochemical & Photobiological Sciences: Official Journal of the European Photochemistry Association and the European Society for Photobiology 10 (12), 1945–1953. Available at https://doi.org/10.1039/c1pp05270e.

Parsons, J.R., S´aez, M., Dolfing, J., De Voogt, P., 2002. Biodegradation of perfluorinated compounds. Food and Chemical Toxicology: an International Journal Published for the British Industrial Biological Research Association 23 (10), 950. Available at https://doi.org/10.1016/0278-6915(85)90116-4.

Paul, A.G., Jones, K.C., Sweetman, A.J., 2009. A first global production, emission, and envi ronmental inventory for perfluorooctane sulfonate. Environmental Science & Technology 43, 386–392.

Pflieger, R., Nikitenko, S.I., Cair´os, C., Mettin, R., 2019. Characterization of Cavitation Bubbles and Sonoluminescence. Springer Springer. Available at https://doi.org/10.1007/978-3-030-11717-7.

Pirsaheb, M., Moradi, N., 2020. Sonochemical degradation of pesticides in aqueous solution: investigation on the influence of operating parameters and degradation pathway-a systematic review. RSC Advances 10 (13), 7396–7423. Available at https://doi.org/10.1039/c9ra11025a.

Prevedouros, K., Cousins, I.T., Buck, R.C., Korzeniowski, S.H., 2006. Sources, fate and trans port of perfluorocarboxylates. Environmental Science & Technology 40, 32–44.

Qingguo, J.H., 2013. Remediation of perfluoroalkyl contaminated aquifers using an in situ two-layer barrier: laboratory batch and column study, SERDP project ER-2127.

Rahman, M.F., Peldszus, S., Anderson, W.B., 2014. Behaviour and fate of perfluoroalkyl and polyfluoroalkyl substances (PFASs) in drinking water treatment: a review. Water Research 50, 318–340. Available at https://doi.org/10.1016/j.watres.2013.10.045.

Rayne, S., Forest, K., 2009. Perfluoroalkyl sulfonic and carboxylic acids: a critical review of physicochemical properties, levels and patterns in waters and wastewaters, and treatment methods. Journal of Environmental Science and Health 44, 1145–1199.

Rodriguez-Freire, L., Balachandran, R., Sierra-Alvarez, R., Keswani, M., 2015. Effect of sound frequency and initial concentration on the sonochemical degradation of perfluorooctane sulfonate (PFOS). Journal of Hazardous Materials 300 6Cheng, J.

Ross, I., Hurst, J., Burdick, J., Kalve, E., Houtz, E., Pancras, T., 2016. Remediation of poly- and perfluoroalkyl substances: new remediation technologies for emerging challenges.

Ross, I., McDonough, J., Miles, J., Storch, P., Kochunarayanan, P.T., Kalve, E., et al., 2018. A review of emerging technologies for remediation of PFASs. Remediation 28, 101–126.

Rusinek, C.A., 2019. PFAS remediation at MSU-Fraunhofer: electrochemical destruction using boron-doped diamond electrodes. Urbana, Michigan.

Schröder, H.F., Meesters, R.J., 2005. Stability of fluorinated surfactants in advanced oxida tion processes-a follow up of degradation products using flow injection-mass spec trometry, liquid chromatography-mass spectrometry and liquid chromatography multiple stage mass spectrometry. Journal of Chromatography. A 1082, 110–119.

Sekiguchi, K., Kudo, T., Sankoda, K., 2017. Combined sonochemical and short-wavelength UV degradation of hydrophobic perfluorinated compounds. Ultrasonics Sonochemistry 39 (January), 87–92. Available at https://doi.org/10.1016/j.ultsonch.2017.04.002.

Shende, T., Andaluri, G., Suri, R.P.S., 2019. Kinetic model for sonolytic degradation of non-volatile surfactants: perfluoroalkyl substances. Ultrasonics Sonochemistry 51, 359–368. Available at https://doi.org/10.1016/j.ultsonch.2018.08.028.

Shen, M., Ma, H., 2022. Metal-organic frameworks (MOFs) and their derivative as electrode materials for lithium-ion batteries. Coordination Chemistry Reviews 470, 214715.

Sidnell, T., et al., 2022. Sonolysis of per- and polyfluoroalkyl substances (PFAS): A review of reaction mechanisms, kinetics, and applications. Science of the Total Environment 806, 150597.

Singh, R.K., Fernando, S., Fakouri Baygi, S., Multari, N., Mededovic Thagard, S., Holsen, T.M., 2019a. Breakdown products from perfluorinated alkyl substances (PFAS) degradation in a plasma-based water treatment process. Environmental Science & Technology 53. Available at https://doi.org/10.1021/acs.est.8b07031 acs.est.8b07031.

Singh, R.K., Multari, N., Nau-Hix, C., et al., 2019b. Rapid removal of poly- and perfluorinated compounds from investigation-derived waste (IDW) in a pilot- scale plasma reactor. Environmental Science & Technology. Available at https://doi.org/10.1021/acs.est.9b02964.

Singh, R.K., Brown, E., Mededovic Thagard, S., Holsen, T.M., 2020a. Treatment of PFAS- containing landfill leachate using an enhanced contact plasma reactor. Journal of Hazardous Materials 2021 (408). Available at https://doi.org/10.1016/j.jhazmat.2020.124452.

Singh, R.K., Multari, N., Nau-Hix, C., et al., 2020b. Removal of poly- and per-fluorinated compounds from ion exchange regenerant still bottom samples in a plasma reactor. Environmental Science & Technology 54 (21), 13973–13980. Available at https://doi.org/10.1021/acs.est.0c02158.

Song, Z., Tang, H., Wang, N., Zhu, L., 2013. Reductive defluorination of perfluorooctanoic acid by hydrated electrons in a sulfite-mediated UV photochemical system. Journal of Hazardous Materials 262, 332–338. Available at https://doi.org/10.1016/j.jhazmat.2013.08.059.

Sostaric, J.Z., Riesz, P., 2001. Sonochemistry of surfactants in aqueous solutions: an EPR spin-trapping study. Journal of the American Chemical Society 123 (44), 11010–11019. Available at https://doi.org/10.1021/ja010857b.

Stahl, T., Mattern, D., Brunn, H., 2011. Toxicology of perfluorinated compounds. Environmental Sciences Europe 23, 1–52.

Stratton, G.R., Dai, F., Bellona, C.L., Holsen, T.M., Dickenson, E.R.V., Mededovic, T.S., 2017. Plasma-based water treatment: efficient transformation of perfluoroalkyl substances in prepared solutions and contaminated groundwater. Environmental Science & Technology 51 (3), 1643–1648. Available at https://doi.org/10.1021/acs.est.6b04215.

Sukeesan, S., Boontanon, N., Boontanon, S.K., 2021. Improved electrical driving current of electrochemical treatment of per- and polyfluoroalkyl substances (PFAS) in water using boron-doped diamond anode. Environmental Technology & Innovation 23. Available at https://doi.org/10.1016/j.eti.2021.101655.

Suslick, K., Doktycz, S., Flint, E., 1990. On the origin of sonoluminescence and sonochemistry. Ultrasonics 28, 280–290.

Suslick, K.S., Flannigan, D.J., 2008. Inside a collapsing bubble: sonoluminescence and the conditions during cavitation. Annual Review of Physical Chemistry 59, 659–683.

Taylor, P.H., Yamada, T., Striebich, R.C., Graham, J.L., Giraud, R.J., 2014. Investigation of waste incineration of fluorotelomer-based polymers as a potential source of PFOA in the environment. Chemosphere 110, 17–22. Available at https://doi.org/10.1016/j.chemosphere.2014.02.037.

Thangavadivel, K., Okitsu, K., Owens, G., Lesniewski, P.J., Nishimura, R., 2013. Influence of sonochemical reactor diameter and liquid height on methyl orange degradation under 200 kHz indirect sonication. Journal of Environmental Chemical Engineering 1 (3), 275–280. Available at https://doi.org/10.1016/j.jece.2013.05.005.

Thi, L.A.P., et al., 2014. Enhancing decomposition rate of perfluorooctanoic acid by carbonate radical oxidants. Environmental Science & Technology 48 (2), 1232–1239.

Trudel, D., Horowitz, L., Wormuth, M., Scheringer, M., Cousins, I.T., Hungerbühler, K., 2008. Estimating consumer exposure to PFOS and PFOA. Risk Analysis: An Official Publication of the Society for Risk Analysis 28, 251–269.

Tseng, N., 2012. Feasibility of Biodegradation of Polyfluoroalkyl and Perfluoroalkyl Sub stances. Civil Engineering. UCLA (Electronic Thesis and Dissertation).

Turner, L.P., Kueper, B.H., Weber, K.P., 2019. Remediation of perfluorooctane sulfonate (PFOS) and perfluorooctanoic acid (PFOA) spiked sands and aqueous film forming foam (AFFF) impacted soils by ball milling. 21, 10908.

US EPA, 2016. Drinking water health advisories for PFOA and PFOS. Available at https://www.epa.gov/ground-water-and-drinking-water/drinking-water-health-advisoriespfoa-and-pfos.

US EPA, 2017. Emerging contaminants-perfluorooctane sulfonate (PFOS) and perfluorooctanoic acid(PFOA). Available at https://www.epa.gov/sites/production/files/2017-12/documents/ffrrofactsheet_contaminants_pfos_pfoa_11-20-17_508_0.pdf.

US EPA, 2018. Edition of the drinking water standards and health advisories, EPA 822-F 18-00. Available at https://www.epa.gov/sites/production/files/2018-03/documents/dwtable2018.pdf.

Uwayezu, J.N., Carabante, I., Lejon, T., et al., 2021. Electrochemical degradation of per- and poly-fluoroalkyl substances using boron-doped diamond electrodes. Journal of Environmental Management 290 (February). Available at https://doi.org/10.1016/j.jenvman.2021.112573.

Vecitis, C.D., Park, H., Cheng, J., Mader, B.T., Hoffmann, M.R., 2008a. Kinetics and mechanism of the sonolytic conversion of the aqueous perfluorinated surfactants, perfiuorooctanoate (PFOA), and perfluorooctane sulfonate (PFOS) into inorganic products. The Journal of Physical Chemistry. A 112 (18), 4261–4270. Available at https://doi.org/10.1021/jp801081y.

Vecitis, C.D., Park, H., Cheng, J., Mader, B.T., Hoffmann, M.R., 2008b. Enhancement of perfluorooctanoate and perfluorooctanesulfonate activity at acoustic cavitation bubble interfaces. The Journal of Physical Chemistry C 112, 16850–16857. Available at https://doi.org/10.1021/jp804050p.

Vecitis, C.D., Park, H., Mader, B.T., Hoffmann, M.R., 2010a. Sonochemical degrada tion of perfluorooctane sulfonate (PFOS) and perfluorooctanoate (PFOA) in ground water: kinetic effects of matrix inorganics. Environmental Science & Technology 44, 445–450 62–669.

Vecitis, C.D., Wang, Y.J., Cheng, J., Park, H., Mader, B.T., Hoffmann, M.R., 2010b. Sonochemical degradation of perfluorooctane sulfonate in aqueous film-forming foams. Environmental Science & Technology 44 (1), 432–438. Available at https://doi.org/10.1021/Es902444r.

Vecitis, C.D., et al., 2010c. Sonochemical degradation of perfluorooctane sulfonate and perfluorooctanoate in aqueous film-forming foams. Environmental Science & Technology 44 (4), 1288–1293.

Verma, S., Varma, R.S., Nadagouda, M.N., 2021. Remediation and mineralization processes for per-and polyfluoroalkyl substances (PFAS) in water: a review. Science of the Total Environment 794, 148987.

Vo, H.N.P., Ngo, H.H., Guo, W., Nguyen, T.M.H., Li, J., Liang, H., et al., 2020. Poly-and perfluoroalkyl substances in water and wastewater: a compre hensive review from sources to remediation. Journal of Water Process Engineering 36, 101393.

Wang, G., Cheng, H., 2023. Application of photocatalysis and sonocatalysis for treatment of organic dye wastewater and the synergistic effect of ultrasound and light. Molecules (Basel, Switzerland) 28, 3706.

Wang, Y., Zhang, P., Pan, G., Chen, H., 2008. Ferric ion mediated photochemical decomposition of perfluorooctanoic acid (PFOA) by 254 nm UV light. Journal of Hazardous Materials 160 (1), 181–186. Available at https://doi.org/10.1016/j.jhazmat.2008.02.105.

Wang, F., Lu, X., Li, X.-Y., Shih, K., 2015. Effectiveness and mechanisms of defluorination of perfluorinated alkyl substances by calcium compounds during waste thermal treat ment. Environmental Science & Technology 499, 5672–5680.

Weinfurtner, K., Kördel, W., Bücking, M., 2008. Studies on the transfer of PFT from con taminated soils to plants. Soil Protection 03, 88–92 (Edition).

Wen, Y., Mirji, N., Irudayaraj, J., 2020. Epigenetic toxicity of PFOA and GenXin HepG2 cells and their role in lipid metabolism. Toxicology In Vitro: An International Journal Published in Association with BIBRA 65, 104797.

Winchell, L.J., Ross, J.J., Wells, M.J.M., Fonoll, X., Norton, J.W., Bell, K.Y., 2021. Per- and polyfluoroalkyl substances thermal destruction at water resource recovery facilities: a state of the science review. Water Environment Research: A Research Publication of the Water Environment Federation 93 (6), 826–843. Available at https://doi.org/10.1002/wer.1483.

Woodard, S., Berry, J., Newman, B., 2017. Ion exchange resin for PFAS removal and pilot test comparison to GAC. Remediation 27 (3), 19–27. Available at https://doi.org/10.1002/rem.21515.

Wood, R.J., Lee, J., Bussemaker, M.J., 2017. A parametric review of sonochemistry: control and augmentation of sonochemical activity in aqueous solutions. Ultrasonics Sonochemistry 38, 351–370. Available at https://doi.org/10.1016/j.ultsonch.2017.03.030.

Wood, R.J., Sidnell, T., Ross, I., et al., 2019. Ultrasonic degradation of perfluorooctane sulfonic acid (PFOS) correlated with sonochemical and sonoluminescence characterisation. Ultrasonics Sonochemistry 2020 (68). Available at https://doi.org/10.1016/j.ultsonch.2020.105196.

Dewulf, J., Van Langenhove, H., De Visscher, A., Sabbe, S., 2001. Ultrasonic degradation of trichloroethylene and chlorobenzene at micromolar concentrations: kinetics and modelling. Ultrasonics Sonochemistry 8, 143–146.

Xu, B., Liu, S., Zhou, J.L., Zheng, C., Weifeng, J., Chen, B., et al., 2021. PFAS and their substitutes in groundwater: occurrence, transformation and remediation. Journal of Hazardous Materials 412, 125159.

Yamada, T., Taylor, P.H., Buck, R.C., Kaiser, M.A., Giraud, R.J., 2005. Thermal degradation of f luorotelomer treated articles and related materials. Chemosphere 61, 974–984.

Yamamoto, T., Noma, Y., Sakai, S.I., Shibata, Y., 2007. Photodegradation of perfluorooctane sulfonate by UV irradiation in water and alkaline 2-propanol. Environmental Science & Technology 41 (16), 5660–5665. Available at https://doi.org/10.1021/es0706504.

Yang, S., Sun, J., Hu, Y., Cheng, J., Liang, X., 2013. Effect of vacuum ultraviolet on ultrasonic defluorination of aqueous perfluorooctanesulfonate. Chemical Engineering Journal 234, 106–114.

Yasui, K., Tuziuti, T., Lee, J., Kozuka, T., Towata, A., Iida, Y., 2008. The range of ambient radius for an active bubble in sonoluminescence and sonochemical reactions. The Journal of Chemical Physics 128 (18). Available at https://doi.org/10.1063/1.2919119.

Yasuoka, K., Sasaki, K., Hayashi, R., 2011. An energy-efficient process for decomposing perfluorooctanoic and perfluorooctane sulfonic acids using dc plasmas generated within gas bubbles. Plasma Sources Science and Technology 20 (3). Available at https://doi.org/10.1088/0963-0252/20/3/034009.

Ye, Y.-F., Zhu, Y., Lu, N., Wang, X., Su, Z., 2021. Treatment of rhodamine B with cavitation technology: comparison of hydrodynamic cavitation with ultrasonic cavitation. RSC Advances 11, 5096–5106.

Yusof, N.S.M., Anandan, S., Sivashanmugam, P., Flores, E.M., Ashokkumar, M., 2022. A correlation between cavitation bubble temperature, sonoluminescence and interfacial chemistry–a minireview. Ultrasonics Sonochemistry 85, 105988.

Yu, G., Huang, J., Lu, M., Cagnetta, G., Zhang, K., 2017. Mechanochemical mineralization of "very persistent" fluorocarbon surfactants – 6:2 fluorotelomer sulfonate (6: 2FTS) as an example. Scientific Reports 7 (1), 1–10. Available at https://doi.org/10.1038/s41598-017-17515-7.

Zeidabadi, F.A., Esfahani, E.B., Mohseni, M., 2023. Effects of water matrix on per-and poly-fluoroalkyl substances (PFAS) treatment: Physical-separation and degradation processes–a review. Journal of Hazardous Materials Advances 10, 100322.

Zhang, K., Huang, J., Yu, G., et al., 2013. Destruction of perfluorooctane sulfonate (PFOS) and perfluorooctanoic acid (PFOA) by ball milling. Environmental Science & Technology 47 (12), 6471–6477.

Zhang, K., Cao, Z., Huang, J., Deng, S., Wang, B., Yu, G., 2016. Mechanochemical destruction of Chinese PFOS alternative F-53B. Chemical Engineering Journal 286, 387–393. Available at https://doi.org/10.1016/j.cej.2015.10.103.

Zhao, Z., Cheng, X., Hua, X., Jiang, B., Tian, C., Tang, J., et al., 2020. Emergingandlegacyper-andpolyfluoroalkylsubstancesinwater, sediment, and air of the Bohai Sea and its surrounding rivers. Environmental Pollution (Barking, Essex: 1987) 263, 114391.

Zheng, Z., Yu, H., Geng, W.C., Hu, X.Y., Wang, Y.Y., Li, Z., et al., 2019. Guanidinocalix [5] arene for sensitive fluorescence detection and magnetic removal of perfluorinated pollutants. Nature Communications 10, 1–9.

Zupanc, M., Pandur, Ž., Perdih, T.S., Stopar, D., Petkovšek, M., Dular, M., 2019. Effects of cavitation on different microorganisms: the current understanding of the mecha nisms taking place behind the phenomenon. a review and proposals for further re search. Ultrasonics Sonochemistry 57, 147–165.

CHAPTER

# 8

# Electrochemical oxidation of per- and polyfluoroalkyl substances: treatment mechanisms, electrode materials, and future directions

*Sajjad Hussain*[1,2], *Saima Gul*[3], *Hammad Khan*[1], *Arslan Maqbool*[1], *Nadeem Khan*[4], *Farrukh Faheem*[4], *Josiel Viana E.Viana*[2], *and Sabir Khan*[5]

[1]Faculty of Materials and Chemical Engineering, GIK Institute of Engineering Sciences and Technology, Swabi, Khyber Pakhtunkhwa, Pakistan [2]Programa de Pós-Graduação em Química (PPGQ), Universidade Federal Do Amazonas (UFAM), Manaus, Amazonas (AM), Brazil [3]Department of Chemistry, Islamia College Peshawar, Peshawar, Khyber Pakhtunkhwa, Pakistan [4]São Carlos Institute of Chemistry, University of São Paulo, São Carlos, SP, Brazil [5]Technological Development Center-CDTec, Postgraduate Program in Materials Science and Engineering, PPGCEM, Federal University of Pelotas- UFPel, Pelotas, RS, Brazil

## 8.1 Introduction

Our planet is undoubtedly vulnerable to potential threats by virtue of increasing energy demand and environmental pollution induced by hasty industrialization and urbanization. Our ecosystem has been brutally contaminated with a variety of hazardous toxins, which are mostly anthropogenic in origin. These pollutants not only spoil the quality of air and soil but also pose potential threats to aquatic flora and fauna. The index of water contamination is elevating simultaneously with the proportionate increase in scientific and technological developments, particularly in the fields of chemistry, chemical/ applied engineering, and technology (Hussain et al., 2020a). One such invention that creates incredibly valuable goods for the automotive, electronics, aero-space, military, and construction industries is per- and polyfluoroalkyl substances (PFAS). However, these chemicals are remarkably inert and manifest potential resistance to heat, water, oil, and even corrosion. These contaminants are

*Per- and Polyfluoroalkyl Substances (PFAS) as Environmental Contaminants*
DOI: https://doi.org/10.1016/B978-0-443-43856-1.00010-1

nonbiodegradable and can persist in the environment. There are over a thousand different types of synthetic or man-made organic substances in this class (Sultan et al., 2024). By virtue of their incredible stability, these PFAS are extensively used in commercial goods, military tools, and industrial setups. These molecules are polar, and their head groups make them hydrophilic; however, their perfluorinated tail renders them hydrophobic. The widespread use of PFAS in paints, insecticides, soap, hardwood floors, packaging, spraying foams, as well as cleaning products, gives rise to extensive contamination of groundwater, potable water, and soil (Per - and Polyfluoroalkyl Substance (PFAS) - Potential Risk to Rural Groundwater Systems, 2025). Moreover, PFAS are also widely utilized in water and stain-resistant textiles. PFAS are crowned with a designation of "forever chemicals" because of their extreme stability and ubiquitous presence in nature. The sulfonate and carboxylate-based PFAS, for instance, perfluorooctanoic acid (PFOA) and perfluorooctane sulfonic acid (PFOS) have drawn much attention of the scientific community globally owing to their adverse effects on human health (Habib et al., 2024; Yang et al., 2020). The aquatic ecology, particularly the marine life, is at higher risk from PFOA and PFOS due to their toxicity, bioaccumulation, and environmental persistence. Large plumes may emerge because of PFAS's great mobility, which allows them to migrate over long distances through soil and groundwater transport from the site of their origin. These substances have been found in landfills and groundwater all around the world (Rekik et al., 2024; Urtiaga et al., 2022). PFAS concentrations and detection rate in the aquatic environment evaluated in Reykjavik varied, according to the general range of results (Table 8.1). The capital's drinking water supply, the groundwater near Heidmörk, has the lowest PFAS detection frequency and is in good condition. While the PFAS concentration and detection frequency in wastewater at the Klettagardar WWT plant were higher, the runoff water from residential and industrial areas was almost comparable. With a PFAS concentration more than 500 times greater than the other discharge sites in the three groups of PFAS, the sites with a history of using firefighting foams had the largest detection frequency and were particularly extensively contaminated at Reykjavík Airport (Gunnarsdóttir et al., 2025).

When the global dissemination of certain PFAS was investigated in 2000, the use of PFAS increased dramatically. PFAS are present in practically everybody of water and are released into the environment because of their widespread use, raising serious environmental concerns. PFAS were recently found in snow and meltwater samples taken from the Khumbu Glacier on Mount Everest, indicating that PFAS are present even in the inaccessible environments (Saidi et al., 2025).

Despite being present in small amounts, PFAS pose plausible risks to aquatic life and humans because of their environmental persistence and water solubility. Because long-chain PFAS like PFOA and PFOS were wreaking havoc in the environment due to their persistence, this raised concerns and stimulated investigations (Abbasian Chaleshtari and Foudazi, 2022). PFAS are fully and partially fluorinated alkyl compounds that may be short-chain or long-chain compounds. These compounds have two parts: one hydrophobic carbon chain is totally fluorinated with the terminal end, which has a polar functional group such as carboxylate ($COO^-$), sulfonate ($SO_3^-$) or phosphate ($OPO_3^-$). The number of fluorinated carbons represented by n, which may be the most common, is $n = 4$ to 8; the short-chain PFAS are $n < 7$, like trifluoroacetic acid (TFA) , which is a very short-chain PFCA ($n = 1$), with a -COOH group. In polyfluoroalkyl, some carbon atoms are partially fluorinated as

**TABLE 8.1** SUM PFAS concentration, range of concentration, count determined, and frequency of detection in each category for industrial wastewater, surface water, and drinking water at various locations.

| Type of site | Matrix | No. PFAS analyzed | No. of samples tested | Median ng $L^{-1}$ (without <MDL) | Range of sum of PFAS ng $L^{-1}$ | Count detected >MDL/sum chemicals tested | Detection frequency % |
|---|---|---|---|---|---|---|---|
| Drinking water WS | 3 intake zones | PFAS-4[a] | 8 | 0.14 | <MDL–0.14 | 1/32 | 3% |
| | | 18 PFAS[b] | 8 | 0.18 | <MDL–0.41 | 11/144 | 8% |
| | | 6 PFAS[c] | 5 | 0.17 | <MDL–0.19 | 4/30 | 13% |
| Wastewater WWT | Klettagardar | PFAS-4 | 8 | 2.31 | 1.57–4.21 | 29/44 | 91% |
| | | 18 PFAS | 8 | 9.00 | 7.28–14.11 | 81/144 | 56% |
| | | 6 PFAS | 6 | 2.28 | 0.84–24.4 | 23/32 | 64% |
| Surface water | Residential | PFAS-4 | 7 | 0.34 | 0.03–0.53 | 7/16 | 44% |
| | | 18 PFAS | 4 | 4.07 | 3.11–5.69 | 23/72 | 32% |
| | | 6 PFAS | 2 | 0.11 | 0.10–0.11 | 4/12 | 33% |
| | Industry | PFAS-4 | 9 | 0.35 | <MDL–0.96 | 14/36 | 39% |
| | | 18 PFAS | 9 | 4.57 | 3.23–7.48 | 51/162 | 31% |
| | | 6 PFAS | 7 | 0.32 | 0.11–1.50 | 15/42 | 33% |
| | AFFFs | PFAS-4 | 4 | 796 | 2.13–1622 | 16/16 | 100% |
| | | 18 PFAS | 4 | 1039 | 6.38–2152 | 49/72 | 68% |
| | | 6 PFAS | 2 | 971 | 500–1442 | 11/12 | 92% |

[a] *PFAS-4 = PFHxS, PFOA, PFNA, PFOS.*
[b] *18 PFAS = PFBA, PFBS, PFHxA, PFHpA, PFHxS, PFOA, PFNA, PFOS, PFDA, PFUnDA, PFDoDA, PFTrDA, PFTeDA, PFPeA, PFPeS, PFHpS, PFNS, PFDS.*
[c] *Additional 6 PFAS = FBSA, FHxSA, 4:2 FTS, 6:2 FTS, 8:2 FTS, FOSA.*

seen in 6:2 fluorotelomer alcohol (6:2) $F(CF_2)_6$–$CH_2CH_2OH$, as 6 represents the fluorinated Carbon and 2 represents non fluorinated carbon, and it is bound to a hydrogen atom instead of fluorine. Fig. 8.1 shows a variety of PFAS types (n = 1–8) with different function groups and long chains. The types of PFAS that combine with surfactants are the most common in water and wastewater. In this sense, new and efficient remediation methods for PFAS from an aquatic environment are required because of the implementation of stricter regulatory restrictions. Wastewater has been effectively treated for more than a century using a variety of techniques and technologies, including adsorption, membrane separation, chemical coagulation, advanced oxidation, biological, electrochemical, and bio-electrochemical processes. Almost every strategy has its inherent advantages as well as associated concerns. For example, biological remediation is thought to be the most cost-effective method for treating organic molecules, but it is less effective when it comes to refractory contaminants (Pierpaoli et al., 2021).

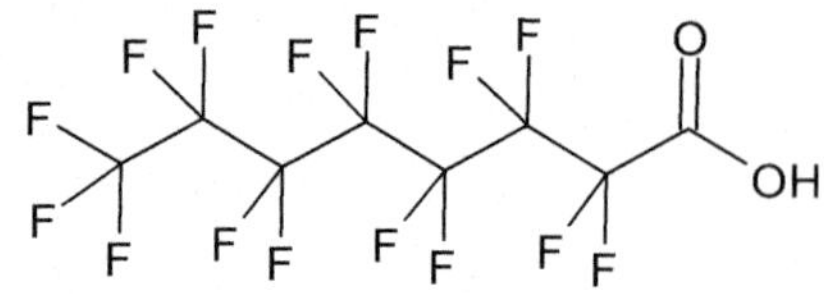

**PFOA** *(Perfluorooctanoic acid) n=7*

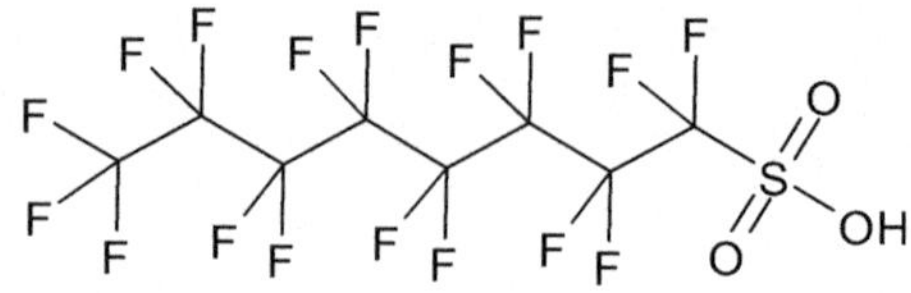

**PFOS** *(Perfluorooctanesulfunic acid) n=8*

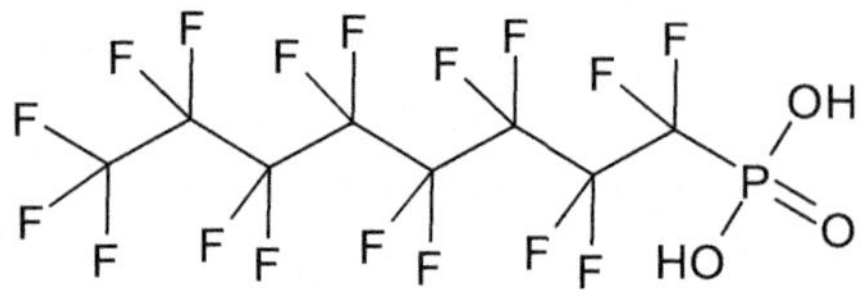

**PFOPA** *(Perfluorooctylphosphonic acid) n=8*

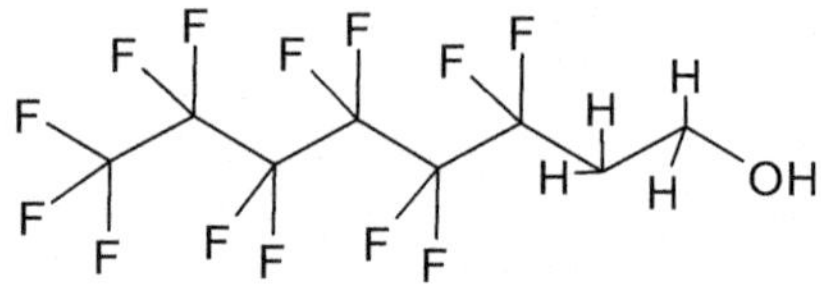

**6:2 FTOH** *(Fluorotelomer alcohol) or 2H,2H-Perfluorooctanoic acid*

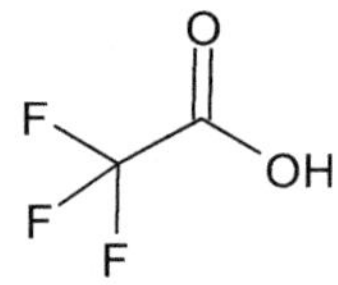

**TFA** *(Trifluoroacetic acid) n=1*

**FIGURE 8.1** The figure demonstrates the chemical structure and bonding of polyfluoroalkyl substances.

Similarly, incineration may result in great removal efficiency, but it may also produce hazardous gaseous pollutants and require high operating costs in harsh environments (Zhang et al., 2023). For refractory pollutants, chemical oxidation and coagulation may work well, but they need a constant supply of reagents, which could result in the generation of unwanted intermediates. Wastewater containing recalcitrant organic components may benefit greatly from the application of advanced oxidation processes (AOPs). The process of oxidation of organic compounds (RH) is initiated by $^{\bullet}OH$ radicals generated in wastewater, as per Eq. (8.1) and Eq. (8.2), which illustrates how the $^{\bullet}OH$ radicals would improve oxidized organic molecules through hydrogen abstraction, producing highly reactive organic radicals ($R^{\cdot}$), which can carry out further oxidation. They can break down almost every type of organic compound without discrimination, mostly by generating highly oxidizing ($^{\bullet}OH$) as shown below:

$$COF_2 + H_2O \rightarrow CO_2 + 2HF \tag{8.1}$$

$$RH + HO^{\bullet} \rightarrow H_2O + R\cdot \rightarrow \text{oxidizedproducts} \tag{8.2}$$

AOPs are thought to be more environmentally benign than many physical separation-driven treatment systems since they chemically break down pollutants, frequently achieving partial or total mineralization rather than just moving them to another phase (Manna and Sen, 2023). Furthermore, sonolysis, microwave, ultrasonication, photolysis, electrochemistry, advanced oxidation process, boiling, high-energy E-beam, high- temperature steam, plasma, hydrothermal treatment, and supercritical water oxidation process (SCWO) are some of the current technologies incorporated to eradicate PFAS (Xu et al., 2024). It is an undeniable fact that technology used to destroy PFAS may consume a lot of energy. For this reason, reverse osmosis, membrane separation, and foam fractionation (FF) are the preferred methods for targeting highly concentrated, low-volume waste (Awoyemi et al., 2024).

Fig. 8.2 summarizes and displays the energy consumption of PFOS and PFOA annihilation by these technologies. Electrochemical oxidation process (EAOP) has several advantages over the others, including lower equipment requirements, ambient operation, and energy efficiency. The energy consumption of PFAS destruction by E-beam cannot be calculated due to insufficient data (Xu et al., 2024). According to Leung et al. (2022), PFOA and PFOS degradations follow a pseudo-first-order kinetic model. At a pH of 13 and a dosage of 500 kGy (0.139 kWh $L^{-1}$), the degradation rates of PFOA and PFOS in a nitrogen environment were 95.7% and 85.9%, respectively. The rates of defluorination were 71.4% and 46.8%, respectively. These findings suggest that PFOA and PFOS in anoxic alkaline solutions can be efficiently broken down by electron beam irradiation. SCWO was unable to determine the precise energy required to destroy PFAS (Leung et al., 2022). According to reports, when the chemical oxygen demand (COD) concentration rises above 60,000 mg $L^{-1}$, the SCWO system can become energy self-sufficient (Xu et al., 2024).

Due to their great interest in treating both organic and inorganic contaminants from a range of wastewaters, electrochemical oxidation (EO) processes contribute to a complementary subset of the many AOPs. EO and electrocoagulation (EC) are promising approaches to treat a variety of industrial wastewater effluents. The EO techniques have manifested

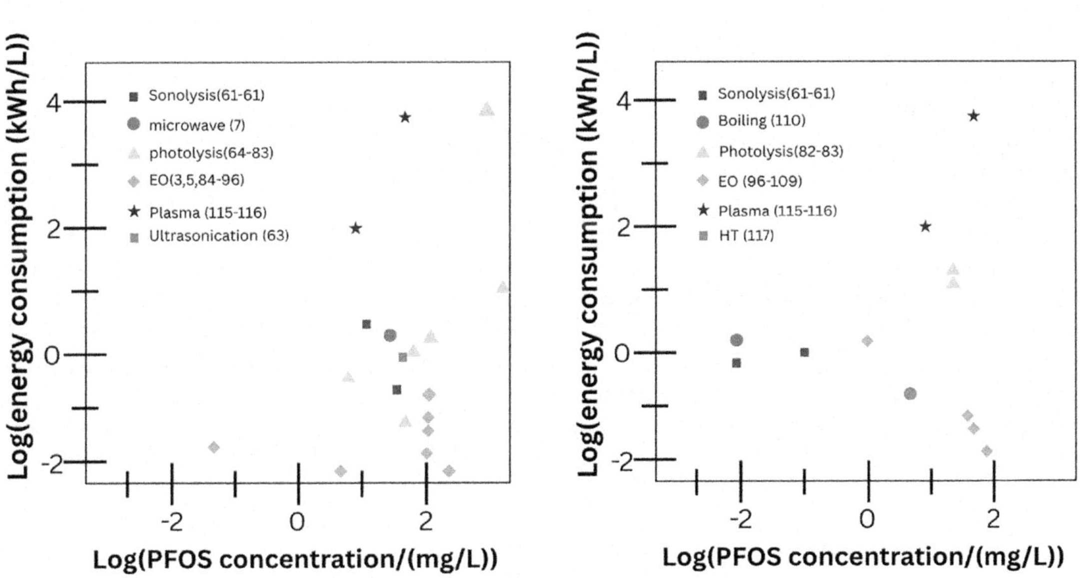


**FIGURE 8.2** Energy consumption of per- and polyfluoroalkyl substances destruction (A) perfluorooctanoic acid (PFOA); (B) perfluorooctane sulfonic (PFOS). *Adopted from Xu, X., Li, Y., Vo, P.H.N., Shukla, P., Ge, L. & Zhao, C.X. (2024). Electrochemical advanced oxidation of per- and polyfluoroalkyl substances (PFASs): Development, challenges and perspectives. Chemical Engineering Journal, 500. https://doi.org/10.1016/j.cej.2024.157222.*

themselves as a reliable method that encompasses in situ electro-generation of potent oxidants, especially the hydroxyl ($^{\bullet}OH$) radical, to oxidize contaminants. In contrast to several other traditional treatment technologies, EO offers an environmentally favorable option because the electron is the sole reagent used during treatment (Yuan et al., 2023; Yanagida et al., 2022). In the meantime, EC relies on the sacrificial metal anode's electro dissolution to produce flocs by forming insoluble metal hydroxides and other hydroxy metallic compounds. With prominent advantages of the ease of applicability, less dependence on outside chemicals, and the production of a relatively smaller amount of sludge, these technologies have offered themselves as the best substitute for chemical coagulation. Nevertheless, EO and EC have a number of drawbacks, including the need for salt to maintain solution conductivity and electrode passivation. In this scenario, the degradation efficacy can be further enhanced by coupling or integrating the advanced EAOPs with conventional treatment procedures. The in situ generations of potent oxidants and a reduced need for chemicals have turned integrated EAOPs into cutting-edge modern technological approaches to cope with PFAS contamination. Notably, PFAS are chemically and physiologically inert, hence using EAOPs in combination with other therapeutic options has lowered the drawbacks of each EAOP alone while offering the facility of synergistic benefits (Rahimi et al., 2024).

### 8.1.1 Source, properties, and toxicity of per- and polyfluoroalkyl substances

According to a typical estimate, there are 4700 different forms of PFAS in the world. Based on the number of carbon atoms in the chain, they are categorized as either polymers or nonpolymers and further subdivided into short- and long-chain compounds. The various functional groupings at the longer or shorter chains serve as the basis for additional classification. These substances have a Per- and polyfluoroalkyl perfluoroalkyl chain that is both hydrophobic and lipophobic, as well as a sulfonic acid functional group that imparts polarity to the chains. Because of the incredible polarity and subsequent stability of the carbon–fluoride bond (C–F), they are very stable and are present everywhere in soil, water, and in the air. PFOA and PFOS are the two most often utilized PFAS (L. Shi et al., 2024). These two organic, completely fluorinated substances belong to the ubiquitous polyfluoroalkyl acids (PFAAs) found in the environment. Because PFOS is a liquid that dissolves somewhat in water, it can have a negative impact on a person's health, including raising their risk of thyroid and cholesterol issues. Since its chain has eight carbons, PFOA, a fluoroalkanoic acid, is often referred to as C8. A surfactant is a material that possesses a carbohydrate "head group" and a perfluorinated, n-octyl “tail group” (Shi et al., 2024). Because of their strong carbon-fluoride linkages, (1) PFOS and (2) PFOA are very stable molecules. Both groups are relatively dissolvable in water, with molecular weights that are 414 $g\,mol^{-1}$ for PFOA and 500 $g\,mol^{-1}$ for PFOS. PFOS is comparatively nonvolatile at 0.002 mmHg, but PFOA is also known to have minimal volatility at 0.54 mmHg. The ratio of a chemical's concentration in the octanol phase to its concentration in the aqueous phase of a two-phase octanol/water system is known as the octanol/water partition coefficient ($K_{ow}$) (Corviseri et al., 2025). It is a helpful indicator of a compound's propensity to bind with either hydrophilic or hydrophobic substances. Nevertheless, the exact value of $K_{ow}$ in (1) PFOS and (2) PFOA cannot be determined. This is because (1) PFAS groups develop several layers in an octanol and water mixture, and (2) they behave differently from other hydrophobic substances. In comparison to PFOA, PFOS has a greater value for boiling point and organic carbon partition coefficient ($K_{oc}$) (Zhu et al., 2022).

The majority of PFAS's effects on the environment are not well understood or explored yet. Research has been conducted for years to investigate the effects of PFAS exposure on aquatic species, humans, vegetation, and fauna. They have been found in all environmental matrices due to their distinct chemical characteristics and persistence, and people and biota can be exposed to them through a variety of routes. Epidemiological research has demonstrated that exposure to small amounts of PFAS would cause major health problems in both humans and animals, including immunotoxicity, endocrine disruption, cancer, and developmental abnormalities (Morales et al., 2024).

#### 8.1.1.1 *Per- and polyfluoroalkyl substances sources from water*

PFAS are so common that even in the Arctic, PFAS can be found in water across the globe in a variety of environmental matrices. The polytetrafluoroethylene (PTFE) coating found on nonstick cookware is one of the most prevalent sources of PFAS. When the pan overheats, Teflon, a brand of PTFE, may leak PFAS into the air. Teflon-coated nonstick cookware that is misused and tossed in the trash can either be recycled or burned, depending on convenience. As a result, it discharges harmful gases and/or leachate of PFAS into the environment as demonstrated in Fig. 8.3 (A) and (B) (Sharma et al., 2022).

High-hazard flammable liquid fires can be effectively put out with aqueous films-forming foam (AFFF). While PFAS are the active components utilized for surfactant characteristics,

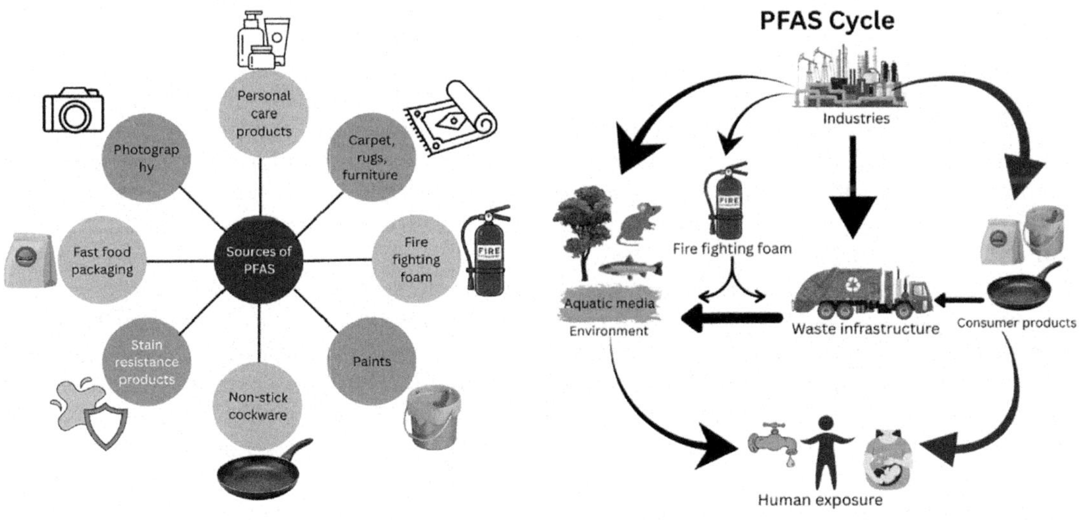


**FIGURE 8.3** **Sources of PFAS** (A) Different sources of PFAS in day-to-day life. (B) PFAS cycle illustrating their ubiquitous presence. *Adopted from Sharma, S., Shetti, N.P., Basu, S., Nadagouda, M.N. & Aminabhavi, T.M. (2022). Remediation of per- and polyfluoroalkyls (PFAS) via electrochemical methods. Chemical Engineering Journal, 430. https://doi.org/10.1016/j.cej.2021.132895.*

AFFF is often made by mixing fluorinated surfactants with foaming compounds. Because of their nonstick qualities and resistance to water, oil, and grease, PFAS compounds are used as grease-proofing agents. PFAS are the ideal foil for greasy dishes because of their ability to repel grease. According to a recent study conducted in the United States, prevalent mosquito repellents in the area are laden with PFAS. Additionally, PFAS have been widely found in shampoos and paints. Aqueous firefighting foams that contain perfluoroalkyl carboxylates (PFCAs), perfluoroalkyl sulfonates (PFSAs), and perfluorooctane-sulfonamide may also be a source of PFAS in water (Cornelsen et al., 2021). They are released in greater quantities over a shorter period, which contaminates the water in the surrounding areas. Additionally, one of the main sources of PFAS in water and wastewater is the electrical and electronic manufacturing sectors. The amount of PFAS in water was also influenced by food packaging materials that were more water-repellent and oil-proof. Every year, millions of metric tons of garbage are created, and PFAS may leak into water systems if they are disposed of in landfills. Three distinct PFAS (PFBS, PFHxS, and PFOS) were measured by Satriaji et al. (2025) in Taiwanese rivers that received the discharge of electronic industrial wastewater (Satriaji et al., 2025).

#### 8.1.1.2 *Toxicity of per- and polyfluoroalkyl substances*

Due to their high stability, PFAS have a propensity to persist in the environmental water bodies and even in human organisms for extended periods of time. Humans in developed nations have high serum concentrations of PFCAs and PFSAs; they are so widespread that almost everyone has been confronted with them and has PFAS traces in their blood. Health risks may arise from PFAS's propensity to build up in human tissues, such as the liver and kidneys (Table 8.2). Over time, correlations have also been discovered between higher blood serum PFAS levels and elevated levels of uric acid, cholesterol, and liver enzyme activity, as well as pregnancy-induced hypertension (Liao et al., 2023). Low birth weight was also found to be strongly linked to PFAS-contaminated blood serum levels. Additionally, it has been demonstrated that PFAS negatively impacts newborns' and older children's behavior, learning, and growth. Children with "attention deficit hyperactivity disorder" have higher levels of PFHxS and PFOS in their serum samples. US-EPA has determined that both PFOA and PFOS may cause cancer.

Despite extensive research on the toxicity of the most common PFAS, statistics about the detrimental impacts on humans are still being developed, and the majority of this research has been carried out on animals. There is no proof that changing to short-chained PFAS would mitigate the problem, though, as taking a substance off the market and substituting it with another of the same class has rarely worked in the past. The toxicity and bio-accumulative potential of the available fluorinated substitutes are little understood, and for all those fluorinated experiments, gathering the missing data is expensive and time-consuming (Garg et al., 2023). Table 8.3 presents a brief account of PFAS and their adverse effects at various localities of the world. Perfluorooctanesulphonyl fluoride is represented by POSF, tetrafluoroethylene by TFE, perfluorooctanoate by PFDA, perfluorohexane sulfonic acid by PFHxS, perfluo-rononanoic acid by PFNA, and perfluorooctanoate by PFOA.

**TABLE 8.2** Impacts of various PFAS on various creatures, including humans, and the circumstances surrounding their exposure.

| PFAS | Organism | Toxicity | Exposure | Region | References |
|---|---|---|---|---|---|
| POSF | Males | Cancer leading to death | Exposure at a factory that produces fluorochemicals | USA | Alexander et al. (2024) |
| PFOA and PFOS | Males | Hyperuricemia risk | Long-term exposure to a chemical facility | USA | Liu et al. (2023) |
| TFE | Males/females | Leukemia | Polytetrafluoroethylene production sites | Europe | Ozben et al. (2024) |
| PFDA and PFOA | Children <13 years | Diphtheria antibody concentrations | Postnatal exposure | Faroe Islands | Timmermann et al. (2022) |
| PFOS, PFOA, PFHxS, and PFNA | Female | British boys' birth sizes | Exposure to PFA during pregnancy | Britain | Marks et al. (2021) |
| PFOA and PFOS | Japanese medaka | Modification of the reproductive impact | Exposure to chemicals directly | Korea | Dasmahapatra et al. (2023) |
| PFOA, PFDA, and PFOS | Fish | Biomagnification in the intestine | Long-term exposure to aquatic environments | Belgium | Cara et al. (2022) |
| PFOS | Zebrafish | Functional disruption of the intestine | Long-term exposure in lab settings | China | Huang et al. (2021) |
| PFOA, PFNA, and PFHxS | Males > age 50 years | Unknown | Mostly with drinking water in the field study (through food, consumer goods, household dust, or soil) | USA | Garnick et al. (2021) |
| PFOA, PFOS, PFNA, PFHxS, and GenX | C57BL/6 J wild-type male and female mice | Elevated bile acids, sterol metabolites, and cholesterol circulation | Dietary atherogenic exposure to water | USA | Sivagami et al. (2023) |

**TABLE 8.3** A brief account of PFAS remediations incorporating electrochemical oxidations (EOs) techniques.

| Removal technique | Target contaminant | Anode/cathode | Parameters | % Degradation/ mineralization | Reference |
|---|---|---|---|---|---|
| EO | Perfluorooctanoic acid (PFOA) | BDD/titanium | Concentration = 10–100 μg $L^{-1}$<br>pH = 3–11<br>power = 49.4–109 kWh $m^{-3}$ | 91%–99% | Samuel et al. (2024) |
| EO | PFOA, PFOS, PFBA | Boron-doped graphene-based sponge | Concentration = 10–100 μg $L^{-1}$<br>pH = 7–7.2<br>power = 10.1 kWh $m^{-3}$ | 74%–87% | Duinslaeger and Radjenovic (2022) |
| EO | PFOA, PFOS, PFBA | Graphene sponge anode functionalized with borophene | Concentration = 0.2 μM<br>pH = 5.6–7.3<br>current density = 115–460 $Am^{-2}$ | 32%–77% | Duinslaeger et al. (2023) |
| Foam fractionation (FF) and EO | Different PFAS | $Ti_4O_7$/stainless steel | Environmental sample + 0.02 g $L^{-1}$ HPMCpH = 5.6–7.3 current density = 20 mA $cm^{-2}$ | 50%–70% | Wang et al. (2023) |
| EO | Hexafluoro propylene oxide dimer acid | $CeO_2$+ $Ti_4O_7$+graphene oxide, UV/KI-$SO_3^{-2}$ | pH = variable concentrations = 20 mM KI and $Na_2SO_3$, 245 nm UV lamp | 70%–90% | Liang et al. (2025) |
| Electrocoagulation (EC) and EO | Ten different PFAS | $Ti_4O_7$ ceramic plate (EO) and Zn anode (EC) | PFAS concentration =0.1 μM<br>pH = 3.8<br>electrolyte = 20 mM $Na_2SO_4$<br>current density =10 mA $cm^{-2}$ | More than 90% | Shi et al. (2021) |
| EO | 53 individual PFAS compounds | BDD anode/stainless steel cathode | 5 mM ammonium acetate<br>sample Volume for LC = 10 μL | ≦90% | Rehnstam et al. (2024) |
| EO | PFOA and PFOS | BDD-persulphate | PFAS concentration = 50 μM<br>pH = 3.8<br>Current density = 40 mA $cm^{-2}$ | 33%–64% defluorination | Zhao et al. (2025) |
| EO | 29 PFAS compounds | Ti/BDD and Ti/$IrO_2$ anodes and a Pt cathode | PFAS concentration = 0.1 μM<br>pH = 2.5–03<br>electrolyte = NaCl 2 g $L^{-1}$<br>current density = 15 mA $cm^{-2}$ | 54%–98% | Hieu Nhu et al. (2025) |
| EO | PFOA | $CeO_2$ electrodes | Concentration of POFA = 1 mg $L^{-1}$<br>applied potential = 1.4 V<br>power = 0.62 kWh $m^{-3}$ | 94% removal<br>73% defluorination | Liu et al. (2025) |

## 8.1.2 Per- and polyfluoroalkyl substances transport in groundwater and surface water

Leachate from industrial effluents and municipal solid waste is the primary source of the elevated PFAS concentrations in both groundwater as well as surface systems. The abundance of PFAS found in water sources, waterways, wastewater, rivers, lakes, and seas has been the subject of more investigation in recent years. Even samples taken from Khumbu Glacier, Mount Everest, contained PFAS.

PFOS, PFOA, and PFHxA were found to be present in meltwater and snow samples at levels of 26.14 and 10.34 ng $L^{-1}$ of PFOS, respectively, out of the 14 chemicals that were tested for presence (Saidi et al., 2025). The kinds of sources, site features, and environmental circumstances all affect the movement of PFAS in water. Chain length, molecular charge, fluorination level, and functional group are some of the characteristics of PFAS. The environmental site circumstances, on the other hand, can be listed as follows: the duration and extent of active pollution; the flow rates of surface water and groundwater; the existence of medium interfaces such as air and nonaqueous phase liquids (NAPL); the level of groundwater; infiltrate or precipitation; and the presence of cocontaminants. Sediment and particle properties, together with solution properties, once PFAS reach water bodies, their fate and transportation will be influenced by factors such as redox-potential, pH, and the existence of cocontaminants (Cara et al., 2022).

Because PFAS are hydrophobic, they can: (1) sorb organic carbon molecules and become suspended, which leads to their partitioning into solids such as metal oxides, organic materials, sediments, and soil particles. (2) electronic contact with both organic and inorganic materials. Uncertainty regarding the extent of the connection between partitioning coefficients and soil characteristics, as well as the coefficients' implied disdain for electrostatic interactions, could result in inaccurate transport predictions. The partitioning behavior of PFAS is also significantly influenced by electrostatic contact. Anionic PFAS have more mobility than cationic or zwitterionic PFAS because of interactions with negatively charged sediments and particles (Kabiri et al., 2023). Fluid interactions, such as those between air and water or NAPL and water, PFAS preferentially accumulate because they function as surfactants.

Self-assemblages of PFAS as micelles, hemi-micelles, in addition to bilayer structures, are common if their concentration surpasses the critical micelle concentration (CMC). PFOS has a CMC within the 500–5000 mg $L^{-1}$ range, but studies show that hemi-micelles can develop at quantities as low as 0.5–5 mg $L^{-1}$ (Lebouachera et al., 2024). The movement of PFAS, particularly in aquatic bodies, is significantly influenced by migration. It has been discovered that PFAS can spread widely through soil, groundwater, surface water, and even the atmosphere. Advection and dispersion are the main mechanisms, as is the case with most chemical pollutants. Research has also demonstrated that leaching is a significant factor in the movement of PFAS from the soil's surface to groundwater and surface water, indicating migration in response to irrigation operations or natural events like flooding or precipitation. Local PFAS transmission in the air is one of the main causes of deposits on water bodies' surfaces. Numerous investigations have proven the transformation of polyfluorinated PFAS. It has been demonstrated that PFAS tend to bioaccumulate in higher trophic level bodies. Ingestion and inhalation are two possible routes of entry into human bodies, where enterohepatic circulation keeps it there. Though it can also enter through consumer goods, the primary source of entry is up for discussion (Lebouachera et al., 2024; Lu et al., 2024).

## 8.2 Key concepts

### 8.2.1 Electrochemical oxidation for the degradation of per- and polyfluoroalkyl substances

Over the past two decades, EO has emerged as a highly effective and operationally simple method for wastewater treatment. Its remarkable efficiency and versatility have placed it at the forefront of advanced environmental remediation technologies. The growing interest in this field is reflected in a wealth of scholarly books and comprehensive reviews dedicated to the application of electrochemistry in environmental protection (Hussain et al., 2020b). Since EO can completely mineralize both biodegradable and nonbiodegradable organic materials, it has proven to be a reliable and adaptable method for the degradation of wastewater pollutants. High amenability, reduced chemical requirements, increased environmental compatibility, and high adaptability are characteristics of these procedures (Bellanthudawa et al., 2023). This electrooxidation has consistently demonstrated its capability to mineralize a wide array of organic pollutants across various water matrices, ranging from synthetic solutions to real wastewater samples. However, the overall efficiency of EO in achieving effective disinfection and decontamination is strongly influenced by both the intrinsic properties of the electrocatalytic material, particularly the anode, and the specific operational conditions under which the process is carried out (Martínez-Huitle et al., 2023; Sharma et al., 2022). EO has been widely used in recent years not only to treat wastewater and disinfect the portable water but also to help with soil remediation. EO of organic pollutants in wastewater generally proceeds through two main pathways: direct oxidation and indirect oxidation. The distinction between these pathways is closely linked to the nature of the anode material- specifically, whether it is an active or nonactive-anode and how it interacts with the pollutant molecule (Nair et al., 2023).

#### *8.2.1.1 Direct electro-oxidation (active anodes)*

Direct oxidation primarily occurs at active anodes like (Pt, $RuO_2/Ti$, $IrO_2/Ti$), which possess electrocatalytically active sites that allow for the chemisorption of organic molecules. These anodes can reach high oxidation states, typically above the standard potential for oxygen evolution, enabling the formation of surface-bound oxidizing species such as EO (electrode metal oxides or superoxides).

Literature reveals that water ($H_2O$) molecules discharged on the surface of electrode (E), it produces hydroxyl radicals ($^{\bullet}OH$) represented as E[$^{\bullet}OH$] (Eq. 8.3). These surface-bound radicals can either react with organic compounds (R) to form oxidized products (Eq. 8.4) or convert into oxygen-containing products through an oxygen transfer process (Eq. 8.5) as shown in Fig. 8.4. After that, it takes part as mediator via (Eq. 8.6) in the selective oxidation of pollutants, which occurs in parallel with oxygen evolution (Eq. 8.7), as side reaction, which may influence the performance of EO process.

$$E + H_2O \rightarrow E(^{\cdot}OH) + H + +e- \quad (8.3)$$

$$E(\cdot OH) + R \rightarrow E + mCO_2 + nH_2O + nH + +ne- \quad (8.4)$$

$$E(\cdot OH) \rightarrow EO + H^{+} + e- \quad (8.5)$$

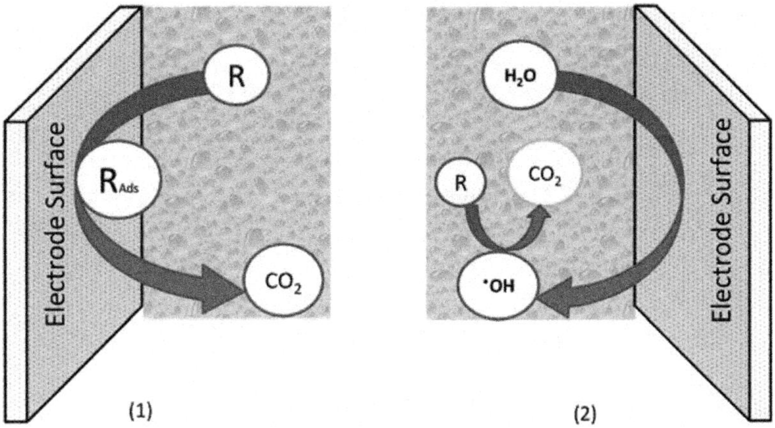


**FIGURE 8.4** Direct (1) and indirect (2) electrochemical degradation process diagrammatic representation of the intricate processes that take place throughout the electro-oxidation process. *From Hussain, Sajjad, Ullah Khan, Saad & Gul, Saima. (2020). Electrochemical Treatment of Antibiotics in Wastewater (pp. 355–394). Springer Science and Business Media LLC. https://doi.org/10.1007/978-3-030–40422-2_17.*

$$EO + R \rightarrow E + RO \tag{8.6}$$

$$EO \rightarrow E + \frac{1}{2}O_2 \tag{8.7}$$

In this mechanism, the organic pollutant is adsorbed directly onto the anode surface and undergoes oxidation through electron transfer reactions. Physisorption and chemisorption are the two ways that radicals interact with the electrode because of their production. The competitive nature of both processes' participation is mostly determined by the anode material (Fig. 8.4). "When physisorption is the predominant mechanism (Eq. (8.4)), where (E) and (R) stand for the electrode and contaminant, respectively, this is the primary process that influences the oxidation of pollutants." Nevertheless, if ($^{\bullet}$OH) is chemisorbed, it is limited to reacting with contaminants close to the anode in Eq. (8.5) and Eq. (8.6), generating superoxide or higher state oxides (Fitch et al., 2022).

### *8.2.1.2 Indirect electro-oxidation (nonactive anodes)*

During Indirect EO, the anode (typically nonactive anodes like boron-doped diamond [BDD], $SnO_2$ [e.g., Sb-doped $SnO_2$], Ti/$PbO_2$, Ni/Sb–$SnO_2$, Ti/$IrO_2$, Ti/ $SnO_2$) does not interact strongly with pollutants. Instead, oxidation occurs through reactive oxygen species (ROS) generated from water discharge or electrolyte oxidation (Tasca et al., 2025). Different studies demonstrated that a variety of oxidizing species can be formed from water discharge and other oxidant sources, like ROS ($O_3$ and $H_2O_2$) perxo-compounds like (perox-odisulfate [$S_2O_8^{2-}$], peroxodicarbonate $C_2O_6^{2-}$), peroxodiphosphate ($P_2O_8^{4-}$), and active chlorine. The most common oxidant produced by the anodic oxidation of chloride is chlorine. The production of active chlorine species, such as hypochlorous acid (HOCI) and hypochlorite ions, further facilitated the oxidation (Somashekara and Mulky, 2023). The in situ produced oxidants quickly oxidize the contaminants in the bulk of the solution, as shown in the following equations (Eq. 8.8–8.16):

$$2Cl^- \rightarrow Cl_2 + 2e^- \tag{8.8}$$

$$2H_2O + 2e^- \rightarrow H_2 + 2OH^- \tag{8.9}$$

$$Cl_2 + H_2O \rightarrow HOCl + H^+ + Cl^- \tag{8.10}$$

$$HOCl \rightarrow H + +OCl^- \tag{8.11}$$

$$R + OCl^- \rightarrow CO_2 + H_2O + Cl^- \tag{8.12}$$

$$ClO^- + H_2O \rightarrow ClO_2 - +2H + +2e^- \tag{8.13}$$

$$ClO_2^- + H_2O \rightarrow ClO_3^- + 2H^+ + 2e^- \tag{8.14}$$

$$ClO_3^- + H_2O \rightarrow ClO_4^- + 2H^+ + 2e^- \tag{8.15}$$

It is also possible that the loss of $ClO^-$ by reduction to chloride ion at the cathode:

$$OCl^- + H_2O + 2e^- \rightarrow C^{1-} + 2OH- \tag{8.16}$$

Furthermore, direct EO via a direct electron transfer to the anode results in partial contamination of wastewater. Indirect or mediated oxidation by the electro-generation of oxidizing species produced from water discharge or supporting electrolyte oxidation at elevated current can be observed a total decontamination.

However, electro-oxidation offers numerous benefits over many traditional methods. The electro-oxidation technique is renowned for being environmentally benign and producing very little waste throughout the treatment process. Despite being heralded as a technology of the future and being environmentally friendly, electro-oxidation has limited application in the industry due to certain trade-offs (Brosler et al., 2023). Because they offer a high surface area for improved reaction kinetics, nanomaterial- based electrocatalysts could be used to overcome the restrictions.

As Table 8.3 elaborate various electrodes used for degradation and mineralization of PFAS in electrochemical processes. Researchers have frequently discovered that acidic environments maximize removal efficiency. The ideal pH, according to a study by Loc et al. (2024), is acidic because of increased ($^{\bullet}OH$) production and a larger oxygen overpotential. Degradation efficiency is improved at lower pH levels because the oxygen generation process is suppressed; ($^{\bullet}OH$) production would decrease if oxygen is produced more quickly in an alkaline solution. However, the ideal pH may also be influenced by electrode material (Loc et al., 2024). Low OEP readings indicate that the production of oxygen by $H_2O$ may be limited in acidic environments. A more thorough evaluation of the type of pollutant and removal process, as well as other aspects, could be taken into consideration, even though most of the research concurred that the ideal starting pH tends to be neutral. Although research indicates that the current density supplied to the cell is proportionate to the electrical consumption, the clearance rate is likewise related to current density. As a result, energy consumption and removal rates must be traded off. Temperature was also discovered to have an impact on EO treatment, and the outcomes were found to be contradictory (Morales et al., 2024). $TiO_2$ nanotube electrodes were used to execute degradation, and the effects of temperature on lignin oxidation were investigated at temperatures between 0°C and 30°C. Here, it was found that the reaction rate increased as the temperature rose. The kind of electrode that was utilized determined how temperature affected things. A few PFOS carboxyl groups and $SO_4$ radical dot groups may be to blame for this. When both processes take place at the same time, the overall effectiveness of treatment is improved (Leung et al., 2022; Li et al., 2025).

### 8.2.2 Per- and polyfluoroalkyl substances degradation mechanisms during electrochemical oxidation

EO is an effective way for the degradation of PFAS. As it produces strong oxidants, which can break the exceptionally strong carbon–fluorine (C–F) bonds that differentiate PFAS compounds. Understanding the underlying mechanisms of EO is important to optimizing system performance and tailoring treatment strategies for real-world applications. The researchers are still working to fully understand how PFAS are broken down during EO. However, most agree that the first step, where PFAS molecules lose an electron directly at the electrode surface, is the slowest rate-dependent step. This idea is backed by both computer simulations and lab experiments, which also suggest that PFAS aren't very reactive to hydroxyl radicals ($^{\bullet}OH$), meaning other pathways are likely to be more critical in their breakdown (Barisci and Suri, 2023).

Radjenovic et al. and Xu et al. (Radjenovic et al., 2020; Xu et al., 2024) explained the EO mechanism and used $^{\bullet}OH$ (Cycles I and II), $O_2$ (Cycle III), and $H_2O$ (Cycle IV) to summarize four different PFAS degradation pathways, as shown in Fig. 8.5 **(A)**. This was accomplished by combining experimental results with the outcomes of multiple density functional theory (DFT) investigations (Li et al., 2025).

The PFAS radicals are produced by the direct electron transfer of PFAS at the anode surface, $C_nF_{2n+1}COO\cdot$ or $C_nF_{2n+1}SO_3\cdot$, which is typically credited with initiating degradation because $\cdot OH$ alone has little reactivity to degrade PFAS. This was further supported by the fact that tert-butyl alcohol, a hydroxyl radical scavenger, had no discernible impact on the rates of PFOA and PFOS elimination or the creation of intermediates of small-chain PFAS, predicting that the BDD region was the main location of degradation. As seen in Cycles I–IV in A, these radicals are indirectly reduced to generate the perfluoroalkyl radicals $C_nF_{2n+1}\cdot$ by Kolbe decarboxylation or desulfonation processes. Cycle I is predicated on research demonstrating that PFOS degrades quickly without producing intermediate small-chain PFAS. Cycle II considers research in which PFOS forms short-chain PFCAs and studies where the $CF_2$ unzipping mechanism causes the production of small-chain PFAS during degradation. As seen in Cycle III, the perfluoroalkyl radicals $C_nF_{2n+1}$ may potentially react with $O_2$ in water and serve as the main reaction in processes with high $^{\bullet}OH$ competitions or low $^{\bullet}OH$ concentrations. In Cycle IV, the H atoms are finally abstracted from $H_2O$ by $C_nF_{2n+1}$, which subsequently reacts with $^{\bullet}OH$ to create intermediates of shorter-chain PFAS (Bani-Melhem et al., 2025). In another study by Liu et al. (2019) (Liu et al., 2019) demonstrated the possible route for PFOA transformation by $SO_4^{\bullet-}$ and $^{\bullet}OH$ radicals, Fig. 8.5 (**B**), is proposed based on the intermediate analysis. For $SO_4^{\bullet-}$ oxidation, direct electron transfer happens between $SO_4^{\bullet-}$ and the carboxylic group of PFOA ($C_7F_{15}COOH$) due to its electrophilic

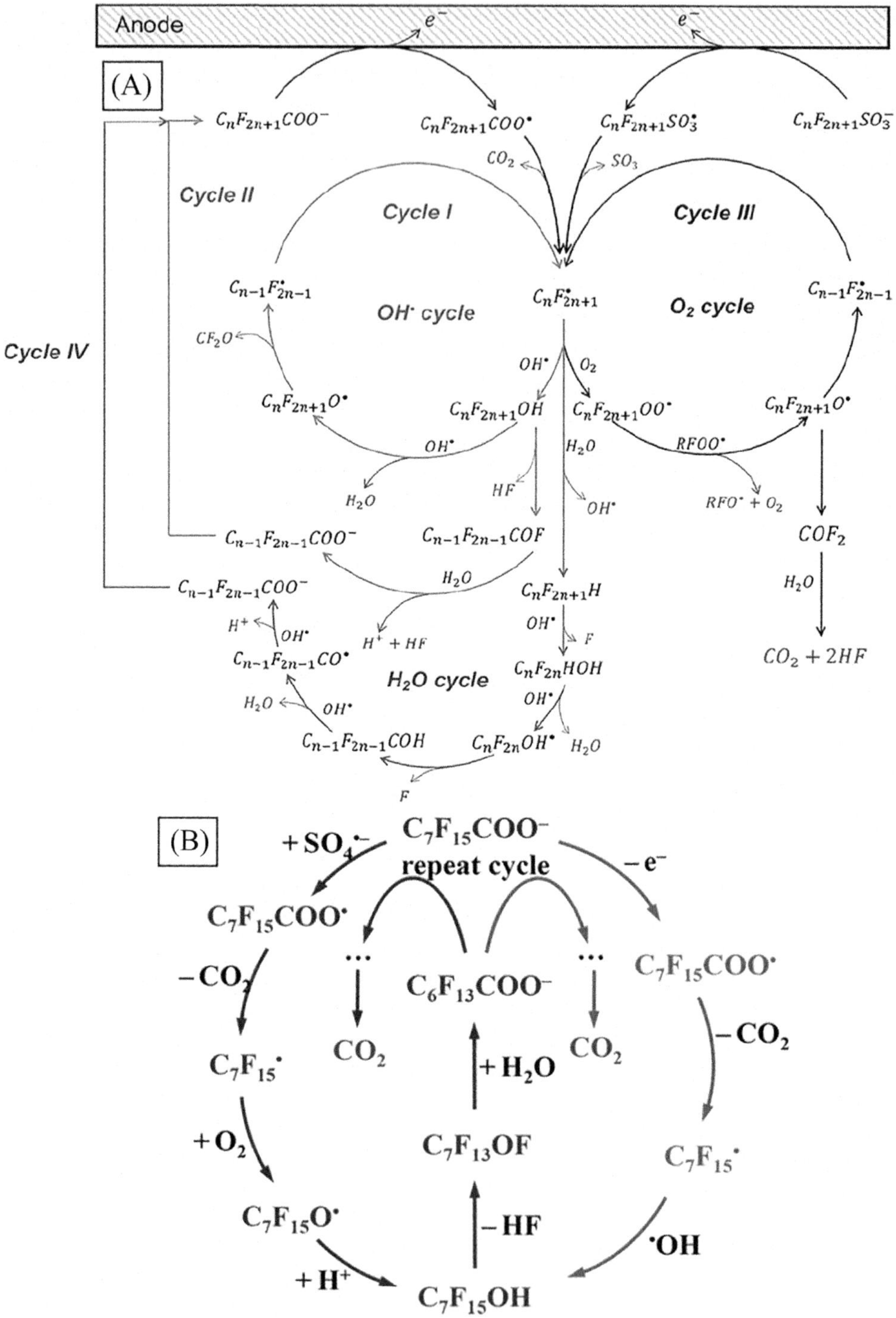
Anode
(A)
Cycle I
Cycle II
Cycle III
Cycle IV
OH˙ cycle
O₂ cycle
H₂O cycle
(B)
C₇F₁₅COO⁻
repeat cycle
C₆F₁₃COO⁻
C₇F₁₃OF
C₇F₁₅OH

behavior, making it possible to form $C_7F_{15}COO^{\bullet}$ and a subsequent Kolbe decarboxylation reaction, which converts $C_7F_{15}COO^{\bullet}$ to $C_7F_{15}^{\bullet}$. The $C_7F_{15}^{\bullet}$ reacts with oxygen and a proton in the reaction medium and transfers to $C_6F_{13}CF_2OH$. As it is reactive, it will transfer to $C_6F_{13}COF$ by intramolecular rearrangement, which will precede further hydrolysis and then form $C_6F_{13}COOH$. In the meantime, PFOA can be degraded by $^{\bullet}OH$, where pollutant release at anode and subsequently by decarboxylation reaction, and then the formed $C_7F_{15}^{\bullet}$ reacts with $^{\bullet}OH$, resulting in $C_7F_{15}OH$ formation, which is finally produce $C_6F_{13}COOH$ via similar rout as mentioned above. It was recently proposed that a distinct degradation route governs the oxidation process during EO treatment when hydroxyl and sulfate radicals are generated electrochemically, as shown in Fig. 8.5 **(B)**. Degradation of PFOA is initiated by direct oxidation at the anode surface as well as indirect oxidation of $SO_4^{\bullet-}$ and carboxylic group in PFOA. Perfluoroalkyl radicals $C_nF_{2n+1}$ can be generated by the former. These radicals then constantly undergo $CF_2$ degradation to create intermediates of shorter-chain PFAS until they completely degrade into $CO_2$ and HF. Findings from EO treatments with electrochemically generated $SO_4^{\bullet-}$ PFOA and hexafluoropropylene oxide dimer acid (HFPO-DA) support this reaction, which was initially noted in PFOA treatments employing heat-activated $SO_4^{\bullet-}$. Higher PFAS removal, faster reaction rates, and a reduced system energy need were the outcomes of the combination of both processes that occurred when sulfate radicals were present during therapy. However, as other PFAS have not been used to study the use of electrochemically produced sulfate radicals, this process has only been documented for PFOA and HFPO-DA thus far (Bani-Melhem et al., 2025).

Recently, a study conducted by Liang et al. (2025) demonstrated the possible oxidation mechanism of different PFAS in the $Ti_4O_7$ Heterojunction composite electrode system, as shown in Fig. 8.6. The EO of a short-chain (HFPO-DA so-called GenX) initiate at electrode surface by direct electron transfer from carboxylate group of compounds. It forms a carbon-oriented radical, which quickly decarboxylates and releases $CO_2$ and produces a reactive intermediate radical. The ether bond in this process undergoes hemolytic bond rapture and generate $CF_3COF$ (trifluoro-acetyl fluoride) and a $C_3F_7^{\bullet}$ (perfluoroalkyl) radical. TFA and HF are also produced from acyl fluoride ($CF_3COF$) hydrolysis in a solution medium. Subsequently, the hydroxyl radicals ($^{\bullet}OH$) electrode surface attack the $C_3F_7^{\bullet}$ and transfer into $C_3F_7OH$, which is further hydroxylated and abstracts the hydrogen, and generating another radical $C_3F_7O^{\bullet}$. This reactive radical proceeds by bond cleavage and is transformed into $C_2F_5^{\bullet}$ and $COF_2$ (carbonyl fluoride). These radicals are further hydrolyzed and converted into $CO_2$ and HF. There is also some rearrangement reaction of $C_3F_7OH$ to $C_2F_5COF$, which hydrolyzes

**FIGURE 8.5** Electrochemical oxidation mechanism of PFAS: Electrochemical oxidation of PFAS broke down in two ways: (A) By combining direct-oxidation on the anode surface along with indirectoxidation with ($^{\bullet}OH$) radicals, i-(Cycles I and II), dissolved ($O_2$), ii-(Cycle III), and the creation of volatile fluorinated organic compounds, iii-(Cycle IV) and (B) Possible routes for PFOA oxidation via sulfate solution activation and $^{\bullet}OH$ radicals. *(A) Adopted from Radjenovic, J., Duinslaeger, N., Avval, S.S. & Chaplin, B.P. (2020). Facing the Challenge of Poly And Perfluoroalkyl Substances in Water: Is Electrochemical Oxidation the Answer? Environmental Science and Technology, 54(23), 14815–14829. https://doi.org/10.1021/acs.est.0c06212, with permission, Copyright © 2020, American Chemical Society; B) Liu, Y., Fan, X., Quan, X., Fan, Y., Chen, S., & Zhao, X. (2019). Enhanced Perfluorooctanoic Acid Degradation by Electrochemical Activation of Sulfate Solution on B/N Codoped Diamond. Environmental Science & Technology, 53(9), 5195–5201. https://doi.org/10.1021/acs.est.8b06130, with permission, Copyright © 2019, American Chemical Society.*

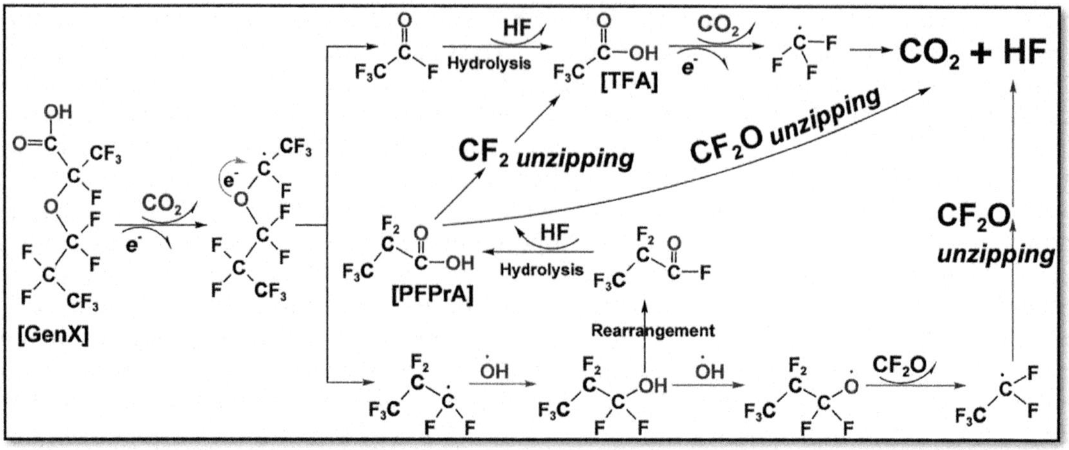


**FIGURE 8.6** Electro-oxidation pathway of hexafluoropropylene oxide dimer acid (HFPO-DA or GenX) (Eqs. 8.18–8.21): possible electrochemical mineralization mechanisms of (HFPO-DA or GenX). *From Liang, Y., Wang, A., Liang, S., Sun, K., Xie, R., Zheng, C., Zhang, S., Tang, C., Cheng, D., Wang, J., Huang, Q. & Lin, H. (2025). Durable Ti4O7 Heterojunction Composite Membrane Encapsulating N-Doped Graphen Nanosheets for Efficient Electro-Oxidation of GenX and Other PFAS in Fluorochemical Wastewater. Environmental Science and Technology, 59(9), 47454755. https://doi.org/10.1021/acs.est.4c09423.*

into perfluoropropionic acid and HF. Finally, all fragments of PFAS are oxidized into $CO_2$ and HF.

These suggested pathways highlight the stepwise oxidative degradation and defluorination of HFPO-DA, which essentially proceeds by radical formation, hydrolysis of acyl fluorides, and $CF_2$-unit chain cleavage at the electrode solution interface. Key reactions in this process are represented as the overall mechanistic route given in Fig. 8.6.

1. **Initial electron transfer and decarboxylation:**

$$C_3F_7O(CF_3)CFCO_2^- - e^- \rightarrow C_3F_7O(CF3)CFCO_2 \rightarrow C_3F_7O(CF_3)CF^\bullet + CO_2 \quad (8.17)$$

2. **C–O bond cleavage:**

$$C_3F_7O(CF_3)CF\bullet \rightarrow C_3F_7^\bullet + CF_3COF \quad (8.18)$$

3. **Hydrolysis of acyl fluoride:**

$$CF_3COF + H_2O \rightarrow CF_3COOH(TFA) + HF \quad (8.19)$$

4. **Hydroxyl radical reaction and ether radical cleavage:**

$$C_3F_7^\cdot + \dot{O}H \rightarrow C_3F_7OH + \dot{O}H \rightarrow C_3F_7OH\cdots OH \rightarrow C_3F_7O^\cdot \rightarrow COF_2 + C_2F_5^\cdot \quad (8.20)$$

**5. $COF_2$ hydrolysis:**

$$COF_2 + H_2O \rightarrow CO_2 + 2HF \tag{8.21}$$

## 8.3 Methods and applications of electrochemical oxidation

### 8.3.1 Experimental setup

The EO of PFAS is typically performed in a well-controlled laboratory-scale batch or continuous reactors. A representative EO reactor consists of three main components: an electrochemical cell, electrode materials, and an electrolyte solution. Most of the lab-scale experiments were performed either in a batch reactor (undivided or divided) or a continuous flow cell, depending on the kinetics of PFAS degradation and the complexity of the water matrix necessary to avoid interference or reformation of byproducts. Undivided cells are commonly used for simplicity, where both the anode and cathode are placed in the same compartment. Divided cells (using a membrane or diaphragm) are employed when separation of anodic and cathodic reactions is necessary to avoid interference or reformation of byproducts (Xu et al., 2024).

### 8.3.2 Electrode materials and their applications for per- and polyfluoroalkyl substances removal mechanism

In EO technology, the electrode is very important since it determines the cost and effectiveness of pollution removal. These parameters are directly influenced by the electrode's performance characteristics and material composition. In the EO process, the anode is primarily where the oxidation reaction occurs. The key factors affecting oxidation capacity and operating lifespan of the EO process are the anode's stability and catalytic activity. Therefore, creating anode materials that are more effective, affordable, and suitable for practical uses is the main goal of current research initiatives. This is accomplished by improving the preparation procedure, designing advantageous structures, and carefully choosing the right materials. A thorough analysis of current research developments is then presented, followed by measurements of the BDD electrode and the stable metal oxide electrode dimensionally stable anodes (DSA) (Li et al., 2024).

#### *8.3.2.1 Graphite and pure metal electrodes*

Graphite and all-metal electrodes were the most widely employed in the early phases of EO treatment research, as illustrated in Fig. 8.7. Because of their remarkable electrical conductivity and inexpensive preparation costs, they are frequently utilized in the application of wastewater treatment. It has been demonstrated that 98.0% pigment removal in wastewater is possible under the given operating parameters, A pH of 2.0, an electrolyte concentration of $0.1\ mol/dm^{-3}$, a current density of $13\ mA/cm^{-2}$, and an electrolysis duration of 20 min are among them. This outcome suggests that the removal procedure

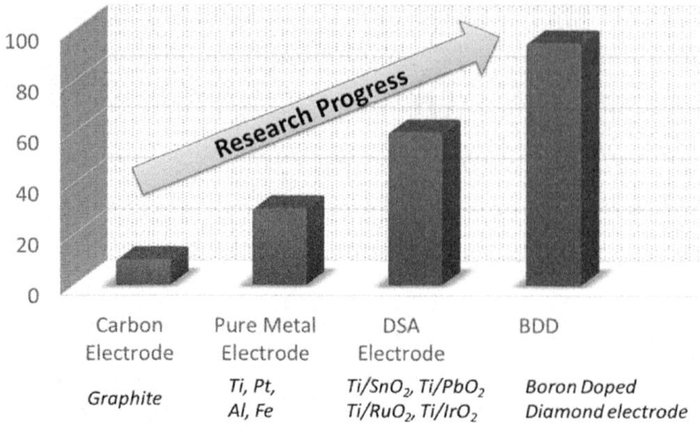


**FIGURE 8.7 Different types of electrodes for electrochemical oxidation of PFAS:** common electrodes used for electrochemical oxidation of PFAS. *From Shi, Lifeng, Leng, Chunpeng, Zhou, Yunlong, Yuan, Yue, Liu, Lin, Li, Fuping & Wang, Hao. (2024). A review of electrooxidation systems treatment of poly-fluoroalkyl substances (PFAS): electrooxidation degradation mechanisms and electrode materials. Environmental Science and Pollution research, 31(30), 42593–42613. https://doi.org/10.1007/s11356-024-34014-1.*

was effective (Shi et al., 2024). The findings show that $Cl_2$ and HClO are the primary oxidants in the electrochemical process, which is primarily indirect oxidation. The change in reaction rate during electrolysis was ascertained using reaction kinetics. The process's kinetic rate constant, which ranges from 0.15 to 0.07 $min^{-1}$ when the pH is between 3 and 7, indicates that lower pH values promote the breakdown of CBN. After 95 min of electrolysis, the removal efficiency of CBN reached to 60% at a pH of 3 and a current density of 10 mA $cm^{-2}$ (Camcioglu et al., 2022). A variety of metal electrodes, including those from the platinum group, Al, Fe, and Ti, are frequently utilized in EO procedures. Al and Fe can oxidize organic contaminants when employed as anodes, generating colloidal particles in the original electrolyte solution.

These anodes use electrical flocculation that is accomplished by coagulation, precipitation, or air flotation to influence the removal of colloidal pollutants from the wastewater (Ansari et al., 2022). Using aluminum as the anode material, Mao et al. (2023) (Mao et al., 2023) examined the chemical process involved in treating paracetamol and naturally occurring organic debris in river water. The researchers discovered that 60 min reaction time, 2 cm electrode spacing, and 9 V of voltage produced the best therapy outcomes. The efficiency of iron and aluminum anodes in the removal process was evaluated by a comparative study. The investigation's findings demonstrated that aluminum anodes were more effective than iron anodes. The creation of hydrogen bonds between paracetamol and aluminum hydroxide is the main mechanism of degradation, according to posttreatment X-ray diffraction data (Mao et al., 2023; Shi et al., 2024).

#### *8.3.2.2 Dimensionally stable anodes electrodes for environmental applications*

Mostly made of metal oxides like $PbO_2$, $SnO_2$, $IrO_2$, and $RuO_2$, DSAs have been widely used in industrial processes, most notably in the manufacturing of chlor-alkali, where they make up around 90% of the electrodes (Hussain et al., 2015). Their superiority over graphite or pure metal electrodes stems from their excellent conductivity and resilience to corrosion. Although titanium, a common DSA substrate, can produce a stable oxide layer that impairs performance, its advantageous conductivity and stability have made it a popular choice for

Ti-based DSAs. EO systems have been used to treat complicated industrial effluents, including wastewater from coke ovens, using these DSAs (Mörttinen, 2023). For instance, under ideal circumstances, a Ti/$RuO_2$ anode and graphite cathode combination removed 90% of organics, demonstrating outstanding endurance and surface integrity across several cycles. Doping with metal and nonmetal elements, which creates surface defects including crystal steps and dislocations, has been the subject of recent developments aimed at improving DSA performance. These flaws promote more effective electrochemical reactions by expanding electron routes, enhancing conductivity, and increasing catalytic activity. Higher oxygen evolution potential and better total organic carbon (TOC) removal kinetics were shown by Yb-doped Ti/$SnO_2$-$RuO_2$ electrodes. Like this, sol-gel produced $RuO_2$-$IrO_2$-$ZrO_2$ trimers exhibited improved catalytic activity and porosity with higher Ru levels, albeit at the expense of some stability (Khan et al., 2020). Notably, titanium-doped DSA electrodes have demonstrated better long-chain PFAS defluorination, highlighting the need for customized electrode design for dealing with persistent environmental pollutants (Qiu et al., 2024).

#### 8.3.2.3 *Boron-doped diamond electrode for advanced oxidation*

Due to their extremely poor electrical conductivity and very large bandgap (>5 eV), diamonds have intrinsic limits as an electrode material, despite their extraordinary hardness, chemical inertness, and outstanding thermal conductivity. But in the late 1990s, boron atoms were added to the diamond lattice to create BDD, which retained the beneficial qualities of diamond while greatly increasing conductivity. BDD electrodes have outstanding corrosion resistance, chemical stability, mechanical strength, and long service life. They may effectively generate hydroxyl radicals ($^{\cdot}OH$), which are potent oxidants for EO, thanks to their broad potential window and high oxygen evolution overpotential (Khan et al., 2023). Furthermore, BDD's low adsorption properties and inert surface allow for self-cleaning and minimizing electrode fouling, which enhances operational stability and lowers energy usage (Qiu et al., 2024). The strong oxidative potential of BDD electrodes, particularly in wastewater treatment applications, has been confirmed by recent investigations employing DFT and experimental analysis. According to comparative analyses, BDD performs better than conventional DSA electrodes such as Ti/$RuO_2$-$IrO_2$ in eliminating COD and TOC from difficult-to-treat effluents, including landfill leachate and chlorinated aromatic compounds. For example, under comparable circumstances, BDD's COD elimination efficiency reached up to 92%, whilst DSA only reached roughly 35%. Furthermore, BDD electrodes have demonstrated encouraging outcomes in the breakdown of persistent contaminants like PFOA and PFOS, with substrate composition affecting efficiency. These results establish BDD as an extremely efficient and successful anode material for EO, particularly when dealing with organic pollutants that are difficult to manage (Mörttinen, 2023).

### 8.3.3 Supporting electrolytes in electrochemical oxidation of PFAS

During any electrochemical degradation process, the choice of an appropriate supporting electrolyte is also indispensable, not only for providing sufficient ionic conductivity but also to directly impact the formation of reactive species and thereby the degradation efficiency of PFAS. Commonly, NaCl, $Na_2SO_4$, and $NaClO_4$ are used as electrolytes because of their reactivity with the electrode and the matrix water. NaCl has two major roles in this process: it improves conductivity and favors the production of reactive chlorine species like chlorine radicals ($Cl^•$), HOCl, and chlorine gas ($Cl_2$), which could be beneficial for the indirect PFAS oxidation. During the process, $Cl^•$ interacts with $^•OH$ (which comes from anodic surface) and produces perchlorate ions ($ClO_4^-$). As $ClO_4^-$ is a very stable and unreactive species, it may minimize the overall degradation performance and defluorination efficiency. On the other hand, $Na_2SO_4$ provides a preferred condition for sulfate radicals ($SO_4^{•-}$) generation in certain electrochemical conditions. Such radicals are strong oxidizing agents, and thus are important for PFAS degradation, especially when operated with high performing electrodes such as BDD or Magnéli phase titanium suboxide ($Ti_4O_7$). While this gives a competitive advantage, the sulfate-based systems usually have lower conductivities than chloride electrolytes, requiring higher cell voltages or longer treatment times, particularly for less reactive electrodes. The behavior of each electrolyte may vary significantly depending on the electrode material. For example, research using $Ti/RuO_2$ anodes has indicated that NaCl can be a more effective electrolyte than $Na_2SO_4$ in removing PFOS and PFOA, likely because the NaCl has a greater potential drop (and hence conductivity) at a given voltage. This increased resistance per Ohm's law translates to higher oxidation potentials at the anode surface, leading to higher reactivity for reaction intermediates, and thus to faster PFAS degradation. In contrast, with high-oxidation potential electrodes such as BDD or $Ti_4O_7$, $Na_2SO_4$ is observed to be the leading solution for the degradation of PFOA, and this could be ascribed to the efficient production of $SO_4^{•-}$ radicals. On the other hand, PFOS is more persistent under the same conditions, which indicates the impact of supporting electrolyte reactivity on PFAS, and the PFAS structure should be discussed accordingly when developing treatment protocols. Last but not least, the chemical inertness and high solubility of $NaClO_4$ do not make any significant contribution to the oxidation chemistry, because of the inherent stability of $ClO_4^-$. Therefore, it is frequently described as a control or model electrolyte in mechanistic studies and is less suitable for application-based conditions for PFAS destruction. Owing to the complex interplay between electrolyte species, PFAS molecules, and the electrode surface, additional investigations are needed to elucidate the underlying reaction pathways, especially as they pertain to chlorine-based radicals and their role in modulating the defluorination process.

### 8.3.4 Applications of electrochemical oxidation for per- and polyfluoroalkyl substances removal

EO is a highly customized method and could be operated on a lab scale and on a large scale. At the lab scale, it enables controlled treatment and mechanism investigation using advanced anode materials. Similarly, its integration into pilot-scale systems often highlights its practical potential for real-world remediation of PFAS-contaminated water matrices.

#### 8.3.4.1 *Laboratory-scale: graphene sponge electrodes*

Duinslaeger and Radjenovic (2022) present a boron-doped graphene sponge anode for POFA degradation in a flow-through bench-top system. The reactor was operated at a current density of 230 A $m^2$, which obtained PFAS removal efficiencies ranging from 17%–67% for C4-C8 molecules, with electrical energy consumption around 10 kWh $m^3$. The electrochemical system collectively electro sorption (7%–35%) and electrooxidation (9%–32%), and achieved 74%–87% defluorination of the electrooxidized fraction. The reactor also demonstrates that no chloride-harmful chlorinated oxidants were generated, making it promising for intricate and brackish waters (Xu et al., 2024). It highlights how advanced electrode materials, such as graphene sponges, can effectively cleave PFAS at a lab scale with controlled energy use and effective C–F bond disruption.

#### 8.3.4.2 *Pilot/field-scale: foam fractionation coupled electrochemical oxidation*

A recent study by Wang et al. demonstrated a hybrid treatment assay, coupling FF followed by EO degrade PFAS waste groundwater and landfill leachate using BDD or Ti-based electrodes (Xu et al., 2024; Wang et al., 2023). This system achieved an average 50% PFAS removal, with long-chain degradation up to 86% and short-chain up to 31%. Interestingly, the treatment reduced biological toxicity (via bacterial assays), and organic fluorine dropped by ~44%. FF-EO method was more energy-efficient as compared to EO, especially for longer-chain PFAS and waters with higher organic matter, confirming its promise for scaling up. Maldonado et al. (2021) explored scale-up of EO by treating PFAS waste containing brine from ion-exchange resin regeneration using BDD electrodes at both laboratory and semipilot scales. In the lab (2 L reactor), applying 50 mA $cm^2$ achieved >99% total PFAA removal within 8 h. Moving to the semipilot scale (~14 L), degradation slowed somewhat (~0.8× the rate), yet still delivered >94% removal, validating EO's robustness across scales. It is worth noting that combining ion-exchange concentration with EO reduces energy consumption by more than 90% compared to treating raw water, an important insight for cost-effective implementation (Maldonado et al., 2021).

#### 8.3.4.3 *High-pressure membrane couple electrochemical oxidation*

More than 90% of PFAS in polluted water can be eliminated by nanofiltration and reverse osmosis, but the concentrate that is produced requires additional treatment. According to Mastropietro et al. (2021), some research has investigated the use of EO and established optimization methodologies for potential commercial applications to address higher levels of PFAS in RO as well as NF concentrates (Mastropietro et al., 2021). RO concentrates from contaminated groundwater with soluble organic carbon of 38 mg $L^{-1}$ and PFAS values ranging from 13 to 440 μg $L^{-1}$ were successfully treated by EO. While the quantity of PFOA stayed constant, the concentrations of short-chain PFCAs, such as PFBA, PFPeA, PFHxA, and PFHpA, increased by 21, 6.8, and 4.8 times, respectively. Following 18 h of treatment, PFSA removals were seen at 83% PFBS, 92% PFHxS, 96% PFOS, and 60% 6:2 FTSA. There was 47% DOC removal and 42% defluorination. Another study that dealt with Gen-X in a synthetic freshwater with a starting level of 1 mg $L^{-1}$ was able to reduce the total volume by

80% and upconcentrate the desired component by a factor of five by using NF-90 membranes as a preliminary concentration step. With 99.5% Gen -X eradication, energy usage dropped by 83%, from 1470 kWh $m^3$ with EO treatment alone to 237 kWh $m^3$. Comparably, a study by def that used an NF-270 flat membrane was able to reduce the volume by a factor of five and raise the PFHxA concentration up to five times prior to EO treatment. With an energy usage of 15.2 kWh $m^3$, 90% PFHxA was removed, falling within the smaller bound of the declared EE/O values for single EO treatment. According to a different study that used mathematical models to assess the usage of NF-90 and NF-270 for the NF-EO mix treatment of PFHxA, the NF-90–EO combination would be capable of achieving 99% eradication at 11.5 kWh $m^3$ (Rekik et al., 2024; Veciana et al., 2022).

#### *8.3.4.4 Zwitterionic hydrogel-assisted electrochemical oxidation*

Sulfobetaine methacrylate (SBMA)-based hydrogel integrated into a flow- throughEO reactor is used in a comprehensive case study investigating the synergy between zwitterionic hydrogels and EO for PFAS removal. Using ethylene glycol dimethacrylate as a cross-linker, SBMA was polymerized to create the hydrogel used in this investigation, creating a three-dimensional, highly hydrophilic, and porous network. This zwitterionic hydrogel demonstrated remarkable resistance to fouling and selective adsorption of PFAS molecules, especially PFOA, because of its natural equilibrium of negatively charged sulfonate groups and positively charged quaternary ammonium groups. A mix of hydrophobic and electrostatic interactions underpinned the adsorption mechanism: the fluorinated tail of PFOA entrenched in the hydrophobic regions of the hydrogel matrix, while the negatively charged carboxylate head of the compound connected with the quaternary ammonium sites (Mayakaduwage et al., 2022). The hydrogel served as a preconcentration medium, enriching PFAS close to the anode surface when PFOA-contaminated water was run through the hydrogel-packed column next to a BDD anode. Because of the near proximity, which allowed for better contact between PFAS molecules and the hydroxyl radicals produced during anodic oxidation, lowering mass transfer restrictions and speeding up degradation rates, this spatial localization greatly increased the efficiency of EO (Lee and Drobn, 2023).

More than 90% of PFOA was broken down in 2 h with the hydrogel in place, at a comparatively low current density (10 mA $cm^2$). This occurred according to pseudo-first-order reaction kinetics, with a rate constant that was roughly 2.5 times greater than that of the system without hydrogel support. Because of more thorough and prolonged radical interaction, the zwitterionic matrix not only accelerated degradation but also reduced the production of harmful byproducts, such as shorter-chain PFAS intermediates (Berhanu et al., 2023). Crucially, the technique was both efficient and financially feasible because the hydrogel could be renewed and used again for up to 10 treatment cycles with little performance loss. This case study demonstrates how important constraints, including poor PFAS accessibility, fouling by organic matter, and inefficiencies in mass transfer, can be addressed by incorporating zwitterionic hydrogels into EO systems. Because of the chemical resilience of PFAS and the complexity of real-world water matrices, the hybrid approach, which combines selective adsorption and reactive degradation, offers a promising pathway for treating PFAS-contaminated water in both engineered and natural systems (Restivo et al., 2024).

## 8.4 Challenges and scientific issues associated with per- and polyfluoroalkyl substances electro-oxidation

Unlike other AOPs, the electro-oxidation of PFAS, has emerged as a robust system to target PFAS contaminants without generating secondary pollutants if the electrolytic parameters are tuned appropriately. EO has so far offered a promising candidate to address the decontamination of a broad spectrum of pollutants in water. It deems a great decontamination promise by virtue of its scalability, ease of application, nonselectivity, modulation, automation and efficiency. PFAS are often termed "forever chemicals" due to their extreme stability and persistence, which is why most of the conventional water treatment strategies are ineffective in their removal, potentially. In this scenario, so far, the EO techniques have demonstrated superior mineralization of PFAS via direct, indirect, or mediated EO. Nevertheless, the EO of PFAS is prone to several challenges and issues, which are indispensably worth mentioning here. The first but most serious challenge encountered by EO for PFAS degradation is passivity and poisoning of the electrode surface, which leads to a potential decline in catalytic activity and its overall service life as well. This activity decrease of the catalyst surface depends upon the inherent properties of the catalyst itself as well as on the organic species prevailing in the electrolyte solution. This issue of catalytic poisoning/passivity can be minimized by incorporating inert surface electrodes, for instance, BDD and Dimensionally Stable Anodes (DSA). Furthermore, catalytic poisoning can also be pacified by executing nano-catalyst materials owing to their exceptionally large surface areas, where surface poisoning renders meaningless. The nano-based catalytic materials, by virtue of their huge surface areas, diligently trigger the degradation kinetics to such an extent that the surface poisoning can be conveniently overwhelmed (Shi et al., 2024).

The second challenge associated with the use of the EO technique for PFAS abatement is its capital and operational costs. The capital costs are more pronounced when we rely upon noble metals (like Pt and its counterparts) for the fabrication of electrode materials, and the skyrocketing operational costs are due to large energy consumption. The former issue can be appropriately addressed by launching earth's abundant metals (like Fe, Co, Ni, etc) for fabrication of electrode material, while the latter problem of high energy consumption can be moderated promptly by installation of a renewable solar energy system. Similarly, sometimes the byproducts of electrochemical degradation are chlorinated compounds (like chlorinated hydrocarbons, which are known carcinogens), especially when NaCl is used as electrolyte. Thus, if degradation parameters are not adequately optimized and tuned, then these chlorinated compounds pose additional threats. Occasionally, EO techniques demand highly conductive effluents to execute degradations, which further contributes to the complexity of the degradation protocols in question (Hussain et al., 2020b).

The prevailing pH of the electrolytic solution is of pivotal significance as it not only controls the concentration and kinetics of $^{\bullet}OH$ radicals, but also oxygen evolution overpotential and ultimately the service life of the electrode materials. A broad spectrum of research has demonstrated that degradation efficiency is optimal in acidic protocols of electrolytes, since at lower pH, the generation of $^{\bullet}OH$ radicals is more likely and presents a higher overpotential for oxygen evolution reaction (OER), which remarkably enhances the degradation efficiency of EO. Conversely, in an alkaline medium, where the OER overpotential is low, which severely declines the kinetics of $^{\bullet}OH$ radical generation and hence retards EO efficacy. Thus in basic medium, the EO confronts a critical challenge (Niu et al., 2016). Some authors have linked the efficiency of PAFSs' electrooxidation as a function of chloride concentration in the electrolyte. During anodic production of chlorine at the active anode material, the active sites become preoccupied and drastically decrease the accessibility of PFAS molecules to the electrode surface, which in turn tremendously hampers the electro-oxidation of PFAS. Furthermore, it is quite challenging to distinguish between interconnected aspects of the solution, such as pH, ionic strength, electrolyte, and initial PFAS strength, as they vary from one PFAS to another. Similarly, EO efficiency is also much influenced by the presence of organic materials existing in the background in addition to PFAS target materials, which is why it is most likely to assess and scale up the laboratory-scale EO experiments to real samples from effluents. Extensive research devoted to PFAS and PFOA has revealed that the short-chain PFAS depicts an exceptional reluctance and stability towards electrooxidation and hence poses a critical challenge to the EO design and systems (Radjenovic et al., 2020).

## 8.5 Future directions

In the future, the development of novel electrode materials will hold great promise for the abatement of PFAS pollution. But these electrode materials must fulfill the demands like corrosion resistant, stable, cost- effective, efficient, scalable, nontoxic, and could easily be fabricated using low-cast technologies like spray coating, self-assembly, thermal treatments, etc. Here, a transition from noble metal-based catalysts to earth's abundant metals for electrode fabrication can be of crucial significance as far as cost- effectiveness is concerned. Nevertheless, they must present a high overpotential for OER to ensure an effective electrooxidation of PFAS. Fortunately, as far as cost-effectiveness is concerned, considerable research has been accomplished, and it has turned into an emerging approach for electrochemists and engineers to probe deeper insights. For instance, with respect to selectively targeting PFAS, several materials have been developed, like molecularly imprinted polymers (MIPs). Similarly, MIP conductive coatings equipped with binding sites that are dimensionally more compatible with PFAS constitute a novel approach to electrode design via a bottom-up mode of fabrication. Among the 2D materials, the graphene-based electrodes are gaining popularity with respect to selective degradations of PFAS materials; this will open a new window for further research. Similarly, the transition from 2D to 3D electrodes as well as electrochemical reactor design can play a pivotal role and has escalated the efficacy of EO of PFAS to an appreciable extent, which would be compatible with the demands of future pollution control. These 3D electrodes and reactor assemblies enhance mass transport and ensure the accessibility of PFAS to the electrode surface, which leads to high degradation kinetics and efficiency of EO, which decreases energy consumption promptly.

Last but not least, the role of twin technologies like EO/EC, EO/UV, EO/$O_3$, EO/Fenton can also be considered if a singular EO approach is not enough to achieve the desired goals of PFAS pollution abatement. This may be ascribed to incomplete degradation of PAFS contaminants, the presence of short- chain PFAS molecules (which are less vulnerable to degradation) and the interference of other ionic species in the water matrix, hence impeding the overall degradation process. The best solution to these issues is likely the adoption of amalgamated or twin/hybrid approaches, which have showcased excellent degradation outputs (Rahimi et al., 2024). Moreover, it is worth mentioning here that the efficiencies of such EO processes are the function of mass transport phenomenon, which is of paramount significance here as it takes the pollutant molecules to site of degradation, that is the electrode surface, from the bulk of electrolyte. Thus, a combined synergistic effect of an amalgamated approach like EO/UV with twin simultaneous approaches of irradiation accompanied by electrochemical treatment is superior to EO alone to achieve effective degradation goals. Thus, an integrated strategic approach for PFAS pollution abatement has emerged as a promising sector of scholarly interest. In the same way, nano- based electrocatalysts have also been tested by researchers and have presented remarkable PFAS degradations owing to their large surface areas and their potential resistance to getting poisoned. Hence, the nanoworld is expected to explore new horizons in the research domain of catalysis.

## 8.6 Conclusion

Innovative remediation strategies are needed because PFAS are chemically stable and resistant to conventional treatment methods. EO has been examined in this chapter as a potent and versatile method for PFAS degradation, with a focus on its benefits, including ease of use, low levels of secondary pollution, and the possibility of full mineralization. EO exhibits notable efficacy in degrading long- and short-chain PFAS chemicals via both direct and indirect pathways, especially when using sophisticated anode materials such as dimensionally stable anodes (DSAs) and BDD. Despite its potential, EO has drawbacks that need to be resolved for wider use, including energy consumption and electrode passivation. Combining EO with complementary techniques like adsorption onto new materials like zwitterionic hydrogels, EC, or UV irradiation provides synergistic solutions that improve performance while reducing drawbacks. Scaling EO techniques to real-world circumstances still requires the development of selective, long-lasting, and reasonably priced electrode materials in addition to reactor configurations that are optimized.

## References

Abbasian Chaleshtari, Z., Foudazi, R., 2022. A review on per- and polyfluoroalkyl substances (PFAS) remediation: separation mechanisms and molecular interactions. In ACS ES and T Water 2 (12). Available at https://doi.org/10.1021/acsestwater.2c00271.

Alexander, B.H., Ryan, A., Church, T.R., Kim, H., Olsen, G.W., Logan, P.W., 2024. Mortality and cancer incidence in perfluorooctanesulfonyl fluoride production workers. American Journal of Industrial Medicine 67 (4). Available at https://doi.org/10.1002/ajim.23568.

Ansari, K., Shrikhande, A., Malik, M.A., Alahmadi, A.A., Alwetaishi, M., Alzaed, A.N., et al., 2022. Optimization and operational analysis of domestic greywater treatment by electrocoagulation filtration using response surface methodology. Sustainability (Switzerland) 14 (22). Available at https://doi.org/10.3390/su142215230.

Awoyemi, O.S., Naidu, R., Fang, C., 2024. Advancements on ultrasonic degradation of per- and polyfluoroalkyl substances (PFAS): toward hybrid approaches. Environments - MDPI 11 (9). Available at https://doi.org/10.3390/environments11090187.

Bani-Melhem, K., Alnaief, M., Al-Qodah, Z., Al-Shannag, M., Elnakar, H., AlJbour, N., et al., 2025. On the performance of electrocoagulation treatment of high-loaded gray water: kinetic modeling and parameters optimization via response surface methodology. Applied Water Science 15 (5), 1–21. Available at https://doi.org/10.1007/s13201-025-02451-z.

Barisci, S., Suri, R., 2023. Degradation of emerging per- and polyfluoroalkyl substances (PFAS) using an electrochemical plug flow reactor. Journal of Hazardous Materials 460, 132419. Available at https://doi.org/10.1016/j.jhazmat.2023.132419 June.

Bellanthudawa, B.K.A., Nawalage, N.M.S.K., Handapangoda, H.M.A.K., Suvendran, S., Wijayasenarathne, K.A.S.H., Rathnasuriya, M.L.D., et al., 2023. A perspective on biodegradable and non-biodegradable nanoparticles in industrial sectors: applications, challenges, and future prospects. In Nanotechnology for Environmental Engineering 8 (4). Available at https://doi.org/10.1007/s41204-023-00344-7.

Berhanu, A., Mutanda, I., Taolin, J., Qaria, M.A., Yang, B., Zhu, D., 2023. A review of microbial degradation of per- and polyfluoroalkyl substances (PFAS): biotransformation routes and enzymes. Science of the Total Environment 859, 160010. Available at https://doi.org/10.1016/J.SCITOTENV.2022.160010.

Brosler, P., Girão, A.V., Silva, R.F., Tedim, J., Oliveira, F.J., 2023. In-house vs. commercial boron-doped diamond electrodes for electrochemical degradation of water pollutants: a critical review. In Frontiers in Materials 10. Available at https://doi.org/10.3389/fmats.2023.1020649.

Camcioglu, S., Özyurt, B., Oturan, N., Trellu, C., Oturan, M.A., 2022. Fast and complete destruction of the anti-cancer drug cytarabine from water by electrocatalytic oxidation using electro-fenton process. Catalysts 12 (12). Available at https://doi.org/10.3390/catal12121598.

Cara, B., Lies, T., Thimo, G., Robin, L., Lieven, B., 2022. Bioaccumulation and trophic transfer of perfluorinated alkyl substances (PFAS) in marine biota from the Belgian North Sea: distribution and human health risk implications. Environmental Pollution 311, 119907. Available at https://doi.org/10.1016/J.ENVPOL.2022.119907.

Cornelsen, M., Weber, R., Panglisch, S., 2021. Minimizing the environmental impact of PFAS by using specialized coagulants for the treatment of PFAS polluted waters and for the decontamination of firefighting equipment. Emerging Contaminants 7, 63–76. Available at https://doi.org/10.1016/J.EMCON.2021.02.001.

Corviseri, M.C., Polidoro, A., De Poli, M., Stevanin, C., Chenet, T., D'Anna, C., et al., 2025. Targeted determination of volatile fluoroalkyl pollutants and non-targeted screening for environmental monitoring. Talanta 292, 127944. Available at https://doi.org/10.1016/J.TALANTA.2025.127944.

Dasmahapatra, A.K., Williams, C.B., Myla, A., Tiwary, S.K., Tchounwou, P.B., 2023. A systematic review of the evaluation of endocrine-disrupting chemicals in the Japanese medaka (*Oryzias latipes*) fish. In Frontiers in Toxicology 5. Available at https://doi.org/10.3389/ftox.2023.1272368.

Duinslaeger, N., Doni, A., Radjenovic, J., 2023. Impact of supporting electrolyte on electrochemical performance of borophene-functionalized graphene sponge anode and degradation of per- and polyfluoroalkyl substances (PFAS). Water Research 242, 120232. Available at https://doi.org/10.1016/j.watres.2023.120232.

Duinslaeger, N., Radjenovic, J., 2022. Electrochemical degradation of per- and polyfluoroalkyl substances (PFAS) using low-cost graphene sponge electrodes. Water Research 213, 118148. Available at https://doi.org/10.1016/j.watres.2022.118148.

Fitch, A., Balderas-Hernandez, P., Ibanez, J.G., 2022. Electrochemical technologies combined with physical, biological, and chemical processes for the treatment of pollutants and wastes: a review. Journal of Environmental Chemical Engineering 10 (3), 107810. Available at https://doi.org/10.1016/j.jece.2022.107810.

Garg, A., Shetti, N.P., Basu, S., Nadagouda, M.N., Aminabhavi, T.M., 2023. Treatment technologies for removal of per- and polyfluoroalkyl substances (PFAS) in biosolids. In Chemical Engineering Journal 453. Available at https://doi.org/10.1016/j.cej.2022.139964.

Garnick, L., Massarsky, A., Mushnick, A., Hamaji, C., Scott, P., Monnot, A., 2021. An evaluation of health-based federal and state PFOA drinking water guidelines in the United States. Science of the Total Environment 761, 144107. Available at https://doi.org/10.1016/J.SCITOTENV.2020.144107.

Gunnarsdóttir, M.J., Andradóttir, H., Ólafsdóttir, K., Hlödversdóttir, Á.Ó., Kallenborn, R., Ræder, E.M., et al., 2025. PFAS in drinking water, wastewater and surface water in Reykjavik, Iceland. Environmental Science: Advances 4 (9), 1427–1443. Available at https://doi.org/10.1039/d5va00054h.

Habib, Z., Song, M., Ikram, S., Zahra, Z., 2024. Overview of per- and polyfluoroalkyl substances (PFAS), their applications, sources, and potential impacts on human health. In Pollutants 4 (1). Available at https://doi.org/10.3390/pollutants4010009.

Hieu Nhu, H.D., Tri, D.V., Luu, T.L., Trippel, J., Wagner, M., 2025. Degradation of 29 per- and poly-fluoroalkyl substances (PFAS) in water using fenton-assisted electrochemical oxidation process. Separation and Purification Technology 362, 131908. Available at https://doi.org/10.1016/j.seppur.2025.131908.

Huang, J., Wang, Q., Liu, S., Zhang, M., Liu, Y., Sun, L., et al., 2021. Crosstalk between histological alterations, oxidative stress and immune aberrations of the emerging PFOS alternative OBS in developing zebrafish. Science of the Total Environment 774, 145443. Available at https://doi.org/10.1016/J.SCITOTENV.2021.145443.

Hussain, S., Gul, S., Steter, J.R., Miwa, D.W., Motheo, A.J., 2015. Route of electrochemical oxidation of the antibiotic sulfamethoxazole on a mixed oxide anode. Environmental Science and Pollution Research 22 (19), 15004–15015. Available at https://doi.org/10.1007/s11356-015-4699-9.

Hussain, S., Khan, N., Gul, S., Khan, S., Khan, H., 2020a. Contamination of water resources by food dyes and its removal technologies. Water Chemistry. IntechOpen. Available at https://doi.org/10.5772/intechopen.90331.

Hussain, S., Ullah Khan, S., Gul, S., 2020b. Electrochemical Treatment of Antibiotics in Wastewater (pp. 355–394). https://doi.org/10.1007/978-3-030-40422-2_17.

Kabiri, S., Monaghan, C.L., Navarro, D., McLaughlin, M.J., 2023. Hydrophobic interaction is the dominant mechanism of zwitterionic PFAS adsorption to carbon-based sorptive materials in water and soil. Environmental Science: Water Research and Technology. Available at https://doi.org/10.1039/d3ew00550j.

Khan, H., Hussain, S., Ud Din, M.A., Arshad, M., Wahab, F., Hassan, U., et al., 2023. Multiple design and modelling approaches for the optimisation of carbon felt electro-Fenton treatment of dye laden wastewater. Chemosphere 338, 139510. Available at https://doi.org/10.1016/j.chemosphere.2023.139510.

Khan, S.U., Perini, J.A.L., Hussain, S., Khan, H., Khan, S., Boldrin Zanoni, M.V., 2020. Electrochemical preparation of Nb2O5 nanochannel photoelectrodes for enhanced photoelectrocatalytic performance in removal of RR120 dye. Chemosphere 257. Available at https://doi.org/10.1016/j.chemosphere.2020.127164.

Lebouachera, S.E.I., Balamane-Zizi, O., Boublia, A., Ghriga, M.A., Hasanzadeh, M., Hadri, H.E., et al., 2024. Understanding the factors affecting the adsorption of surface-active agents onto reservoir rock in chemical enhanced oil recovery applications: a comprehensive review. Chemistry Africa 7 (5), 2283–2306. Available at https://doi.org/10.1007/s42250-024-00931-4.

Lee, H., Drobn, I., 2023. Proceedings of APCE-CECE-ITP-IUPAC 2022.

Leung, S.C.E., Shukla, P., Chen, D., Eftekhari, E., An, H., Zare, F., et al., 2022. Emerging technologies for PFOS/PFOA degradation and removal: A review. In Science of the Total Environment 827. Available at https://doi.org/10.1016/j.scitotenv.2022.153669.

Li, G., Peng, M., Huang, Q., Huang, C.-H., Chen, Y., Hawkins, G., et al., 2025. A review on the recent mechanisms investigation of PFAS electrochemical oxidation degradation: mechanisms, DFT calculation, and pathways. Frontiers in Environmental Engineering 4, 1–15. Available at https://doi.org/10.3389/fenve.2025.1568542 April.

Li, S., Jiang, B., Liu, G., Shi, C., Yu, H., Lin, Y., 2024. Recent progress of particle electrode materials in three-dimensional electrode reactor: synthesis strategy and electrocatalytic applications. In Environmental Science and Pollution Research 31 (8). Available at https://doi.org/10.1007/s11356-023-31807-8.

Liang, Y., Wang, A., Liang, S., Sun, K., Xie, R., Zheng, C., et al., 2025. Durable $Ti_4O_7$ heterojunction composite membrane encapsulating N-doped graphene nanosheets for efficient electro-oxidation of GenX and other PFAS in fluorochemical wastewater. Environmental Science & Technology 59 (9), 4745–4755. Available at https://doi.org/10.1021/acs.est.4c09423.

Liao, Q., Tang, P., Fan, H., Song, Y., Liang, J., Huang, H., et al., 2023. Association between maternal exposure to per- and polyfluoroalkyl substances and serum markers of liver function during pregnancy in China: a mixture-based approach. Environmental Pollution 323, 121348. Available at https://doi.org/10.1016/J.ENVPOL.2023.121348.

Liu, D., Yan, S., Wang, P., Chen, Q., Liu, Y., Cui, J., et al., 2023. Perfluorooctanoic acid (PFOA) exposure in relation to the kidneys: a review of current available literature. In Frontiers in Physiology 14. Available at https://doi.org/10.3389/fphys.2023.1103141.

Liu, X., Shu, Y., Pan, Y., Zeng, G., Zhang, M., Zhu, C., et al., 2025. Electrochemical destruction of PFAS at low oxidation potential enabled by $CeO_2$ electrodes utilizing adsorption and activation strategies. Journal of Hazardous Materials 486, 137043. Available at https://doi.org/10.1016/j.jhazmat.2024.137043.

Liu, Y., Fan, X., Quan, X., Fan, Y., Chen, S., Zhao, X., 2019. Enhanced perfluorooctanoic acid degradation by electrochemical activation of sulfate solution on B/N codoped diamond. Environmental Science & Technology 53 (9), 5195–5201. Available at https://doi.org/10.1021/acs.est.8b06130.

Loc, N.X., Bao, T.T., Thi, P., Tuyen, T., Phong, L.H., Thi, D., et al., 2024. International Journal of Recycling Organic Waste Enhanced removal of $Pb^{2+}$ from aqueous solutions by iron / manganese binary oxide loaded on Eichhornia crassipes stem biochar. 10, 1–17.

Lu, Y., Guan, R., Zhu, N., Hao, J., Peng, H., He, A., et al., 2024. A critical review on the bioaccumulation, transportation, and elimination of per- and polyfluoroalkyl substances in human beings. In Critical Reviews in Environmental Science and Technology 54 (2). Available at https://doi.org/10.1080/10643389.2023.2223118.

Maldonado, V.Y., Becker, M.F., Nickelsen, M.G., Witt, S.E., 2021. Laboratory and semi-pilot scale study on the electrochemical treatment of perfluoroalkyl acids from ion exchange still bottoms. Water 13 (20), 2873. Available at https://doi.org/10.3390/w13202873.

Manna, M., Sen, S., 2023. Advanced oxidation process: a sustainable technology for treating refractory organic compounds present in industrial wastewater. In Environmental Science and Pollution Research 30 (10). Available at https://doi.org/10.1007/s11356-022-19435-0.

Mao, Y., Zhao, Y., Cotterill, S., 2023. Examining current and future applications of electrocoagulation in wastewater treatment. In Water (Switzerland) 15 (8). Available at https://doi.org/10.3390/w15081455.

Marks, K.J., Howards, P.P., Smarr, M.M., Flanders, W.D., Northstone, K., Daniel, J.H., et al., 2021. Prenatal exposure to mixtures of persistent endocrine-disrupting chemicals and birth size in a population-based cohort of british girls. Epidemiology (Cambridge, Mass.) 32 (4). Available at https://doi.org/10.1097/EDE.0000000000001351.

Martínez-Huitle, C.A., Rodrigo, M.A., Sirés, I., Scialdone, O., 2023. A critical review on latest innovations and future challenges of electrochemical technology for the abatement of organics in water. Applied Catalysis B: Environmental 328, 122430. Available at https://doi.org/10.1016/j.apcatb.2023.122430.

Mastropietro, T.F., Bruno, R., Pardo, E., Armentano, D., 2021. Reverse osmosis and nanofiltration membranes for highly efficient PFASs removal: overview, challenges and future perspectives. Dalton Transactions 50 (16). Available at https://doi.org/10.1039/d1dt00360g.

Mayakaduwage, S., Ekanayake, A., Kurwadkar, S., Rajapaksha, A.U., Vithanage, M., 2022. Phytoremediation prospects of per- and polyfluoroalkyl substances: a review. Environmental Research 212, 113311. Available at https://doi.org/10.1016/J.ENVRES.2022.113311.

Morales, M., Arp, H.P.H., Castro, G., Asimakopoulos, A.G., Sørmo, E., Peters, G., et al., 2024. Eco-toxicological and climate change effects of sludge thermal treatments: pathways towards zero pollution and negative emissions. Journal of Hazardous Materials 470, 134242. Available at https://doi.org/10.1016/J.JHAZMAT.2024.134242.

Mörttinen, J., 2023. Dimensionally stable anodes and their possibilities in neutral electrolytic pickling.

Nair, G., Soni, B., Shah, M., 2023. A comprehensive review on electro-oxidation and its types for wastewater treatment. Groundwater for Sustainable Development 23, 100980. Available at https://doi.org/10.1016/J.GSD.2023.100980.

Niu, J., Li, Y., Shang, E., Xu, Z., Liu, J., 2016. Electrochemical oxidation of perfluorinated compounds in water. Chemosphere 146, 526–538. Available at https://doi.org/10.1016/j.chemosphere.2015.11.115.

Ozben, T., Fragão-Marques, M., Tomasi, A., 2024. A comprehensive review on PFAS including survey results from the EFLM Member Societies. Clinical Chemistry and Laboratory Medicine 62 (6). Available at https://doi.org/10.1515/cclm-2023-1418.

Per - and Polyfluoroalkyl Substance (PFAS) - Potential Risk to Rural Groundwater Systems, 2025. January, 1–10.

Pierpaoli, M., Szopińska, M., Wilk, B.K., Sobaszek, M., Łuczkiewicz, A., Bogdanowicz, R., et al., 2021. Electrochemical oxidation of PFOA and PFOS in landfill leachates at low and highly boron-doped diamond electrodes. Journal of Hazardous Materials 403. Available at https://doi.org/10.1016/j.jhazmat.2020.123606 July 2020.

Qiu, F., Wang, L., Fan, Y., Pan, Y., Song, H., Ye, Z., et al., 2024. Review on structural adjustment strategies of titanium-based metal oxide dimensionally stable anodes in electrochemical advanced oxidation technology. Advanced Engineering Materials 26 (8), 1–22. Available at https://doi.org/10.1002/adem.202400122.

Radjenovic, J., Duinslaeger, N., Avval, S.S., Chaplin, B.P., 2020. Facing the challenge of poly- and perfluoroalkyl substances in water: is electrochemical oxidation the answer? Environmental Science & Technology 54 (23), 14815–14829. Available at https://doi.org/10.1021/acs.est.0c06212.

Rahimi, F., Nasiri, A., Hashemi, M., Rajabi, S., Abolghasemi, S., 2024. Advances in three-dimensional electrochemical degradation: a comprehensive review on pharmaceutical pollutants removal from aqueous solution. Chemosphere 362, 142620. Available at https://doi.org/10.1016/J.CHEMOSPHERE.2024.142620.

Rehnstam, S., Smith, S.J., Ahrens, L., 2024. Suspect and non-target screening of per- and polyfluoroalkyl substances (PFAS) and other halogenated substances in electrochemically oxidized landfill leachate and groundwater. Journal of Hazardous Materials 480, 136316. Available at https://doi.org/10.1016/j.jhazmat.2024.136316.

Rekik, H., Arab, H., Pichon, L., El Khakani, M.A., Drogui, P., 2024. Per-and polyfluoroalkyl (PFAS) eternal pollutants: sources, environmental impacts and treatment processes. Chemosphere 358, 142044. Available at https://doi.org/10.1016/j.chemosphere.2024.142044 January.

Restivo, J., Orge, C.A., Soares, O.S.G.P., Pereira, M.F.R., 2024. A review of current and prospective catalytic routes for the management of PFAs contamination in water. Journal of Environmental Chemical Engineering 12 (3), 112859. Available at https://doi.org/10.1016/J.JECE.2024.112859.

Saidi, M., Reguig, B.A., El Amine Monir, M., Althagafi, T.M., Fatmi, M., Remil, A., et al., 2025. Kinetics thermodynamics and adsorption study of raw treated diatomite as a sustainable adsorbent for crystal violet dye. Scientific Reports 15 (1), 1–18. Available at https://doi.org/10.1038/s41598-025-05787-3.

Samuel, M.S., Kadarkarai, G., Ryan, D.R., McBeath, S.T., Mayer, B.K., McNamara, P.J., 2024. Enhanced perfluorooctanoic acid (PFOA) degradation by electrochemical activation of peroxydisulfate (PDS) during electrooxidation for water treatment. Science of the Total Environment 942, 173736. Available at https://doi.org/10.1016/j.scitotenv.2024.173736.

Satriaji, F.V., Le Tong, C.T., Marlina, N., Lin, Y., Dat, N.D., Bui, H.M., et al., 2025. Comprehensive risk evaluation of perfluoroalkyl substance pollution in urban riverine systems: ecotoxicological and human health perspectives. Toxics 13 (6), 435. Available at https://doi.org/10.3390/toxics13060435.

Sharma, S., Shetti, N.P., Basu, S., Nadagouda, M.N., Aminabhavi, T.M., 2022. Remediation of per- and polyfluoroalkyls (PFAS) via electrochemical methods. Chemical Engineering Journal 430, 132895. Available at https://doi.org/10.1016/j.cej.2021.132895.

Shi, H., Chiang, S.-Y.(D.), Wang, Y., Wang, Y., Liang, S., Zhou, J., et al., 2021. An electrocoagulation and electrooxidation treatment train to remove and degrade per- and polyfluoroalkyl substances in aqueous solution. Science of the Total Environment 788, 147723. Available at https://doi.org/10.1016/j.scitotenv.2021.147723.

Shi, L., Leng, C., Zhou, Y., Yuan, Y., Liu, L., Li, F., et al., 2024. A review of electrooxidation systems treatment of poly-fluoroalkyl substances (PFAS): electrooxidation degradation mechanisms and electrode materials. Environmental Science and Pollution Research 31 (30), 42593–42613. Available at https://doi.org/10.1007/s11356-024-34014-1.

Sivagami, K., Sharma, P., Karim, A.V., Mohanakrishna, G., Karthika, S., Divyapriya, G., et al., 2023. Electrochemical-based approaches for the treatment of forever chemicals: Removal of perfluoroalkyl and polyfluoroalkyl substances (PFAS) from wastewater. Science of the Total Environment 861, 160440. Available at https://doi.org/10.1016/J.SCITOTENV.2022.160440.

Somashekara, D., Mulky, L., 2023. Sequestration of contaminants from wastewater: a review of adsorption processes. ChemBioEng Reviews 10 (4), 491–509. Available at https://doi.org/10.1002/cben.202200050.

Sultan, M.B., Anik, A.H., Rahman, M.M., 2024. Emerging contaminants and their potential impacts on estuarine ecosystems: are we aware of it? In Marine Pollution Bulletin 199. Available at https://doi.org/10.1016/j.marpolbul.2023.115982.

Tasca, A.L., Uwayezu, J.N., Carabante, I., Kumpiene, J., 2025. Electrochemical remediation of PFAS by boron-doped diamond electrodes: a review. Journal of Environmental Chemical Engineering 13 (3), 117044. Available at https://doi.org/10.1016/j.jece.2025.117044.

Timmermann, C.A.G., Pedersen, H.S., Weihe, P., Bjerregaard, P., Nielsen, F., Heilmann, C., et al., 2022. Concentrations of tetanus and diphtheria antibodies in vaccinated Greenlandic children aged 7–12 years exposed to marine pollutants, a cross sectional study. Environmental Research 203. Available at https://doi.org/10.1016/j.envres.2021.111712.

Urtiaga, A., Gómez-Lavín, S., Soriano, A., 2022. Electrochemical treatment of municipal landfill leachates and implications for poly- and perfluoroalkyl substances (PFAS) removal. Journal of Environmental Chemical Engineering 10 (3). Available at https://doi.org/10.1016/j.jece.2022.107900.

Veciana, M., Bräunig, J., Farhat, A., Pype, M.L., Freguia, S., Carvalho, G., et al., 2022. Electrochemical oxidation processes for PFAS removal from contaminated water and wastewater: fundamentals, gaps and opportunities towards practical implementation. Journal of Hazardous Materials 434, 128886. Available at https://doi.org/10.1016/J.JHAZMAT.2022.128886.

Wang, Y., Ji, Y., Li, K., Huang, Q., 2023. Foam fractionation and electrochemical oxidation for the treatment of per- and polyfluoroalkyl substances (PFAS) in environmental water samples. Chemosphere 339, 139615. Available at https://doi.org/10.1016/j.chemosphere.2023.139615.

Xu, X., Li, Y., Vo, P.H.N., Shukla, P., Ge, L., Zhao, C.-X., 2024. Electrochemical advanced oxidation of per- and polyfluoroalkyl substances (PFASs): development, challenges and perspectives. Chemical Engineering Journal 500, 157222. Available at https://doi.org/10.1016/j.cej.2024.157222.

Yanagida, A., Webb, E., Harris, C.E., Christenson, M., Comfort, S., 2022. Using electrochemical oxidation to remove PFAS in simulated investigation-derived waste (IDW): laboratory and pilot-scale experiments. Water (Switzerland) 14 (17). Available at https://doi.org/10.3390/w14172708.

Yang, J.S., Lai, W.W.P., Panchangam, S.C., Lin, A.Y.C., 2020. Photoelectrochemical degradation of perfluorooctanoic acid (PFOA) with GOP25/FTO anodes: intermediates and reaction pathways. Journal of Hazardous Materials 391. Available at https://doi.org/10.1016/j.jhazmat.2020.122247 February.

Yuan, Q., Qu, S., Li, R., Huo, Z.Y., Gao, Y., Luo, Y., 2023. Degradation of antibiotics by electrochemical advanced oxidation processes (EAOPs): performance, mechanisms, and perspectives. In Science of the Total Environment 856. Available at https://doi.org/10.1016/j.scitotenv.2022.159092.

Zhang, J., Gao, L., Bergmann, D., Bulatovic, T., Surapaneni, A., Gray, S., 2023. Review of influence of critical operation conditions on by-product/intermediate formation during thermal destruction of PFAS in solid/biosolids. In Science of the Total Environment 854. Available at https://doi.org/10.1016/j.scitotenv.2022.158796.

Zhao, L., Pu, R., Deng, S., Lin, L., Mantzavinos, D., Naidu, R., et al., 2025. Enhanced electrochemical degradation of per- and polyfluoroalkyl substances (PFAS) by activating persulfate on boron-doped diamond (BDD) anode. Separation and Purification Technology 359, 130459. Available at https://doi.org/10.1016/j.seppur.2024.130459.

Zhu, J., Chen, Y., Gu, Y., Ma, H., Hu, M., Gao, X., et al., 2022. Feasibility study on the electrochemical reductive decomposition of PFOA by a Rh/Ni cathode. Journal of Hazardous Materials 422, 126953. Available at https://doi.org/10.1016/j.jhazmat.2021.126953 August 2021.

CHAPTER

# 9

# Remediation of PFAS-contaminated soils: Current and emerging technologies

*Vasiliki Kinigopoulou*

**Department of Environmental Engineering, School of Engineering, International Hellenic University, Sindos, Thessaloniki, Greece**

## 9.1 Introduction

Per- and polyfluoroalkyl substances (PFAS) are a diverse family of thousands of synthetic chemicals engineered for their resistance to oil and water (National Institute of Environmental Health Sciences Perfluoroalkyl and Polyfluoroalkyl Substances, n.d.; EPA, U.S.E.P.A (n.d.)). Their defining feature is a chain of carbon atoms fully or partially substituted with fluorine, forming some of the strongest chemical bonds in organic chemistry, which renders them thermally and chemically stable. Because of these properties, PFAS have been used since the 1940s in numerous industrial and consumer applications, including firefighting foams, nonstick cookware, stain-resistant textiles, food packaging, and electronics. The widespread use and persistence of PFAS have led to their global distribution: PFAS are now detected in air, water, soil, and biota, and surveys find measurable concentrations in the blood of most Americans.

Soils are a critical environmental sink for PFAS because they accumulate contamination from multiple sources, including atmospheric deposition, irrigation with PFAS-laden water, land application of biosolids and industrial sludge, use of aqueous film-forming foams for firefighting, pesticide formulations, unlined landfills, and accidental spills (Evich et al., 2022). Once deposited, PFAS can persist in the solid matrix for decades due to their low reactivity and affinity for organic matter and mineral surfaces. Contaminated soils pose a long-term risk by slowly releasing PFAS to the hydrosphere through leaching and runoff, and by providing an exposure pathway to terrestrial organisms and humans via dust inhalation, ingestion of crops or livestock raised on contaminated sites.

The term PFAS encompasses several structural classes—perfluoroalkyl carboxylic acids (PFCAs), perfluoroalkyl sulfonic acids (PFSAs), sulfonamides, and precursors—with chain

*Per- and Polyfluoroalkyl Substances (PFAS) as Environmental Contaminants*
DOI: https://doi.org/10.1016/B978-0-443-43856-1.00008-3

lengths typically between four and sixteen carbon atoms (National Institute of Environmental Health Sciences Perfluoroalkyl and Polyfluoroalkyl Substances, (n.d.)). Physicochemical behavior varies across these classes: short-chain PFAS (≤C8) are more water soluble and mobile, whereas long-chain PFAS exhibit greater sorption to soils and bioaccumulation potential (Evich et al., 2022). The remarkable diversity of PFAS (thousands of compounds) makes hazard assessment challenging and underscores the need for robust remediation strategies (EPA, U.S.E.P.A, (n.d.)). Effective management requires technologies capable of addressing both mobile and strongly sorbed PFAS fractions and of adapting to different soil types and contamination scenarios.

The behavior of PFAS in soil is strongly influenced by carbon chain length and functional group chemistry, which controls sorption strength, mobility, and bioavailability. Long-chain PFAS are preferentially adsorbed onto soil organic matter and mineral surfaces, often displacing shorter-chain compounds through competitive sorption, whereas short-chain PFAS remain more mobile and prone to leaching. This heterogeneous sorption behavior complicates remediation efforts and partly explains the limited effectiveness of single-technology approaches in complex soil systems (Higgins and Luthy, 2006; Maimaiti et al., 2018; Fabregat-Palau et al., 2021; Abraham et al., 2022).

This chapter reviews the current state of PFAS soil remediation technologies with a focus on five representative approaches: thermal treatment, extraction, oxidation, bioremediation, and phytoremediation. Thermal treatment encompasses high-temperature incineration, pyrolysis, thermal desorption, and smoldering combustion. Extraction covers soil washing and chemical extraction to mobilize PFAS into an aqueous phase. Oxidation explores heat-activated persulfate, electrochemical methods, and other destructive chemical processes. Bioremediation focuses on microbial processes, including bioaugmentation and biostimulation, which aim to transform PFAS compounds or their precursors through biological activity in soil systems. Phytoremediation discusses plant-based and hybrid techniques for risk management. By examining the mechanisms, operating conditions, efficiencies, and practical challenges of each technology, the chapter aims to guide the selection of appropriate remediation strategies and to highlight emerging opportunities such as renewable energy integration.

## 9.2 Thermal treatment

### 9.2.1 Overview and mechanism

Thermal treatment is one of the most well-established methods for destroying PFAS in soil matrices due to its ability to reach temperatures high enough to break the strong carbon–fluorine (C–F) bonds. The thermal degradation of PFAS involves volatilization followed by oxidative or pyrolytic destruction. The stability of PFAS, particularly long-chain compounds such as perfluorooctanoic acid (PFOA) and perfluorooctane sulfonate (PFOS), requires extremely high temperatures, typically >1000°C, for complete mineralization (Hurst et al., 2024; Vargette et al., 2023). These high temperatures can be achieved via incineration, rotary kilns, or in situ thermal desorption. During heating, PFAS molecules may undergo defluorination, releasing hydrogen fluoride (HF), $CO_2$, CO, and fluorinated hydrocarbons as byproducts. Careful control of residence time and oxygen availability is crucial to prevent the formation of incomplete combustion products such as perfluoroalkenes and fluoroaromatics (Brusseau, 2018).

## 9.2.2 Incineration and vitrification technologies

Incineration is a widely used thermal treatment technology for the destruction of hazardous organic waste. The process involves the oxidation of organic compounds at elevated temperatures in the presence of oxygen, converting contaminants primarily into carbon dioxide, water vapor, and inorganic residues. Typical combustion systems operate at temperatures between approximately 800°C and 1200°C, depending on the waste composition and reactor design. Efficient incineration requires adequate oxygen supply, sufficient residence time, and effective mixing within the combustion chamber to ensure complete oxidation of target compounds (Urciuolo et al., 2025).

For PFAS-contaminated materials, high-temperature incineration in dedicated hazardous-waste combustion facilities is considered one of the most effective treatment options. Due to the exceptional stability of the C–F bond, temperatures exceeding 1000°C–1200°C are typically required to achieve effective degradation and mineralization of PFAS molecules. Under optimized operating conditions, incinerators can achieve destruction efficiencies (DE) exceeding 99.99%. However, concerns regarding incomplete combustion, the formation of intermediate PFAS compounds, and the management of fluorinated off-gas emissions have led to increased regulatory scrutiny and the development of strict operational guidelines for PFAS incineration facilities (USEPA Interim Guidance on the Destruction and Disposal of PFAS and Materials Containing PFAS, 2020) (see Figs. 9.1 and 9.2).

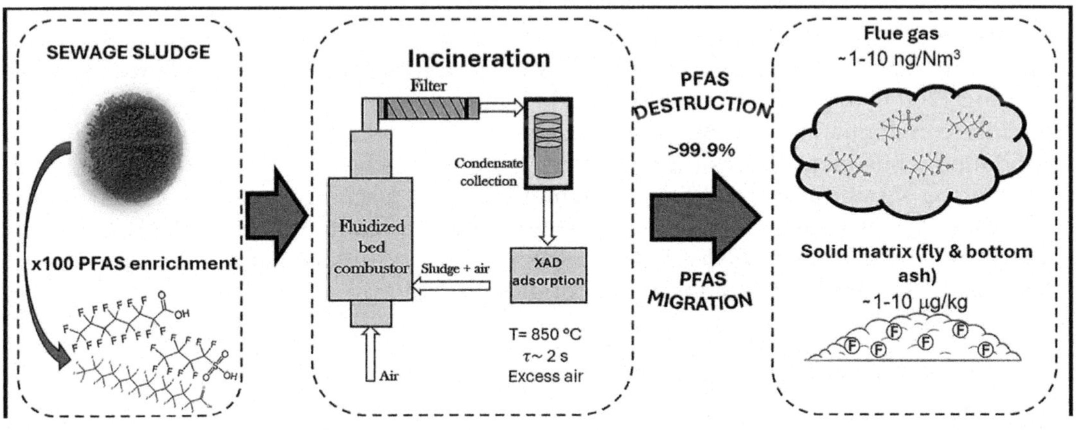


**FIGURE 9.1** Schematic representation of per- and polyfluoroalkyl substance (PFAS) degradation pathways during high-temperature incineration. Oxidative thermal treatment promotes defluorination of PFAS and formation of inorganic fluorine species, while fluorinated gaseous byproducts may form and require advanced off-gas treatment. *Adapted from Urciuolo, M., Migliaccio, R., Ciccone, B., Ruoppolo, G., de Folly d'Auris, A., Frisario, S., et al., 2025. Experimental investigation on PFAS degradation through municipal sludge combustion processes. Applied Thermal Engineering 276, 127006, licensed under CC-BY.*

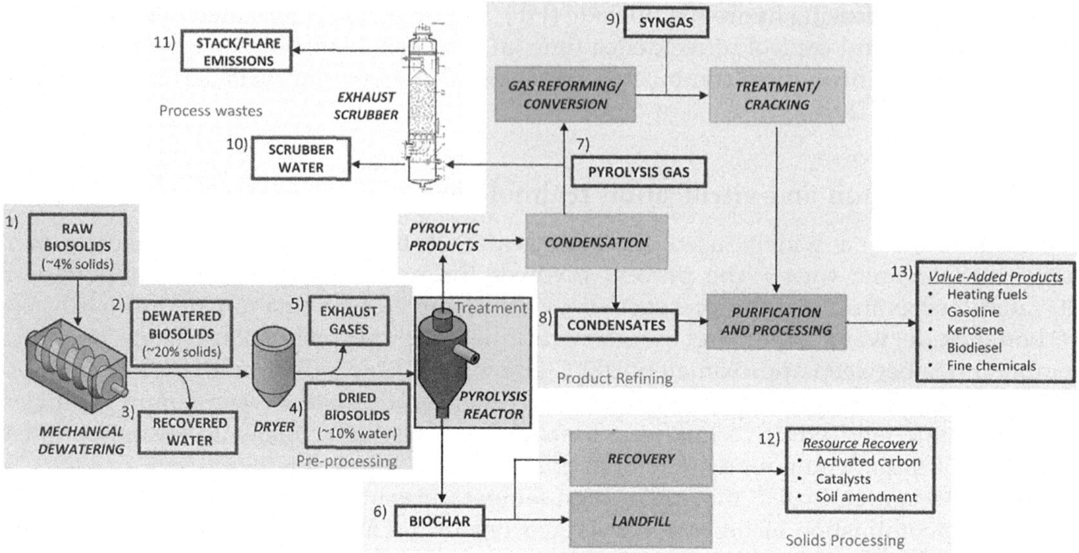


**FIGURE 9.2** Schematic representation of a pyrolysis treatment system showing process compartments and critical sampling locations. Feedstock drying and devolatilization occur in stages (1–5), followed by thermal decomposition producing biochar (6), pyrolysis gas (7), and condensable liquids such as bio-oils and tars (8). Noncondensable gases may be combusted to supply heat to the reactor (9), while downstream streams and products (10–13) represent potential pathways for contaminant transformation or release. *Adapted from Wallace, J.S., Edirisinghe, D., Seyedi, S., Noteboom, H., Blate, M., Balci, D.D., et al., 2023. Burning questions: current practices and critical gaps in evaluating removal of per-and polyfluoroalkyl substances (PFAS) during pyrolysis treatments of biosolids. Journal of Hazardous Materials Letters. 4, 100079.*

In addition to conventional incineration, vitrification technologies have emerged as an alternative thermal approach for the treatment of contaminated soils and mixed hazardous wastes. Vitrification involves melting soil and waste matrices at temperatures typically exceeding 1400°C to produce a stable glass-like material that immobilizes both organic and inorganic contaminants. During this process, organic pollutants may undergo thermal degradation while metals and other inorganic components become encapsulated within the glass matrix, significantly reducing their mobility and leaching potential. Navarro et al. demonstrated that solar-powered thermal vitrification of contaminated soils could immobilize heavy metals such as lead and copper while simultaneously degrading organic pollutants under concentrated solar flux conditions (Navarro et al., 2013).

### 9.2.3 Pyrolysis and ermerging solar thermal systems

Pyrolysis is a thermochemical treatment process that decomposes organic materials at elevated temperatures (typically 500°C–1000°C) under inert or oxygen-deficient conditions (Winchell et al., 2021). Unlike incineration, pyrolysis breaks down organic matter into three main product streams: a carbon-rich solid residue (biochar), condensable liquids (bio-oils and tars), and noncondensable gases such as CO, $CO_2$, $H_2$, and light hydrocarbons (Wallace et al., 2023). Prior to thermal decomposition, contaminated materials such as biosolids or soils typically undergo drying and devolatilization at temperatures below approximately 300°C, releasing moisture, $CO_2$, and volatile organic compounds. As the temperature increases to around 350°C–500°C, primary pyrolysis reactions generate combustible gases including carbon monoxide, hydrogen, alcohols, aldehydes, and organic acids, while leaving behind a carbon-rich char matrix (Demirbas and Arin, 2002). At higher temperatures (>500°C), secondary reactions promote the formation of condensable liquids and additional gaseous products. The resulting pyrolysis gas can be captured and combusted to provide heat for the reactor, partially sustaining the pyrolysis process.

The fate of PFAS during pyrolysis remains an active area of research. Bench-scale studies of PFAS-contaminated biosolids report removals exceeding 80% for several PFAS species at temperatures between 500°C and 800°C; however, certain short-chain PFAS may persist or become enriched in some pyrolysis products (Wallace et al., 2023; McNamara et al., 2023).

Recent innovations explore the use of solar thermal energy as a renewable heat source for thermochemical treatment systems. Solar concentrators can generate temperatures exceeding 1000°C, enabling decentralized treatment systems with reduced carbon footprint (Navarro (n.d.)). Navarro et al. highlighted the potential of solar-driven thermal confinement and evaporation processes for contaminant extraction and partial degradation, suggesting that similar approaches could be adapted for PFAS treatment when combined with suitable posttreatment technologies (Wu et al., 2023).

### 9.2.4 Thermal desorption and hybrid systems

Thermal desorption is a nondestructive technology that volatilizes PFAS from soil matrices rather than destroying them. Operating at moderate temperatures (typically 350°C–550°C), it contrasts with high-temperature incineration and relies on off-gas scrubbers to collect vaporized compounds (Hurst et al., 2024). Ex situ systems may use direct or indirect-fired rotary dryers or thermal screw conveyors; modular thermopiles provide mobile options for remote sites. Although the approach has been demonstrated only at bench and pilot scales, it offers a means to treat soils contaminated with PFAS by transferring them to a gaseous stream that can be processed separately.

Performance depends on operating conditions and the chemical structure of the target PFAS. Studies show that optimal removal occurs between 350°C and 450°C with residence times of 15–45 min (Hurst et al., 2024; Sörengård et al., 2020). Sulfonic and sulfonamide PFAS are more readily volatilized than carboxylic PFCAs, while adding calcium hydroxide can lower the volatilization temperature of sulfonates when the molar Ca/F ratio exceeds unity. Desorption is more challenging from clay-rich or organic-rich soils, where PFAS are strongly adsorbed, and pretreatment to remove debris or moisture may be required to improve heat transfer and reduce energy consumption.

Thermal desorption is generally applied to soils with high contaminant concentrations but limited volumes, such as shallow source zones impacted by firefighting foam. The high capital and operational costs of thermal technologies make them most cost-effective when treatment plants are located near the contaminated site (Hurst et al., 2024). Mobile thermopile systems can reduce transport distances but typically have lower throughput and longer treatment times. Careful management of vapor-phase emissions is necessary; recovered off-gas often requires further treatment by carbon adsorption or catalytic oxidation to prevent secondary pollution.

### 9.2.5 Smoldering combustion

Smoldering combustion is a flameless, self-sustaining oxidation process that occurs on the surface of a fuel within a porous medium. Unlike flaming combustion, smoldering involves heterogeneous oxidation reactions in which oxygen directly reacts with the fuel surface, producing heat that propagates through the porous matrix (Duchesne et al., 2020). When sufficient heat is generated to offset thermal losses, the reaction can become self-sustaining and continue without additional external energy input after ignition. This process has been widely investigated as a remediation strategy for contaminated soils, wastewater sludge, and other organic wastes due to its ability to destroy organic contaminants while leaving behind primarily inert mineral residues.

In PFAS-contaminated soils, smoldering combustion is typically initiated by adding a supplemental fuel, most commonly granular activated carbon (GAC), to ensure that the oxidation reactions generate sufficient heat to reach temperatures required for PFAS degradation (approximately 900°C) (Fournie et al., 2023). Once ignition occurs, the smoldering front propagates through the soil matrix, transferring heat and oxygen to the contaminated material. During this process, perfluoroalkyl acids (PFAAs) and other PFAS compounds can undergo thermal decomposition, producing inorganic fluoride species along with carbon dioxide and water.

Laboratory and pilot-scale experiments have demonstrated that smoldering combustion can reduce both short- and long-chain PFAS concentrations by more than 99.9% under optimized conditions (Duchesne et al., 2020; Major, 2019). In column experiments using GAC-amended soils, most of the fluorine was retained in the residual ash, while vapor-phase emissions accounted for less than one percent of the initial PFAS mass. The addition of calcium oxide (CaO) can further reduce hydrogen fluoride emissions by binding fluoride species in the combustion zone. Field demonstrations treating contaminated soil volumes up to approximately 10 $m^3$ have confirmed the potential of smoldering as an effective remediation approach where soil permeability allows adequate airflow and distribution of the surrogate fuel.

Fig. 9.3 illustrates the conceptual mechanism of PFAS destruction during smoldering combustion. GAC acts as a surrogate fuel that sustains the exothermic oxidation reactions, while calcium oxide may be added to capture fluoride species generated during PFAS degradation.

Despite its high removal efficiency and relatively low external energy demand, smoldering combustion remains at the pilot-scale stage for PFAS remediation. Process performance depends strongly on soil permeability, fuel distribution, air injection rates, and temperature control. Consequently, careful system design and emission management are required to ensure complete contaminant destruction and minimize potential secondary releases of fluorinated compounds (Duchesne et al., 2020; Major, 2019).

## 9.3 Extraction

### 9.3.1 Overview and mechanism

Extraction strategies for PFAS-contaminated soils aim to mobilize and transfer the contaminants from the solid matrix to a liquid phase where further treatment can occur.

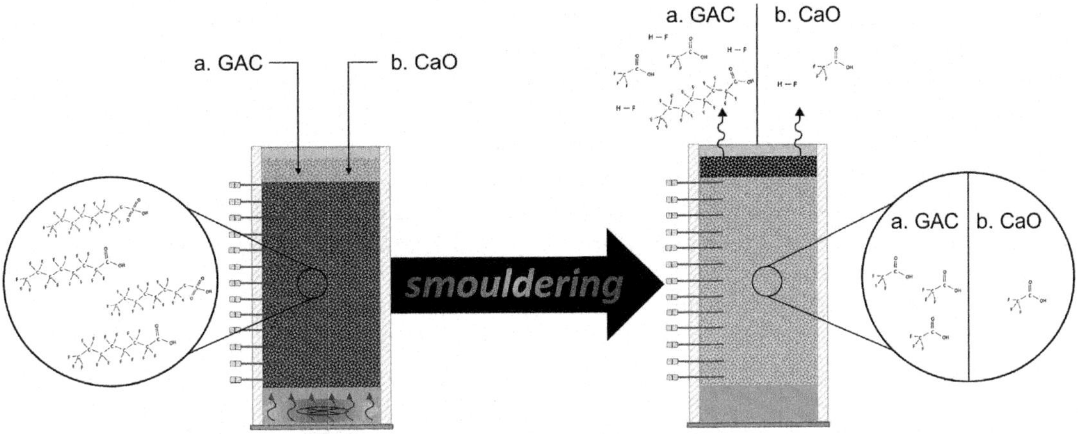


**FIGURE 9.3** Conceptual representation of per- and polyfluoroalkyl substance (PFAS) destruction during smoldering combustion in contaminated soils. Granular activated carbon (GAC) acts as a surrogate fuel that sustains a self-propagating smoldering front, generating high temperatures capable of degrading PFAS compounds. Calcium oxide (CaO) may be added to capture fluorine species formed during the process and reduce hydrogen fluoride emissions. *Adapted from Fournie, T., Rashwan, T.L., Switzer, C., Gerhard, J.I., 2023. Smouldering to treat PFAS in sewage sludge. Waste Management. 164, 219–227.*

PFAS molecules, particularly the ionic forms such as PFOA and PFOS, adsorb strongly to soil particles via electrostatic and hydrophobic interactions, especially in high-organic or clay-rich soils (Wang et al., 2017). Short-chain PFAS exhibit higher mobility due to their lower molecular weight and weaker sorption, making them more amenable to aqueous-phase extraction. The efficiency of extraction depends on soil properties (e.g., pH, organic matter content), PFAS chain length, and the chemistry of the extracting solution. Typically, water-based extraction enhanced with surfactants or complexing agents improves PFAS desorption, though removal rates vary considerably across compounds (Meegoda et al., 2022).

## 9.3.2 Soil washing and surfactant-enhanced techniques

Soil washing is an ex situ extraction process that uses water, often amended with surfactants, chelating agents, or pH adjustment, to transfer PFAS from the solid matrix into a process water stream (Hurst et al., 2024; Uwayezu et al., 2024). Excavated soils are fed into washing units comprising log washers, hydrocyclones, screens, vibrating tables, and other mechanical separators to disaggregate soil aggregates and separate coarse particles from fine-grained fractions. The wash water is recycled through treatment units such as GAC and ion-exchange resins to remove PFAS before being reused or discharged; however, fine particles and residuals with elevated PFAS concentrations may require further treatment or disposal.

The efficiency of soil washing depends on both the properties of the PFAS and the soil. Laboratory and pilot observations indicate that between 71% and 90% of the total PFAS mass in excavated soils is mobilized into the water phase. Removal efficiencies exceeding

85%–97% have been reported for short-chain PFAS (C4–C8) in pilot plants, whereas full-scale systems have achieved 90%–99.7% reduction across a broad range of PFAS (Hurst et al., 2024; Grimison et al., 2023). Longer-chain compounds (>C8) exhibit stronger hydrophobic interactions and electrostatic binding to soil surfaces; their removal may require multiple washing cycles, mechanical shredding to increase contact, or a combination with other technologies.

Soil washing is most effective for granular, sandy soils with low clay and organic matter content. High fines content reduces the efficiency of separation and increases the volume of fine-grained residue (filter cake) that must be treated or disposed of (Hurst et al., 2024; Quinnan et al., 2022). Traditional guidelines suggest applying soil washing to soils containing less than 40% fines; however, recent applications to cohesive soils have achieved removals of 85%–97% for C4–C8 PFAS, demonstrating that the technology can be adapted to more clay-rich matrices with appropriate process optimization. Adjusting the pH of the process water can alter PFAS speciation and improve desorption, while surfactants or cosolvents can enhance mobilization but require careful selection to prevent secondary contamination.

Management of the fine-grained residue and process water is critical for sustainable soil washing operations. Filter cakes rich in PFAS are commonly stabilized and sent to landfills or subjected to destructive treatments such as thermal desorption or smoldering (Hurst et al., 2024; Quinnan et al., 2022; Pennell et al., 2023; Nguyen et al., 2023). The addition of binders, such as Portland cement, during stabilization can improve the mechanical properties of the fines and enable their reuse in construction, provided that leachable PFAS concentrations meet regulatory criteria. The choice between on-site mobile washing units and off-site fixed facilities depends on the volume of soil and proximity to treatment infrastructure; mobile systems reduce transport costs for large projects but have lower throughput, whereas small volumes or sites near existing plants may favor off-site treatment.

Case studies of large-scale soil washing projects demonstrate its viability for remediating PFAS-impacted soils. Facilities processing thousands of tons of soil have reported removals above 90%, with treated soils meeting reuse criteria and a significant reduction in waste volume (Hurst et al., 2024; Quinnan et al., 2022; Londhe and Venkatesan, 2025). These examples, combined with ongoing improvements in surfactant formulations and process design, support the role of soil washing as a key technology within integrated PFAS remediation strategies.

Soil washing, often conducted ex situ, uses a liquid solution to dissolve and separate PFAS from soil particles. Surfactants such as Tween 20, Triton X-100, and SDS can enhance desorption by forming micelles that encapsulate PFAS molecules and carry them into the aqueous phase (Du et al., 2014). However, the resulting PFAS-laden wash water requires further treatment, often via GAC or high-pressure membrane filtration. Recent studies have evaluated the use of biosurfactants and biodegradable chelating agents to reduce secondary pollution risk (Mahinroosta and Senevirathna, 2020). Some success has been observed with cosolvent extraction using ethanol or methanol-water mixtures, though these methods introduce handling and safety challenges. The scalability of soil washing depends heavily on the volume of soil, extraction kinetics, and treatment logistics of the washate.

The efficiency of soil washing is closely linked to PFAS partitioning behavior, with reported partition coefficients following the order PFCAs < PFSAs < fluorotelomer sulfonates < sulfonamides. This hierarchy explains the preferential removal of short-chain PFAS during washing and the persistence of strongly sorbed long-chain compounds in fine soil fractions. Consequently, soil washing is increasingly implemented as a pretreatment step prior to advanced degradation technologies, particularly when treating large volumes of heterogeneous soil (Hubert et al., 2023; Pang et al., 2022).

### 9.3.3 Electrokinetic remediation

Electrokinetic remediation (EKR) is a promising in situ technique that applies a low-voltage direct current across electrodes inserted into the soil, mobilizing contaminants through electromigration, electroosmosis, and electrophoresis (Reddy and 2013, 2013). EKR has shown high removal efficiencies (up to 90%) for charged PFAS species, particularly when combined with enhancing agents such as surfactants or complexing ions. Zhou et al. (2024) demonstrated that solar-powered EKR systems, using natural saponins as mobilizers, could significantly enhance metal removal, suggesting similar adaptations for PFAS (Zhou et al., 2024). Advantages include minimal soil excavation and targeted treatment zones. Limitations include high energy requirements (though solar assistance mitigates this), pH front management, and variable removal efficiency for long-chain PFAS.

### 9.3.4 Interfacial solar evaporation and hybrid systems

A novel concept introduced by Navarro et al. (2023) involves interfacial solar evaporation—using solar-absorbing materials placed at the soil–water interface to generate localized heat and accelerate evaporation. This approach creates a thermocapillary flow that concentrates contaminants in the upper soil layers for subsequent removal (Wu et al., 2023). While originally applied for heavy metal extraction, theoretical modeling and early-stage trials suggest this method may also be effective for concentrating surface-active PFAS. Such systems, when combined with membrane distillation or foam fractionation, could serve as integrated treatment solutions. Key research directions include material optimization (e.g., photothermal coatings), field-scale validation, and postevaporation treatment of condensed phases.

While extraction techniques effectively reduce PFAS mass, residual contamination often remains in the soil matrix, necessitating complementary approaches such as immobilization to control mobility and exposure.

## 9.4 Immobilization and soil stabilization

Immobilization is widely applied as a cost-effective strategy to mitigate PFAS leaching and plant uptake in contaminated soils, particularly where excavation or destructive treatment is impractical. The approach involves mixing sorbent materials into soil to increase PFAS sorption capacity and reduce permeability, thereby limiting transport to groundwater. Carbon-based materials such as GAC and biochar promote PFAS retention through hydrophobic interactions, electrostatic attraction, and ion exchange, with activated carbon generally outperforming biochar due to its higher surface area and structural stability (Mahinroosta and Senevirathna, 2020; Hale et al., 2017; Li et al., 2023).

Composite amendments combining activated carbon with mineral binders (e.g., aluminum hydroxide, clay minerals, or cement) have demonstrated enhanced immobilization efficiency and long-term stability, achieving PFAS leaching reductions exceeding 99% in laboratory and pilot-scale studies (Sorengard et al., 2019; Barth et al., 2021; Kabiri and McLaughlin, 2021). While immobilization does not destroy PFAS, it plays a critical role in exposure control and risk management, especially in agricultural soils where plant uptake represents a significant pathway.

## 9.5 Oxidation

### 9.5.1 Principles and oxidant selection

Oxidation-based remediation seeks to chemically degrade PFAS by introducing highly reactive species such as sulfate or hydroxyl radicals. Due to the exceptional stability of PFAS, traditional oxidants like chlorine or ozone are largely ineffective against the C–F backbone. Instead, advanced oxidation processes, including persulfate activation, electrochemical oxidation (EO), and photolysis, are required to initiate degradation (Ross et al., 2018). The choice of oxidant and activation method depends on the target PFAS, soil conditions (e.g., redox potential, moisture), and operational constraints. Activation of persulfate ($S_2O_8^{2-}$) to generate $SO_4{\bullet}^-$ or hydroxyl radicals is typically achieved via heat, base, or transition metals (Travar et al., 2021; Riegel et al., 2020).

### 9.5.2 Heat-activated persulfate oxidation

Heat-activated persulfate oxidation (HAP) is a widely studied AOP in PFAS remediation. Under elevated temperatures (~80°C–90°C), persulfate decomposes to form sulfate radicals that can attack and cleave PFAS molecules. Horst et al. noted that PFOA removal in aqueous matrices exceeded 90% after 24 h of HAP treatment, although results varied significantly for PFOS and other long-chain PFAS (Riegel et al., 2020). The process is most efficient in slurry systems where uniform heating can be maintained. However, sulfate radical reactivity is nonselective and can be consumed by natural organic matter or cocontaminants. Thus pretreatment to reduce background organic load is often necessary. HAP can also be coupled with electrokinetic heating or solar thermal input to improve sustainability.

### 9.5.3 Electrochemical oxidation

EO involves the use of electrodes, typically boron-doped diamond (BDD), to generate strong oxidants in situ via water electrolysis. The anode generates hydroxyl radicals and direct oxidative potential that can degrade PFAS chains (Tasca et al., 2025). Cai et al., 2024 demonstrated that solar thermoelectric systems integrated with EO achieved effective chromium removal in soil—a model system for PFAS adaptation (Cai et al., 2024). EO can proceed via direct electron transfer at the anode surface or indirectly through the in situ generation of reactive species such as hydroxyl radicals, persulfate, and chlorine-based oxidants. The choice of electrode material critically influences degradation pathways, service life, and by-product formation. Studies have shown that BDD electrodes may promote the formation of short-chain PFAS during degradation, whereas iron-based electrodes exhibit lower secondary PFAS formation, highlighting the importance of electrode selection in process design (Öhberg, 2022; Sivagami et al., 2023). EO systems offer excellent control and modularity, but they face challenges including high energy consumption, electrode fouling, and limited field-scale data. Improvements in electrode materials (e.g., titanium–suboxide coatings) and current efficiency are areas of active research.

### 9.5.4 Photocatalysis and combined techniques

Photocatalytic degradation of PFAS, particularly using $TiO_2$ or carbon-doped variants under UV or solar light, has shown limited success. However, in combination with oxidants or sensitizers, photocatalysis can enhance PFAS defluorination rates. UV/sulfite and UV/persulfate systems have been tested in soil slurries with mixed results (Yates, 2022). A key limitation is the need for transparency and exposure, which may restrict application in deep or opaque soils. Hybrid systems combining photocatalysis with EO or HAP offer synergistic benefits and may become viable with renewable energy integration. These systems also offer promise in semisaturated environments such as wetlands or retention basins.

## 9.6 Bioremediation

### 9.6.1 Overview and mechanism

Bioremediation refers to the use of microorganisms to transform, detoxify, or mineralize contaminants in soil through metabolic or cometabolic processes. Compared with physicochemical remediation technologies, microbial approaches are generally considered environmentally friendly and cost-effective because they operate under natural environmental conditions and typically require lower energy inputs. In contrast to many physicochemical treatments that rely on extreme temperatures, high pressures, or aggressive chemicals, microbial remediation can potentially occur in situ under moderate environmental conditions such as ambient temperature and atmospheric pressure, reducing operational risks and infrastructure requirements (Bhatt et al., 2021; Yu et al., 2022). However, the application of bioremediation to PFAS remains particularly challenging because the C–F bond is among the strongest bonds in organic chemistry, making PFAS highly resistant to conventional microbial degradation pathways (Bala et al., 2022; Grgas et al., 2023).

Despite these limitations, increasing evidence suggests that certain microbial communities can participate in the transformation of PFAS or their precursor compounds through mechanisms such as reductive defluorination, cometabolism, or enzymatic transformation.

Microbial activity may lead to gradual shortening of the perfluoroalkyl chain or transformation of fluorinated precursors into terminal PFAAs. Although complete mineralization is rarely achieved, microbial transformation processes can reduce PFAS toxicity, alter their mobility, and potentially enhance the effectiveness of subsequent physicochemical treatment processes (Yu et al., 2022; Grgas et al., 2023; Noori et al., 2025; Beškoski et al., 2024; Wackett, 2022; Hnatko et al., 2023).

Recent experimental studies have demonstrated that partial PFAS degradation may occur under specific biochemical conditions. For example, Sun et al. (2021) reported partial defluorination of PFOS by an anaerobic microbial community in the presence of vitamin $B_{12}$ as a cofactor under reductive conditions involving sulfide ($S^{2-}$) (Sun et al., 2021). Such findings suggest that microbial metabolism may play a role in initiating PFAS transformation pathways, although these processes remain slow and incomplete under natural environmental conditions.

Given the inherent resistance of PFAS molecules to biodegradation, recent research has increasingly focused on biotechnological approaches aimed at enhancing microbial degradation capabilities. In particular, genetic and metabolic engineering strategies are being explored to modify microorganisms so that they can express enzymes capable of weakening or cleaving the C–F bond (LaFond et al., 2023). Advances in synthetic biology and gene-editing technologies—such as CRISPR-based genome engineering—have opened new possibilities for designing microbial strains with improved PFAS transformation potential (Smorada et al., 2024). Engineered microorganisms may be capable of producing specific enzymes that initiate defluorination reactions or transform PFAS molecules into less toxic intermediates or ultimately mineralize them into inorganic fluoride and carbon dioxide (Webster et al., 2024).

### 9.6.2 Bioaugmentation

Bioaugmentation is a remediation strategy that involves the introduction of specialized microbial strains or microbial consortia into contaminated environments to enhance biodegradation processes. The objective of this approach is to provide metabolic capabilities that may be absent or insufficient within the indigenous microbial community, thereby promoting the transformation or degradation of specific contaminants (Lyon and Vogel, 2012). In soil and groundwater systems, bioaugmentation may be applied when native microorganisms are unable to degrade the target pollutants efficiently or when their populations are too small to sustain effective biodegradation rates.

In the context of PFAS-contaminated soils, bioaugmentation focuses on introducing microbial strains capable of initiating transformation pathways for fluorinated compounds. Several bacterial genera—including *Pseudomonas*, *Rhodococcus*, and *Acidimicrobium* sp. A6—have been investigated for their potential role in PFAS transformation. Laboratory studies have reported partial degradation or transformation of compounds such as PFOA and PFOS under both aerobic and anaerobic conditions (Grgas et al., 2023; Savvidou et al., 2024; Chiriac et al., 2023). Although complete mineralization remains rare, these microbial processes may contribute to the gradual transformation of PFAS or their precursors into less persistent intermediates.

In many cases, microbial consortia have demonstrated greater effectiveness than individual strains because synergistic metabolic interactions can facilitate multi-step transformation pathways. Some microorganisms may degrade fluorinated precursor compounds, while others participate in subsequent transformation reactions, leading to the formation of shorter-chain PFAS or partially defluorinated products. Consequently, bioaugmentation strategies often employ mixed microbial cultures designed to enhance overall PFAS biotransformation through complementary metabolic functions.

For bioaugmentation to be successful, the introduced microorganisms must be able to survive, proliferate, and maintain their degradative capabilities within the environmental conditions of the contaminated site. However, several factors can limit the effectiveness of this approach. Indigenous microbial populations may outcompete the introduced strains, injected microorganisms may lose their metabolic activity after introduction, and the transport and distribution of microbial inocula within subsurface environments can be difficult to control (Lyon and Vogel, 2012). Additionally, distinguishing the introduced microorganisms from native microbial communities during monitoring can be challenging, although advances in molecular techniques are improving the ability to track specific microbial populations in environmental systems (Underwood et al., 2022).

Despite these challenges, bioaugmentation remains an attractive remediation strategy because it allows the deliberate selection of microorganisms with desired metabolic activities. This targeted approach can significantly accelerate contaminant degradation compared with relying solely on natural microbial adaptation, which may take considerable time for certain pollutants to occur (Franzmann et al., 2000). Recent developments in molecular biology and microbial genetic engineering have further expanded the potential of bioaugmentation strategies, including the concept of gene bioaugmentation, where engineered microorganisms or genetic elements are introduced to confer specific degradation pathways for otherwise recalcitrant contaminants (Janssen and Stucki, 2020). Although most applications remain at laboratory or pilot scale, these advances highlight the potential role of bioaugmentation within integrated PFAS remediation strategies.

### 9.6.3 Biostimulation

Biostimulation is a widely applied strategy in environmental bioremediation that aims to enhance the activity of indigenous microbial communities by modifying environmental conditions to favor microbial metabolism. Rather than introducing external microorganisms, this approach focuses on stimulating the native microbial populations already present in contaminated soils. This is typically achieved through the addition of limiting nutrients or by adjusting environmental parameters such as pH, moisture, and redox conditions, thereby creating conditions that promote microbial growth and enzymatic activity (Bala et al., 2022).

In practice, biostimulation often involves optimizing the availability of essential nutrients—particularly carbon, nitrogen, and phosphorus—to maintain a suitable carbon–nitrogen–phosphorus (C:N:P) ratio that supports microbial proliferation and metabolic activity. Additional amendments may include electron donors, electron acceptors, biosurfactants, or oxygen-releasing compounds that stimulate microbial respiration and increase the biodegradation potential of the indigenous microbial community (Gomez and Sartaj, 2020). These amendments can enhance the metabolic capacity of microorganisms capable of transforming a variety of organic contaminants in soil and groundwater systems.

Biostimulation has demonstrated considerable potential for the remediation of contaminated soils and aquifers, particularly in situations where microbial degradation is limited by nutrient deficiencies or unfavorable redox conditions (Rajput et al., 2021). By improving microbial growth conditions, the technique can accelerate the biodegradation of organic pollutants and increase the bioavailability of contaminants to microbial enzymes.

Recent developments in biostimulation strategies have explored the use of advanced nutrient delivery systems, including slow-release fertilizers and nanomaterials designed to maintain sustained microbial stimulation over extended periods. These approaches can help regulate nutrient availability, reduce nutrient losses, and enhance contaminant bioavailability, thereby improving the efficiency of microbial remediation processes (Javaid et al., 2023). Field-scale applications have also demonstrated that biostimulation can provide a cost-effective and environmentally compatible alternative to more invasive remediation methods, supporting its use within integrated and sustainable soil remediation frameworks (Alonso-Gutiérrez et al., 2024).

In the context of PFAS-contaminated soils, biostimulation may also support cometabolic transformation processes. The addition of organic substrates or other amendments can stimulate microbial communities capable of indirectly transforming PFAS while metabolizing other primary substrates. Although complete PFAS mineralization remains rare, optimized environmental conditions may enhance microbial activity and improve the potential for PFAS transformation within soil systems.

### 9.6.4 Limitations and future perspectives

Although microbial remediation offers a promising, sustainable approach, several limitations currently restrict its large-scale application. PFAS biodegradation rates are typically slow, and many microbial processes result in partial transformation rather than complete mineralization. Additionally, intermediate products may include shorter-chain PFAS compounds, which can be more mobile in soil and groundwater systems.

Consequently, bioremediation is increasingly viewed as a complementary technology within integrated remediation strategies. Combining microbial processes with physicochemical treatment, such as EO, thermal activation, or soil washing—may enhance overall remediation efficiency while maintaining the environmental benefits of biological systems. Continued research into microbial metabolic pathways, enzyme systems capable of C–F bond cleavage, and engineered microbial consortia will be essential for advancing the role of bioremediation in PFAS-contaminated soil management (Grgas et al., 2023; Beškoski et al., 2024; Chiriac et al., 2023).

## 9.7 Phytoremediation

### 9.7.1 Mechanism and plant selection

Phytoremediation utilizes plants to extract, degrade, or stabilize contaminants in soil and groundwater. In the case of PFAS, the dominant mechanisms are phytoextraction and phytostabilization, with limited evidence of enzymatic degradation. Uptake is influenced by PFAS chain length, functional group, and hydrophobicity. Short-chain compounds (e.g., Perfluorobutanoic acid (PFBA), Perfluorobutanesulfonic acid (PFBS)) are more likely to be absorbed and translocated to plant tissues than long-chain PFAS (Gobelius et al., 2017; Shi et al., 2023). Ideal plant candidates are fast-growing, deeply rooted species with high transpiration rates. Studies have highlighted *Phragmites australis*, *Populus deltoides*, and *Brassica juncea* as effective accumulators under field conditions (Evangelou and Robinson, 2022; Ambaye et al., 2022).

## 9.7.2 Constructed wetlands and rhizofiltration

Constructed wetlands (CWs) simulate natural wetland processes, enhancing PFAS retention through sedimentation, sorption, and biological activity. Plants facilitate rhizofiltration and improve hydraulic conditions. In pilot CWs, PFAS removal efficiencies between 30%–70% have been reported depending on the hydraulic residence time and PFAS composition (Savvidou et al., 2024; Awad et al., 2022). The use of organic amendments (e.g., compost, biochar) within CW substrates can further improve retention. However, long-term accumulation in biomass raises questions about safe harvesting and disposal. CWs are best suited for passive systems with stable flow conditions and low PFAS concentrations.

## 9.7.3 Rhizosphere microbiome interactions

The rhizosphere, or root–soil interface, is a hotspot of microbial activity that can influence PFAS mobility. Though direct microbial degradation of PFAS is rare, some studies suggest cometabolic defluorination by anaerobic microbes under specific conditions (Noori et al., 2025). Enhancing microbial consortia through inoculation or nutrient amendment is an emerging strategy to improve PFAS transformation in phytoremediation systems. Additionally, plant root exudates can alter PFAS bioavailability by changing pH and ionic strength, indirectly affecting sorption-desorption dynamics.

Beyond plant uptake, microbial activity in soils also plays a complementary role in PFAS transformation through biodefluorination and biotransformation processes. Several bacterial and fungal species, including *Pseudomonas*, *Rhodococcus*, and *Acidimicrobium* sp. A6 and white-rot fungi have demonstrated partial degradation of PFAS under aerobic and anaerobic conditions. These processes are typically slow and often result in the formation of short-chain PFAS intermediates, but they offer environmentally benign pathways that may be integrated with physicochemical treatments to enhance overall remediation performance (Grgas et al., 2023; Chiriac et al., 2023; Zhang et al., 2022; Sanzana et al., 2025).

## 9.7.4 Limitations and research needs

While phytoremediation and associated rhizosphere and microbial processes offer environmentally sustainable and low-cost options for managing PFAS-contaminated soils, their practical application remains constrained by slow remediation rates and incomplete degradation. Microbial biotransformation and biodefluorination pathways frequently result in the formation of short-chain PFAS intermediates, whose mobility and ecotoxicological implications are not yet fully understood. Moreover, treatment performance is highly site-specific and influenced by soil properties, microbial community composition, PFAS speciation, and environmental conditions, limiting the predictability and scalability of biological approaches. Consequently, phytoremediation is currently best suited for long-term risk reduction and containment rather than rapid or complete PFAS removal.

Future research should focus on elucidating the metabolic pathways and enzymatic mechanisms involved in PFAS biodegradation, as well as on identifying the fate and persistence of transformation products in soil–plant systems. Integrating biological processes with physicochemical treatments, such as electrokinetic enhancement or oxidative posttreatment, may improve degradation efficiency while maintaining the sustainability advantages of phytoremediation (Beškoski et al., 2024; Shi et al., 2023). Long-term field studies and life-cycle assessment (LCA) approaches are also required to evaluate ecological trade-offs and to support informed decision-making for large-scale implementation.

## 9.8 Conclusion

Remediating PFAS-contaminated soils is one of the most pressing environmental challenges due to the extraordinary persistence, mobility, and bioaccumulation potential of these substances. Unlike conventional organic pollutants, PFAS resist most biological and chemical degradation pathways, requiring specialized, often energy-intensive or hybridized treatment approaches. Five major remediation strategies - thermal treatment, extraction, oxidation, remediation, and phytoremediation - each present unique mechanisms, advantages, and limitations. Thermal methods, particularly at high temperatures (>1000°C), remain the most effective destructive option, with emerging innovations such as solar-driven systems and vitrification offering both destruction and immobilization. Extraction techniques, including surfactant-enhanced flushing and electrokinetic methods, mobilize PFAS—especially short-chain compounds—but necessitate subsequent treatment of concentrated leachate. Oxidation processes, such as heat-activated persulfate and electrochemical systems, are advancing toward overcoming PFAS' chemical resilience, with combined thermal, electrical, and chemical activation showing promising results in laboratory and pilot studies. Bioremediation, based on microbial transformation processes such as bioaugmentation and biostimulation, offers a sustainable and low-energy complementary approach, although current applications are generally limited by slow degradation rates and incomplete mineralization of PFAS compounds. Phytoremediation, though nondestructive, provides a low-cost and environmentally friendly strategy for risk reduction and containment, particularly when integrated with CWs or electrokinetic systems. Across all methods, renewable energy—especially solar—plays a growing role in improving performance, reducing operational costs, and supporting sustainable remediation.

No single technology can be considered universally applicable; instead, hybrid or sequential treatment trains that integrate separation, concentration, and destruction steps are increasingly recognized as the most viable pathways to balance efficiency with sustainability. Beyond technical performance, the selection of a remediation strategy must also consider regulatory frameworks, site-specific constraints, energy requirements, economic feasibility, and social acceptance. High-temperature approaches, for example, may be effective but face sustainability barriers unless coupled with renewable inputs and advanced emission controls, while soil washing and other separation-based methods require optimization alongside robust water treatment.

Future research should prioritize integrated, site-specific solutions, with a focus on PFAS-specific catalytic and electrochemical materials, the application of LCA to evaluate trade-offs, and continued investment in field trials and long-term monitoring. Strengthening interdisciplinary collaboration across environmental engineering, policy, and public health will be critical to ensuring that remediation strategies are not only technically sound but also practical, cost-effective, and socially acceptable. Ultimately, overcoming the PFAS soil contamination challenge requires a holistic framework that combines technological innovation with regulatory, economic, and societal dimensions to achieve effective and sustainable outcomes (see Table 9.1).

**TABLE 9.1** Comparative overview of per- and polyfluoroalkyl substances (PFAS) soil remediation technologies.

| Technology | Operating conditions (range) | PFAS removal efficiency | Suitability and considerations |
|---|---|---|---|
| **Thermal treatment (desorption and incineration)** | • Thermal desorption: 350°C–550°C, residence time 15–45 min<br>• Ex situ rotary dryers or thermopiles<br>• High-temperature incineration: >900°C–1100°C with extended residence time<br>• Smoldering combustion: self-sustaining oxidation >900°C using surrogate fuel (e.g., spent GAC, rubber) | • Thermal desorption: transfers 70%–90% of PFAS mass into off-gas stream<br>• Incineration: >99% destruction efficiency<br>• HF is typically captured using CaO sorbents<br>• Smoldering: >99.9% PFAS reduction with <1% vaporized fraction | • Suitable for highly contaminated soils and source-zone "hotspots"<br>• Smoldering has lower energy demand but remains mostly at pilot scale<br>• High capital and operational costs<br>• Requires strict off-gas treatment and regulatory permitting |
| **Extraction and soil washing** | • Ex situ treatment using water, surfactants, or cosolvents<br>• Modular systems with log washers, hydrocyclones, and attrition scrubbers<br>• Process water treated with GAC or ion-exchange resins<br>• Operates at ambient temperature and pressure | • Transfers 70%–97% of PFAS into process water<br>• Full-scale systems report 90%–99.7% removal of C4–C8 PFAS from sandy fractions<br>• Short-chain PFAS are more easily mobilized than long-chain compounds<br>• Efficiency decreases in clay-rich or organic soils | • Suitable for large soil volumes with low–moderate contamination<br>• Generates PFAS-rich fines and process water requiring further treatment<br>• Considered a mass-reduction technology, not destructive<br>• More cost-effective when combined with on-site water polishing |

*(Continued)*

**TABLE 9.1** (Continued)

| Technology | Operating conditions (range) | PFAS removal efficiency | Suitability and considerations |
|---|---|---|---|
| **Oxidation (chemical and electrochemical)** | • Heat-activated persulfate (40°C–90°C; sometimes >100°C with catalysts)<br>• Fenton-like systems and UV-activated oxidation<br>• Electrochemical oxidation using boron-doped diamond or Ti/Ru electrodes<br>• Often applied to soil slurries or extracted leachates | • Highly variable removal efficiency<br>• Persulfate oxidation: 30%–80% PFAS reduction in soils<br>• C–S bond cleavage more feasible than C–F bond breakage<br>• Electrochemical processes can achieve near complete defluorination in aqueous phases | • More effective for treatment of extracted leachates or slurries<br>• High energy demand and limited field-scale applications<br>• Efficiency improved with thermal or electrical activation<br>• Active research area: PFAS-specific catalysts and hybrid systems |
| **Bioremediation (bioaugmentation and biostimulation)** | • In situ or ex situ microbial treatment of contaminated soils<br>• Bioaugmentation: introduction of specialized microbes (e.g., *Pseudomonas, Rhodococcus, Acidimicrobium*)<br>• Biostimulation: addition of nutrients (C, N, P), electron donors or acceptors<br>• Microbial activity is influenced by pH, moisture, temperature, and redox conditions | • Generally, a partial transformation rather than complete mineralization<br>• Microbial processes may convert PFAS precursors into terminal PFAAs<br>• Limited defluorination observed in laboratory studies<br>• Removal efficiency strongly site-dependent and typically lower than physicochemical methods | • Environmentally sustainable and low-energy approach<br>• Suitable for long-term management of low-to-moderate contamination<br>• Transformation products may include more mobile short-chain PFAS<br>• Often integrated with oxidation, electrochemical, or soil washing technologies |
| **Phytoremediation** | • In situ use of plants (e.g., reeds, willow, grasses)<br>• Applied in wetlands or engineered systems<br>• Timescale ranges from seasons to years<br>• Often combined with electrokinetic enhancement or constructed wetlands | • Nondestructive process<br>• PFAS removal through uptake, stabilization, and rhizosphere interactions<br>• Short-chain PFAS more easily taken up by plants<br>• Reported reductions typically <50% over multiple years | • Low-cost and environmentally friendly approach<br>• Suitable for large areas with diffuse contamination<br>• Functions primarily as risk management or containment strategy<br>• Often combined with other remediation technologies |

# References


National Institute of Environmental Health Sciences. Perfluoroalkyl and polyfluoroalkyl substances (PFAS). Available at https://www.niehs.nih.gov/health/topics/agents/pfc (accessed on 7 May 2026).

U.S. Environmental Protection Agency. Per- and polyfluoroalkyl substances (PFAS). Available at https://www.epa.gov/pfas (accessed on 7 May 2026).

Abraham, J.E.F., Mumford, K.G., Patch, D.J., Weber, K.P., 2022. Retention of PFOS and PFOA mixtures by trapped gas bubbles in porous media. Environmental Science & Technology 56, 15489–15498.

Alonso-Gutiérrez, J., Vilchez-Vargas, R., Leiva, C., 2024. Field-scale biostimulation of hydrocarbon degradation in a marine sediment chronically impacted by oil pollution. Marine Pollution Bulletin 200, 115806.

Ambaye, T.G., Vaccari, M., Prasad, S., Rtimi, S., 2022. Recent progress and challenges on the removal of per-and poly-fluoroalkyl substances (PFAS) from contaminated soil and water. Environmental Science and Pollution Research 29, 58405–58428.

Awad, J., Brunetti, G., Juhasz, A., Williams, M., Navarro, D., Drigo, B., et al., 2022. Application of native plants in constructed floating wetlands as a passive remediation approach for PFAS-impacted surface water. Journal of Hazardous Materials 429, 128326.

Bala, S., Garg, D., Thirumalesh, B.V., Sharma, M., Sridhar, K., Inbaraj, B.S., et al., 2022. Recent strategies for bioremediation of emerging pollutants: a review for a green and sustainable environment. Toxics 10, 484.

Barth, E., McKernan, J., Bless, D., Dasu, K., 2021. Investigation of an immobilization process for PFAS contaminated soils. Journal of Environmental Management 296, 113069.

Beškoski, V., Lješević, M., Jiménez, B., Muñoz-Arnanz, J., Colomer-Vidal, P., Inui, H., et al., 2024. Bioremediation of soils contaminated with PFAS: an update on available techniques, pilot studies, challenges, and future directions. Soil Remediation Science and Technology. Springer, pp. 331–366.

Bhatt, P., Gangola, S., Bhandari, G., Zhang, W., Maithani, D., Mishra, S., et al., 2021. New insights into the degradation of synthetic pollutants in contaminated environments. Chemosphere 268, 128827.

Brusseau, M.L., 2018. Assessing the potential contributions of additional retention processes to PFAS retardation in the subsurface. The Science of the Total Environment 613, 176–185.

USEPA Interim Guidance on the Destruction and Disposal of PFAS and Materials Containing PFAS, 2020.

Cai, D., Hu, R., Guo, M., Wang, D., Zhu, Y., Sun, C., et al., 2024. A solar thermoelectric system by temperature difference for efficient removal of chromium (VI) in water and soil. Journal of Hazardous Materials 480, 136309.

Chiriac, F.L., Stoica, C., Iftode, C., Pirvu, F., Petre, V.A., Paun, I., et al., 2023. Bacterial biodegradation of perfluorooctanoic acid (PFOA) and perfluorosulfonic acid (PFOS) using pure Pseudomonas strains. Sustainability 15, 14000.

Demirbas, A., Arin, G., 2002. An overview of biomass pyrolysis. Energy Sources 24, 471–482.

Du, Z., Deng, S., Bei, Y., Huang, Q., Wang, B., Huang, J., et al., 2014. Adsorption behavior and mechanism of perfluorinated compounds on various adsorbents—a review. Journal of Hazardous Materials 274, 443–454.

Duchesne, A.L., Brown, J.K., Patch, D.J., Major, D., Weber, K.P., Gerhard, J.I., 2020. Remediation of PFAS-contaminated soil and granular activated carbon by smoldering combustion. Environmental Science & Technology 54, 12631–12640.

Evangelou, M.W.H., Robinson, B.H., 2022. The phytomanagement of PFAS-contaminated land. International Journal of Environmental Research and Public Health 19, 6817.

Evich, M.G., Davis, M.J.B., McCord, J.P., Acrey, B., Awkerman, J.A., Knappe, D.R.U., et al., 2022. Per-and polyfluoroalkyl substances in the environment. Science 80, 375 eabg9065.

Fabregat-Palau, J., Vidal, M., Rigol, A., 2021. Modelling the sorption behaviour of perfluoroalkyl carboxylates and perfluoroalkane sulfonates in soils. The Science of the Total Environment 801, 149343.

Fournie, T., Rashwan, T.L., Switzer, C., Gerhard, J.I., 2023. Smouldering to treat PFAS in sewage sludge. Waste Management 164, 219–227.

Franzmann, P.D., Zappia, L.R., Tilbury, A.L., Patterson, B.M., Davis, G.B., Mandelbaum, R.T., 2000. Bioaugmentation of atrazine and fenamiphos impacted groundwater: laboratory evaluation. Bioremediation Journal 4, 237–248.

Gobelius, L., Lewis, J., Ahrens, L., 2017. Plant uptake of per-and polyfluoroalkyl substances at a contaminated fire training facility to evaluate the phytoremediation potential of various plant species. Environmental Science & Technology 51, 12602–12610.

Gomez, D.E., Sartaj, M., 2020. Biostimulation for in situ bioremediation of petroleum hydrocarbon-contaminated soils: a review. Journal of Environmental Management 255, 109896.

Grgas, D., Petrina, A., Štefanac, T., Bešlo, D., Landeka Dragičević, T., 2023. A review: per-and polyfluoroalkyl substances—biological degradation. Toxics 11, 446.

Grimison, C., Knight, E.R., Nguyen, T.M.H., Nagle, N., Kabiri, S., Bräunig, J., et al., 2023. The efficacy of soil washing for the remediation of per-and poly-fluoroalkyl substances (PFASs) in the field. Journal of Hazardous Materials 445, 130441.

Hale, S.E., Arp, H.P.H., Slinde, G.A., Wade, E.J., Bjørseth, K., Breedveld, G.D., et al., 2017. Sorbent amendment as a remediation strategy to reduce PFAS mobility and leaching in a contaminated sandy soil from a Norwegian firefighting training facility. Chemosphere 171, 9–18.

Higgins, C.P., Luthy, R.G., 2006. Sorption of perfluorinated surfactants on sediments. Environmental Science & Technology 40, 7251–7256.

Hnatko, J.P., Liu, C., Elsey, J.L., Dong, S., Fortner, J.D., Pennell, K.D., et al., 2023. Microbial reductive dechlorination by a commercially available dechlorinating consortium is not inhibited by perfluoroalkyl acids (PFAAs) at field-relevant concentrations. Environmental Science & Technology 57, 8301–8312.

Hubert, M., Arp, H.P.H., Hansen, M.C., Castro, G., Meyn, T., Asimakopoulos, A.G., et al., 2023. Influence of grain size, organic carbon and organic matter residue content on the sorption of per-and polyfluoroalkyl substances in aqueous film forming foam contaminated soils-Implications for remediation using soil washing. The Science of the Total Environment 875, 162668.

Hurst, J., Hale, S., Miles, J., Dal, E., Gevaerts, W., Burdick, J., 2024. *PFAS Soil Treatment Processes: A Review of Operating Ranges and Constraints*; Brussels.

Janssen, D.B., Stucki, G., 2020. Perspectives of genetically engineered microbes for groundwater bioremediation. Environmental Science: Processes & Impacts 22, 487–499.

Javaid, S., Mushtaq, S., Mumtaz, M.Z., Rasool, G., Naqqash, T., Afzal, M., et al., 2023. Mineral solubilizing rhizobacterial strains mediated biostimulation of rhodes grass seedlings. Microorganisms 11, 2543.

Kabiri, S., McLaughlin, M.J., 2021. Durability of sorption of per-and polyfluorinated alkyl substances in soils immobilized using common adsorbents: 2. Effects of repeated leaching, temperature extremes, ionic strength and competing ions. The Science of the Total Environment 766, 144718.

LaFond, J.A., Hatzinger, P.B., Guelfo, J.L., Millerick, K., Jackson, W.A., 2023. Bacterial transformation of per-and poly-fluoroalkyl substances: a review for the field of bioremediation. Environmental Science: Advances 2, 1019–1041.

Li, Q., Wang, L., Fu, Y., Lin, D., Hou, M., Li, X., et al., 2023. Transformation of soil organic matter subjected to environmental disturbance and preservation of organic matter bound to soil minerals: a review. Journal of Soils and Sediments 23, 1485–1500. Available at https://doi.org/10.1007/s11368-022-03381-y.

Londhe, K., Venkatesan, A.K., 2025. Air bubbling assisted soil washing to treat PFAS in high organic content soils. Environments 12, 20.

Lyon, D.Y., Vogel, T.M., 2012. Bioaugmentation for groundwater remediation: an overview. Bioaugmentation for Groundwater Remediation 1–37.

Mahinroosta, R., Senevirathna, L., 2020. A review of the emerging treatment technologies for PFAS contaminated soils. Journal of Environmental Management 255, 109896.

Maimaiti, A., Deng, S., Meng, P., Wang, W., Wang, B., Huang, J., et al., 2018. Competitive adsorption of perfluoroalkyl substances on anion exchange resins in simulated AFFF-impacted groundwater. Chemical Engineering Journal 348, 494–502.

Major, D., 2019. Demonstration of Smoldering Combustion Treatment of PFAS Impacted Investigation-Derived Waste.

McNamara, P., Samuel, M.S., Sathyamoorthy, S., Moss, L., Valtierra, D., Lopez, H.C., et al., 2023. Pyrolysis transports, and transforms, PFAS from biosolids to py-liquid. Environmental Science: Water Research & Technology 9, 386–395.

Meegoda, J.N., Bezerra de Souza, B., Casarini, M.M., Kewalramani, J.A., 2022. A review of PFAS destruction technologies. International Journal of Environmental Research and Public Health 19, 16397.

Navarro, A., n.d. A review on the Solar Thermal Treatment of Contaminated soils. (n.d.).

Navarro, A., Cardellach, E., Cañadas, I., Rodríguez, J., 2013. Solar thermal vitrification of mining contaminated soils. International Journal of Mineral Processing 119, 65–74.

Nguyen, D., Schaefer, C.E., Bamer, J.T., Lanza, H.A., Wintle, D., Maynard, K.G., et al., 2023. Bench-scale testing of a novel soil PFAS treatment train for informed remedial planning and decision-making. Remediation Journal 33, 309–321.

Noori, M.T., Gupta, P., Hellgardt, K., Min, B., 2025. Per-and polyfluoroalkyl substances (PFAS): an emerging environmental challenge and (microbial) bioelectrochemical treatment strategies. Current Opinion in Environmental Science & Health. 43, 100588.

Pang, H., Dorian, B., Gao, L., Xie, Z., Cran, M., Muthukumaran, S., et al., 2022. Remediation of poly-and perfluoroalkyl substances (PFAS) contaminated soil using gas fractionation enhanced technology. The Science of the Total Environment 827, 154310.

Pennell, K.D., Capiro, N.L., Fortner, J.D., 2023. Development of coupled physicochemical and biological systems for in situ remediation of mixed perfluorinated chemical and chlorinated solvent groundwater plumes.

Quinnan, J., Morrell, C., Nagle, N., Maynard, K.G., 2022. Ex situ soil washing to remove PFAS adsorbed to soils from source zones. Remediation Journal 32, 151–166.

Rajput, M.S., Kamboj, M.L., Kumar, N., 2021. Effect of biostimulation on some behavioral aspects and energy balance of sahiwal cows during peri-estrual period. Indian Journal of Animal Research 55.

Reddy, K.R., 2013. Electrokinetic remediation of soils at complex contaminated sites: technology status, challenges, and opportunities. *Coupled Phenom. Environ. Geotech. al (Eds).-2013. CRC Press. Taylor Fr. Gr*, pp. 131–147.

Riegel, M., Egner, S., Sacher, F., Scholtissek, B., Hjort, M., Beuthe, B., et al., 2020. Review of Water Treatment Systems for PFAS Removal. *TZW, Brussels Aug*.

Ross, I., McDonough, J., Miles, J., Storch, P., Thelakkat Kochunarayanan, P., Kalve, E., et al., 2018. A review of emerging technologies for remediation of PFASs. Remediation Journal 28, 101–126.

Sanzana, S., Fenti, A., Iovino, P., Panico, A., 2025. A review of PFAS remediation: separation and degradation technologies for water and wastewater treatment. Journal of Water Process Engineering 74, 107793.

Savvidou, P., Dotro, G., Campo, P., Coulon, F., Lyu, T., 2024. Constructed wetlands as nature-based solutions in managing per-and poly-fluoroalkyl substances (PFAS): evidence, mechanisms, and modelling. The Science of the Total Environment 934, 173237.

Shi, Q., Kaur, P., Gan, J., 2023. Harnessing the potential of phytoremediation for mitigating the risk of emerging contaminants. Current Opinion in Environmental Science & Health 32, 100448.

Sivagami, K., Sharma, P., Karim, A.V., Mohanakrishna, G., Karthika, S., Divyapriya, G., et al., 2023. Electrochemical-based approaches for the treatment of forever chemicals: removal of perfluoroalkyl and polyfluoroalkyl substances (PFAS) from wastewater. The Science of the Total Environment 861, 160440.

Smorada, C.M., Sima, M.W., Jaffé, P.R., 2024. Bacterial degradation of perfluoroalkyl acids. Current Opinion in Biotechnology 88, 103170.

Sorengard, M., Kleja, D.B., Ahrens, L., 2019. Stabilization of per-and polyfluoroalkyl substances (PFASs) with colloidal activated carbon (PlumeStop®) as a function of soil clay and organic matter content. Journal of Environmental Management 249, 109345.

Sun, Z., Geng, D., Zhang, C., Chen, J., Zhou, X., Zhang, Y., et al., 2021. Vitamin B12 (CoII) initiates the reductive defluorination of branched perfluorooctane sulfonate (br-PFOS) in the presence of sulfide. Chemical Engineering Journal 423, 130149.

Sörengård, M., Lindh, A.S., Ahrens, L., 2020. Thermal desorption as a high removal remediation technique for soils contaminated with per-and polyfluoroalkyl substances (PFASs). PLoS ONE 15, e0234476.

Tasca, A.L., Uwayezu, J.N., Carabante, I., Kumpiene, J., 2025. Electrochemical remediation of PFAS by Boron-Doped Diamond electrodes: a review. Journal of Environmental Chemical Engineering 13, 117044. Available at https://doi.org/10.1016/J.JECE.2025.117044.

Travar, I., Uwayezu, J.N., Kumpiene, J., Yeung, L.W.Y., 2021. Challenges in the PFAS remediation of soil and landfill leachate: a review. Advances in Environmental Engineering Research 2, 1–40.

Underwood, J.C., Akob, D.M., Lorah, M.M., Imbrigiotta, T.E., Harvey, R.W., Tiedeman, C.R., 2022. Microbial community response to a bioaugmentation test to degrade trichloroethylene in a fractured rock aquifer, Trenton, NJ. FEMS Microbiol. Ecol. 98. pp. fiac077.

Urciuolo, M., Migliaccio, R., Ciccone, B., Ruoppolo, G., de Folly d'Auris, A., Frisario, S., et al., 2025. Experimental investigation on PFAS degradation through municipal sludge combustion processes. Applied Thermal Engineering 276, 127006.

Uwayezu, J.N., Ren, Z., Sonnenschein, S., Leiviskä, T., Lejon, T., van Hees, P., et al., 2024. Combination of separation and degradation methods after PFAS soil washing. The Science of the Total Environment 907, 168137.

Vargette, L.D.S., De Coensel, N., De Ras, K., Van de Vijver, R., Voorspoels, S., Van Geem, K.M., 2023. Prospects of complete mineralization of per-and polyfluoroalkyl substances by thermal destruction methods. Current Opinion in Chemical Engineering 42, 100954.

Wackett, L.P., 2022. Nothing lasts forever: understanding microbial biodegradation of polyfluorinated compounds and perfluorinated alkyl substances. Microbial Biotechnology 15, 773–792.

Wallace, J.S., Edirisinghe, D., Seyedi, S., Noteboom, H., Blate, M., Balci, D.D., et al., 2023. Burning questions: current practices and critical gaps in evaluating removal of per-and polyfluoroalkyl substances (PFAS) during pyrolysis treatments of biosolids. Journal of Hazardous Materials Letters 4, 100079.

Wang, Z., DeWitt, J.C., Higgins, C.P., Cousins, I.T., 2017. A never-ending story of per- and polyfluoroalkyl substances (PFASs)?

Webster, L.J., Villa-Gomez, D., Brown, R., Clarke, W., Schenk, P.M., 2024. A synthetic biology approach for the treatment of pollutants with microalgae. Frontiers in Bioengineering and Biotechnology 12, 1379301.

Winchell, L.J., Wells, M.J.M., Ross, J.J., Fonoll, X., Norton Jr, J.W., Kuplicki, S., et al., 2021. Analyses of per-and polyfluoroalkyl substances (PFAS) through the urban water cycle: toward achieving an integrated analytical workflow across aqueous, solid, and gaseous matrices in water and wastewater treatment. The Science of the Total Environment 774, 145257.

Wu, P., Wu, X., Wang, Y., Zhao, J., Xu, H., Owens, G., 2023. Boosting extraction of Pb in contaminated soil via interfacial solar evaporation of multifunctional sponge. Green Energy Environ 8, 1459–1468.

Yates, C., 2022. UV Sulfite Advanced Reduction Process for PFOS Transformation in Wastewater Treatment. University of New Hampshire ISBN 9798819395134.

Yu, Y., Che, S., Ren, C., Jin, B., Tian, Z., Liu, J., et al., 2022. Microbial defluorination of unsaturated per-and polyfluorinated carboxylic acids under anaerobic and aerobic conditions: a structure specificity study. Environmental Science & Technology 56, 4894–4904.

Zhang, Y., Qv, Z., Wang, J., Yang, Y., Chen, X., Wang, J., et al., 2022. Natural biofilm as a potential integrative sample for evaluating the contamination and impacts of PFAS on aquatic ecosystems. Water Research 215, 118233.

Zhou, M., Ding, Y., Li, Q., Wei, X., 2024. Electrokinetic remediation of cadmium-spiked soil using tea saponins powered by solar energy. Environmental Engineering Science 41, 381–390.

Öhberg, A., 2022. Electrochemical oxidation of PFAS in soil conditions: using boron-doped diamond electrodes and iron electrodes.

CHAPTER

# 10

# Advanced oxidation processes for per- and polyfluoroalkyl substances remediation in contaminated soils

*Ganeshkumar Govindasamy*[1], *Mohammed Rizwan*[1], *Sunantha Ganesan*[2], *and Krishnasamy Sivagami*[1]

[1]School of Chemical Engineering, Vellore Institute of Technology, Vellore, Tamil Nadu, India
[2]Department of Biomedical Sciences, School of Biosciences and Technology, Vellore Institute of Technology, Vellore, Tamil Nadu, India

## 10.1 Introduction

The need for more effective remediation approaches is driven by the alarming environmental concerns of the Industrial Revolution. To restore the soil and water environment, advanced treatment and remediation technologies have to be employed. Perfluoroalkyl substances (PFAS) compounds were first made by two specific companies (DuPont and 3M) about 70 or 80 years ago. At the time, scientists thought they were amazing because they were "inert," meaning they didn't react with other things or rot away. Because they lasted so long and resisted heat and water, they were put into everything from nonstick pans to waterproof clothes. PFAS includes thousands of compounds; among them, perfluorooctanoic acid (PFOA) and perfluoro octane sulfonate (PFOS) have been widely studied for their potential health effects. Existing research is actively investigating the health effects of PFAS. PFAS creates significant environmental challenges due to its persistent nature and long-term damage. Aquatic organisms and wildlife are affected through the disruption of hormonal and reproductive systems and through ingestion. There is a need for advanced remediation technology to be designed to neutralize and degrade the contaminants to prevent their spread in the ecosystem. Based on the application, the mechanism and technology of remediation methods are classified as following conventional or physical, chemical, and biological approaches, differentiated into three categories. The conventional remediation methods, such as thermal desorption, excavation, and containment washing, remain ineffective. Biological remediation tends to degrade, absorb, or neutralize the toxic contaminant substance, and the microorganisms, such as fungi, bacteria, and microbes, or

*Per- and Polyfluoroalkyl Substances (PFAS) as Environmental Contaminants*
DOI: https://doi.org/10.1016/B978-0-443-43856-1.00025-3

plants leverage the process. It's an eco-friendly approach to cleaning up soil impurities and contaminants with hydrocarbon oils, pesticides, heavy metals, and other hazardous substances. However, this technique has limitations, as it requires a long time for complete degradation. Environmental conditions such as temperature, pH, and oxygen level are highly influenced by the efficiency of the biodegradation process. Chemical remediation is effective for the decomposition of PFAS, which includes advanced treatment processes such as chemical soil washing, oxidation, reduction, and defluorination. This process involves major chemical alteration in the PFAS structure, which includes chain scission, breaking down the molecules into harmless compounds. In addition to these methods, absorbent materials such as activated carbon, biochar, and zeolites are the specialized classes of materials used to capture pollutants from the soil to increase the efficiency of the remediation process. Likely, polyurethane-foam materials have high durability and chemical resistance used to encapsulate the hazardous compound. Most research on PFAS oxidation focuses on water-based systems, but this study investigates how PFAS bound to soil can be broken down.

By considering complex environmental factors, such as competition from organic matter, the neutralization of reactive species, and the rates at which PFAS detach from surfaces. This soil-focused approach helps connect controlled laboratory experiments with real-world cleanup efforts in the field. Using advanced oxidation methods is essential for cleaning up soils contaminated with PFAS. This need arises from the compounds' high toxicity, mobility in the environment, and exceptional persistence. The strong carbon–fluorine bonds make PFAS extremely resistant to natural breakdown, leading to their long-term buildup in soils. In turn, soil becomes a secondary reservoir of contamination, gradually releasing PFAS into groundwater. This movement spreads contamination throughout surface waters and allows these chemicals to be absorbed by crops. Since PFAS continue to leach from soil over time, they ultimately enter drinking water and the food supply, increasing the likelihood of human exposure. Traditional methods such as landfilling and adsorption only shift the contamination rather than eliminate it. Because the PFAS compounds remain unchanged, these strategies do not offer a lasting solution and instead pose ongoing environmental risks. Advanced oxidation processes (AOPs) provide a powerful alternative by producing highly reactive radicals that can break the strong carbon–fluorine bonds. This enables the stepwise degradation of PFAS, often resulting in substantial or complete defluorination of the original compounds. As a result, AOP-based treatments mark a shift in soil remediation—from simply containing pollutants to destroying them at the molecular level. This approach offers a more sustainable solution by permanently reducing PFAS toxicity rather than just delaying the spread of contamination.

## 10.2 Technologies for soil oxidation remediation

Perfluoro carboxylic acid, a subset of per- and polyfluoroalkyl substance (PFAS), has been widely detected in environmental matrices such as water and soil due to their extensive industrial and commercial application. These compounds are used as lubricants, surfactants, additives, nonstick cookware, food packaging, stain-resistant fabrics, and firefighting aqueous film-forming foams (Lin et al., 2015). Perfluoroalkyl carboxylic acids (PFCAs) with PFOA consist of a perfluorinated chain which is attached to a carbonyl functional group ($C_nF_{2n+1}COOH$). The carbon–fluorine (C–F) bond is a polar covalent bond with high bond dissociation energy. This extreme strength stabilizes the polymer and defies standard environmental breakdown. Consequently, these characteristics label the compounds as persistent organic pollutants (Huang et al., 2016). The degradation rate of PFCAs is slow, which results in an adverse effect on the environment by enhancing the life period. The long-term stability of PFCA leads them to accumulate in surface and groundwater, further worsening their environmental impact. Other than environmental persistence, PFCAs' toxicological and bioaccumulation effects have raised health concerns in living beings (Qi et al., 2022). It has been investigated that the presence of PFOA compounds in wildlife and human blood serum, other than that it found in remote ecosystems over long range because of its long-range transport nature in water and air. Because of their toxicity and persistence, there is an alarming concern to develop efficient and economical degradation technology to regulate their health and environmental impact (Ehsan et al., 2024). The chemical stability and mobility of PFAS compounds make them accumulate in soil and water over time and create substantial ecological concerns, as shown in the Fig. 10.1.

Biological oxidation is a conventional oxidation process which capable of breaking C–C bonds, but it cannot break C–F persistent bonds due to their chemical stability (Yin et al., 2025). Defluorination of PFAS has a limited option, short-chain PFAS can be decomposed using microorganisms. Many other processes have been investigated to enhance the efficiency of defluorinated PFAS and destroy stable PFAS compounds. A process like photocatalytic ozonation, UV/sulfite, advanced oxidation and reduction methods (Zelekew et al., 2024). The stability and toxicity of PFAS are ascribed to the fluorine atom; scissoring of carbon skeleton C–C without cleavage of C–F bonds makes shorter chain PFAS that still

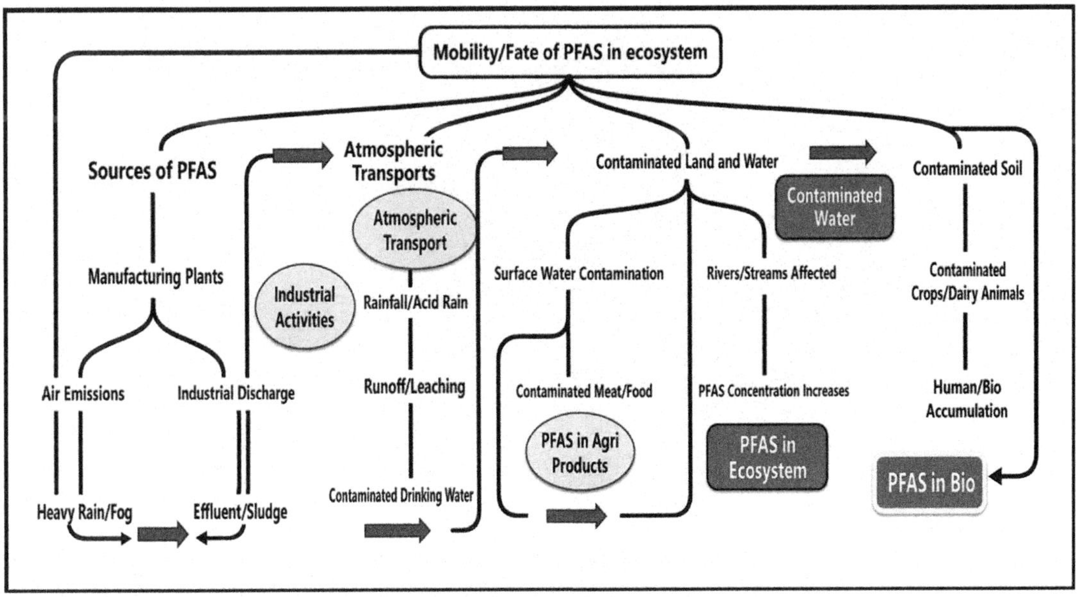


FIGURE 10.1 Schematic of perfluoroalkyl substances transport in the ecosystem.

harms the environment by its persistent nature (Alalm and Boffito, 2022). Although numerous studies have reported the degradation of PFAS and defluorination, other than the oxidation pathway, researchers have introduced advanced oxidation and reduction processes based on speculation screening of spectroscopy analysis techniques (Li et al., 2021). The PFAS treatment by various types of AOPs is shown in Fig. 10.2.

## 10.3 Advanced oxidation processes

AOPs offer a promising approach for breaking down PFAS in contaminated soils; they functions by producing reactive oxygen species (ROS) that can break the strong carbon–fluorine bonds in PFAS molecules. The most successful remediation technique is AOPs, which use ROS

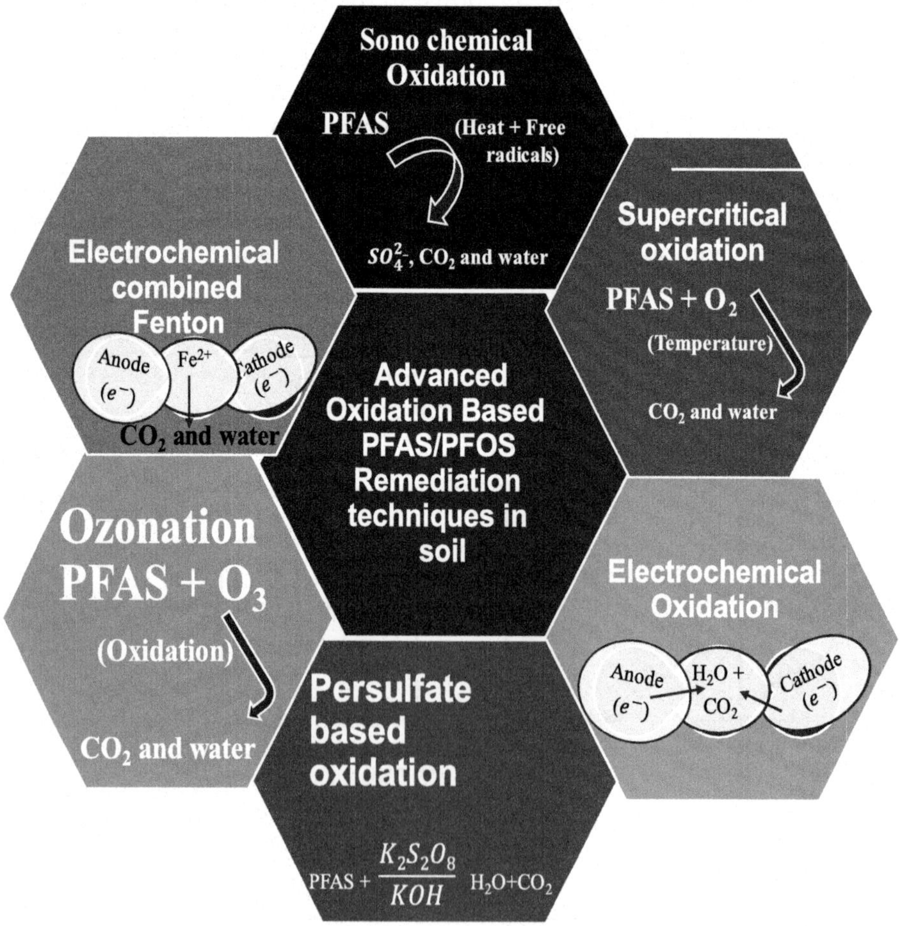


FIGURE 10.2 Perfluoroalkyl substances treatment by various types of advanced oxidation processes.

such as hydroxyl radicals ($^{\bullet}OH$) and sulfate radicals ($SO_4^{\bullet-}$) to cleave the carbon–fluorine bonds in PFAS, resulting in smaller and less toxic byproducts (Ehsan et al., 2024). These radicals are created by several processes, such as the Fenton reaction, in which $Fe^{2+}$ reacts with $H_2O_2$ to produce hydroxyl radicals; persulfate activation, which uses heat or ultraviolet light to produce sulfate radicals; and electrochemical oxidation, which uses high-voltage electrodes (Xu et al., 2020). Chain scission of PFAS molecules results in degradation reaction by producing fluoride ions ($F^-$), $CO_2$, and other organic chemicals. AOPs' reaction of PFAS generates three types of radical species, hydroxyl radical ($OH^{\bullet}$), which initiates the degradation reaction, and these radical exhibits strong oxidation properties can attack PFAS chains, sulphate radical ($SO_4^{\bullet-}$) has strong oxidative properties which are generated from persulfate activation, and superoxide radical ($O_2^{\bullet-}$) tends to break down the PFAS precursors into the transitional products and initiate the mineralization process (Belkouteb et al., 2020). The degradation rate and pathway of PFAS highly influence and dominate free radicals and ROS generation. In homogeneous oxidation reactions, operating conditions such as PFAS concentration and pH are the major factors influencing the oxidation reaction by generating alternative types and quantities of ROS formation (Sharma et al., 2022). Degradation occurs through the breakdown of PFCA and PFSA molecules by eliminating a sulfur trioxide or carboxyl group to release $SO_3$ or $CO_2$ (decarboxylation) and replacing hydrogen atoms in place of fluorine atoms and chain scission of C–C bonds. In the case of heterogeneous catalyst oxidation reactions, the degradation pathway and defluorination rate are influenced by photo and catalysts (Ren et al., 2021). The heterogeneous APS process includes photocatalytic and Fenton processes; most of the interaction between the ROS and contaminants will occur in the bulk solution or on the catalyst surface. Fig. 10.3 illustrates the interaction between the catalyst and PFAS compounds and ions transferring mechanism. The activity of the degradation reaction depends on the charge potential between existing functional groups of organic molecules and existing catalysis. The zero charge point zone in the catalyst has maximum efficiency in the degradation reaction (Wang et al., 2019).

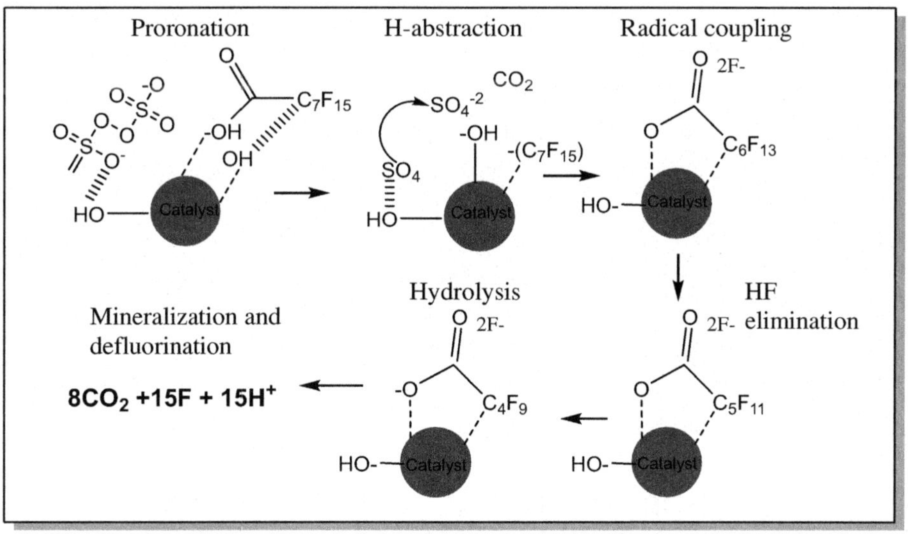


FIGURE 10.3 Perfluorooctanoic acid degradation mechanism with the catalyst.

## 10.3.1 Electrochemical degradation

The carbon–halogen bond (C–X) chain scission mechanism is widely explored in organic electrochemistry and halogenated organic containment, becoming a common pollutant in the environment, and is particularly relevant for environmental remediation (Han et al., 2023). Dissociative electron transfer (DET) is used to explain the pathway of the direct electron transfer mechanism, which is used to understand electrochemical reduction. PFAS has a specific group of alkyl-halides with C–F bonds (Su et al., 2019). There are two possible pathways of the DET model, which seem applicable to aryl halides (R–X) and alkyl (Scheme 10.1) stepwise mechanism, a two-electron process is implicated. In the first step, an intermediate anion radical $[R–X]^{\bullet}$ is yielded, then it further decomposes to produce a free radical $R^{\bullet}$ and X. In Scheme 10.2, bonding splitting and radical transfer happen both in a single step to yield $R^{\bullet}$ free radical (Yin et al., 2025). The next step of the reduction process is carbonation. Once $R^{\bullet}$ is generated, it quickly undergoes a second electro-transfer process. Further chemical reactions are carried out by radical species, which are coupled to the electron transfer process, such as deprotonations and nucleophilic attack (Pierpaoli et al., 2021). In electro-oxidation processes, the electric potential is applied in a slurry of contaminants using electrodes, where an oxidation reaction occurs on the anode and a reduction in the cathode. As a result, high concentrations of reactive radicals are generated at the anode, and it has a high tendency to degrade the organic pollutant (Qiao et al., 2023). The following reactions will take the path in PFAS compounds.

$$C_7F_{15}CO_2^- + M \rightarrow C_7F_{15}CO_2 + e^- \tag{10.1}$$

$$C_7F_{15}CO_2^{-\cdot} \rightarrow C_7F_{15} + CO_2 \tag{10.2}$$

$$C_7F_{15}^{\cdot} + OH^{\cdot} \rightarrow C_7F_{15}OH \tag{10.3}$$

$$C_7F_{15}OH \rightarrow C_6F_{13}COF + HF \tag{10.4}$$

$$C_7F_{15} + O_2 \rightarrow C_6F_{15}O_2 \tag{10.5}$$

$$C_7F_{15}O_2^{\cdot} + C_7F_{15}O_2^{\cdot} \rightarrow 2C_7F_{15}^{\cdot}O^{\cdot} + O_2 \tag{10.6}$$

$$C_6F_{15} + CO_2 \rightarrow C_6F_{13} + COF_2 \tag{10.7}$$

The EO process has a high potential to degrade the complex organic pollutant; the outcomes from the mineralization process are $CO_2$ and $H_2O$ (Karatas et al., 2023). Likely in

$$C_7F_{15}CO_2^- + M \rightarrow C_7F_{15}CO_2^{\square} + e^- \quad (1)$$

$$C_7F_{15}CO_2^{-\square} \rightarrow C_7F_{15} + CO_2 \quad (2)$$

$$C_7F_{15}^{\square} + OH^{\square} \rightarrow C_7F_{15}OH \quad (3)$$

$$C_7F_{15}OH \rightarrow C_6F_{13}COF + HF \quad (4)$$

**SCHEME 10.1** Degradation pathway via dissociative electron transfer 1.

$$C_7F_{15} + O_2 \rightarrow C_6F_{15}^{\square}O_2 \tag{5}$$

$$C_7F_{15}O_2^{\square} + C_7F_{15}O_2^{\square} \rightarrow 2\ C_7F_{15}^{\square}O^{\square} + O_2 \tag{6}$$

$$C_6F_{15}^{\square} + CO_2^{\square} \rightarrow C_6F_{13}^{\square} + COF_2 \tag{7}$$

**SCHEME 10.2** Degradation pathway via dissociative electron transfer 2.

the following process, there is no need for additional chemicals and catalysts, and it eliminates secondary by-product pollutants. The produced $C_7F_{15}O_2$ are highly reactive and unstable, which undergo a decarboxylation process (Kutsuna and Hori, 2007). The degradation pathway is not standalone; the mentioned one may follow a different pathway and mechanisms.

## 10.3.2 Thermal degradation

PFAS-contaminated sewage spreads through sewage systems and can travel far away from where it was first used. Using very high temperatures to "break" the chemical bonds. To turn a permanent, toxic chemical into something that isn't harmful. In this, the containment is heated to high temperatures (600 to 1000) to mineralize the PFAS and destroy the harmful compounds. Meantime there is a risk of releasing volatile gas from polymer compounds to the surrounding environment. The incineration process is applied to evaluate the destiny of PFAS; the combustion of PFAS yields toxic gas substances, such as hexafluoroethane and tetrafluoromethane ($CF_4$) (Watanabe et al., 2016). The thermal decomposition of PFCA, perfluoroalkyl ether carboxylic acid, and perfluoroalkyl sulfonic acids (PFAS) is performed under ($N_2$, $O_2$, $CO_2$) atmosphere conditions for thermal reactivation to spent granular activated carbon (Xiao et al., 2020). Since heating the PFAS compound is not sufficient to degrade the compound, combustion is carried out in controlled atmosphere conditions, there is a necessity to use an inline gas chromatography apparatus to monitor the process, also there is a need to optimize the combustion process and to understand the kinetic of degradation process to improve efficient of yield and to attain complete degradation of PFAS compounds (Wang et al., 2022).

## 10.3.3 Ozone degradation

The efficiency of the PFAS degradation in the soil is enhanced by combining the soil washing and ozonation During the soil washing process, the PFAS compound is transported to the aqueous medium, which stays as a suspension. In the aqueous medium, the removal efficiency reached up to 85%–100% (Thomas et al., 2020). Moreover, the efficiency of the process high influenced by pH; several studies state that pH ranges from 4–5 or 10–11 are maintained for the ozonization process. $O_3$ is introduced in the bubble form into a contaminant-containing solution for pretreatment up to 15 min, and the contaminated system is maintained between 4–5 pH, then is gradually increased to 11 pH by adding

NaOH solution, and $O_2$ gas is introduced in the contamination chamber (Lin et al., 2012). To make these methods work better, researchers varied the amount of ozone and oxygen used. This technique is effective at breaking down different types of PFAS, including both short and long versions of the chemicals. Additionally, it is a reliable and eco-friendly way to remove these pollutants from the environment (Barisci and Suri, 2023). Research shows that adding oxygen and ozone directly, rather than using ozone as a pretreatment, creates a high amount of active free radicals. This reaction causes the chemicals to break apart effectively (D'Agostino and Mabury, 2014). Table 10.1 shows the literature survey of various types of PFAS and describes the short-chain products degradation through AOPs.

## 10.4 Technologies for soil oxidation remedies

Research has firmly established a connection between PFAS exposure and a higher likelihood of developing cancer, specifically highlighting a direct link between elevated PFOA concentrations and a risk of kidney and testicular malignancies (Seyyed Salehi and Boffetta, 2023). Although Skeen land et al. noted that the evidence linking PFOA or PFOS levels to bladder, liver, pancreatic, or prostate cancers remains inconclusive, the International Agency for Research on Cancer has formally categorized PFOA as a "human carcinogen." In contrast, an official evaluation for PFOS has yet to be established (Steenland and Winquist, 2022; Chang et al., 2014). The hazards of PFAS exposure have created a critical demand for reliable cleanup solutions (Baldwin et al., 2025). The pervasive spread of PFAS "forever chemicals" creates a major hurdle for meeting the 2030 sustainable development goals. Addressing this crisis requires a two-pronged approach: aggressive remediation of existing pollutants and a dramatic reduction in current usage to safeguard both public health and ecological integrity. Moving forward, the global outlook for PFAS management centers on massive financial investment and strict regulatory frameworks designed to curb their long-term impact. This financial shift, coupled with a sharp rise in academic and industrial studies, reflects a massive surge of interest in developing technologies to neutralize these "forever chemicals" (Luche et al., 2026; Berthold et al., 2023).

Remediating PFAS-contaminated soil is exceptionally difficult due to the following primary factors, such as, their robust chemical and thermal bonds make them resistant to traditional destruction methods, contaminated sites rarely contain a single type of PFAS, usually presenting a complicated variety of different compounds, their simultaneous water- and oil-repellent properties (hydrophobic and oleophobic) complicate standard separation and filtration techniques, because they do not biodegrade, these "forever chemicals" remain in the environment indefinitely without intervention (Zhu et al., no date). While multiple approaches exist for treating PFAS-contaminated soil, many current technologies struggle with high costs, secondary waste production, or the inability to capture short-chain compounds. Remediation approaches are generally categorized into three main frameworks: in situ, ex situ, and AOPs. According to Qutob et al., these methods utilize specific chemical reactions to neutralize soil contaminants. In situ methods are helpful to treat the soil in place using techniques such as Fenton's reagent, ozone treatment, and the application of permanganate or persulfate. Ex situ methods require soil excavation for treatment,

**TABLE 10.1** A literature survey of perfluoroalkyl substances and other short-chain products degradation from various types of wastewaters.

| Pollutants | Methods | Efficiency | Kinetics used | Conditions | Advanced oxidation processes (AOP) type | $R^2$ values | References |
|---|---|---|---|---|---|---|---|
| Perfluorooctanoate (PFOA) | Electro-Fenton/boron doped diamond/ membrane distillation (EF–BDD–MD) | 95.0% | Pseudo–first order | Current density = 50 to 500 A $m^{-2}$ $H_2O_2$ = 0.25 to 2.3 mM | Combined AOP | 0.98532 | Yousefi et al. (no date) |
| Perfluoro butanoic acid (PFBA), perfluoro-n-pentatonic acid (PFPeA), Perfluoro hexanoic acid (PFHxA), and perfluoroheptanoic acid (PFHpA) | Electrochemical oxidation | 31.8% ± 4.6%<br>41.4% ± 4.1%<br>78.2% ± 3.2%<br>97.9% ± 4.6% | Pseudo–first order | Wastewater = 140.8 –1386.9 ppm, and $PBO_2$ = film electrode | Single AOP | 0.996<br>0.999<br>0.994<br>0.997 | Niu et al. (2012) |
| PFOA and perfluoro octane sulfonate (PFOS) | Electron–beam degradation | 95.7% and 85.9% | Pseudo–first order | Initial concentration of wastewater = 10–40 ppm, ambient temperature = 1.8 MeV, and irradiation = 100–500 kGy. | Single AOP | >0.90 | Ma et al. (2017) |
| PFOA and PFOS | Electrochemical oxidation using $Ti_4O_7$ reactive electrochemical membrane anode | 79.6% and 73.0% | Pseudo–first order | pH = 7, electrode surface area = 0.33 $cm^2$ | Combined AOP | Not mentioned | Le et al. (2019) |
| PFOA | Photodecomposition (VUV light) | 100% | Pseudo–first order | VUV = 5 W lamp, wavelength = 185 to 254 nm, and 20 µm initial dosage of PFOA, | Single AOP | Not mentioned | Liang et al. (2016) |

(*Continued*)

TABLE 10.1 (Continued)

| Pollutants | Methods | Efficiency | Kinetics used | Conditions | Advanced oxidation processes (AOP) type | $R^2$ values | References |
|---|---|---|---|---|---|---|---|
| PFOA | Hg lamp using (500 W), $Pt/In_2O_3$ photocatalyst | 98% | Pseudo–first order | PFOA = 200 ppm, catalyst dosage = 0.4 g $L^{-1}$ | Single AOP | Not mentioned | Chenmin et al. (2017) |
| PFOA | Photodegradation using a (200 W) xenon lamp | 61.7% | Pseudo–first order | VUV light irradiation time = 120 min., | Single AOP | 0.9691 | Chen et al. (2007) |

primarily utilizing thermal oxidation or wet-air oxidation. AOPs represent specialized or hybrid approaches, including electrochemical oxidation, persulfate-based AOPs, or integrated systems that combine thermal treatment with adsorption, as shown in Table 10.2 (Qutob et al., 2025). The comparison of different soil PFAS removal technologies are compiled in Table 10.3. To figure out which sources are most responsible for PFAS pollution, need to map out where these chemicals come from and how they spread. Crucial to the design of mitigation frameworks is the identification of PFAS sources and the assessment of their severity, ensuring that intervention efforts are both prioritized and effective.

### 10.4.1 Reaction mechanism of perfluoroalkyl substances removal from soil

Soil oxidation remediation involves the use of highly reactive radical species called cleaner radicals, which dismantle the stubborn pollutant molecules. These chemicals specifically sulfate, hydroxyl, and superoxide atoms, which attack the organic pollutant molecules and break down into smaller, as well as nontoxic substances. The process starts by introducing the oxidants into the soil to produce highly reactive radicals. Then, these radicals function as chemical shears, aggressively targeting the complex organic pollutant bonds. After the chemical bonds are broken, the pollutants are converted into harmless products, typically small fragments of water and carbon dioxide ($CO_2$). Common oxidants, such as hydrogen peroxide, permanganate, persulfate, or ozone, neutralize the pollutants by stripping away their electrons, attacking them with reactive radicals, or physically splitting their chemical bonds. These reactions destroy complex organic molecules by converting them into intermediate products and mineralizing them into carbon dioxide, water, and inorganic ions. The general reaction mechanism of PFAS removal from soil from AOPs is shown in Fig. 10.4. The efficiency of the process depends on soil properties, oxidant dosage, pH, and contaminant type.

### 10.4.2 Synergistic importance of energy and cost effects of advanced oxidation processes-based soil remediation

The effectiveness of hybrid AOPs in PFAS cleanup comes from their enhanced performance, achieved by activating multiple complementary pathways at the same time (Silva et al., 2025). In advanced oxidation and soil remediation, the following four points are important: chemical reactions, mass transfer, intermediate byproducts formation, and the

**TABLE 10.2** Classification of soil oxidation/remediation techniques

| Category | Primary methods |
|---|---|
| In situ | Fenton's reagent, persulfate, permanganate, and ozone. |
| Ex situ | Thermal oxidation, wet-Air oxidation. |
| Advanced (AOPs) | Electrochemical, persulfate oxidation, hybrid of thermal + adsorption approach. |

**TABLE 10.3** Comparison of perfluoroalkyl substances removal from distinct types of contaminated soil

| Methods | Conditions | In situ/ ex situ | Pollutant removed | Used materials | Single/hybrid techniques | Efficiency | References |
|---|---|---|---|---|---|---|---|
| Electrochemical oxidation | Current: 5 mA $cm^{-2}$, with the pore size of is greater than 2.6 μm and the porosity of 21.6% | In situ | PFASs | Anode: magneli ($Ti_4O_7$) cathode: stainless steel | Single system | 99.0% PFOA 93.4% PFOS | Lin et al. (no date) |
| Ultrasound/ persulfate (US/PS) | For 1 L of slurry containing 200 g of contaminated soil, time duration: 6 hrs. | In situ | PFASs | Ultrasound capacity: 20 to 43 kHz, of a titanium probe of 25.4 mm diameter, | Integrated system | 71.0% | Lie et al. (2020) |
| Soil washing | pH: 7, conductivity of sample (1 L): 200 μS/ cm. secondary treatment is required. | Ex situ | PFOSs | Soil samples stored at 4°C, in a polyamide sampling bag. | Single technique | 73.0% | Hoisaeter et al. (2021) |
| Vacuum ultraviolet (VUV)/sulfite system | Time duration: 4 hrs., initial concentration of pollutant at the pH level of 10. | Ex situ | PFOSs | 185 nm photons are used through the VUV system. | Hybrid system | PFOS: 97.3% defluorination removal: 75.4% | Gu et al. (2017) |
| Ozonation | pH: 4 to 5, inlet concentration: 50 μg $L^{-1}$, time duration: 4 hrs. | Ex situ | PFAAs, PFOA, and PFOS | Ozone producer with a capacity of 4.2 L $min^{-1}$. magnetic stirrer with water bath. | Single system | ≈ 85%–100% | Lin et al. (2012) |
| Photodegradation | $TiO_2$ combined with peroxymono-sulfate (PMS), pH: 3, at the inlet concentration of 50 mg $L^{-1}$, $TiO_2$: 0.25 g $L^{-1}$, PMS: 0.75 g $L^{-1}$. | Ex situ | PFOA | Visible light (300 W): 400 to 700 nm source, and UV light (32 W): 185 to 254 nm sources. | Hybrid phenomenon. | 100% | Xu et al., (no date) |
| 3D electrochemical advanced oxidation process (3D-EAP) | Time duration: 120 min., inlet concentration of 100 μg $L^{-1}$, in a 200 ml | Both can be done. | PFOA | Electrode: ($Fe_3O_4$@ MHCS) used. | Single system | 96.0% | Annamalai et al. (2026) |

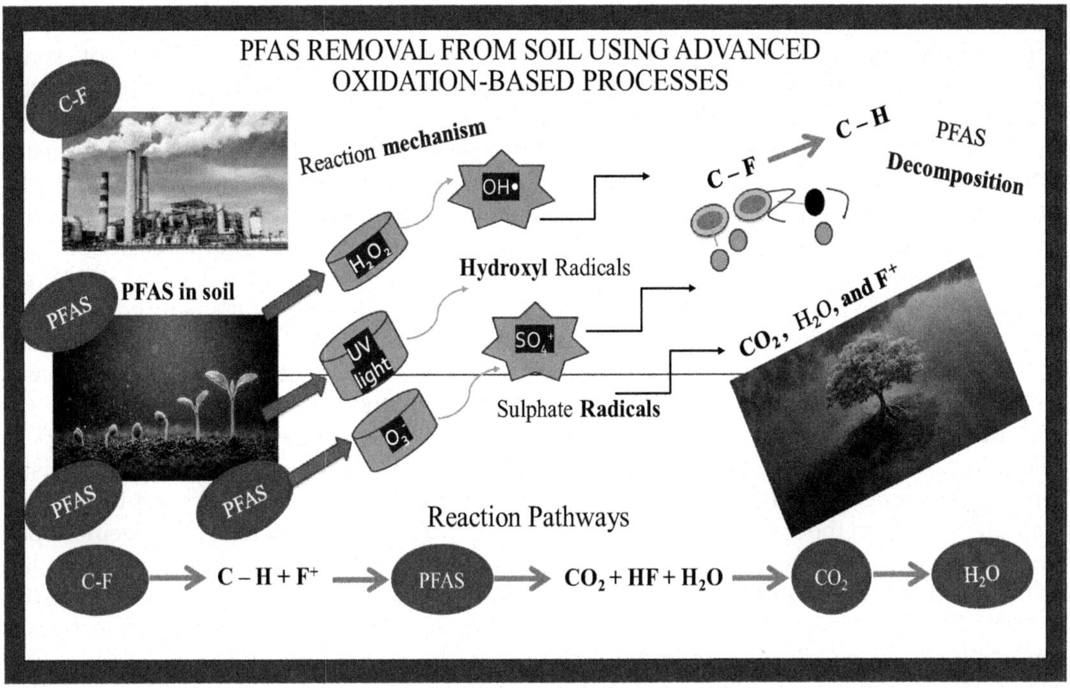


**FIGURE 10.4** Perfluoroalkyl substances (PFAS) removal reaction mechanism of contaminated soil by advanced oxidation processes.

improved soil recovery function, like a relay race. The physical processes (mixing) ensure the chemicals meet the pollutants, while the removal of inhibitors ensures that the "finish line" is called as clean soil is reached without the reaction being stalled by competing molecules (Xiang et al., 2022). Unlike traditional chemical oxidation, electrochemical processes can produce potent radicals and high-energy intermediates specifically tailored for carbon–fluorine bond activation (Li et al., 2024). The constructive collaboration between these species and photocatalytic systems creates a catalytic loop that accelerates reaction rates and optimizes the efficiency of photon utilization (Fang et al., 2025). Quantifying the effectiveness of hybrid systems often involves calculating synergy factors. This metric represents the ratio of the hybrid system's performance to the combined results of its individual components when operated independently.

By combining electrochemical and photocatalytic methods to remove PFAS, studies achieved with great synergy factors, proving the hybrid system is far more effective than the two methods acting independently (Tshangana et al., 2025). The degree of synergistic improvement is heavily influenced by operational parameters. Achieving peak performance typically involves a two-stage approach: first, optimizing the variables for each individual process, followed by a focused optimization of the integrated system's parameters. To streamline this complex process, factorial design experiments are employed, allowing researchers to identify ideal operating windows efficiently without the need for exhaustive, repetitive testing (Jankovic et al., 2025). Compared to individual operations achieving equivalent removal, hybrid configurations demonstrate enhanced energy efficiency. This reduction in overall energy intensity per unit of PFAS is a direct result of shared infrastructure and accelerated degradation kinetics (Arun et al., 2026). Research indicates that integrating gold and silver nanoparticles into fiber-gel electrodes significantly enhances efficiency compared to traditional standalone systems. By utilizing these nanoparticle-embedded hydrogel electrodes, energy consumption can be slashed by 18% to 20%. At a current density of 10 mA $cm^{-2}$, the system achieved the following results within a two-hour window. The transition to these advanced electrodes allows for a dramatic reduction in the specific energy required for PFAS remediation, dropping from a high of 164.9 kWh $m^{-3}$ down to just 90 kWh $m^{-3}$. Economic assessments reveal that hybrid AOPs can reduce overall costs by 15% to 20% compared to using individual treatment methods (Hwang et al., 2022). The strength of the applied voltage is a pivotal driver in the breakdown of PFAS. Increasing the voltage amplitude intensifies the electric field within the reactor, which in turn accelerates electron-impact reactions. This surge in energy significantly ramps up the generation of the reactive species required to successfully dismantle these highly resistant "forever chemicals." Crucially, the impact of voltage on treatment success is not universal; it fluctuates based on the specific pollutant being targeted. Researchers tested a specific type of plasma technology called a reverse vortex gliding arc (GAP). They used air, nitrogen, and oxygen to see how well this system removes different types of PFAS from liquids. The study focused on three specific chemical groups: PFCAs, PFSAs, and FTSs. All tests were conducted using a one-litre treatment container. In this study, PFAS levels are used between 0.05 and 1 g $L^{-1}$. They found that this method successfully removed many of these chemicals. The success of the treatment depended mostly on the length of the chemical chains. Specifically, compounds with eight or more carbon atoms saw a removal rate of over 90%. In just 1 h, these longer chains also experienced a 25% reduction in fluorine levels using electrical energy 23.2 kW h $m^{-3}$ only (Lewis et al., 2020).

This disparity is driven by the inherent chemical structures of the molecules: the sulfonate group in PFOS is more susceptible to plasma-induced breakdown than the carboxylate group in PFOA. While UV/electrochemical hybrid systems offer superior performance, they involve a higher initial capital investment compared to standard methods. Specifically, establishing a hybrid system requires an estimated \$100,000 to \$150,000 for every 1000 $m^3$ $day^{-1}$ of treatment capacity. Despite the higher initial capital requirements, hybrid AOP systems offer substantial operational savings that lead to a short payback period of 12 to 24 months (Ghosh and Sahu, no date). This rapid return on investment makes these advanced systems a financially viable solution for medium- to large-scale PFAS remediation projects.

### 10.4.3 Life cycle assessment of perfluoroalkyl substances-degradation in soil by advanced oxidation processes

Unlike AOPs, conventional methods typically fail to achieve destruction and often require secondary treatment. For instance, while incineration can fully break down PFAS, it still necessitates complex air emission controls to prevent secondary pollution. While techniques such as soil washing and stabilization only relocate PFAS or change its physical state, AOPs provide a definitive destructive pathway. By breaking down these "forever chemicals" entirely, AOPs offer a superior solution for lowering long-term ecotoxicity and protecting biodiversity. While AOP efficiency depends on energy sources, scale, and chemical optimization, a hybrid approach is often best. Combining AOPs with conventional methods can outperform individual treatments, potentially reducing the overall energy footprint within a life cycle assessment (LCA).

Uncertainty analysis involves navigating a complex web of variables and assumptions. In the context of PFAS remediation, the specific soil composition and the intensity of the contamination significantly dictate the process. Furthermore, chemical oxidation strategies rely on precise dosing to trigger reactive soil radicals; these strategies are often influenced by the nuances of electrical mixing speeds and the resulting reaction kinetics. To ensure these findings are robust and reliable, researchers utilize advanced simulations, such as Monte Carlo scenarios, to account for potential variability. In AOPs, optimizing LCA performance depends on identifying and managing specific "hotspots" or high-impact variables (Ribeiro et al., 2024). Key factors include the carbon footprint of electricity generation, the chemical grade and volume of persulfates used, and the integration of UV-based photocatalysis. By transitioning toward renewable energy sources and refining oxidant dosages at these critical points, the overall environmental and economic sustainability of the treatment process can be significantly enhanced. LCA highlights several critical pillars for modern environmental management, emphasizing the integration of renewable energy with in situ treatment methods. By applying the 3R principles (reduce, reuse, and recycle), and combining AOPs with biological remediation, practitioners can move toward a circular economy—specifically through the recovery and repurposing of by-products such as fluoride. These strategies shift operational standards toward the core tenets of green chemistry, ensuring that chemical handling is both sustainable and environmentally responsible.

LCA serves as a vital evaluative tool for measuring the environmental footprint of PFAS removal in soil remediation. Because AOPs typically require significant chemical reagents and high-energy intensity, the LCA focuses on three primary pillars: system design, oxidant dosage management, and total energy consumption (Sanzana et al., 2025). When these processes are integrated with low-carbon energy sources, AOPs achieve superior efficiency, effectively transitioning the remediation-strategy from simple pollutant transfer to a "zero containment" or complete destruction model. To ensure that PFAS destruction technologies provide authentic environmental benefits, future studies must bridge the gap between laboratory results and real-world applications. This requires a shift toward pilot-scale data collection, rigorous long-term toxicity monitoring, and the development of hybrid treatment systems. Without this holistic approach, there is a risk of "burden shifting"—where solving a contamination issue in one area inadvertently creates a new environmental problem elsewhere in the life cycle.

## 10.5 Conclusion

Application of AOPs for PFAS soil remediation involves a fundamental trade-off between destruction efficacy, cost, and complexity. It offers complete elimination of contaminant mass and liability; they are high-cost, high-energy operations. The most powerful destructive technologies (EO and SCWO) are best applied to concentrated waste streams, requiring substantial initial capital investment in separation infrastructure (e.g., soil washing, IXR, FF) and high ongoing operating expenditure for energy. Furthermore, EO necessitates management of regulated toxic byproducts, primarily chlorate, which complicates regulatory approval and requires an additional treatment step. For soil remediation, destruction must occur ex situ to minimize the required oxidant dose and increase the destruction efficiency of the AOP. Mitigating the demonstrated ability of EO to reduce overall toxic potency, even of unidentified compounds, confirms the necessity of moving beyond simple target compound quantification. Concentrates should be isolated and stored or immediately routed to a high-energy destructive AOP to manage or destroy the residues formed during the extraction and treatment of PFAS. Insufficient data from pilot- and full-scale AOP applications for PFAS soil remediation limits optimization and advancement. Future remediation projects should focus on rigorous field-scale evaluation of integrated soil washing and AOP trains to refine operational parameters, particularly concerning short-chain degradation and the practical management of byproduct formation.

## References

Alalm, M.G., Boffito, D.C., 2022. Mechanisms and pathways of PFAS degradation by advanced oxidation and reduction processes: a critical review. Chemical Engineering Journal 450 (4), 138352. Available at https://doi.org/10.1016/j.cej.2022.138352.

Annamalai, S., Kim, N., Hwang, S., 2026. Performance of a 3D-particle electrode for PFOA degradation by persulfate-activated electrochemical advanced oxidation processes. Journal of Hazardous Materials: Organics 2, 100014. Available at https://doi.org/10.1016/j.hazmo.2026.100014.

Arun, K.K., Yasi, H.S.A., Wada, O.Z., AlMomani, F., Mahmoud, K.A., 2026. Comprehensive review of PFAS occurrence, regulatory frameworks, analytical, and remediation methodologies: emphasis on the Arabian Gulf and arid regions. Environmental Science: Water Research & Technology 12, 121. Available at https://doi.org/10.1039/d5ew00570a.

Baldwin, W.S., Bloom, M.S., Chung, K.W., Dasgupta, S., DeLorenzo, M.E., Hunt, K.J., et al., 2025. Toxicological, chemical, social, and economic challenges associated with PFAS and replacement aqueous film-forming foams (AFFF). Toxics 13 (9), 732. Available at https://doi.org/10.3390/toxics13090732.

Barisci, S., Suri, R., 2023. Degradation of emerging per- and polyfluoroalkyl substances (PFAS) using an electrochemical plug flow reactor. Journal of Hazardous Materials 460, 132419. Available at https://doi.org/10.1016/j.jhazmat.2023.132419.

Belkouteb, N., Franke, V., McCleaf, P., Köhler, S., Ahrens, L., 2020. Removal of per- and polyfluoroalkyl substances (PFASs) in a full-scale drinking water treatment plant: Long-term performance of granular activated carbon (GAC) and influence of flowrate. Water Research 182, 115913. Available at https://doi.org/10.1016/j.watres.2020.115913.

Berthold, T.A., McCrary, A., Villeneuve, S.D., Schramm, M., 2023. Let's talk about PFAS: inconsistent public awareness about PFAS and its sources in the United States. PLoS One 18 (11), e0294134 https://doi.org/10.1371/journal. pone.0294134.

Chang, E.T., Adami, H.O., Boffetta, P., Cole, P., Starr, T.B., Mandel, J.S., 2014. A critical review of perfluorooctanoate and perfluorooctanesulfonate exposure and cancer risk in humans. Critical Reviews in Toxicology 44, 1–81. Available at https://doi.org/10.3109/10408444.2014.905767.

Chen, J., Zhang, P.Y., Liu, J., 2007. Photodegradation of perfluorooctanoic acid by 185 nm vacuum ultraviolet light. Journal of Environmental Science 19 (4), 387–390. Available at https://doi.org/10.1016/S1001-0742(07)60064-3.

Chenmin, X., Pingxiang, Q., Huan, C., Fang, J., 2017. Platinum modified indium oxide nanorods with enhanced photocatalytic activity on degradation of perfluorooctanoic acid (PFOA. Journal of the Taiwan Institute of Chemical Engineers 80, 761–768. Available at https://doi.org/10.1016/j.jtice.2017.09.018.

D'Agostino, L.A., Mabury, S.A., 2014. Identification of novel fluorinated surfactants in aqueous film forming foams and commercial surfactant concentrates. Environmental Science & Technology 48, 121. Available at https://doi.org/10.1021/es403729e.

Ehsan, M.N., Riza, M., Pervez, M.N., Li, C.W., Zorpas, A.A., Naddeo, V., 2024. PFAS contamination in soil and sediment: contribution of sources and environmental impacts on soil biota. Case Studies in Chemical and Environmental Engineering 9, 100643. Available at https://doi.org/10.1016/j.cscee.2024.100643.

Fang, R., Xing, Z., Yang, Q., Li, X., Wang, Q., Feng, F., et al., 2025. Spin polarization and barrier synergy: new paradigms in catalytic science. ChemSusChem 18, 1164. Available at https://doi.org/10.1002/cssc.202501164.

Ghosh, S., Sahu, M., n.d. Ultrasound for the degradation of endocrine disrupting compounds in aqueous solution: a review on mechanisms, influence of operating parameters and cost estimation. Chemosphere, 349, 140864. https://doi.org/10.1016/j.chemosphere.2023.140864.

Gu, Y., Liu, T., Wang, H., Han, H., Dong, W., 2017. Hydrated electron-based decomposition of perfluoro octane sulfonate (PFOS) in the VUV/sulphite system. Science of the Total Environment 607 – 608, 541–548 https://doi.org/10.1016/j.scitotenv.2017.06.197.

Han, J., Zhao, M., Wu, K., Hong, Y., Huang, T., Gu, X., et al., 2023. A bifunctional single-atom catalyst assisted by $Fe_2O_3$ for efficiently electrocatalytic perfluorooctanoic acid degradation by integrating reductive and oxidative processes. Chemical Engineering Journal 470, 144270. Available at https://doi.org/10.1016/j.cej.2023.144270.

Hoisaeter, A., Arp, H.P.H., Slinde, G., Knutsen, H., Hale, S.E., Breedveld, G.D., et al., 2021. Excavated vs novel in situ soil washing as a remediation strategy for sandy soils impacted with per- and polyfluoroalkyl substances from aqueous film forming foams. Science of the Total Environment 794, 148763. Available at https://doi.org/10.1016/j.scitotenv.2021.148763.

Huang, D., Yin, L., Niu, J., 2016. Photoinduced hydro defluorination mechanisms of perfluorooctanoic acid by the SiC/graphene catalyst. Environmental Science & Technology 50 (11), 5857–5863. Available at https://doi.org/10.1021/acs.est.6b00652.

Hwang, J.H., Sip, Y.Y.L., Kim, K.T., Han, G., Rodriguez, K.L., Fox, D.W., et al., 2022. Nanoparticle-embedded hydrogel synthesized electrodes for electrochemical oxidation of perfluorooctanoic acid (PFOA) and perfluorooctanesulfonic acid (PFOS). Chemosphere 296, 134001. Available at https://doi.org/10.1016/j.chemosphere.2022.134001.

Jankovic, A., Chaudhary, G., Goia, F., 2025. Optimization through classical design of experiments (DOE): an investigation on the performance of different factorial designs for multi-objective optimization of complex systems. Journal of Building Engineering 102, 111931. Available at https://doi.org/10.1016/j.jobe.2025.111931.

Karatas, O., Khataee, A., Kobya, M., Yoon, Y., 2023. Electrochemical oxidation of perfluorooctanesulfonate (PFOS) from simulated soil leachate and landfill leachate concentrate. Journal of Water Process Engineering 56, 104292. Available at https://doi.org/10.1016/j.jwpe.2023.104292.

Kutsuna, S., Hori, H., 2007. Rate constants for aqueous-phase reactions of $SO_4^-$ with $C_2F_5C(O)O^-$ and $C_3F_7C(O)O^-$ at 298 K. International Journal of Chemical Kinetics 39 (5), 276–288. Available at https://doi.org/10.1002/kin.20239.

Le, T.X.H., Haflich, H., Shah, A.D., Chaplin, B.P., 2019. Energy-efficient electrochemical oxidation of perfluoroalkyl substances using a $Ti_4O_7$ reactive electrochemical membrane anode. Environmental Science & Technology Letters 6, 504–510 https://10.1021/acs.estlett.9b00397.

Lewis, A.J., Joyce, T., Hadaya, M., Ebrahmim, F., Dragiev, I., Giardetti, N., et al., 2020. Rapid degradation of PFAS in aqueous solutions by reverse vortex flow gliding arc plasma. Environmental Science: Water Research & Technology 6, 1044–1057. Available at https://doi.org/10.1039/C9EW01050E.

Li, M., Zhang, H., Cheng, J., Song, J., Han, W., Zhou, S., et al., 2024. Unveiling the decomposing and mineralizing mechanism of novel perfluoroalkyl acids via hydroxyl radical dominated electrochemical oxidation. Applied Catalysis B: Environment and Energy 351, 123983. Available at https://doi.org/10.1016/j.apcatb.2024.123983.

Li, N., Lu, X., He, M., Duan, X., Yan, B., Chen, G., et al., 2021. Catalytic membrane-based oxidation-filtration systems for organic wastewater purification: a review. Journal of Hazardous Materials 414, 125478. Available at https://doi.org/10.1016/j.jhazmat.2021.125478.

Liang, X., Cheng, J., Yang, C., Yang, S., 2016. Factors influencing aqueous perfluorooctanoic acid (PFOA) photodecomposition by VUV irradiation in the presence of ferric ions. Chemical Engineering Journal 298, 291–299 https://10.10.1016/j.cej.2016.03.150.

Lie, J.J., Tian, Y., Sobhani, Z., Naidu, R., Fang, C., 2020. Synergistic degradation of PFAS in water and soil by dual-frequency ultrasonic activated persulfate. Chemical Engineering Journal 388, 124215. Available at https://doi.org/10.1016/j.cej.2020.124215.

Lin, A.Y.C., Panchangam, S.C., Chang, C.Y., Hong, P.K.A., Hsueh, H.F., 2012. Removal of perfluorooctanoic acid and perfluoro octane sulfonate via ozonation under alkaline condition. Journal of Hazardous Materials 243, 272–277. Available at https://doi.org/10.1016/j.jhazmat.2012.10.029.

Lin, H., Niu, J., Liang, S., Wang, C., Wang, Y., Jin, F., et al., n.d. Development of microporous Magneli phase Ti4O7 ceramic materials: as an efficient anode for mineralization of poly- and perfluoroalkyl substances. Chemical Engineering Journal, 354, 1058 – 1067. https://doi.org/10.1016/j.cej.2018.07.210.

Lin, H., Wang, Y., Niu, J., Yue, Z., Huang, Q., 2015. Efficient sorption and removal of perfluoroalkyl acids (PFAAs) from aqueous solution by metal hydroxides generated in situ by electrocoagulation. Environmental Science & Technology 49 (17), 10562–10569. Available at https://doi.org/10.1021/acs.est.5b02092.

Luche, S., Fanelli, A., Buschini, A., Ricci, A., Maccari, C., Andreoli, R., et al., 2026. PFOA and PFOS accumulation induces genotoxic damage and proteomic alterations in *Cannabis sativa* shoots. Journal of Hazardous Materials 501, 140686. Available at https://doi.org/10.1016/j.jhazmat.2025.140686.

Ma, S.H., Wu, M.H., Tang, L., Sun, R., Zang, C., Xiang, J.J., et al., 2017. EB degradation of perfluorooctanoic acid and perfluoro octane sulfonate in aqueous solution. Nuclear Science & Techniques 28 (9), 137 https://10.1007/s41365-017-0278-8.

Niu, J., Lin, H., Xu, J., Wu, H., Li, Y., 2012. Electrochemical mineralization of perfluoro carboxylic acids (PFCAs) by ce-doped modified porous nanocrystalline $PbO_2$ film electrode. Environmental Science & Technology 46 (18), 10191–10198 https://10.1021/es302148z.

Pierpaoli, M., Szopińska, M., Wilk, B.K., Sobaszek, M., Łuczkiewicz, A., Bogdanowicz, R., et al., 2021. Electrochemical oxidation of PFOA and PFOS in landfill leachates at low and highly boron-doped diamond electrodes. Journal of Hazardous Materials 403, 123606. Available at https://doi.org/10.1016/j.jhazmat.2020.123606.

Qi, Y., Cao, H., Pan, W., Wang, C., Liang, Y., 2022. The role of dissolved organic matter during Per- and Polyfluorinated Substance (PFAS) adsorption, degradation, and plant uptake: a review. Journal of Hazardous Materials 436, 129139. Available at https://doi.org/10.1016/j.jhazmat.2022.129139.

Qiao, Q., Singh, S., Patidar, R., Wang, L., Li, Y., Shi, J., et al., 2023. Contribution of electrolyte in parametric optimization of perfluorooctanoic acid during electro-oxidation: active chlorinated and sulfonated by-products formation and distribution. Chemosphere 312, 137202. Available at https://doi.org/10.1016/j.chemosphere.2022.137202.

Qutob, M., Rafatullah, M., Flafel, H.M., 2025. Advancement into the integration of chemical oxidation and bioremediation for the removal of hydrocarbon-contaminated soil. Journal of Environmental Chemical Engineering 13, 116846. Available at https://doi.org/10.1016/j.jece.2025.116846.

Ren, Z., Bergmann, U., Leiviska, T., 2021. Reductive degradation of perfluorooctanoic acid in complex water matrices by using the UV/sulphite process. Water Research 205, 117676. Available at https://doi.org/10.1016/j.watres.2021.117676.

Ribeiro, J.P., Sarinho, L., Nunes, M.I., 2024. Application of life cycle assessment to Fenton processes in wastewater treatment – a review. Journal of Water Process Engineering 57, 104692. Available at https://doi.org/10.1016/j.jwpe.2023.104692.

Sanzana, S., Fenti, A., Iovino, P., Panico, A., 2025. A review of PFAS remediation: separation and degradation technologies for water and wastewater treatment. Journal of Water Process Engineering 74, 107793. Available at https://doi.org/10.1016/j.jwpe.2025.107793.

Seyyed Salehi, M.S., Boffetta, P., 2023. Per-and poly-fluoroalkyl substances (PFAS) exposure and risk of kidney, liver, and testicular cancers: a systematic review and meta-analysis. La Medicina del Lavoro 114 (5), e2023040. Available at https://doi.org/10.23749/mdl.v114i5.15065.

Sharma, S., Shetti, N.P., Basu, S., Nadagouda, M.N., Aminabhavi, T.M., 2022. Remediation of per- and polyfluoroalkyls (PFAS) via electrochemical methods. Chemical Engineering Journal 430, 132895. Available at https://doi.org/10.1016/j.cej.2021.132895.

Silva, J.R., Quinta-Ferreira, R.M., Castro, L.M., 2025. Biological treatments for VOC-contaminated off-gas: advances, challenges, and energetic valorisation opportunities. sustainability 17, 4802. Available at https://doi.org/10.3390/su17114802.

Steenland, K., Winquist, A., 2022. PFAS and cancer, a scoping review of the epidemiologic evidence. Environmental Research 194, 110690. Available at https://doi.org/10.1016/j.envres.2020.110690.

Su, Y., Rao, U., Khor, C.M., Jensen, M.G., Teesch, L.M., Wong, B.M., et al., 2019. Potential-driven electron transfer lowers the dissociation energy of the C–F bond and facilitates reductive defluorination of perfluoro octane sulfonate (PFOS). ACS Applied Materials & Interfaces 11 (37), 33913–33922. Available at https://doi.org/10.1021/acsami.9b10449.

Thomas, R., Jenkins, K., Landale, B., Trigger, G., Holsen, T.M., Dore, S., et al., 2020. Evaluation of PFAS treatment technology: alkaline ozonation. Remediation Journal 30, 27–37. Available at https://doi.org/10.1002/rem.21654.

Tshangana, C.S., Nhlengethwa, S.T., Glass, S., Denison, S., Kuvarega, A.T., Nkambule, T.T.I., et al., 2025. Technology status to treat PFAS contaminated water and limiting factors for their effective full-scale application. npj Clean Water 8 (41), 4573. Available at https://doi.org/10.1038/s41545-025-00457-3.

Wang, J., Lin, Z., He, X., Song, M., Westerhoff, P., Doudrick, K., et al., 2022. Critical review of thermal decomposition of per- and polyfluoroalkyl substances: mechanisms and implications for thermal treatment processes. Environmental Science & Technology 56 (9), 5355–5370. Available at https://doi.org/10.1021/acs.est.2c02251.

Wang, N., Lv, H., Zhou, Y., Zhu, L., Hu, Y., Majima, T., et al., 2019. Complete defluorination and mineralization of perfluorooctanoic acid by a mechanochemical method using alumina and persulfate. Environmental Science & Technology 53 (14), 8302–8313. Available at https://doi.org/10.1021/acs.est.9b00486.

Watanabe, N., Takemine, S., Yamamoto, K., Haga, Y., Takata, M., 2016. Residual organic fluorinated compounds from thermal treatment of PFOA, PFHxA and PFOS adsorbed onto granular activated carbon (GAC). Journal of Material Cycles and Waste Management 18, 625–630. Available at https://doi.org/10.1007/s10163-016-0532-x.

Xiang, I., Harindintwali, J.D., Wang, F., Redmile-Gordon, M., Chang, S.X., Fu, Y., et al., 2022. Integrating biochar, bacteria, and plants for sustainable remediation of soils contaminated with organic pollutants. Environmental Science & Technology 56, 16546–16566. Available at https://doi.org/10.1021/acs.est.2c02976.

Xiao, F., Sasi, P.C., Yao, B., Kubátová, A., Golovko, S.A., Golovko, M.Y., et al., 2020. Thermal stability and decomposition of perfluoroalkyl substances on spent granular activated carbon. Environ. Science & Technology Letters 7 (5), 343–350. Available at https://doi.org/10.1021/acs.estlett.0c00114.

Xu, B., Ahmed, M.B., Zhou, J.L., Altaee, A., n.d. Visible and UV photocatalysis of aqueous perfluorooctanoic acid by $TiO_2$ and peroxy mono-sulphate: process kinetics and mechanistic insights. Chemosphere, 243, 125366. https://doi.org/10.1016/j.chemosphere.2019.125366.

Xu, B., Zhou, J.L., Altaee, A., Ahmed, M.B., Johir, M.A.H., Ren, J., et al., 2020. Improved photocatalysis of perfluorooctanoic acid in water and wastewater by $Ga_2O_3$/UV system assisted by peroxymonosulfate. Chemosphere 239, 124722. Available at https://doi.org/10.1016/j.chemosphere.2019.124722.

Yin, S., Calvillo Solís, J.J., Sandoval-Pauker, C., Puerto-Diaz, D., Villagran, D., 2025. Advances in PFAS electrochemical reduction: mechanisms, materials, and future perspectives. Journal of Hazardous Materials 491, 137943. Available at https://doi.org/10.1016/j.jhazmat.2025.137943.

Yousefi, A., Omi, F.R., Yang, L., Ganiyu, S.O., Ullah, A., El-Din, M.G., et al., n.d. Innovative hybrid approach for enhanced PFAS degradation and removal: integrating membrane distillation, cathodic electro-Fenton, and anodic oxidation. Journal of Environmental Management, 379, 124818. https://doi.org/10.1016/j.jenvman.2025.124818.

Zelekew, O.A., Wang, Z., Gu, Y., Wu, Y., 2024. Recent progress on photoactive heterogeneous photocatalysts for the degradation of per-and polyfluoroalkyl substances (PFAS): mechanisms, DFT calculations, limitations, and prospects. Journal of Environmental Chemical Engineering 12 (5), 113323. Available at https://doi.org/10.1016/j.jece.2024.113323.

Zhu, W., Liang, Q., Gu, W., Wang, H., Xu, Y., Li, Y., n.d. Hydra's scenes of per- and polyfluoroalkyl substances (PFAS): occurrence and removal across environments. Advanced Industrial and Engineering Polymer Research, 12, 000752. https://doi.org/10.1016/j.aiepr.2025.12.007.

CHAPTER

# 11

# Phytoremediation of PFAS contaminated soils

*Harshavardhan M., Nandini I.V., and Vaishakh Nair*

**Department of Chemical Engineering, National Institute of Technology Karnataka (NITK), Mangalore, Karnataka, India**

## 11.1 Introduction

Per- and polyfluoroalkyl substances (PFAS) constitute a broad class of synthetic organic chemicals applied in numerous industrial processes and consumer products. Their widespread use is attributed to high chemical stability, strong oil- and water-repellent characteristics, and resistance to heat and environmental degradation. Commonly found in firefighting foams, nonstick cookware, water-resistant fabrics, and food packaging, PFAS have become ubiquitous in the environment (Zahra et al., 2025). They have been detected in water sources, soils, and even in human and animal bloodstreams, raising significant health concerns (Yogarathinam et al., 2025). Due to their extremely strong carbon–fluorine (C–F) bonds, PFAS have remarkable resistance to thermal, chemical, and biological degradation. A major reason for concern is that these compounds have been detected worldwide, primarily due to their continued use in various industrial and consumer products (Hussain et al., 2025).

PFAS molecules mostly consist of a hydrophilic "head" and a hydrophobic fluorinated tail. The hydrophilic "head" often is a carboxylate or sulfonate group. This molecule structure closely mimics the natural fatty acids, which are essential for plant cell membranes and energy storage (Fawad et al., 2026). Moreover, it is observed that the PFAS, especially long-chain molecules, can interact with plant proteins and lipids in the roots, thereby mimicking the way natural fatty acids interact with these plant components. Similarly, short-chain PFAS are highly water-soluble and can be transported along with water through aquaporins during the transpiration stream. These compounds can enter the food chain when plants absorb them from contaminated irrigation water, leading to their presence in vegetables, fruits, cereals, milk, and meat. In aquatic environments, PFAS tend to accumulate in the benthic zone, posing serious risks to bottom-dwelling organisms. Notably, elevated levels of perfluorooctanesulfonic acid (PFOS), a long-chain type of PFAS, have been

*Per- and Polyfluoroalkyl Substances (PFAS) as Environmental Contaminants*
**DOI: https://doi.org/10.1016/B978-0-443-43856-1.00023-X**

identified in marine species, including polar bears, dolphins, seals, whales, and porpoises, with concentrations reaching up to 1900 ng $g^{-1}$. These findings emphasize the persistence of PFAS and their widespread toxic effects across the food web, causing bioaccumulation, thereby making them a major pollutant of global concern (Bhattacharya et al., 2025).

### 11.1.1 Chemical structure and classification

The chemical structure of PFAS is stated to be an alkyl chain of carbon atoms where all of the hydrogen atoms are replaced by fluorine atoms. PFAS feature at least one perfluoroalkyl moiety, $C_nF_{2n+1}$–, which is made up of one or more carbon atoms on which all hydrogen substituents have been replaced by fluorine atoms (Buck et al, 2011). For more than 50 years, around 4730 compounds belonging to the PFAS family have been widely manufactured for a range of home and industrial uses (Arslan and Gamal El-Din, 2021). Fig. 11.1 shows PFAS compounds were classified into three primary groups: per- and polyfluoroalkyl ether acids (PFEA), Per- and Polyfluoroalkyl acids (PFAA) precursors, and other types of PFAS (Yogarathinam et al, 2025).

### 11.1.2 Toxicity studies

The chemical inertness of PFAS, arising from the high bond energy of the C–F bond, under engineered conditions, partial degradation and defluorination have been reported,

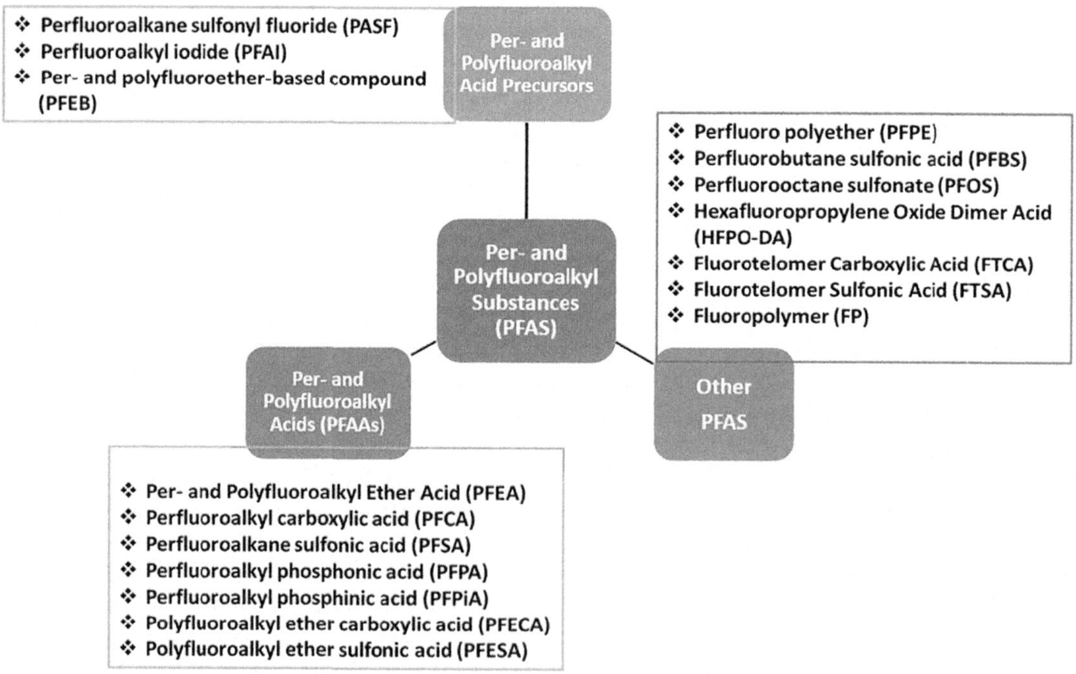


FIGURE 11.1 Classification of per- and polyfluoroalkyl substances (PFAS).

but the resulting products are often PFAAs. PFAAs display strong environmental mobility and are known to bioaccumulate, with evidence indicating their transfer and concentration within food chains (Fair et al., 2019).

PFAS contamination occurs across air, water, and soil, and exposure through these media has been associated with adverse health effects in both humans and wildlife. Documented toxicological impacts include reproductive toxicity, endocrine disruption, neurotoxicity, immunotoxicity, hepatotoxicity, and pulmonary toxicity (Bhattacharya et al., 2025). In humans, exposure has been linked to developmental impairments, liver injury, immune system dysfunction, and increased risk of carcinogenesis (Yogarathinam et al., 2025).

The International Agency for Research on Cancer has classified perfluorooctanoic acid (PFOA) as carcinogenic to humans and PFOS as possibly carcinogenic to humans, based on evaluations of epidemiological, animal, and mechanistic evidence (Drury et al., 2024). PFAS exposure in aquatic organisms has been shown to cause reproductive toxicity, behavioral alterations, and bioaccumulation in higher trophic levels, with reports of accumulation in top predators such as dolphins and polar bears (Houde et al., 2011).

In soil, elevated PFAS concentrations are associated with microbial population decline, cellular damage, and reduced enzymatic activity (Cai et al., 2020). The findings indicate that PFAS contamination negatively affects both the abundance and diversity of soil microorganisms, with uncontaminated soil showing significantly greater bacterial sequence diversity than contaminated soil (Karimi Douna and Yousefi, 2023). Bioaccumulation studies indicate that PFAAs accumulate in soil invertebrates such as earthworms, with equilibrium concentrations reached within 28 days (Rich et al., 2015). Increasing soil concentrations of PFOS and PFOA have been linked to DNA damage, reduced survival, and, in some cases, mortality in earthworms (Zheng et al., 2016; Zareitalabad et al., 2013).

### 11.1.3 Phytoremediation

Phytoremediation is a type of bioremediation which specifically uses that down to focus on the capabilities of plants to reduce contaminant concentrations in the environment (Sharma, 2021). It includes a range of eco-friendly and cost-effective techniques that utilize plant-based processes to clean-up contaminated environments. Phytoremediation has been applied to the removal of a wide range of contaminants, including heavy metals, pesticides, and hydrocarbons. Five different species of *Brassica* (*Brassica juncea, Brassica campestris, Brassica carinata, Brassica napus, and Brassica nigra*) have been shown to accumulate zinc, copper, lead, and nickel from contaminated soils, demonstrating effectiveness in phytoextraction processes. The accumulation varies according to species type (Purakayastha et al., 2008). In aquatic systems, *Eichhornia crassipes* (water hyacinth) and *Lemna minor* (duckweed) have been employed to reduce concentrations of nutrients, dyes, and organic pollutants in wastewater, with significant decreases in contaminant levels reported (Srivastava and Gupta, 2021). In terrestrial environments, *Vetiveria zizanioides* (vetiver grass) has been used to stabilize soils and enhance the degradation of petroleum hydrocarbons, contributing to the recovery of contaminated sites (Dudai et al., 2018).

### 11.1.4 Economic and practical considerations for per- and polyfluoroalkyl substances removal

Conventional remediation technologies, such as ion exchange resins and granular activated carbon, offer high removal efficiencies but are hindered by high operation and maintenance costs and the requirement for frequent regeneration. Similarly, advanced filtration methods, such as nanofiltration and reverse osmosis, provide consistent results; they face significant practical challenges regarding the disposal of highly concentrated retentate and spent membranes. Phytoremediation offers an environmentally benign, low-maintenance alternative, yet it is often impractical for urgent needs due to remediation timeframes that can span decades or centuries (Kafle et al., 2022).

## 11.2 Phytoremediation of per- and polyfluoroalkyl substances

For persistent organic pollutants such as PFAS, plants achieve remediation through several mechanisms in both soil and water medium as shown in the Figs. 11.2 and 11.3. It is observed that the pollutants are absorbed through the root system and either translocated to above-ground tissues or retained within root structures. Some contaminants are transformed into less toxic forms through enzymatic or metabolic processes within the plant

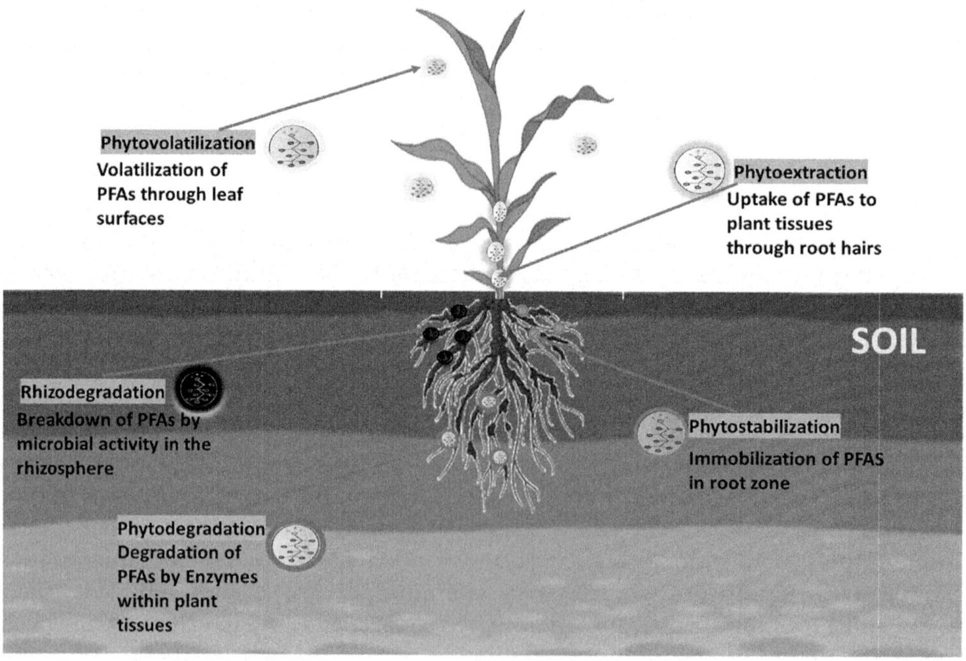


FIGURE 11.2 Mechanisms of per- and polyfluoroalkyl substances (PFAS) phytoremediation in contaminated soils.

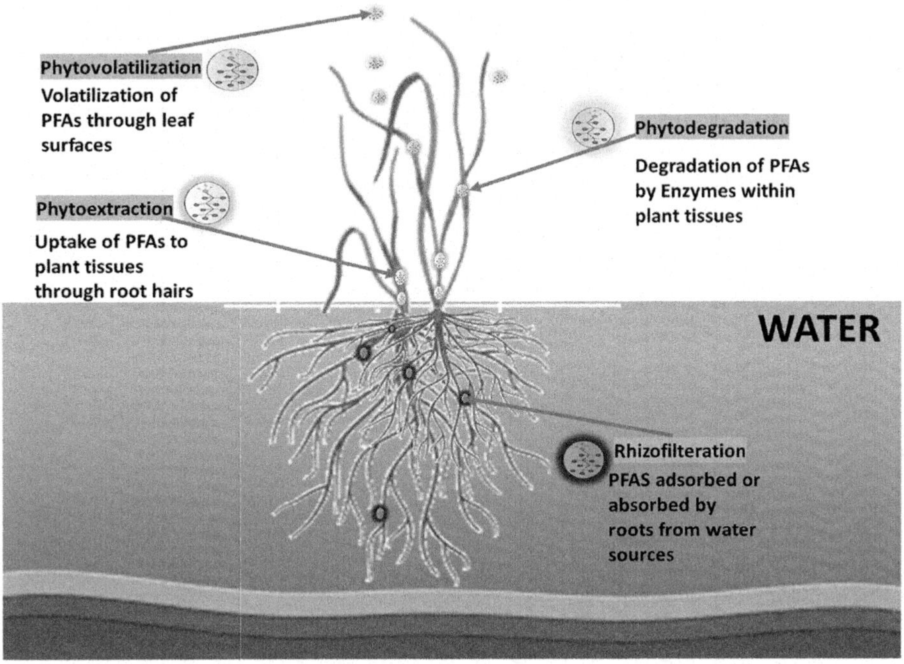


**FIGURE 11.3** Mechanisms of per- and polyfluoroalkyl substances (PFAS) removal from aquatic systems via phytoremediation.

tissues. In certain cases, plants stabilize pollutants in the rhizosphere by reducing their mobility and bioavailability, thereby limiting environmental dispersion. These mechanisms function simultaneously or independently, depending on plant species, pollutant characteristics, and environmental conditions. Six key phytoremediation strategies have been employed for addressing PFAS-contaminated soils and waterbodies: phytodegradation, phytostabilization, phytoextraction, phytovolatilization, phytofiltration (rhizofiltration), and rhizodegradation (Mayakaduwage et al., 2022).

## 11.2.1 Phytodegradation

Phytodegradation is the process by which plants take up organic pollutants and enzymatically break them down into less toxic, low-molecular-weight compounds through internal metabolic activity or via enzymes secreted by their roots (Kavusi et al., 2023; Susarla et al., 2002). Enzymes such as dehalogenases, oxygenases, and reductases produced by plant tissues are responsible for catalyzing these degradation pathways (Grzegórska et al., 2020). This mechanism involves the absorption of hydrophobic organic contaminants by roots, followed by their transformation into less harmful substances within plant tissues (Bhattacharya et al., 2025). The phytoremediation technique is useful for the remediation of different organic contaminants, including chlorinated solvents, pesticides, and insecticides across contaminated environments such as soil, sediments, sludge, surface water, and

**TABLE 11.1** Various types of phytoremediation studies for the treatment of per- and polyfluoroalkyl substances (PFAS).

| S. no. | Type | Plants used | PFAs types | Removal efficiency (%) | References |
|---|---|---|---|---|---|
| 1. | Phytoextraction | Weeds (*Phyllanthus urinaria, Aster indicus*) | All PFAS | 41.40 | He et al. (2023) |
| 2. | Rhizofiltration | Wetland plants (*Eriophorum angustifolium, Carex rostrata,* etc.) | PFOS, PFOA | 90 | Greger and Landberg (2024) |
| 3. | Rhizofiltration | *Phragmites australis* | PFAA | 30–50 | Ferrario et al. (2022) |
| 4. | Phytoextraction | *Schedonorus arundinaceus* | PFAS | 42 | Scearce et al. (2025) |
| 5. | Phytostabilization | *Lemna minor* (duckweed) | PFHpA, PFHxS, PFOA, PFOS | >95 | Zhang and Liang (2020) |
| 6. | Phytoextraction | Woody plant species | PFPeA, PFHxA, PFOA, PFBS, PFHxS, PFOS | 42 | Huff et al. (2020) |
| 7. | Phytoextraction | Hemp (*Cannabis sativa*) | PFAS | 2 | Niarchos et al. (2023) |
| 8. | Phytoextraction | Hemp (*Cannabis sativa*) | PFAS | 45 | Wright et al. (2025) |
| 9. | Phytostabilization | *Juncus effusus* | PFAA | 11.40 | Zhang et al. (2019) |
| 10. | Phytoextraction | Hemp (*Cannabis sativa*) | PFOA, PFOS | ——— | Nason et al. (2024) |

groundwater (Susarla et al., 2002). Studies have reported PFAS uptake and partial transformation within plant systems; however, these effects are generally linked to microbial activity in the rhizosphere or plant-assisted transport mechanisms rather than direct enzymatic degradation by the plants themselves (Mayakaduwage et al., 2022). Recent studies, as shown in Table 11.1 have demonstrated that plant species such as *Phragmites australis, Helianthus annuus,* and *Brassica juncea* can effectively remove PFAAs from contaminated soil and water, achieving removal efficiencies ranging from 30% to 60% (Ferrario et al., 2022; Nassazzi et al., 2023). Moreover, plants can absorb and store PFAS but do not possess the enzymatic pathways required to dismantle these bonds (Huang et al., 2021). Naturally autotrophic plants lack the specialized catabolic enzymes required to cleave such resilient bonds and mineralize these chemicals into carbon dioxide and water. While plants can successfully degrade other organics, any "degradation" observed with PFAS is typically a biotransformation of precursor compounds rather than true mineralization. For example, plants may transform precursors such as perfluorooctanesulfonamide into PFOS, a terminal product that is even more resistant to further biological breakdown (Zareitalabad et al., 2013).

### 11.2.2 Phytostabilization

Phytostabilization is a key mechanism within phytoremediation that uses plant-root interactions to immobilize contaminants in soil and groundwater. Phytostabilization involves two main processes: stabilization and immobilization. Stabilization occurs when plant roots form a physical barrier that reduces erosion and leaching. Root systems bind soil particles, restrict contaminant movement by wind or water, and limit percolation into groundwater, thereby confining pollutants within the root zone. Immobilization takes place through chemical and biological interactions in the rhizosphere. In the rhizosphere, root exudates, which include organic acids, amino acids, sugars, and phenolic compounds, can modify soil pH, redox potential, and microbial activity. These changes affect contaminant speciation, leading to precipitation as insoluble salts, adsorption onto root surfaces, or binding to soil organic matter. As a result, contaminants become less mobile and less available for uptake (Kängsepp et al., 2025). This approach is especially beneficial for soils contaminated with PFAS located near waterbodies, as it anchors them within the rhizosphere and reduces the risk of leaching into groundwater (Bolan et al., 2021). In most cases, long-chain PFAS (e.g., PFOS, PFDA) are sequestered within plant-root tissues. The presence of fibrous roots also improves soil stability and reduces erosion, which further limits the movement of contaminants. Grasses such as *Vetiveria zizanioides* (vetiver), *Lolium perenne* (ryegrass), and *Panicum virgatum* (switchgrass) are commonly used for phytostabilization. These species develop dense root systems that maintain soil structure and moisture, creating conditions that are favorable for retaining PFAS. Their adaptability to various soil types makes them suitable for field applications (Bolan et al., 2021).

### 11.2.3 Phytoextraction

Phytoextraction is a phytoremediation technique where plants absorb pollutants from the soil or water through their roots and move them to their above-ground biomass.

This technique has proven effective in remediating heavy metal pollution and may also aid in the uptake of organic pollutants such as PFAS from soil into plant tissues (Bauddh et al., 2017). It is particularly relevant for short-chain PFAS such as perfluorobutanoic acid and perfluorohexanoic acid, which are more mobile and water-soluble. Because these compounds can be more easily transported from the roots to the shoots of plants, they can be effectively removed by harvesting the plant biomass (Nason et al., 2024). The phytoextraction process starts with the passive or facilitated uptake of PFAS from soil pore water into the roots, driven by transpiration. PFAS molecules are transported through either apoplastic (cell wall) or symplastic (cytoplasmic) pathways, depending on their chain length and functional group. Once inside the root cells, short-chain PFAS can enter the xylem and be transported to above-ground tissues through mass flow (Huff et al., 2020). The efficiency of this translocation is much higher for short-chain compounds compared to long-chain compounds. Factors such as root exudates, membrane transporters, and rhizosphere chemistry all play a role in influencing the uptake and mobility of PFAS (Gobelius et al., 2017). Several plant species have been recognized as effective candidates for the phytoextraction of PFAS due to their rapid growth, high-biomass production, and deep root systems. For instance, hemp (*Cannabis sativa*) has been effectively used to absorb PFAS from soils amended with biosolids. Similarly, wetland species such as *Phragmites australis* and *Alisma plantago-aquatica* have demonstrated significant PFAS uptake capabilities. Supporting these findings, Brusseau et al. (2023) reported that high-biomass crops, including sunflower (*Helianthus annuus*), Indian mustard (*Brassica juncea*), and hemp (*Cannabis sativa*), could remove between 0.2%–33% of PFAS from soil within a single cropping cycle (Nason et al., 2024; Brusseau et al., 2023). A key drawback of phytoextraction is the potential accumulation of PFAS in edible plant parts, which may pose health risks upon consumption. However, there are evidences that plant-associated microbes, including endophytes and rhizospheric bacteria, can initiate the biotransformation of certain PFAS, particularly shorter-chain precursors (Fawad et al., 2026).

### 11.2.4 Phytovolatilization

Phytovolatilization is the process by which plants absorb contaminants, followed by their transformation and release into the atmosphere as gases through transpiration or metabolic conversion (Bhat et al., 2022; Ali et al., 2013). This technique is commonly used for elements such as mercury and selenium, as well as for volatile organic compounds such as chlorinated solvents. However, its effectiveness for PFAS is highly limited due to the unique chemical stability and low volatility of most PFAS compounds (Sharma et al., 2024). Plant species such as *Populus deltoides* (Eastern cottonwood) and *Salix alba* (white willow) have been extensively studied for their ability to phytovolatilize organic pollutants due to their high transpiration rates and rapid growth (Burken and Schnoor, 1998). However, the effectiveness of this technique for PFAS remains limited. Some studies suggest that short-chain PFAS or related precursors could potentially undergo volatilization in specific environmental conditions, particularly in saturated soils or wetland systems. Although PFAS are generally considered nonvolatile due to their high molecular weight and the strength of their C–F bonds, the volatility varies among different PFAS compounds. Long-chain PFAS exhibit extremely low volatility; however, certain short-chain PFAS and volatile precursors may undergo physicochemical transformations that facilitate their movement through plant tissues and potential release into the atmosphere via transpiration. This theoretical possibility highlights the need for further investigation into phytovolatilization as a potential pathway for PFAS dissipation in plant-based remediation systems. Kavusi et al. (2023) emphasized that current research is insufficient to confirm the efficiency or safety of PFAS phytovolatilization.

### 11.2.5 Rhizofiltration

In the rhizofiltration technique, the roots of aquatic or semiaquatic plants are utilized to remove contaminants from water (Pratas et al., 2014). As shown in Fig. 11.3, the plant roots absorb, adsorb, or concentrate pollutants from aqueous environments such as contaminated groundwater, surface water, or wastewater. In this process, the plant roots act as natural biofilters, taking up dissolved contaminants through root uptake (absorption) or binding them to root surfaces (adsorption). For removal of PFAS through various physical and biological processes, systems such as wetlands or hydroponics are particularly effective (Sharma et al., 2024). PFAS in water are mostly found as anionic species (e.g., $PFOS^-$, $PFOA^-$) with high solubility and stability. In hydroponic systems, plants grown in water-based environments filter PFAS through root interactions. Sörengård et al. (2023) reported that *Carex rostrata* (bottle sedge) and *Eriophorum angustifolium* (common cotton grass) were among the most effective species for the removal of PFAS. In the study, *Carex rostrata* removed 42% of PFOA and 64% of PFOS within 12 days. In the same study, *Elodea canadensis* (Canadian pondweed) achieved removal rates of 14% for PFOA and 19% for PFOS in just 3 days. Similarly, floating wetland systems utilizing Australian native species such as *Phragmites australis, Baumea articulata,* and *Juncus kraussii* achieved 42%–53% removal of legacy PFAS at 10 $\mu g\ L^{-1}$ concentrations, as demonstrated by researchers at the University of South Australia (Awad et al., 2022). Felizeter et al. (2012) observed that in hydroponic conditions, crops such as tomato, cabbage, and zucchini primarily accumulated long-chain PFAAs in the roots, whereas shorter-chain PFAAs were translocated to the aerial parts of the plants. In another study, Zhang et al. (2023) found that Chinese cabbage accumulated the highest concentrations of PFAS (up to 13,456 $ng\ g^{-1}$), followed by lettuce, cucumber, and chrysanthemum. These findings collectively suggest that a variety of aquatic plants hold promise for PFAS remediation through rhizofiltration. According to Idris and Erasmus (2024), the hydrophobicity and lipophobicity of PFAS vary with chain length. Short-chain PFASs are found in all above-ground tissues, but long-chain PFASs are concentrated in the root. The hydrophobicity of long-chain PFAAs allows them to bioaccumulate largely in the root due to the electron-drawing effect of fluorine atoms and functional groups.

### 11.2.6 Rhizodegradation

Rhizodegradation is a promising, low-cost plant-assisted bioremediation approach where microorganisms in the plant-root zone (rhizosphere) help to alter or degrade PFAS (Kuiper et al., 2004; Pathak et al., 2024). This process involves the plant roots, which release various organic compounds, such as enzymes, exudates, and sloughed-off cells, which act as a carbon source that stimulates bacteria and fungi populations in the rhizosphere, leading to the breakdown of contaminants (Evich et al., 2022; Grzegórska et al., 2020). The rhizodegradation of PFAS primarily involves three different mechanisms, which can be aerobic, enzyme-catalyzed oxidative degradation, or anaerobic reductive defluorination (Niu et al., 2024). *Acidimicrobium sp. A6* bacterial strain found in the plant-root zone has demonstrated the ability to defluorinate PFOA and PFOS under anaerobic conditions by reducing iron (Huang et al., 2021). Moreover, plant species such as *Betula pendula* (silver birch) and *Juncus effusus* (soft rush) support active microbial communities in the root zone that can accumulate or partially degrade various PFAS compounds (Zhang et al., 2019).

## 11.3 Strategies for enhancement of phytoremediation

The major drawback of phytoremediation is that it is a biological process relying on the natural growth rate of plants, which results in longer clean-up time to achieve adequate decontamination levels compared to other methods (Kafle et al., 2022). In addition, the mismanagement of phytoremediation techniques increases the significant risk of transferring these toxic, bioaccumulative compounds to animals and humans through the food chain (Wani et al., 2023). Moreover, the choice and the build-up of the phytoremediation system for PFAS removal needs to be tailored primarily based on the type and chain size of the PFAS. Different strategies (see Fig. 11.4) are used to enhance the phytoremediation of PFAS, which involves the use of biosystems involving external microorganisms, biomaterials, and bio-electrochemical systems (BES). The combination of phytoremediation and BES is an emerging, sustainable strategy for removing PFAS, which addresses the high energy demands of purely electrochemical methods and the slow kinetics of traditional bioremediation (Yang and Liu, 2025). Constructed wetland-microbial fuel cells are an integrated approach that synergizes the biological capabilities of plants with the electrochemical potential of electroactive microbes to

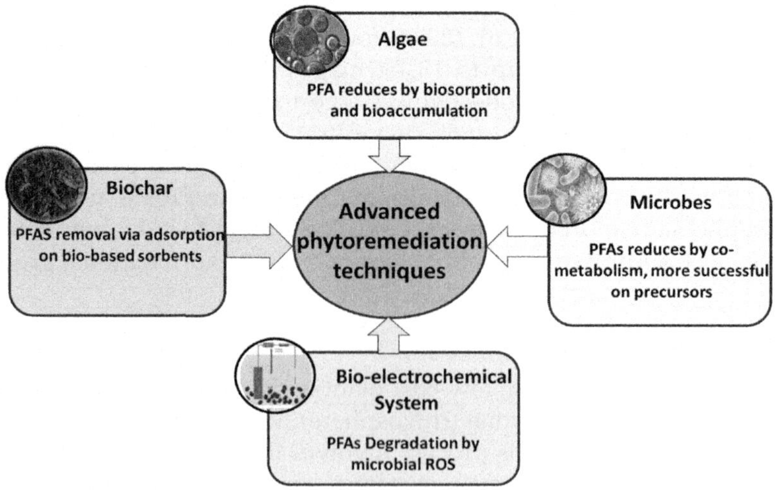


**FIGURE 11.4** Strategies for enhancement of phytoremediation.

overcome the extreme stability of the C–F bond (Kang et al., 2025). Moreover, such a system, along with remediation of wastewater and soil, can also generate small amounts of bioelectricity, thereby providing a "green" renewable energy byproduct.

The combination of phytoremediation and biochar is another sustainable method to effectively manage the PFAS removal. Plants such as *Miscanthus* and *Salix* efficiently take up (ultra) short-chain PFAS, which are otherwise poorly adsorbed by biochar. Biochar simultaneously immobilizes long-chain PFAS (such as PFOS and PFOA) through hydrophobic interactions and pore filling (Sørmo et al., 2021). Moreover, biochar improves soil quality by increasing water retention, regulating pH, and supplying nutrients for rhizosphere microorganisms (Mayakaduwage et al., 2022). Similarly, instead of biochar, algae can be used for the removal of PFAS by biosorption and bioaccumulation. Moreover, algae can be effectively used for the removal of both short- and long-chain PFAS, with removal rates up to 88% observed for PFOS in lab studies (Marchetto et al., 2021). Combining them allows for a broader range of PFAS chain lengths to be targeted simultaneously. Algae, through photosynthesis, produce oxygen, which can enhance the aerobic bacterial degradation potential of PFAS precursors. Plants release root exudates that foster a rich microbial community in the rhizosphere. These associated microorganisms, along with those in the algal-bacteria symbiotic systems, can actively participate in the biotransformation or even defluorination (breaking the strong C–F bond) of some PFAS, a process that is otherwise very difficult (Bhattacharya et al., 2025).

Microbial-assisted phytoremediation is an emerging, sustainable strategy for removing PFAS by utilizing the synergistic relationship between plants and specialized microbes. Recent research highlights this "plant-microbe mechanism" as a promising path toward partial mineralization and effective immobilization of these persistent contaminants. Pseudomonas species are also widely used for their aerobic degradation capabilities. White-rot fungi, such as *Phanerochaete chrysosporium,* use ligninolytic enzymes to transform PFAS precursors into shorter-chain, more manageable metabolites (Sanzana et al., 2025). Additional care should be taken to enrich the bacterial population by maintaining the soil conditions, such as pH and nutrient concentration.

Genetic engineering can be used for developing transgenic species to overcome natural biological limitations, such as slow growth rates, low biomass production, and limited toxic tolerance in highly contaminated environments for PFAS removal. Overexpression of genes in transgenic plants regulates the uptake and translocation of contaminants from roots to harvestable tissues. Furthermore, genetic modification offers a path toward internal mineralization by incorporating microbial genes from efficient degraders, such as *Pseudomonas* or *Acidimicrobium*, which produce enzymes such as dehalogenases (FAcD) and monooxygenases capable of defluorination (Huang and Jaffe, 2019). Transgenic approaches can enhance environmental adaptability, addressing challenges related to site-specificity and vulnerability to pests or high phytotoxicity.

## 11.4 Impact on multimetal pollution on phytoremediation of per- and polyfluoroalkyl substances

Multimetal pollution, specifically the cooccurrence of heavy metals such as lead, copper, and zinc with PFAS, significantly complicates phytoremediation efforts, generally acting as an opposing factor that reduces the efficiency of plant uptake. While heavy metals can enhance the sorption of long-chain PFAS to soil components via cation bridging and by neutralizing the negative charge on soil surfaces, this increased retention often binds the PFAS too tightly, limiting its bioavailability to plant roots. Furthermore, the combined toxicity of heavy metals and PFAS often exceeds the tolerance of many plant species, which leads to reduced biomass production, stunted root growth, and severe chlorosis, which in turn lowers the total transpiration-driven accumulation of contaminants (Mayakaduwage et al., 2022). In some cases, hyperaccumulators are forced into a trade-off where heavy metals will be efficiently removed at the expense of lower PFAS remediation efficiency, necessitating specialized amendments to enhance phytoavailability in complex industrial sites.

## 11.5 Conclusion

Phytoremediation offers a sustainable approach to reducing the environmental load of PFAS, a group of contaminants characterized by high persistence, mobility, and bioaccumulative behavior. Six phytoremediation strategies have been reported for the removal of various pollutants, however extensive studies on PFAS removal using these strategies are required. Techniques such as phytoextraction and phytostabilization have demonstrated effectiveness for certain PFAS compounds. In contrast, approaches such as phytovolatilization are constrained by the inherently low volatility of these substances, limiting their broader applicability. Recent advancements include microbial-assisted phytoremediation, bio-electrochemical treatment, and the incorporation of biochar and algae, along with the phytoremediation techniques, can improve PFAS retention and transformation prove sufficiently than the conventional phytoremediation approach. These enhancements rely on interactions among microbial communities, plant growth-promoting rhizobacteria, and soil amendments to increase contaminant bioavailability, promote enzymatic activity, and support plant resilience under chemical stress.

## References

Ali, H., Khan, E., Sajad, M.A., 2013. Phytoremediation of heavy metals—concepts and applications. Chemosphere 91, 869–881.

Arslan, M., Gamal El-Din, M., 2021. Removal of per- and poly-fluoroalkyl substances (PFASs) by wetlands: prospects on plants, microbes and the interplay. The Science of the Total Environment 800, 149570.

Awad, J., Brunetti, G., Juhasz, A., Williams, M., Navarro, D., Drigo, B., et al., 2022. Application of native plants in constructed floating wetlands as a passive remediation approach for PFAS-impacted surface water. Journal of Hazardous Materials 429, 128326.

Bauddh, K., Singh, B., Korstad, J. (Eds.), 2017. Phytoremediation Potential of Bioenergy Plants. Springer Singapore, Singapore.

Bhat, S.A., Bashir, O., Haq, S.A.U., Amin, T., Rafiq, A., Ali, M., et al., 2022. Phytoremediation of heavy metals in soil and water: an eco-friendly, sustainable and multidisciplinary approach. Chemosphere 303, 134788.

Bhattacharya, A., Fathima, J., Varghese, S., Chatterjee, P., Gadhamshetty, V., 2025. Advances in bioremediation strategies for PFAS-contaminated water and soil. Soil Science and Environmental. 3, 100126.

Bolan, N., Sarkar, B., Yan, Y., Li, Q., Wijesekara, H., Kannan, K., et al., 2021. Remediation of poly- and perfluoroalkyl substances (PFAS) contaminated soils – To mobilize or to immobilize or to degrade? Journal of Hazardous Materials 401, 123892.

Brusseau, M.L., Anderson, R.H., Guo, B., Yan, N.R., Ma, R., 2023. Greenhouse experiments on phytoextraction of PFAS from soil using high biomass crop plants. The Science of the Total Environment 873, 162232.

Buck, R.C., Franklin, J., Berger, U., Conder, J.M., Cousins, I.T., de Voogt, P., et al., 2011. Perfluoroalkyl and polyfluoroalkyl substances in the environment: terminology, classification, and origins. Integrated Environmental Assessment and Management 7, 513–541.

Burken, J.G., Schnoor, J.L., 1998. Predictive relationships for uptake of organic contaminants by hybrid poplar trees. Environmental Science & Technology 32 (21), 3379–3385.

Cai, Y., Chen, H., Yuan, R., Wang, F., Chen, Z., Zhou, B., 2020. Metagenomic analysis of soil microbial community under PFOA and PFOS stress. Environmental Research 188, 109838.

Drury, N.L., Prueitt, R.L., Beck, B.D., 2024. Commentary: understanding IARC's PFOA and PFOS carcinogenicity assessments. Regulatory Toxicology and Pharmacology 154, 105726.

Dudai, N., Tsion, I., Shamir, S.Z., Nitzan, N., Chaimovitsh, D., Shachter, A., et al., 2018. Agronomic and economic evaluation of Vetiver grass (*Vetiveria zizanioides* L.) as means for phytoremediation of diesel polluted soils in Israel. Journal of Environmental Management 211, 247–255.

Evich, M.G., Davis, M., Weber, E.J., Tebes-Stevens, C., Acrey, B., Henderson, W.M., et al., 2022. Environmental fate of Cl-PFPECAs: predicting the formation of PFAS transformation products in New Jersey soils. Environmental Science & Technology 56, 7779–7788.

Fair, P.A., Wolf, B., White, N.D., Arnott, S.A., Kannan, K., Karthikraj, R., et al., 2019. Perfluoroalkyl substances (PFASs) in edible fish species from Charleston Harbor and tributaries, South Carolina, United States: exposure and risk assessment. Environmental Research 171, 266–277.

Fawad, M., Wu, T., Xu, X., Wang, J., Zhou, X., Rajput, V.D., et al., 2026. Mechanisms of PFAS uptake and bioaccumulation in plants. Ecotoxicology and Environmental Safety 311, 119888.

Felizeter, S., McLachlan, M.S., de Voogt, P., 2012. Uptake of perfluorinated alkyl acids by hydroponically grown lettuce (*Lactuca sativa*). Environmental Science & Technology 46, 4487–4493.

Ferrario, C., Peruzzi, C., Cislaghi, A., Polesello, S., Valsecchi, S., Lava, R., et al., 2022. Assessment of reed grasses (*Phragmites australis*) performance in PFAS removal from water: a phytoremediation pilot plant study. Water 14, 946.

Gobelius, L., Lewis, J., Ahrens, L., 2017. Plant uptake of per- and polyfluoroalkyl substances at a contaminated fire training facility to evaluate the phytoremediation potential of various plant species. Environmental Science & Technology 51, 12602–12610.

Greger, M., Landberg, T., 2024. Removal of PFAS from water by aquatic plants. Journal of Environmental Management 351, 119895.

Grzegórska, A., Rybarczyk, P., Rogala, A., Zabrocki, D., 2020. Phytoremediation—from environment cleaning to energy generation—current status and future perspectives. Energies 13, 2905.

He, Q., Yan, Z., Qian, S., Xiong, T., Grieger, K.D., Wang, X., et al., 2023. Phytoextraction of per- and polyfluoroalkyl substances (PFAS) by weeds: effect of PFAS physicochemical properties and plant physiological traits. Journal of Hazardous Materials 454, 131492.

Houde, M., Martin, J.W., Letcher, R.J., Solomon, K.R., Muir, D.C.G., 2011. Biological monitoring of polyfluoroalkyl substances: a review. Environmental Science & Technology 45, 7962–7973.

Huang, D., Xiao, R., Du, L., Zhang, G., Yin, L.S., Deng, R., et al., 2021. Phytoremediation of poly- and perfluoroalkyl substances: a review on aquatic plants, influencing factors, and phytotoxicity. Journal of Hazardous Materials 418, 126314.

Huang, S., Jaffe, P.R., 2019. Defluorination of perfluorooctanoic acid (PFOA) and perfluorooctane sulfonate (PFOS) by *Acidimicrobium* sp. Strain A6. Environmental Science & Technology 53, 11410–11419.

Huff, D.K., Morris, L.A., Sutter, L., Costanza, J., Pennell, K.D., 2020. Accumulation of six PFAS compounds by woody and herbaceous plants: potential for phytoextraction. International Journal of Phytoremediation. 22, 1538–1550.

Hussain, H.N., Jilani, M.I., Imtiaz, F., Ahmed, T., Arshad, M.B., Mudassar, M., et al., 2025. Advances in the removal of Polyfluoroalkyl Substances (PFAS) from water using destructive and non-destructive methods. Green Analytical Chemistry 12, 100225.

Idris, O.A., Erasmus, M., 2024. Degradation pathways of perfluoroalkyl and polyfluoroalkyl compounds: Removal in water and soil using fungi and plant-based remediation. Environmental Advances 18, 100598.

Kafle, A., Timilsina, A., Gautam, A., Adhikari, K., Bhattarai, A., Aryal, N., 2022. Phytoremediation: mechanisms, plant selection and enhancement by natural and synthetic agents. Environmental Advances 8, 100203.

Kang, C., Zhao, Y., Tang, C., 2025. First study to explore the feasibility of simultaneously applying microbial fuel cell and microbial electrolysis cell in constructed wetland. Chemical Engineering Journal 514, 163430.

Kängsepp, P., Svensson, B.-M., Djerf, H., Roos, S., 2025. Phytoremediation with *Salix* spp.: a sustainable approach to PFAS remediation at firefighting training sites. IOP Conference Series: Earth and Environmental Science 1474, 12015.

Karimi Douna, B., Yousefi, H., 2023. Removal of PFAS by biological methods. Asian Pacific Journal of Environmental Cancer 6, 53–68.

Kavusi, E., Ansar, B.S.K., Ebrahimi, S., Sharma, R., Ghoreishi, S.S., Nobaharan, K., et al., 2023. Critical review on phytoremediation of polyfluoroalkyl substances from environmental matrices: need for global concern. Environmental Research 217, 114844.

Kuiper, I., Lagendijk, E.L., Bloemberg, G.V., Lugtenberg, B.J., 2004. Rhizoremediation: a beneficial plant-microbe interaction. Molecular Plant-Microbe Interactions MPMI 17, 6–15.

Marchetto, F., Roverso, M., Righetti, D., Bogialli, S., Filippini, F., Bergantino, E., et al., 2021. Bioremediation of per- and poly-fluoroalkyl substances (PFAS) by *Synechocystis* sp. PCC 6803: a chassis for a synthetic biology approach. Life (Chicago, Ill. 1978) 11 (12), 1300.

Mayakaduwage, S., Ekanayake, A., Kurwadkar, S., Rajapaksha, A.U., Vithanage, M., 2022. Phytoremediation prospects of per- and polyfluoroalkyl substances: a review. Environmental Research 212, 113311.

Nason, S.L., Thomas, S., Stanley, C., Silliboy, R., Blumenthal, M., Zhang, W., et al., 2024. A comprehensive trial on PFAS remediation: hemp phytoextraction and PFAS degradation in harvested plants. Environmental Science: Advances 3, 304–313.

Nassazzi, W., Wu, T.C., Jass, J., Lai, F.Y., Ahrens, L., 2023. Phytoextraction of per- and polyfluoroalkyl substances (PFAS) and the influence of supplements on the performance of short–rotation crops. Environmental Pollution (Barking, Essex 1987) 333, 122038.

Niarchos, G., Ahrens, L., Kleja, D.B., Leonard, G., Forde, J., Bergman, J., et al., 2023. In-situ application of colloidal activated carbon for PFAS-contaminated soil and groundwater: a Swedish case study. Remediation 33, 101–110.

Niu, Q., Lin, X., Zheng, X., Wu, Y., Long, M., Chen, Y., 2024. Aerobic or anaerobic? Microbial degradation of per- and polyfluoroalkyl substances: a review. Journal of Hazardous Materials 480, 136173.

Pathak, H.K., Chauhan, P.K., Shekhar, C., Dubey, G., Upadhyay, K., 2024. Mechanistic and future prospects in rhizospheric engineering for agricultural contaminants removal, soil health restoration, and management of climate change stress. The Science of the Total Environment 927, 172116.

Pratas, J., Paulo, C., Favas, P.J.C., Venkatachalam, P., 2014. Potential of aquatic plants for phytofiltration of uranium-contaminated waters in laboratory conditions. Ecological Engineering 69, 170–176.

Purakayastha, T.J., Viswanath, T., Bhadraray, S., Chhonkar, P.K., Adhikari, P.P., Suribabu, K., 2008. Phytoextraction of zinc, copper, nickel and lead from a contaminated soil by different species of Brassica. International Journal of Phytoremediation 10, 61–72.

Rich, C.D., Blaine, A.C., Hundal, L., Higgins, C.P., 2015. Bioaccumulation of perfluoroalkyl acids by earthworms (*Eisenia fetida*) exposed to contaminated soils. Environmental Science & Technology 49, 881–888.

Sanzana, S., Fenti, A., Iovino, P., Panico, A., 2025. A review of PFAS remediation: separation and degradation technologies for water and wastewater treatment. Journal of Water Process Engineering 74, 107793.

Scearce, A.E., MacRae, J.D., Goossen, C.P., Zhang, Y.-J., Holt, K.P., Schattman, R.E., 2025. Uptake of per- and polyfluoroalkyl substances (PFAS) into lettuce (*Lactuca sativa*), tall fescue (*Schedonorus arundinaceus*) and tomato (*Solanum lycopersicum*): a greenhouse experiment evaluating bioconcentration factors and testing the effect of intercropping. Environmental Advances 20, 100629.

Sharma, I., 2021. Bioremediation techniques for polluted environment: concept, advantages, limitations, and prospects. In: Murillo-Tovar, M.A., Saldarriaga-Noreña, H., Saeid, A. (Eds.), Trace Metals in the Environment - New Approaches and Recent Advances. IntechOpen, London, pp. 221–236.

Sharma, M., Rawat, S., Rautela, A., 2024. Phytoremediation in sustainable wastewater management: an eco-friendly review of current techniques and future prospects. AQUA – Water Infrastructure, Ecosystems and Society 73, 1946–1975.

Sörengård, J., Khodadoust, M., Benskin, J.P., Radke, M., 2023. Evaluation of PFAS phytoremediation potential in Nordic aquatic plants. The Science of the Total Environment 895, 164961.

Sørmo, E., Silvani, L., Bjerkli, N., Hagemann, N., Zimmerman, A.R., Hale, S.E., et al., 2021. Stabilization of PFAS-contaminated soil with activated biochar. The Science of the Total Environment 763, 143031.

Srivastava, N.K., Gupta, S., 2021. Phytoremediation of water pollutants using *Eichhornia crassipes* and Lemna minor. Journal of Water Pollution and Purification Research 8, 39–44.

Susarla, S., Medina, V.F., McCutcheon, S.C., 2002. Phytoremediation: an ecological solution to organic chemical contamination. Ecological Engineering 18, 647–658.

Wani, Z.A., Ahmad, Z., Asgher, M., Bhat, J.A., Sharma, M., Kumar, A., et al., 2023. Phytoremediation of potentially toxic elements: role, status and concerns. Plants 12 (3), 429.

Wright, T., Crompton, M., Bishop, D., Currell, G., Suwal, L., Turner, B.D., 2025. Phytoremediation evaluation of forever chemicals using hemp (*Cannabis sativa* L.): pollen bioaccumulation and the risk to bees. Chemosphere 370, 143859.

Yang, H., Liu, J., 2025. Recent advances in microbial and bioelectrochemical strategies for degradation of per- and polyfluoroalkyl substances: mechanisms, limitations, and research opportunities. Biotechnology Letters 47 (3), 48.

Yogarathinam, L.T., Baig, U., Waheed, U.A., Aljundi, I.H., 2025. Recent progress and challenges in nanocomposite membranes for efficient removal of hazardous per-/polyfluoroalkyl substances from water sources. Journal of Environmental Chemical Engineering 13, 115470.

Zahra, Z., Song, M., Habib, Z., Ikram, S., 2025. Advances in per- and polyfluoroalkyl substances (PFAS) detection and removal techniques from drinking water, their limitations, and future outlooks. Emerging Contaminants 11 (1), 100434.

Zareitalabad, P., Siemens, J., Hamer, M., Amelung, W., 2013. Perfluorooctanoic acid (PFOA) and perfluorooctanesulfonic acid (PFOS) in surface waters, sediments, soils and wastewater – a review on concentrations and distribution coefficients. Chemosphere 91, 725–732.

Zhang, W., Liang, Y., 2020. Removal of eight perfluoroalkyl acids from aqueous solutions by aeration and duckweed. The Science of the Total Environment 724, 138357.

Zhang, W., Zhang, D., Zagorevski, D.V., Liang, Y., 2019. Exposure of Juncus effusus to seven perfluoroalkyl acids: uptake, accumulation and phytotoxicity. Chemosphere 233, 300–308.

Zhang, X., Zhao, Y., Liu, Y., Zhu, L., Zhang, Y., 2023. Hydroponic accumulation and translocation of per- and polyfluoroalkyl substances in leafy vegetables. Environmental Pollution (Barking, Essex 1987) 314, 120243.

Zheng, X., Shi, Y., Lu, Y., Xu, X., 2016. Growth inhibition and DNA damage in the earthworm (*Eisenia fetida*) exposed to perfluorooctane sulphonate and perfluorooctanoic acid. Chem. Ecol. 32, 103–116.

PART 3

# Materials for PFAS remediation

CHAPTER

# 12

# Per- and polyfluoroalkyl substances removal through separation of wastewater foams

*Angel Chyi En We*[1], *Sanne J. Smith*[2], *and Stefano Freguia*[1]
[1]Department of Chemical Engineering, The University of Melbourne, Parkville, VIC, Australia
[2]Department of Water Management, Delft University of Technology, Delft, The Netherlands

## 12.1 Introduction

Foams can act as a medium for separating surface-active substances (surfactants), including per- and polyfluoroalkyl substances (PFAS). Naturally occurring environmental foams, such as sea, river, and lake foams, have been found to contain relatively high concentrations of PFAS (Madsen et al., 2025; Schwichtenberg et al., 2020; Sherman-Bertinetti et al., 2024). Certain PFAS species exhibit concentrations several orders of magnitude higher in these foams than in the corresponding bulk water. This phenomenon is driven by differences in surface activity, which cause PFAS (and other surfactants) to preferentially adsorb at the air–water interfaces of bubbles formed during turbulence events that introduce air into the water. Such turbulence events include wind, waves, and flow disturbances from hydraulic structures (e.g., weir drop and cascade). The presence of natural and/or anthropogenic foam-stabilizing agents can stabilize the air bubbles, enhancing foam formation. As a result, foams with low interstitial liquid content are formed, in which PFAS are enriched compared to the bulk water.

While the above describes the unintended accumulation of PFAS in naturally occurring environmental foams, the same principle can be exploited as a PFAS removal strategy at wastewater treatment plants (WWTPs), which are known to be major PFAS discharge points. In aerated bioreactors at WWTPs, aeration is primarily provided to enable the biological oxidation of organic carbon and ammonia, while also promoting the formation of air–water interfaces through rising bubbles. Additionally, spontaneous foam formation in aerated bioreactors, possibly due to the presence of foam-forming bacteria with hydrophobic cell surfaces and other (bio)surfactants (Collivignarelli et al., 2020), is not uncommon. These conditions in aerated bioreactors present an opportunity for PFAS removal through foam separation at WWTPs.

*Per- and Polyfluoroalkyl Substances (PFAS) as Environmental Contaminants*
DOI: https://doi.org/10.1016/B978-0-443-43856-1.00018-6

## 12.2 Key concepts

### 12.2.1 Adsorptive bubble separation technique—foam fractionation

The removal or recovery of surfactants via foam separation is not a new technology; rather, it has long been established and well-studied as an adsorptive bubble separation technique known as foam fractionation. In essence, foam fractionation relies on the formation and separation of foam to remove dissolved or colloidal substances with sufficient surface activity from the bulk liquid, after their adsorption at the air–water interfaces. Typical historical applications of this process include protein separation from aquaculture wastewater (Chen et al., 1993) and agro-food processing wastewater (Li et al., 2014).

Certain PFAS with high surface activity can also be effectively separated and removed from water through foam fractionation. The amphiphilic nature of PFAS (Fig. 12.1A) promotes their adsorption at the air–water interfaces formed during bubble generation, with the hydrophobic moiety oriented toward the air phase and the hydrophilic moiety facing the water phase (Fig. 12.1B).

Studies on the feasibility of foam fractionation for PFAS removal from contaminated water have been extensively carried out, spanning laboratory, pilot, and commercial scales. Overall, previous studies consistently show that the effectiveness of foam fractionation varies depending on the specific PFAS species (McCleaf et al., 2021; Burns et al., 2022; Smith et al., 2022). For instance, among perfluoroalkyl acids (PFAAs) species, those with a perfluorinated carbon number (or perfluorinated chain length) of six or more have been shown to be effectively removed by foam fractionation, whereas those with a perfluorinated carbon number of five and fewer are only marginally or negligibly removed.

The effectiveness of foam fractionation for various PFAS species is governed by their affinity for interfacial adsorption, which is commonly expressed using the interfacial

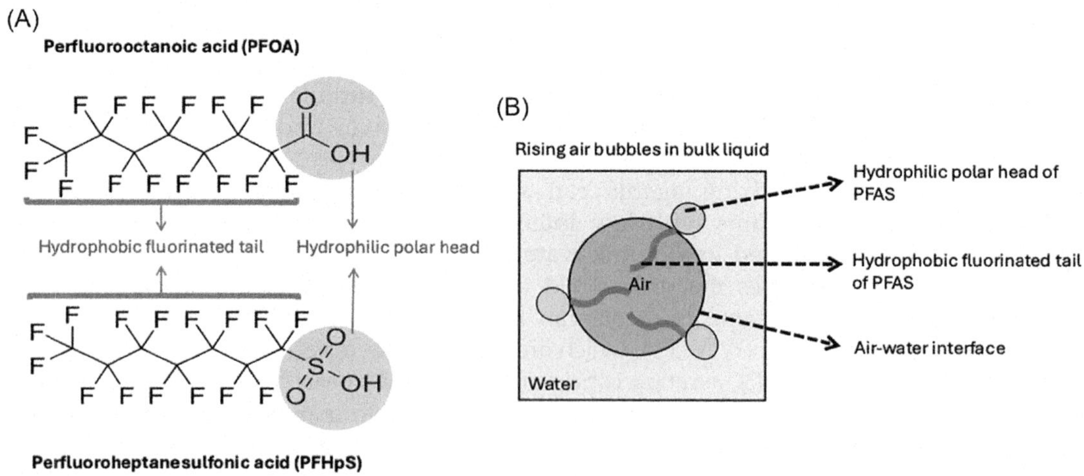


FIGURE 12.1 (A) Amphiphilic structure of perfluorooctanoic acid (PFOA) and perfluoroheptanesulfonic acid (PFHpS), both with a perfluorinated carbon number of seven. (B) Orientation of the hydrophobic and hydrophilic moieties of per- and polyfluoroalkyl substances (PFAS) at the air–water interfaces of rising air bubbles.

adsorption coefficient ($K_i$), or more precisely in this context, the air–water interfacial adsorption coefficient ($K_{aw}$). A larger $K_{aw}$ indicates a stronger tendency of a PFAS species for accumulation at the air–water interfaces. According to Brusseau (2019), the molar volume of PFAS can serve as a reliable single descriptor for predicting log $K_{aw}$ across a wide range of PFAS with varying headgroup charges in an air-deionised water system. In general, PFAS with larger molar volumes exhibit larger $K_{aw}$ values, indicating that molar volume effectively captures the influence of molecular size on hydrophobic interactions, which are the primary driving force for PFAS interfacial adsorption (Brusseau, 2019).

However, this does not apply to nonionic PFAS with very large headgroups—specifically, those with a headgroup-to-tail molar-volume ratio exceeding 1.3. A follow-up study by Brusseau (2023) showed that both molar volume and the headgroup-to-tail molar-volume ratio are necessary for predicting $K_{aw}$ for these compounds. This is because, unlike most PFAAs and cationic PFAS, the hydrophilic headgroups in nonionic PFAS are often as large as, or larger than, their hydrophobic tails. Consequently, increases in molar volume in nonionic PFAS are primarily driven by the increase in the proportional size of the hydrophilic headgroup, rather than the hydrophobic tail, making molar volume alone an unreliable predictor of their behavior for air–water interfacial adsorption (Brusseau, 2023).

While molar volume can be the most precise factor to predict $K_{aw}$ of most common PFAS species, the perfluorinated carbon number and the type of functional headgroup are more commonly used in the literature to compare removal performance in foam fractionation. This is likely due to the common practice of classifying PFAS into short- or long-chain categories based on their perfluorinated carbon number and functional headgroup type—particularly those PFAS of greatest concern and research interest. In general, $K_{aw}$ increases as the perfluorinated carbon number increases. For compounds with the same perfluorinated carbon number (e.g., perfluorooctanoic acid [PFOA] and perfluoroheptanesulfonic acid [PFHpS], both with seven perfluorinated carbons), PFHpS has a higher $K_{aw}$ due to its sulfonic acid functional headgroup, whereas PFOA has a carboxylic acid headgroup.

In addition to PFAS molar volume, several other factors that impact $K_{aw}$, including solution ion composition, ionic strength, pH, and the presence of cosolutes, were investigated by Brusseau and Van Glubt (2019). The authors concluded that while all the factors investigated have some degree of impact on $K_{aw}$, their effect is small when PFAS concentrations are below 1–10 mg $L^{-1}$, which is commonly the case in real environments.

### 12.2.2 Removal and enrichment of per- and polyfluoroalkyl substances in foam fractionation

Foam fractionation for PFAS separation is commonly carried out in engineered foam fractionators, as reported in previous studies (Burns et al., 2022; Smith et al., 2022). Bubble generation and foam collection systems are essential components of the foam fractionation process (Fig. 12.2). In the context of PFAS remediation of contaminated water using foam fractionation, both removal (or concentration reduction) and enrichment are key performance indicators (Eqs. 12.1–12.4). Achieving a high removal percentage with low PFAS concentrations remaining in the treated bulk (also known as the bottom) is crucial to ensure that the treated bulk can be safely discharged into the receiving environment. At the same

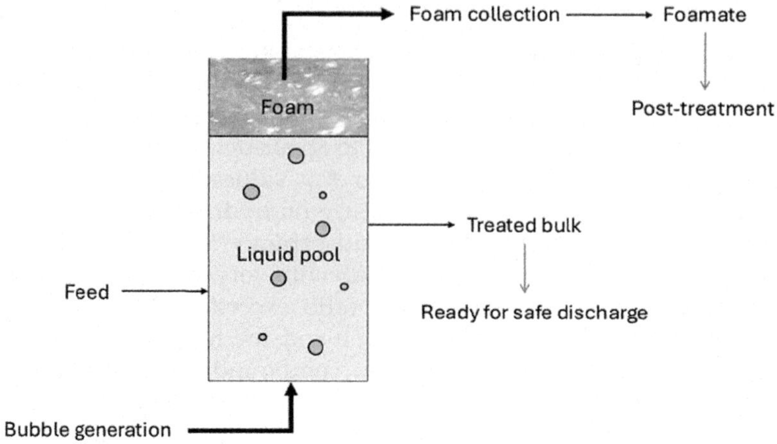


**FIGURE 12.2** Foam fractionator with a simple mode.

time, a high enrichment factor (EF) is preferred, as it indicates a greater degree of concentration into a smaller volume of waste stream, manifested as the foamate (i.e., collapsed foams). Thus both the treated bulk and the foamate (Fig. 12.2), which are the product streams of the foam fractionation process for PFAS treatment, are of interest.

$$\textit{Mass-based removal}\ (\%),\ \textbf{\textit{batch}}\ \textit{operation mode} = \left(1 - \frac{\textit{PFAS concentrations in treated bulk}\left(\frac{ng}{L}\right) \times \textit{Volume of treated bulk}\ (L)}{\textit{PFAS concentrations in feed}\left(\frac{ng}{L}\right) \times \textit{Volume of feed}\ (L)}\right) \times 100 \tag{12.1}$$

$$\textit{Mass-based removal}\ (\%),\ \textbf{\textit{continuous}}\ \textit{operation mode} = \left(1 - \frac{\textit{PFAS concentrations in treated bulk}\left(\frac{ng}{L}\right) \times \textit{Volumetric flowrate of treated bulk}\left(\frac{L}{min}\right)}{\textit{PFAS concentrations in feed}\left(\frac{ng}{L}\right) \times \textit{Volumetric flowrate of feed}\left(\frac{L}{min}\right)}\right) \times 100 \tag{12.2}$$

$$\textit{Concentration reduction}\ (\%) = \left(1 - \frac{\textit{PFAS concentrations in treated bulk}\left(\frac{ng}{L}\right)}{\textit{PFAS concentrations in feed}\left(\frac{ng}{L}\right)}\right) \times 100 \tag{12.3}$$

$$\text{Enrichment factor} = \frac{\text{PFAS concentrations in foamate } \left(\frac{ng}{L}\right)}{\text{PFAS concentrations in feed } \left(\frac{ng}{L}\right)} \tag{12.4}$$

In addition, the extent of volume reduction in the PFAS-contaminated stream may also be worth reporting. This can be expressed in terms of the foamate fraction (Eqs. 12.5–12.6) or volume reduction factor (Eqs. 12.7–12.8). Such information is valuable for evaluating and designing the process, as it is relevant to the downstream handling of the foamate, which often involves destructive PFAS treatment as part of a complete treatment train. Standalone foam fractionation, with its capability to only separate and concentrate PFAS, does not constitute a comprehensive PFAS treatment strategy. The generation of foamate from the foam fractionation process, which is highly enriched in PFAS, requires downstream destructive treatment to convert harmful PFAS species into benign and safe final products. Given the higher energy requirement and operational complexity of destructive treatments such as thermal treatment and advanced oxidation/reduction processes (Meegoda et al., 2022), a lower foamate fraction or, equivalently, a higher volume reduction factor is always favorable for improving the efficiency of these treatments.

$$\text{Foamate fraction (\%), } \textbf{batch} \text{ operation mode} = \frac{\text{Volume of foamate (L)}}{\text{Volume of feed (L)}} \times 100 \tag{12.5}$$

$$\text{Foamate fraction (\%), } \textbf{continuous} \text{ operation mode} = \frac{\text{Volumetric flowrate of foamate } \left(\frac{L}{\text{min}}\right)}{\text{Volumetric flowrate of feed } \left(\frac{L}{\text{min}}\right)} \times 100 \tag{12.6}$$

$$\text{Volume reduction factor, } \textbf{batch} \text{ operation mode} = \frac{\text{Volume of feed (L)}}{\text{Volume of foamate (L)}} \tag{12.7}$$

$$\text{Volume reduction factor, } \textbf{continuous} \text{ operation mode} = \frac{\text{Volumetric flowrate of feed } \left(\frac{L}{\text{min}}\right)}{\text{Volumetric flowrate of foamate } \left(\frac{L}{\text{min}}\right)} \tag{12.8}$$

As mentioned earlier, both a high removal percentage and a high EF (which often also corresponds to a low foamate fraction or a high volume reduction factor) are desirable in foam fractionation. However, it is important to note that there is often a trade-off between these two indicators in a conventional foam fractionation process (Burns et al., 2022). The opposing relationships of removal and enrichment are illustrated in Fig. 12.3, which shows scenarios with different foam harvesting locations along a column, where changes in bubble size distribution are observed with height, along with the distortion of spherical bubbles (wet foam) into polyhedral bubbles (dry foam). For instance, harvesting foam with a higher wetness (i.e., higher liquid content) can enhance PFAS removal as a larger amount of PFAS

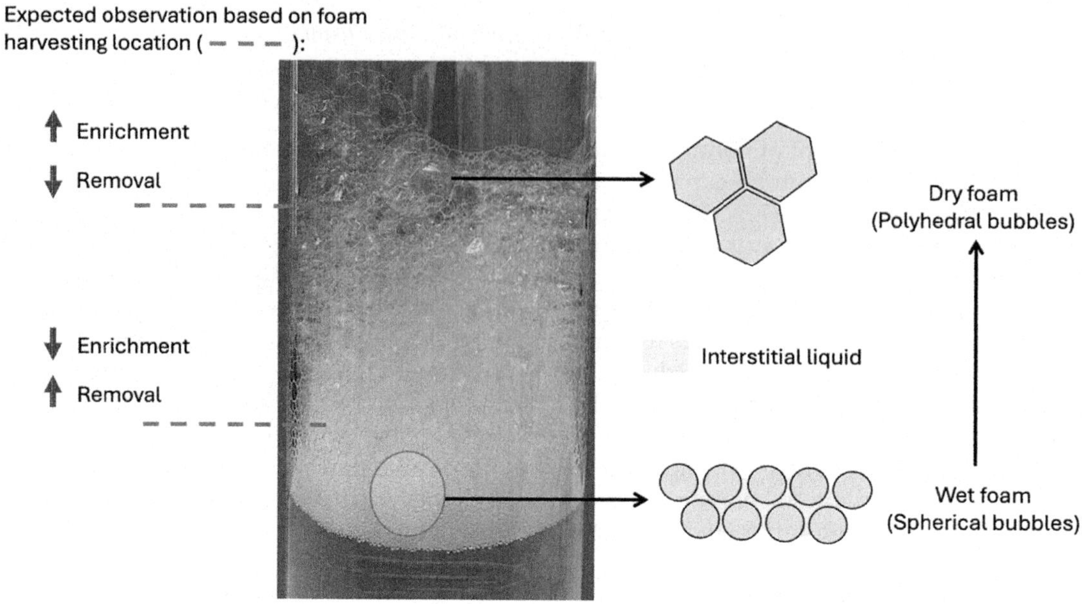


FIGURE 12.3 Foam structure variation due to changes in bubble size distribution with height.

is removed with the foamate; however, this leads to a lower EF. This is because PFAS adsorption primarily occurs at the air–water interfaces of bubbles, whereas wetter foam contains more interstitial liquid, which generally has a significantly lower PFAS concentration. As a result, collecting more of this liquid "dilutes" the PFAS concentration in the foamate, thereby increasing the foamate volume and lowering the EF.

Adjustments to key operating parameters, such as bubble size and gas flow rate, also have competing effects on removal and enrichment. For example, a smaller bubble size, which provides a larger specific surface area for PFAS adsorption, can enhance PFAS removal. However, this can lead to a lower PFAS enrichment, as smaller bubble sizes result in a wetter foam. Similarly, increasing gas flow rate tends to increase both surface area flux for air–liquid interfacial adsorption of PFAS (enhancing removal) and liquid flux carried by the foam (reducing EF).

Several other operating parameters, including but not limited to the addition of external cosurfactants, foam height, and temperature, can also affect both performance indicators, with their relationships described by Sochacki et al. (2024) and We et al. (2024a). Hence, optimizing the operating parameters of foam fractionation for PFAS remediation should focus on balancing removal and enrichment to achieve optimal performance. Design interventions, such as multistage (demonstrated by Burns et al., 2022) and counterflow foam fractionators (illustrated by Lemlich, 1972), may be adopted to help reduce the tension between removal and enrichment by providing more than one equilibrium stage. However, it is important to note that implementing such design interventions at WWTPs usually requires additional treatment units (e.g., external multistage foam fractionators) beyond the existing aerated bioeactors.

### 12.2.3 Proposed integration of foam fractionation into aerated bioreactors at wastewater treatment plants

WWTPs inevitably receive PFAS-contaminated water from domestic and industrial discharges. Due to the persistent nature of PFAS and the fact that conventional wastewater treatment trains are not designed to remove these compounds, treated effluent and sewage sludge become conduits for PFAS release into the environment (Behnami et al., 2024; Nguyen et al., 2022). To address the challenges posed by PFAS at WWTPs, source-control measures, such as restricting nonessential PFAS production and use, substituting toxic PFAS with safer alternatives, and providing on-site treatment for PFAS-contaminated water, should be prioritized. However, as these actions may take time to be fully implemented and domestic-source PFAS discharges may continue indefinitely thereafter, treatment approaches within WWTPs could serve as complementary mitigation strategies, where necessary.

The growing concern over emerging contaminants (not limited to PFAS) may drive the implementation of treatment approaches at WWTPs. For example, with the introduction of Directive (EU) 2024/3019, all wastewater treatment facilities serving a population equivalent of 150,000 or more are expected to implement quaternary treatment to reduce pharmaceuticals in effluents by at least 80% relative to the influent load (European Parliament, Council of the European Union, 2024). Where necessary, treatment approaches may also be implemented within WWTPs to specifically target PFAS.

The ultimate goal of PFAS remediation is the degradation of these compounds into products that can be safely discharged, which can be achieved through destructive treatment. However, destructive treatment is not always practical as a single-stage treatment for WWTPs treating large volumes of water, as it is often associated with operational complexity, high energy consumption, and harsh conditions. Therefore a two-step treatment train combining separation/concentration and destructive treatment approaches is suggested as a more feasible strategy.

Activated carbon, ion exchange resin, nanofiltration, and reverse osmosis (RO) are among the separation/concentration approaches that have been extensively studied (Abbasian Chaleshtari and Foudazi, 2022). However, these methods present several drawbacks, such as the need to manage spent solid media (e.g., regeneration, replacement, and/or incineration), high sensitivity to matrix effects, and the generation of large volumes of RO retentate (20%–50% of the influent volume) (Horst et al., 2018). Foam fractionation, by contrast, has the potential to overcome or minimize many of these limitations.

Integrating foam fractionation into aerated bioreactors could serve as a suitable separation/concentration method, particularly with regard to PFAS treatment at WWTPs. Unlike conventional foam fractionation, which is typically performed in dedicated engineered foam fractionators, this integrated approach takes advantage of existing operational conditions in aerated bioreactors. The aeration process, primarily intended to support biological oxidation of organic carbon and ammonia, inherently generates air–water interfaces and often results in spontaneously occurring foams. These conditions may facilitate PFAS removal via foam fractionation, provided that suitable foam harvesting systems are in place. This integration offers several advantages: (1) no additional process is required to generate air–water interfaces, as aeration is already provided, thereby reducing

**FIGURE 12.4** Photos showing spontaneous foaming in aerated bioreactors at wastewater treatment plants in Australia.

energy consumption and operational costs, and (2) it eliminates the need for additional infrastructure, reducing capital investment and land requirements.

Foaming, though varying in extent, is commonly observed on the surface of aerated bioreactors during aeration at WWTPs (Fig. 12.4). A foam structure forms when gas bubbles are injected faster than the liquid drains from the films between these bubbles (Schramm and Wassmuth, 1994). Some foams collapse rapidly, while others can remain stable for longer periods. The stability of foam depends on the amount and nature of foam-stabilizing agents, such as (bio)surfactants, foam-forming (filamentous) biomass with hydrophobic cell surfaces, and oil and grease, which are the most likely contributors in aerated bioreactors (Collivignarelli et al., 2020).

When foam-stabilizing agents are abundantly present, they can lead to the formation of highly stable foam, which poses undesired operational challenges at WWTPs. These challenges include reduced oxygen transfer efficiency in the aerated bioreactors, biomass loss, odor issues, safety concerns due to foam overflowing onto walkways, and potential exposure to pathogens via wind-blown foam (Collivignarelli et al., 2020; Heard et al., 2008; de Ios Reyes, 2010). The cause of stable foam can be preliminarily identified through the appearance of the foam. Whitish, lightweight foam is typically associated with poorly biodegradable surfactants, commonly observed during the start-up phase of a WWTP (Pujol et al., 1991) or following the inflow of excess chemical surfactants from industrial discharge (Collivignarelli et al., 2020). In contrast, brownish, viscous, and stable foam, commonly termed biological foam, is usually caused by a combination of certain (filamentous) bacteria with hydrophobic cell surfaces (e.g., mycolic acid-containing *Actinobacteria*) and the presence of surfactants (de Ios Reyes, 2010).

Relying on the erratic foams at WWTPs discussed above for PFAS removal is impractical, due to the unpredictability of their occurrence and the associated operational and safety concerns. Such foaming should always be controlled and suppressed. The type of foaming that supports the concept of PFAS removal through foam separation in aerated bioreactors is more likely to occur naturally under typical operational conditions and to be mild, as illustrated in Fig. 12.4. A dedicated foam separation device can then be engineered and integrated to collect those PFAS-enriched foams.

While the concept of integrating of foam fractionation into aerated bioreactors for PFAS removal appears promising, pilot- and full-scale demonstrations are, to the best of our knowledge, still lacking. Current efforts remain at the conceptual stage, involving field-based investigations through foam sampling and EF evaluation, as well as lab-scale foam fractionation experiments using real wastewater, specifically mixed liquor suspension or activated sludge.

## 12.3 Methods and applications

### 12.3.1 Field-based conceptual studies

Preliminary conceptual investigations into the integration of foam fractionation within aerated bioreactors have begun through field-based studies (Schaefer et al., 2023; Smith et al., 2023; We et al., 2024b). In these studies, foam and other water samples were collected from WWTPs, and PFAS enrichment was evaluated by comparing concentrations in the aerated bioreactors' foam to those in the effluent of the aerated bioreactors (Schaefer et al., 2023), influent to the aerated bioreactors (Smith et al., 2023), or underlying mixed liquor suspension in the aerated bioreactors (We et al., 2024b). This comparison was typically expressed as an EF, which quantifies the extent of PFAS separation based on concentration differences between the foam and the reference water samples. Prior to PFAS analysis, foam samples were collapsed into liquid form (i.e., foamate), and both foamate and reference water samples were either filtered or centrifuged to remove solids. As a result, the PFAS analysis and the calculated PFAS EF primarily reflected the aqueous phase of the samples, except in We et al. (2024b), where the centrifuged solids were also examined.

Fig. 12.5 illustrates the aqueous PFAS EFs by combining data from Smith et al. (2023) and We et al. (2024b), focusing only on samples collected from conventional activated sludge (CAS)-based aerated bioreactors. Overall, an obvious trend of increasing EF is observed with increasing perfluorinated carbon number of PFAS. Short-chain PFAS, particularly those with fewer than six perfluorinated carbon numbers, exhibit markedly low median EFs of four or below, indicating a low tendency to adsorb at the air–water interfaces due to their lower hydrophobicity.

It is noteworthy that even for the same PFAS species, the EF can vary by several orders of magnitude. For instance, the EF of PFOA and perfluorooctane sulfonate (PFOS) ranged from approximately 1 to 550 and 1 to 5500, respectively (Fig. 12.5). Possible factors contributing to the wide variation in PFAS EF include: (1) challenges and uncertainties associated with sampling foam and reference water samples; (2) variations in the definition of EF used between these studies; and (3) differences in the physical, chemical, and biological characteristics of foam and reference water samples, as well as (4) the variations in the operating conditions of CAS-based aerated bioreactors across different WWTPs.

Undoubtedly, the varying configurations and accessibility of aerated bioreactors may require adjustments to sample collection procedures, potentially introducing uncertainties

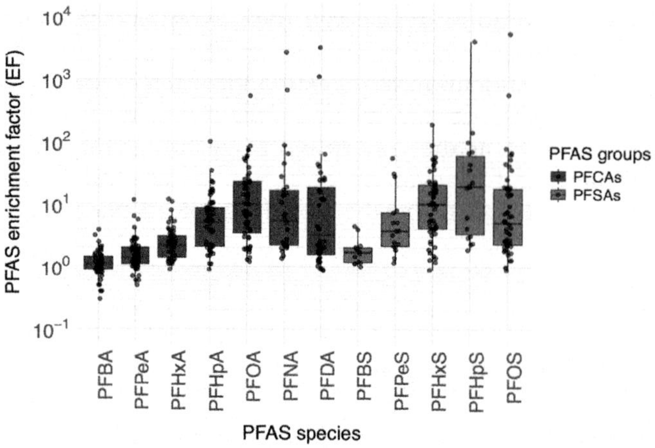


**FIGURE 12.5** Per- and polyfluoroalkyl substances (PFAS) enrichment factors (EF) data. The *box* represents the interquartile range (IQR) from the 25th percentile to the 75th percentile, with the *middle line* in the *box* indicating the median value. The *whiskers* represent the maximum and minimum values within 1.5 times the IQR. *Note*: Only data with concentrations above the limit of quantification or method reporting limit for both foam and their corresponding water samples were used to generate this figure. Smith et al. (2023) defined EF as the ratio between PFAS concentrations in foam and influent to the aerated bioreactors, whereas We et al. (2024b) defined EF as the ratio between PFAS concentrations in foam and underlying mixed liquor suspension in the aerated bioreactors. Source: *From Smith, S.J., Keane, C., Ahrens, L., Wiberg, K., 2023. Integrated treatment of per- and polyfluoroalkyl substances in existing wastewater treatment plants—scoping the potential of foam partitioning. ACS ES&T Engineering. https://doi.org/10.1021/acsestengg.3c00091 (using only data from activated sludge-based wastewater treatment plants) and We, A.C.E., Stickland, A.D., Clarke, B.O., Freguia, S., 2024. The role of suspended biomass in PFAS enrichment in wastewater treatment foams. Water Research. 254, 121349. https://doi.org/10.1016/j.watres.2024.121349 (aqeuous phase EF data).*

in the evaluation of PFAS EF. Developing robust sampling methods for field-based studies can indeed help reduce such uncertainties. To improve the reproducibility of the sampling methods, Coffin et al. (2024) developed a novel foam sampling device for collecting cascade-derived foam at WWTPs, aiming to replace grab sampling and hand-scooping methods previously used by other studies. Additionally, the device enables the collection of data on the foam's liquid content, providing useful insights into foam characteristics. While the use of the foam sampling device appears promising for cascade-derived foam collection, its applicability to foam collection in aerated bioreactors, such as those illustrated in Fig. 12.4, remains untested.

Furthermore, as discussed by We et al. (2024b), inconsistencies in the definition of the EF across studies—foam:influent in Smith et al. (2023) and foam:underlying bulk in We et al. (2024b)—may have contributed to some of the variations in reported values. These differences, combined with the limitations of grab sampling, which capture PFAS concentrations at a single time point and overlook the differing retention times of the foam and reference water samples, introduce additional uncertainty. Nevertheless, despite these constraints, a consistent finding across these studies is that foam in aerated bioreactors generally contains significantly higher (long-chain) PFAS concentrations compared to all types of water samples, supporting its potential as a medium for PFAS separation and concentration at WWTPs.

Attempts have been made by We et al. (2024c) to elucidate the factors contributing to PFAS EF variability in their dataset presented in We et al. (2024b), by examining several mixed liquor suspension and foam characteristics through statistical modeling. Foam half-life, filament content, total suspended solids (TSS) concentration, and the surface tension of the mixed liquor suspension were found to have statistically significant negative relationships with the enrichment of certain PFAS species, particularly those with perfluorinated carbon numbers of six or above (We et al., 2024c). In brief, increased foam half-life, filament content, and TSS concentrations may indicate/cause increased foam wetness, contributing to lower EF. Meanwhile, the modestly lower surface tension in mixed liquor suspension (65–71 $mN\,m^{-1}$ in We et al. (2024c)) compared to pure water at room temperature (~72 $mN\,m^{-1}$) may have helped enhance foam formation, leading to higher enrichment in the foams.

Findings by Smith et al. (2023) and We et al. (2024b) found insufficient evidence to support the relationships between parameters such as conductivity, pH, total organic carbon, PFAS concentrations, and metal ion concentrations in the influent or mixed liquor suspension and PFAS EF. These findings differ from lab-scale studies (e.g., Meng et al., 2018; Wang et al., 2023), where, on the contrary, such parameters were found to affect PFAS EF. This discrepancy may be attributed to the relatively narrow and lower ranges of these parameters typically found at WWTPs, which may not be sufficient to produce observable effects as seen under the controlled conditions and wider parameter value ranges of laboratory experiments. Moreover, there could be multiple interacting parameters in the field-based studies that mask the effects of individual parameters on PFAS EF, warranting further investigation.

In general, liquid content (i.e., the foam wetness) could be the most direct factor in governing the PFAS EF, with lower liquid content contributing to higher long-chain PFAS EF (Coffin et al., 2024). However, the liquid content of the foam is considered a secondary factor, influenced by a range of interacting parameters (i.e., physicochemical and biological characteristics of water, as well as the operational parameters of aerated bioreactors). These parameters primarily affect foam formation and stability, thereby contributing to variations in the liquid content of the foam.

An interesting finding by Smith et al. (2023) suggests that PFAS EF may potentially be influenced by the treatment processes used in aerated bioreactors. As shown in Fig. 12.6, higher EFs for PFOA and PFOS were generally observed in foam samples collected from moving-bed biofilm reactor (MBBR) compared to those from CAS systems. Smith et al. (2023) speculated that the absence (or minimal presence) of suspended solids in MBBRs may be beneficial to PFAS enrichment. In contrast, the higher suspended solids content in CAS systems may have reduced PFAS enrichment in the aqueous phase of the foamate, due to the interplay of PFAS sorption onto the solids, as well as hindered foam buildup, which prevented the formation of dry foam.

This was further supported by We et al. (2024b), who reported that PFAS enrichment occurs in both the aqueous and solid fractions of the foamate collected from CAS systems. This dual-phase distribution of the enriched PFAS could explain the lower aqueous phase PFAS EFs observed in CAS compared to MBBRs, where the foamate collected primarily consists of the aqueous phase. These findings raise the question of whether foam fractionation may be more suitable for aerated bioreactors with lower suspended solids concentrations, such as MBBRs or other systems that facilitate the generation of air–water interfaces and foaming at WWTPs.

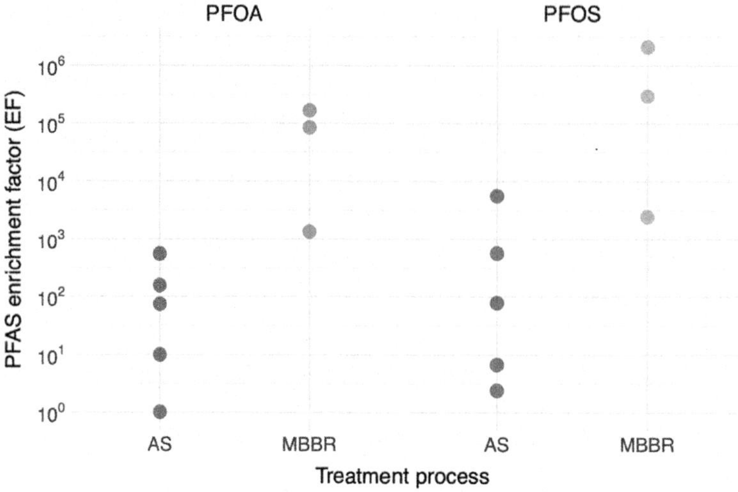


**FIGURE 12.6** PFAS enrichment factor (EF) for samples collected from two different treatment processes—conventional activated sludge (AS) and moving-bed biofilm reactors (MBBRs). Source: *From Smith, S.J., Keane, C., Ahrens, L., Wiberg, K., 2023. Integrated treatment of per- and polyfluoroalkyl substances in existing wastewater treatment plants—scoping the potential of foam partitioning. ACS ES&T Engineering. https://doi.org/10.1021/acsestengg.3c00091.*

Coffin et al. (2024) investigated PFAS enrichment in foams derived from cascades at WWTPs after tertiary treatment, where minimal solids were reasonably expected. In their study, high EFs were observed for large molar volume PFAS; for instance, PFOS exhibited EFs ranging from 295 to 143,000. Based on estimations, effective foam collection at the cascade could potentially remove a substantial amount of PFAS (>99 %) with molar volume equal to or greater than that of PFOS (Coffin et al., 2024). Unintentional foam fractionation in WWTP cascades may offer several advantages over aerated bioreactors. For example, it does not require electricity to generate air–water interfaces, occurs over a relatively smaller surface area that enables simpler foam collection system design, and may be less affected by interference from solids. However, the applicability of this approach may be constrained by the need for specific hydraulic structures (i.e., cascades) and by the relatively uncommon formation of dense foam at these locations, which may differ from observations reported in Coffin et al. (2024).

### 12.3.2 Lab-scale foam fractionation experiments

Field-based studies generally indicate significantly high enrichment of long-chain PFAS in the foams of aerated bioreactors. However, high enrichment does not necessarily correlate with a high removal percentage. Given that the primary objective of PFAS separation at WWTPs is to ensure that PFAS levels in the treated water are within safe limits before discharge into receiving water bodies, achieving high enrichment in foam is by itself insufficient. Lab-scale experiments are critical for further assessing the removal percentage, particularly through the application of foam harvesting techniques, which are difficult to evaluate through field-based studies.

Unlike foam fractionation applied to other contaminated streams, special consideration should be given to suspended solids in wastewater from aerated bioreactors, as these are present in substantial amounts—typically ranging from 2 to 5 g $L^{-1}$. These solids can act as a

PFAS retention source, particularly for long-chain PFAS (Arvaniti et al., 2014), and thus may need to be accounted for to provide a more comprehensive assessment. Depending on the choice of data interpretation, PFAS removal can be reported either as a combined value from both the aqueous and solid phases, reflecting the overall system removal, or separately for each phase. The latter option recognizes that the aqueous and solid phases may represent distinct WWTP products. In practice, mixed liquor suspension typically undergoes further treatment, namely secondary sedimentation, to enable phase separation. Subsequently, the two phases follow different treatment pathways—the aqueous phase is either discharged or further treated with tertiary or advanced treatment, whereas a fraction of the solid phase is removed as waste activated sludge (WAS) and further treated into a biosolids stream.

Malovanyy et al. (2025) and We et al. (2025) investigated PFAS removal from actual activated sludge samples (i.e., mixed liquor suspension) collected from WWTPs employing the CAS process, using a semibatch foam column. In both studies, influent was introduced into the foam column at the start of the experiments, followed by continuous aeration and vacuum-assisted foam harvesting throughout the experimental period. The foamate fraction collected after 30 min of aeration was 2% and 8%, respectively, in We et al. (2025) and Malovanyy et al. (2025). Overall, aqueous-phase removal of approximately 85% ± 5% for PFOA and 75% ± 5% for PFOS was reported. For the solid phase, removal ranged from 40%–60% for PFOA and 6%–35% for PFOS. However, when both aqueous and solid phases were considered together, the overall removal percentage decreased significantly to approximately 70% for PFOA and 15% for PFOS in both studies, compared to analyzing only the aqueous phase. This illustrates how the interpretation of treatment performance can vary depending on the phases considered.

The presence of high suspended solids content introduces additional factors that may affect foam fractionation performance, specifically the solid–water adsorption coefficient, alongside the previously discussed air–water adsorption coefficient ($K_{aw}$). The findings by Malovanyy et al. (2025) and We et al. (2025) suggest that solid-phase removal is less pronounced compared to aqueous-phase removal for the majority of long-chain PFAS. Although solid-phase desorption of PFAS occurs and the desorbed PFAS can subsequently adsorb onto the air–water interfaces of the rising bubbles during foam fractionation, this process may be limited to some extent. A factor that could limit this process is the high binding affinity of PFAS to the organic matter in the suspended solids, which makes desorption challenging—particularly for PFAS with higher perfluorinated carbon numbers. The low removal of PFAS from the solid phase during foam fractionation may be considered more of an additional benefit, rather than a primary solution for reducing PFAS in the solid fraction, based on the findings of these semibatch experiments.

In contrast to long-chain PFAS (i.e., those with a perfluorinated carbon number of ≥6), which showed good removal from the aqueous phase and marginal to moderate removal from the solid phase, short-chain PFAS tend to remain predominantly in the aqueous phase and appear to be less influenced by suspended solids retention. Consistent with field-based conceptual studies reporting low EFs for short-chain PFAS (i.e., those with a perfluorinated carbon number of <6), their removal from the aqueous phase was generally poor, with values typically around 20% or lower (Malovanyy et al., 2025; We et al., 2025). This underscores a key limitation of foam fractionation in effectively removing short-chain PFAS.

In addition to CAS, Malovanyy et al. (2025) also tested other types of wastewater samples collected from WWTPs, including wastewater after screening, after primary sedimentation, and treated wastewater, as feed for the foam column. Overall, the highest removal from the aqueous phase was observed when treated wastewater was used as feed, followed by wastewater after primary sedimentation, wastewater after screening, and lastly, the CAS. The authors speculated that this trend was linked to the decreasing concentrations of suspended solids and organic substances across these samples. Although the treated wastewater showed the highest PFAS removal from the aqueous phase, it also exhibited the lowest foaming potential. As a result, the authors suggested that the addition of a surfactant might be necessary to facilitate foam formation in treated wastewater, in contrast to the wastewater after sedimentation and screening. These findings highlight that while applying foam fractionation at WWTP stages other than aerated bioreactors may enhance PFAS removal, it also introduces practical challenges. These challenges include increased capital and operational costs due to the absence of existing air–water interface generation processes and, in some cases, the need for external foaming agents to promote foam formation.

Depending on the final treatment goal, it is important to note that the removal percentage alone may not provide sufficient information for evaluation. The key consideration is likely whether PFAS can be reduced to a "safe" level after treatment, recognizing that this varies depending on context. The initial concentration of PFAS at WWTPs typically ranges from $ng\,L^{-1}$ to $\mu g\,L^{-1}$, which may result in a lower driving force for interfacial adsorption. Additionally, where regulations or guidelines exist, they are typically based on a defined concentration of PFAS considered to be safe. Therefore it is crucial to report and consider the final concentration of PFAS in the treated bulk, in addition to the removal percentage during performance assessment. For example, in We et al. (2025), both PFOA and PFOS, with initial aqueous concentrations in the feed of approximately $10\,ng\,L^{-1}$, were reduced to $2\,ng\,L^{-1}$ or below in the treated bulk after a 30-min semibatch operation, corresponding to a removal percentage of approximately 75%–85%.

### 12.3.3 Capturing aerosols as a pathway for per- and polyfluoroalkyl substances removal

Building on the understanding of preferential air–water interfacial adsorption of PFAS and generation of aerosols during the aeration process (aerosolisation), the capture of aerosols as a medium for PFAS removal has been proposed and explored in Nguyen et al. (2024) and Schaefer et al. (2025). This approach may be particularly beneficial for water that does not foam well. A proof-of-concept bench-scale study showed that about 70%–75% removal was achieved for long-chain PFAS such as perfluorohexane sulfonic acid (PFHxS), PFOA, and PFOS with less than 1% of the total water volume in the vessel being recovered as aerosol in sorbent material (i.e., paper towels) placed 2.5 cm from the water surface. Increasing the aeration flow rate has been shown to increase PFAS removal through aerosolisation. For instance, PFOS removal increased from 75% to 97% at aeration flow rates of 20 and $30\,mL\,min^{-1}$, respectively (Nguyen et al., 2024), in a column with a diameter of 8.9 cm and a working volume of 1.7 L.

Similarly, Schaefer et al. (2025) estimated that between 64% and 92% removal of long-chain PFAS could be achieved with near-surface aerosol capture (within centimeters of the water surface). The authors suggest that near-surface capture, with a reduced aerosol-capturing distance above the wastewater surface to maximize aerosol collection, may be a more effective approach than increasing aeration flow rates, which would otherwise lead to higher energy consumption. Although this approach avoids reliance on foam formation and the need for foam harvesting devices, harvesting PFAS-enriched aerosols presents its own challenges. Capturing aerosols likely requires sorbent materials, which would need to be regenerated or replaced periodically, and properly managed as solid waste upon disposal.

## 12.4 Challenges

The concept of utilizing the existing conditions in aerated bioreactors for the implementation of foam fractionation offers advantages in terms of minimizing capital and operational costs, making it an attractive option. However, several challenges may be anticipated and warrant further investigation. First, it is important to note that the aerated bioreactors primarily serve the biological treatment process and are not specifically designed for PFAS removal. Any attempt to facilitate and optimize the implementation of foam fractionation must not compromise the biological removal performance of ammonia and organics. Therefore constraints exist compared to a dedicated engineered foam fractionator.

It is not difficult to recognize that thick and dense foam formation would not be generated in a healthy activated sludge system; instead, the foam would likely be relatively transient and mild, thereby presenting challenges for full-scale foam harvesting. Additionally, as foam is harvested from the aerated bioreactors, it may lead to a reduced extent of foam formation, and optimized, intermittent harvesting may be required. In fact, Schaefer et al. (2023, 2025) estimated that the mass flow of PFAS into the foam of a typical aerated bioreactor is generally insufficient to achieve a meaningful reduction of PFAS mass and concentration, even if the foam is aggressively harvested. Their estimation is based on a mass fraction of $10^{-6}$ of the total wastewater passing through a WWTP being harvested as foam, along with a PFAS EF of 1000, which contributes to only 0.1% PFAS mass removal. Stimulation of foam formation has been proposed as one approach to enhance the removal (Schaefer et al., 2023; Smith et al., 2023), but this often contradicts the established operational norms of WWTPs, and thus requires careful consideration. Also, the increased foam mass being harvested may lead to a higher foamate volume, which could be adverse to downstream destructive treatment processes.

In addition, a fraction of the biomass (i.e., activated sludge) is expected to be removed from the system during foam harvesting. Depending on the amount of sludge removed, it may be necessary to separate the solids fraction from the foamate and recycle it to maintain the intended WAS mass and ensure proper biological treatment performance. However, this is generally undesirable, as it could reintroduce PFAS into the aerated bioreactors due to the high PFAS concentration in the solids fraction of the foamate. Also, undesirable filamentous biomass is likely to accumulate in the foam, and purposefully recycling these microorganisms could lead to other operational issues such as sludge bulking.

The vacuum-assisted foam harvesting approach demonstrated in lab-scale studies may face challenges during full-scale implementation, particularly considering the large surface area of aerated bioreactors, as discussed by Malovanyy et al. (2025). Coffin and Reeves (2025) attempted to develop and evaluate continuous cascade foam collection methods at a WWTP. An estimation based on the PFAS EF of the foamate collected suggests that foam harvesting using a specially designed mesh-belt skimmer could potentially achieve over 90% removal of most long-chain PFAS, with only approximately 0.26% of the wastewater being recovered as foamate (Coffin and Reeves, 2025). The mesh-belt skimmer enhances PFAS enrichment by allowing foam drainage and collapse, which would not otherwise be achieved by passive overflow foam collection methods (e.g., traditional skimmers or vacuum extraction). Based on their calculations, approximately three times more foamate volume (i.e., 0.87% of the wastewater) would need to be collected to achieve 90% removal of PFOS, 20% of PFOA, and 22% of PFHxS using the passive foam collection methods, making it a less efficient approach. While the mesh-belt skimmer appears promising, its suitability for application in aerated bioreactors remains unknown, given the differences in operational conditions and foam characteristics compared to cascade-derived foam. Also, the actual PFAS removal performance with this foam harvesting method remains to be confirmed through long-term continuous operation, alongside PFAS measurements in the effluent.

Consistent findings indicate that the ineffective removal of short-chain PFAS (those with perfluorinated carbon numbers $<6$) is a key challenge in foam fractionation. The addition of cosurfactants, particularly those with counterions corresponding to the PFAS of interest, has been proposed to enhance the removal percentage of short-chain PFAS. For example, cetyltrimethylammonium bromide (CTAB; a cationic cosurfactant) has been shown to improve the removal of short-chain anionic PFAS by acting as a collector and allowing electrostatic pairing, thereby forming a new surface-active moiety with a higher affinity for air–water interfacial adsorption (Buckley et al., 2023; Vo et al., 2023). However, the potential toxicity, increased operational cost, the competition between PFAS and other constituents in the mixed liquor suspension for pairing with the added cosurfactants, and the availability of treatment methods to manage the residual concentration of these cosurfactants are among the factors that should be considered.

## 12.5 Future directions

Based on the limitations and challenges identified in previous studies, three main future directions are suggested to advance the implementation of foam fractionation in aerated bioreactors at WWTPs:

1. **Conducting studies that closely simulate the operational mode and setup of aerated bioreactors at WWTPs**
   The majority of studies were conducted in semibatch mode, which differs from the continuous operation of aerated bioreactors. A continuous mode experiment should be conducted to better analyze performance through a comprehensive PFAS mass-balance perspective. Given that the studies typically used foam columns with relatively smaller

diameters and lower heights compared to full-scale aerated bioreactors, the hydrodynamics of bubbles during aeration may differ, potentially affecting PFAS adsorption and overall performance. Moreover, long-term studies could be carried out to monitor the effects of integrating foam fractionation into the aerated bioreactors on biological performance, including the potential for biomass loss through foam harvesting. Aerosolisation could also be explored as an alternative approach. Finally, pilot-scale studies could follow to further assess the feasibility and effectiveness.

2. **Improving short-chain PFAS removal through sustainable alternatives**
   Short-chain PFAS (e.g., PFBA, PFBS, and PFHxA), which exhibit low surface activity, are commonly challenging to remove effectively using foam fractionation and other adsorptive methods. As long-chain PFAS are gradually phased out and restricted, and the short-chain variants become more widely used, the need for effective removal methods for these compounds is becoming increasingly important. While previous studies have suggested the use of cationic cosurfactants to enhance the removal of anionic short-chain PFAS, this approach may not be sustainable due to its potential environmental impact and associated costs. Therefore there is a need to identify cheaper and safer alternative methods that can effectively remove these compounds.
3. **Developing suitable foam harvesting devices and downstream foamate management**
   To facilitate effective PFAS removal through foam separation, it is crucial to design and implement efficient foam harvesting devices using advanced engineering approaches. The design should account for different aerated bioreactor configurations and be robust against varying weather and flow conditions. As part of establishing a comprehensive PFAS treatment train, downstream foamate management using destructive treatment processes should also be considered. Future studies could be conducted to identify suitable destructive treatments, considering compatibility, cost-effectiveness, and the formation of any transformation products other than full mineralization. Given the large volumetric throughput of water at WWTPs, the 2% to 8% of foamate collected in lab-scale experiments, as demonstrated in Malovanyy et al. (2025) and We et al. (2025), still represents a significant volume of the waste stream. Therefore introducing a secondary concentration step or implementing process interventions may be necessary to further reduce the foamate volume collected from the aerated bioreactors, highlighting the need for further studies.

## 12.6 Conclusion

PFAS removal through foam separation from aerated bioreactors at WWTPs leverages the surface-active properties of PFAS and their preferential adsorption at the air–water interfaces. This approach shows significant potential, offering several advantages over conventional adsorptive PFAS separation methods. Effective PFAS removal has been demonstrated at commercial scale, in purpose-built foam fractionators for other types of PFAS-contaminated water. Conversely, integrating foam fractionation into aerated bioreactors is still in the conceptual development stage, with few lab-scale studies conducted thus far. Compared to purpose-built foam fractionators, this integration reduces capital and operational costs by utilizing the existing infrastructure at WWTPs. However, it also introduces challenges in optimizing and designing the process for PFAS removal, which would otherwise be easier to achieve with specialized foam fractionators. Further studies are needed to focus on the practicality and feasibility of this integration, building on the existing knowledge that foam fractionation is an effective method for PFAS separation and concentration.

## References

Abbasian Chaleshtari, Z., Foudazi, R., 2022. A review on per- and polyfluoroalkyl substances (PFAS) remediation: separation mechanisms and molecular interactions. ACS ES&T Water 2 (12), 2258–2272. Available at https://doi.org/10.1021/acsestwater.2c00271.

Arvaniti, O.S., Andersen, H.R., Thomaidis, N.S., Stasinakis, A.S., 2014. Sorption of perfluorinated compounds onto different types of sewage sludge and assessment of its importance during wastewater treatment. Chemosphere 111, 405–411. Available at https://doi.org/10.1016/j.chemosphere.2014.03.087.

Behnami, A., Zoroufchi Benis, K., Pourakbar, M., Yeganeh, M., Esrafili, A., Gholami, M., 2024. Biosolids, an important route for transporting poly- and perfluoroalkyl substances from wastewater treatment plants into the environment: A systematic review. Science of The Total Environment 925, 171559. Available at https://doi.org/10.1016/j.scitotenv.2024.171559.

Brusseau, M.L., 2019. The influence of molecular structure on the adsorption of PFAS to fluid-fluid interfaces: Using QSPR to predict interfacial adsorption coefficients. Water Research 152, 148–158. Available at https://doi.org/10.1016/j.watres.2018.12.057.

Brusseau, M.L., 2023. QSPR-based prediction of air-water interfacial adsorption coefficients for nonionic PFAS with large headgroups. Chemosphere 340, 139960. Available at https://doi.org/10.1016/j.chemosphere.2023.139960.

Brusseau, M.L., Van Glubt, S., 2019. The influence of surfactant and solution composition on PFAS adsorption at fluid-fluid interfaces. Water Research 161, 17–26. Available at https://doi.org/10.1016/j.watres.2019.05.095.

Buckley, T., Karanam, K., Han, H., Vo, H.N.P., Shukla, P., Firouzi, M., et al., 2023. Effect of different co-foaming agents on PFAS removal from the environment by foam fractionation. Water Research 230, 119532. Available at https://doi.org/10.1016/j.watres.2022.119532.

Burns, D.J., Hinrichsen, H.M., Stevenson, P., Murphy, P.J.C., 2022. Commercial-scale remediation of per- and polyfluoroalkyl substances from a landfill leachate catchment using Surface-Active Foam Fractionation (SAFF®). Remediation Journal 32 (3), 139–150. Available at https://doi.org/10.1002/rem.21720.

Chen, S., Timmons, M.B., Bisogni, J.J., Aneshansley, D.J., 1993. Protein and its removal by foam fractionation. The Progressive Fish-Culturist 55 (2), 76–82. Available at https://doi.org/10.1577/1548-8640(1993)055<0076:PAIRBF>2.3.CO;2.

Coffin, E.S., Reeves, D.M., 2025. Application of novel continuous foam collection methods for removal of PFAS from cascade generated foam at wastewater treatment plants. Journal of Hazardous Materials 491, 137888. Available at https://doi.org/10.1016/j.jhazmat.2025.137888.

Coffin, E.S., Reeves, D.M., Cassidy, D.P., Danielson, N.D., Henry, M.A., 2024. PFAS distribution in cascade derived foam at wastewater treatment plants: the role of non-linear drainage, collapse induced enrichment, and implications for removal. Journal of Hazardous Materials 477, 135334. Available at https://doi.org/10.1016/j.jhazmat.2024.135334.

Collivignarelli, M.C., Baldi, M., Abbà, A., Caccamo, F.M., Carnevale Miino, M., Rada, E.C., et al., 2020. Foams in wastewater treatment plants: from causes to control methods. Applied Sciences 10 (8). Available at https://doi.org/10.3390/app10082716 Article 8.

de los Reyes III, F.L., 2010. Foaming. In: Seviour, R., Nielsen, P.H. (Eds.), Microbial Ecology of Activated Sludge. IWA Publishing, London, UK, pp. 215258.

European Parliament, Council of the European Union, 2024. Directive (EU) 2024/3019 of 27 November 2024 on urban wastewater treatment (recast) (Text with EEA relevance). Official Journal of the European Union. Available at https://eur-lex.europa.eu/legal-content/EN/TXT/?uri=CELEX:32024L3019.

Heard, J., Harvey, E., Johnson, B.B., Wells, J.D., Angove, M.J., 2008. The effect of filamentous bacteria on foam production and stability. Colloids and Surfaces B: Biointerfaces 63 (1), 21–26. Available at https://doi.org/10.1016/j.colsurfb.2007.10.011.

Horst, J., McDonough, J., Ross, I., Dickson, M., Miles, J., Hurst, J., et al., 2018. Water treatment technologies for PFAS: the next generation. Groundwater Monitoring & Remediation 38 (2), 13–23. Available at https://doi.org/10.1111/gwmr.12281.

Lemlich, R., 1972. Principles of foam fractionation and drainage. In: Lemlich, R. (Ed.), Adsorptive Bubble Separation Techniques. Academic Press, INC, pp. 33–51.

Li, R., Wu, Z., Wang, Y., Liu, W., 2014. Pilot study of recovery of whey soy proteins from soy whey wastewater using batch foam fractionation. Journal of Food Engineering 142, 201–209. Available at https://doi.org/10.1016/j.jfoodeng.2014.05.004.

Madsen, M.H., Møller, J.J., Ebbehøj, N.E., Nielsen, F., Severinsen, M.T., Jensen, J.F., et al., 2025. PFAS concentrations in the blood of Danish surfers. International Journal of Hygiene and Environmental Health 264, 114522. Available at https://doi.org/10.1016/j.ijheh.2025.114522.

Malovanyy, A., Forsén, E., Lihammar, R., 2025. Removal of per- and polyfluoroalkyl substances (PFAS) from municipal wastewater by foam fractionation. Water Research 268, 122660. Available at https://doi.org/10.1016/j.watres.2024.122660.

McCleaf, P., Kjellgren, Y., Ahrens, L., 2021. Foam fractionation removal of multiple per- and polyfluoroalkyl substances from landfill leachate. AWWA Water Science 3 (5). Available at https://doi.org/10.1002/aws2.1238.

Meegoda, J.N., Bezerra De Souza, B., Casarini, M.M., Kewalramani, J.A., 2022. A review of PFAS destruction technologies. International Journal of Environmental Research and Public Health 19 (24), 16397. Available at https://doi.org/10.3390/ijerph192416397.

Meng, P., Deng, S., Maimaiti, A., Wang, B., Huang, J., Wang, Y., et al., 2018. Efficient removal of perfluorooctane sulfonate from aqueous film-forming foam solution by aeration-foam collection. Chemosphere 203, 263–270. Available at https://doi.org/10.1016/j.chemosphere.2018.03.183.

Nguyen, H.T., McLachlan, M.S., Tscharke, B., Thai, P., Braeunig, J., Kaserzon, S., et al., 2022. Background release and potential point sources of per- and polyfluoroalkyl substances to municipal wastewater treatment plants across Australia. Chemosphere 293, 133657. Available at https://doi.org/10.1016/j.chemosphere.2022.133657.

Nguyen, D., Stults, J., Devon, J., Novak, E., Lanza, H., Choi, Y., et al., 2024. Removal of per- and polyfluoroalkyl substances from wastewater via aerosol capture. Journal of Hazardous Materials 465, 133460. Available at https://doi.org/10.1016/j.jhazmat.2024.133460.

Pujol, R., Duchene, P., Schetrite, S., Canler, J.P., 1991. Biological foams in activated sludge plants: characterization and situation. Water Research 25 (11), 1399–1404. Available at https://doi.org/10.1016/0043-1354(91)90118-A.

Schaefer, C.E., Hooper, J.L., Strom, L.E., Wu, K., Guelfo, J.L., 2023. Per- and polyfluoroalkyl substances in foam and dewatering streams at wastewater treatment plants. AWWA Water Science 5 (4), e1349. Available at https://doi.org/10.1002/aws2.1349.

Schaefer, C.E., Novak, E., Nguyen, D., 2025. Estimating potential for nonvolatile PFAS removal in aeration basins via near-surface aerosol capture. AWWA Water Science 7 (2), e70021. Available at https://doi.org/10.1002/aws2.70021.

Schramm, L.L., Wassmuth, F., 1994. Foams: basic principles. In: Schramm, L.L. (Ed.), Foams: Fundamentals and Applications in the Petroleum Industry Advances in Chemistry. American Chemistry Society, Washington, DC, pp. 3–45.

Schwichtenberg, T., Bogdan, D., Carignan, C.C., Reardon, P., Rewerts, J., Wanzek, T., et al., 2020. PFAS and dissolved organic carbon enrichment in surface water foams on a Northern U.S. Freshwater Lake. Environmental Science & Technology 54 (22), 14455–14464. Available at https://doi.org/10.1021/acs.est.0c05697.

Sherman-Bertinetti, S.L., Kostelnik, E.G., Gruber, K.J., Balgooyen, S., Remucal, C.K., 2024. Preferential partitioning of per- and polyfluoroalkyl substances (PFAS) and dissolved organic matter in freshwater surface microlayer and natural foam. Environmental Science & Technology 58 (29), 13099–13109. Available at https://doi.org/10.1021/acs.est.4c02285.

Smith, S.J., Wiberg, K., McCleaf, P., Ahrens, L., 2022. Pilot-scale continuous foam fractionation for the removal of per- and polyfluoroalkyl substances (PFAS) from Landfill leachate. ACS ES&T Water 2 (5), 841–851. Available at https://doi.org/10.1021/acsestwater.2c00032.

Smith, S.J., Keane, C., Ahrens, L., Wiberg, K., 2023. Integrated treatment of per- and polyfluoroalkyl substances in existing wastewater treatment plants—scoping the potential of foam partitioning. ACS ES&T Engineering. Available at https://doi.org/10.1021/acsestengg.3c00091.

Sochacki, M., Michorczyk, P., Vogt, O., 2024. Foam fractionation as an efficient method for the separation and recovery of surfactants and surface-inactive agents: state of the art. ACS Omega. Available at https://doi.org/10.1021/acsomega.4c08413 acsomega.4c08413.

Vo, P.H.N., Buckley, T., Xu, X., Nguyen, T.M.H., Rudolph, V., Shukla, P., 2023. Foam fractionation of per- and polyfluoroalkyl substances (PFASs) in landfill leachate using different cosurfactants. Chemosphere 310, 136869. Available at https://doi.org/10.1016/j.chemosphere.2022.136869.

Wang, Y., Ji, Y., Li, K., Huang, Q., 2023. Foam fractionation and electrochemical oxidation for the treatment of per- and polyfluoroalkyl substances (PFAS) in environmental water samples. Chemosphere 339, 139615. Available at https://doi.org/10.1016/j.chemosphere.2023.139615.

We, A.C.E., Zamyadi, A., Stickland, A.D., Clarke, B.O., Freguia, S., 2024a. A review of foam fractionation for the removal of per- and polyfluoroalkyl substances (PFAS) from aqueous matrices. Journal of Hazardous Materials 465, 133182. Available at https://doi.org/10.1016/j.jhazmat.2023.133182.

We, A.C.E., Stickland, A.D., Clarke, B.O., Freguia, S., 2024b. The role of suspended biomass in PFAS enrichment in wastewater treatment foams. Water Research 254, 121349. Available at https://doi.org/10.1016/j.watres.2024.121349.

We, A.C.E., Stickland, A.D., Clarke, B.O., Freguia, S., 2024c. Analysis of wastewater treatment plant data identifies the drivers of PFAS enrichment in foams. Water Research 266, 122397. Available at https://doi.org/10.1016/j.watres.2024.122397.

We, A.C.E., Stickland, A.D., Clarke, B.O., Freguia, S., 2025. PFAS removal through foam harvesting during wastewater aeration. Journal of Hazardous Materials 491, 137936. Available at https://doi.org/10.1016/j.jhazmat.2025.137936.

CHAPTER

# 13

# Emerging approaches for Per and polyfluoroalkyl substances removal by using clay minerals

*Monisha Manogharan*[1], *Sivagami Krishnasamy*[2], *and Sunantha Ganesan*[3]

[1]VIT School of Agricultural Innovations and Advanced Learning (VAIAL), Vellore Institute of Technology, Vellore, Tamil Nadu, India [2]School of Chemical Engineering, Vellore Institute of Technology, Vellore, Tamil Nadu, India [3]Department of Biomedical sciences, School of BioSciences and Technology, Vellore Institute of Technology, Vellore, Tamil Nadu, India

## 13.1 Introduction

Per- and polyfluoroalkyl substances (PFAS) are a substantial category of organofluorine synthetic aliphatic compounds, distinguished by their unique physicochemical characteristics (resistance to water, heat, oil, and microbial degradation). They can be widely used in many industrial sectors, such as firefighting foams, paints, cosmetics, nonstick cookware, textiles, and food packaging materials. PFAS are widely detected in several environmental matrices (air, water, and soil) (Glüge et al., 2020; Jian et al., 2018). PFAS are highly persistent in the environment because of the presence of C–F bonds, which offer remarkable stability and resistance to various degradation (Dias et al., 2024). PFAS are broadly classified as long-chain PFAS and short-chain PFAS based on carbon chain length. Long-chain PFAS, which contain eight or more carbons, include perfluorooctanoic acid (PFOA), perfluorooctane sulfonate (PFOS), and perfluorononanoic acid (PFNA). Short-chain PFAS consist of 4–7 carbon atoms, such as perfluorobutane sulfonic acid, perfluoroheptanoic acid, and GenX (Gebbink et al., 2017; Heydebreck et al., 2015).

PFAS have been used in various industrial applications since the 1950s; however, because of their persistence, these substances are widely detected in various environmental matrices like water, air, soil, and biota (ITRC, 2023). PFAS can enter the environment through various direct and indirect sources. According to Jala et al. (2023), PFAS may bioaccumulate in living organisms (Jala et al., 2023). This leads to biomagnification across the food chain

*Per- and Polyfluoroalkyl Substances (PFAS) as Environmental Contaminants*
DOI: https://doi.org/10.1016/B978-0-443-43856-1.00006-X

(Glaser et al., 2021). As a result, humans, particularly those who consume contaminated fish or livestock products like meat, eggs, and milk products, may experience increased exposure levels (Domingo and Nadal, 2019). Exposure to these chemicals leads to adverse health effects like cancer, liver and kidney damage, endocrine disruption, thyroid disease, fertility issues, neurological disease, asthma, and cardiovascular diseases (US EPA, 2024).

To date, most of the removal techniques have mainly focused on the removal of long-chain PFAS such as PFOS and PFOA due to their regulatory measures; however, the removal of short-chain PFAS is equally important and increasingly necessary. Nowadays, many treatment technologies are used to remediate PFAS in aquatic systems, which include biodegradation, advanced oxidation processes, reverse osmosis, adsorption, UV radiation, and nanofiltrations (Smith, 2023; Sörengård et al., 2020; Chen et al., 2020). However, these techniques are used to remove PFAS and also have some drawbacks, which include high cost and energy consumption, low removal efficiency, limited specificity, and difficulty in large-scale and real world applications (Lee et al., 2012). Adsorption is the most cost effective method for removing PFAS from polluted water, with selectivity and efficacy suitable for large-scale field applications (Schroeder et al., 2021).

Recently, clay minerals have gained attention in pollutant removal due to their eco-friendly, low-cost, and widespread availability. However, in their natural state, they are frequently ineffective at removing huge levels of contaminants from water. To enhance clay minerals' adsorption capacity, these minerals are treated with organic or inorganic chemicals that increase adsorption efficiency, selectivity, and surface characteristics (Mukhopadhyay et al., 2019; Sarkar et al., 2018). This chapter provides an overview of various types of clay minerals, properties, and applications. Furthermore, it highlights various clay mineral -based adsorbents, their adsorption mechanisms, the life cycle of used/spent clay mineral-based adsorbents, and the associated challenges.

## 13.2 Occurrence of per- and polyfluoroalkyl substances in various environment matrices

PFAS enter into the environment mainly via various direct and indirect sources such as wastewater discharge from fluorochemical manufacturing industries and the use of PFAS -containing products, such as aqueous film-forming foam (Sture et al., 2026). Indirect sources involve landfill leachates, agricultural activities, stormwater runoff, atmospheric runoff, and sewage sludge into the environment (Huset et al., 2011; Lang et al., 2017) (Fig. 13.1).

PFAS have become prevalent in environmental matrices such as air, water, soil, and biota (Bolan et al., 2021a; Zhang et al., 2020). PFAS compounds such as PFOA, PFOS, perfluorohexane sulfonate (PFHxS), perfluorohexanoic acid (PFHxA), and PFNA are frequently detected in drinking water (Koulini and Nambi, 2024), surface water (Koulini and Nambi, 2024; Stefano et al., 2023), groundwater (Koulini and Nambi, 2024; McMahon et al., 2022; Stefano et al., 2023), air, soil (Jia et al., 2023; Zhu et al., 2022), and sediments (Ahmadireskety et al., 2022). The concentrations of PFAS in various countries are presented in Table 13.1.

The bioaccumulative nature of PFAS poses adverse effects on human health and ecosystems, since these substances may accumulate in organisms and cause various issues (De Silva et al., 2020; Pandey et al., 2022). Human exposure to PFAS from contaminated water, food, and skin

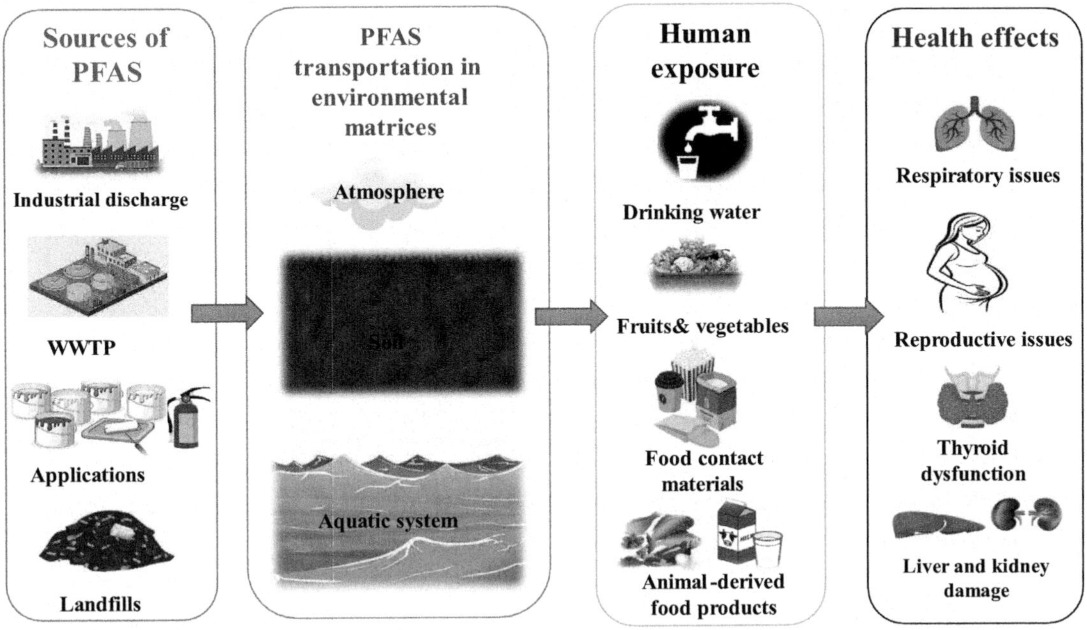


**FIGURE 13.1** Sources and transportation of per- and polyfluoroalkyl substances (PFAS) in various environmental matrices.

contact leads to severe health issues. The complexity of PFAS chemicals highlights the urgent need for comprehensive monitoring, regulatory actions, and efficient removal techniques (Alazaiza et al., 2025). Various remediation methods, including advanced oxidation, reverse osmosis, bioremediation, and filtration, are employed to eliminate PFAS, with PFAS with clay minerals proving effective due to their unique properties.

## 13.3 Properties of clay minerals

Clay minerals, such as montmorillonite, kaolinite, illite, etc., have some unique characteristics such as ion exchange capacities, electrical conductivity, and swelling capacity, which have been discussed below.

### 13.3.1 Ion exchange capacity

The exposure of clay minerals to environmental changes (erosion, deposition, and transport) causes chemical reactions, such as the creation of new novel clay minerals, ion exchange capacity, and mineral restructuring. One of the major properties of clay minerals is the cation exchange capacity (CEC), which enables such minerals to remain in liquid solution with their negative charge so that they can reverse this feature. The

**TABLE 13.1** Concentration level of per- and polyfluoroalkyl substances (PFAS) in soil and aquatic environment.

| Samples | Country | PFAS | Concentration level | References |
|---|---|---|---|---|
| Surface water and groundwater | India | PFHxA<br>Perfluorobutane sulfonic acid (PFBS)<br>perfluorobutanoic acid (PFBA)<br>Perfluorooctanoic acid (PFOA)<br>Perfluorooctane sulfonate (PFOS)<br>L-PFBS<br>PFOA | 0.40–4.70 ng $L^{-1}$<br><MQL–10.20 ng $L^{-1}$<br><MQL–9.20 ng $L^{-1}$<br><1.50–24.80 ng $L^{-1}$<br><1.20–13.90 ng $L^{-1}$<br>136.27 ng $L^{-1}$<br>77.61 ng $L^{-1}$ | Hariharan et al. (2023), Koulini and Nambi (2024), Sharma et al. (2016) |
| Lake water (Lake Victoria Gulf) | Kenya | PFOS<br>PFOA | <0.40–2.53 ng $L^{-1}$<br><0.40–11.70 ng/L | Orata et al. (2009) |
| Surface water (Little Arch Creek) | North America | PFOA<br>PFOS | 21.93 ng $L^{-1}$<br>2.64 ng $L^{-1}$ | Li et al. (2022) |
| Surface water and groundwater | Brazil | Perfluoroheptanoic acid (PFHpA) | 11–718 ng $L^{-1}$ | Stefano et al. (2023) |
| Freshwater lake (Kallinge) | Sweden | PFBA<br>PFBS<br>PFHpA<br>PFHxS | 110,000 ng $L^{-1}$<br>78,000 ng $L^{-1}$<br>63,000 ng $L^{-1}$<br>92,000 ng $L^{-1}$ | Greger and Landberg (2024) |
| Taihu Lake | China | PFOS<br>PFOA | 9.7–26.5 ng $L^{-1}$<br>21.7–25.4 ng $L^{-1}$ | Ma et al. (2024) |
| Sediment, military bases, carpet industrial sites, and airport areas | USA (Pensacola) | $\Sigma_{51}$PFAS | 0.04–0.48 ng $g^{-1}$ | Ahmadireskety et al. (2022) |
| Soil, aqueous film-forming foam-polluted sites | China (Shanghai) | $\Sigma_{11}$PFAS | 3.4–531.7 ng $g^{-1}$ | Cao et al. (2022) |
| Landfill | China (Hangzhou city) | $\Sigma_{17}$PFAS | 5.31–108 ng $g^{-1}$ | Xu et al. (2021) |

reversible exchange of cations between the clay surfaces and the solutions surrounding is known as ion exchange, which is determined by the negative charge of a clay and is measured by the CEC. CEC depends on the type of clay used and is 3–15 meq per 100 g with kaolinite and 100–150 meq per 100 g with vermiculite (Carroll, 1959). Swelling clays such as smectite have higher CEC since they have large surface areas and are able to retain water. Ionic strength, temperature, and pH have an effect on the cation exchange ability of ions such as $Ca^{2+}$, $Mg^{2+}$, and K (Yakovleva and Tu, 2012). Marine-based systems may have a reduced exchange capacity of their clays at higher salinity, and internal chemistry, such as the interlayer water content and layer charge, may be of greater significance in the ion adsorption and fixation processes.

### 13.3.2 Swelling behavior

Swelling clays, such as Na-MMT, show significant changes in volume when hydrated due to their electrostatic interactions between water molecules and clay layers mediated by sodium ions (Faisal et al., 2024). The swelling mechanism of Na-MMT is categorized into two types: crystalline swelling and osmotic swelling. Crystalline swelling occurs through stepwise formation of water molecules (1 W, 2 W, 3 W) in the interlayer space of the clay, and this process is stimulated by the exchangeable sodium cations located between the clay layers. Osmotic swelling occurs in dilute salt solution or pure water and follows initial crystalline swelling, leading to significant expansion caused by the difference in ion concentrations between the interlayer space and the bulk solution (Li et al., 2024). The extent of swelling is influenced by several factors, for example, in the presence of alkalies or the addition of byproducts such as ground granulated slag and fly ash have been shown significant reduce swelling in alkali-transformed clays (Sruthi et al., 2022). The adsorption of surfactants affects swelling in organoclays, showing a linear relationship between d-spacing and the organic-to-clay mass ratio. While hydration and electrochemical interactions primarily drive the swelling behavior of clay minerals, irreversible alterations may occur under certain conditions (Zeng et al., 2007). Understanding these dynamics is critical for geotechnical engineering applications and the production of polymer nanocomposites, which frequently need controlled swelling.

### 13.3.3 Adsorptive and low permeability

Clay minerals possess high adsorbent capacity and low permeability, allowing them to effectively interact with various substances, especially water and heavy metals, through ion exchange and surface complexation (Gu et al., 2019; Schoonheydt and Johnston, 2011). As the content of clay in the hydrocarbon reservoirs increases, the porosity and permeability decrease. They have a high specific surface area and CEC, which makes them helpful in the removal of heavy metals from wastewater (Jiang et al., 2013; Yang et al., 2024). Furthermore, clay minerals can be used as efficient adsorbents because of their low permeability, which can be a barrier to waste remediation. This shows the need to be cautious about using clay minerals with regard to environmental remediation and resource development.

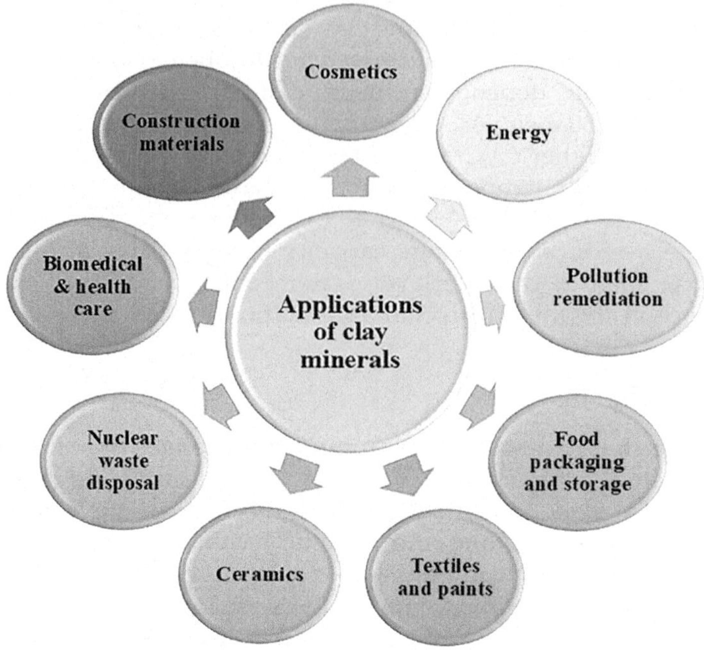


FIGURE 13.2 Various applications of clay minerals.

## 13.4 Applications of clay minerals

Clays and clay minerals are crucial industrial materials due to their unique physicochemical characteristics and diverse applications across various industries, as shown in Fig. 13.2. These minerals are used in various industrial sectors such as manufacturing ceramics, biomedical, health care, paper, cement, construction, soil fertility enhancement, paints, etc. Their large surface, high porosity, and layered structure make them ideal for creating various composites for a variety of applications (Nomicisio et al., 2023; Selvasudha et al., 2021). Clay minerals are also used as catalysts in many chemical reactions due to their altered surface and high surface area, which increases their catalytic activity (Cecilia and Jiménez-Gómez, 2021). Moreover, clay minerals are also used in construction materials, such as cement, to improve concrete strength and reduce $CO_2$ emissions. These diverse and innovative applications highlight the versatility and industrial significance of clay minerals. Clay minerals are also used in the removal of various environmental pollutants. However, the environmental impact of clay mining and processing continues to be a source of concern, emphasizing sustainable and eco-friendly use methods.

## 13.5 Classification of clay minerals

Clay minerals are a common class of materials recognized for their unique properties and diverse applications among various scientific disciplines. These minerals are mainly

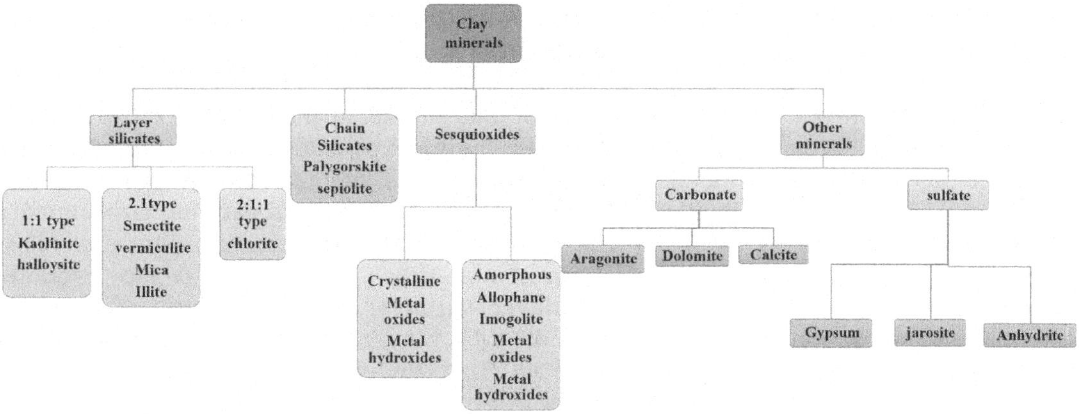


**FIGURE 13.3** Types of clay minerals.

formed through chemical weathering of rocks, particularly metamorphic and igneous rocks, and composed of hydrous phyllosilicates and characterized by a layered structure, such as tetrahedral and octahedral sheets. Clay minerals are distinguished by their layered structure, which consists of two or three sheets of tetrahedral and octahedral units. The main groups of clay minerals are illustrated in Fig. 13.3.

Clay minerals are a broad group of hydrous aluminosilicates distinguished by their layered structures, which are composed mostly of silica tetrahedra and alumina octahedra. These minerals are classified into four categories: kaolinite (1:1 layer structure), illite (2:1 layer structure), smectite (2:1 layer structure with expandable capabilities), and vermiculite (similar to smectite but with a higher degree of hydration) (Bibi et al., 2016; Kumari and Mohan, 2021).

Attapulgite and sepiolite are fibrous clay minerals from the palygorskite group. Sepiolite, with the chemical formula Mg 8($H_2O$) $_4[Si_4O_{10}]_3(OH)4{\cdot}8$ $H_2O$, is a magnesium-rich silicate with a 2:1-type trioctahedral layer chain structure. Sepiolite, a hydrous magnesium phyllosilicate clay mineral, is closely linked to palygorskite and belongs to the sepiolite-palygorskite group. Its fibrous appearance and channel-like crystal structure identify it as a one-dimensional functional clay mineral. (Wang, 2025). Sesquioxide clays arise in areas with significant rainfall as a result of severe leaching that removes silica and alumina from alumino–silicate clays, leaving less readily soluble iron and aluminum oxides and hydroxides (e.g., $Fe_2O_3$, $Fe(OH)_3$, Al $(OH)_3$). Sesquioxides, such as $Al_2O_3$ and $La_2O_3$, are compounds composed of two metal atoms (+3 oxidation state) and three oxygen atoms. Traditional clays, are nonadhesive, do not expand when exposed to water, and exhibit a low CEC. Other minerals, like carbonate and sulfate, are inorganic soil components. The major carbonate (calcite, dolomite, and aragonite) and sulfate (gypsum, jarosite, and anhydrite) minerals are most prevalent in arid and semiarid regions.

## 13.6 Removal of Per and polyfluoroalkyl substances by using clay minerals and modified clays

Naturally occurring clay minerals are a popular adsorbent material for PFAS removal because of their extensive availability, high CEC, and inexpensive cost, and the adsorption efficiency of various clay minerals has been shown in Table 13.2. Montmorillonite clay, especially calcium montmorillonite (CM), has been shown to bind to various environmental pollutants through active interlayer surfaces, basal planes, and edge sites (Orr et al., 2020; Wang et al., 2019; Wei et al., 2019; Zhang et al., 2014; Gu et al., 2010). However, the hydrophilic and negatively charged interlayer surfaces of natural phyllosilicate clay minerals, such as kaolinite, palygorskite, and MMT, hinder their ability to adsorb anionic PFAS via electrostatic attraction (Bolan et al., 2021b). The hydroxyl groups on the edges of aluminum-octahedral and silicon-tetrahedral layers play an important role in PFAS adsorption on minerals such as kaolinite and MMT (Wei et al., 2019). CM clays have been chemically treated with L-carnitine and choline to enhance their ability to adsorb hydrophobic and anionic pollutants, resulting in modified clays that exhibit increa sed sorption of chemicals, such as PFOA and PFOS, especially at acidic pH 2. These modifications have been deemed safe for humans and animals in multiple clinical trials (Wang et al., 2021; Phillips et al., 2019; Darlington et al., 2018).

Modified clay minerals (MC) demonstrate effective PFAS removal due to their strong adsorption capacity, and their enhanced removal efficiency arises from treatments, such as surfactants and ionic liquids (ILs) that increase interactions with PFAS. Surfactant-modified clay minerals removed more than 90% of PFAS from the snowmelt, which is better than traditional adsorbents like powder activated carbon (Jiang et al., 2024). The ability of MC to be regenerated and reused shows an effective alternative for real-world applications. Zhang et al. (2014) found that the PFOS adsorption was greater on kaolinite versus montmorillonite, where outer sphere complexes formed quickly, followed by slower ligand exchange. Humic acid coatings reduced PFOS adsorption due to competitive site occupation and increased electrostatic repulsion, emphasizing the importance of surface modifications in enhancing clay-based PFAS adsorption (Zhang et al., 2014).

Xiao et al. (2011) found that PFOS adsorbed more strongly than PFOA on kaolinite surfaces (pH 7.5, 22.2 °C), probably due to differences in functional groups and interactions with mineral surfaces. Because anionic PFAS have low pKa values, they become negatively charged under environmental pH circumstances, and their adsorption behaviour is affected by the electronegative head group structure and fluorinated carbon chain (Xiao et al., 2011). Kaolinite effectively adsorbs PFAS, with adsorption kinetics influenced by the molecular structure, including chain length and functional groups. Longer chain PFAS exhibit greater sorption due to surface aggregation on kaolinite, while shorter chains are less restricted (Loganathan and Wilson, 2022). Furthermore, it has been shown that it can be supplemented with activated carbon to increase its removal efficiency, such as a mixture of electrokinetic treatment and an iron-coated activated carbon permeable reactive barrier was used to eliminate PFOA in kaolinite soil and reached up to 59.55 % (Ganbat et al., 2024). Moreover, the partition coefficient of different PFAS compounds is related to kaolinite's exchange capacity of cations, suggesting that its mineralogical characteristics are critical to adsorption (Ahmad et al., 2023). In general, kaolinite is an attractive and affordable substitute for PFAS

**TABLE 13.2** Removal of per- and polyfluoroalkyl substances (PFAS) in aqueous systems by using clay and clay minerals.

| Types of clay and clay minerals used | Modifiers used | PFAS | Initial dosage of PFAS | Clay and clay mineral dose level (g $L^{-1}$) | Partition coefficient | pH | Adsorption or removal efficiency | Reference |
|---|---|---|---|---|---|---|---|---|
| Montmorillonite | - | Perfluorooctane sulfonate (PFOS) | 1.0 mg $L^{-1}$ | 5.0 | 0.07 L $g^{-1}$ | 7.0 | 0.05 mg $g^{-1}$ | Zhang et al. (2014) |
| | – | PFOS | 10–1200 μg $L^{-1}$ | 0.4 | 0.17 L $g^{-1}$ | 7.0 | 0.29–0.31 mg-$g^{-1}$ | Zhao et al. (2014) |
| | HDTMA | PFOS | 50 mg $L^{-1}$ | 4.0 | 16.25 L $g^{-1}$ | 3.0 | 890 mg $g^{-1}$ | Zhou et al. (2013) |
| | Quaternary ammonium | Perfluorooctanoic acid (PFOA) and PFOS | 0–500 μg $L^{-1}$ | 0.5 | – | 7.6 | – | Ray et al. (2019) |
| | HDTMA and 3-indole-acetic acid | PFOA | 10 mg $L^{-1}$ | 2.2 | – | 3.0 | 100% | Tian et al. (2016) |
| | HDTMA and poly-4-vinylpyridine-co-styrene mixture | PFOA | 10 mg $L^{-1}$ | 2.2 | – | – | >90% defluori-nation | Chen et al. (2019) |
| Kaolinite | – | PFOS | 1.0 mg $L^{-1}$ | 5.0 | 0.13 L $g^{-1}$ | 7.0 | 0.08 mg $g^{-1}$ | Zhang et al. (2014) |
| | – | PFOA | 10–1200 μg $L^{-1}$ | 0.4 | 0.54 L $g^{-1}$ | 7.0 | 0.10–0.11 mg-$g^{-1}$ | Zhao et al. (2014) |
| Na-mica | Octadecylamine | PFOS | 10 μg $mL^{-1}$ | 6.25 | – | – | 100% | Martín et al. (2018) |
| | Octadecylamine | PFOA | 10 μg $mL^{-1}$ | 6.25 | 1.61 L $g^{-1}$ | – | 91% | |
| Nnao-$Fe_3O_4$ | Starch stabilized | PFOA | 0–4000 μg $L^{-1}$ | 8 mg $L^{-1}$ | 85.66 L $g^{-1}$ | 6.8 | 62.5 mg $g^{-1}$ | Gong et al. (2016) |
| Goethite | – | PFOS | 0.12–8 mg $L^{-1}$ | – | 2.8–8.9 L $kg^{-1}$ | 4.0 | $1.4 \times 10^{-3}$ mole-cules $nm^{-2}$ | Johnson et al. (2007) |
| | – | PFOS | 5–1000 μg $L^{-1}$ | 0.417 | 0.011 L $m^{-2}$ | 7.0 | 2.4 μg $m^{-2}$ | Tang et al. (2010) |

(*Continued*)

TABLE 13.2 (Continued)

| Types of clay and clay minerals used | Modifiers used | PFAS | Initial dosage of PFAS | Clay and clay mineral dose level ($g\ L^{-1}$) | Partition coefficient | pH | Adsorption or removal efficiency | Reference |
|---|---|---|---|---|---|---|---|---|
| Boehmite | – | PFOA | 50–1000 $\mu g\ L^{-1}$ | 5.0 | 0.002 $L\ m^{-2}$ | 7.0 | 0.633 $\mu g\ m^{-2}$ | Wang et al. (2012) |
| Alumina | – | PFOA | 40–400 $\mu g\ L^{-1}$ | 10.0 | 0.001 $L\ m^{-2}$ | 4.3 | 0.16 $\mu g\ m^{-2}$ | Wang and Shih. (2011) |
| Iranian bentonite | Arquad® 2HT-75 (IBA)<br>IBA modified with palmitic acid (IBAP) | PFOS<br>PFOS | 5–300 $mg\ L^{-1}$ | 2.5 | 1.49 $L\ g^{-1}$<br>0.89 $L\ g^{-1}$ | 7.8<br>7.8 | 99.9 $mg\ g^{-1}$<br>78.1 $mg\ g^{-1}$ | Khodabakhshloo et al. (2021) |

treatment in polluted sites. However, the effectiveness of kaolinite can be different when it comes to specific PFAS chemicals and the nature of the environmental circumstances, which implies that additional research should be conducted to ensure the maximum use of this material in different scenarios.

Zhao et al. (2014) investigated the adsorption of four perfluoroalkyl acids (PFOS, PFHxS, PFOA, and PFHxA) on minerals MMT, kaolinite, and hematite. The study found that PFOS and PFHxS exhibited greater sorption compared to PFOA and PFHxA, with hematite demonstrating the highest sorptive capacity, followed by kaolinite and montmorillonite (Zhao et al., 2014). Sorption declined with increasing ionic strength and pH, and especially in the sorption of PFOS. The paper demonstrates that surface complexation and hydrogen bonding play a crucial role in influencing the PFOS sorption, which is applicable in the interpretation of the movement and persistence of PFAAs in samples such as sediment and groundwater. PFAS adsorbed on IL-modified clays and was affected by the CEC of the clay and the structure of PFAS (Dong et al., 2024). Long-chain ILs enhance the kinetics and efficiency of PFOA and PFOS in different chemistries of water (Dong et al., 2021). The silylated clay (QAG-ZrMT) is more able to adsorb less hydrophobic PFAS compared to traditional organoclays and highly regenerates, which is an improvement over existing PFAS-polluted water treatment methods (Dong et al., 2024).

PFAS can be removed by using halloysite minerals, a type of naturally occurring clay mineral, because of their nanotubular shape, which has an enormous surface area and numerous adsorption sites (Matusik, 2016). Alterations in halloysite could increase its adsorption level, and it is likely to be more efficient than conventional adsorbents. Although no research has investigated the ability of halloysite to eliminate PFAS, its performance in eliminating other contaminants, such as ammonium and phosphates, implies the same (Leszczyński, 2021; Saki et al., 2019). The way in which halloysite can be optimized to be used in the removal of PFAS, as well as specific modifications to its interaction with PFAS and integration with advanced remediation techniques, should be a subject of future research (Dudarko et al., 2024; Kim et al., 2024). Magnetic surfactant-modified clay effectively absorbed more than 90% of PFAS from the snowmelt. The study also emphasized the modified clay's regenerability and reusability, indicating its potential for industrial uses (Jiang et al., 2024). Gianni et al. (2025) found that nanotubular clay minerals such as halloysite and imogolite can effectively adsorb both PFAS and herbicides, demonstrating their versatility (Gianni et al., 2025). Molecular simulations revealed that stable PFAS-mineral complexes form due to electrostatic interactions. While these experiments demonstrate the efficacy of several clay minerals in PFAS removal, questions remain about their scalability and long-term stability in real-world applications. More study is needed to optimize these technologies for larger-scale environmental remediation efforts.

## 13.7 Mechanisms involved in the removal of per- and polyfluoroalkyl substances by using clay-based adsorbents

Clay-based adsorbents effectively remove various pollutants like PFAS from water through various mechanisms, such as electrostatic attraction, hydrophobic interactions,

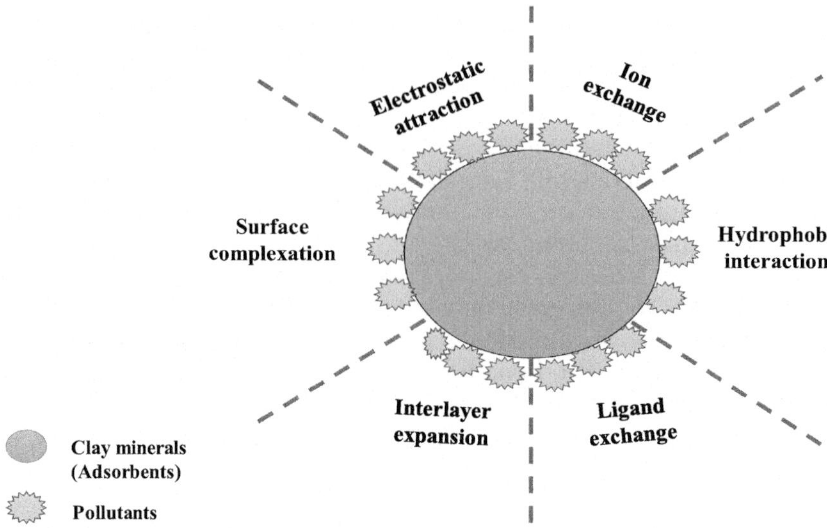


**FIGURE 13.4** Pollutants removal mechanisms.

and ligand exchange, as shown in Fig. 13.4. The surface alterations of clay minerals improve their efficiency as PFAS removals. The adsorption procedure depends on the properties of adsorbents, where PFAS has been modified as lipophobic and hydrophobic owing to the low polarizability of fluorine atoms. The longer the carbon chain of PFAS compounds, the more lipophilic they are and the greater the adsorbent affinity, resulting in the ability of the negatively charged PFAS to interact with the negatively charged surfaces of the mineral in conditions where the hydrophobicity is sufficient. Dong et al. (2024) explain that the longer carbon chains of PFAS compounds demonstrated high hydrophilicity, which leads to the ability of the negatively charged PFAS to interact with the negatively charged surfaces of the mineral in the longer carbon chains (Dong et al., 2024). Organoclays modified with fluorophilic amide groups and long alkyl-chain surfactant, which enhance these interactions and increase long-chain PFAS adsorption efficiency. Adsorbent interactions can also be promoted in cases involving negatively charged mineral surfaces adsorbing negatively charged PFAS compounds, provided that they are hydrophobic (Riess, 2005). The length of carbon chains increases their hydrophobicity. Long alkyl-chain surfactants of organoclay minerals are hydrophobic because of the presence of fluorophilic amide groups, which promote the adsorption of short-chain PFAS by hydrophobic effects (Stebel et al., 2019). However, the electrostatic interactions have a strong impact on the PFAS adsorption onto clay-based materials, particularly when they are functionalized with ILs and assisted by the cation exchange properties of the clays (Dong et al., 2024).

Tang et al. (2010) discovered that the sorption of PFAS on mineral surfaces is influenced by solution properties, such as pH, as well as ionic strength. Acidification of the surface raises the electrostatic attraction of anionic PFOS (Tang et al., 2010). The role that ion exchange plays is important, particularly in modified clays, where they possess anion-

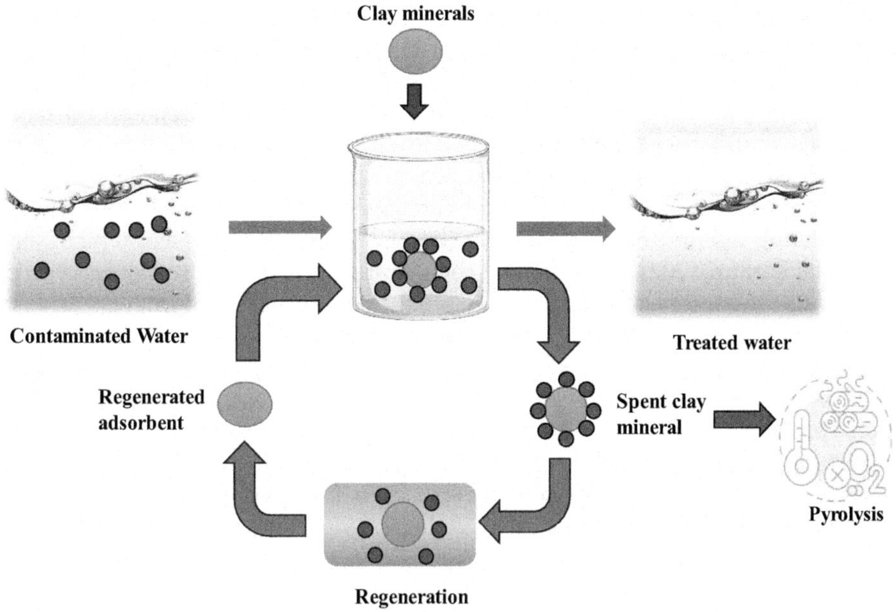


FIGURE 13.5 Regeneration and reusability of used/spent clay minerals.

exchange abilities that boost the removal of PFAS. Pseudo-second-order models can indicate chemisorption when adsorption kinetics are used with pseudo-second-order models (Liu et al., 2024; Valencia et al., 2023), and the intraparticle diffusion model can be used to describe PFAS adsorption in clay-based adsorbents, implying that internal diffusion is a rate-limiting factor (Valencia et al., 2023).

## 13.7.1 Life cycle of spent clay mineral adsorbents

The regeneration and reusability of spent clay mineral adsorbents for contaminants in water treatment are crucial for sustainable environmental management. Regeneration and reusability of used/spent clay mineral-based adsorbents for the removal of pollutants are shown in Fig. 13.5. Recent studies highlight various regeneration methods, including thermal degradation, chemical treatment, supercritical extraction, photocatalytic activity, and biodegradation for these adsorbent materials, ensuring their efficacy in removing pollutants while minimizing waste (Momina et al., 2018). Thermal regeneration of natural clay minerals like montmorillonite can be thermally regenerated at high temperature (e.g., 600°C) successfully mineralizing adsorbed contaminants into harmless byproducts such as $CO_2$ and water vapor with over 85% efficiency after multiple cycles (Dawood et al., 2025). Pyrolysis at 600°C is more effective and environmentally friendly than heat treatment at 1000°C in treating PFAS-loaded adsorbents, allowing clay materials to be regenerated. The literature on PFAS elimination during pyrolysis is limited. Kim et al. (2015) discovered that biochar produced at 300°C and 700°C did not significantly affect PFAS concentrations, however, pyrolysis at 850°C resulted in undetectable PFAS levels (Kim et al., 2015). Kundu et al. (2021) reported more than 90% removal of PFOA and PFOS from a biosolid pyrolysed at 600°C under specific circumstances, with the resulting biochar adsorbing considerable amounts of long- and short-chain PFAS from contaminated water (Kundu et al., 2021). Furthermore, organoclay/metal-based nanoparticles as photocatalysts could assist in PFAS breakdown under light irradiation, providing an alternative to high-temperature incineration. Chemical regeneration of used clay mineral adsorbents improves their pollutant removal capacity. This method involves utilizing acids, alkalies, or chelating agents to remove impurities from the clay surface, which allows reuse (Rezai and Allahkarami, 2023). While the regeneration of clay adsorbents is a promising approach for water treatment, it is crucial to assess the environmental impacts associated with the regeneration processes. Techniques such as thermal and microwave regeneration, although effective, may require considerable energy inputs that could negate some ecological benefits. Future study is needed to optimize these methods to improve their sustainability. Chemical regeneration, on the other hand, considerably aids in the restoration of clay adsorbent functionality, but there are still worries regarding the environmental impact of the regeneration chemicals used. Future research should also seek to improve these processes in order to decrease potential negative consequences while increasing operating efficiency.

## 13.8 Challenges and factors affecting per- and polyfluoroalkyl substances adsorption by clay-based adsorbents

There are several challenges with the removal of PFAS with the help of clay minerals because of the high water solubility of PFAS, and stable C–F bonds make them persistent in the environment. Although ILs have been used to treat clay minerals such as MMT to improve their adsorption, this process is difficult and very expensive. The effects of water pH and competing ions, and the molecular structure of PFAS, on adsorption efficiency affect the rate of adsorption. IL clays show potential PFAS removal by altering their surface and expanding the interlayer, but their effectiveness depends on PFAS chain length and water chemistry conditions (Dong et al., 2021; Zhou et al., 2024). Moreover, it is difficult to regenerate and reuse clay adsorbents because of the possible saturation of adsorbent sites, which will result in wastage. Though clay minerals provide a cost-effective solution to the problem of PFAS, their drawbacks suggest that other materials, such as metal–organic frameworks (MOFs) with enhanced adsorption properties and reuse, can be used with clay in tackling the issue of PFAS contamination (Gomri et al., 2024; Zhang et al., 2024). The efficacy of PFAS removal using clay-based adsorbing agents depends on several factors, such as the clay composition, the chemical structure of PFAS, and environmental factors such as pH, ionic strength, natural organic matter, competing ions, and temperature. Short-chain PFAS are difficult to eliminate compared to long-chain because they are more mobile and soluble. Modified clay-based adsorbents, particularly those containing ILs or surfactants, have high potential for removing PFAS by increasing the number of electrostatic and hydrophobic interactions. The loading of modifying agents and the adsorption capacity of clay in general are also influenced by the CEC of the clay.

### 13.8.1 Temperature

The temperature in the case of removal of PFAS using clay-based adsorbents significantly influences the efficacy of PFAS removal, as rising temperatures raise the energy of the molecular kinetics, therefore raising the adsorption rate and capacity. It has been found that ionically modified liquid clays show temperature-dependent adsorption behaviors, particularly for long-chain PFAS, because of stronger hydrophobic affinities and low water viscosity (Dong et al., 2021). Besides, the temperature may influence the structural characteristics of clays, including the CEC, as well as surface area, which in turn has an additional effect on adsorption performance (Jiang et al., 2024). Future studies on the thermodynamics of PFAS removal by using natural or organically modified clays are required due to the limited data.

### 13.8.2 pH

PFAS removal efficiency using clay-based adsorbents is affected by the pH factor, which affects the surface charge and ionization condition of the adsorbent. It has been shown that the adsorption capacity of the modified clays depends on the pH, with the highest capacity increasing electrostatic and hydrophobic interactions (Higgins and Luthy, 2006). PFAS compounds are highly adsorbed at low pH through the protonation or deprotonation of functional groups. In a study conducted by Deng et al. (2012), it was found that carbon nanotubes becomes negatively charged at pH above 3.4, and hence the adsorption performance becomes poor. The same outcome was noticed with clay-based adsorbents such as goethite, in which low pH forms a positively charged surface, enhancing electrostatic interactions (Deng et al., 2012; Tang et al., 2010).

### 13.8.3 Organic matter

The literature shows that the existence of natural organic matter (NOM) has a significant effect on the elimination of PFAS on clay-based adsorbents. NOM is able to compete with PFAS when adsorbing on clay surfaces and decreases the overall removal (Jiang et al., 2024), and the surface charge and hydrophobicity of clay adsorbents alter the adsorption of PFAS (Dong et al., 2024). As an example, dissolved organic matter (OM) can cause short-chain PFAS adsorption to decrease due to the presence of dissolved OM in aqueous environments, since both PFAS and dissolved OM are hydrophobic interacting agents. OM is a negatively charged substance that has hydrophobic compounds like aromatic groups, which cause electrostatic repulsion with PFAS. Although NOM is capable of adsorbing PFAS, the hydrophobic interactions of OM–PFAS in aqueous conditions are scarcely comprehended (Campos Pereira et al., 2018).

### 13.8.4 Impacts of competitive ions and ionic strength

The ionic strength and competing ions have a significant effect on the adsorption of PFAS by clay minerals. Competitive adsorption takes place when ions such as sodium and hydrogen change the electrostatic forces between the PFAS and clay surfaces, affecting adsorption performance. Long chains usually win over short chains because they have stronger hydrophobic interactions, especially when monovalent cations are present, which minimizes the repulsion effect of electrostatic forces. Furthermore, high ionic strength can improve the adsorption of short chain PFAS by weakening repulsive forces and facilitating their attachment to clay surfaces (Xiao et al., 2011). Also, competing ions such as $Mg^{2+}$ impact certain adsorption at clay edges, which underscores the significance of the geochemical conditions to forecast PFAS retention (Orucoglu et al., 2022). However, while ionic strength and competing ions can enhance adsorption, they can also generate higher rates of desorption and complicate the dynamics of retention in polluted areas (Umeh et al., 2023).

## 13.9 Conclusion and future aspects

Emerging pollutants such as PFAS contamination are becoming a serious concern throughout the world. Because of their physicochemical properties, PFAS can be used in various industrial sectors. The widespread nature of these substances in environmental matrices, such as soil, air, water, and biota, causes serious issues to both humans and wildlife. PFAS are extremely stable and resistant to water, oil, heat, and microbial degradation, leading to their long-term presence in water, air, and soil. This contributes to widespread contamination. PFAS enter the human body through various sources, such as contaminated water, food, and food packaging materials, and through inhalation and dermal contact. Exposure to these compounds leads to several harmful health effects, including cancer, thyroid disease, liver damage, kidney damage, cardiovascular disease, neurotoxicity, and reproductive issues. Natural and engineered or modified clay minerals showed significant potential for the adsorption of PFAS in an aqueous environment. Their layered structure, high surface area, ion exchange capacities, swelling behavior, and surface charge characteristics, nucleophilic and electrophilic sites help them to effectively bind with long-chain PFAS, such as PFOA and PFOS. Among various case studies related to clay minerals, montmorillonite and kaolinite have shown promising removal efficiency towards PFAS, which can be enhanced through physical and chemical modification, such as thermal activation, polymer functionalization, and surface intercalation. Moreover, when compared to natural clay minerals, modified clay minerals showed high removal efficiency of PFAS. However, various molecular structure of PFAS, environmental conditions such as high temperature, OM, ionic strength, pH, and competing ions, and physicochemical properties of clay minerals affects the removal efficiency of clay-based adsorbents towards PFAS.

- Additionally, desorption, regeneration, and long-term stability of clay-based adsorbents need more attention to ensure sustainable remediation.
- Compared to long-chain PFAS removal, short-chain PFAS is more difficult because of its high mobility and desorption efficiency. Therefore future research should focus on the development of engineered clay-based materials specifically designed for the removal of PFAS.
- The development of engineered clay factionalized with tailored hybrid material (organo modifiers), for example, MOF–clay hybrid materials, biochar–clay composites, and

chitosan–clay composites, may improve the specificity and adsorption capacity for the removal of diverse groups of PFAS, including short-chain PFAS and emerging analogs.

- Advanced spectroscopic and modeling approaches, for example, synchrotron-based research and molecular dynamics simulations helps to understand the interactions between PFAS and clay minerals, and also allow for the development of more selective and efficient adsorbents.
- Future research is needed to understand how factors such as competing ions, temperature, and OMs that affect the removal of PFAS by using clay minerals, mainly for optimizing large-scale wastewater treatment and real-world applications.
- Uses of chemical modifiers (e.g., surfactants) to enhance the performance of clay minerals may be toxic to microbes and aquatic life, and future research must focus on the development of nontoxic and biobased, eco-friendly, and cost-effective alternatives.

## References

Ahmad, A., Tian, K., Tanyu, B., Foster, G.D., 2023. Effect of clay mineralogy on the partition coefficients of perfluoroalkyl substances. ACS ES&T Water 3 (9), 2899–2909. Available at https://doi.org/10.1021/acsestwater.3c00105.

Ahmadireskety, A., Da Silva, B.F., Robey, N.M., Douglas, T.E., Aufmuth, J., Solo-Gabriele, H.M., et al., 2022. Per- and polyfluoroalkyl substances (PFAS) in street sweepings. Environmental Science & Technology 56 (10), 6069–6077. Available at https://doi.org/10.1021/acs.est.1c03766.

Alazaiza, M.Y.D., Alzghoul, T.M., Ramu, M.B., Abu Amr, S.S., Abushammala, M.F.M., 2025. PFAS contamination and mitigation: a comprehensive analysis of research trends and global contributions. Case Studies in Chemical and Environmental Engineering 11, 101127. Available at https://doi.org/10.1016/j.cscee.2025.101127.

Bibi, I., Icenhower, J., Niazi, N.K., Naz, T., Shahid, M., Bashir, S., 2016. Clay minerals. Environmental Materials and Waste. Elsevier, pp. 543–567. Available at https://doi.org/10.1016/B978-0-12-803837-6.00021-4.

Bolan, N., Sarkar, B., Vithanage, M., Singh, G., Tsang, D.C.W., Mukhopadhyay, R., et al., 2021a. Distribution, behaviour, bioavailability and remediation of poly- and per-fluoroalkyl substances (PFAS) in solid biowastes and biowaste-treated soil. Environment International 155, 106600. Available at https://doi.org/10.1016/j.envint.2021.106600.

Bolan, N., Sarkar, B., Yan, Y., Li, Q., Wijesekara, H., Kannan, K., et al., 2021b. Remediation of poly- and perfluoroalkyl substances (PFAS) contaminated soils – to mobilize or to immobilize or to degrade? Journal of Hazardous Materials 401. Available at https://doi.org/10.1016/j.jhazmat.2020.123892.

Campos Pereira, H., Ullberg, M., Kleja, D.B., Gustafsson, J.P., Ahrens, L., 2018. Sorption of perfluoroalkyl substances (PFASs) to an organic soil horizon – effect of cation composition and pH. Chemosphere 207. Available at https://doi.org/10.1016/j.chemosphere.2018.05.012.

Cao, L., Xu, W., Wan, Z., Li, G., Zhang, F., 2022. Occurrence of PFASs and its effect on soil bacteria at a fire-training area using PFOS-restricted aqueous film-forming foams. IScience 25 (4), 104084. Available at https://doi.org/10.1016/j.isci.2022.104084.

Carroll, D., 1959. Ioon exchange in clays and other minerals. Bulletin of the Geological Society of America 70 (6). Available at https://doi.org/10.1130/0016-7606(1959)70[749:IEICAO]2.0.CO;2.

Cecilia, J.A., Jiménez-Gómez, C.P., 2021. Catalytic applications of clay minerals and hydrotalcites. Catalysts 11 (1), 68. Available at https://doi.org/10.3390/catal11010068.

Chen, Z., Li, C., Gao, J., Dong, H., Chen, Y., Wu, B., et al., 2020. Efficient reductive destruction of perfluoroalkyl substances under self-assembled micelle confinement. Environmental Science and Technology 54 (8). Available at https://doi.org/10.1021/acs.est.9b06599.

Chen, Z., Tian, H., Li, H., Li, J.S., Hong, R., Sheng, F., et al., 2019. Application of surfactant modified montmorillonite with different conformation for photo-treatment of perfluorooctanoic acid by hydrated electrons. Chemosphere 235. Available at https://doi.org/10.1016/j.chemosphere.2019.07.032.

Darlington, R., Barth, E., McKernan, J., 2018. The challenges of PFAS remediation. Military Engineer. Vol. 110, Number 712.

Dawood, I., Zyoud, A.H., Zyoud, S., Amireh, A., Zyoud, S.H., Kim, T.W., 2025. Sustainable water treatment using thermally stable natural clay: dual adsorption–thermolysis approach for organic pollutants and nitrate removal. Scientific Reports 15 (1). Available at https://doi.org/10.1038/s41598-025-12148-7.

De Silva, A.O., Armitage, J.M., Bruton, T.A., Dassuncao, C., Heiger-Bernays, W., Hu, X.C., et al., 2020. PFAS exposure pathways for humans and wildlife: a synthesis of current knowledge and key gaps in understanding. Environmental Toxicology and Chemistry 40 (3), 631–657. Available at https://doi.org/10.1002/etc.4935.

Deng, S., Zhang, Q., Nie, Y., Wei, H., Wang, B., Huang, J., et al., 2012. Sorption mechanisms of perfluorinated compounds on carbon nanotubes. Environmental Pollution 168, 138–144. Available at https://doi.org/10.1016/j.envpol.2012.03.048.

Dias, D., Bons, J., Kumar, A., Kabir, M., Liang, H., 2024. Forever chemicals, per-and polyfluoroalkyl substances (PFAS), in lubrication. Lubricants 12 (4), 114. Available at https://doi.org/10.3390/lubricants12040114.

Domingo, J.L., Nadal, M., 2019. Human exposure to per- and polyfluoroalkyl substances (PFAS) through drinking water: a review of the recent scientific literature. Environmental Research 177, 108648. Available at https://doi.org/10.1016/j.envres.2019.108648.

Dong, Q., Min, X., Huo, J., Wang, Y., 2021. Efficient sorption of perfluoroalkyl acids by ionic liquid-modified natural clay. Chemical Engineering Journal Advances 7, 100135. Available at https://doi.org/10.1016/j.ceja.2021.100135.

Dong, Q., Min, X., Zhao, Y., Wang, Y., 2024. Adsorption of per- and polyfluoroalkyl substances (PFAS) by ionic liquid-modified clays: effect of clay composition and PFAS structure. Journal of Colloid and Interface Science 654, 925–934. Available at https://doi.org/10.1016/j.jcis.2023.10.112.

Dudarko, O., Budnyak, T., Tkachenko, O., Agback, T., Agback, P., Bonnet, B., et al., 2024. Removal of poly- and perfluoroalkyl substances from natural and wastewater by tailored silica-based adsorbents. ACS ES&T Water 4 (4), 1303–1314. Available at https://doi.org/10.1021/acsestwater.3c00408.

Faisal, H.M.N., Gaikwad, H.K., Katti, K.S., Katti, D.R., 2024. In-silico investigation of mechanical behavior of hydrated Na-montmorillonite tactoid. Colloids and Surfaces A: Physicochemical and Engineering Aspects 694, 134117. Available at https://doi.org/10.1016/j.colsurfa.2024.134117.

Ganbat, N., Altaee, A., Hamdi, F.M., Zhou, J., Chowdhury, M.H., Zaidi, S.J., et al., 2024. PFOA remediation from kaolinite soil by electrokinetic process coupled with activated carbon/iron coated activated carbon - permeable reactive barrier. Journal of Contaminant Hydrology 267, 104425. Available at https://doi.org/10.1016/j.jconhyd.2024.104425.

Gebbink, W.A., van Asseldonk, L., van Leeuwen, S.P.J., 2017. Presence of emerging per- and polyfluoroalkyl substances (PFASs) in river and drinking water near a fluorochemical production plant in the Netherlands. Environmental Science & Technology 51 (19), 11057–11065. Available at https://doi.org/10.1021/acs.est.7b02488.

Gianni, E., Scholtzová, E., Tyrologou, P., Couto, N., Pospíšil, M., Papoulis, D., et al., 2025. Nanotubular clay minerals for simultaneous sorption of pesticides and PFCAs: a molecular simulation study. Integrated Environmental Assessment and Management 21 (6), 1403–1415. Available at https://doi.org/10.1093/inteam/vjae038.

Glaser, D., Lamoureux, E., Opdyke, D., LaRoe, S., Reidy, D., Connolly, J., 2021. The impact of precursors on aquatic exposure assessment for PFAS: insights from bioaccumulation modeling. Integrated Environmental Assessment and Management 17 (4), 705–715. Available at https://doi.org/10.1002/ieam.4414.

Glüge, J., Scheringer, M., Cousins, I.T., DeWitt, J.C., Goldenman, G., Herzke, D., et al., 2020. An overview of the uses of per- and polyfluoroalkyl substances (PFAS). Environmental Science: Processes & Impacts 22 (12), 2345–2373. Available at https://doi.org/10.1039/D0EM00291G.

Gomri, C., Benkhaled, B.T., Cretin, M., Semsarilar, M., 2024. Adsorbent material used for the treatment of per- and poly-fluoroalkyl substances (PFAS): a short review. Macromolecular Chemistry and Physics 225 (11). Available at https://doi.org/10.1002/macp.202400012.

Gong, Y., Wang, L., Liu, J., Tang, J., Zhao, D., 2016. Removal of aqueous perfluorooctanoic acid (PFOA) using starch-stabilized magnetite nanoparticles. Science of the Total Environment 562. Available at https://doi.org/10.1016/j.scitotenv.2016.03.100.

Greger, M., Landberg, T., 2024. Removal of PFAS from water by aquatic plants. Journal of Environmental Management 351, 119895. Available at https://doi.org/10.1016/j.jenvman.2023.119895.

Gu, S., Kang, X., Wang, L., Lichtfouse, E., Wang, C., 2019. Clay mineral adsorbents for heavy metal removal from wastewater: a review. Environmental Chemistry Letters 17 (2), 629–654. Available at https://doi.org/10.1007/s10311-018-0813-9.

Gu, X., Evans, L.J., Barabash, S.J., 2010. Modeling the adsorption of Cd (II), Cu (II), Ni (II), Pb (II) and Zn (II) onto montmorillonite. Geochimica et Cosmochimica Acta 74 (20). Available at https://doi.org/10.1016/j.gca.2010.07.016.

Hariharan, G., S, G., R, R.S., R, D.R., P, R., R, R., 2023. Early detection of emerging persistent perfluorinated alkyl substances (PFAS) along the east coast of India. Science of the Total Environment 902, 166155. Available at https://doi.org/10.1016/j.scitotenv.2023.166155.

Heydebreck, F., Tang, J., Xie, Z., Ebinghaus, R., 2015. Alternative and legacy perfluoroalkyl substances: differences between European and Chinese river/estuary systems. Environmental Science & Technology 49 (14), 8386–8395. Available at https://doi.org/10.1021/acs.est.5b01648.

Higgins, C.P., Luthy, R.G., 2006. Sorption of perfluorinated surfactants on sediments. Environmental Science & Technology 40 (23), 7251–7256. Available at https://doi.org/10.1021/es061000n.

Huset, C.A., Barlaz, M.A., Barofsky, D.F., Field, J.A., 2011. Quantitative determination of fluorochemicals in municipal landfill leachates. Chemosphere 82 (10), 1380–1386. Available at https://doi.org/10.1016/j.chemosphere.2010.11.072.

Interstate Technology & Regulatory Council, 2023. PFAS — per and polyfluoroalkyl substances.

Jala, A., Adye, D.R., Borkar, R.M., 2023. Occurrence and risk assessments of per- and polyfluoroalkyl substances in tea bags from India. Food Control 151, 109812. Available at https://doi.org/10.1016/j.foodcont.2023.109812.

Jia, X., Li, X., Zhou, L., Hui, Y., Li, W., Cai, Y., et al., 2023. Variations of the level, profile, and distribution of PFAS around POSF manufacturing facilities in China: an overlooked source of PFCA. Environmental Science & Technology 57 (13), 5264–5274. Available at https://doi.org/10.1021/acs.est.2c08995.

Jian, J.-M., Chen, D., Han, F.-J., Guo, Y., Zeng, L., Lu, X., et al., 2018. A short review on human exposure to and tissue distribution of per- and polyfluoroalkyl substances (PFASs). Science of the Total Environment 636, 1058–1069. Available at https://doi.org/10.1016/j.scitotenv.2018.04.380.

Jiang, T., Pervez, M.N., Ilango, A.K., Ravi, Y.K., Zhang, W., Feldblyum, J.I., et al., 2024. Magnetic surfactant-modified clay for enhanced adsorption of mixtures of per- and polyfluoroalkyl substances (PFAS) in snowmelt: improving practical applicability and efficiency. Journal of Hazardous Materials 471, 134390. Available at https://doi.org/10.1016/j.jhazmat.2024.134390.

Jiang, Y., Zhang, C., Deng, H., Wang, M., Luo, M., 2013. Influences of clay minerals on physical properties of low permeability and tight sandstones. Xinan Shiyou Daxue Xuebao/Journal of Southwest Petroleum University 35 (6). Available at https://doi.org/10.3863/j.issn.1674-5086.2013.06.005.

Johnson, R.L., Anschutz, A.J., Smolen, J.M., Simcik, M.F., Lee Penn, R., 2007. The adsorption of perfluorooctane sulfonate onto sand, clay, and iron oxide surfaces. Journal of Chemical and Engineering Data 52 (4). Available at https://doi.org/10.1021/je060285g.

Khodabakhshloo, N., Biswas, B., Moore, F., Du, J., Naidu, R., 2021. Organically functionalized bentonite for the removal of perfluorooctane sulfonate, phenanthrene and copper mixtures from wastewater. Applied Clay Science 200. Available at https://doi.org/10.1016/j.clay.2020.105883.

Kim, J.H., Ok, Y.S., Choi, G.H., Park, B.J., 2015. Residual perfluorochemicals in the biochar from sewage sludge. Chemosphere 134. Available at https://doi.org/10.1016/j.chemosphere.2015.05.012.

Kim, N., Elbert, J., Shchukina, E., Su, X., 2024. Integrating redox-electrodialysis and electrosorption for the removal of ultra-short- to long-chain PFAS. Nature Communications 15 (1), 8321. Available at https://doi.org/10.1038/s41467-024-52630-w.

Koulini, G.V., Nambi, I.M., 2024. Occurrence of forever chemicals in Chennai waters, India. Environmental Sciences Europe 36 (1), 60. Available at https://doi.org/10.1186/s12302-024-00881-1.

Kumari, N., Mohan, C., 2021. Basics of clay minerals and their characteristic properties. Clay and Clay Minerals. IntechOpen. Available at https://doi.org/10.5772/intechopen.97672.

Kundu, S., Patel, S., Halder, P., Patel, T., Hedayati Marzbali, M., Pramanik, B.K., et al., 2021. Removal of PFASs from biosolids using a semi-pilot scale pyrolysis reactor and the application of biosolids derived biochar for the removal of PFASs from contaminated water. Environmental Science: Water Research and Technology 7 (3). Available at https://doi.org/10.1039/d0ew00763c.

Lang, J.R., Allred, B.M., Field, J.A., Levis, J.W., Barlaz, M.A., 2017. National estimate of per- and polyfluoroalkyl substance (PFAS) release to U.S. Municipal Landfill Leachate. Environmental Science & Technology 51 (4), 2197–2205. Available at https://doi.org/10.1021/acs.est.6b05005.

Lee, Y., Lo, S., Kuo, J., Hsieh, C., 2012. Decomposition of perfluorooctanoic acid by microwaveactivated persulfate: effects of temperature, pH, and chloride ions. Frontiers of Environmental Science and Engineering in China 6 (1). Available at https://doi.org/10.1007/s11783-011-0371-x.

Leszczyński, J., 2021. Removal of ammonium ions from aqueous solutions using weathered halloysite. Materials 14 (16), 4359. Available at https://doi.org/10.3390/ma14164359.

Li, B., Li, C., Gui, Y., Zhan, H., Gu, Y., Yu, M., et al., 2024. Understanding structural anisotropy and mechanical properties of Na-montmorillonite with crystalline swelling and uniaxial deformation under different hydration degrees. Computers and Geotechnics 169. Available at https://doi.org/10.1016/j.compgeo.2024.106200.

Li, X., Fatowe, M., Cui, D., Quinete, N., 2022. Assessment of per- and polyfluoroalkyl substances in Biscayne Bay surface waters and tap waters from South Florida. Science of The Total Environment 806, 150393. Available at https://doi.org/10.1016/j.scitotenv.2021.150393.

Liu, F., Pignatello, J.J., Sun, R., Guan, X., Xiao, F., 2024. A comprehensive review of novel adsorbents for per- and polyfluoroalkyl substances in water. ACS ES&T Water 4 (4), 1191–1205. Available at https://doi.org/10.1021/acsestwater.3c00569.

Loganathan, N., Wilson, A.K., 2022. Adsorption, structure, and dynamics of short- and long-chain PFAS molecules in kaolinite: molecular-level insights. Environmental Science and Technology 56 (12). Available at https://doi.org/10.1021/acs.est.2c01054.

Ma, K., Lu, Y., Zhang, Y., Zhang, Y., 2024. Trend of PFAS concentrations and prediction of potential risks in Taihu Lake of China by AQUATOX. Environmental Research 251, 118707. Available at https://doi.org/10.1016/j.envres.2024.118707.

Martín, J., Orta, M.del M., Medina-Carrasco, S., Santos, J.L., Aparicio, I., Alonso, E., 2018. Removal of priority and emerging pollutants from aqueous media by adsorption onto synthetic organo-funtionalized high-charge swelling micas. Environmental Research 164. Available at https://doi.org/10.1016/j.envres.2018.03.037.

Matusik, J., 2016. Halloysite for Adsorption and Pollution Remediation, pp. 606–627. https://doi.org/10.1016/B978-0-08-100293-3.00023-6.

McMahon, P.B., Tokranov, A.K., Bexfield, L.M., Lindsey, B.D., Johnson, T.D., Lombard, M.A., et al., 2022. Perfluoroalkyl and polyfluoroalkyl substances in groundwater used as a source of drinking water in the Eastern United States. Environmental Science & Technology 56 (4), 2279–2288. Available at https://doi.org/10.1021/acs.est.1c04795.

Momina, Shahadat, M., Isamil, S., 2018. Regeneration performance of clay-based adsorbents for the removal of industrial dyes: a review. RSC Advances Vol. 8 (Number 43). Available at https://doi.org/10.1039/c8ra04290j.

Mukhopadhyay, R., Adhikari, T., Sarkar, B., Barman, A., Paul, R., Patra, A.K., et al., 2019. Fe-exchanged nano-bentonite outperforms Fe3O4 nanoparticles in removing nitrate and bicarbonate from wastewater. Journal of Hazardous Materials 376. Available at https://doi.org/10.1016/j.jhazmat.2019.05.025.

Nomicisio, C., Ruggeri, M., Bianchi, E., Vigani, B., Valentino, C., Aguzzi, C., et al., 2023. Natural and synthetic clay minerals in the pharmaceutical and biomedical fields. Pharmaceutics 15 (5), 1368. Available at https://doi.org/10.3390/pharmaceutics15051368.

Orata, F., Quinete, N., Werres, F., Wilken, R.-D., 2009. Determination of perfluorooctanoic acid and perfluorooctane sulfonate in Lake Victoria Gulf Water. Bulletin of Environmental Contamination and Toxicology 82 (2), 218–222. Available at https://doi.org/10.1007/s00128-008-9543-1.

Orr, A.A., He, S., Wang, M., Goodall, A., Hearon, S.E., Phillips, T.D., et al., 2020. Insights into the interactions of bisphenol and phthalate compounds with unamended and carnitine-amended montmorillonite clays. Computers and Chemical Engineering 143. Available at https://doi.org/10.1016/j.compchemeng.2020.107063.

Orucoglu, E., Grangeon, S., Gloter, A., Robinet, J.C., Madé, B., Tournassat, C., 2022. Competitive adsorption processes at clay mineral surfaces: a coupled experimental and modeling approach. ACS Earth and Space Chemistry 6 (1). Available at https://doi.org/10.1021/acsearthspacechem.1c00323.

Pandey, K., Rai, S., Kuril, S., Singh, P., Rajput, M.S., Pilli, S., et al., 2022. Environmental effects of per- and polyfluoroalkyl substances exposure. Current Developments in Biotechnology and Bioengineering. Elsevier, pp. 15–33. Available at https://doi.org/10.1016/B978-0-323-99906-9.00002-4.

Phillips, T.D., Wang, M., Elmore, S.E., Hearon, S., Wang, J.-S., 2019. NovaSil Clay for the Protection of Humans and Animals from Aflatoxins and Other Contaminants. Clays and Clay Minerals 67, 99–110. Available at 10.1007/s42860-019-0008-x.

Ray, J.R., Shabtai, I.A., Teixidó, M., Mishael, Y.G., Sedlak, D.L., 2019. Polymer-clay composite geomedia for sorptive removal of trace organic compounds and metals in urban stormwater. Water Research 157. Available at https://doi.org/10.1016/j.watres.2019.03.097.

Rezai, B., Allahkarami, E., 2023. Desorption/regeneration of adsorbents and their performance. Sustainable Remediation Technologies for Emerging Pollutants in Aqueous Environment. Available at https://doi.org/10.1016/B978-0-443-18618-9.00021-8.

Riess, J.G., 2005. Understanding the fundamentals of perfluorocarbons and perfluorocarbon emulsions relevant to in vivo oxygen delivery. Artificial Cells, Blood Substitutes, and Immobilization Biotechnology Vol. 33 (Number 1). Available at https://doi.org/10.1081/BIO-200046659.

Saki, H., Alemayehu, E., Schomburg, J., Lennartz, B., 2019. Halloysite nanotubes as adsorptive material for phosphate removal from aqueous solution. Water 11 (2), 203. Available at https://doi.org/10.3390/w11020203.

Sarkar, B., Rusmin, R., Ugochukwu, U.C., Mukhopadhyay, R., Manjaiah, K.M., 2018. Modified clay minerals for environmental applications. Modified Clay and Zeolite Nanocomposite Materials: Environmental and Pharmaceutical Applications. Available at https://doi.org/10.1016/B978-0-12-814617-0.00003-7.

Schoonheydt, R.A., Johnston, C.T., 2011. The surface properties of clay minerals. Layered Mineral Structures and their Application in Advanced Technologies. Mineralogical Society of Great Britain and Ireland, pp. 335–370. Available at https://doi.org/10.1180/EMU-notes.11.10.

Schroeder, T., Bond, D., Foley, J., 2021. PFAS soil and groundwater contamination: via industrial airborne emission and land deposition in SW Vermont and Eastern New York State, USA. Environmental Science: Processes and Impacts 23 (2). Available at https://doi.org/10.1039/d0em00427h.

Selvasudha, N., Dhanalekshmi, U.-M., Krishnaraj, S., Harish Sundar, Y., Sri Durga Devi, N., Sarathchandiran, I., 2021. Multifunctional clay in pharmaceuticals. Clay Science and Technology. IntechOpen. Available at https://doi.org/10.5772/intechopen.92408.

Sharma, B.M., Bharat, G.K., Tayal, S., Larssen, T., Bečanová, J., Karásková, P., et al., 2016. Perfluoroalkyl substances (PFAS) in river and ground/drinking water of the Ganges River basin: emissions and implications for human exposure. Environmental Pollution 208, 704–713. Available at https://doi.org/10.1016/j.envpol.2015.10.050.

Smith, S., 2023. Innovative Treatment Technologies for PFAS-Contaminated Water - Utilizing Foam Partitioning and Exploring Electrochemical Oxidation. Swedish University of Agricultural Sciences. Available at https://doi.org/10.54612/a.6p8i5lbs8l vol. 53.

Sörengård, M., Östblom, E., Köhler, S., Ahrens, L., 2020. Adsorption behavior of per- and polyfluoralkyl substances (PFASs) to 44 inorganic and organic sorbents and use of dyes as proxies for PFAS sorption. Journal of Environmental Chemical Engineering 8 (3). Available at https://doi.org/10.1016/j.jece.2020.103744.

Sruthi, P.L., Reddy, P.H.P., Moghal, A.A.B., 2022. Swelling behavior of alkali transformed kaolinitic clays treated with flyash and ground granulated blast furnace slag. Indian Geotechnical Journal 52 (1), 145–160. Available at https://doi.org/10.1007/s40098-020-00489-1.

Stebel, E.K., Pike, K.A., Nguyen, H., Hartmann, H.A., Klonowski, M.J., Lawrence, M.G., et al., 2019. Absorption of short-chain to long-chain perfluoroalkyl substances using swellable organically modified silica. Environmental Science: Water Research and Technology 5 (11). Available at https://doi.org/10.1039/c9ew00364a.

Stefano, P.H.P., Roisenberg, A., D'Anna Acayaba, R., Roque, A.P., Bandoria, D.R., Soares, A., et al., 2023. Occurrence and distribution of per-and polyfluoroalkyl substances (PFAS) in surface and groundwaters in an urbanized and agricultural area, Southern Brazil. Environmental Science and Pollution Research 30 (3), 6159–6169. Available at https://doi.org/10.1007/s11356-022-22603-x.

Sture, M., Valdersnes, S., Boitsov, S., Grøsvik, B.E., Ali, A., 2026. Source and fate of ultra-short-chain PFAS in water and biota from an AFFF impacted site. Journal of Chromatography B 1268, 124848. Available at https://doi.org/10.1016/j.jchromb.2025.124848.

Tang, C.Y., Shiang Fu, Q., Gao, D., Criddle, C.S., Leckie, J.O., 2010. Effect of solution chemistry on the adsorption of perfluorooctane sulfonate onto mineral surfaces. Water Research 44 (8), 2654–2662. Available at https://doi.org/10.1016/j.watres.2010.01.038.

Tian, H., Gao, J., Li, H., Boyd, S.A., Gu, C., 2016. Complete defluorination of perfluorinated compounds by hydrated electrons generated from 3-indole-acetic-acid in organomodified montmorillonite. Scientific Reports 6. Available at https://doi.org/10.1038/srep32949.

Umeh, A.C., Hassan, M., Egbuatu, M., Zeng, Z., Al Amin, M., Samarasinghe, C., et al., 2023. Multicomponent PFAS sorption and desorption in common commercial adsorbents: kinetics, isotherm, adsorbent dose, pH, and index ion and ionic strength effects. Science of the Total Environment 904. Available at https://doi.org/10.1016/j.scitotenv.2023.166568.

US EPA, 2024. Our Current Understanding of the Human Health and Environmental Risks of PFAS. US EPA.

Valencia, A., Ordonez, D., Sadmani, A.H.M.A., Reinhart, D., Chang, N.B., 2023. Comparing the removal and fate of long and short chain per- and polyfluoroalkyl substances (PFAS) during surface water treatment via specialty adsorbents. Journal of Water Process Engineering 56. Available at https://doi.org/10.1016/j.jwpe.2023.104345.

Wang, F., Shih, K., 2011. Adsorption of perfluorooctanesulfonate (PFOS) and perfluorooctanoate (PFOA) on alumina: influence of solution pH and cations. Water Research 45 (9). Available at https://doi.org/10.1016/j.watres.2011.03.007.

Wang, F., Liu, C., Shih, K., 2012. Adsorption behavior of perfluorooctanesulfonate (PFOS) and perfluorooctanoate (PFOA) on boehmite. Chemosphere 89 (8). Available at https://doi.org/10.1016/j.chemosphere.2012.06.071.

Wang, L., 2025. Sepiolite. Manual of Mineral Material Science. Springer Nature Singapore, pp. 482–487. Available at https://doi.org/10.1007/978-981-97-7558-3_108.

Wang, M., Orr, A.A., He, S., Dalaijamts, C., Chiu, W.A., Tamamis, P., et al., 2019. Montmorillonites can tightly bind glyphosate and paraquat reducing toxin exposures and toxicity. ACS Omega 4 (18). Available at https://doi.org/10.1021/acsomega.9b02051.

Wang, M., Orr, A.A., Jakubowski, J.M., Bird, K.E., Casey, C.M., Hearon, S.E., Tamamis, P., Phillips, T.D., 2021. Enhanced adsorption of per- and polyfluoroalkyl substances (PFAS) by edible, nutrient-amended montmorillonite clays. Water Research 188, 116534. Available at 10.1016/j.watres.2020.116534.

Wei, Z., Xu, T., Zhao, D., 2019. Treatment of per- and polyfluoroalkyl substances in landfill leachate: status, chemistry and prospects. Environmental Science: Water Research and Technology Vol. 5 (Number 11). Available at https://doi.org/10.1039/c9ew00645a.

Xiao, F., Zhang, X., Penn, L., Gulliver, J.S., Simcik, M.F., 2011. Effects of monovalent cations on the competitive adsorption of perfluoroalkyl acids by kaolinite: experimental studies and modeling. Environmental Science and Technology 45 (23). Available at https://doi.org/10.1021/es202524y.

Xu, C., Song, X., Liu, Z., Ding, X., Chen, H., Ding, D., 2021. Occurrence, source apportionment, plant bioaccumulation and human exposure of legacy and emerging per- and polyfluoroalkyl substances in soil and plant leaves near a landfill in China. Science of The Total Environment 776, 145731. Available at https://doi.org/10.1016/j.scitotenv.2021.145731.

Yakovleva, A.A., Tu, V.D., 2012. Ion exchange on clay minerals from some deposits of Irkutsk region. Russian Journal of Applied Chemistry 85 (3). Available at https://doi.org/10.1134/S1070427212030044.

Yang, X., Zhou, Y., Hu, J., Zheng, Q., Zhao, Y., Lv, G., et al., 2024. Clay minerals and clay-based materials for heavy metals pollution control. Science of the Total Environment 954, 176193. Available at https://doi.org/10.1016/j.scitotenv.2024.176193.

Zeng, Q.H., Paul, D.R., Yu, A.B., Lu, G.Q.M., 2007. The swelling behavior of organoclays. Key Engineering Materials 353–358. Available at https://doi.org/10.4028/www.scientific.net/kem.353-358.2239.

Zhang, B., He, Y., Huang, Y., Hong, D., Yao, Y., Wang, L., et al., 2020. Novel and legacy poly- and perfluoroalkyl substances (PFASs) in indoor dust from urban, industrial, and e-waste dismantling areas: the emergence of PFAS alternatives in China. Environmental Pollution 263. Available at https://doi.org/10.1016/j.envpol.2020.114461.

Zhang, K., Cheng, P., Liu, Y., Xia, S., 2024. Efficient removal of per- and polyfluoroalkyl substances by a metal-organic framework membrane with high selectivity and stability. Water Research 265, 122276. Available at https://doi.org/10.1016/j.watres.2024.122276.

Zhang, R., Yan, W., Jing, C., 2014. Mechanistic study of PFOS adsorption on kaolinite and montmorillonite. Colloids and Surfaces A: Physicochemical and Engineering Aspects 462. Available at https://doi.org/10.1016/j.colsurfa.2014.09.019.

Zhao, L., Bian, J., Zhang, Y., Zhu, L., Liu, Z., 2014. Comparison of the sorption behaviors and mechanisms of perfluorosulfonates and perfluorocarboxylic acids on three kinds of clay minerals. Chemosphere 114. Available at https://doi.org/10.1016/j.chemosphere.2014.03.098.

Zhou, H., Jiang, L., Yang, Y., Zhang, H., Zhang, S., Wang, J., et al., 2024. Efficient microwave regeneration of iron-modified Na-based montmorillonite and Na-based attapulgite for enhanced adsorption of tetracycline in water. Journal of Environmental Chemical Engineering 12 (4). Available at https://doi.org/10.1016/j.jece.2024.113229.

Zhou, Q., Pan, G., Shen, W., 2013. Enhanced sorption of perfluorooctane sulfonate and Cr(VI) on organo montmorillonite: influence of solution pH and uptake mechanism. Adsorption 19 (2–4), 709–715. Available at https://doi.org/10.1007/s10450-013-9496-5.

Zhu, Q., Qian, J., Huang, S., Li, Q., Guo, L., Zeng, J., et al., 2022. Occurrence, distribution, and input pathways of per- and polyfluoroalkyl substances in soils near different sources in Shanghai. Environmental Pollution 308, 119620. Available at https://doi.org/10.1016/j.envpol.2022.119620.

CHAPTER

# 14

# Zeolites for adsorptive and catalytic per- and polyfluoroalkyl substances remediation

*Foteini Zormpa and Erisa Saraçi*

**Institute of Catalysis Research and Technology, Karlsruhe Institute of Technology, Eggenstein-Leopoldshafen, Germany**

## 14.1 Introduction

Per- and polyfluoroalkyl substances (PFAS) are a large group of synthetic chemicals that contain fluorine atoms bonded to carbon chains, which present extreme resistance to heat, water, and oil. In perfluoroalkyl substances, all hydrogen atoms in the carbon chain are replaced by fluorine, while polyfluoroalkyl substances contain partially fluorinated chains (Buck et al., 2011). These properties have made PFAS valuable in a wide range of industrial and consumer products, including nonstick cookware, water-repellent fabrics, firefighting foams, and food packaging. However, their presence in the environment has raised serious concerns due to their widespread occurrence in drinking water, food, wildlife, and even in the human body (Evich et al., 2022). PFAS exposure poses significant health hazards, including increased risks of liver and kidney cancers, as well as potential impacts on fetal development and childhood growth (Brown et al., 2020; Blake et al., 2018; Anderko and Pennea, 2020). As a result, global regulatory actions have sought to restrict PFAS production and set maximum permissible levels in both air and water. These regulations focus mainly on two of the most frequently detected PFAS compounds: perfluorooctanoic acid (PFOA) and perfluorooctanesulfonic acid (PFOS) (Abunada et al., 2020), whose chemical structures are presented in Fig. 14.1.

In recent years, research on PFAS remediation has grown, encompassing a variety of emerging technologies (Rahman et al., 2014; Lott et al., 2025). In aqueous media, the majority of these studies focus on PFAS adsorption (Park et al., 2020; Gagliano et al., 2020), photocatalytic conversion (Xu et al., 2017; Li et al., 2020), electrocatalytic oxidation (Gomez-Ruiz et al., 2017; Lin et al., 2021), persulfate-based oxidation (Lei et al., 2020; Yang et al., 2020), as well as membrane technologies (Tsai et al., 2010; Murray et al., 2019).

*Per- and Polyfluoroalkyl Substances (PFAS) as Environmental Contaminants*
DOI: https://doi.org/10.1016/B978-0-443-43856-1.00011-3

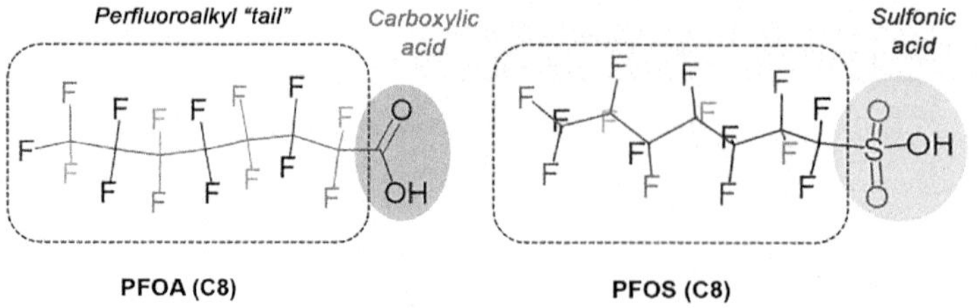


FIGURE 14.1 Chemical structure of perfluorooctanoic acid (PFOA) and perfluorooctanesulfonic acid (PFOS) molecules.

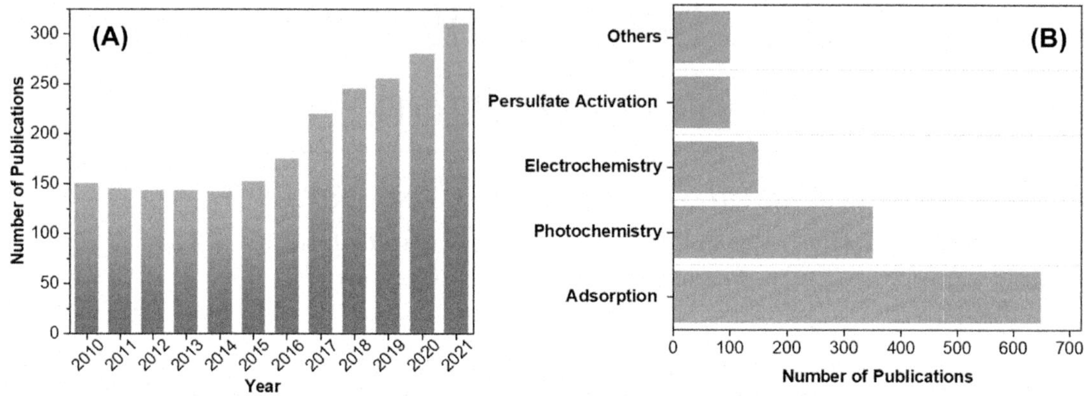


FIGURE 14.2 Number of per- and polyfluoroalkyl substances remediation publications by (A) year from 2010–21, and (B) by technology type used during recent years. Source: *Adapted from Schrama, F.R.A., Massimi, S.E., Dooley, M.R., Trewyn, B.G., Vyas, S., Richards, R.M., 2024. Light-driven interfaces for PFAS detection and destruction. RSC Applied Interfaces. 1, 833–845. https://doi.org/10.1039/D1034LF00171K, licensed under CC BY 3.0. © The Royal Society of Chemistry.*

According to a recent review by Schrama et al. (2024), adsorption and photocatalysis are the predominant approaches discussed in the current literature on PFAS remediation. The statistical data illustrating this trend are presented in Fig. 14.2A and B.

Among the commonly studied adsorbents are carbon materials (activated carbon [AC], carbon nanotubes, and graphene-based materials), ion exchange resins, and zeolites. Zeolites are exceptionally distinguished by their highly tunable structures, chemical and thermal stability, regeneration potential, and cost-effective synthesis. Their pore size and surface chemistry can be optimized for different PFAS chain lengths, enhancing both adsorption capacity and degradation efficiency. For example, silica-rich zeolites with high Si/Al ratios form hydrophobic surfaces that interact favorably with the perfluorinated tails of perfluoroalkyl carboxylic acids (PFCAs), such as PFOA and PFOS (Van den Bergh et al., 2020). Additionally, surface functionalization with metals such as iron (Fe) can catalyze the degradation of PFAS into smaller molecules when exposed to light (Qian et al., 2020).

This chapter examines the role of zeolitic materials in PFAS remediation within aqueous media. In the first part, the main features of zeolites, such as structure, topologies, and acidity/basicity, are described, offering an overview of the materials' functionalities. Specific examples of important applications where zeolites are used either as adsorbents or catalysts are mentioned, highlighting the importance of these materials in industrial and environmental applications. The main focus of the chapter is the role of zeolites in PFAS remediation via adsorption and photocatalytic degradation technologies. Details about the adsorption and photocatalytic reaction mechanisms are discussed, as well as the importance of surface chemistry, pore structure, and active metal species. The chapter concludes with a critical perspective on the current challenges and future directions of zeolite-based PFAS remediation technologies.

## 14.2 Zeolites: structural and functional overview

### 14.2.1 Structure and composition

Zeolites are crystalline, microporous materials composed of $SiO_4$ and $AlO_4$ tetrahedra as primary building units, which are interconnected through bridging oxygen atoms to form secondary building units (SBUs). These SBUs assemble into more complex structures, including sodalite cages and hexagonal prisms, ultimately resulting in three-dimensional frameworks (Fig. 14.3) (Kulprathipanja, 2010; Kordala and Wyszkowski, 2024). The term "zeolite" was first introduced by Axel F. Cronstedt, who observed that when a mineral sample was heated, it resembled the action of water boiling. As a result, he named the mineral "zeolite," derived from the Greek words zein (to boil) and lithos (stone). This boiling effect was later attributed to the release of water molecules, trapped within the material's pores, upon heating (Colella and Gualtieri, 2007). Zeolite frameworks vary in pore sizes, shapes, and channel connectivity. Conventional zeolites with 8- to 12-membered ring pores typically have pore sizes ranging from 0.3 to 1.2 nm, and surface areas from 300 to 800 $m^2 g^{-1}$ (Moliner and Martínez, 2015; Bae et al., 2010; Möller and Bein, 2013). Each zeolite framework (topology) is described by a unique three-letter code, generated by International Zeolite Association. Common topologies include chabazite (CHA), zeolite A (LTA), ZSM-5 (MFI), mordenite (MOR), ferrierite (FER), zeolites Y and X (FAU), and Beta (BEA) (Colella and Wise, 2014).

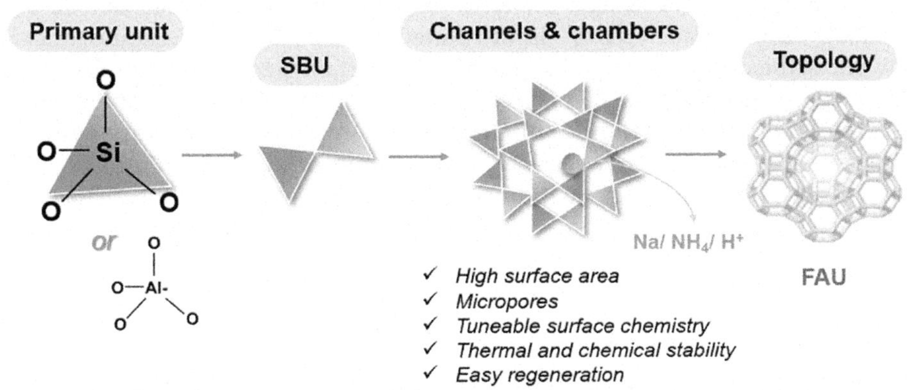


FIGURE 14.3 Schematic representation of zeolitic structure.

All-silica zeolites consist of only $SiO_4$ tetrahedra, have a well-defined structure, and a highly hydrophobic surface. Silicalite-1 is the most well-known all-silica version of ZSM-5 zeolite, characterized by high thermal and chemical stability, features that make it effective in many catalytic and adsorption processes (Zhang et al., 2017; Serrano et al., 2007; Chen et al., 2022). Beta zeolite is available in both all-silica and aluminum-containing forms, characterized by a 12-membered ring pore structure with relatively large pore openings. These structural features make it well-suited for various catalytic applications, such as alkylation (Jin et al., 2022; Rastegar et al., 2020), cracking (Khalil et al., 2021; Mensah et al., 2021), and the transformation of aromatic hydrocarbons (Yan et al., 2023). Substitution of $Si^{4+}$ by $Al^{3+}$ in the framework results in a negative charge, usually compensated by cations, such as $Na^+$ and $K^+$, or $NH_4^+$, or $H^+$. The presence of protons forms Si–OH–Al hydroxyl groups, which behave as Brønsted acid sites (Lukyanov et al., 2014). In addition to Brønsted acid sites, these zeolites also have Lewis acidity, which can originate from the framework Al species and/or framework modifications with other metals (Ravi et al., 2020; Luo et al., 2016).

### 14.2.2 Adsorptive properties

Adsorption is mainly a reversible process in which atoms, ions, or molecules from a gas, liquid, or dissolved solid accumulate on the surface of a solid or liquid at specific sites. The accumulated molecules are called adsorbates (or sorbates). Adsorption can be physical (physisorption) or chemical (chemisorption), depending on the nature of interactions (ranging from weak van der Waals forces to strong chemical bonds) between the adsorbate and the adsorbent (Masel, 1996). An effective adsorbent should have a high surface area due to the existence of porous networks (to enhance adsorption capacity), while also being selective, regenerable, stable under operating conditions, and cost-effective. Zeolites, due to their well-defined microporous structures, ion-exchange capability, thermal and chemical stability, and low-cost synthesis, are among the most widely used adsorbents in separation applications (Pérez-Botella et al., 2022).

Damour (1840) and Friedel (1896) were among the first to study the reversible adsorption of molecules by zeolites. Later, in the 1930s McBain demonstrated their potential for highly selective adsorption and introduced the term "molecular sieves" (Pérez-Botella et al., 2022; Szostak, 1989). "Molecular sieving" is attributed to the ability of zeolite micropores to selectively allow the diffusion of molecules below a certain size, while excluding larger ones (Suib et al., 2023). Since then, zeolites have been widely adopted for commercial separation processes involving gas and liquid mixtures. Key applications include pressure swing adsorption systems for $O_2/N_2$ separation in air (Chin et al., 2023; Ruthven and Farooq, 1990), natural gas and biogas purification (Alonso-Vicario et al., 2010), as well as treatment and desiccant systems (Zheng et al., 2014; Wang et al., 2013; Mahdavi Far et al., 2022; Kallo, 2001).

### 14.2.3 Catalytic properties and applications

Zeolites possess key catalytic properties that make them highly effective in a wide range of industrial processes, especially in the petrochemical and refining industries. Zeolites are widely used in processes such as fluid catalytic cracking for the conversion of heavy hydrocarbons into gasoline (Tonetto et al., 2004; Vogt and Weckhuysen, 2015), hydroisomerization and alkylation for the production of high-octane fuels (Pastvova et al., 2017; Deldari, 2005), aromatization of alkanes (Bhan and Nicholas Delgass, 2008), methanol-to-hydrocarbons and dimethyl ether-to-hydrocarbons conversions (Paunović et al., 2022; Niethammer et al., 2025, 2023), and in environmental applications such as selective catalytic reduction of $NO_x$ emissions (Zengel et al., 2021; Doronkin et al., 2014).

Their unique crystalline structure, acid-base properties, and shape selectivity enable zeolites to serve as highly efficient, selective, and stable solid catalysts (Weitkamp, 2000; Primo and Garcia, 2014). In addition to acting as standalone catalysts, zeolites can host active metal species either in their framework or on their surface. Isomorphous substitution of Si or Al atoms with other metals such as Ti, Sn, V, Zr, Hf, Nb, Ga, B, and Fe can be performed without disrupting the topological structure, resulting in metal-containing zeolite frameworks with new properties (Wang et al., 2023; Peeters et al., 2022; Ni et al., 2022; Bols et al., 2021; Zhou et al., 2021). Extra-framework metal species, in the form of single metal sites, clusters, and nanoparticles, can also be generated on the internal or external surfaces via methods including impregnation, deposition, or ion exchange, resulting in bifunctional materials (Zhang et al., 2023a; Sun et al., 2021).

## 14.3 Zeolites as adsorbents for per- and polyfluoroalkyl substances remediation

Adsorption is considered a simple, efficient, low-energy, cost-effective, and environmentally friendly method for PFAS remediation in aqueous media (Kucharzyk et al., 2017). The most commonly used adsorbents are ACs and resins. However, the large scale use of resins is limited due to their relatively high cost and challenging regeneration (Nassi et al., 2014). Other materials that have been tested for PFAS adsorption are polymers and biomaterials, which also possess the disadvantages of using pricey precursor materials and complicated synthesis procedures (Vu and Wu, 2022). During recent years, many works have also focused on metal–organic frameworks (MOFs) and covalent organic frameworks for PFAS remediation. Especially MOFs, which own both organic and inorganic counterparts, possess large specific surface areas, and flexibility regarding their surface hydrophobicity, thus enhancing the adsorption capacity of PFAS remarkably. Despite the performance advantages, there are still limitations in terms of their economic competitiveness (Liu et al., 2025). Regarding inorganic adsorbents, layered double hydroxides (LDH), such as Cu–Al, Zn–Al, Mg–Al, and Mg–Fe LDH, are promising for PFAS remediation due to their high anion-exchange capacity, tunable composition, and stability, making them effective for long-chain PFAS removal. However, they face challenges with short-chain PFAS, competition from other anions, and limited field-scale validation (Kim et al., 2025; Zhang et al., 2025; Min et al., 2022; Yan et al., 2024; Huo et al., 2022).

Thanks to their high surface area, tunable pore size and surface chemistry, ease of regeneration, and low cost, zeolites have emerged as very promising adsorbents for PFAS remediation, even at a large scale. Zeolites with pore sizes in the range of 0.7 to 0.8 nm are particularly well-suited for most PFAS molecules, which, according to the literature, have an average molecular width of about 0.5 nm (Zhang et al., 2023b). Therefore the correlation between the zeolite's network and the chain length of the adsorbed PFAS is very important. Additionally, factors such as the Si/Al ratio of the zeolite (which regulates the hydrophilicity/hydrophobicity) and surface modifications with metals or organic agents can enhance the adsorption and photocatalytic performance.

## 14.3.1 Influence of Si/Al ratio and hydrophobicity

As discussed in section 2, the hydrophobicity governed by the Si/Al ratio in zeolites plays a central role in PFAS adsorption. Van den Bergh et al. (2020) performed an informative study regarding PFOA and PFOS adsorption on a hydrophobic, all-silica Beta zeolite (with a $SiO_2/Al_2O_3$ ratio 300). Their findings revealed that high adsorption capacities (350 mg $g^{-1}$) and removal efficiencies (21%–37%) were achieved for both PFOA and PFOS, at an initial concentration ($C_e$) of 150 mg $L^{-1}$, outperforming the industrial benchmark adsorbent AC. Even in the presence of organic competitors, such as phenol, benzoic acid, adipic acid, caprylic acid, and sodium dodecyl sulfate, PFOA adsorption remained highly selective, demonstrating the superior affinity of the molecule with the surface of this defect-free, silica-rich zeolite. When an Al-containing zeolite Beta was used (with a $SiO_2/Al_2O_3$ ratio of 22), the adsorption capacity was much lower, probably due to the presence of Al and silanol defects, which increased the adsorption enthalpy for PFOA. Density functional theory calculations provided a more clear overview of the interaction between PFOA molecules as well as their position within the zeolite's pores, revealing that head-to-head interactions are more preferable and stable due to the formation of carboxylic acid dimers, driven by H-bonding. PFOA molecules are located within the a- and b-channels, and the helical conformation of their perfluorinated chains enables an extremely tight packing, achieving the maximum sterically possible packing density.

## 14.3.2 Role of per- and polyfluoroalkyl substances chain length and functional groups

In addition to the Si/Al ratio, the chain length of PFAS can also affect adsorption capacity. Lauwers et al. (2023) suggested that Beta zeolite is an ideal sorbent for the removal of PFAS with eight carbons. They studied the adsorption of four PFCAs and four perfluorosulfonic acids (PFSAs) with carbon chain lengths ranging from C4 for perfluorobutanoic acid (PFBA) and perfluorobutanesulfonic acid (PFBS), to C6 for perfluorohexanoic acid (PFHxA) and perfluorohexanesulfonic acid (PFHxS), C8 for PFOA and PFOS, and C10 for perfluorodecanoic acid (PFDA). Adsorption isotherms showed increasing capacity with PFAS chain lengths: PFBA (25 mg $g^{-1}$) < PFHxA (200 mg $g^{-1}$) < PFOA (360 mg $g^{-1}$) < PFDA (380 mg $g^{-1}$) (Fig. 14.4). However, when the adsorption capacity was expressed in mmol of PFCA $g^{-1}$ dry matter, the capacity for PFOA was the highest (0.87 mmol $g^{-1}$). As described above, for Beta zeolite, PFOA molecules fit perfectly between the channel intersections of this zeolite topology, compared to PFDA, which can be too long for ideal packing. For PFAS with equal carbon numbers, such as PFOA/PFOS and PFHxA/PFHxS, adsorption was lower for sulfonic acids, possibly due to their higher polarity. Additional studies regarding

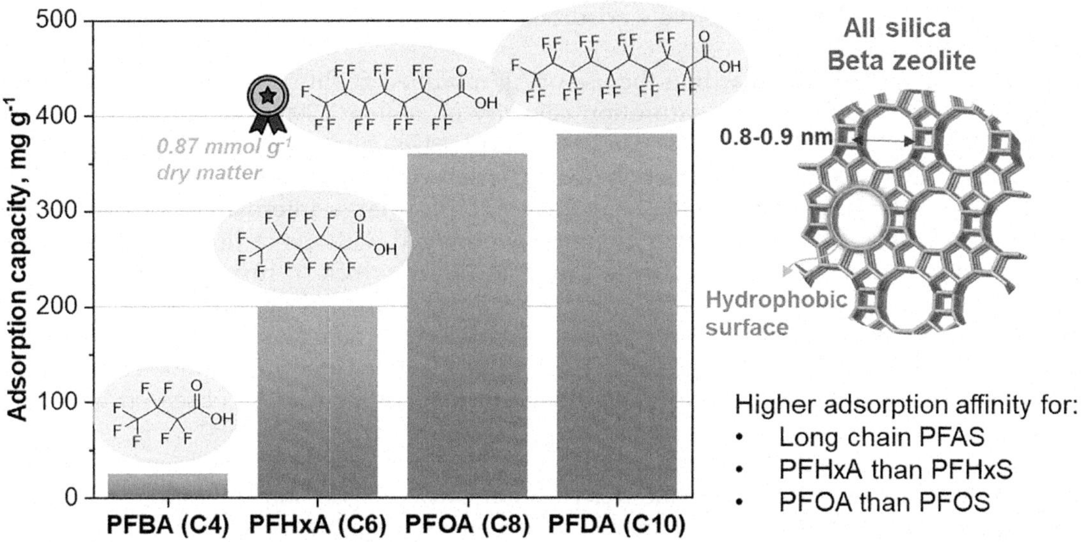


FIGURE 14.4 Adsorption capacity of all-silica Beta zeolite for per- and polyfluoroalkyl substances (PFAS) of varying chain lengths. Graph based on the data from Lauwers et al. (2023).

the water matrix revealed that the solution pH and the addition of cations or organic substances (such as NOM, humic acid, or fulvic acid) had minimal influence on PFOA/PFOS removal. Similar observations regarding the influence of water matrix variations were also made by Sheehan et al. (2025). Adsorption experiments of PFOA and PFOS showed that zeolite Beta is able to treat four times more of the PFOA solution with an efficiency higher than that of AC, thus making it a great alternative to the other sorbents, also for larger-scale applications. Regarding adsorbent regeneration, the thermal treatment at 550°C for 5 h was sufficient for maintaining high PFOS uptake even after three cycles (Lauwers et al., 2023).

### 14.3.3 Modified and composite zeolites

In a similar study, where a mixture of 18 PFAS, including $C_3$–$C_{13}$ PFCAs, $C_4$, $C_6$, and $C_8$ PFSAs, three fluorotelomer sulfonate, and perfluorooctane sulfonamide was used in batch sorption experiments with Al-containing Beta zeolite ($SiO_2/Al_2O_3$ ratio 25), 99.5% removal efficiency was achieved (Mancinelli et al., 2023a). Equilibrium was reached within 10 min, outperforming even AC with 98.5% removal in 15 min. Other promising materials were found to be Beta-360 (with a $SiO_2/Al_2O_3$ ratio of 360) and mordenite, as well as silver-functionalized zeolites (AgY390 and AgY760), which also demonstrated potential catalytic activity for PFAS degradation. Silver can provide antibiofouling and bactericidal properties to zeolites, thereby increasing material lifetime, allowing a constant permeation flow, and preventing pore clogging. In general, adsorption efficiency can be influenced by PFAS molecular characteristics (e.g., chain length and functional groups), as well as adsorbent properties such as pore size and hydrophobicity (i.e., Si/Al ratio). As already demonstrated, long-chain PFAS are more effectively adsorbed than short-chain variants (Mancinelli et al., 2023a; Ponge et al., 2024).

A more detailed work regarding the use of Ag-modified zeolite Y390 (with a $SiO_2/Al_2O_3$ ratio of 500) revealed that the saturation capacity of zeolite Y390 and the Ag-exchanged sample can reach ~43% and ~62% for PFOA, and ~17% and ~32% for PFOS in weight, respectively. X-ray powder diffraction (XRPD) rietveld refinements showed that PFAS are located at the center of zeolite Y supercages, typical large internal cavities within the zeolite Y structure. These supercages are linearly arranged with the carboxylic acid group in PFOA and with the sulfonyl hydroxide group in PFOS, which in turn were H-bonded to both coadsorbed water molecules and framework oxygen atoms (Mancinelli et al., 2022). For the Ag-exchanged Y zeolite, further studies regarding the effect of thermal treatment for adsorbent regeneration were conducted to monitor all the structural parameter variations occurring during both dehydration and PFAS burning processes. As observed also here, PFAS sites could become empty in the temperature range of 550°C–650°C, in correspondence with the thermal degradation of PFOA and PFOS. Upon heating, the crystallinity of the samples was not affected, while dehydration caused silver ion migration into supercages. The dimensions of the channels remained comparable to those of the pristine materials, and Ag-exchanged Y could readsorb PFAS in amounts comparable to those adsorbed in the first cycle (Mancinelli et al., 2023b).

Surface interactions between zeolites and PFAS are important for achieving a high removal degree. As already presented, the Si/Al ratio as well as metal modifications can have a strong effect on the adsorption capacity of the materials. Another type of modification tested involves cationic surfactants, such as cetrimonium bromide (CTAB), and cationic polyelectrolytes such as poly(diallyldimethylammonium chloride) (PDADMAC). These modifiers are supposed to change the electrostatic and hydrophobic interactions, thus enhancing the adsorption capacity. According to Hossain et al. (2023), PDADMAC surface coatings reduced PFOA adsorption capacity from 91.87 to 77.68 $mg\ g^{-1}$ but enhanced the adsorption capacity of PFBA by up to 51.3%. Similarly, CTAB coatings decreased PFOA adsorption capacity by 58.7%. The decreased adsorption capacity was attributed to the lower specific surface area following surface coatings. On the contrary, the pristine Beta zeolite (with $SiO_2/Al_2O_3$ ratio 300) demonstrated the highest PFOA uptake (91.87 $mg\ g^{-1}$). For this material, the equilibrium adsorption uptake was also modeled using the Langmuir and the Freundlich adsorption isotherms, concluding that the Langmuir model (with $R^2$ value 0.994) fit slightly better to the experimental data than the Freundlich model (with $R^2$ value 0.909).

Although powdered adsorbents have been extensively researched for PFAS remediation, their application in water treatment plants is still limited. This is primarily due to difficulties in removing them before discharge, the risk of flow obstruction due to their low mass, and their typically higher costs and limited regeneration potential. Therefore Licato et al. (2022) investigated a method of simultaneous pharmaceuticals and personal care products (PPCPs) and PFAS removal utilizing zeolite–sodium silicate composite materials. Composites were synthesized by using sodium silicate and synthetic zeolite powder, which after heating and water removal, formed a cohesive material. These zeolite composites, along with a natural zeolite, were tested for a laboratory-controlled mixture of 21 PPCPs and PFAS and for wastewater treatment facility effluent samples containing 84 PPCPs and 24 PFASs. Treatment with the zeolite composites resulted in an average mass decrease of 72% for

PFAS. The authors found that the novel zeolite composite materials were effective in removing PPCPs and PFASs simultaneously under both controlled and real-world conditions. However, their adsorption effectiveness could not be attributed to any single property of the materials, indicating that multiple factors likely influence performance. In Table 14.1, further information about the studies dealing with PFAS adsorption in zeolites is summarized.

Despite their many advantages, zeolites also face certain limitations compared to conventional adsorbents, such as AC and ion-exchange resins. For instance, zeolites may exhibit decreased performance in the presence of natural organic matter (NOM), which can competitively adsorb onto their surfaces and block access to micropores (Lei et al., 2023). Additionally, their relatively rigid framework can lead to lower adsorption capacities for some short-chain or branched PFAS compared to more flexible sorbents (Leung et al., 2023). While ion-exchange resins naturally offer strong electrostatic affinity for anionic PFAS, zeolites require surface modification to achieve similar selectivity. Moreover, fouling and pore clogging can still occur, particularly in real wastewater matrices, necessitating careful pretreatment or regeneration strategies (Lauwers et al., 2023). These factors should be considered when evaluating zeolites for large-scale implementation.

Nevertheless, when tailored appropriately, zeolites remain highly effective adsorbents for PFAS remediation, with materials such as Beta zeolite and zeolite Y demonstrating excellent performance due to their well-structured channels and favorable surface properties. In summary, zeolites are highly effective adsorbents for PFAS remediation, with Beta zeolite and zeolite Y being identified as the most promising materials due to their well-structured channels and large pore size. All-silica zeolites, as a result of their hydrophobic nature, exhibit greater adsorption capacity for less hydrophilic PFAS, such as PFOA, especially those with longer carbon chains, while shorter-chain PFAS and sulfonic variants like PFOS are less efficiently adsorbed. Zeolites generally maintain consistent performance, even under varying pH conditions or in the presence of competing organic species in water (Van den Bergh et al., 2020; Lauwers et al., 2023). Adsorbents can be easily regenerated by calcination, and modification with metals, such as silver, enhances antibacterial properties (Mancinelli et al., 2023a, 2023b). Additionally, customizing zeolites with cationic agents can selectively improve adsorption for certain PFAS, though it may decrease the efficiency for others like PFOA (Hossain et al., 2023). Zeolite composites also offer the advantage of simultaneously removing PFAS and other organic contaminants, such as PPCPs (Licato et al., 2022).

## 14.4 Zeolites as photocatalysts for per- and polyfluoroalkyl substances remediation

Beyond adsorption, catalytic remediation technologies are acknowledged as promising methods to decrease PFAS concentrations in water. These techniques primarily target the dehalogenation of alkyl chains, aiming to replace fluorine atoms with hydrogen (a process known as defluorination or hydrodefluorination). Depending on the functional groups of the original compound, partial dehalogenation can yield carboxylic acids, sulfonic acids, saturated alkanes, or other intermediates (Restivo et al., 2024). In aqueous media, photocatalysis has emerged as an important tool for PFAS remediation (Leonello et al., 2021).

**TABLE 14.1** Summary of per- and polyfluoroalkyl substances (PFAS) adsorption studies using zeolites.

| Adsorbate | Adsorbent | Conditions | Removal | References |
|---|---|---|---|---|
| Perfluorooctanoic acid (PFOA), perfluorooctanesulfonic acid (PFOS) | All-silica Beta | 0.1–500 mg $L^{-1}$, 20°C, 24 h | 37% for PFOA (350 mg $g^{-1}$ for Ce 150 mg $L^{-1}$), 21% for PFOS | Van den Bergh et al. (2020) |
| PFSAs, perfluoroalkyl carboxylic acids (PFCAs) | All-silica Beta | 0.1–250 mg $L^{-1}$ for PFSAs, 500 mg $L^{-1}$ for PFCAs, 20°C–25°C, 24 h | PFOA ($C_8$): 350 for Ce 150 mg $L^{-1}$, PFDA ($C_{10}$): ~400 for Ce 100 mg $L^{-1}$ | Lauwers et al. (2023) |
| 18 PFAS | Beta-25 | 0.1 mg $mL^{-1}$ for each PFAS, 24 h | 99.5% PFAS uptake | Mancinelli et al. (2023a) |
| PFOA, PFOS | Zeolite Y390 | 0.1–35 mg $L^{-1}$ for PFOS, 0.1–100 mg $L^{-1}$ for PFOA, 25.3°C ± 0.5°C, 24 h | 43% for PFOA, 17% for PFOS | Mancinelli et al. (2022) |
| | Ag–zeolite Y390 | | 62% for PFOA, 32% for PFOS | |
| Pharmaceuticals and personal care products and PFAS | Sodium silicate–zeolite composite | 1 mg $L^{-1}$ for each component | 72% | Licato et al. (2022) |

Among the most common photocatalytic materials are titanium dioxide ($TiO_2$), zinc oxide (ZnO), graphitic carbon nitride ($g\text{-}C_3N_4$), as well as bismuth- and iron-based catalysts (Xu et al., 2020a; Navidpour et al., 2021, 2024; Xu et al., 2020b; Rather et al., 2024). Zeolites, although lacking intrinsic photocatalytic activity, can serve as supports for hosting other photoactive metals, providing high surface areas and well-defined pore channels that enhance PFAS adsorption and the subsequent degradation under ultraviolet (UV) light. Two major classes can be distinguished: $TiO_2$-modified zeolites and Fe-modified zeolites.

### 14.4.1 $TiO_{2}$_-zeolite composites

Titanium dioxide ($TiO_2$) is a highly efficient semiconductor that generates charge carriers ($e^-/h^+$) under irradiation, enabling redox reactions, with widespread use in applications such as organic pollutant degradation and biomass conversion (Gaya and Abdullah, 2008; Giannakoudakis et al., 2023; Colmenares et al., 2013; Belessi et al., 2007). The benchmark commercial $TiO_2$ P25 material completes PFOA degradation within ~4 h under 254 nm UV, though its performance declines markedly at longer wavelengths due to surface charge effects and band-gap limitations (Leonello et al., 2021). Additionally, the defluorination efficiency of $TiO_2$ is usually low (<30%), even for some other titanium oxide structures such as titanate nanotubes. Higher removal grades and defluorination degree can be achieved through $TiO_2$ modification with other metals (e.g., Pd, Pt, Ce, and Co) (Li et al., 2016; Samuel et al., 2022), or composites of $TiO_2$ with porous materials such as carbons and zeolites, which can increase the interaction between the photogenerated reactive species and PFAS degradation (Juve et al., 2024). Composites of $TiO_2$ with graphene oxide (GO) and reduced GO have shown very high activity in PFOA defluorination, achieving 80%–90% removal and 40% defluorination under 254 nm UV after 5 h (Bouteh et al., 2023). Recently, a $TiO_2$–zeolite nanocomposite was used for pentafluoropropionic acid (PFPA) photodegradation under UV light (185 nm), exhibiting a notably high defluorination degree of 54.5% after 8 h. The composite, prepared using a microwave-assisted sol-gel technique, displayed high $TiO_2$ grain dispersion due to the porous zeolite carrier, thereby enhancing the material's photocatalytic performance (Halder et al., 2024). $TiO_2$–zeolite systems function via photocatalytic oxidation driven by electron-hole pair generation (see Section 14.4.4 for full mechanism). Compared to $TiO_2$-graphene composites, zeolitic materials are more cost-effective, making them attractive for large-scale environmental applications (Armaković and Armaković, 2025).

### 14.4.2 Fe- zeolite systems

Fe-zeolites exploit ligand-to-metal charge transfer (LMCT) mechanisms for PFAS oxidation under light irradiation (Juve et al., 2024). Qian et al. (2020). used zeolite Beta-35 ($SiO_2/Al_2O_3$ 35) as a support for iron deposition, and this catalyst was utilized for PFOA degradation under UV-A irradiation (315–400 nm). After 24 h of equilibration, the majority of molecules were in the adsorbed state, resulting in 16.8 mg $g^{-1}$ loading of PFOA on the zeolite. After 6 h under light irradiation, Fe–BEA-35 resulted in up to 90% of the initial PFOA decomposition and complete degradation (>99%) within 24 h. The same Fe-modified zeolite Beta was also used for PFOS degradation under irradiation with 254 nm UV light, achieving 99% degradation within 96 h under acidic conditions (Qian et al., 2021). The main products were fluoride, sulfate, and short-chain perfluoroalkyl acids. Another member of per- and polyfluoroalkyl ether acids, which has been detected globally in river water and is considered potentially more toxic than legacy PFAS, is the ammonium salt of hexafluoropropylene oxide dimer acid, named GenX. In their study, Wen et al. (2023) compared the photocatalytic degradability of GenX with that of PFOA and PFOS using Fe–Beta-360 ($SiO_2/Al_2O_3$ ratio 360) photocatalyst. After 7 h of irradiation, GenX showed lower removal (79%) and defluorination (33%) compared to PFOA (100% removal and 69% defluorination) and PFOS (100% removal and 51% defluorination). These outcomes correlate with GenX's lower hydrophobicity, steric effects, and formation of persistent transformation byproducts, leading to reduced adsorption (40%) compared to PFOA (99%) and PFOS (87%).

According to the literature, the superior activity of Fe-zeolites is attributed to the monomeric iron species (e.g., $Fe^{3+}$) on the internal surface of the zeolite, which enhance PFAS photodegradation more effectively than larger iron clusters and nanoparticles (Qian et al., 2024). Active $Fe^{3+}$ species can be introduced into the zeolite framework via ion exchange (Arana Juve et al., 2024). The superior activity of monomeric Fe species over other iron species has also been observed in catalytic applications such as the oxidative cleavage of C–C bonds of vicinal diols (Treu et al., 2022).

### 14.4.3 Performance comparison and optimization strategies

Titania–zeolite ($TiO_2$–zeolite) and iron-exchanged zeolite (Fe–Beta) catalysts both show promise for PFAS photocatalytic remediation, despite differences in performance, cost, stability, and synthetic complexity. The photocatalytic activity of $TiO_2$ can be further enhanced through structural modifications (e.g., engineered morphologies, titanate nanotubes), metal modifications (Pd, Pt, Ce, and Co), and combinations of $TiO_2$ with porous materials such as carbonaceous and zeolites, which result in composite formation (Giannakoudakis et al., 2021; Safajou et al., 2017; Bai et al., 2009; Eldoma et al., 2024; Huang et al., 2008). These composites offer better oxide grain dispersion, high surface areas, increased adsorption capacity, and can be easily activated by UV-visible radiation, thereby enhancing PFAS degradation. $TiO_2$–zeolite nanocomposites can achieve a high degree of PFOA defluorination (54.5%) under standard conditions (at 85°C, after 8 h exposure to radiation). However, during composite synthesis via sol-gel or hydrothermal methods, care should be taken to prevent pore blockage (Liao et al., 2021).

Although slower, Fe-zeolites demonstrate substantial defluorination: ~69 % PFOA defluorination in 7 h at 365 nm and pH 4.4, outperforming unmodified zeolite under similar conditions. PFOS degradation follows similar trends, while even for the more challenging pollutant GenX, high removal degrees and approximately 33% defluorination can be achieved. The water matrix composition significantly influences the performance, as acidic conditions ($pH < 5.5$) enhance PFOA degradation, while under neutral and basic conditions, the catalytic activity was negligible (Qian et al., 2020, 2021). Remarkably, the catalyst preserved 80%–90% of its initial activity even after six successive runs, highlighting its stability and reusability. Fe-zeolites are prepared via simple ion exchange from aqueous $Fe^{3+}$ solutions, followed by mild calcination, resulting in uniform active sites while preserving zeolite porosity (Treu et al., 2022). This enhanced activity, even after many cycles, demonstrates the importance of iron-modified zeolites as photocatalysts for large-scale applications. Currently, compared to large-scale PFAS removal technologies such as adsorption, ion exchange, and reverse osmosis, PFAS remediation via Fe-zeolite-based photocatalysis has similar energy demands ($\geq 1$ € $m^{-3}$) in terms of total operation costs, while there is substantial room for further optimization (Qian et al., 2021). An overview of the remediation of PFAS using metal-modified zeolites is shown in Table 14.2.

### 14.4.4 Mechanistic insights

A clear understanding of the photocatalytic mechanisms of zeolite-based systems is essential for the rational design of effective PFAS degradation technologies. The two principal mechanisms, corresponding to $TiO_2$ and Fe-modified zeolites, function via distinct yet complementary pathways. These mechanisms are schematically illustrated in Fig. 14.5.

In $TiO_2$–zeolite composites, UV light excites electrons, thus generating electron-hole pairs ($e^-/h^+$). Photogenerated holes ($h^+$) promote water oxidation and hydroxyl radicals (•OH) formation, while the electrons ($e^-$) promote dissolved oxygen reduction to superoxide ($\bullet O_2^-$). These reactive oxygen species initiate PFAS degradation via decarboxylation and cleavage of C–F bonds (Figure 14.5A). In this system, the zeolite matrix aids the process by dispersing $TiO_2$ particles and enhancing the adsorption of PFAS (Halder et al., 2024; Zelekew et al., 2024).

Fe-modified zeolites operate via a LMCT mechanism (Wen et al., 2023). Isolated $Fe^{3+}$ ions in tetrahedral and octahedral coordination inside the pores form PFOA–$Fe^{3+}$ complexes, which, under UV-A irradiation in the presence of $O_2$, enhance PFOA oxidation. Decarboxylation results in fluoride release and formation of perfluoroalkyl radicals. Subsequent oxidation by molecular oxygen regenerates $Fe^{3+}$ ions, while producing additional radicals (e.g., •OH) that sustain the degradation cycle. Iron species in tetrahedral and octahedral coordination on the zeolite's internal surface enhance PFAS photodegradation relative to larger iron clusters and nanoparticles, due to the hydrophobic interactions between O-Si-O groups and the PFOA tail, and electrostatic interactions between Fe species and the carboxyl group (Qian et al., 2020). The reaction mechanism is depicted in Figure 14.5B. In essence, Ti-based systems rely on semiconductor photocatalysis with enhanced charge dynamics, while Fe-based systems utilize redox-active centers and adsorption-driven activation. Both systems benefit from the structural features of zeolites that promote interactions between PFAS and reactive sites.

To conclude, beyond adsorption, catalytic degradation, particularly using metal-modified zeolite catalysts, has emerged as a promising strategy for PFAS remediation, with $TiO_2$ and Fe-based zeolites showing significant photocatalytic activity under UV irradiation. $TiO_2$–zeolite composites enhance defluorination by preventing $TiO_2$ grain growth, while Fe–Beta-35 demonstrates excellent degradation of PFOA and PFOS under UV-A irradiation, driven by strong adsorption and LMCT processes. These catalysts convert PFAS into less harmful products, such as fluoride, $CO_2$, and short-chain PFCAs. The effectiveness is linked to the presence of isolated $Fe^{3+}$ species in tetrahedral and octahedral coordination, which improves both adsorption and reactivity.

**TABLE 14.2** Application of zeolites in the photocatalytic remediation of per- and polyfluoroalkyl substances (PFAS).

| Reactant | Catalyst | Conditions | Degradation | References |
|---|---|---|---|---|
| Pentafluoropropionic acid (PFPA) | $TiO_2$–zeolite nanocomposite | $C_{0,\ PFPA} = 10$ mg $L^{-1}$, $C_{TiO2\text{–}zeolite} = 0.5$ g $L^{-1}$, 85°C, pH 13, 8 h | 54.5% defluorination | Halder et al. (2024) |
| Perfluorooctanoic acid (PFOA) | Fe–Beta-35 ($SiO_2/Al_2O_3$ ratio 35) | 300 ml water, $C_{0,\ PFOA} = 48$ μM, 0.1 $g_{catalyst}$, $pH_0 = 5.5$, 315–400 nm, 24 h | >99% | Qian et al. (2020) |
| Perfluorooctanesulfonic acid (PFOS) | Fe–Beta-35 ($SiO_2/Al_2O_3$ ratio 35) | $C_{0,\ PFOS} = 20$ μM, 0.5 $g_{catalyst}$ $L^{-1}$, $pH_0 = 3.0$, 254 nm, 96 h | 99% | Qian et al. (2021) |
| GenX, PFOA, and PFOS | Fe-Beta-360 ($SiO_2/Al_2O_3$ ratio 360) | $C_{initial} = 10$ μM, 1 $g_{catalyst}$ $L^{-1}$, $pH_0 = 4.4$, 254 nm, for 0–7 h without stirring | GenX: 79% removal, 33% defluorination. PFOA: 100% removal, 69% defluorination. PFOS: 100% removal, 51% defluorination. | Wen et al. (2023) |

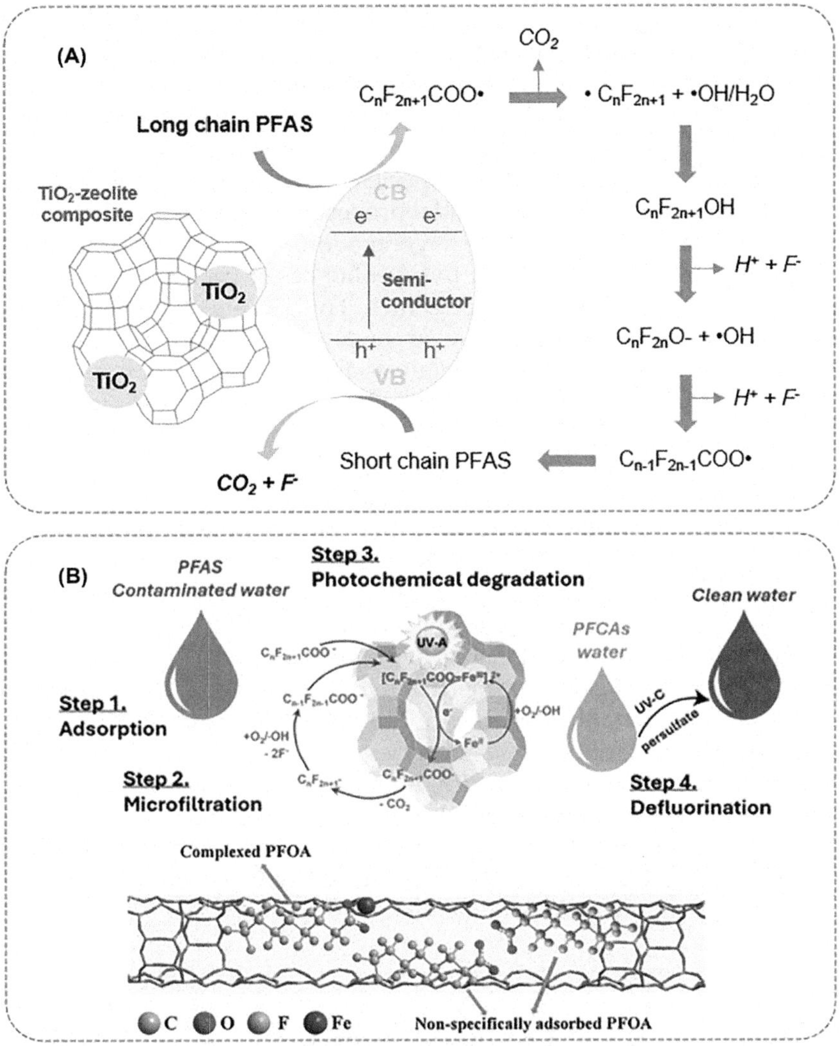


**FIGURE 14.5** Proposed reaction mechanism for per- and polyfluoroalkyl substances (PFAS) degradation via photocatalysis by utilizing (A) $TiO_2$–zeolite nanocomposites, based on Halder et al. (2024), Zelekew et al. (2024), and (B) Fe-zeolite catalyst, based on Qian et al. (2020).

## 14.5 Conclusions

Zeolites are important materials for both adsorption and catalytic applications due to their porous structure, with well-defined channels, high surface area, hydrophobic/hydrophilic properties, thermal and chemical stability, as well as low-cost synthesis. In the context of PFAS remediation, zeolites have emerged as highly efficient adsorbents for removing these compounds from aqueous environments. In particular, all-silica zeolite Beta, with its 12-membered ring pores, long channels, and highly hydrophobic surface, has been established as a highly effective adsorbent for PFOA and PFOS removal. The Si/Al ratio, and thus hydrophobicity/hydrophilicity, influences adsorption capacity, as the presence of Al and silanol defects increases the PFOA adsorption enthalpy, thereby decreasing overall uptake. PFAS chain length also plays an important role, with shorter-chain compounds (4–6 carbons) showing lower adsorption efficacy than longer-chain PFAS (8–10 carbon atoms). Surface modification of zeolites with metal species, such as Ag, can enhance PFAS removal while providing antibiofouling and bactericidal properties. Thermal treatment of PFAS-loaded zeolites at 550°C–600°C allows complete regeneration, while preserving their adsorption efficiency. In photocatalytic applications, two zeolite-based catalysts can be distinguished: $TiO_2$–zeolite composites and Fe-modified zeolites. For $TiO_2$–zeolite composites, the zeolite matrix enhances $TiO_2$ dispersion, improving both photodegradation and defluorination activity. Fe-based zeolites are promising catalysts for the degradation of a range of PFAS (GenX, PFOA, and PFOS) under UV irradiation, often achieving near-complete removal and a high defluorination degree. Effective adsorption and degradation of short-chain PFAS can be achieved by selecting zeolites with smaller pore sizes and optimized surface-pollutant interactions.

Looking ahead, future research should focus on improving the practical viability of zeolite-based technologies for PFAS remediation. Key directions include the use of hierarchical zeolites with micro and mesoporosity, which can be synthesized either via postmodification methods (e.g., desilication) or bottom-up approaches (e.g., templated hydrothermal synthesis), which could enhance the simultaneous adsorption of PFAS with both short and longer chain lengths. Additionally, the development of hybrid materials that integrate photocatalytic and electrochemical functionalities enables more complete mineralization of PFAS under ambient conditions. Moreover, improved strategies for regeneration, resistance to fouling, and real-water matrix compatibility will be critical for scale-up. Life-cycle assessments and techno-economic analyzes are also needed to benchmark zeolite performance against conventional technologies. Finally, exploring the synergistic effects of multifunctional modifications, such as combining metal and structural tuning, may unlock next-generation materials with both high efficiency and environmental sustainability.

## List of abbreviations

| | |
|---|---|
| **AC** | Activated carbon |
| **CTAB** | Cetrimonium Bromide |
| **COFs** | Covalent organic frameworks |
| **DFT** | Density functional theory |
| **DTH** | Dimethyl ether-to-hydrocarbons |
| **FCC** | Fluid catalytic cracking |
| **FOSA** | Perfluorooctane sulfonamide |
| **FTSAs** | Fluorotelomer sulfonates |
| **GO** | Graphene oxide |
| **IZA** | International Zeolite Association |
| **LDH** | Layered double hydroxides |
| **MTH** | Methanol-to-hydrocarbons |
| **MOFs** | Metal–organic frameworks |

| | |
|---|---|
| **PDADMAC** | Poly(diallyldimethylammonium chloride) |
| **PFAS** | Per- and polyfluoroalkyl substances |
| **PFBA** | Perfluorobutanoic acid |
| **PFBS** | Perfluorobutanesulfonic acid |
| **PFCAs** | Perfluoroalkyl carboxylic acids |
| **PFDA** | Perfluorodecanoic acid |
| **PFHxA** | Perfluorohexanoic acid |
| **PFHxS** | Perfluorohexanesulfonic acid |
| **PFOA** | Perfluorooctanoic acid |
| **PFOS** | Perfluorooctanesulfonic acid |
| **PFPA** | Pentafluoropropionic acid |
| **PFSAs** | Perfluorosulfonic acids |
| **PPCPs** | Pharmaceuticals and personal care products |
| **SBUs** | Secondary building units |
| **UV** | Ultraviolet |
| **WWTF** | Wastewater treatment facility |

# References

Abunada, Z., Alazaiza, M.Y.D., Bashir, M.J.K., 2020. An overview of per- and polyfluoroalkyl substances (PFAS) in the environment: source, fate, risk and regulations. Water 12, 3590. Available at https://doi.org/10.3390/w12123590.

Alonso-Vicario, A., Ochoa-Gómez, J.R., Gil-Río, S., Gómez-Jiménez-Aberasturi, O., Ramírez-López, C.A., Torrecilla-Soria, J., et al., 2010. Purification and upgrading of biogas by pressure swing adsorption on synthetic and natural zeolites. Microporous and Mesoporous Materials 134, 100–107. Available at https://doi.org/10.1016/j.micromeso.2010.05.014.

Anderko, L., Pennea, E., 2020. Exposures to per-and polyfluoroalkyl substances (PFAS): potential risks to reproductive and children's health. Current Problems in Pediatric and Adolescent Health Care 50, 100760. Available at https://doi.org/10.1016/j.cppeds.2020.100760.

Arana Juve, J.-M., Baami González, X., Bai, L., Xie, Z., Shang, Y., Saad, A., et al., 2024. Size-selective trapping and photocatalytic degradation of PFOA in Fe-modified zeolite frameworks. Applied Catalysis B: Environment and Energy 349, 123885. Available at https://doi.org/10.1016/j.apcatb.2024.123885.

Armaković, S.J., Armaković, S., 2025. Zeolite-supported $TiO_2$ for enhanced photocatalytic performance in environmental applications: a review. Catalysts 15, 174. Available at https://doi.org/10.3390/catal15020174.

Bae, Y.-S., Yazaydın, A.Ö., Snurr, R.Q., 2010. Evaluation of the BET method for determining surface areas of MOFs and zeolites that contain ultra-micropores. Langmuir: The ACS Journal of Surfaces and Colloids 26, 5475–5483. Available at https://doi.org/10.1021/la100449z.

Bai, Y., Li, W., Liu, C., Yang, Z., Feng, X., Lu, X., et al., 2009. Stability of Pt nanoparticles and enhanced photocatalytic performance in mesoporous Pt-(anatase/$TiO_2$(B)) nanoarchitecture. Journal of Materials Chemistry 19, 7055–7061. Available at https://doi.org/10.1039/B910240J.

Belessi, V., Lambropoulou, D., Konstantinou, I., Katsoulidis, A., Pomonis, P., Petridis, D., et al., 2007. Structure and photocatalytic performance of $TiO_2$/clay nanocomposites for the degradation of dimethachlor. Applied Catalysis B: Environmental 73, 292–299.

Van den Bergh, M., Krajnc, A., Voorspoels, S., Tavares, S.R., Mullens, S., Beurroies, I., et al., 2020. Highly selective removal of perfluorinated contaminants by adsorption on all-silica zeolite beta. Angewandte Chemie 132, 14190–14194. Available at https://doi.org/10.1002/anie.202002953.

Bhan, A., Nicholas Delgass, W., 2008. Propane aromatization over HZSM-5 and Ga/HZSM-5 catalysts. Catalysis Reviews 50, 19–151. Available at https://doi.org/10.1080/01614940701804745.

Blake, B.E., Pinney, S.M., Hines, E.P., Fenton, S.E., Ferguson, K.K., 2018. Associations between longitudinal serum perfluoroalkyl substance (PFAS) levels and measures of thyroid hormone, kidney function, and body mass index in the Fernald Community Cohort. Environmental Pollution 242, 894–904. Available at https://doi.org/10.1016/j.envpol.2018.1007.1042.

Bols, M.L., Devos, J., Rhoda, H.M., Plessers, D., Solomon, E.I., Schoonheydt, R.A., et al., 2021. Selective Formation of α-Fe (II) sites on Fe-zeolites through one-pot synthesis. Journal of the American Chemical Society 143, 16243–16255. Available at https://doi.org/10.11021/jacs.16241c07590.

Bouteh, E., Bentel, M.J., Cates, E.L., 2023. Semiconductor-hydrophobic material interfaces as a new active site paradigm for photocatalytic degradation of perfluorocarboxylic acids. Journal of Hazardous Materials 453, 131437 https://doi.org/10.1016/j.jhazmat.132023.131437.

Brown, J.B., Conder, J.M., Arblaster, J.A., Higgins, C.P., 2020. Assessing human health risks from per- and polyfluoroalkyl substance (PFAS)-impacted vegetable consumption: a tiered modeling approach. Environmental Science & Technology 54, 15202–15214. Available at https://doi.org/10.1021/acs.est.15200c03411.

Buck, R.C., Franklin, J., Berger, U., Conder, J.M., Cousins, I.T., de Voogt, P., et al., 2011. Perfluoroalkyl and polyfluoroalkyl substances in the environment: Terminology, classification, and origins. Integrated Environmental Assessment and Management 7, 513–541. Available at https://doi.org/10.1002/ieam.1258.

Chen, D., Tang, Q., Deng, W., Chaianansutcharit, S., Guo, L., 2022. Comparative studies on the toluene sorption performance over silicalite-1 zeolites with different morphologies. Microporous and Mesoporous Materials 346, 112275. Available at https://doi.org/10.1016/j.micromeso.2022.112275.

Chin, C., Kamin, Z., Bahrun, M.H.V., Bono, A., 2023. The production of industrial-grade oxygen from air by pressure swing adsorption. International Journal of Chemical Engineering 2023, 2308227. Available at https://doi.org/10.1155/2023/2308227.

Colella, C., Gualtieri, A.F., 2007. Cronstedt's zeolite. Microporous and Mesoporous Materials 105, 213–221. Available at https://doi.org/10.1016/j.micromeso.2007.1004.1056.

Colella, C., Wise, W.S., 2014. The IZA handbook of natural zeolites: a tool of knowledge on the most important family of porous minerals. Microporous and Mesoporous Materials 189, 4–10. Available at https://doi.org/10.1016/j.micromeso.2013.1008.1028.

Colmenares, J.C., Lisowski, P., D., Ł., 2013. omot, A novel biomass-based support (Starbon) for $TiO_2$ hybrid photocatalysts: a versatile green tool for water purification. RSC Advances 3, 20186–20192. Available at https://doi.org/10.21039/C20183RA43673J.

Deldari, H., 2005. Suitable catalysts for hydroisomerization of long-chain normal paraffins. Applied Catalysis A: General 293, 1–10. Available at https://doi.org/10.1016/j.apcata.2005.1007.1008.

Doronkin, D.E., Casapu, M., Günter, T., Müller, O., Frahm, R., Grunwaldt, J.-D., 2014. Operando spatially- and time-resolved XAS study on zeolite catalysts for selective catalytic reduction of NOx by $NH_3$. The Journal of Physical Chemistry C 118, 10204–10212. Available at https://doi.org/10.1021/jp5028433.

Eldoma, M.A., Alaswad, S.O., Mahmoud, M.A., Qudsieh, I.Y., Hassan, M., Bakather, O.Y., et al., 2024. Enhancing photocatalytic performance of Co-$TiO_2$ and Mo-$TiO_2$-based catalysts through defect engineering and doping: a study on the degradation of organic pollutants under UV light. Journal of Photochemistry and Photobiology A: Chemistry 446, 115164. Available at https://doi.org/10.111016/j.jphotochem.2023.115164.

Evich, M.G., Davis, M.J.B., McCord, J.P., Acrey, B., Awkerman, J.A., Knappe, D.R.U., et al., 2022. Per- and polyfluoroalkyl substances in the environment. Science 375. Available at https://doi.org/10.1126/science.abg9065.

Gagliano, E., Sgroi, M., Falciglia, P.P., Vagliasindi, F.G.A., Roccaro, P., 2020. Removal of poly- and perfluoroalkyl substances (PFAS) from water by adsorption: Role of PFAS chain length, effect of organic matter and challenges in adsorbent regeneration. Water Research 171, 115381. Available at https://doi.org/10.111016/j.watres.2019.115381.

Gaya, U.I., Abdullah, A.H., 2008. Heterogeneous photocatalytic degradation of organic contaminants over titanium dioxide: a review of fundamentals, progress and problems. Journal of Photochemistry and Photobiology C: Photochemistry Reviews 9, 1–12. Available at https://doi.org/10.1016/j.jphotochemrev.2007.1012.1003.

Giannakoudakis, D.A., Qayyum, A., Barczak, M., Colmenares-Quintero, R.F., Borowski, P., Triantafyllidis, K., et al., 2023. Mechanistic and kinetic studies of benzyl alcohol photocatalytic oxidation by nanostructured titanium (hydro)oxides: do we know the entire story? Applied Catalysis B: Environmental 320, 121939–https://doi.org/10.121016/j.apcatb.2022.121939.

Giannakoudakis, D.A., Qayyum, A., Nair, V., Khan, A., Pradhan, S.R., Prekodravac, J., et al., 2021. Ultrasound-assisted decoration of CuOx nanoclusters on $TiO_2$ nanoparticles for additives free photocatalytic hydrogen production and biomass valorization by selective oxidation. Molecular Catalysis 514, 111664. Available at https://doi.org/10.111016/j.mcat.2021.111664.

Gomez-Ruiz, B., Gómez-Lavín, S., Diban, N., Boiteux, V., Colin, A., Dauchy, X., et al., 2017. Boron doped diamond electrooxidation of 6:2 fluorotelomers and perfluorocarboxylic acids. Application to industrial wastewaters treatment. Journal of Electroanalytical Chemistry 798, 51–57. Available at https://doi.org/10.1016/j.jelechem.2017.1005.1033.

Halder, S., Chou, F.-C., Wan, T.-K., Tsai, Y.-T., Cheng, P.-C., Lin, Y.-C., 2024. Enhanced photodegradation of pentafluoropropionic acid (PFPA) via TiO2-synthetic zeolite composites: a sustainable approach for effective defluorination. Journal of Water Process Engineering 62, 105296. Available at https://doi.org/10.101016/j.jwpe.2024.105296.

Hossain, M.S., Stuart, T.D., Ramarao, B.V., VanLeuven, C.C., Wriedt, M., Kiemle, D., et al., 2023. Investigation into cationic surfactants and polyelectrolyte-coated β-zeolites for rapid and high-capacity adsorption of short- and long-chain PFAS. Industrial & Engineering Chemistry Research 62, 8373–8384. Available at https://doi.org/10.1021/acs.iecr.8373c00468.

Huang, M., Xu, C., Wu, Z., Huang, Y., Lin, J., Wu, J., 2008. Photocatalytic discolorization of methyl orange solution by Pt modified $TiO_2$ loaded on natural zeolite. Dyes and Pigments 77, 327–334. Available at https://doi.org/10.1016/j.dyepig.2007.1001.1026.

Huo, J., Min, X., Dong, Q., Xu, S., Wang, Y., 2022. Comparison of Zn–Al and Mg–Al layered double hydroxides for adsorption of perfluorooctanoic acid. Chemosphere 287, 132297. Available at https://doi.org/10.131016/j.chemosphere.2021.132297.

Jin, M., Oh, M., Choi, M., 2022. Hierarchical BEA zeolite with trimodal micro-/meso-/macroporosity as a selective and long-lived catalyst for isobutane/2-butene alkylation. ACS Catalysis 12, 4067–4077. Available at https://doi.org/10.1021/acscatal.1c05786.

Juve, J.-M.A., Donoso Reece, J.A., Wong, M.S., Wei, Z., Ateia, M., 2024. Photocatalysts for chemical-free PFOA degradation – what we know and where we go from here? Journal of Hazardous Materials 462, 132651. Available at https://doi.org/10.131016/j.jhazmat.2023.132651.

Kallo, D., 2001. Applications of natural zeolites in water and wastewater treatment. Reviews in mineralogy and geochemistry 45, 519–550. Available at https://doi.org/10.2138/rmg.2001.2145.2115.

Khalil, U., Liu, Z., Peng, C., Hikichi, N., Wakihara, T., García-Martínez, J., et al., 2021. Ultrafast surfactant-templating of *BEA zeolite: an efficient catalyst for the cracking of polyethylene pyrolysis vapours. Chemical Engineering Journal 412, 128566. Available at https://doi.org/10.121016/j.cej.2021.128566.

Kim, K.H., Chung, Y., Kenyon, P., Tran, T.N., Rees, N.H., Choi, S.J., et al., 2025. Regenerable water remediation platform for ultrafast capture and mineralization of per-and polyfluoroalkyl substances. Advanced Materials. Available at https://doi.org/10.01002/adma.202509842e09842.

Kordala, N., Wyszkowski, M., 2024. Zeolite properties, methods of synthesis, and selected applications. Molecules (Basel, Switzerland) 29, 1069. Available at https://doi.org/10.3390/molecules29051069.

Kucharzyk, K.H., Darlington, R., Benotti, M., Deeb, R., Hawley, E., 2017. Novel treatment technologies for PFAS compounds: a critical review. Journal of Environmental Management 204, 757–764. Available at https://doi.org/10.1016/j.jenvman.2017.1008.1016.

Kulprathipanja, S., 2010. Zeolites in Industrial Separation and Catalysis. John Wiley & Sons.

Lauwers, A., Vercammen, J., Vos, D.De, 2023. Adsorption of PFAS by all-silica zeolite β: insights into the effect of the water matrix, regeneration of the material, and continuous PFAS adsorption. ACS Applied Materials & Interfaces 15, 52612–52621. Available at https://doi.org/10.51021/acsami.52613c12321.

Lei, X., Lian, Q., Zhang, X., Karsili, T.K., Holmes, W., Chen, Y., et al., 2023. A review of PFAS adsorption from aqueous solutions: current approaches, engineering applications, challenges, and opportunities. Environmental Pollution 321, 121138.

Lei, Y.-J., Tian, Y., Sobhani, Z., Naidu, R., Fang, C., 2020. Synergistic degradation of PFAS in water and soil by dual-frequency ultrasonic activated persulfate. Chemical Engineering Journal 388, 124215. Available at https://doi.org/10.121016/j.cej.2020.124215.

Leonello, D., Fendrich, M.A., Parrino, F., Patel, N., Orlandi, M., Miotello, A., 2021. Light-induced advanced oxidation processes as PFAS remediation methods: a review. Applied Sciences 11, 8458. Available at https://doi.org/10.3390/app11188458.

Leung, S.C.E., Wanninayake, D., Chen, D., Nguyen, N.-T., Li, Q., 2023. Physicochemical properties and interactions of perfluoroalkyl substances (PFAS) - challenges and opportunities in sensing and remediation. Science of the Total Environment 905, 166764. Available at https://doi.org/10.161016/j.scitotenv.2023.166764.

Li, F., Wei, Z., He, K., Blaney, L., Cheng, X., Xu, T., et al., 2020. A concentrate-and-destroy technique for degradation of perfluorooctanoic acid in water using a new adsorptive photocatalyst. Water Research 185, 116219. Available at https://doi.org/10.111016/j.watres.2020.116219.

Li, M., Yu, Z., Liu, Q., Sun, L., Huang, W., 2016. Photocatalytic decomposition of perfluorooctanoic acid by noble metallic nanoparticles modified $TiO_2$. Chemical Engineering Journal 286, 232–238. Available at https://doi.org/10.1016/j.cej.2015.1010.1037.

Liao, G., She, A., Chu, W., Zuo, J., Yao, W., 2021. Zeolite-loaded titanium dioxide photocatalytic cement-based materials for efficient degradation of drinking water disinfection byproduct trichloroacetamide. Frontiers in Materials Volume 8 (2021), 674287. Available at https://doi.org/10.673389/fmats.2021.674287.

Licato, J.J., Foster, G.D., Huff, T.B., 2022. Zeolite composite materials for the simultaneous removal of pharmaceuticals, personal care products, and perfluorinated alkyl substances in water treatment. ACS ES&T Water 2, 1046–1055. Available at https://doi.org/10.1021/acsestwater.1042c00024.

Lin, H., Xiao, R., Xie, R., Yang, L., Tang, C., Wang, R., et al., 2021. Defect engineering on a $Ti_4O_7$ electrode by $Ce^{3+}$ doping for the efficient electrooxidation of perfluorooctanesulfonate. Environmental Science & Technology 55, 2597–2607. Available at https://doi.org/10.1021/acs.est.2590c06881.

Liu, Z., Liao, M., Wang, L., Zhuang, S., 2025. Recent advancements in PFAS adsorptive removal using MOFs and COFs: a review. Reviews in Environmental Science and Bio/Technology 24, 63–95. Available at https://doi.org/10.1007/s11157-11024-09715-11151.

Lott, P., Maurer, F., Beck, A., 2025. Roadmap to catalytic abatement of gas phase per- and polyfluoroalkyl substances (PFAS). Angewandte Chemie International Edition 64. Available at https://doi.org/10.202421002/anie.202424718e202424718.

Lukyanov, D.B., Vazhnova, T., Cherkasov, N., Casci, J.L., Birtill, J.J., 2014. Insights into brønsted acid sites in the zeolite mordenite. The Journal of Physical Chemistry C 118, 23918–23929. Available at https://doi.org/10.21021/jp5086334.

Luo, H.Y., Lewis, J.D., Román-Leshkov, Y., 2016. Lewis acid zeolites for biomass conversion: perspectives and challenges on reactivity, synthesis, and stability. Annual review of chemical and biomolecular engineering 7, 663–692. Available at https://doi.org/10.1146/annurev-chembioeng-080615-034551.

Moliner, M., Martínez, C., 2015. Corma multipore zeolites: synthesis and catalytic applications. Angewandte Chemie International Edition 54, 3560–3579. Available at https://doi.org/10.1002/anie.201406344 Cormamultipore zeolites: synthesis and catalytic applications.

Mahdavi Far, R., Van der Bruggen, B., Verliefde, A., Cornelissen, E., 2022. A review of zeolite materials used in membranes for water purification: history, applications, challenges and future trends. Journal of Chemical Technology & Biotechnology 97, 575–596. Available at https://doi.org/10.1002/jctb.6963.

Mancinelli, M., Martucci, A., Ahrens, L., 2023. Exploring the adsorption of short and long chain per- and polyfluoroalkyl substances (PFAS) to different zeolites using environmental samples. Environmental Science: Water Research & Technology 9, 2595–2604. Available at https://doi.org/10.1039/D2593EW00225J.

Mancinelli, M., Martucci, A., Salani, G.M., Bianchini, G., Gigli, L., Plaisier, J.R., et al., 2023. High temperature behaviour of Ag-exchanged Y zeolites used for PFAS sequestration from water. Physical Chemistry Chemical Physics 25, 20066–20075. Available at https://doi.org/10.21039/D20063CP01584J.

Mancinelli, M., Stevanin, C., Ardit, M., Chenet, T., Pasti, L., Martucci, A., 2022. PFAS as emerging pollutants in the environment: a challenge with FAU type and silver-FAU exchanged zeolites for their removal from water. Journal of Environmental Chemical Engineering 10, 108026. Available at https://doi.org/10.101016/j.jece.2022.108026.

Masel, R.I., 1996. Principles of Adsorption and Reaction on Solid Surfaces. John Wiley & Sons.

Mensah, J., Yan, P., Kennedy, E., Drewery, M., Stockenhuber, M., 2021. Novel hierarchical core-shell BEA@NanoZSM-5 zeolite for improved cracking performance for 1,3,5-triisopropylbenzene and n-hexadecane. Microporous and Mesoporous Materials 328, 111399. Available at https://doi.org/10.111016/j.micromeso.2021.111399.

Min, X., Huo, J., Dong, Q., Xu, S., Wang, Y., 2022. Enhanced sorption of perfluorooctanoic acid with organically functionalized layered double hydroxide. Chemical Engineering Journal 446, 137019. Available at https://doi.org/10.131016/j.cej.2022.137019.

Murray, C.C., Vatankhah, H., McDonough, C.A., Nickerson, A., Hedtke, T.T., Cath, T.Y., et al., 2019. Removal of per- and polyfluoroalkyl substances using super-fine powder activated carbon and ceramic membrane filtration. Journal of Hazardous Materials 366, 160–168. Available at https://doi.org/10.1016/j.jhazmat.2018.1011.1050.

Möller, K., Bein, T., 2013. Mesoporosity – a new dimension for zeolites. Chemical Society Reviews 42, 3689–3707. Available at https://doi.org/10.1039/C3683CS35488A.

Nassi, M., Sarti, E., Pasti, L., Martucci, A., Marchetti, N., Cavazzini, A., et al., 2014. Removal of perfluorooctanoic acid from water by adsorption on high surface area mesoporous materials. Journal of Porous Materials 21, 423–432. Available at https://doi.org/10.1007/s10934-10014-19788-10935.

Navidpour, A.H., Hosseinzadeh, A., Zhou, J.L., Huang, Z., 2021. Progress in the application of surface engineering methods in immobilizing $TiO_2$ and ZnO coatings for environmental photocatalysis. Catalysis Reviews 65, 822–873. Available at https://doi.org/10.1080/01614940.2021.01983066.

Navidpour, A.H., Safaei, J., Johir, M.A.H., Ni, B.-J., Dashti, A., Li, X., et al., 2024. Zinc oxide@ citric acid-modified graphitic carbon nitride nanocomposites for adsorption and photocatalytic degradation of perfluorooctanoic acid. Advanced Composites and Hybrid Materials 7, 53. Available at https://doi.org/10.1007/s42114-42024-00867-w.

Ni, L., Khare, R., Bermejo-Deval, R., Zhao, R., Tao, L., Liu, Y., et al., 2022. Highly active and selective sites for propane dehydrogenation in zeolite Ga-BEA. Journal of the American Chemical Society 144, 12347–12356. Available at https://doi.org/10.11021/jacs.12342c03810.

Niethammer, B., Arnold, U., Sauer, J., 2023. Suppressing the aromatic cycle of the dimethyl ether to hydrocarbons reaction on zeolites. Applied Catalysis A: General 651, 119021. Available at https://doi.org/10.111016/j.apcata.2023.119021.

Niethammer, B., Zormpa, F., Hoang, G.T., Slaby, N.A., Zevaco, T.A., Karakoulia, S., et al., 2025. Conversion of dimethyl ether to hydrocarbons catalyzed by Pd-loaded *MRE zeolites. Catalysis Today 453, 115258. Available at https://doi.org/10.111016/j.cattod.2025.115258.

Park, M., Wu, S., Lopez, I.J., Chang, J.Y., Karanfil, T., Snyder, S.A., 2020. Adsorption of perfluoroalkyl substances (PFAS) in groundwater by granular activated carbons: roles of hydrophobicity of PFAS and carbon characteristics. Water Research 170, 115364. Available at https://doi.org/10.111016/j.watres.2019.115364.

Pastvova, J., Kaucky, D., Moravkova, J., Rathousky, J., Sklenak, S., Vorokhta, M., et al., 2017. Effect of enhanced accessibility of acid sites in micromesoporous mordenite zeolites on hydroisomerization of n-hexane. ACS Catalysis 7, 5781–5795. Available at https://doi.org/10.1021/acscatal.5787b01696.

Paunović, V., Sushkevich, V., Rzepka, P., Artiglia, L., Hauert, R., Sik Lee, S., et al., 2022. Reactivation of catalysts for methanol-to-hydrocarbons conversion with hydrogen. Journal of Catalysis 407, 54–64. Available at https://doi.org/10.1016/j.jcat.2022.1001.1018.

Peeters, E., Calderon-Ardila, S., Hermans, I., Dusselier, M., Sels, B.F., 2022. Toward industrially relevant Sn-BETA zeolites: synthesis, activity, stability, and regeneration. ACS Catalysis 12, 9559–9569. Available at https://doi.org/10.1021/acscatal.9552c02527.

Ponge, C.A., Sheehan, N.P., Corbin, D.R., Peltier, E., Hutchison, J.M., Shiflett, M.B., 2024. Zeolites for sorption of PFAS from water. Industrial & Engineering Chemistry Research 63, 12102–12112. Available at https://doi.org/10.11021/acs.iecr.12104c00541.

Primo, A., Garcia, H., 2014. Zeolites as catalysts in oil refining. Chemical Society Reviews 43, 7548–7561. Available at https://doi.org/10.1039/C7543CS60394F.

Pérez-Botella, E., Valencia, S., Rey, F., 2022. Zeolites in adsorption processes: state of the art and future prospects. Chemical Reviews 122, 17647–17695. Available at https://doi.org/10.11021/acs.chemrev.17642c00140.

Qian, L., Georgi, A., Gonzalez-Olmos, R., Kopinke, F.-D., 2020. Degradation of perfluorooctanoic acid adsorbed on Fe-zeolites with molecular oxygen as oxidant under UV-A irradiation. Applied Catalysis B: Environmental 278, 119283. Available at https://doi.org/10.111016/j.apcatb.2020.119283.

Qian, L., Kopinke, F.-D., Georgi, A., 2021. Photodegradation of perfluorooctanesulfonic acid on Fe-zeolites in water. Environmental Science & Technology 55, 614–622. Available at https://doi.org/10.1021/acs.est.1020c04558.

Qian, L., Zhao, H., Schierz, A., Mackenzie, K., Georgi, A., 2024. A deep insight into perfluorooctanoic acid photodegradation using metal ion-exchanged zeolites. ACS ES&T Engineering 4, 748–757. Available at https://doi.org/10.1021/acsestengg.1023c00462.

Rahman, M.F., Peldszus, S., Anderson, W.B., 2014. Behaviour and fate of perfluoroalkyl and polyfluoroalkyl substances (PFASs) in drinking water treatment: a review. Water Research 50, 318–340. Available at https://doi.org/10.1016/j.watres.2013.1045.

Rastegar, S.F., Sadovska, G., Pilar, R., Moravkova, J., Kaucky, D., Brabec, L., et al., 2020. Analysis of decisive structural parameters of zeolites for alkylation of benzene with ethylene. Applied Catalysis A: General 591, 117379. Available at https://doi.org/10.111016/j.apcata.2019.117379.

Rather, R.A., Xu, T., Leary III, R.N., Zhao, D., 2024. Aqueous and solid phase photocatalytic degradation of perfluorooctane sulfonate by carbon-and Fe-modified bismuth oxychloride. Chemosphere 346, 140585. Available at https://doi.org/10.141016/j.chemosphere.2023.140585.

Ravi, M., Sushkevich, V.L., van Bokhoven, J.A., 2020. Towards a better understanding of Lewis acidic aluminium in zeolites. Nature Materials 19, 1047–1056. Available at https://doi.org/10.1038/s41563-41020-40751-41563.

Restivo, J., Orge, C.A., Soares, O.S.G.P., Pereira, M.F.R., 2024. A review of current and prospective catalytic routes for the management of PFAs contamination in water. Journal of Environmental Chemical Engineering 12, 112859. Available at https://doi.org/10.111016/j.jece.2024.112859.

Ruthven, D.M., Farooq, S., 1990. Air separation by pressure swing adsorption. Gas Separation & Purification 4, 141–148. Available at https://doi.org/10.1016/0950-4214(1090)80016-E.

Safajou, H., Khojasteh, H., Salavati-Niasari, M., Mortazavi-Derazkola, S., 2017. Enhanced photocatalytic degradation of dyes over graphene/Pd/$TiO_2$ nanocomposites: $TiO_2$ nanowires versus $TiO_2$ nanoparticles. Journal of Colloid and Interface Science 498, 423–432. Available at https://doi.org/10.1016/j.jcis.2017.1003.1078.

Samuel, M.S., Shang, M., Niu, J., 2022. Photocatalytic degradation of perfluoroalkyl substances in water by using a duo-functional tri-metallic-oxide hybrid catalyst. Chemosphere 293, 133568. Available at https://doi.org/10.131016/j.chemosphere.2022.133568.

Schrama, F.R.A., Massimi, S.E., Dooley, M.R., Trewyn, B.G., Vyas, S., Richards, R.M., 2024. Light-driven interfaces for PFAS detection and destruction. RSC Applied Interfaces 1, 833–845. Available at https://doi.org/10.1039/D1034LF00171K.

Serrano, D.P., Calleja, G., Botas, J.A., Gutierrez, F.J., 2007. Characterization of adsorptive and hydrophobic properties of silicalite-1, ZSM-5, TS-1 and Beta zeolites by TPD techniques. Separation and Purification Technology 54, 1–9. Available at https://doi.org/10.1016/j.seppur.2006.1008.1013.

Sheehan, N.P., Ponge, C.A., Pankratz, A., Hutchison, J.M., Laird, B.B., Nguyen, N.P., et al., 2025. Interference of PFAS sorption on zeolites from natural water characteristics. Chemosphere 378, 144414. Available at https://doi.org/10.141016/j.chemosphere.2025.144414.

Suib, S.L., Přech, J., Szaniawska, E., Čejka, J., 2023. Recent advances in tetra- (Ti, Sn, Zr, Hf) and pentavalent (Nb, V, Ta) metal-substituted molecular sieve catalysis. Chemical Reviews 123, 877–917. Available at https://doi.org/10.1021/acs.chemrev.1022c00509.

Sun, Q., Wang, N., Yu, J., 2021. Advances in catalytic applications of zeolite-supported metal catalysts. Advanced Materials 33, 2104442. Available at https://doi.org/10.2101002/adma.202104442.

Szostak, R., 1989. Molecular sieves, 1 ed., Springer Dordrecht.

Tonetto, G., Atias, J., de Lasa, H., 2004. FCC catalysts with different zeolite crystallite sizes: acidity, structural properties and reactivity. Applied Catalysis A: General 270, 9–25. Available at https://doi.org/10.1016/j.apcata.2004.1003.1042.

Treu, P., Sarma, B.B., Grunwaldt, J.-D., Saraçi, E., 2022. Oxidative cleavage of vicinal diols catalyzed by monomeric Fe-sites inside MFI zeolite. ChemCatChem 14. Available at https://doi.org/10.202201002/cctc.202200993.

Tsai, Y.-T., Yu-Chen Lin, A., Weng, Y.-H., Li, K.-C., 2010. Treatment of perfluorinated chemicals by electro-microfiltration. Environmental Science & Technology 44, 7914–7920. Available at https://doi.org/10.1021/es101964y.

Vogt, E.T.C., Weckhuysen, B.M., 2015. Fluid catalytic cracking: recent developments on the grand old lady of zeolite catalysis. Chemical Society Reviews 44, 7342–7370. Available at https://doi.org/10.1039/C5CS00376H.

Vu, C.T., Wu, T., 2022. Recent progress in adsorptive removal of per- and poly-fluoroalkyl substances (PFAS) from water/wastewater. Critical Reviews in Environmental Science and Technology 52, 90–129. Available at https://doi.org/10.1080/10643389.2020.11816125.

Wang, B., Guo, Y., Zhu, J., Ma, J., Qin, Q., 2023. A review on titanosilicate-1 (TS-1) catalysts: research progress of regulating titanium species. Coordination Chemistry Reviews 476, 214931. Available at https://doi.org/10.211016/j.ccr.2022.214931.

Wang, W., Luoming, W., Zhong, L., Yutang, F., Jing, D., Xiao, J., 2013. An overview of adsorbents in the rotary desiccant dehumidifier for air dehumidification. Drying Technology 31, 1334–1345. Available at https://doi.org/10.1080/07373937.2013.07792094.

Weitkamp, J., 2000. Zeolites and catalysis. Solid State Ionics 131, 175–188. Available at https://doi.org/10.1016/S0167-2738(1000)00632-00639.

Wen, J., Li, H., Ottosen, L.D.M., Lundqvist, J., Vergeynst, L., 2023. Comparison of the photocatalytic degradability of PFOA, PFOS and GenX using Fe-zeolite in water. Chemosphere 344, 140344. Available at https://doi.org/10.141016/j.chemosphere.2023.140344.

Xu, B., Ahmed, M.B., Zhou, J.L., Altaee, A., 2020. Visible and UV photocatalysis of aqueous perfluorooctanoic acid by $TiO_2$ and peroxymonosulfate: process kinetics and mechanistic insights. Chemosphere 243, 125366. Available at https://doi.org/10.121016/j.chemosphere.2019.125366.

Xu, B., Ahmed, M.B., Zhou, J.L., Altaee, A., Wu, M., Xu, G., 2017. Photocatalytic removal of perfluoroalkyl substances from water and wastewater: mechanism, kinetics and controlling factors. Chemosphere 189, 717–729. Available at https://doi.org/10.1016/j.chemosphere.2017.1009.1110.

Xu, T., Zhu, Y., Duan, J., Xia, Y., Tong, T., Zhang, L., et al., 2020. Enhanced photocatalytic degradation of perfluorooctanoic acid using carbon-modified bismuth phosphate composite: effectiveness, material synergy and roles of carbon. Chemical Engineering Journal 395, 124991. Available at https://doi.org/10.1016/j.cej.2020.124991.

Yan, H., Chen, Q., Lin, Y., Zu, Y., Zhang, J., Feng, T., et al., 2024. New insights into removal efficacy of perfluoroalkyl substances by layered double hydroxide and its composite materials. Chemical Engineering Journal Advances 19, 100636. Available at https://doi.org/10.101016/j.ceja.2024.100636.

Yan, P., Kennedy, E.M., Zhang, H., Stockenhuber, M., 2023. Catalytic hydropyrolysis of lignocellulosic biomass to BTX and biofuels over zeolite beta based catalysts. Fuel 332, 125946. Available at https://doi.org/10.121016/j.fuel.2022.125946.

Yang, L., He, L., Xue, J., Ma, Y., Xie, Z., Wu, L., et al., 2020. Persulfate-based degradation of perfluorooctanoic acid (PFOA) and perfluorooctane sulfonate (PFOS) in aqueous solution: review on influences, mechanisms and prospective. Journal of Hazardous Materials 393, 122405. Available at https://doi.org/10.121016/j.jhazmat.2020.122405.

Zelekew, O.A., Wang, Z., Gu, Y., Wu, Y.-n, 2024. Recent progress on photoactive heterogeneous photocatalysts for the degradation of Per-and Polyfluoroalkyl Substances (PFAS): mechanisms, DFT calculations, limitations, and future prospects. Journal of Environmental Chemical Engineering 12, 113323. Available at https://doi.org/10.111016/j.jece.2024.113323.

Zengel, D., Barth, S., Casapu, M., Grunwaldt, J.-D., 2021. The impact of pressure and hydrocarbons on NOx abatement over Cu- and Fe-zeolites at pre-turbocharger position. Catalysts 11, 336. Available at https://doi.org/10.3390/catal11030336.

Zhang, J., Zhu, X., Wang, G., Wang, P., Meng, Z., Li, C., 2017. The origin of the activity and selectivity of silicalite-1 zeolite for toluene methylation to para-xylene. Chemical Engineering Journal 327, 278–285. Available at https://doi.org/10.1016/j.cej.2017.06.114.

Zhang, M., Sicard, F., Erkal, T.S., Bowers, G.M., Yazaydin, A.O., 2023. The role of surface thermodynamics and kinetics in the removal of PFOA from aqueous solutions. Surfaces and Interfaces 41, 103271. Available at https://doi.org/10.101016/j.surfin.2023.103271.

Zhang, Q., Gao, S., Yu, J., 2023. Metal sites in zeolites: synthesis, characterization, and catalysis. Chemical Reviews 123, 6039–6106. Available at https://doi.org/10.1021/acs.chemrev.2c00315.

Zhang, Y., Wei, C., Song, X., Tang, Z., Li, Y., Zhang, M., 2025. Rapid and effective removal of PFOS from water using ultrathin MgAl-LDH nanosheets with defects. Chemical Engineering Journal 505, 158998. Available at https://doi.org/10.151016/j.cej.2024.158998.

Zheng, X., Ge, T.S., Wang, R.Z., 2014. Recent progress on desiccant materials for solid desiccant cooling systems. Energy 74, 280–294. Available at https://doi.org/10.1016/j.energy.2014.07.027.

Zhou, H., Yi, X., Hui, Y., Wang, L., Chen, W., Qin, Y., et al., 2021. Isolated boron in zeolite for oxidative dehydrogenation of propane. Science 372, 76–80. Available at https://doi.org/10.1126/science.abe7935.

CHAPTER

# 15

# Metal–organic frameworks for the remediation of per- and polyfluoroalkyl substances

*Stephen Sunday Emmanuel*[1,2] *and Ademidun Adeola Adesibikan*[1]

[1]Department of Industrial Chemistry, Faculty of Physical Sciences, University of Ilorin, Ilorin, Kwara, Nigeria [2]New Life Research Foundation, Atlanta, GA, United States

## 15.1 Introduction

Metal–organic frameworks (MOFs) are a rapidly expanding class of crystalline porous materials constructed from metal ions or metal clusters interconnected by organic ligands through coordination bonds (Oggianu et al., 2026; Yuldashboyevna et al., 2026). Following the pioneering work of Richard Robson on porous coordination networks and the subsequent development of the first crystalline metal–organic frameworks with strong metal–ligand bonds by Yaghi and co-workers in 1995, MOF materials have attracted increasing attention from the scientific community (Oggianu et al., 2026; Liu et al., 2026). The scientific significance of these materials was further highlighted by the 2025 Nobel Prize in Chemistry awarded to Richard Robson, Susumu Kitagawa, and Omar M. Yaghi for their contributions to the development of MOFs. Their modular, highly ordered structures and unique properties distinguished them from traditional porous materials such as activated carbon, zeolites, and polymeric resins. Over the years, MOFs have been positioned as versatile materials for a wide range of applications, including gas storage and separation, catalysis, sensing, and, more importantly, environmental remediation (Sharma et al., 2026a; Emmanuel et al., 2026a). Specifically, MOFs have found application in the remediation of various environmental pollutants such as dyes, pharmaceuticals, pesticides, per- and polyfluoroalkyl substances (PFAS), etc. (Kumar et al., 2026; Lee et al., 2025; Bayode et al., 2025).

Among these pollutants, PFAS are pollutants of particular concern owing to their hazardousness, stubbornness, and persistence because of their stable C–F bonds (Bayode et al., 2024; Chaix et al., 2025). In fact, they have been nicknamed "forever chemicals," and thus their increasing presence in the ecosystem, especially aquatic environments have

*Per- and Polyfluoroalkyl Substances (PFAS) as Environmental Contaminants*
DOI: https://doi.org/10.1016/B978-0-443-43856-1.00021-6

emerged as a critical global concern (Bayode et al., 2024; Chaix et al., 2025). PFAS are continuously introduced into the water bodies through industrial discharges, domestic wastewater, nonstick cookware, cosmetics, fire-fighting foams, textiles, agricultural runoff, aerospace, electronics, etc. (Chaix et al., 2025; Peter and Lee, 2025; Ismail et al., 2025). Conventional wastewater treatment plants are often ineffective at fully removing these forever chemicals, leading to their persistence in surface water, groundwater, drinking water, and even the environment at large. The continuous exposure of aquatic ecosystems and human beings to PFAS raises concerns related to endocrine disruption, cancer, and long-term ecological toxicity (Bayode et al., 2024).

In response to these challenges, advanced materials capable of selectively capturing and removing PFAS contaminants are increasingly being explored, and MOFs have attracted significant attention in this context due to their versatility, exceptional porosity, and high surface area. Moreover, through rational design of metal nodes, organic linkers, and framework architecture, MOFs can be engineered to promote strong host–guest interactions, enhance adsorption selectivity, and facilitate degradation (Ismail et al., 2025; Grancha et al., 2025).

Notably, as shown in Fig. 15.1, an extensive literature survey revealed that many studies have been carried out on the use of MOFs for PFAS remediation, and given the growing interest, there is a need to continuously consolidate and update recent/existing knowledge in this area. This book chapter primarily focuses on the removal of PFAS using MOFs.

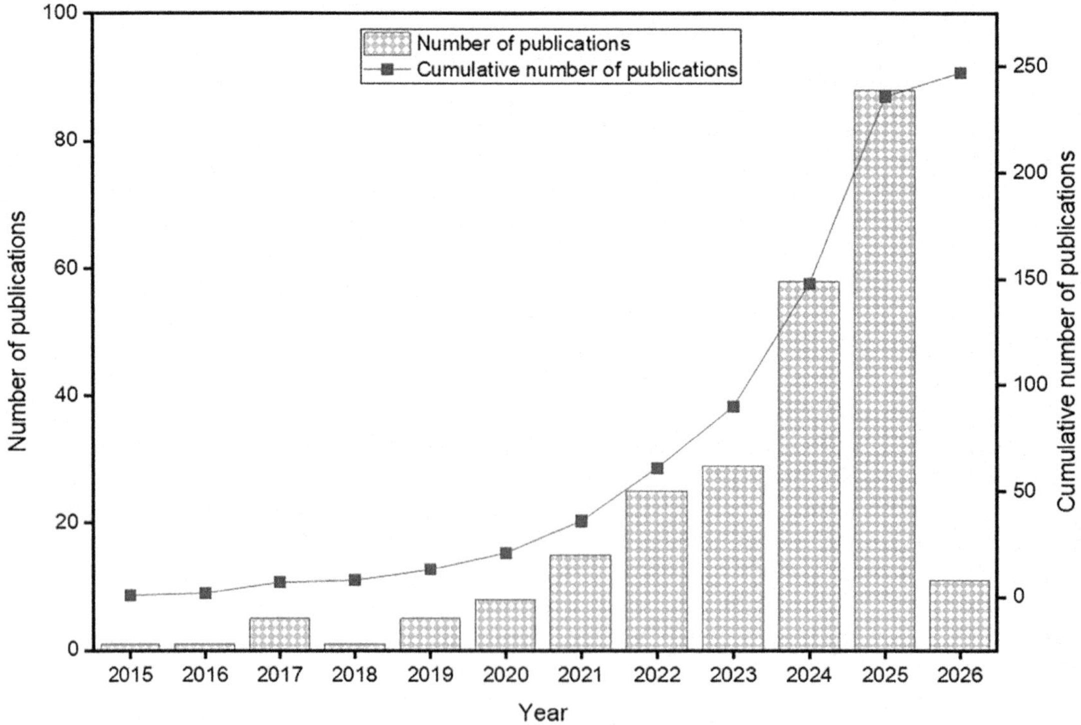


**FIGURE 15.1** Research/publication growth of metal–organic frameworks-per- and polyfluoroalkyl substances remediation-related studies.

The aim was to explore the recent trend of MOFs' performance. To set the study tone, this chapter firstly presents an overview of MOFs' unique structure, properties, and synthesis strategies. Thereafter, special attention was devoted to the performance evaluation, reusability efficiency/stability dynamics, and the potential for real PFAS wastewater treatment was discussed. At the end, current practical problematic issues and future research hotspots are also discussed to highlight pathways for translating MOF-based technologies from laboratory-scale studies to sustainable and effective PFAS wastewater treatment solutions on an industrial scale.

## 15.2 Fundamental structure and unique properties of metal–organic frameworks

MOFs are crystalline porous materials comprising metal ions/metal clusters and organic linkers connected by coordination bonds to form 1D, 2D, or 3D structures (Pala et al., 2023; Ahmadi et al., 2025; Shi et al., 2025). An overview of their general architecture is illustrated in Fig. 15.2 alongside their unique properties. Owing to their tunable pore sizes, exceptionally high surface areas, and highly tailorable structures, MOFs have attracted

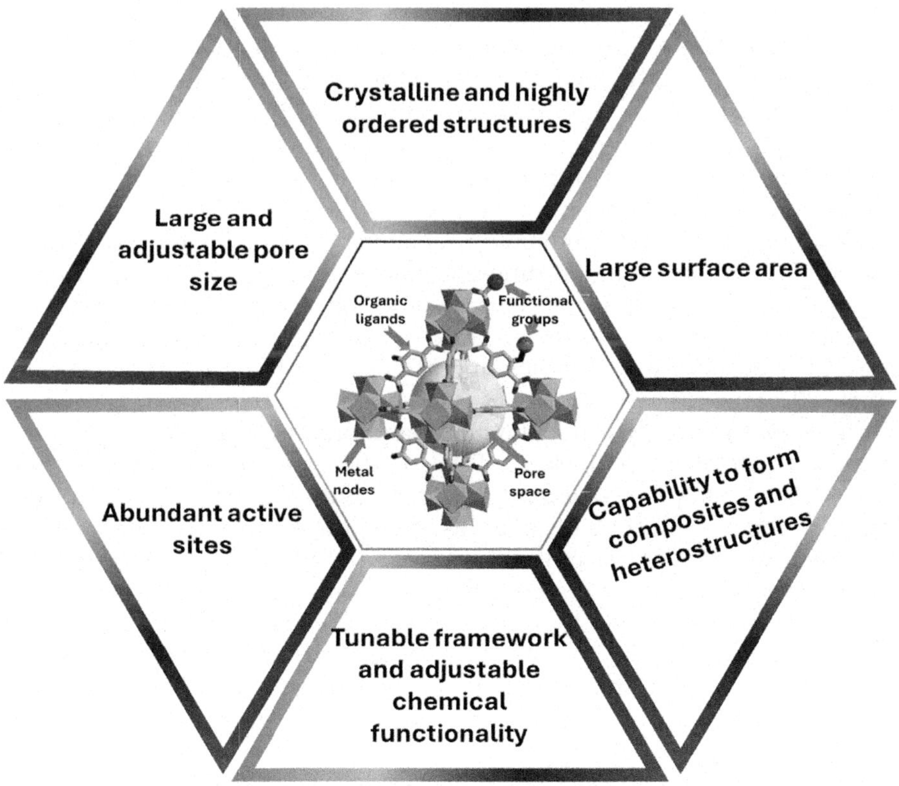


FIGURE 15.2 Unique properties of metal–organic frameworks.

considerable attention as advanced materials for environmental remediation, particularly for applications in water treatment (Jiao et al., 2019; Su et al., 2025).

In recent years, MOFs have emerged as promising adsorbents for the removal of various organic pollutants, including PFAS, from contaminated water (Shi et al., 2025; Su et al., 2025). Similar to other porous materials, the efficiency of PFAS adsorption by MOFs is governed largely by their physicochemical characteristics. Enhancing the adsorption performance of MOFs typically involves three main strategies: modifying the metal ions or clusters that serve as framework nodes, tailoring the organic ligands that construct the framework backbone, and optimizing the coordination mode to regulate pore structure. Through these structural adjustments, both adsorption capacity and selectivity toward PFAS can be significantly improved (Su et al., 2025; Liu et al., 2025).

Metal nodes form the structural core of MOFs and play a critical role in determining framework stability and adsorption behavior. MOFs based on chromium and zirconium are particularly stable in aqueous environments due to their strong metal–ligand bonds, high coordination numbers, and pronounced Lewis acidity, which explains their frequent use in PFAS adsorption studies. Beyond these, a wide range of metal ions-including zinc, nickel, cobalt, iron, copper, and magnesium-have also been incorporated into MOF structures. In such systems, the negatively charged functional head groups of PFAS molecules can interact strongly with the positively charged inorganic nodes through electrostatic attraction (Shi et al., 2025; Liu et al., 2025; Sharma et al., 2026b).

Organic ligands constitute the second key structural component of MOFs, forming the scaffold that connects metal nodes into extended networks (Ahmadi et al., 2025). Ligands bearing multiple functional groups generally provide a higher density of active sites for adsorbate binding, while specific chemical functionalities can enhance affinity toward particular PFAS molecules through surface interactions. The type and quantity of functional groups introduced into MOFs are largely controlled by synthesis and postsynthetic modification strategies. Fluorine-containing functional groups have been widely explored for MOF functionalization, although different fluorination approaches can lead to varying adsorption efficiencies in aqueous systems. In addition, the presence of defects within the organic linkers can be beneficial, as they increase accessible surface area, expose unsaturated metal sites, and create additional binding opportunities for PFAS molecules (Liu et al., 2025; Raghav et al., 2025).

The coordination mode between metal nodes and organic linkers strongly influences the overall framework architecture of MOFs. This coordination geometry determines key pore characteristics, such as pore size and channel configuration, which in turn govern adsorption behavior. For effective PFAS uptake, the pore dimensions of MOFs must exceed the kinetic diameter of PFAS molecules to allow their entry into the framework. Furthermore, the internal channel structure influences molecular diffusion and residence time within the pores. 2D MOFs typically possess 1D tubular channels that enable rapid molecular transport, whereas 3D MOFs exhibit interconnected pore networks with larger surface areas. This reticular architecture provides a greater number of adsorption sites, making 3D MOFs particularly effective for capturing a broad range of contaminants (Liu et al., 2025; Rezvani et al., 2025).

Additional structural parameters, including surface area and particle size, also affect PFAS adsorption performance. A large surface area generally enhances interactions between

MOFs and PFAS molecules, leading to improved removal efficiency. However, it has also been observed that some MOFs with relatively low surface areas can outperform high-surface-area counterparts, indicating that surface chemistry and binding site accessibility may be more critical than surface area alone. Particle size further influences remediation kinetics by controlling diffusion pathways. Smaller MOF particles shorten diffusion distances, accelerate mass transfer, and allow rapid occupation of available adsorption sites, resulting in faster removal rates (Liu et al., 2025; Ilić et al., 2025). Overall, while these structural properties of MOFs dictate their end use, these properties/functionalities are typically defined by the synthesis process, which is underscored in the succeeding section.

## 15.3 Common methods for metal–organic frameworks synthesis

Different methods for the fabrication of MOFs are available. However, this section concisely focuses on methods that have seen the light of scalability or have scalability potential. This includes hydrothermal, electrochemical, microwave, and solvothermal processes as summarized in Fig. 15.3, as per their merits and demerits (Raghav et al., 2025).

Notably, among these methods, solvothermal synthesis is the most widely used method for preparing MOFs. It involves dissolving metal salts and organic ligands into organic solvents and heating the mixture in a sealed vessel (e.g., Teflon-lined autoclave), typically

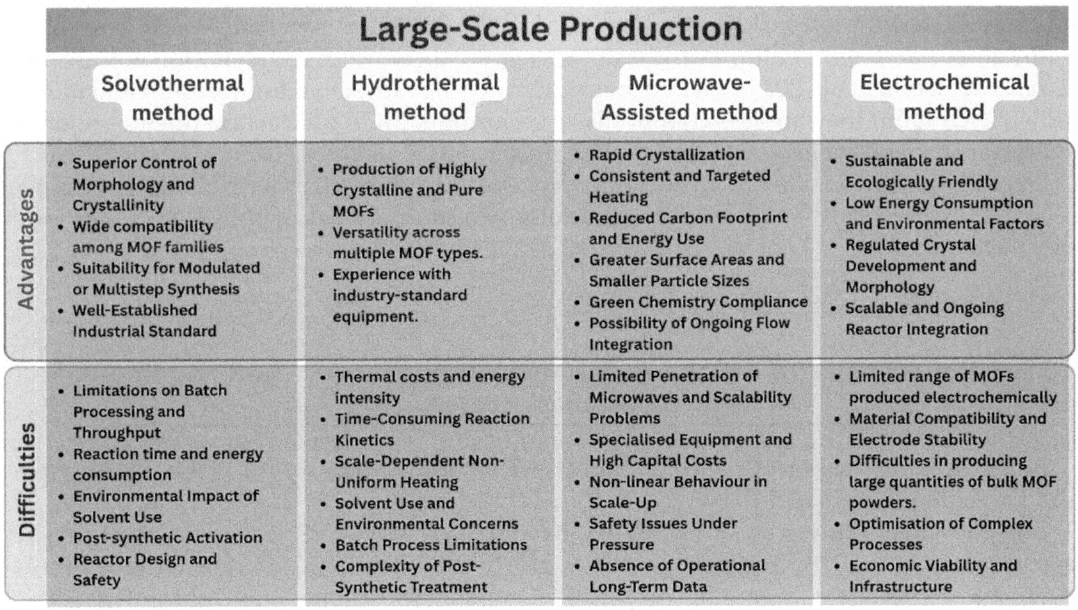


**FIGURE 15.3** Merits and demerits of common large-scale production metal–organic frameworks (MOFs) fabrication techniques (Raghav et al., 2025). *Reproduced with permission from Elsevier (Authorization code: 6185360225564).*

under autogenous pressure. This approach allows precise control over crystal growth, morphology, and framework topology by adjusting temperature, solvent type, reaction time, and modulators. Solvothermal methods generally produce highly crystalline MOFs with well-defined pore structures, although they often require long reaction times and large volumes of solvent (Gao et al., 2025; Main et al., 2026; Nguyen et al., 2026).

Microwave-assisted synthesis uses microwave irradiation to rapidly heat the reaction mixture, resulting in fast and uniform nucleation. Compared to conventional solvothermal routes, this method significantly reduces synthesis time from hours or days to minutes while maintaining good crystallinity. Microwave heating can also improve particle size uniformity and reduce energy consumption (Nguyen et al., 2026; Mahmoud, 2025). However, scalability and control over framework defects can be more challenging than in traditional solvothermal synthesis.

Hydrothermal synthesis is similar to solvothermal synthesis but uses water as the primary solvent. Reactions are carried out in sealed autoclaves at elevated temperatures and pressures. Precise control over reaction conditions is often realized in this method by heating the reaction mixture below the solvent's boiling point. This method is particularly attractive due to its environmental friendliness, low cost, and suitability for producing water-stable MOFs. Hydrothermal synthesis is commonly used for frameworks containing high-valent metal ions like MIL-88, although the limited solubility of some organic ligands in water can restrict its applicability (Gao et al., 2025; Ozdemir and Unlu, 2026; Chen et al., 2026).

Electrochemical synthesis involves the generation of metal ions directly from a metal electrode under an applied electric potential, which then coordinates with organic ligands in solution to form MOFs. This method eliminates the need for metal salts and allows continuous, room-temperature synthesis with precise control over metal ion release (Lian et al., 2025; Dong et al., 2025). Electrochemical routes are attractive for thin-film and surface-grown MOFs, but they are less commonly used for bulk MOF production due to limitations in crystal size and yield.

Overall, the various MOFs synthesis approaches highlighted above directly govern their behavior in practical remediation applications like the removal of PFAS discussed in the next sections.

## 15.4 Performance of metal–organic frameworks for per- and polyfluoroalkyl substances removal

In this section, the performance of various MOFs for the removal of PFAS was discussed and presented in Table 15.1.

In a particular study, MIL-101(Cr)@AC, 85% PFOS removal efficiency was accomplished, in contrast to MIL-101(Cr), which gave 80%. However, MIL-101(Cr) had faster kinetics (Pala et al., 2023). Similarly, in another experiment, under environmentally relevant conditions ($pH = 7$, dose $= 25$ mg), the investigated MOF demonstrated outstanding PFAS removal efficiency. Specifically, 80%–100% removal efficiencies were achieved for long-chain PFAS ($C_7$–$C_{12}$), such as PFDA, PFOS, PFUnDA, and PFDoDA. Notably, the MOF material also

**TABLE 15.1** Summary of metal–organic frameworks (MOFs)' performance for per- and polyfluoroalkyl substances (PFAS) removal.

| MOFs | PFAS | Optimal conditions | Removal efficiency (%) | Time (min.) | References |
|---|---|---|---|---|---|
| MIL-101(Cr)@AC | PFOS | Dose = 4 mg, temp. = 25°C, pH = 4 | 85 | 120 | Pala et al. (2023) |
| MIL-101(Cr) | PFOS | Dose = 4 mg, temp. = 25°C, pH = 4 | 80 | 15 | Pala et al. (2023) |
| Zr-MOF | PFBS | pH = 3, dose = 500 mg | >69 | 10 | Zhang et al. (2024) |
| Zr-MOF | PFHxS | pH = 3, dose = 500 mg | 71 | 10 | Zhang et al. (2024) |
| Zr-MOF | PFOS | pH = 3, dose = 500 mg | 77 | 10 | Zhang et al. (2024) |
| MOF | PFBA | pH = 7, dose = 25 mg | 70 | 1800 | Grancha et al. (2025) |
| MOF | PFBS | pH = 7, dose = 25 mg | 86 | 1800 | Grancha et al. (2025) |
| MOF | PFPeS | pH = 7, dose = 25 mg | 99.7 | 1800 | Grancha et al. (2025) |
| MOF | PFHxA | pH = 7, dose = 25 mg | 78.2 | 1800 | Grancha et al. (2025) |
| MOF | PFHxS | pH = 7, dose = 25 mg | 100 | 1800 | Grancha et al. (2025) |
| MOF | PFHpA | pH = 7, dose = 25 mg | 96.7 | 1800 | Grancha et al. (2025) |
| MOF | PFHpS | pH = 7, dose = 25 mg | 100 | 1800 | Grancha et al. (2025) |
| MOF | PFOA | pH = 7, dose = 25 mg | 99.7 | 1800 | Grancha et al. (2025) |
| MOF | PFOS | pH = 7, dose = 25 mg | 100 | 1800 | Grancha et al. (2025) |
| MOF | PFNA | pH = 7, dose = 25 mg | 99.7 | 1800 | Grancha et al. (2025) |
| MOF | PFDA | pH = 7, dose = 25 mg | 100 | 1800 | Grancha et al. (2025) |
| MOF | PFUnDA | pH = 7, dose = 25 mg | 100 | 1800 | Grancha et al. (2025) |
| MOF | PFDoDA | pH = 7, dose = 25 mg | 100 | 1800 | Grancha et al. (2025) |
| MIL-101(Fe)-$NH_2$@ MgCo-LDH | PFOA, PFOS, PFHxS, and PFBS | pH = 3, dose = 15 mg | >95 | 240 | Abouzied et al. (2026) |
| MIL-101(Cr) | PFOA | pH = 7, dose = 5 mg | >98 | - | Ismail et al. (2025) |
| Chitin-Zr-based MOF-808 | PFDA | pH = 6.8, dose = 5 mg, temp. = 25°C | ~86 | 30 | Dutta et al. (2025) |
| Chitin-Zr-based MOF-808 | PFOA | pH = 6.8, dose = 5 mg, temp. = 25°C | ~95 | 30 | Dutta et al. (2025) |

(*Continued*)

**TABLE 15.1** (Continued)

| MOFs | PFAS | Optimal conditions | Removal efficiency (%) | Time (min.) | References |
|---|---|---|---|---|---|
| ZIF-8 carbon | PFCAs, PFSAs, 6:2 FTS, and N-EtFOSAA | Dose = 5 mg | >99 | 1440 | Ilango et al. (2025) |
| UiO-67 | PFAS | Dose = 10 mg | 99 | 30 | Hua et al. (2025) |
| UiO-66-$N(CH_3)_3^+$ | PFOA | Dose = 500 mg | >99 | 5 | Dalapati et al. (2025) |
| ZIF-67@$C_3N_4$ | PFOA | Dose = 50 mg | 79.2 | 360 | Su et al. (2023) |
| MIL-101(Fe) | PFOA | pH = 3, temp. = 25°C, dose = 300 mg | ~100 | 30 | Qi et al. (2025) |
| UiO-66 | PFOA | NA | 73.6 | NA | Luo et al. (2025) |
| UiO-66 | PFNA | NA | 90.1 | NA | Luo et al. (2025) |
| UiO-66 | PFDA | NA | 96.2 | NA | Luo et al. (2025) |
| UiO-66 | PFOS | NA | 99.4 | NA | Luo et al. (2025) |
| UiO-66 | PFNS | NA | 78.5 | NA | Luo et al. (2025) |
| UiO-66 | PFOA | NA | 96 | NA | Di Palma et al. (2025) |

*NA*, Not available.

exhibited substantial affinity toward short-chain PFAS, attaining removal efficiencies of 70% for PFBA and 86% for PFBS. This level of performance, particularly for short-chain analogs that are typically difficult to remove by conventional materials like activated carbon, highlights the superior adsorption capability of MOFs and underscores their potential as next-generation materials for effective PFAS remediation (Grancha et al., 2025).

In another study, an optimized MIL-101(Fe)-$NH_2$@MgCo-LDH composite exhibited exceptional PFAS removal performance, achieving >98% elimination of PFOA, PFOS, PFHxS, and PFBS. The MOF composite delivered a remarkably enhanced adsorption capacity (747.5 mg $g^{-1}$) and faster PFOA uptake relative to pristine MIL-101(Fe) and LDH counterparts, underscoring the benefits of interfacial hybridization. The PFAS remediation kinetics data were well described by the pseudo-second-order, indicating a chemisorption-dominated, heterogeneous surface interactions remediation process. Notably, the removal efficiency increased systematically with perfluoroalkyl chain length ($C_4$–$C_9$), rising from 82.5% to 98.9%, which is suggested to be due to strengthened hydrophobic and fluorophilic interactions. In addition, the MOF composite maintained >95% PFAS removal in the presence of $Cl^-$ and $NO_3^-$, while competitive suppression was observed for $F^-$ and $SO_4^{2-}$. Mechanistic analyses attributed PFAS capture to synergistic Lewis acid–base coordination, electrostatic attraction, and hydrophobic/fluorophilic partitioning (Abouzied et al., 2026).

Recently, bio-derived functionalization strategies have been shown to enhance PFAS remediation by MOFs. In a particular study, a chitin-modified MOF-808 achieved rapid PFAS removal at a low adsorbent dosage (5 mg), reaching approximately 86% removal of PFDA and 95% removal of PFOA within 30 min. The mechanistic investigations revealed that the interfacial synergy between the chitin matrix and the MOF framework created site-specific binding environments with elevated PFAS binding energies, thereby accelerating uptake kinetics and improving overall removal efficiency. These findings further corroborate the growing claim that functionalization of MOFs with biopolymers can substantially amplify their removal performance for PFAS (Dutta et al., 2025).

Another recent emerging trend is in the area of the application of heterojunction MOF composite for the photocatalytic degradation of PFAS. In one study, ZIF-67 was grown in situ on g-$C_3N_4$ via a hydrothermal route to construct a heterostructure photocatalyst for PFOA degradation. Compared with bare g-$C_3N_4$, the resulting ZIF-67@g-$C_3N_4$ composite exhibited a substantially increased surface area, which promoted PFAS adsorption and facilitated interfacial contact. Moreover, the formation of a MOF-semiconductor heterojunction enhanced visible-light absorption and suppressed electron-hole recombination, thereby improving photocatalytic activity. Specifically, the hybrid MOF material achieved a PFOA removal efficiency of approximately 79%, remarkably outperforming pristine g-$C_3N_4$ under identical conditions. These findings highlight the synergetic promise of MOF-enabled heterojunction engineering for PFAS photocatalytic degradation (Su et al., 2023). The performance of MOFs in other PFAS removal studies is summarized in Table 15.1. Notably, as presented in Table 15.1, various MOF materials (with a dose range of 4–500 mg), especially composites, displayed excellent remediation efficiency (>70%–100% in 5–1800 min) for different PFAS. This is remarkable, as it points to the level of advancement in MOF as coordinated porous materials for PFAS removal.

## 15.5 Reusability performance and stability potential

This section is devoted to the evaluation of the reusability and stability of MOFs for the removal of PFAS. Generally, the reusability performance and stability of a material speak volumes about its industrial applicability potential (Luo et al., 2025). This evaluation usually involves the recovery of the spent material, regeneration, or elution prior to reuse (Emmanuel and Adesibikan, 2025; Emmanuel et al., 2026b).

For example, in a particular reusability study, following the removal of PFOS by MIL-101(Cr)@AC, EtOH and $H_2O$ (50:50) was employed as the regenerating agent to restore the performance capacity of the spent MOF material. The findings revealed that after four cycles, PFOS removal efficiency was still over 70%. This diminution in removal efficiency was suggested to be a result of an alteration in the crystallinity of MIL-101 (Cr) @AC (Pala et al., 2023). A similar reusability efficiency of 69.1% to 100% at the 15th cycle was recorded using MOF for the removal of PFBS, PFBA, PFPeS, PFHxS, PFHxA, PFHpS, PFHpA, PFOA, PFOS, PFNA, PFDA, PFUnDA, and PFDoDA. Additionally, as established by ICP-MS analysis, no metal leaching was noticed for the MOF material during the PFAS remediation process. This was corroborated by the PXRD patterns of both reused and pristine MOF, which were similar in terms of crystallinity, confirming its excellent structural stability and industrial applicability potential (Grancha et al., 2025). In a related study, a mixed eluent consisting of 30% NaCl and 70% MeOH enabled near-complete PFAS desorption (>99.6%), thereby ensuring effective regeneration of the UiO-66 framework. According to the study findings, the high polarity of MeOH facilitated PFAS transfer from the MOF's surface into the liquid phase, while the presence of NaCl attenuated electrostatic binding interactions, collectively promoting efficient desorption. Interestingly, UiO-66 maintained excellent chemical stability in methanol, preserving its structural integrity throughout repeated cycling. As a result, the material demonstrated strong recyclability over five adsorption–desorption cycles, retaining removal efficiencies above 69.8% for PFOA and 98.5% for PFOS. These findings further emphasize the importance of solvent–salt synergy in designing effective regeneration protocols for MOF-based PFAS treatment systems (Luo et al., 2025).

In another reusability study, the operational stability of the MIL-101(Fe)-$NH_2$@MgCo-LDH was evaluated through repeated adsorption–regeneration cycles. Across five successive runs, PFOA removal efficiency showed only a modest decline, decreasing from 96.8% initially to 92.0% by the fifth cycle. This marginal performance loss (<5%) suggests limited adsorptive site degradation or partial pore obstruction, while indicating that the primary PFAS uptake interactions remain largely reversible under the applied regeneration conditions. In parallel, potential secondary contamination was assessed by analyzing treated effluents for metal leaching using ICP-OES. Detected concentrations of Fe (0.0498 mg $L^{-1}$), Co (0.0135 mg $L^{-1}$), and Mg (0.2024 mg $L^{-1}$) were well below commonly accepted discharge thresholds. Collectively, these findings demonstrate strong chemical stability and minimal metal release, reinforcing MIL-101(Fe)-$NH_2$@MgCo-LDH suitability for repeated use in real PFAS-contaminated water treatment (Abouzied et al., 2026). Likewise, the stability and reusability of ZIF-8 carbon were examined through repeated adsorption–regeneration cycles. In this study, the performance of the MOF was evaluated four times for the removal of a complex mixture comprising 11 PFAS (PFHxA,

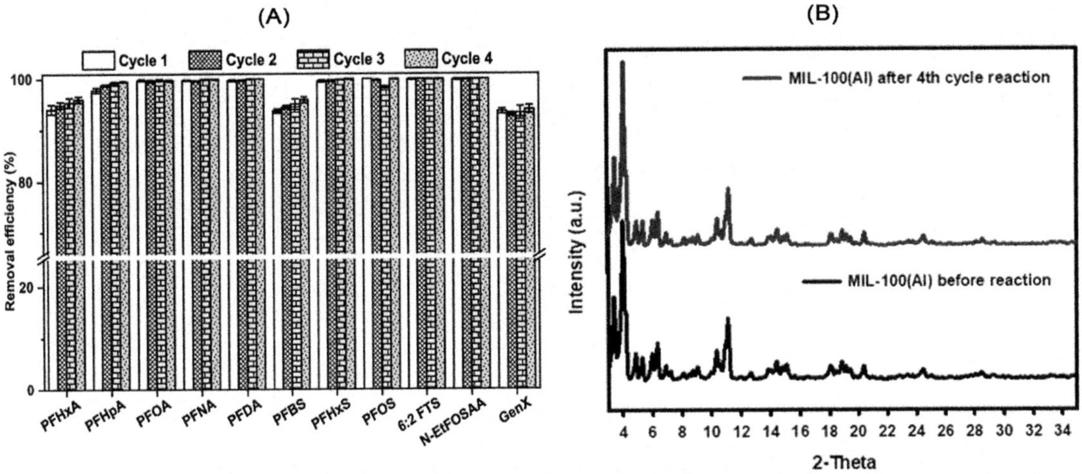


**FIGURE 15.4** (A) Performance of ZIF-8 carbon for per- and polyfluoroalkyl substances removal for four reusability–regeneration rounds (Ilango et al., 2025). a) Reproduced with permission from Elsevier (Authorization code: 6186581417093). (b) XRD analysis of MIL-100(Al) before and after the recycling test for the removal of PFOA (Yoon et al., 2025). *Reproduced with permission from Elsevier (Authorization code: 6186601328578).*

PFHpA, PFOA, PFNA, PFDA, PFBS, PFOS, 6:2 FTS, N-EtFOSAA, and GenX). Following regeneration with MeOH, the MOF retained high uptake efficiencies for both long- and short-chain PFAS, although a minor decline was observed after the first regeneration operation. Specifically, as displayed in Fig. 15.4A, removal efficiencies were still >90% after four cycles, and the minor reduction is suspected to be due to partial desorption or incomplete recovery during the regeneration process. Overall, the results highlight the strong regenerative potential of ZIF-8 carbon, its cost-effectiveness, and suitability for real-world application while underscoring the challenges associated with fully restoring adsorption sites for highly mobile PFAS (Ilango et al., 2025). Comparable stability and reusability performance were also reported for the removal of PFOA using MIL-100(Al) in four consecutive rounds. Moreover, as displayed in Fig. 15.4B, the XRD examination showed that there were no notable alterations in the crystal structure following the fourth cycle, demonstrating MIL-100(Al) structural stability throughout the recycling test and potential for pilot-scale application. The reusability performance of other various MOFs for PFAS removal is summarized in Table 15.2.

Notably, as noted from the recyclability data presented in Table 15.2 that the least number of times that MOF can be reused was 4, whereas the highest number of times was 15, the average PFAS removal efficiency after nth cycle (last cycle) of application is greater than 70% in most cases, and different desorbing/regenerating agents that have been explored to restore the integrity of spent MOF materials include MeOH, EtOH, NaCl, HCl, $HNO_3$, and deionized/distilled water. Among these, EtOH and water are recommended due to availability, low cost, eco-friendliness, and safety.

**TABLE 15.2** Summary of recyclability potential of metal–organic frameworks (MOFs) for the removal of per- and polyfluoroalkyl substances (PFAS).

| MOFs | PFAS | Eluting/regenerating agent | No. of cycles | Removal efficiency in the last cycle (%) | Reference |
|---|---|---|---|---|---|
| MIL-101(Cr)@AC | PFOS | EtOH and $H_2O$ (50:50) | 4 | >70 | Pala et al. (2023) |
| Zr-MOF | PFOS | 0.1 M HCl and MeOH (30/70) | 5 | >95 | Zhang et al. (2024) |
| MOF | PFBA | MeOH | 15 | 69.1 | Grancha et al. (2025) |
| MOF | PFBS | MeOH | 15 | 78.3 | Grancha et al. (2025) |
| MOF | PFPeS | MeOH | 15 | 98.8 | Grancha et al. (2025) |
| MOF | PFHxA | MeOH | 15 | 81.3 | Grancha et al. (2025) |
| MOF | PFHxS | MeOH | 15 | 99.5 | Grancha et al. (2025) |
| MOF | PFHpA | MeOH | 15 | 94.1 | Grancha et al. (2025) |
| MOF | PFHpS | MeOH | 15 | 100 | Grancha et al. (2025) |
| MOF | PFOA | MeOH | 15 | 97.7 | Grancha et al. (2025) |
| MOF | PFOS | MeOH | 15 | 100 | Grancha et al. (2025) |
| MOF | PFNA | MeOH | 15 | 99.7 | Grancha et al. (2025) |
| MOF | PFDA | MeOH | 15 | 100 | Grancha et al. (2025) |
| MOF | PFUnDA | MeOH | 15 | 100 | Grancha et al. (2025) |
| MOF | PFDoDA | MeOH | 15 | 100 | Grancha et al. (2025) |
| MIL-101(Fe)-$NH_2$@MgCo-LDH | PFOA | 70% MeOH + 1% NaCl | 5 | 92.0 | Abouzied et al. (2026) |
| MIL-101(Cr) | PFOA | NA | 7 | >75 | Ismail et al. (2025) |
| Chitin-Zr-based MOF-808 | PFDA | MeOH + 1% NaCl | 5 | ~90 | Dutta et al. (2025) |
| Chitin-Zr-based MOF-808 | PFOA | MeOH + 1% NaCl | 5 | ~85 | Dutta et al. (2025) |

| | | | | | |
|---|---|---|---|---|---|
| ZIF-8 carbon | PFHxA, PFHpA, PFOA, PFNA, PFDA, PFBS, PFHxS, PFOS, 6:2 FTS, N-EtFOSAA, GenX | MeOH and deionized water | 4 | >90 | Ilango et al. (2025) |
| MIL-100(Al) | PFOA | $HNO_3$ | 4 | >80 | Yoon et al. (2025) |
| UiO-66-$N(CH_3)_3^+$ | PFOA | 0.2 M HCl and MeOH (30:70 v/v) | 5 | 93 | Dalapati et al. (2025) |
| MIL-101(Fe) | PFOA | MeOH | 4 | >90 | Qi et al. (2025) |
| UiO-66 | PFOS | 30% NaCl + 70% MeOH | 5 | 98.5 | Luo et al. (2025) |
| UiO-66 | PFOA | 30% NaCl + 70% MeOH | 5 | 69.8 | Luo et al. (2025) |

*NA*, Not available.

## 15.6 Real per- and polyfluoroalkyl substances wastewater remediation and pilot-scale applicability potential

In recent times, research on PFAS removal has advanced to the applicability of MOF-based materials under environmentally relevant conditions using real water matrices from lakes and surface water. This is to evaluate the industrial application potential of MOFs and the possibility for scale-up.

For example, in a particular study, the performance of MOF for the removal of various long- and short-chain PFAS from the Turia River in Valencia, Spain, was evaluated. It was shown that 72.9%–100% removal efficiency was achieved by the MOF in contrast to activated carbon, that achieve far less for PFBS, PFBA, PFPeS, PFHxS, PFHxA, PFHpS, PFHpA, PFOA, PFOS, PFNA, PFDA, PFUnDA, and PFDoDA. This degree of success emphasizes the greater adsorption capacity of MOFs and shows their promise as next-generation materials for efficient PFAS remediation, especially for short-chain analogs that are usually challenging to remove by conventional materials like activated carbon (Grancha et al., 2025). In a similar case study, MIL-101(Fe)-$NH_2$@MgCo-LDH achieved excellent PFAS removal effectiveness in river water. Specifically, under optimized conditions, 96.5% to 78.4% of PFDA, PFNA, PFOA, PFHpA, PFHxA, PFPeA, and PFBA were removed in that order. The observed chain-length effect emphasizes the role of hydrophobic/fluorophilic partitioning in the mesoporous framework, promoting enhanced capture of longer perfluoroalkyl moieties (Abouzied et al., 2026). In a comparable work, the selectivity of Chitin-Zr-based MOF-808 was systematically evaluated using multicomponent PFAS solutions. In this study, Chitin-Zr-based MOF-808 exhibited substantially higher adsorption performance than the pristine chitin cryogel for an equimolar mixture of four PFAS (100 ppm each). The MOF material displayed pronounced selectivity toward carboxylate-terminated PFAS, achieving preferential removal of ~90% for PFOA and ~95% for PFDA relative to sulfonated analogs such as PFOS (~82%) and PNBS (~73%). Furthermore, the applicability of the MOFs composite under realistic water matrices was confirmed using PFAS-spiked lake water (Lake *Embalse de Gorostiza,* located on the west coast of Spain, in Bizkaia province) containing a broad spectrum of competing dissolved metal ions. Despite this chemical complexity, Chitin-Zr-based MOF-808 maintained appreciable removal efficiencies, following the trend PNBS < PFOS < PFOA < PFDA, and this is in very good agreement with the static binding energy pattern obtained using density functional theory PFDA > PFOA > PFOS > PNBS as displayed in Fig. 15.5. Notably, high adsorption capacities for PFOA (~300 $mg\ g^{-1}$) and PFDA (~320 $mg\ g^{-1}$) were sustained under these conditions, reinforcing the potential of Chitin-Zr-based MOF-808 as practical and selective PFAS sorbents in real-world water treatment scenarios (Dutta et al., 2025). In a related investigation, the performance of UiO-66 was assessed in municipal wastewater and surface water collected from Chaohu Lake to better approximate realistic PFAS contamination scenarios. Despite the complex background of the water matrices, the MOF material demonstrated a strong affinity toward long-chain PFAS spiked at environmentally relevant levels (1.0 $\mu g\ L^{-1}$). Notably, removal efficiencies exceeding 88% for PFOA and 97% for PFOS were achieved, underscoring the robustness and selectivity of UiO-66 in complex aqueous environments (Luo et al., 2025). These results further support the practical viability of MOFs for PFAS remediation beyond idealized laboratory conditions.

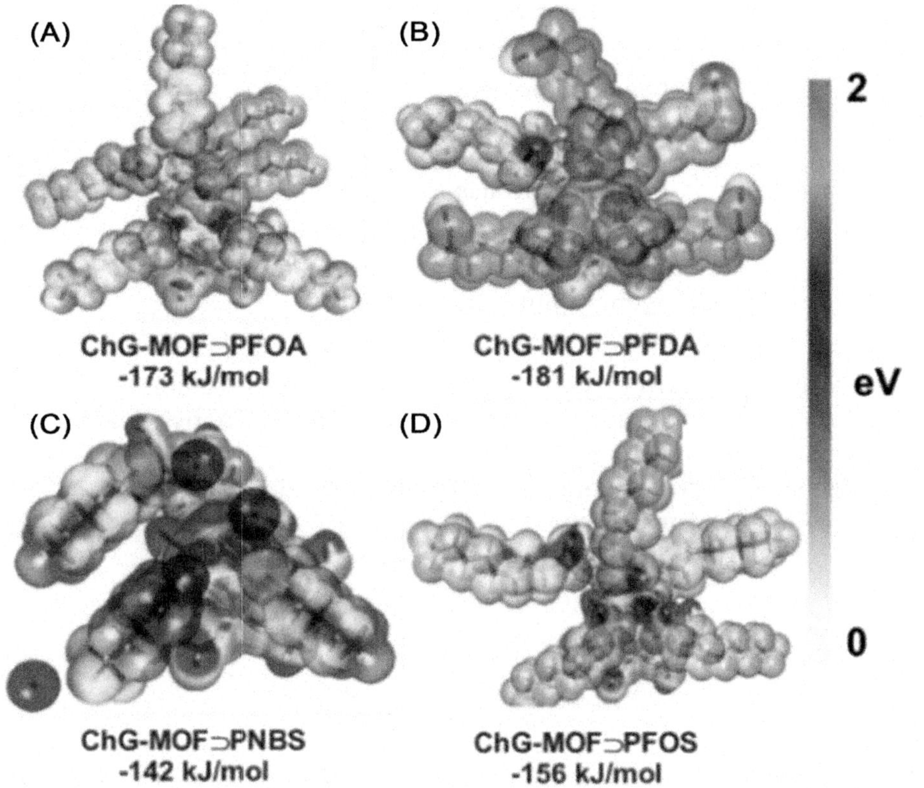


**FIGURE 15.5** DFT optimization-based energy-optimized binding sites (A) PFOA, (B) PFDA, (C) PNBS, (D) PFOS (Dutta et al., 2025). *Reproduced with permission from Elsevier (Authorization code: 6186541497090).*

## 15.7 Challenges and future research directions

Despite the remarkable performance of MOFs for the removal of PFAS, some scientific problematic issues and frontiers remain, as presented in Fig. 15.6 and outlined below. Some of these challenges have been recently highlighted by Raghav's research group (Raghav et al., 2025).

## 15.8 Synthesis efficiency and economic viability

**Challenges:** Conventional MOF synthesis methods rely on high temperatures, long reaction times, and hazardous organic solvents, resulting in high energy consumption and limited scalability. The cost of organic linkers and metal precursors further restricts large-scale application.

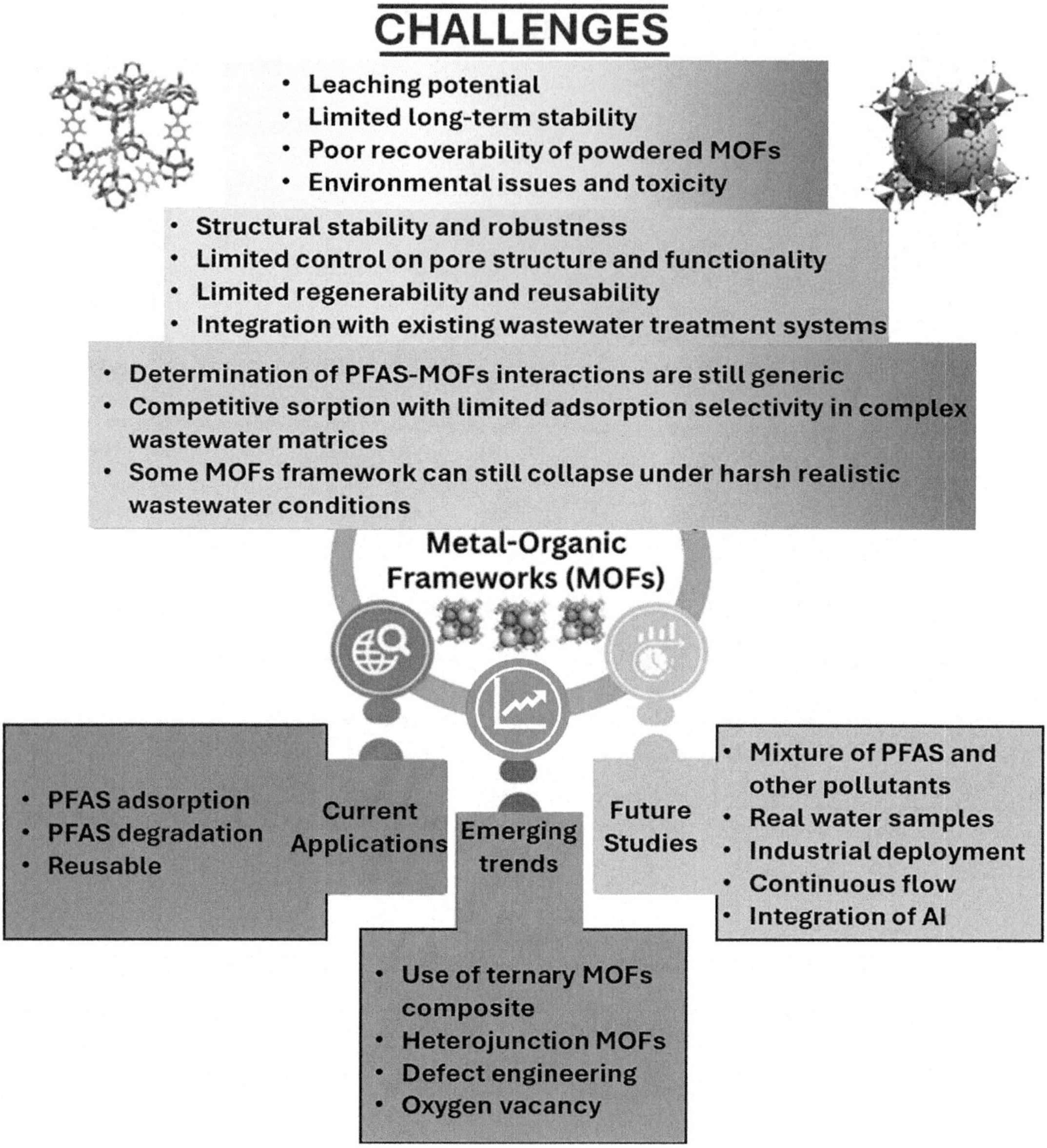


FIGURE 15.6 Overview of challenges and future research directions (Raghav et al., 2025). *Modified with permission from Elsevier (Authorization code: 6185360225564).*

**Future perspectives:** Development of green, low-energy synthesis routes such as mechanochemical, microwave-assisted, electrochemical, and continuous-flow methods is essential. Water-based, solvent-free, and scalable synthesis strategies will be critical for cost-effective PFAS remediation.

## 15.9 Hydrolytic and chemical stability in aqueous media

**Challenges:** Many MOFs suffer from hydrolysis and framework collapse under realistic wastewater conditions, particularly across varying pH ranges and in the presence of competing ions.

**Future perspectives:** Future research should focus more on water-stable MOFs based on strong metal–ligand bonds (e.g., Zr–O, Ti–O, Cr–N), hydrophobic surface modifications, and postsynthetic stabilization strategies to ensure long-term durability.

## 15.10 Toxicity, leaching, and environmental safety

**Challenges:** The potential toxicity of MOFs remains a critical concern for PFAS removal applications. Many MOFs contain transition metals or organic ligands that may leach into treated water during operation, regeneration, or long-term exposure, posing risks to aquatic ecosystems and human health. Framework degradation under variable pH, ionic strength, and oxidative conditions can release metal ions and ligand fragments, potentially generating secondary pollutants that persist in the environment.

**Future perspectives:** Future research should prioritize the development of nontoxic, low-leaching MOFs constructed from environmentally benign metals and biodegradable or bio-derived ligands. Comprehensive toxicity assessments, including leaching studies, ecotoxicological evaluations, and life-cycle analyses, are essential before large-scale deployment. Designing inherently stable or self-passivating MOFs that prevent metal release, as well as integrating toxicity screening into early-stage material design, will be crucial for safe and sustainable PFAS wastewater treatment.

## 15.11 Control of pore structure and active sites

**Challenges:** Precise control over defects and functional site distribution remains difficult. Poorly controlled modification can lead to pore blockage, reduced crystallinity, and diminished adsorption or photocatalytic performance.

**Future perspectives:** Rational pore engineering, defect control, and molecularly imprinted MOFs designed for PFAS-specific recognition can enhance selectivity and adsorption efficiency while maintaining structural integrity.

## 15.12 Selectivity in complex wastewater matrices

**Challenges:** In real wastewater, PFAS compete with dyes, pesticides, natural organic matter, and inorganic ions for adsorption sites, significantly reducing MOF efficiency.

**Future perspectives:** Selective functionalization using charged or hydrophobic groups and tailoring host–guest interactions can improve PFAS specificity. Computational screening can guide the design of MOFs with enhanced selectivity under competitive conditions.

## 15.13 Regeneration, reusability, and secondary pollution

**Challenges:** Repeated adsorption–desorption cycles often lead to irreversible adsorption, pore blockage, structural damage, and incomplete regeneration. Metal and ligand leaching raise environmental safety concerns.

**Future perspectives:** Low-energy regeneration techniques such as photoregeneration and electrochemical regeneration should be prioritized. Designing nonleaching, biodegradable, or self-regenerating MOFs can minimize secondary contamination risks.

## 15.14 Technoeconomic considerations in metal–organic framework-based per- and polyfluoroalkyl substances treatment

**Challenges:** Despite promising laboratory-scale performance, the practical deployment of MOF-based systems for PFAS removal is constrained by limited technoeconomic viability. High synthesis costs, reliance on expensive organic linkers and metal precursors, energy-intensive fabrication processes, and scalability issues significantly increase overall treatment costs. Additionally, the lack of standardized cost models and limited pilot-scale data further complicates economic assessment and comparison with conventional technologies such as activated carbon and ion exchange resins.

**Future perspectives:** Integrating technoeconomic analysis into the development of MOF-based treatment systems is essential to bridge the gap between laboratory research and industrial application. In the future, it is recommended that research efforts should be directed to low-cost, green synthesis routes, utilization of abundant or waste-derived precursors, and process optimization to reduce energy and material inputs. Coupling technoeconomic analysis with life-cycle assessment will provide a more holistic understanding of economic and environmental trade-offs. Moreover, pilot-scale demonstrations and process modeling can help identify cost drivers, optimize regeneration strategies, and evaluate long-term feasibility, ultimately guiding the design of economically competitive and scalable PFAS remediation technologies.

## 15.15 Mechanical stability

**Challenges:** Powdered MOFs suffer from poor mechanical strength, difficult recovery, pressure drops, and clogging in continuous-flow systems, limiting industrial implementation.

**Future perspectives:** Advanced shaping strategies, including pelletization, granulation, electrospinning, 3D printing, and membrane immobilization, can improve mechanical stability and enable seamless integration into existing PFAS wastewater treatment systems.

## 15.16 Real-world validation, regulation, and predictive design

**Challenges:** Most studies rely on simulated wastewater, lack standardized testing protocols, and face regulatory uncertainty regarding long-term safety and ecotoxicity.

**Future perspectives:** Future work should emphasize real wastewater testing, life-cycle assessment, and regulatory alignment. Artificial intelligence, machine learning, and computational modeling can enhance MOF design, performance prediction, and transition from laboratory research to practical PFAS remediation systems.

## 15.17 Conclusion

This chapter provides an overview of MOFs as emerging materials for the removal of PFAS contaminants from water systems. Owing to their tunable porosity, high surface area, and versatile coordination chemistry, MOFs have displayed strong adsorption and photocatalytic capabilities toward a wide range of PFAS pollutants. Notably, numerous studies report high removal efficiencies, often exceeding 70%–90% under optimized conditions, highlighting the effectiveness of MOFs in both adsorption- and degradation-based treatment. Studies show that structural modification, functionalization, and composites with materials such as activated carbon, MgCo-LDH, g-$C_3N_4$, etc., significantly enhance PFAS removal efficiency and stability. In particular, ternary MOFs composite outperform binary and pristine counterparts due to synergistic activities and functionalities. Also, several MOFs demonstrate promising reusability, maintaining substantial performance over multiple regeneration cycles (>70% PFAS removal efficiency in 4–15 cycles), which indicates their potential for practical and economically viable water treatment applications. However, challenges related to hydrolytic stability, toxicity, regeneration efficiency, long-term stability, scalability, and performance under realistic wastewater conditions remain key barriers to full-scale deployment. Future research should therefore prioritize the development of water-stable and environmentally benign MOFs, green and scalable synthesis routes, and standardized testing in complex wastewater matrices. The integration of artificial intelligence and computational tools for material design/screening, life-cycle assessment for environmental safety, and system-level integration with existing treatment technologies will be essential for translating MOF-based PFAS removal from laboratory studies to real-world implementation.

## References


Abouzied, A.S., Al, A.B.M., Alsaiari, N.S., Aldaghfag, S.A., Ismail, M.A., Karimov, M., et al., 2026. Enhanced removal of per- and polyfluoroalkyl substances (PFAS) by $NH_2$-functionalized MOF with ex-situ grown layered double hydroxides: mechanistic insights and adsorption behavior. Separation and Purification Technology 382, 136051. Available at https://doi.org/10.1016/j.seppur.2025.136051.

Ahmadi, P., Ahmadi, A., Tabibiazar, M., Hamidi, S., Ramezani, S., 2025. Current advances in the synthesis of beta-cyclodextrin-based metal organic frameworks and their potential applications in the adsorption/removal of pollutants: a comprehensive review. Carbohydrate Polymers 353, 123241. Available at https://doi.org/10.1016/j.carbpol.2025.123241.

Bayode, A.A., Emmanuel, S.S., Akinyemi, A.O., Ore, O.T., Akpotu, S.O., Koko, D.T., et al., 2024. Innovative techniques for combating a common enemy forever chemicals: a comprehensive approach to mitigating per- and polyfluoroalkyl substances (PFAS) contamination. Environmental Research 261, 119719. Available at https://doi.org/10.1016/j.envres.2024.119719.

Bayode, A.A., Emmanuel, S.S., Terlanga, K.D., 2025. Chapter 19 - Nanomaterials and their derivative biocomposites for dye adsorption. In: Jagaba, A.H., Mohamed Kutty, S.R., Isa, M.H., Birniwa, A.H.B.T.-EB for DA (Eds.), Engineering Biocomposites for Dye Adsorption. Elsevier, pp. 313–337. Available at https://doi.org/10.1016/B978-0-443-29877-6.00019-6.

Chaix, A., Gomri, C., Benkhaled, B.T., Habib, M., Dupuis, R., Petit, E., et al., 2025. Efficient PFAS removal using reusable and non-toxic 3D printed porous trianglamine hydrogels. Advanced Materials 37, 2410720.

Chen, W., Yao, M., Zheng, W., Li, J., Zheng, Y., Dong, L., 2026. A bimetallic MOF-based electrochemical sensor for detecting histamine in spiked wine samples. Food Analytical Methods 19, 1–12.

Dalapati, R., Shi, J., Hunter, M., Zang, L., 2025. Dual-functional metal–organic framework for efficient removal and fluorescent detection of perfluorooctanoic acid (PFOA) from water. Journal of Materials Chemistry C 13, 16753–16762.

Dong, H., Zheng, L., Jiang, Z., Wang, T., Li, X., Ma, Y., et al., 2025. In situ electrosynthesis of bimetallic CuNiMOF nanocomplexes on graphite paper for ratiometric electrochemical detection of maltol. Microchemical Journal 211, 113134.

Dutta, S., Aretxabaleta, M.A., Shirolkar, M.M., Wuttke, S., Lizundia, E., 2025. Fungal chitin nanofibril-MOF hybrids for PFAS remediation in water: mechanism and performance-sustainability trade-off. Chemical Engineering Journal 526, 170898. Available at https://doi.org/10.1016/j.cej.2025.170898.

Emmanuel, S.S., Adesibikan, A.A., 2025. Nanoremediation of chemical pollutants using green carbon nanomaterials. Sustainable Carbon Nanomaterials and Their Applications 5–29. Available at https://doi.org/10.1088/978-0-7503-6325-9ch5.

Emmanuel, S.S., Alanazi, A.K., Adesibikan, A.A., Olawoyin, C.O., Abimbola, E.T., Oluwole, O.J., 2026a. Graphene/graphitic-based metal-organic frameworks (MOFs) for photocatalytic degradation of pharmaceutical pollutants: a review. Journal of Organometallic Chemistry 1043, 123915. Available at https://doi.org/10.1016/j.jorganchem.2025.123915.

Emmanuel, S.S., Suleiman, A.I., Itopa, A.A., Adesibikan, A.A., Idris, M.O., 2026b. Zero-valent iron and its derivative composites for heavy metal adsorption. Composites for Biocomposites and Heavy Metal Adsorption. Elsevier, pp. 73–88.

Gao, Q., Sheng, Q., Zhang, S., Tang, Y., 2025. Recent trends on MIL-88 (Fe) metal–organic frameworks: synthesis and applications in pollutant removal and detection. RSC Advances 15, 26184–26200.

Grancha, T., García-Atienza, P., García, L., Armenta, S., Herrero-Martínez, J.M., Armentano, D., et al., 2025. Efficient capture of short-and long-chain PFAS from water by a metal–organic framework. Small (Weinheim an der Bergstrasse, Germany) 21, e10000.

Hua, L., Solomon, M.B., D'Alessandro, D.M., Donald, W.A., 2025. Dual-functional metal-organic frameworks for adsorptive removal and ultra-trace quantitation of 50 per- and polyfluoroalkyl substances in water. Journal of Hazardous Materials 494, 138679. Available at https://doi.org/10.1016/j.jhazmat.2025.138679.

Ilango, A.K., Mekkat, R., Jeyalakshmi, V., Pervez, M.N., Jiang, T., Chand, P., et al., 2025. Enhanced removal of PFAS in water using activated ZIF-8 carbons: high adsorption efficiency, repeatable regenerability and reusability. Chemical Engineering Journal 507, 160192. Available at https://doi.org/10.1016/j.cej.2025.160192.

Ilić, N., Tan, K., Mayr, F., Hou, S., Aumeier, B.M., Morales, E.M.C., et al., 2025. Trace adsorptive removal of PFAS from water by optimizing the UiO-66 MOF interface. Advanced Materials 37, 2413120.

Ismail, M.A., Taki, A.G., Kumar, S., Sammen, S.S., Amari, A., Bongale, A., et al., 2025. Effectiveness of waste-derived MIL type MOFs in removing PFOA and PFAS pollutants for environmental remediation. Scientific Reports 15, 9439.

Jiao, L., Seow, J.Y.R., Skinner, W.S., Wang, Z.U., Jiang, H.-L., 2019. Metal–organic frameworks: structures and functional applications. Materials Today 27, 43–68. Available at https://doi.org/10.1016/j.mattod.2018.10.038.

Kumar, B., Uttam, A., Agrahari, G.K., Sistla, V.S., 2026. Effect of distinct carboxylic acid modulators on Ni-Zn MOF synthesis and their application in "hydroxy naphthol blue" dye removal from aqueous solutions. Materials Science and Engineering B 325, 119074.

Lee, D., Jun, B.-M., Park, J., Jang, M., Lee, S., Snyder, S.A., et al., 2025. A review of removal of per-and polyfluoroalkyl substances using metal–organic framework-based nanoadsorbents. Nano Convergence 12, 56.

Lian, H., Wang, Q., Bai, X., Hua, J., Han, Q., Pan, Y., 2025. Electrochemical synthesis of ZIF-8 membrane on commercial ceramic support for propylene/propane separation. Journal of Membrane Science 719, 123733.

Liu, J.-Q., Chen, R., He, Y., Chen, Z., Ruan, H., Pan, Y., et al., 2026. Machine learning and AI empowering metal-organic frameworks: synthesis, performance prediction and therapeutic application. Dalton Transactions.

Liu, Z., Liao, M., Wang, L., Zhuang, S., 2025. Recent advancements in PFAS adsorptive removal using MOFs and COFs: a review. Reviews in Environmental Science and Bio/Technology 24, 63–95.

Luo, J., Luo, F., Li, H., Mao, C., Pan, Y., Fang, Z., et al., 2025. Structure-oriented metal-organic framework activation via proximal oligoalkyl quaternary ammonium grafting enhances long-chain PFAS sorption. Angewandte Chemie 137, e202514746.

Mahmoud, M.M., 2025. Microwave-assisted fast synthesis of MOF-801. Next Materials 6, 100316. Available at https://doi.org/10.1016/j.nxmate.2024.100316.

Main, R.M., Rainer, D.N., Bauzá, M., Ettlinger, R., Kelly, N.L., Coles, S.J., et al., 2026. A flexible copper MOF as a carboxylate-specific crystalline sponge for structure solution using X-ray and electron diffraction. Chemical Science.

Nguyen, M.B., Van Dung, L., Doan, T.L.H., Son, P.T., Nayak, S., Pham, T.H.Y., et al., 2026. Linker-driven tuning of pore structure and acidity in Zr-MOFs for paraoxon-methyl hydrolysis and efficient oxidative desulfurization. Journal of the Taiwan Institute of Chemical Engineers 182, 106588. Available at https://doi.org/10.1016/j.jtice.2025.106588.

Oggianu, M., Manna, F., Mameli, V., Galan-Mascaros, J.R., Capelo, S., de Oliveira, R.T., et al., 2026. A 3D polymorphic Cu-based ultramicroporous MOF capable of CO2 uptake and conversion. Journal of Materials Chemistry A.

Ozdemir, R., Unlu, D., 2026. Synergistic effect of graphene oxide and MOF-235 for efficient hydrogen generation from sodium borohydride hydrolysis. International Journal of Hydrogen Energy 204, 153217. Available at https://doi.org/10.1016/j.ijhydene.2025.153217.

Pala, J., Le, T., Kasula, M., Esfahani, M.R., 2023. Systematic investigation of PFOS adsorption from water by metal organic frameworks, activated carbon, metal organic framework@ activated carbon, and functionalized metal organic frameworks. Separation and Purification Technology 309, 123025.

Di Palma, G., Banerjee, P., Enemark-Rasmussen, K., Talebi Deylamani, S., Jinschek, J.R., 2025. Highly efficient removal of perfluorooctanoic acid from water using zirconium terephthalate (UiO-66) gel. Advanced Materials Interfaces 2500166.

Peter, L.G., Lee, L.S., 2025. Sources and pathways of PFAS occurrence in water sources: relative contribution of land-applied biosolids in an agricultural dominated watershed. Environmental Science & Technology 59, 1344–1353.

Qi, Y., Li, F., Zhang, C., Tang, X., Wang, C., Liang, Y., et al., 2025. Enhancement of the coadsorption of per- and polyfluoroalkyl substances onto Fe-MOF derivative in surface water through strengthening hydrophobic interactions. ACS ES&T Engineering 5, 1405–1416. Available at https://doi.org/10.1021/acsestengg.4c00932.

Raghav, S., Chavda, V., Yadav, A., Mansi, G., Raghav, J., Jain, P., et al., 2025. Advanced MOF-based materials for adsorptive and photocatalytic removal of emerging organic pollutants from wastewater. Applied Materials Today 46, 102865. Available at https://doi.org/10.1016/j.apmt.2025.102865.

Rezvani, M.A., Mousavi, S.A., Rafiei, Z., Hemmatzadeh, A., 2025. Synthesis of new nanocomposite based on Keggin-type heteropolyanion and metal-organic framework (MOF) as a new and effective nanocatalyst in removing environmental pollutants. Polyhedron 281, 117724. Available at https://doi.org/10.1016/j.poly.2025.117724.

Sharma, M., Dixit, V., Sharma, P., Gupta, R., 2026a. Economical synthesis of MOF from CHNS analyzer waste CuO and PET bottles for Congo red sequestration: a pathway towards dual mitigation. Reaction Chemistry & Engineering.

Sharma, G., Verma, A., Zhi, J., Wang, T., Verma, Y., Dhiman, P., et al., 2026b. Heteroatom-doped MOFs as hybrid materials for the removal of organic pollutants. Coordination Chemistry Reviews 549, 217260. Available at https://doi.org/10.1016/j.ccr.2025.217260.

Shi, Y., Yang, Y., Ge, S., Wu, M., Jiang, J., Fan, W., et al., 2025. Progress in advanced MOF-derived materials for the removal of organic and inorganic pollutants. Coordination Chemistry Reviews 530, 216474. Available at https://doi.org/10.1016/j.ccr.2025.216474.

Su, R., Yao, H., Wang, H., Chen, Y., Huang, S., Luo, Y., et al., 2025. Metal-organic frameworks for removing emerging organic pollutants: a review. Journal of Water Process Engineering 70, 107096.

Su, P., Zhang, C., Liu, Y., Zhang, J., Djellabi, R., Wang, R., et al., 2023. Boosting PFOA photocatalytic removal from water using highly adsorptive and sunlight-responsive ZIF67/MIL-100(Fe) modified C3N4. Journal of Environmental Chemical Engineering 11, 110765. Available at https://doi.org/10.1016/j.jece.2023.110765.

Yoon, S., Park, J., Police, A.K.R., Choe, J.K., Bae, S., 2025. Enhanced removal of perfluorooctanoic acid by aluminum-based metal–organic frameworks prepared by bauxite residue. Journal of Hazardous Materials 483, 136687. Available at https://doi.org/10.1016/j.jhazmat.2024.136687.

Yuldashboyevna, D.S., Bakhtiyarovna, B.G., Yakubzhanova, Z.B., Kumar, A., Yusupboevich, Y.Y., Suleimanovich, T.K., et al., 2026. Synthesis, crystal structure, and multifunctional adsorptive properties of a copper (II) 4-acetamidobenzoate MOF. Adsorption 32 (1).

Zhang, Y., Kong, K., Wu, Q., Ma, T., Liang, J., Wang, R., 2024. A porphyrinic metal-organic framework with cooperative adsorption domains for PFAS removal from water. ChemSusChem 17, e202400069.

CHAPTER

# 16

# Activated carbon for per- and polyfluoroalkyl substances remediation: properties, mechanisms, challenges, and future directions

*Mohamed Zbair*[1,2] *and Jocelyne Brendle*[1,2]

[1]Institut de Science des Mat´eriaux de Mulhouse CNRS UMR 7361, Université de Haute Alsace, Mulhouse, Grand Est, France [2]Université de Strasbourg, Strasbourg, Grand Est, France

## Content tags

- Adsorption Technologies
- Carbon-based materials
- Water and wastewater treatment
- Environmental engineering
- Environmental chemistry
- Emerging contaminants
- PFAS remediation
- Advanced materials for environmental applications
- Sustainable remediation strategies

## 16.1 Introduction

Per- and polyfluoroalkyl substances (PFASs) are synthetic chemicals known for their carbon–fluorine bonds, which give them extreme persistence in the environment, giving them the nickname of eternal chemicals (Buck et al., 2011). In addition to the fluorinated carbon chain, PFAS features a terminal functional group, which may include carboxylic, sulfonic, phosphonic, phosphinic, sulfonamide, ether carboxylic, or ether sulfonic

*Per- and Polyfluoroalkyl Substances (PFAS) as Environmental Contaminants*
DOI: https://doi.org/10.1016/B978-0-443-43856-1.00004-6

functionalities (Nakayama et al., 2019). Hence, this distinctive combination gives PFASs unique chemical and thermal stability, as well as a hydrophilic–lipophilic balance that influences their use in fire-fighting foams, stain-repellent treatments, waterproof textiles, carpets, food packaging, nonstick cookware, cleaning agents, pesticides, metal plating, and electrical wire coatings (Buck et al., 2011; Nakayama et al., 2019; Phong Vo et al., 2020). Due to their widespread use in many applications, PFAS are becoming more and more prevalent in the world's ecosystems (Pauletto and Bandosz, 2022). Their resistance to degradation and bioaccumulation properties present significant health risks, including cancer, developmental toxicity, and immune dysfunction, requiring urgent regulatory action (Boiteux et al., 2012). Regulatory actions reflect these dangers:

**a.** 2009/2019: listed under the Stockholm Convention on Persistent Organic Pollutants (Stockholm Convention, no date).
**b.** EU Drinking Water Directive: 100 ng $L^{-1}$ (hazardous PFAS sum), 500 ng $L^{-1}$ (total PFAS) (Official Journal of the European Union, no date).
**c.** U.S. EPA: 70 ng $L^{-1}$ health advisory for perfluorooctanoic acid (PFOA)/perfluorooctanesulfonic acid (PFOS) (EPA– United States Environmental Protection Agency Drinking Water Health Advisory for Perfluorooctanoic Acid PFOA, no date; EPA-United States Environmental Protection Agency Drinking Water Health Advisory for Perfluorooctanoic Acid PFOS, no date).

High levels of PFAS have been identified in water sources in various parts of the world, often exceeding recommended limits for safe consumption (Pauletto and Bandosz, 2022). There is therefore an urgent need to develop and apply effective methods for removing PFAS from water and soil in order to protect the environment and human health.

Activated carbons (ACs) are carbon-rich materials characterized by a high surface area, tunable porosity, and diverse surface functional groups. They are among the most effective and widely used adsorbents for environmental purification. Their physicochemical properties, such as hydrophobicity, pore size distribution, and surface chemistry, make them particularly suitable for the adsorption of PFAS, especially long-chain variants that exhibit strong affinity toward carbon surfaces (Wu et al., 2020; Wang et al., 2019; Yuan et al., 2020; Mohd Azmi et al., 2021; Saeidi et al., 2020).

Among the available technologies for PFAS removal, adsorption using ACs stands out due to its high adsorption capacity, cost-effectiveness, and scalability in water treatment applications (Wu et al., 2020; Wang et al., 2019; Yuan et al., 2020; Mohd Azmi et al., 2021; Saeidi et al., 2020).

This chapter offers a detailed analysis of the potential of ACs for the adsorptive elimination of PFAS. After a description of the physicochemical properties of AC materials and the fundamental mechanisms that govern PFAS adsorption, the influence of the key parameters governing adsorption efficiency and outlines current limitations will be presented, along with future perspectives for the practical application of these materials in environmental remediation. Ultimately, this work aims to contribute to the development of more effective and sustainable strategies for PFAS decontamination in response to an urgent global environmental challenge.

## 16.2 Key concepts: activated carbon for per- and polyfluoroalkyl substances remediation

### 16.2.1 Activated carbon fundamentals

ACs are one of the most widely used and efficient adsorbents in a variety of environmental applications particularly in water treatment (materials for adsorption-based remediation) (González-García, 2018; Choi et al., 2016; Mamun et al., 2016; Sethia and Sayari, 2016; Zbair et al., 2023), owing to its high surface area, tunable pore structure, and versatile surface chemistry (e.g., phenols, carboxyls, quinones, lactones, amines, pyridinic nitrogen), which contribute to its strong adsorption performance (Gayathiri et al., 2022; Zbair et al., 2018a). This carbon-rich material is produced by the thermal or chemical activation of carbon-containing precursors such as coal, coconut shells, wood, or nutshells (Zbair et al., 2018a,b; Ouasfi et al., 2019). The activation process results in a highly porous structure with surface areas often exceeding $1000\,m^2\,g^{-1}$, making it an ideal compound for capturing organic pollutants through a variety of adsorption mechanisms (Gayathiri et al., 2022). The interplay between surface functional groups, pore structure (especially microporosity), and hydrophobic domains plays a crucial role in the removal of persistent contaminants like PFAS, as highlighted in a recent comparative review of ACs and metal–organic frameworks (MOFs) (Pauletto and Bandosz, 2022).

Thanks to these properties, AC is widely applied in fields such as food processing (Nieto-Delgado and Rangel-Mendez, 2011), pharmaceuticals (Li et al., 2015; Ouasfi et al., 2019), textiles (Ahsaine et al., 2018), gas purification and storage (Mamun et al., 2016; Sethia and Sayari, 2016; Zbair et al., 2023; Bennici et al., 2022), and above all, water and air treatment (Choi et al., 2016; Zietzschmann et al., 2016). The application of ACs in the tertiary treatment phase targets the extraction of micropollutants. Given the vast heterogeneity in ACs characteristics (influenced by precursor type, pyrolysis temperature/duration, and activation process) and the diverse range of waterborne contaminants, it is imperative to select a carbon whose adsorptive properties are specifically tailored for the intended pollutants.

The production route of ACs from biomass involves multiple thermochemical conversion pathways such as pyrolysis, gasification, hydrothermal treatment, torrefaction, and flash carbonization (Anastopoulos et al., 2024). These methods differ significantly in terms of operating temperatures, heating rates, residence times, and yields, which in turn influence the physicochemical properties of the final adsorbent. The resulting (bio)char is then subjected to either physical or chemical activation (using agents like steam, $CO_2$, KOH, or $ZnCl_2$) to enhance porosity and surface functionality. A concise overview of these processes is provided in Fig. 16.1, highlighting how varying process conditions and activating agents lead to the formation of porous AC.

Among its numerous applications, AC has emerged as one of the most effective and widely used materials for PFAS removal from aqueous environments (Pauletto and Bandosz, 2022; Yuan et al., 2020; Liu et al., 2020). The potential hydrophobic nature of long-chain PFAS compounds, characterized by strong carbon–fluorine bonds, allows them to bind strongly to the nonpolar carbon surface of AC (Appleman et al., 2014). This makes

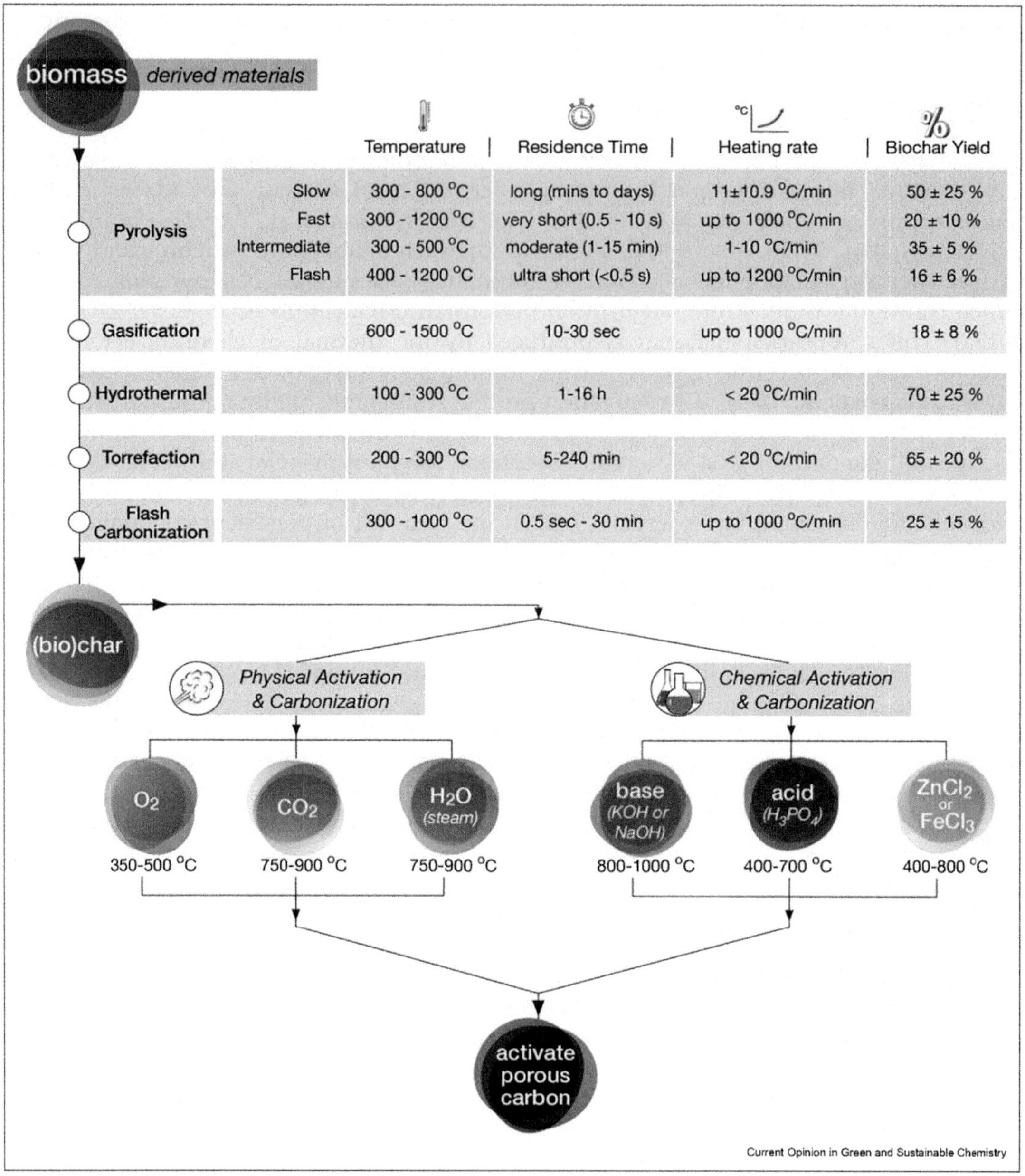


**FIGURE 16.1** Overview of biomass-derived activated carbon (AC) production pathways, showing temperature ranges, heating rates, and biochar yields for different thermochemical processes. Final activation of biochar into porous carbon can be achieved via physical ($O_2$, $CO_2$, steam) or chemical (acid/base or metal salts) methods. *Adapted from Anastopoulos, I., Giannakoudakis, D.A., Frontistis, Z., Zorpas, A.A., Pashalidis, I., Triantafyllidis, K., 2024. Biomass-derived adsorbents: universal database development for their synthesis and remediation efficiency as a necessary step to move from laboratory- to pilot-scale applications. Curr Opin Green Sustain ChemCurrent Opinion in Green and Sustainable Chemistry. 47, 100902. https://doi.org/10.1016/j.cogsc.2024.100902.*

ACs particularly efficient in removing PFAS from water, especially for longer-chain variants that exhibit a strong affinity for the carbon surface.

In practice, three main forms of AC are employed for water treatment: granular activated carbon (GAC) (Yu et al., 2009), powdered activated carbon (PAC) (Hansen et al., 2010), and activated carbon fibers (ACFs) (Wang et al., 2015). GAC is typically used in fixed-bed filters, where contaminated water flows through a packed column, and in this case, PFAS molecules can be retained within the pores. PAC, on the other hand, is added directly to water during treatment and then removed by sedimentation or filtration. ACFs, although less common in full-scale applications, have gained attention in research due to their high surface area and uniform microporosity, which may offer enhanced adsorption performance. All three formats are modular and scalable, making them suitable for applications ranging from household systems to large-scale municipal facilities.

The key advantages of AC for PFAS remediation include:

**a.** High adsorption capacity: for organic pollutants, particularly for long-chain PFAS.
**b.** Scalability and flexibility: AC can be used effectively in various configurations, from small-scale filters to large municipal treatment systems.
**c.** Cost-effectiveness: compared to advanced oxidative or electrochemical methods, AC requires relatively low infrastructure costs and energy input, making it a cost-efficient solution.
**d.** Environmental friendly: the use of AC does not generate harmful by-products, and spent carbon can be repurposed in applications such as fuel or soil remediation, subject to local regulations.

## 16.2.2 Mechanism of per- and polyfluoroalkyl substances adsorption onto activated carbon

The adsorption of PFAS onto ACs is governed by multiple mechanisms, including potential hydrophobic nature, electrostatic forces, pore filling, and surface complexation. The efficiency of PFAS removal depends on parameters such as PFAS chain length, functional groups, water chemistry (pH, ionic strength), and the physicochemical properties of the ACs themselves (Gagliano et al., 2020).

### *16.2.2.1 Potential hydrophobic nature*

The primary mechanism driving PFAS adsorption onto ACs is potential hydrophobic nature. PFAS molecules consist of a fluorinated carbon tail, which is highly hydrophobic, and a hydrophilic functional head group (e.g., sulfonate or carboxylate). The nonpolar, hydrophobic regions of PFAS tend to adsorb onto the hydrophobic domains of ACs, which include both graphitic surfaces and amorphous carbon phases within the pores (Yu et al., 2009). Long-chain PFAS (such as PFOS and PFOA) exhibit stronger adsorption than short-chain variants (e.g., PFBA, PFBS) due to their greater hydrophobicity (Appleman et al., 2014). This is consistent with Traube's rule, which states that adsorption affinity increases with alkyl chain length in surfactant molecules (Rosen, 2004).

#### 16.2.2.2 Electrostatic interactions

The surface charge of ACs and the ionization state of PFAS play a crucial role in adsorption. ACs typically carry a negative surface charge at neutral to high pH due to the deprotonation of acidic oxygen-containing functional groups on their surface. The primary functional groups responsible for this pH-dependent charge development are carboxylic groups (-COOH, pKa ~3–6), which deprotonate most readily, and phenolic hydroxyl groups (-OH), which contribute significantly at higher pH (Moreno-Castilla, 2004; Radovic et al., 2000). This deprotonation (-COOH → $-COO^-$ + $H^+$ and -OH → $-O^-$ + $H^+$) releases $H^+$ ions, consuming acidity and increasing the local pH near the carbon surface (Qiu et al., 2023; Vervald et al., 2022). However, at lower pH (below the point of zero charge, pH ~7–9), the ACs' surface becomes more positively charged, enhancing the adsorption of anionic PFAS (e.g., PFOA, PFOS) through electrostatic attraction (Ochoa-Herrera and Sierra-Alvarez, 2008). Conversely, at higher pH, electrostatic repulsion between negatively charged ACs and anionic PFAS can reduce adsorption efficiency (Wang and Shih, 2011). The presence of divalent cations (e.g., $Ca^{2+}$, $Mg^{2+}$) can further enhance adsorption by forming cation bridges between the negatively charged PFAS and AC's surface (Zhao et al., 2014; Liu et al., 2022; Cantoni et al., 2021; Park et al., 2020).

#### 16.2.2.3 Pore filling and size exclusion

The microporous structure of ACs, typically defined by pore sizes smaller than 2 nm, plays a key role in the physical adsorption of PFAS, primarily through a pore filling mechanism. A crucial aspect of this interaction is the compatibility between the molecular dimensions of PFAS and the pore architecture of the adsorbent. For example, the molecular size of PFOA ($CF_3C_6F_{12}COOH$) is approximately 1.20 nm in length, 0.80 nm in width, and 0.50 nm in thickness. Similarly, PFOS ($CF_3C_7F_{14}SO_3H$) has dimensions around $1.32 \times 1.00 \times 0.50$ nm (Punyapalakul et al., 2013; Selmi et al., 2024). These values highlight the relatively large and elongated structure of these molecules, largely driven by their rigid fluorinated carbon backbones and polar head groups. Given these dimensions, it becomes evident that ultramicropores (<0.7 nm) are generally too small to accommodate PFOS and PFOA. Even smaller micropores (<1 nm) may restrict molecular diffusion, thereby limiting adsorption efficiency, particularly for larger molecules such as PFOS. While micropores are critical for providing a large internal surface area, they must also be sufficiently accessible to ensure effective interaction with PFAS. Mesopores (2–50 nm) and macropores (>50 nm), although contributing less directly to adsorption capacity, are important for facilitating the transport of PFAS molecules toward internal adsorption sites (Yu et al., 2009). An optimal pore structure should therefore include a balance between micropores that can accommodate PFAS molecules and mesopores that enhance diffusion. This concept is supported by several studies indicating that ACs with a well-developed micropore volume and a complementary mesopore network tend to exhibit superior removal performance for long-chain PFAS (Gagliano et al., 2020; Cantoni et al., 2021; Park et al., 2020).

#### 16.2.2.4 Surface complexation and functional group interactions

Oxygen-containing functional groups (e.g., carboxyl, hydroxyl, carbonyl) on ACs surfaces can interact with PFAS through weak chemical interactions, such as hydrogen bonding or ligand exchange (Liu et al., 2022; Cantoni et al., 2021; Park et al., 2020). While these interactions are secondary to hydrophobic and electrostatic effects, they can enhance adsorption under certain conditions. For example, ACs modified with amine groups have shown increased PFAS uptake due to stronger electrostatic and complexation effects (Ateia et al., 2019).

The main PFAS adsorption mechanisms on ACs have been highlighted in studies by (Gagliano et al., 2020; Du et al., 2014):

**a.** Electrostatic interactions between anionic PFAS and positively charged ACs surfaces (at $pH < pH_{PZC}$).
**b.** Hydrogen bonding involving oxygen or nitrogen groups on ACs and PFAS head groups.
**c.** Hydrophobic effects induced by PFAS fluorinated tails.
**d.** Ionic exchange or bridging effects, where divalent cations (e.g., $Ca^{2+}$) play a binding-mediating role.

π–π interactions play a minor role, probably due to the absence of aromaticity in most PFAS structures. The predominance of hydrophobic and electrostatic mechanisms underlines the importance of tailoring AC surface chemistry.

Following the earlier discussion on molecular size and pore compatibility, this section explores how material characteristics such as surface area and pore volume influence PFAS adsorption capacity across various types of adsorbents (Punyapalakul et al., 2013; Selmi et al., 2024). As previously discussed, the compatibility between PFAS size and pore structure is essential for effective adsorption. In addition to this, a high specific surface area significantly enhances removal performance. Several studies have compared the removal capacities of various adsorbents, including wood and fishery-derived biochars, inorganic porous silicas, Norit GAC 830, and ACFs (Wu et al., 2020; Saeidi et al., 2020; Punyapalakul et al., 2013; Chen et al., 2017; Appleman et al., 2013; Inyang and Dickenson, 2017). A consistent trend emerges: as specific surface area increases—from 6.4 to 413 $m^2 g^{-1}$ for porous silicas and biochars to 924 $m^2 g^{-1}$ for ACs,—PFOA removal efficiency improves significantly, reaching up to 94%. In addition to specific surface area, pore size distribution plays a crucial role in PFAS adsorption. Saeidi et al (Saeidi et al., 2020) studied commercial ACs (VS, FC15, FC10, and WK) with different pore structures (Table 16.1). They found that the increase in BET surface area (from 1100 to 2100 $m^2 g^{-1}$), total pore volume (from 0.53 to 0.99 $cm^3 g^{-1}$), and micropore volume (from 0.36 to 0.80 $cm^3 g^{-1}$) corresponded to an increase in the maximum adsorption capacity of PFOA from 2 to 57 $mg g^{-1}$. Importantly, the maximum adsorption capacity improved with increasing mesopore volume up to 30%–35% of total porosity, highlighting the key role of mesopores in facilitating PFAS diffusion and access to adsorption sites.

**TABLE 16.1** Comparison of activated carbons based on per- and polyfluoroalkyl substances (PFAS) adsorption capacities and surface characteristics.

| ACs | Adsorbates | Adsorption capacity: $Q_{e,max}$ (mg $g^{-1}$) | | | BET surface area ($m^2$ $g^{-1}$) | % volume méso-poreux | $pH_{PZC}$ | Acidic functional groups | Basic functional groups | References |
|---|---|---|---|---|---|---|---|---|---|---|
| 10 ACs | PFOS | 50–650 | | | 903–2341 | - | 4.0–9.2 | 0.07–0.40 μeq $m^{-2}$ | 0.10–0.54 μeq $m^{-2}$ | Zhi and Liu (2015) |
| VS | PFOA; PFOS | 55.6 | 156 | | 2100 | - | 7.3 | 0.22 μmol $m^{-2}$ | - | Saeidi et al. (2020) |
| FC15 | | 35.7 | 108 | | 1600 | - | 7.1 | 0.43 μmol $m^{-2}$ | - | |
| FC10 | | 10.2 | 33.3 | | 1400 | - | 6.5 | 1.09 μmol $m^{-2}$ | - | |
| WK | | 2.08 | 5.0 | | 1100 | - | 5.9 | 2.08 μmol $m^{-2}$ | - | |
| | PFHxA; PFOA; PFOS | $K_F$ (ng $mg^{-1}$)(ng $L^{-1}$)$^{-1/n}$ Freundlich parameter | | | | | | | | Cantoni et al. (2021) |
| vc-AC | | 0.8 | 2.2 | 2.4 | 1002 | 46 | 7.8 | - | - | |
| vb-AC | | 2.3 | 5.4 | 5.9 | 992 | 42 | 9.7 | - | - | |
| rc-AC | | 1.7 | 3.4 | 3.6 | 840 | 59 | 9.7 | - | - | |
| rb-AC | | 2.0 | 4.0 | 4.4 | 897 | 65 | 9.8 | - | - | |
| | PFOS; PFHxA | Breakthrough study | | | | | | | | Appleman et al. (2013) |
| F300 | | 26 | 46 | | - | - | - | - | - | |
| F600 | | 64 | 67 | | - | - | - | - | - | |
| 1240C | | 69 | 72 | | - | - | - | - | - | |

## 16.3 Methods and applications of activated carbon for per- and polyfluoroalkyl substances remediation

The selection of ACs materials is a crucial element in the design of PFAS treatment systems. This section first examines the adsorption capacities of different types of AC, highlighting the influence of their physical form (CAG, PAC, ACF), textural properties, and experimental treatment conditions.

### 16.3.1 Adsorption capacity of activated carbons: influence of material type and experimental conditions

The adsorption capacity is the fundamental parameter that determines the effectiveness and economic feasibility of ACs for PFAS removal. It directly determines treatment system design, operational lifetime, and regeneration frequency, making its accurate assessment essential for engineering design and cost predictions. Capacity reflects the complex interplay between carbon characteristics, PFAS properties, and water chemistry, serving as a key performance indicator for comparing different adsorbents and optimization strategies.

Numerous studies have evaluated the adsorption capacities of various commercial AC materials for PFAS remediation, particularly for PFOA, PFOS, and newer alternatives such as GenX. However, before discussing the adsorption capacities of different AC types, it is important to highlight that the form of GAC, PAC, or ACF, significantly influences its performance. These differences stem from variations in particle size, pore accessibility, and application mode, which directly affect PFAS removal efficiency. The following sections compare adsorption results from representative studies to illustrate the impact of ACs form on performance.

#### 16.3.1.1 *Granular activated carbon*

GACs, commonly used in water treatment systems, show varied performance. For instance, commercial GAC with a surface area of 712 $m^2 g^{-1}$ demonstrated relatively high maximum adsorption capacities for both PFOA (161 $mg\,g^{-1}$) and PFOS (185 $mg\,g^{-1}$) in deionized water at concentrations ranging from 20 to 250 $mg\,L^{-1}$ (Yu et al., 2009). However, other GACs with lower surface areas, such as one with 534 $m^2 g^{-1}$, showed reduced adsorption capacities: 52.8 $mg\,g^{-1}$ for PFOA and 72.2 $mg\,g^{-1}$ for PFOS (Zhang et al., 2016). Interestingly, a GAC with a higher surface area (1050 $m^2 g^{-1}$, Filtrasorb 400) achieved only 0.025 $mg\,g^{-1}$ for PFOA when treating groundwater, reflecting the strong matrix effect of real water systems and the importance of background organic matter and competing ions (McCleaf et al., 2017). Similarly, another commercial GAC (C) yielded 0.396 $mg\,g^{-1}$ for PFOA under trace concentration conditions (Zaggia et al., 2016). These findings underscore the critical influence of water chemistry on adsorption efficiency and highlight the need to assess adsorbent performance under realistic conditions.

#### 16.3.1.2 *Powdered activated carbon*

PACs, due to their smaller particle size and higher accessibility to adsorption sites, often exhibit superior performance compared to GACs. A PAC with 812 $m^2 g^{-1}$ surface area achieved maximum adsorption capacities of 277 mg $g^{-1}$ for PFOA and 520 mg $g^{-1}$ for PFOS under similar experimental conditions (Yu et al., 2009). Another PAC with a lower surface area (764 $m^2 g^{-1}$) still showed relatively high PFOA adsorption (21.8 mg $g^{-1}$), although this was based on equilibrium capacities fitted to kinetic models (Yu et al., 2012). A particularly high-performing PAC from Sinopharm, with a BET surface area of 1539 $m^2 g^{-1}$, showed a remarkable PFOS adsorption capacity of 484 mg $g^{-1}$ at concentrations between 500–1500 mg $L^{-1}$ (Li et al., 2017). These findings highlight the enhanced efficiency of PACs, particularly under a contaminant loading scenario.

#### 16.3.1.3 *Modified and activated carbons*

Beyond conventional GACs and PACs, the development of modified ACs has further expanded the potential for efficient PFAS remediation. These materials are often engineered with specific surface functionalities or chemical modifications aimed at enhancing interactions with PFAS molecules. For instance, ACF from Shanghai Union Soldiers Environmental Technology, with a BET surface area of 1226 $m^2 g^{-1}$, reached a maximum PFOA adsorption capacity of 400.63 mg $g^{-1}$ in batch experiments (Wang et al., 2015). Similarly, magnetic PAC modified with $Fe_3O_4$ achieved 373 mg $g^{-1}$ for PFOA despite having a lower surface area (400 $m^2 g^{-1}$), highlighting the potential of surface functionalization to enhance performance (Meng et al., 2019). Additionally, Jingke Co. produced both GAC and PAC materials that achieved 260 mg $g^{-1}$ adsorption capacities for GenX, a short-chain PFAS, indicating effective treatment potential even for emerging PFAS compounds (Wang et al., 2019).

In summary, the adsorption capacity of AC for PFAS is highly variable and depends on multiple factors, including the type and form of carbon, surface area, surface modifications, and the specific PFAS targeted. Laboratory studies clearly show that powdered and modified ACs can achieve significantly higher adsorption capacities compared to conventional granular forms, especially under controlled conditions and at high contaminant concentrations.

Despite promising adsorption capacities reported under laboratory conditions, carbon-based materials often show significantly inferior performance when applied in real-world scenarios. This discrepancy is particularly evident when it comes to treating groundwater contaminated with PFAS at minute concentrations, typically in the nanogram per liter (ng $L^{-1}$) to low microgram per liter (μg $L^{-1}$) range. Under such realistic conditions, the adsorption behavior of AC is influenced by a series of factors that are absent or minimized in controlled laboratory environments.

For example, Silcarbon TH90 PAC (1100 $m^2 g^{-1}$) adsorbed only 0.020 mg $g^{-1}$ of PFOA when used to treat groundwater at concentrations between 70 ng $L^{-1}$ and 1.4 μg $L^{-1}$ (Hansen et al., 2010). These findings highlight how competing solutes, the presence of natural organic matter (NOM), pH variability, and ionic strength in natural water matrices can significantly reduce the availability of active adsorption sites and alter PFAS-carbon interactions. Moreover, the dynamic conditions encountered in full-scale treatment systems, such as flow rate variations, temperature fluctuations, and intermittent loading, further complicate the translation of laboratory results to field performance. Unlike static batch tests, real systems must operate under continuous flow, where mass transfer limitations and contact time constraints play a critical role in determining the actual treatment efficiency.

These observations underscore the necessity of validating laboratory findings through pilot-scale or field-scale studies before selecting materials for practical implementation. The choice of AC, whether granular, powdered, or modified, must be guided not only by its intrinsic adsorption capacity but also by its robustness under operational conditions, ease of regeneration, cost-effectiveness, and compliance with regulatory discharge limits. While PACs generally outperform GACs in controlled experiments due to their higher surface area and better accessibility to adsorption sites, this advantage may diminish in real applications where handling, dosing precision, and filtration logistics present additional challenges. In conclusion, linking the gap between laboratory and real-world performance requires a thorough understanding of the physicochemical parameters influencing PFAS adsorption, as well as careful consideration of system design and environmental variables. These aspects will be further discussed in the next section, which focuses on the factors that govern adsorption efficiency across different treatment contexts.

## 16.3.2 Role of physicochemical properties of activated carbons in per- and polyfluoroalkyl substances adsorption

The performance of ACs depends not only on their BET surface area or on porosity. The physicochemical properties of the surface, such as functional group chemistry, surface charge, and polarity, also play a fundamental role. This section looks in detail at these factors and their impact on PFAS removal. To gain a comprehensive understanding, it is crucial to explore how these surface characteristics interact with PFAS molecules at the molecular level.

Several studies have shown that simple textural analysis cannot fully explain the variations observed from one material to another. This is particularly the case in the work of Saeidi et al (Saeidi et al., 2020), who proposed a different reading of adsorption mechanisms.

A critical role for surface chemistry in the adsorption of PFAS on ACs has been demonstrated by Saeidi et al (Saeidi et al., 2020) that, while pore size distribution is often considered a key factor in adsorption performance, their work revealed that this parameter alone cannot explain the significant variations observed in PFAS adsorption on different ACs. Instead, they emphasized that surface chemistry, in particular functional groups, charge, and polarity of the carbon surface, plays a dominant role in managing adsorption mechanisms. This is particularly relevant for anionic PFASs such as PFHxA, PFOA, and PFOS, which maintain a persistent negative charge across the entire pH scale (Selmi et al., 2024). The surface charge of ACs, determined by their point of zero charge ($pH_{PZC}$), strongly influences their adsorption capacity. When the solution pH is below $pH_{PZC}$, the AC's surface acquires a positive charge, creating favorable electrostatic interactions with negatively charged PFAS molecules. Therefore to maximize adsorption efficiency, ACs with a $pH_{PZC}$ higher than the operating pH of the water treatment system must be selected. Saeidi et al (Saeidi et al., 2020) have further pointed out that surface polarity adjustments (e.g., reduction of hydrophilic sites) can mitigate competitive adsorption of coexisting organic matter, thus improving PFAS selectivity. As such, surface chemistry emerges not only as a complementary factor but often as a primary determinant in PFAS adsorption. These observations underline the importance of tailoring AC's surface properties rather than just depending on porosity to optimize PFAS removal from complex environmental matrices.

Other research extends this approach by examining the influence of the type of raw material used to produce ACs on their adsorption performance, through variations in $pH_{PZC}$, polarity, or functional group richness. In this context, the raw material used for AC synthesis becomes a decisive variable that directly impacts these surface characteristics. For example, the critical influence of surface chemistry of ACs on PFAS adsorption was also supported by Zhi and Liu (Zhi and Liu, 2015), working systematically to evaluate PFOS and PFOA removal using industrial ACs obtained from different precursor sources. These included ACs-based on coal (F400, GLP), carbon produced from coconut shells (1240C), carbons originating from wood (WVB, BioNC, AquaNC), and carbons synthesized from polymeric materials (Ambersorb, ACF25, ACF20, ACF15). Notably, most carbons exhibited alkaline $pH_{PZC}$ values (7.3–9.3), with the exception of the wood-based variants, which had acidic surfaces with a $pH_{PZC}$ of around 4. Variations in surface chemistry were quantified through acidic functional groups (0.077–0.401 μeq $m^{-2}$) and basic groups (0.103–0.540 μeq $m^{-2}$), revealing marked contrasts in polarity and charge distribution between materials. PFOS adsorption capacities vary considerably: polymer-derived ACF15, ACF20, and ACF25 reach ~50 mg $g^{-1}$, charcoal-based BPL and wood-based AquaNC reach ~70 mg $g^{-1}$, while wood-based WVB and BioNC show exceptional performance (~550 mg $g^{-1}$). The highest capacities (~650 mg $g^{-1}$) were observed for charcoal-based F400, coconut shell-derived 1240C, and polymer-based Ambersorb. These disparities underline the role of raw material provenance in tailoring AC properties. Zhi and Liu (Zhi and Liu, 2015) identified a clear trend: PFAS adsorption increases with the $pH_{PZC}$ and the total basicity of ACs. For example, wood-based ACs, despite their low $pH_{PZC}$, demonstrated superior performance, suggesting that factors such as micropore structure or potential hydrophobic nature can compensate for unfavorable electrostatic conditions. However, the dominant correlation between basicity (high $pH_{PZC}$) and improved adsorption aligns with earlier results by Saeidi et al (Saeidi et al., 2020), strengthening the importance of positively charged surfaces in capturing anionic PFAS. These findings highlight the interaction between ACs' origin, surface chemistry, and affinity for PFAS. While basic ACs with high $pH_{PZC}$ generally favor electrostatic interactions, exceptions such as wood-based ACs imply that synergistic mechanisms (e.g., pore confinement, the hydrophobic partitioning) may also contribute. Overall, these results suggest that an integrated approach, combining textural and chemical descriptors, is needed for rational ACs design. The above information is essential for selecting suitable ACs for specific PFAS contaminants in water treatment systems.

### 16.3.3 Surface chemistry modifications and their impact on per- and polyfluoroalkyl substances removal

Based on the knowledge of the original surface properties, studies also investigated ways of actively modifying ACs' chemistry to further enhance PFAS adsorption. The improvement of the ability of industrially produced ACs to adsorb PFOS and PFOA by modifications aiming at increasing surface basicity has been demonstrated through previous studies (Zhi and Liu, 2015, 2016). These works revealed that annealing ACs (including carbons like

1240C, F400, BioNC, ACF20, and WVB) after thermal treatment at 1000°C and/or nitrogen incorporation through ammonia exposure at 700°C significantly resulted in an elevation of pHpzc values, shifting from 4.50 to 10.50. This change in $pH_{PZC}$ reflected a transition from an acidic to a highly basic surface, which was attributed to two key modifications. Firstly, high-temperature annealing probably removes acidic oxygen-containing functional groups, such as carboxyls and phenolic moieties, thereby reducing the acidity of the surface and increasing the relative abundance of basic sites. Secondly, nitrogen doping introduces nitrogen-based functional groups, such as pyridinic, pyrrolic, or quaternary nitrogen species, which act as Lewis basic sites and contribute to a more positively charged surface. These modifications have simultaneously altered the surface chemistry of ACs, favoring stronger interactions with PFAS. Beyond thermal and chemical treatments, polymer-based surface modifications have also proven effective. It was also reported that by modifying commercial grade known as Filtrasorb 400 (F400) through modification with a cationic polymer, such as polydiallyldimethylammonium chloride, improved its performance. Indeed, the uptake capacity for PFOA and PFOS was enhanced, reaching 185 and 270 mg $g^{-1}$ compared to 160 and 165 mg $g^{-1}$ for the untreated carbon (Ramos et al., 2022). A minor rise in adsorption was also observed for short-chain PFAS (PFHxS), increasing from 167 to 173 mg $g^{-1}$. Significantly, it mitigated the influence of NOM, with humic acids being the main fraction, on the adsorption process, thus prolonging ACs life. However, the magnitude of improvement must be critically assessed to determine its practical relevance. Nevertheless, since the observed difference was around 4%, it is not sufficient to be treated as a meaningful improvement and could instead result from errors inherent to experimental replication.

Surface basicity improvements directly enhanced the adsorption performance of both PFOS and PFOA. For PFOS, uptake capacity was augmented, rising from 60 to 100 mg $g^{-1}$ (~66% increase), while PFOA adsorption showed a moderate increase from 70 to 90 mg $g^{-1}$ (~29% increase) (Zhi and Liu, 2015, 2016). The greater improvement for PFOS is probably due to its sulfonate group ($-SO_3^-$), which has a higher charge density than the carboxylate group ($-COO^-$) of PFOA. The high $pH_{PZC}$ allowed the AC surface to remain positively charged at neutral pH, enhancing electrostatic attraction with the anionic PFOA molecules. In addition, the potential hydrophobic nature, induced by the fluorinated carbon chains of PFOS and PFOA, provided synergy with electrostatic forces, particularly for PFOS, where the hydrophilicity of the sulfonate group could be compensated for by a stronger electrostatic bond. In addition, nitrogen doping may have introduced hydrogen bonding sites or Lewis acid-base interactions between nitrogen groups and PFASs, thereby enhancing adsorption (Zhi and Liu, 2015, 2016). In parallel, other studies have focused on introducing quaternary ammonium groups to promote specific electrostatic interactions. Meanwhile, Yuan et al (Yuan et al., 2020) demonstrated that loading AC with quaternary ammonium/epoxide (QAE-AC) enhanced PFOA adsorption in groundwater. The use of different amounts of QUAB 188 and QUAB 360 affected the carbon structure and adsorption behavior. Low amounts of QAE allowed the molecules to penetrate the pores and form epoxides with the surface groups, thus enhancing adsorption. Higher doses of QAE reduced pore volume, limiting access, while very high amounts created a polymeric layer on the surface, further improving performance. Carbons modified with QUAB 188 had more nitrogen groups and positive charges than those with QUAB 360, enabling better adsorption through the combined electrostatic and potential hydrophobic nature. The best balance between performance and cost was found with the carbon loaded with 1 mL of QUAB 188. Using a three-step process, Sun et al (Sun et al., 2020) developed a quaternary nitrogen-grafted AC to enhance PFOA adsorption in groundwater. Starting with a commercial subbituminous activated carbon (SAC), they oxidized it with $HNO_3$ to add oxygen groups (SO), then heat-treated it with ammonia to introduce amines (SN), and finally performed a Menschutkin reaction with methyl iodide to form quaternary ammonium salts on the surface (SNQ). PFOA adsorption capacities increased accordingly: SNQ ($46.9\ mg\ g^{-1}$) > SN ($34.8\ mg\ g^{-1}$) > SO ($27.5\ mg\ g^{-1}$) > SAC ($23.6\ mg\ g^{-1}$). Micropore volumes and surfaces followed a similar trend. Elemental analysis showed that oxygen content increased after oxidation, nitrogen content increased after ammonia treatment, and quaternized nitrogen increased after the final stage. The $pH_{PZC}$ also increased, with SNQ being the most positively charged at groundwater pH (~7.1), enhancing electrostatic attraction with PFOA anions.

To further consolidate these findings, several investigations have attempted to identify the specific functionalities most responsible for enhanced PFAS uptake.

Previous studies (Saeidi et al., 2020; Punyapalakul et al., 2013; Chen et al., 2017; Zhi and Liu, 2015, 2016) have investigated the surface functionalities and mechanisms governing the adsorption of PFAS onto ACs, identifying specific surface groups that enhance PFAS removal. Among these, hydrophobic groups and nitrogen-containing moieties (e.g., nitriles, pyridines, amides) play a key role. Hydrophobic groups promote interactions with PFAS fluorinated carbon chains via van der Waals forces and hydrophobic effects. In addition, nitrogen-containing functionalities contribute to adsorption via ionic interactions (e.g., attraction between protonated nitrogen groups and anionic PFAS head groups) and hydrogen bonding. For example, pyridine nitrogen can act as a hydrogen bond acceptor, enhancing interactions with PFAS molecules. These synergistic mechanisms enhance both the affinity and adsorption capacity of ACs for PFASs, particularly in complex aqueous environments. Cantoni et al (Cantoni et al., 2021) and Park et al (Park et al., 2020) have further elucidated the factors influencing PFAS adsorption. Their work demonstrated that adsorption capacity correlates strongly with:

**a.** The hydrophobicity of the AC's surface, which leads to PFAS separating from water to the carbon matrix.
**b.** The hydrophobicity of the PFAS compound as longer perfluoroalkyl chains enhance potential hydrophobic nature.
**c.** The $pH_{PZC}$, which enhances electrostatic attraction to anionic PFAS species (e.g., PFOS, PFOA).
**d.** Pore size distribution, where larger mesopore size networks improve the accessibility of larger PFAS molecules.

To quantify the relative impact of each of these parameters on adsorption performance, modeling approaches have also been employed.

A statistical model approach employed by Cantoni et al (Cantoni et al., 2021) quantified these effects, identifying hydrophobicity, $pH_{PZC}$, BET surface area, and functional groups as the most influencing parameters for PFAS adsorption capacity, in order of decreasing influence. For AC's breakthrough performance, this order changed slightly, with hydrophobicity and functional groups dominating, followed by surface area and $pH_{PZC}$ (Table 16.1).

This suggests that while $pH_{PZC}$ is essential for initial adsorption, long-term performance under dynamic conditions relies more on hydrophobic and chemical interactions.

Using column adsorption methods, Appleman et al (Appleman et al., 2013) evaluated PFOS adsorption using commercial GACs—F300, F600, and AquaCarb 1240C. While all three demonstrated efficiencies, F300 showed higher performance, with PFOS breakthrough percentages of 26%, 64%, and 69% for F300, F600, and 1240C, respectively. The authors attributed these differences to variations in material origin and pore structure. Coconut shell-derived ACs, such as 1240C, are predominantly microporous, while bituminous coal-based ACs (F300, F600) exhibit a balanced microporous-mesoporous structure. Although the high iodine value of 1240C, reflects its dense microporosity ideal for small molecules, its limited mesoporosity probably hindered PFOS transport kinetics, reducing its practical efficiency. In contrast, the hybrid porous structure of F300 optimizes both adsorption capacity via micropores and diffusion rates via mesopores, enabling faster access to PFOS molecules.

To achieve maximum PFAS adsorption, ACs should have a high surface area ($\geq$1000 $m^2 g^{-1}$) with a balanced microporous-mesoporous ratio around 65%–70% micropores and 30%–35% mesopores (Table 16.1). Micropores are effective binding sites for small PFAS (e.g., short-chain perfluoroalkyl acids), while mesopores facilitate rapid diffusion and accessibility of larger molecules (e.g., long-chain PFOS/PFOA). Surface modifications, such as nitrogen doping, can further enhance performance by introducing functional groups (e.g., pyridinic N, quaternary amines) that act synergistically with the porous network. These groups enhance electrostatic interactions with anionic PFAS and improve hydrophobicity, thus promoting PFAS separation from aqueous phases.

Overall, the efficiency of ACs for PFAS removal depends on the interaction between surface chemistry (hydrophobicity, charge, functional groups) and pore structure. For example, while the mesopores of F300 enhance PFAS transport, the hydrophobicity and charge of its $pH_{PZC}$ probably enhance adsorption via hydrophobic and electrostatic forces. This dual optimization underlines the need for tailored AC materials must balance physical structure (pore size distribution, surface area) and chemical properties (functionalization, $pH_{PZC}$) to meet the diverse chemistries of PFAS. This is essential to develop adsorbents capable of tackling persistent and emerging contaminants in water treatment systems.

## 16.4 Challenges related to activated carbon in per- and polyfluoroalkyl substances removal

One of the main drawbacks of using ACs (whether granular, powdered, or nanoscale) for PFAS removal is their low reusability. Research shows that the adsorption capacity of ACs drops sharply after only a few regeneration cycles. For example, by the fifth cycle, efficiency often drops by about 30%, mainly because PFAS molecules remain trapped in the pores of the ACs and are not fully removed during regeneration (Dalahmeh et al., 2019). This limitation raises concerns about long-term effectiveness and cost-effectiveness.

The regeneration process is inherently complex due to the multiple mechanisms involved in PFAS adsorption, including potential hydrophobic nature, electrostatic forces, and interstitial diffusion. Breaking these bonds to completely remove PFAS requires harsh conditions, often requiring organic solvents such as alcohols, ethanol, or acetone. However, these solvents generate large volumes of contaminated regenerant, which then requires costly secondary treatment or energy-intensive separation processes for safe disposal (Saleh et al., 2019). This not only complicates the regeneration process but also poses operational and environmental challenges for the management of used adsorbents. After regeneration, a major problem remains: disposing of the extracted PFAS. These chemicals resist natural breakdown, often requiring incineration. However, burning them can release harmful greenhouse gases and toxic by-products if combustion is incomplete (Vecitis et al., 2008). This simply shifts pollution from water to air, highlighting the lack of sustainable solutions for PFAS-contaminated AC.

The economic viability of GAC for PFAS remediation is constrained by significant cost considerations. U.S. EPA data indicate treatment costs of \$0.1 to 1 per 3.7 $m^3$ when utilizing fresh GAC, with demonstrated removal efficiencies of 92%–100% for PFAS contaminants (Stoiber et al., 2020). However, the adsorption capacity shows marked decline upon successive regeneration cycles, particularly for short-chain PFAS compounds due to their enhanced aqueous mobility and reduced potential hydrophobic nature with carbon surfaces. This progressive performance decline necessitates either more frequent GAC replacement or system augmentation, both of which substantially increase life-cycle costs and diminish the long-term economic feasibility of GAC-based treatment systems. In addition to these performance and economic limitations, the durability of ACs in water treatment processes is still under debate. Concerns persist regarding the safe regeneration and disposal of spent ACs. These uncertainties are compounded by the possible adverse health effects arising from the ingestion of, or dermal exposure to, micropollutants that may remain adsorbed on AC particles over prolonged periods of use. Indeed, ACs particles that separate from the filtration media can find their way into the water supply. If accidentally ingested or in contact with skin, these fine particles can act as carriers of residual contaminants, posing a direct risk to human health (Hassan et al., 2020). Furthermore, toxic by-products may form during the regeneration of PFAS-loaded ACs, while some PFAS may be released back into the environment during this process, worsening contamination issues. These limitations call into question the current safety and effectiveness of ACs under real-world conditions, especially when used repeatedly without complete regeneration.

Finally, the release of fine, loaded particles from technical adsorbents is another significant challenge. The extent of particle release depends on several parameters, including the adsorbent's composition, operating conditions, and its structural durability. These tiny particles can be produced when the adsorbent is set up, used, or replaced. Once in potable water systems, they can serve as vectors of adsorbed pollutants, potentially leading to chronic exposure via ingestion or skin absorption. This aspect not only poses a public health risk, but also raises regulatory concerns regarding the safety of using AC in drinking water systems.

## 16.5 Future research directions in activated carbon for per- and polyfluoroalkyl substances remediation

The application of ACs for PFAS removal, while effective for some compounds, presents several scientific and technological challenges that require targeted research. A key research priority is to elucidate the fundamental mechanisms of adsorption at the molecular level, in particular, the interactions between the surface functionalities of ACs and the various structures of PFAS. Advanced characterization techniques combined with molecular dynamics simulations could provide atomic-scale information on binding affinities, particularly for problematic short-chain PFAS, which exhibit lower adsorption efficiencies. Machine learning approaches offer promising tools for predicting adsorption behaviors by analyzing large datasets of ACs' properties and PFAS characteristics, which could accelerate materials optimization.

Material innovation is a critical frontier for addressing PFAS removal. Future research should systematically look into hybrid and composite adsorbents that mix ACs with complementing materials, including biochars, MOFs, and molecularly imprinted polymers. Artificial intelligence could transform this field by allowing for high-throughput virtual screening of prospective material combinations and surface modifications. The development of sustainable, bio-based alternatives with customized pore architectures and surface chemistries should be prioritized, with computational modeling guiding material design to boost both adsorption uptake and regeneration potential.

The critical challenge of spent ACs management requires a multifaceted research approach. Intelligent process optimization using machine learning algorithms could enhance regeneration protocols by predicting optimal conditions for contaminant desorption while minimizing carbon loss. For irreversibly adsorbed PFAS, research must assess advanced destruction technologies such as plasma treatment or supercritical water oxidation, where computational fluid dynamics models could improve reactor designs. Life-cycle assessment models that include AI-powered predictive analytics might be useful for assessing the environmental implications of various treatment and disposal techniques.

Practical research must aim to close the gap between laboratory discoveries and full-scale applications. Machine learning models based on operational data from pilot plants could improve process parameters for various water matrices and pollution conditions. The combination of real-time sensor data and predictive algorithms may enable adaptive treatment systems that respond to changing PFAS concentrations and water quality indices. The use of digital twin technology, which combines multiscale simulations with operational data, could be particularly useful for system design and performance prediction. Emerging computational tools provide great opportunities to advance ACs-based PFAS treatment. Quantum chemical calculations can reveal essential information about adsorption energetics, whereas multiscale modeling approaches can link molecular-level interactions to macroscopic performance. The establishment of open-access databases and established methodologies for computational investigations would improve collaboration and speed up progress in this field of study.

These research paths, which combine experimental investigations with cutting-edge computational and AI-driven methodologies, will be critical in developing next-generation AC technology. A synergistic mix of fundamental research, materials, computer science, and intelligent process engineering will result in more efficient, sustainable, and practical solutions to the global problem of PFAS pollution. Future research should concentrate on developing verified computational frameworks that can accurately forecast ACs' performance across various application scenarios, thereby lowering the time and expense required to design optimum treatment systems.

## 16.6 Conclusion

Adsorption using ACs has been recognized as a crucial technique in PFAS remediation, providing consistent adsorption efficacy, notably for long-chain PFAS chemicals, due to its large surface area and porous structure. Its widespread use in water treatment systems is due to its operational simplicity, scalability, and low-cost relative to many other technologies. However, substantial challenges remain, such as inefficient removal effectiveness for short-chain PFAS, competing adsorption effects in complex water matrices, and the economic and environmental implications of spent carbon regeneration and disposal.

Recent advances in material science are opening the path for next-generation AC technology. Surface-modified and composite adsorbents, such as amine-functionalized ACs, biochar hybrids, and MOF-enhanced carbons, exhibit higher selectivity and adsorption capability, even for challenging short-chain PFAS. Sustainable regeneration approaches, such as electrochemical and solvent-free processes, are also looking promising, with the goal of reducing energy usage and secondary waste generation. The combination of ACs and complementing destructive technologies, such as improved oxidation processes or sonochemical degradation, provides the potential for total PFAS mineralization rather than just phase transfer.

Computational and data-driven approaches are driving an important shift in the field of science. Machine learning algorithms help to accelerate adsorbent design by anticipating structure-performance correlations and optimizing material properties for specific PFAS pollutants. AI-powered process optimization enables real-time modifications to treatment systems, improving efficiency and lowering operational costs. Meanwhile, molecular dynamics simulations reveal atomic-level details about adsorption mechanisms, encouraging the rational development of high-performance materials. These digital tools not only reduce the need for trial-and-error experimentation, but they also help with the shift from laboratory-scale research to full-scale application. The successful implementation of ACs-based PFAS remediation will necessitate a multidisciplinary strategy that combines material innovation, process engineering, and environmental systems analysis. Future research priorities include: (1) developing standardized testing protocols for evaluating ACs' performance under realistic conditions; (2) advancing circular economy approaches to spent carbon management; and (3) establishing robust life-cycle assessment frameworks to assess the sustainability of emerging AC technologies.

Although AC is not an option for every PFAS contamination case, its adaptability and compatibility with both classic treatment methods and cutting-edge technological improvements ensure its continued use. With advancements in material science and digital technology, AC-based solutions will remain critical components of overall PFAS management strategies as regulatory constraints increase and the comprehension of PFAS environmental behavior grows. To successfully address this long-standing environmental challenge, the path forward includes concentrated research, cross-sector collaboration, and smart integration of ACs into larger treatment systems.

# References

Ahsaine, H.A., Zbair, M., Anfar, Z., Naciri, Y., El, R., Alem, N.E., et al., 2018. Cationic dyes adsorption onto high surface area ' almond shell ' activated carbon: kinetics, equilibrium isotherms and surface statistical modeling. Materials Today Chemistry 8, 121–132. Available at https://doi.org/10.1016/j.mtchem.2018.03.004.

Anastopoulos, I., Giannakoudakis, D.A., Frontistis, Z., Zorpas, A.A., Pashalidis, I., Triantafyllidis, K., 2024. Biomass-derived adsorbents: universal database development for their synthesis and remediation efficiency as a necessary step to move from laboratory- to pilot-scale applications. Current Opinion in Green and Sustainable Chemistry 47, 100902. Available at https://doi.org/10.1016/j.cogsc.2024.100902.

Appleman, T.D., Dickenson, E.R.V., Bellona, C., Higgins, C.P., 2013. Nanofiltration and granular activated carbon treatment of perfluoroalkyl acids. Journal of Hazardous Materials 260, 740–746. Available at https://doi.org/10.1016/j.jhazmat.2013.06.033.

Appleman, T.D., Higgins, C.P., Quiñones, O., Vanderford, B.J., Kolstad, C., Zeigler-Holady, J.C., et al., 2014. Treatment of poly- and perfluoroalkyl substances in U.S. full-scale water treatment systems. Water Research 51, 246–255. Available at https://doi.org/10.1016/j.watres.2013.10.067.

Ateia, M., Alsbaiee, A., Karanfil, T., Dichtel, W., 2019. Efficient PFAS removal by amine-functionalized sorbents: critical review of the current literature. Environmental Science & Technology Letters 6, 688–695. Available at https://doi.org/10.1021/acs.estlett.9b00659.

Bennici, S., Dutournié, P., Cathalan, J., Zbair, M., Nguyen, M.H., Scuiller, E., et al., 2022. Heat storage: hydration investigation of MgSO4/active carbon composites, from material development to domestic applications scenarios. Renewable and Sustainable Energy Reviews 158, 112197. Available at https://doi.org/10.1016/j.rser.2022.112197.

Boiteux, V., Dauchy, X., Rosin, C., Munoz, J.-F., 2012. National screening study on 10 perfluorinated compounds in raw and treated tap water in France. Archives of Environmental Contamination and Toxicology 63, 1–12. Available at https://doi.org/10.1007/s00244-012-9754-7.

Buck, R.C., Franklin, J., Berger, U., Conder, J.M., Cousins, I.T., de Voogt, P., et al., 2011. Perfluoroalkyl and polyfluoroalkyl substances in the environment: terminology, classification, and origins. Integrated Environmental Assessment and Management 7, 513–541. Available at https://doi.org/10.1002/ieam.258.

Cantoni, B., Turolla, A., Wellmitz, J., Ruhl, A.S., Antonelli, M., 2021. Perfluoroalkyl substances (PFAS) adsorption in drinking water by granular activated carbon: influence of activated carbon and PFAS characteristics. Science of the Total Environment 795, 148821. Available at https://doi.org/10.1016/j.scitotenv.2021.148821.

Chen, W., Zhang, X., Mamadiev, M., Wang, Z., 2017. Sorption of perfluorooctane sulfonate and perfluorooctanoate on polyacrylonitrile fiber-derived activated carbon fibers: in comparison with activated carbon. RSC Advances 7, 927–938. Available at https://doi.org/10.1039/C6RA25230C.

Choi, Y.-K., Cho, M.-H., Kim, J.-S., 2016. Air gasification of dried sewage sludge in a two-stage gasifier. Part 4: application of additives including Ni-impregnated activated carbon for the production of a tar-free and H2-rich producer gas with a low NH3 content. International Journal of Hydrogen Energy 41, 1460–1467. Available at https://doi.org/10.1016/j.ijhydene.2015.11.125.

Dalahmeh, S.S., Alziq, N., Ahrens, L., 2019. Potential of biochar filters for onsite wastewater treatment: effects of active and inactive biofilms on adsorption of per- and polyfluoroalkyl substances in laboratory column experiments. Environmental Pollution 247, 155–164. Available at https://doi.org/10.1016/j.envpol.2019.01.032.

Du, Z., Deng, S., Bei, Y., Huang, Q., Wang, B., Huang, J., et al., 2014. Adsorption behavior and mechanism of perfluorinated compounds on various adsorbents—a review. Journal of Hazardous Materials 274, 443–454. Available at https://doi.org/10.1016/j.jhazmat.2014.04.038.

EPA-United States Environmental Protection Agency Drinking Water Health Advisory for Perfluorooctanoic Acid (PFOS). Available online: https://www.epa.gov/sites/default/files/2016-05/documents/pfos_health_advisory_final_508.pdf. (no date).

EPA– United States Environmental Protection Agency Drinking Water Health Advisory for Perfluorooctanoic Acid (PFOA), (no date) Available online: https://www.epa.gov/sites/default/files/2016-05/documents/pfoa_health_advisory_final-plain.pdf.

Gagliano, E., Sgroi, M., Falciglia, P.P., Vagliasindi, F.G.A., Roccaro, P., 2020. Removal of poly- and perfluoroalkyl substances (PFAS) from water by adsorption: role of PFAS chain length, effect of organic matter and challenges in adsorbent regeneration. Water Research 171, 115381. Available at https://doi.org/10.1016/j.watres.2019.115381.

Gayathiri, M., Pulingam, T., Lee, K.T., Sudesh, K., 2022. Activated carbon from biomass waste precursors: factors affecting production and adsorption mechanism. Chemosphere 294, 133764. Available at https://doi.org/10.1016/j.chemosphere.2022.133764.

González-García, P., 2018. Activated carbon from lignocellulosics precursors: a review of the synthesis methods, characterization techniques and applications. Renewable and Sustainable Energy Reviews 82, 1393–1414. Available at https://doi.org/10.1016/j.rser.2017.04.117.

Hansen, M.C., Børresen, M.H., Schlabach, M., Cornelissen, G., 2010. Sorption of perfluorinated compounds from contaminated water to activated carbon. Journal of Soils and Sediments 10, 179–185. Available at https://doi.org/10.1007/s11368-009-0172-z.

Hassan, M., Liu, Y., Naidu, R., Du, J., Qi, F., 2020. Adsorption of perfluorooctane sulfonate (PFOS) onto metal oxides modified biochar. Environmental Technology & Innovation 19, 100816. Available at https://doi.org/10.1016/j.eti.2020.100816.

Inyang, M., Dickenson, E.R.V., 2017. The use of carbon adsorbents for the removal of perfluoroalkyl acids from potable reuse systems. Chemosphere 184, 168–175. Available at https://doi.org/10.1016/j.chemosphere.2017.05.161.

Liu, Y., Li, T., Bao, J., Hu, X., Zhao, X., Shao, L., et al., 2022. A review of treatment techniques for short-chain perfluoroalkyl substances. Applied Sciences 12, 1941. Available at https://doi.org/10.3390/app12041941.

Liu, G., Li, C., Stewart, B.A., Liu, L., Zhang, M., Yang, M., et al., 2020. Enhanced thermal activation of peroxymonosulfate by activated carbon for efficient removal of perfluorooctanoic acid. Chemical Engineering Journal 399, 125722. Available at https://doi.org/10.1016/j.cej.2020.125722.

Li, J., Li, Q., Li, L., Xu, L., 2017. Removal of perfluorooctanoic acid from water with economical mesoporous melamine-formaldehyde resin microsphere. Chemical Engineering Journal 320, 501–509. Available at https://doi.org/10.1016/j.cej.2017.03.073.

Li, Q., Qi, Y., Gao, C., 2015. Chemical regeneration of spent powdered activated carbon used in decolorization of sodium salicylate for the pharmaceutical industry. Journal of Cleaner Production 86, 424–431. Available at https://doi.org/10.1016/j.jclepro.2014.08.008.

Mamun, M.R.A., Karim, M.R., Rahman, M.M., Asiri, A.M., Torii, S., 2016. Methane enrichment of biogas by carbon dioxide fixation with calcium hydroxide and activated carbon. Journal of the Taiwan Institute of Chemical Engineers 58, 476–481. Available at https://doi.org/10.1016/j.jtice.2015.06.029.

McCleaf, P., Englund, S., Östlund, A., Lindegren, K., Wiberg, K., Ahrens, L., 2017. Removal efficiency of multiple poly- and perfluoroalkyl substances (PFASs) in drinking water using granular activated carbon (GAC) and anion exchange (AE) column tests. Water Research 120, 77–87. Available at https://doi.org/10.1016/j.watres.2017.04.057.

Meng, P., Fang, X., Maimaiti, A., Yu, G., Deng, S., 2019. Efficient removal of perfluorinated compounds from water using a regenerable magnetic activated carbon. Chemosphere 224, 187–194. Available at https://doi.org/10.1016/j.chemosphere.2019.02.132.

Mohd Azmi, L.H., Williams, D.R., Ladewig, B.P., 2021. Polymer-assisted modification of metal-organic framework MIL-96 (Al): influence of HPAM concentration on particle size, crystal morphology and removal of harmful environmental pollutant PFOA. Chemosphere 262, 128072. Available at https://doi.org/10.1016/j.chemosphere.2020.128072.

Moreno-Castilla, C., 2004. Adsorption of organic molecules from aqueous solutions on carbon materials. Carbon 42, 83–94. Available at https://doi.org/10.1016/j.carbon.2003.09.022.

Nakayama, S.F., Yoshikane, M., Onoda, Y., Nishihama, Y., Iwai-Shimada, M., Takagi, M., et al., 2019. Worldwide trends in tracing poly- and perfluoroalkyl substances (PFAS) in the environment. TrAC Trends in Analytical Chemistry 121, 115410. Available at https://doi.org/10.1016/j.trac.2019.02.011.

Nieto-Delgado, C., Rangel-Mendez, J.R., 2011. Production of activated carbon from organic by-products from the alcoholic beverage industry: surface area and hardness optimization by using the response surface methodology. Industrial Crops and Products 34, 1528–1537. Available at https://doi.org/10.1016/j.indcrop.2011.05.014.

Ochoa-Herrera, V., Sierra-Alvarez, R., 2008. Removal of perfluorinated surfactants by sorption onto granular activated carbon, zeolite and sludge. Chemosphere 72, 1588–1593. Available at https://doi.org/10.1016/j.chemosphere.2008.04.029.

Official Journal of the European Union, Directive (EU) 2020/2184 of the European Parliament and of the Council of 16 December 2020 on the Quality of Water Intended for Human Consumption, (Recast). Available online: http://data.europa.eu/eli/dir/2020/2184/oj (no date).

Ouasfi, N., Zbair, M., Bouzikri, S., Anfar, Z., Bensitel, M., Ait Ahsaine, H., et al., 2019. Selected pharmaceuticals removal using algae derived porous carbon: experimental, modeling and DFT theoretical insights. RSC Advances 9, 9792–9808. Available at https://doi.org/10.1039/C9RA01086F.

Park, M., Wu, S., Lopez, I.J., Chang, J.Y., Karanfil, T., Snyder, S.A., 2020. Adsorption of perfluoroalkyl substances (PFAS) in groundwater by granular activated carbons: roles of hydrophobicity of PFAS and carbon characteristics. Water Research 170, 115364. Available at https://doi.org/10.1016/j.watres.2019.115364.

Pauletto, P.S., Bandosz, T.J., 2022. Activated carbon versus metal-organic frameworks: a review of their PFAS adsorption performance. Journal of Hazardous Materials 425, 127810. Available at https://doi.org/10.1016/j.jhazmat.2021.127810.

Phong Vo, H.N., Ngo, H.H., Guo, W., Hong Nguyen, T.M., Li, J., Liang, H., et al., 2020. Poly-and perfluoroalkyl substances in water and wastewater: a comprehensive review from sources to remediation. Journal of Water Process Engineering 36, 101393. Available at https://doi.org/10.1016/j.jwpe.2020.101393.

Punyapalakul, P., Suksomboon, K., Prarat, P., Khaodhiar, S., 2013. Effects of surface functional groups and porous structures on adsorption and recovery of perfluorinated compounds by inorganic porous silicas. Separation Science and Technology 48, 775–788. Available at https://doi.org/10.1080/01496395.2012.710888.

Qiu, C., Jiang, L., Gao, Y., Sheng, L., 2023. Effects of oxygen-containing functional groups on carbon materials in supercapacitors: a review. Materials and Design 230, 111952. Available at https://doi.org/10.1016/j.matdes.2023.111952.

Radovic, L., Moreno-Castilla, C., Rivera-Utrilla, J., 2000. Carbon materials as adsorbents in aqueous solutions. Chemistry and Physics of Carbon 27, 227–405.

Ramos, P., Singh Kalra, S., Johnson, N.W., Khor, C.M., Borthakur, A., Cranmer, B., et al., 2022. Enhanced removal of per- and polyfluoroalkyl substances in complex matrices by PolyDADMAC-coated regenerable granular activated carbon. Environmental Pollution 294, 118603. Available at https://doi.org/10.1016/j.envpol.2021.118603.

Rosen, M.J., 2004. Adsorption of surface-active agents at interfaces: the electrical double layer. Surfactants and Interfacial Phenomena. John Wiley & Sons, Ltd, pp. 34–104 ISBN 9780471670568.

Saeidi, N., Kopinke, F.-D., Georgi, A., 2020. Understanding the effect of carbon surface chemistry on adsorption of perfluorinated alkyl substances. Chemical Engineering Journal 381, 122689. Available at https://doi.org/10.1016/j.cej.2019.122689.

Saleh, N.B., Khalid, A., Tian, Y., Ayres, C., Sabaraya, I.V., Pietari, J., et al., 2019. Removal of poly- and perfluoroalkyl substances from aqueous systems by nano-enabled water treatment strategies. Environmental Science: Water Research & Technology 5, 198–208. Available at https://doi.org/10.1039/C8EW00621K.

Selmi, T., Gentil, S., Fierro, V., Celzard, A., 2024. Key factors in the selection, functionalization and regeneration of activated carbon for the removal of the most common micropollutants in drinking water. Journal of Environmental Chemical Engineering 12, 113105. Available at https://doi.org/10.1016/j.jece.2024.113105.

Sethia, G., Sayari, A., 2016. Activated carbon with optimum pore size distribution for hydrogen storage. Carbon 99, 289–294. Available at https://doi.org/10.1016/j.carbon.2015.12.032.

Stockholm Convention, All POPs Listed in the Stockholm Convention. Available online: http://chm.pops.int/TheConvention/ThePOPs/ListingofPOPs/tabid/2509/Default.aspx. (no date).

Stoiber, T., Evans, S., Temkin, A.M., Andrews, D.Q., Naidenko, O.V., 2020. PFAS in drinking water: an emergent water quality threat. Water Solutions 1, 49.

Sun, H., Cannon, F.S., He, X., 2020. Effective removal of perfluorooctanoate from groundwater using quaternary nitrogen-grafted granular activated carbon. Journal of Water Process Engineering 37, 101416. Available at https://doi.org/10.1016/j.jwpe.2020.101416.

Vecitis, C.D., Park, H., Cheng, J., Mader, B.T., Hoffmann, M.R., 2008. Kinetics and mechanism of the sonolytic conversion of the aqueous perfluorinated surfactants, perfluorooctanoate (PFOA), and perfluorooctane sulfonate (PFOS) into inorganic products. The Journal of Physical Chemistry A 112, 4261–4270. Available at https://doi.org/10.1021/jp801081y.

Vervald, A., Kozhushnyi, K., Dolenko, T., 2022. Difference in deprotonation for oxygen-containing groups on Sp2 and Sp3 carbons. Fullerenes, Nanotubes and Carbon Nanostructures 30, 106–112. Available at https://doi.org/10.1080/1536383X.2021.1986014.

Wang, W., Maimaiti, A., Shi, H., Wu, R., Wang, R., Li, Z., et al., 2019. Adsorption behavior and mechanism of emerging perfluoro-2-propoxypropanoic acid (GenX) on activated carbons and resins. Chemical Engineering Journal 364, 132–138. Available at https://doi.org/10.1016/j.cej.2019.01.153.

Wang, Y., Niu, J., Li, Y., Zheng, T., Xu, Y., Liu, Y., 2015. Performance and mechanisms for removal of perfluorooctanoate (PFOA) from aqueous solution by activated carbon fiber. RSC Advances 5, 86927–86933. Available at https://doi.org/10.1039/C5RA15853B.

Wang, F., Shih, K., 2011. Adsorption of perfluorooctanesulfonate (PFOS) and perfluorooctanoate (PFOA) on alumina: influence of solution PH and cations. Water Research 45, 2925–2930. Available at https://doi.org/10.1016/j.watres.2011.03.007.

Wu, C., Klemes, M.J., Trang, B., Dichtel, W.R., Helbling, D.E., 2020. Exploring the factors that influence the adsorption of anionic PFAS on conventional and emerging adsorbents in aquatic matrices. Water Research 182, 115950. Available at https://doi.org/10.1016/j.watres.2020.115950.

Yuan, C., Huang, Y., Cannon, F.S., Zhao, Z., 2020. Adsorption mechanisms of PFOA onto activated carbon anchored with quaternary ammonium/epoxide-forming compounds: a combination of experiment and model studies. Journal of Environmental Sciences 98, 94–102. Available at https://doi.org/10.1016/j.jes.2020.05.019.

Yu, J., Lv, L., Lan, P., Zhang, S., Pan, B., Zhang, W., 2012. Effect of effluent organic matter on the adsorption of perfluorinated compounds onto activated carbon. Journal of Hazardous Materials 225–226, 99–106. Available at https://doi.org/10.1016/j.jhazmat.2012.04.073.

Yu, Q., Zhang, R., Deng, S., Huang, J., Yu, G., 2009. Sorption of perfluorooctane sulfonate and perfluorooctanoate on activated carbons and resin: kinetic and isotherm study. Water Research 43, 1150–1158. Available at https://doi.org/10.1016/j.watres.2008.12.001.

Zaggia, A., Conte, L., Falletti, L., Fant, M., Chiorboli, A., 2016. Use of strong anion exchange resins for the removal of perfluoroalkylated substances from contaminated drinking water in batch and continuous pilot plants. Water Research 91, 137–146. Available at https://doi.org/10.1016/j.watres.2015.12.039.

Zbair, M., Ainassaari, K., Drif, A., Ojala, S., Bottlinger, M., Pirilä, M., et al., 2018a. Toward new benchmark adsorbents: preparation and characterization of activated carbon from argan nut shell for Bisphenol A removal. Environmental Science and Pollution Research. Available at https://doi.org/10.1007/s11356-017-0634-6.

Zbair, M., Ainassaari, K., El Assal, Z., Ojala, S., El Ouahedy, N., Keiski, R.L., et al., 2018b. Steam activation of waste biomass: highly microporous carbon, optimization of Bisphenol A, and diuron adsorption by response surface methodology. Environmental Science and Pollution Research 25, 35657–35671. Available at https://doi.org/10.1007/s11356-018-3455-3.

Zbair, M., Nguyen, M.H., Dutournié, P., Bennici, S., 2023. Bi-salts composites to enhance the hydration kinetics and heat storage capacity. Journal of Energy Storage 73, 108862. Available at https://doi.org/10.1016/j.est.2023.108862.

Zhang, D., Luo, Q., Gao, B., Chiang, S.-Y.D., Woodward, D., Huang, Q., 2016. Sorption of perfluorooctanoic acid, perfluorooctane sulfonate and perfluoroheptanoic acid on granular activated carbon. Chemosphere 144, 2336–2342. Available at https://doi.org/10.1016/j.chemosphere.2015.10.124.

Zhao, L., Bian, J., Zhang, Y., Zhu, L., Liu, Z., 2014. Comparison of the sorption behaviors and mechanisms of perfluorosulfonates and perfluorocarboxylic acids on three kinds of clay minerals. Chemosphere 114, 51–58. Available at https://doi.org/10.1016/j.chemosphere.2014.03.098.

Zhi, Y., Liu, J., 2015. Adsorption of perfluoroalkyl acids by carbonaceous adsorbents: effect of carbon surface chemistry. Environmental Pollution 202, 168–176. Available at https://doi.org/10.1016/j.envpol.2015.03.019.

Zhi, Y., Liu, J., 2016. Surface modification of activated carbon for enhanced adsorption of perfluoroalkyl acids from aqueous solutions. Chemosphere 144, 1224–1232. Available at https://doi.org/10.1016/j.chemosphere.2015.09.097.

Zietzschmann, F., Stützer, C., Jekel, M., 2016. Granular activated carbon adsorption of organic micro-pollutants in drinking water and treated wastewater – aligning breakthrough curves and capacities. Water Research 92, 180–187. Available at https://doi.org/10.1016/j.watres.2016.01.056.

CHAPTER

# 17

# Biochar for the remediation of per- and polyfluoroalkyl substances

*Navish Kataria[1], Sangita Yadav[2], and Divya Dhillayan[2]*

[1]Department of Environmental Sciences, J.C. Bose University of Science and Technology, YMCA, Faridabad, Haryana, India [2]Department of Environmental Science and Engineering, Guru Jambheshwar University of Science & Technology, Hisar, Haryana, India

## 17.1 Introduction

The industrial revolution has played an important role in the rapid modernization and the domestic sewage generation. Industrial activities have consequently become a dominant source of emerging contaminants in water (Rani et al., 2024). The release of anthropogenic pollutants into water bodies significantly impacts water quality and affects marine and human health. Consequently, water pollution and the discharge of hazardous chemicals into water bodies must be reduced. The release of different water contaminants like dyes, heavy metals, pharmaceuticals, microbial waste, etc., has led to a decrease in the availability of safe drinking water. Water scarcity has emerged as a significant concern in various countries and is expected to increase as a critical global issue soon. Access to safe and clean drinking water, one of the 17 Sustainable Development Goals, is acknowledged as a fundamental human right. Since the 1940s, PFAS, a category of synthetic organofluorine compounds, have been widely used by the consumer and industrial sectors (Deng et al., 2023). Fig. 17.1 shows several sources of PFASs that contaminate the environment.

PFAS are categorized into two types, namely, long-chain and short-chain PFAS. Long-chain PFAS undergo degradation that results in their conversion into short-chain PFAS (Zhang et al., 2023a). These compounds contain a hydrophobic chain that is either partially or fully saturated with fluorine atoms. PFAS are thermally and chemically stable due to the presence of strong covalent bonds. They are also known as "forever chemicals" (Lenka et al., 2021). They have extensive applications in both industrial and commercial products, textiles, firefighting foams, adhesives, cosmetics, and in the coating of nonstick utensils (Kurwadkar et al., 2022).

PFAS pose a significant risk to both plant and animal life, particularly affecting aquatic species. In the natural food web, the bioaccumulation of PFAS varies across different

*Per- and Polyfluoroalkyl Substances (PFAS) as Environmental Contaminants*
DOI: https://doi.org/10.1016/B978-0-443-43856-1.00005-8

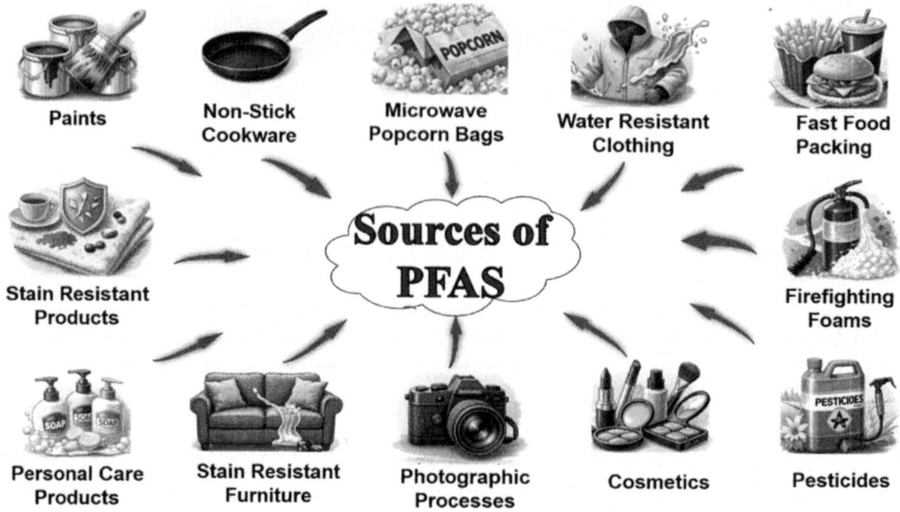


FIGURE 17.1 Sources of per- and polyfluoroalkyl substances (PFASs) in the environment.

species, making it difficult to measure the accurate assessment of ecological risks. This complexity arises from the significant differences in exposure, habitat, and bioaccumulation among various species (Gebbink and van Leeuwen, 2020). Moreover, the extent of PFAS accumulation also varies with the type of internal organs of different species. PFAS tend to accumulate at higher concentrations in the liver than in muscle tissues, with the hierarchy of accumulation being liver > muscle > kidney. Epidemiological studies suggested that exposure to PFAS, chiefly perfluorooctane sulfonate (PFOS) and perfluorooctanoic acid (PFOA), is associated with hormonal imbalance and changes in thyroid function. They are also responsible for the weakening of the immune system, higher cancer risk, cholesterol, and harm to the developing fetus (US Department of Health and Human Services, 1999). PFAS also harm the health of children. Children who are exposed to PFOA have an increased chance of acquiring asthma. Exposure to PFOS is also linked to reduced lung function (Rafiee et al., 2024).

Considering these concerns, effective remediation methods need to be addressed for PFAS contamination in water sources. Conventional wastewater treatment methods like physical separation, chemical oxidation, coagulation with subsequent UV irradiation, aeration, and disinfection are not efficient for removing PFAS. The incapability of conventional treatment processes for the removal of PFAS is due to the presence of a strong carbon–fluorine covalent bond in the molecular structure of PFAS. The hydrophobic nature of PFAS and their low concentrations in water bodies also make the conventional treatment methods ineffective (Ross et al., 2018). The adsorption process has emerged as the most promising and cost-efficient method for the treatment of polluted soil and water. Some of the frequently used adsorbents are biochar, nanoparticles, activated carbon, functionalized clays, ion-exchange resins, and metal organic frameworks (MOFs). At present, anion exchange resins and biochar are the most used adsorbents (Zhang et al., 2022a).

Biochar, a porous carbon rich material, is obtained by the oxygen free carbonization of biomass. It has been extensively used as an alternative adsorbent as it offers a high specific area and highly porous structure, making it a potent adsorbent for the remediation of PFAS. It can be prepared from different waste sources, including kitchen waste, wood, agricultural waste, waste biomass, manure, leaves, and wastewater sludge. It offers both economic and environmental advantages, suggesting a feasible way for scaling up applications. Studies have shown the adsorption of PFAS using modified biochar prepared from a single raw material (Yaashikaa et al., 2020).

This chapter critically assesses recent peer reviewed studies on the use of biochar for the adsorption based removal of both short and long-chain PFAS. The emphasis is on critical elements determining the PFAS removal efficiency of biochar, synthesis techniques, adsorption mechanisms, variables influencing adsorption performance, and adsorbent regeneration methods.

## 17.2 Biochar and its production techniques

Biochar, a carbonaceous material formed through the partial combustion of biomass through pyrolysis at higher temperatures, has gained attention as an effective adsorbent. It can be synthesized from organic waste or residues derived from agricultural and forestry products, offering an environmentally sustainable and cost-effective solution (Sun et al., 2024). It has shown significant results in mitigating environmental pollution caused by both organic and inorganic pollutants. In recent studies, the usage of biochar in the water treatment industry has gained attention and has shown positive outcomes for its application in the removal of PFAS (Wang et al., 2025).

Various thermochemical methods are used to prepare biochar, like pyrolysis, hydrothermal carbonization (HTC), torrefaction, gasification, and flash carbonization (Fig. 17.2). The pyrolysis temperature and the type of initial biomass used to prepare biochar greatly affect the production of biochar. It has been observed that the temperature range between 200°C and 950°C is an optimum temperature range to produce biochar in an oxygen-free atmosphere. The conversion of waste biomass into value-added products, including biochar, biogas, and biooil, characterizes a significant advancement in sustainable resource management (El-Fawal et al., 2025).

HTC is a thermochemical method that converts biomass into biochar within a temperature range of 150°C–300°C and under high pressure conditions. The biomass is combined with water and introduced into a sealed reactor. This process yields various products, including biochar, biooil, and syngas (Shekoohiyan et al., 2025). It is a preferred biochar synthesis method as it operates on low temperature conditions, has fast biochar production rate, and a lower operational cost. Torrefaction is a process carried out in an oxygen limited environment, characterized by a controlled heating rate. The procedure is executed in four distinct phases: heating, dehydrating, torrefying, and condensation (Li et al., 2025). Among others, pyrolysis is often identified as the most efficient and economically viable method for the production of biochar, primarily due to its capacity to generate biochar with properties that can be customized for particular applications. Rodrigo

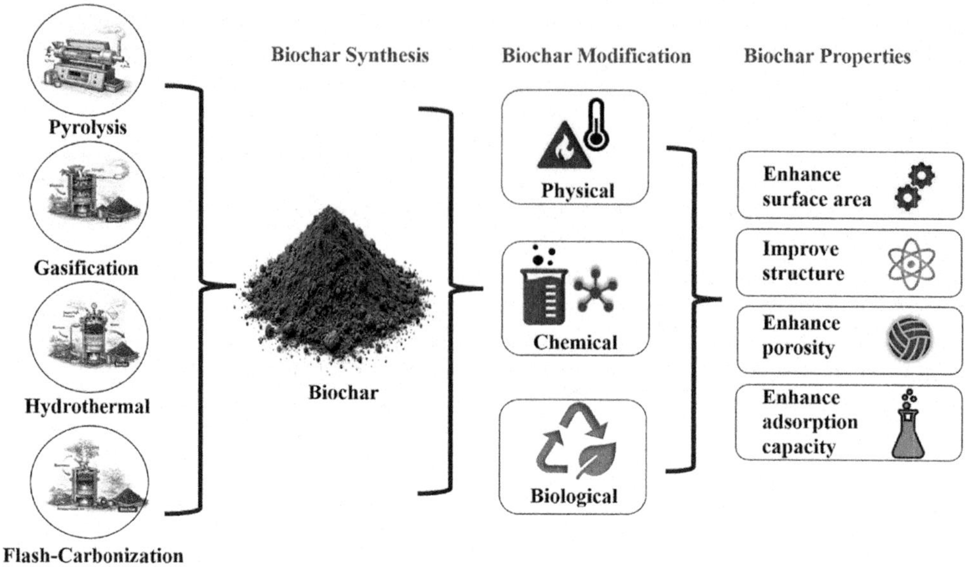


FIGURE 17.2 Representation of various methods of biochar preparation and modification.

et al. (2022) synthesized biochar from Douglas fir at 900°C–1000°C using the gasification technique. Biochar further modified with $Fe_2O_3$ successfully removed PFOS with a 75%–85% removal efficiency. The capacity of biochar to immobilize contaminants for a long time, coupled with its additional advantages such as waste management, climate change mitigation, enhancement of soil fertility, and energy production, makes it for tackling environmental issues, including the removal of PFAS (Kumar et al., 2023a) (see Table 17.1).

## 17.2.1 Modification of biochar for per- and polyfluoroalkyl substances removal

While raw biochar is used for the removal of PFAS, they demonstrate a lower efficiency compared to biochar composites (BCs). This indicates an important pathway for forthcoming advancements: the engineering of biochar to enhance its surface functionality, porosity, and binding sites. BCs are synthesized through the process of impregnating various materials, including metals, metal oxide nanoparticles, clay minerals, carbon nanotubes (CNTs), polysaccharides, or MOFs (Bose et al., 2023). These materials are predominantly introduced prior to pyrolysis, referred to as pretreatment, or subsequent to pyrolysis, known as blending. In this context, biochar serves as a porous material with a functional matrix for the impregnation of the doping with other materials. The primary benefits of BCs encompass the presence of well structured oxygen rich functional groups, an extensive quantity of micro and mesopores, and an elevated surface area tailored for specific applications, indicating a noteworthy advancement in the field. Furthermore, the application of chemical treatments, including impregnation with alkali or acids, can facilitate the addition of functional groups (such as OH, COOH, and $NH_2$ groups) onto the surface of biochar (Kumar et al., 2024) (see Table 17.2).

**TABLE 17.1** Various biochar production techniques.

| Technique | Description | Advantages | Limitation | References |
|---|---|---|---|---|
| Pyrolysis | It is a thermochemical process that generates biochar from biomass under elevated temperatures (350°C–650°C) and pressures. It is an archaic technique for biomass processing, involving the pyrolysis of materials in a sealed reactor with restricted oxygen levels. Based on different characteristics such as residence time, temperature, heating rate, the pyrolysis is divided into three categories: slow, fast, and flash. | Pyrolysis enables for controlled biochar formation with variable surface area, porosity, and chemical characteristics by modifying processing settings. | The process is energy-intensive and complex, with low-temperature pyrolysis producing biochar with fewer functional groups and lesser porosity. | Chi et al. (2021) |
| Gasification | It is an eco-friendly process that transforms fuel at temperatures >700°C under oxygen and/or steam-controlled conditions, resulting in biogas and charcoal as by-products. It is carried out with fixed-bed and fluidized-bed gasifiers. Gasifying factors such as air, oxygen, $CO_2$, and steam affect the composition of syngas. Equipment and operational circumstances have a considerable impact on syngas yield, quality, and biochar properties. | It generates superior syngas and biochar, offering process adaptability depending on gasifying agents, alongside a more straightforward fixed-bed gasifier design that facilitates material handling and reduces gas-phase particulate matter. | The outcome of the process is significantly affected by the type of equipment, raw materials, and operating conditions, resulting in varying biochar properties and necessitating meticulous control for uniform product quality. | Velusamy et al. (2024) |
| Hydrothermal carbonization | It is a thermochemical process that uses water at higher temperatures (180°C–375°C) and pressure conditions to transform biomass, particularly moist or leafy biomass, into hydrochar. HTC modifies biomass via chemical processes, yielding a solid carbon-rich | HTC efficiently transforms wet or green biomass into hydrochar characterized by high carbon content, low ash concentration, superior water retention, and potential applications in energy storage, soil enhancement, and adsorption. | The yield and characteristics of hydrochar are significantly influenced by temperature and reaction duration, while hydrothermal carbonization produces acidic liquid by-products that necessitate | Adeniyi et al. (2024) |

(*Continued*)

**TABLE 17.1** (Continued)

| Technique | Description | Advantages | Limitation | References |
|---|---|---|---|---|
| | product characterized by unique physical and chemical features, including elevated water retention capacity, reduced ash content, and increased energy density. | | meticulous management due to the generation of organic acids. | |
| Torrefaction | It is a mild pyrolysis process conducted at temperatures between 200°C–350°C in an inert or reducing environment, designed to diminish moisture and volatile content in biomass. It degrades hemicellulose, cellulose, and lignin, with a primary focus on hemicellulose, leading to the release of $CO_2$ and $H_2O$. | Torrefaction improves biomass fuel quality by augmenting energy density, diminishing moisture and volatile content, enhancing grindability, and yielding a more uniform, hydrophobic, and stable product. | The process is predominantly designed for energy generation, hence constraining its wider applicability. | Varkolu et al. (2025) |

| Modification technique | Description | Advantage | Limitation | References |
|---|---|---|---|---|
| Physical method | Physical alteration occurs postbiochar generation through techniques such as ball milling, steam treatment, thermal treatment, or gas purging. These processes primarily enhance the porosity and surface area of biochar, thereby influencing its physical structure. | It enhances the porosity and surface area of biochar, hence augmenting its overall adsorption capacity. | It is less efficacious than chemical modification in eliminating certain contaminants, particularly anionic pollutants such as arsenic, fluoride, nitrate, and phosphate. | Kumar et al. (2022b) |
| Chemical method | Chemical modification is a prevalent strategy for improving the surface characteristics of biochar. The process entails the application of acids, alkalis, oxidizing agents, and metal oxides, which can modify the concentration, type, and distribution of surface functional groups and defects, consequently enhancing the sorption capacity of biochar. | It markedly enhances the adsorption capacity of biochar by augmenting its surface with active functional groups and increasing surface reactivity. | The procedure may generate chemical waste and utilize corrosive or toxic substances, hence heightening environmental and safety issues. | Liu et al. (2022) |
| Biological method | Biological modification of biochar entails the integration of advantageous microbes into its structure by techniques such as adsorption, encapsulation, and flocculation. This method enhances microbial activity, stability, and pollutant tolerance, allowing biochar to efficiently eliminate organic contaminants and immobilize heavy metals. | It enhances biochar's efficacy in eliminating organic pollutants and heavy metals, while simultaneously augmenting microbial activity, pollutant resilience, and soil vitality. | The efficacy may be affected by environmental alterations and heavy metal toxicity, which might hinder microbial viability and pollution remediation ability. | Gusiatin and Rouhani (2023) |
| Doping | Mineral doping modification entails the integration of minerals such as clay (montmorillonite, kaolinite), magnesium oxide, birnessite, or calcium compounds into biochar to enhance its structural and chemical characteristics. The process encompasses several doping options, including direct mixing, soaking, or thermal treatment during or subsequent to pyrolysis. | Mineral doped biochars markedly improve sorption capacity, surface functionality, thermal stability, and efficacy in eliminating heavy metals and other contaminants. | It can result in diminished surface area due to pore obstruction and entails intricate manufacturing procedures that may produce byproducts necessitating further processing. | Murtaza et al. (2024) |

The alterations implemented significantly enhance the adsorption capacity of biochar in the adsorption of various organic and inorganic pollutants, including dyes, pharmaceuticals, and heavy metals. It also increases its ability to recollect vital nutrients, including phosphorus and nitrogen, thereby increasing the fertility of soil and supporting plant growth (Patel et al., 2023).

## 17.3 Biochar for per- and polyfluoroalkyl substances removal

Biochar has emerged as an excellent adsorbent, as it is low cost, environmentally sustainable, and is suitable for the adsorption of PFAS from water. From an environmental perspective, it also aligns with the context of a circular economy by promoting waste valorization and resource efficiency. Biochar, produced from a variety of biomass sources, is a versatile substance abundant in aromatic and mineral elements (Kumar et al., 2025). The characterization of biochar is crucial for the analysis of its physical, chemical, and structural characteristics. Different characterization techniques, such as point zero charge ($pH_{pzc}$), Fourier-transform infrared spectroscopy, X-ray diffraction, scanning electron microscopy, Brunauer-Emmett-Teller, and thermogravimetric analysis, are commonly used. They help us in understanding the essential parameters, including surface functional groups, elemental composition, surface area, pH, porosity, surface morphology, and thermal stability. The various characteristics of biochar, including its alkaline pH, elevated porosity, extensive surface area, significant carbon content, water-holding capacity, nutrient and ash composition, and functional groups on the surface, contribute to its effectiveness across a range of applications (Kumar et al., 2022a; 2023a). However, the synthesis conditions and feedstock sources (such as forestry residues or biosolids) result in biochar with different characteristics and have a considerable impact on PFAS adsorption (Liu et al., 2021).

### 17.3.1 Factors affecting the adsorption of per- and polyfluoroalkyl substances by biochar

#### *17.3.1.1 Biochar properties*

Biochar produced through the carbonization of biomass shows excellent efficiency. Biochar is a highly porous structure formed through the volatilization of volatile compounds from biomass, which enhances its adsorption capacity for pollutants (Wu et al., 2021). The physical characteristics of biochar, such as pore distribution, pore size, particle size, and specific surface area, strongly influence the adsorption of PFAS. The elemental composition (carbon, hydrogen, oxygen, nitrogen, and sulfur) and the structural compositions (cellulose, hemicellulose, and lignin) vary with the different raw materials used to prepare biochar (Weber and Quicker 2018). The sludge biochar contains higher nitrogen content, which interacts with the heterocyclic structure to produce pyrrole nitrogen species. The adsorption capacity of sludge biochar for PFAS is enhanced by the presence of these nitrogen atoms through the formation of a weak hydrogen bond with the compound (Zhang et al., 2023b). The hydrophobic interaction is greatly influenced by the presence of oxygen content in carbon-based adsorbents.

It was observed that high-temperature pyrolysis of biochar exhibited an enhanced surface area, increased aromaticity, greater pore volume, and higher micropore formation, enhancing adsorption of PFAS through pore filling (Calisto et al., 2014). It was also observed that at higher pyrolysis temperature, there is a notable reduction in carboxyl and hydroxyl group along with C–O bonds, C–H bonds, and O–H bonds content of biochar. This is due to the increase in mineral and carbonate content, alongside a reduction in acidic volatile products, as the biomass undergoes cracking. This results in a decrease in the polarity and hydrophilicity of biochar, promoting its aromatization, thereby stabilizing its structural integrity, and finally enhancing its efficacy in the adsorption and removal of PFAS (Parthasarathy et al., 2022; Zhang et al., 2022b).

#### 17.3.1.2 *Structure of per- and polyfluoroalkyl substances*

The functional group on the adsorbent and the chain length of the perfluorocarbon of PFAS have a great influence on the removal efficiency of PFAS by biochar from water. The adsorption of PFASs with variable chain lengths has been clearly observed throughout the adsorption process. It was observed that PFAS with the same functional group at the end but a longer C chain show greater hydrophobic interactions. This phenomenon can also be due to the increase in the dissociation constant ($K_d$) due to the elongation of the perfluoroalkyl chain (Kabiri et al., 2023). Du et al. (2015) observed that the hydrophobic interactions of powdered activated carbon (PAC) result in higher adsorption efficiency for long-chain PFASs, specifically PFOA, than short-chain PFASs such as perfluorohexanoic acid (PFHxA) and perfluoroheptanoic acid (PFHpA). In the mixed PFAS system, PAC exhibited limited efficacy in removing PFHxA and PFHpA, while showing a high removal efficiency for PFOA.

The hydrogen bonding and hydrophobic interactions between the adsorbents and PFASs are greatly governed by the nature of functional groups, such as the carboxyl group and sulfonic groups present in PFAC and PFSA, respectively. The reduced size of the carboxyl group in PFCAs could potentially result in a lower hydrophobicity compared to the sulfonic acid group (Vo et al., 2022).

#### 17.3.1.3 *pH of the solution*

The solution pH is a critical parameter that influences the adsorption capacity and efficiency of PFAS and plays an important role in the optimization of adsorption processes. The pH of the solution plays a crucial role in determining the characteristic surface charge of the adsorbent, which in turn influences the ionization and speciation of the adsorbate in the solution. The removal efficiency of PFAS is also influenced by the presence of functional groups under different pH conditions (Gao et al., 2017). It was also observed that pH greatly impacts the sorption of organic anions by regulating the protonation-deprotonation of organic acid functional groups near their pKa value (Salehi, 2025).

When the pH < $pH_{pzc}$ (zero charge point), the $H^+$ ions ionize the surface functional groups of biochar, converting them into positively charged species such as $COH^{2+}$/$COOH^{2+}$/M-$OH^{2+}$/$OH^{2+}$. This influences the electrostatic and ion-exchange interactions, making it easier for biochar to bind to PFOS (Fig. 17.3). On the other hand, when the solution pH exceeds the $pH_{pzc}$, the surface functional group lose their proton and changes into

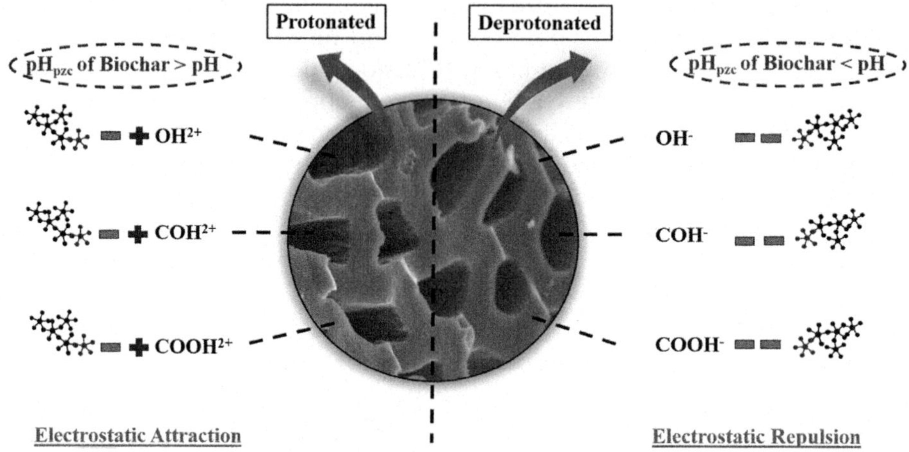


FIGURE 17.3 Effect of pH on the adsorption of per- and polyfluoroalkyl substances by biochar.

negatively charged ($CHO^-/COO^-/OH^-/MO^-$). This results in electrostatic repulsion between the negatively charged biochar surface and the negatively charged PFOS surface (Hassan et al., 2020). Xing et al. (2025) reported that BSA/MBC showed maximum adsorption efficiency at pH 5. When the $pH < 5$, the dissociation of acidic compounds of PFAS into anions favors the maximum electrostatic adsorption. It suggests that hydrophobic interaction dominates the adsorption of PFAS at lower pH.

#### *17.3.1.4 Inorganic ions in the solution*

The presence of various inorganic ions in the solution greatly influences the PFAS adsorption efficiency. It is reported that the PFAS adsorption capacity may either increase or decrease due to the addition of inorganic ions in the solution (Gong et al., 2016; Cao et al., 2017). This variable change in the adsorption efficiency of PFAS by biochar occurs through various mechanisms, such as surface charge neutralization, salting out effects, competition, bridging between the PFAS and OH groups, and electric double layer compression (Du et al., 2014). Du et al. (2015) studied the impact of the presence of inorganic ions in solution on the removal of three PFCAs (PFHxA, PFHpA, and PFOA). It was observed that the inorganic salts affected the adsorption efficiency in the order $CaCl_2 > MgCl_2 > Na_2SO_4 > NaCl > KCl > NaNO_3$. The electrostatic repulsion between the PFAS anions and the adsorbent and the electric double layer of the adsorbents is reduced by the addition of salts, which improves the removal of PFCA. It was also observed that the PFAS adsorption efficiency increases in the presence of divalent cations than monovalent cations in aqueous solution (Arvaniti and Stasinakis, 2015).

The presence of calcium ions ($Ca^{2+}$) in the solution facilitated the adsorption of PFOS and PFOA on biochar. This is due to the charge shielding effect of $Ca^{2+}$, which tends to inhibit the negative charge on the electrical double layer (EDL) of biochar. The presence of $Mg^{2+}$ and $Ca^{2+}$ in the solution neutralizes the negative surface charge of the adsorbent. This tends to reduce the electrostatic repulsion between negatively charged biochar and anionic PFAS and enhances the adsorption capacity of biochar (Rahman et al., 2022). The negatively charged functional groups, such as hydroxyl, carboxyl, or phenolic groups, form bonds with anionic PFAS through a divalent cation bridging mechanism (Issabayeva et al., 2018). There is a decrease in the electrostatic attraction between the negatively charged PFAS and the positively charged adsorbents, such as nanosized $TiO_2$, $Al_2O_3$, and $ZrO_2$, and the negatively charged surfaces of PFAS may be diminished. This phenomenon can be attributed to an increase in ionic strength, which may result in the compression of electric double layers and a subsequent reduction in zeta potential (Kujawa et al., 2014). Nonetheless, the presence of $Na^+$ resulted in a reduction of the adsorption capacity for PFOA and PFBA. This phenomenon can be attributed to the limitations imposed on the formation of PFAS micelles and the compromised ability to facilitate the hydrophobic partitioning of PFASs on biochar (Sheehan et al., 2025). Unlike cations, the inorganic anions in the solution tend to compete for adsorption sites with PFC anions. It was observed that inorganic anions such as $SO_4^{2-}$, $Cl^-$, and $Cr_2O_7^{2-}$ in solution decrease the adsorption capacity of PFC on various adsorbents such as resins, chitosan, and boehmite. This influence on PFC adsorption is complex and is often associated with the concentration and type of inorganic ions in the solution (Du et al., 2014)

#### 17.3.1.5 *Dissolved organic matter in the solution*

It has been observed that the presence of compounds such as dissolved organic carbon (DOC), natural and effluent organic matter tends to decrease the PFAS adsorption capacity. Actually, NOM contains a wide variety of organic compounds, such as humic acid, fulvic acid, proteins, hydrophilic acids, carboxylic acids, and amino acids. The presence of coexisting compounds increases the competition for active sites on the sorbent surface between other pollutants and PFASs. The adsorption kinetics of PFASs exhibit a relatively slow rate, which can be ascribed to phenomena such as blockage of pores and diffusion-controlled mass transport occurring within the pores of adsorbents (Vo et al., 2022). Inyang and Dickenson (2017) reported that PFOA adsorbed by pine wood biochar and hardwood biochar was 25% lower in treated wastewater than in lake water. The elevated DOC concentration ($4.9\ mg\ L^{-1}$) in the treated wastewater preferentially occupied the available adsorption sites on the biochar surface. This competitively inhibits the sorption capacity for PFOA through pore blockage and site saturation mechanisms. Similarly, Wang et al. (2015) reported that with an increase in concentration of HA, there was a gradual decrease in adsorption sites on the surface of boehmite. The removal efficiency is decreased by the deprotonation of HA, turning it negatively charged, which enhances the electrostatic repulsion between the adsorbent and the short chain PFASs (Vo et al., 2022). This can occur due to the competition for adsorption sites between comparable-sized of humus molecules and PFAS molecules (Du et al., 2014).

## 17.4 Adsorption mechanism

The study of the adsorption mechanism of PFAS on biochar provides a theoretical insight of the impact of different components and the complex system. Fig. 17.4 illustrates the

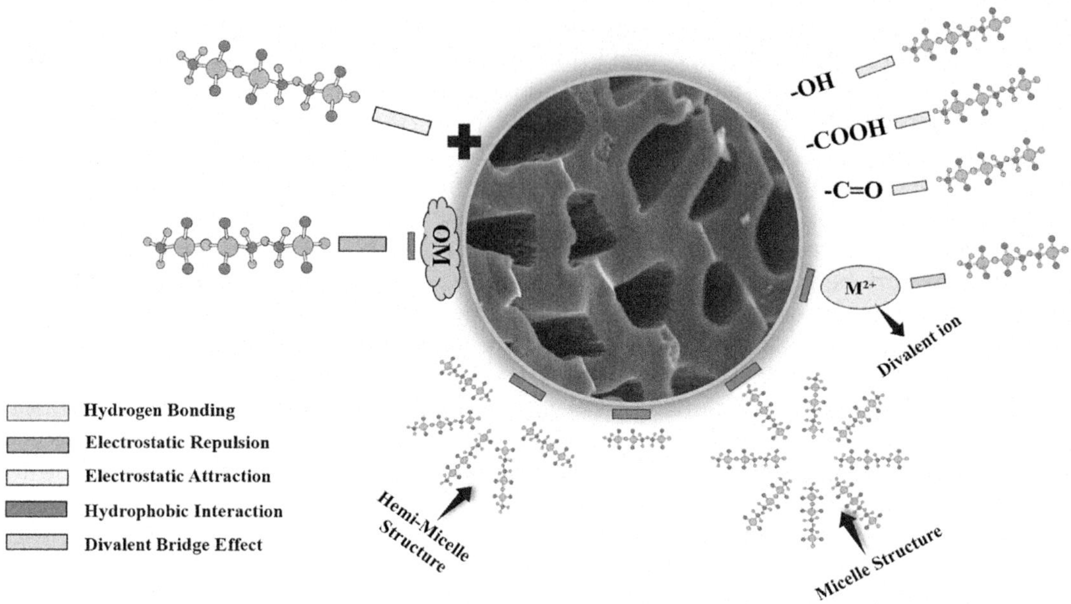


FIGURE 17.4 Adsorption mechanisms for the removal of per- and polyfluoroalkyl substances on biochar.

potential interactions between PFASs and biochar. A variety of interactions, including electrostatic interactions, π–π bonding, H-bonding, hydrophobic effects, ion exchange, and van der Waals forces, have been identified as potentially contributing to the adsorption of PFAS onto different adsorbents. Among these interactions, π–π bonding does not occur during PFAS adsorption because PFAS molecules lack the delocalized electrons and aromatic ring structures required for such interactions. In addition, van der Waals forces contribute minimally to PFAS adsorption because their small molecular size and absence of benzene rings lead to low molecular polarizability, thereby weakening these interactions (Deng et al., 2012).

## 17.4.1 Electrostatic interaction

It is considered one of the main contributors that interact between positively charged adsorbents and anionic PFAS during the adsorption process. The anionic functional group of the PFC molecule is electrostatically attracted by the protonated functional groups of the adsorbent, while the negative surface of the adsorbent repels PFC. A weak ion-dipole interaction can occur between the adsorbent and the PFC molecule in the presence of polar functional groups such as OH and NH (Karoyo and Wilson, 2013). This interaction can be characterized as an electrostatic interaction, which is fundamental in the sorption processes observed in various adsorbents that contain multiple dipole groups. Nonetheless, the existence of inorganic anions, such as $Cl^-$ and $SO^{4-}$, introduces a competitive dynamic with PFAS for adsorption sites (Deng et al., 2010). The pH of the solution plays an important role in the adsorption mechanism as it highly influences the electrostatic attraction between the pollutant and the adsorbent. The functional group of the adsorbent is influenced by the alteration of the solution pH through the protonation deprotonation process. This alteration leads to an impact on the electrostatic attraction and repulsion towards PFAS anions.

Hassan et al. (2020) reported the PFOS adsorption from aqueous solution by magnetically modified bagasse biochar (MBC). It was observed that there was a correlation between the concentration of fluorine in PFOS and the concentration of C, O, Ca, Al, and Si in MBC. The electrostatic interaction was responsible for the adsorption of PFOS on MBC. It was also observed that high temperature pyrolized biochar, CNTs, and graphene favor hydrophobic interaction, while biochar prepared in low temperature conditions facilitates the adsorption of PFOS via electrostatic interactions. This is due to the formation of aromatic functional groups on biochar prepared in high-temperature conditions, enhancing the hydrophobic interaction (Calisto et al., 2014).

## 17.4.2 Hydrophobic interaction

It is the affinity of nonpolar groups such as C–H to aggregate in water, minimizing contact with polar water molecules. PFASs are characterized with a hydrophilic functional group with a carbon–fluorine hydrophobic tail. The electrostatic negativity of PFAS arises due to the functional head group, while the hydrophobic fluorinated C–F chain governs the nonpolar interactions. It appears that electrostatic interactions are predominant in the case of short-chain PFAS. Concurrently, larger PFAS molecules are adsorbed via hydrophobic interactions, while promoting the aggregation of molecules on the active surface (Zaggia et al., 2016). PFASs characterized by long C–F chains are expected to undergo hydrophobic aggregation in aqueous environments, leading to the formation of hemi-micelles and micelles. Due to the dual hydrophobic and hydrophilic nature of PFAS, its molecules self-assemble into structural aggregates called micelles in polar solvents such as water. The core of the micelles or hemi-micelles is formed when the hydrophobic tails of PFAS aggregate with each other. The polar head groups of PFASs are located at the micelle-water interface in contact with and hydrated by several water molecules (Zhang et al., 2019). Multilayer structures, such as micelles and hemi-micelles, may form on adsorbent surfaces when the critical micelle concentration (CMC) is exceeded. These structures may be crucial for PFAS adsorption. Micelles are very difficult to form in solution since the PFAS concentration is much lower than the CMC value. However, highly hydrophobic PFAS with a lower CMC can readily form hemi-micelles in aqueous solution (Rayne and Forest, 2009).

High temperature pyrolysis yields biochar with high aromatic carbon content and low polarity, which often exhibits hydrophobic interaction (Hassan et al., 2022). The polarity of biochar decreases while the amorphous C gradually converts into aromaticity with an increase in pyrolysis temperature. Also, the high temperature degrades the elements of biomass such as O, N, and H. Similarly, Hassan et al. (2020) studied that both hydrophobic and electrostatic interactions are responsible for PFOS adsorption on red mud-modified sawdust biochar, with a notable emphasis on the significance of the former. An increased graphitization of biochar structure was observed due to the presence of protonated metal functional groups in red mud. This could also occur due to the catalytic degradation and conversion of hemicellulose and cellulose. Also, the presence of an aromatic structure may facilitate the PFOS adsorption via a non-ionic interaction mechanism.

### 17.4.3 Hydrogen bonding

The adsorption of PFAS on biochar can also occur via hydrogen bonding interactions. Under alkaline conditions, the electron-rich oxygen atoms in the polar head of PFAS form H-bond with the hydroxyl group of the adsorbent (Xiao et al., 2017). This results in the formation of a strong and stable adsorption interaction (Cheng et al., 2021). The oxygen-containing functional groups (OFGs) such as hydroxyl and carboxyl groups, on the surface of biochar, provide active sites for H-bonding with PFASs, thereby enhancing the adsorption process (Lyu et al., 2022).

Therefore biochar enriched with abundant surface OFGs, including carbonyl, hydroxyl, and carboxyl groups, facilitates strong H-bonding with PFOS, thereby reducing risk and its mobility in the environment. Furthermore, the presence of pyrrole N in biochar also participates in the formation of a weak H-bond with PFAS, enhancing the adsorption process (Cheng et al., 2022). Zhang et al. (2023b) reported that biochar was able to adsorb more PFOS and PFOA from water when the pyrrole N content was higher. This was due to the formation of strong H bonds between hydrochloric acid-modified sludge (H-SL) biochar and the PFASs.

### 17.4.4 Pore diffusion

The adsorption of PFAS is also influenced by the pore size distribution of the adsorbents. The relatively small size of PFOS molecules than the mesopores of biochar enables pore filling, a key mechanism in PFOS adsorption. Krebsbach et al. (2023) reported that biochar with a pore size ~7.5–11 nm is optimal for the adsorption of PFOS via a pore trapping mechanism. This also reduces the subsequent desorption of the adsorbent while shielding it from water turbulence. As the volume of pores in biochar increases, the "trapping" adsorption effect is amplified, thereby increasing the amount of space that is available for PFOS molecules to be adsorbed within the pore structure (the ideal pore: pore volume is approximately 50–150 nm $cm^{-3}$ $g^{-1}$).

## 17.5 Regeneration of biochar

Sustainable usage of the adsorbents depends on the regeneration and recycling of the used adsorbent post adsorption (Kumar et al., 2023b). The potential for PFAS reintroduction during disposal or regeneration is the main environmental and economic concern associated with spent adsorbent management. The cost of the wastewater treatment process decreases with the regeneration of biochar post adsorption. The recycling and reuse of adsorbents are essential for economic efficiency in wastewater control. Various regeneration procedures have been suggested depending on the type and quality of the adsorbent and the nature of the adsorbate. These strategies are essential for minimizing cost (Alsawy et al., 2022). Recent

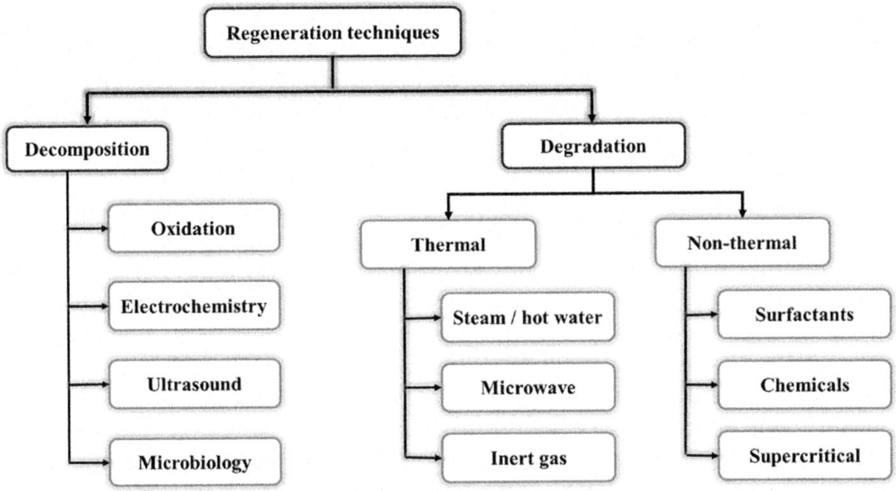


**FIGURE 17.5** Common regeneration techniques and processes used for biochar.

advancements in regeneration methodologies include chemical, thermal, microwave, and hydrothermal treatment methods (Fig. 17.5). These have confirmed significant potential, achieving regeneration efficiencies nearing 99% for carbon-based composites contaminated with PFAS (Vakili et al., 2024).

## 17.5.1 Chemical regeneration

Chemical regeneration is an energy-efficient method for the regeneration of adsorbents saturated with PFAS. Certain chemicals are used in this process to make the desorption process easy by breaking the physicochemical bonds between the adsorbent and adsorbed. Chemical regeneration efficiency depends on the disruption of bonds involving the hydrophilic head of PFAS molecules and the hydrophobic tail. The ability of chemical regeneration to achieve its intended results is dependent upon the use of specialized regenerants, which are intended to disrupt strong bonds that involve the hydrophobic tail and/or the hydrophilic head of PFAS. (Berenguer et al., 2010). PFAS containing adsorbents were treated with different types of regenerant mixtures, such as water, acidic/basic, salt, solvent, and mixed solutions. Among these, the use of water based regenerants is not only inexpensive but it is also easily accessible and beneficial to the environment.

Because of the polar nature of water, it can affect the intermolecular forces that exist between adsorbents and PFAS, which in turn facilitates the release of PFAS into the solution. However, high water temperatures and long contact times are required in water based regenerants to effectively desorb PFAS. The solubility of PFAS increases at high temperature, thereby facilitating the PFAS desorption from the surface of the adsorbent. For instance, the regeneration efficiency went up from 32.7% to 60.8% while increasing the water temperature from 25°C to 92°C (Wang et al., 2019). The pH of the solution also affects the interaction between PFAS and the adsorbents. Therefore various acidic and basic solutions have been used to regenerate absorbents from PFAS. The desorption of the anionic head of PFAS could be achieved using a sodium salt solution, such as NaCl and NaOH. Conventional sodium salt solutions are generally ineffective in regenerating adsorbents saturated with both long and short-chain PFAS due to the lower solubility of PFAS in ionic strength media (Contea et al., 2015).

Electrostatic and ion exchange interactions are the main components of the desorption mechanisms involving acidic or basic regenerants. PFAS molecules become protonated when exposed to highly acidic regenerants like HCl, reducing the electrostatic interactions with the adsorbent surface or functional groups, promoting detachment. Weak base ion-exchange resins (and similar materials) undergo functional group neutralization under basic conditions, whereas the charge on the surface of AC becomes more negative when using NaOH regenerants. The observed outcomes result in the release of PFAS from the adsorbent (Deng et al., 2010). Organic solvents, such as methanol, ethanol, acetone, and acetonitrile, have shown greater efficacy in PFAS desorption. Organic solvents can easily desorb the hydrophobic C–F chain of PFAS. Methanol, typically used at concentrations of 30%–100%, has emerged as a widely adopted and effective solvent for desorbing PFAS from the adsorbents. Usually, 50%–70% ethanol is used, and different adsorbents have been shown to remove PFAS with efficiencies of more than 90%. However, methanol was found to have better regeneration performance, which was explained by the fact that PFAS is less soluble in ethanol than it is in methanol (Du et al., 2016).

### 17.5.2 Thermal regeneration

In thermal regeneration, the adsorbent is subjected to high temperatures, which enables the restoration of spent adsorbents via desorption, release, or thermal destruction of the adsorbed substances, thereby recovering the adsorption performance. It is a three-stage process involving drying, pyrolysis, and oxidation steps. Drying is done at approximately 105°C, while the pyrolysis is done in an inert atmosphere using either argon or nitrogen gas. In the oxidation stage, an oxidizing gas is used to remove the organic residue. Thermal regeneration helps to remove biochar residue and restores the original pores (San Miguel et al., 2001). PFAS saturated adsorbents can be thermally regenerated through a variety of techniques, such as hydrothermal treatment, microwave treatment, and conventional thermal heating.

The pyrolysis method helps to remove volatile compounds from the pore volume, leaving a carbonised char residue. Microwave irradiation (MI) offers a feasible alternative to conventional thermal methods for regeneration. The ability to rapidly and precisely heat materials using MI provides significant benefits, including the reduction of equipment size, the minimum waste generation, reduced energy consumption, and the ability to selectively heat adsorbents (Falciglia et al., 2018). In microwave systems, the regeneration time is reduced compared to conventional thermal regeneration methods due to the fast increase in temperature in the carbon bed (Durán-Jiménez et al., 2019). Operating under high-pressure conditions, hydrothermal treatment helps in the decomposition of PFAS. Sühnholz et al. (2018) used a hydrothermal process under high-pressure conditions to regenerate activated carbon that contained PFOA and PFOS. Hydrothermal regeneration at 200°C for 4 h effectively degraded over 99% of PFOA without producing short-chain PFAS. A recent study focused on the degradation of specific PFAS compounds in wastewater sludge through the application of hydrothermal treatment methods. The system degraded about 99% of PFCAs after 90 min of treatment at 350°C, but it only broke down about 34% of PFOS (Yu et al., 2020).

## 17.6 Conclusion and future perspectives

PFASs are a large group of organic pollutants that present a significant risk to ecosystems and human health, both independently and in conjunction with other environmental contaminants. Among several treatment options, biochar has gained increasing attention as a promising material for PFAS remediation owing to its low cost, high adsorption capacity, simplicity of recovery, and minimum secondary pollution. Biochar-mediated PFAS removal primarily occurs via hydrogen bonding, electrostatic interactions, pore filling processes, and hydrophobic interactions. Moreover, specific types of biochar can promote PFAS degradation through the generation of free radicals, hence diminishing their environmental impact. These processes are affected by various parameters, including the physicochemical characteristics of the biochar, environmental circumstances (e.g., pH, temperature, cocontaminants), and the molecular configuration of the PFASs.

A primary problem in PFAS cleanup is the varying removal effectiveness between long chain and short chain PFAS compounds. Long chain PFASs exhibit enhanced adsorption due to robust hydrophobic interactions, while short-chain PFASs, which depend predominantly on ionic interactions, are typically more challenging to remove. While biochar demonstrates significant promise, its efficacy fluctuates based on the biochar type, modification techniques, and site specific circumstances. Consequently, additional research is required to enhance biochar design and application methodologies. Future research should concentrate on improving biochar's selectivity, regeneration capacity, and applicability at field size for the effective removal of both long- and short-chain PFASs in realistic environmental settings.

## References


Adeniyi, A.G., Iwuozor, K.O., Emenike, E.C., Amoloye, M.A., Adeleke, J.A., Omonayin, E.O., et al., 2024. Leaf-based biochar: a review of thermochemical conversion techniques and properties. Journal of Analytical and Applied Pyrolysis 177, 106352.

Alsawy, T., Rashad, E., El-Qelish, M., Mohammed, R.H., 2022. A comprehensive review on the chemical regeneration of biochar adsorbent for sustainable wastewater treatment. npj Clean Water 5 (1), 29.

Arvaniti, O.S., Stasinakis, A.S., 2015. Review on the occurrence, fate and removal of perfluorinated compounds during wastewater treatment. Science of the Total Environment 524, 81–92.

Berenguer, R., Marco-Lozar, J.P., Quijada, C., Cazorla-Amorós, D., Morallón, E., 2010. Comparison among chemical, thermal, and electrochemical regeneration of phenol-saturated activated carbon. Energy & Fuels 24 (6), 3366–3372.

Bose, B.A., Bhadrapriya, B.C., Kalarikkal, N., 2023. An exploration of some essential inorganic nanofillers apart from metal oxides, metallic particles, and carbon-based nanofillers. Nanofillers. CRC Press, pp. 85–97.

Calisto, V., Ferreira, C.I., Santos, S.M., Gil, M.V., Otero, M., Esteves, V.I., 2014. Production of adsorbents by pyrolysis of paper mill sludge and application on the removal of citalopram from water. Bioresource Technology 166, 335–344.

Cao, F., Wang, L., Tian, Y., Wu, F., Deng, C., Guo, Q., et al., 2017. Synthesis and evaluation of molecularly imprinted polymers with binary functional monomers for the selective removal of perfluorooctanesulfonic acid and perfluorooctanoic acid. Journal of Chromatography. A 1516, 42–53.

Cheng, Y., Wang, B., Shen, J., Yan, P., Kang, J., Wang, W., et al., 2022. Preparation of novel N-doped biochar and its high adsorption capacity for atrazine based on π–π electron donor-acceptor interaction. Journal of Hazardous Materials 432, 128757.

Cheng, N., Wang, B., Wu, P., Lee, X., Xing, Y., Chen, M., et al., 2021. Adsorption of emerging contaminants from water and wastewater by modified biochar: a review. Environmental Pollution (Barking, Essex: 1987) 273, 116448.

Chi, N.T.L., Anto, S., Ahamed, T.S., Kumar, S.S., Shanmugam, S., Samuel, M.S., et al., 2021. A review on biochar production techniques and biochar based catalyst for biofuel production from algae. Fuel 287, 119411.

Contea, L., Fallettia, L., Zaggiaa, A., Milanb, M., 2015. Polyfluorinated organic micropollutants removal from water by ion exchange and adsorption. Chemical Engineering 43.

Deng, J., Han, J., Hou, C., Zhang, Y., Fang, Y., Du, W., et al., 2023. Efficient removal of per-and polyfluoroalkyl substances from biochar composites: cyclic adsorption and spent regenerant degradation. Chemosphere 341, 140051.

Deng, S., Yu, Q., Huang, J., Yu, G., 2010. Removal of perfluorooctane sulfonate from wastewater by anion exchange resins: effects of resin properties and solution chemistry. Water Research 44 (18), 5188–5195.

Deng, S., Zhang, Q., Nie, Y., Wei, H., Wang, B., Huang, J., et al., 2012. Sorption mechanisms of perfluorinated compounds on carbon nanotubes. Environmental Pollution 168, 138–144.

Durán-Jiménez, G., Stevens, L.A., Hodgins, G.R., Uguna, J., Ryan, J., Binner, E.R., et al., 2019. Fast regeneration of activated carbons saturated with textile dyes: textural, thermal and dielectric characterization. Chemical Engineering Journal 378, 121774.

Du, Z., Deng, S., Bei, Y., Huang, Q., Wang, B., Huang, J., et al., 2014. Adsorption behavior and mechanism of perfluorinated compounds on various adsorbents—a review. Journal of Hazardous Materials 274, 443–454.

Du, Z., Deng, S., Chen, Y., Wang, B., Huang, J., Wang, Y., et al., 2015. Removal of perfluorinated carboxylates from washing wastewater of perfluorooctanesulfonyl fluoride using activated carbons and resins. Journal of Hazardous Materials 286, 136–143.

Du, Z., Deng, S., Zhang, S., Wang, B., Huang, J., Wang, Y., et al., 2016. Selective and high sorption of perfluorooctanesulfonate and perfluorooctanoate by fluorinated alkyl chain modified montmorillonite. The Journal of Physical Chemistry C 120 (30), 16782–16790.

El-Fawal, E.M., El Naggar, A.M., El-Zahhar, A.A., Alghandi, M.M., Morshedy, A.S., El Sayed, H.A., 2025. Biofuel production from waste residuals: comprehensive insights into biomass conversion technologies and engineered biochar applications. RSC Advances 15 (15), 11942–11974.

Falciglia, P.P., Roccaro, P., Bonanno, L., De Guidi, G., Vagliasindi, F.G., Romano, S., 2018. A review on the microwave heating as a sustainable technique for environmental remediation/detoxification applications. Renewable and Sustainable Energy Reviews 95, 147–170.

Gao, Y., Deng, S., Du, Z., Liu, K., Yu, G., 2017. Adsorptive removal of emerging polyfluoroalky substances F-53B and PFOS by anion-exchange resin: a comparative study. Journal of Hazardous Materials 323, 550–557.

Gebbink, W.A., van Leeuwen, S.P., 2020. Environmental contamination and human exposure to PFASs near a fluorochemical production plant: review of historic and current PFOA and GenX contamination in the Netherlands. Environment International 137, 105583.

Gong, Y., Wang, L., Liu, J., Tang, J., Zhao, D., 2016. Removal of aqueous perfluorooctanoic acid (PFOA) using starch-stabilized magnetite nanoparticles. Science of the Total Environment 562, 191–200.

Gusiatin, M.Z., Rouhani, A., 2023. Application of selected methods to modify pyrolyzed biochar for the immobilization of metals in soil: a review. Materials 16 (23), 7342.

Hassan, M., Du, J., Liu, Y., Naidu, R., Zhang, J., Ahsan, M.A., et al., 2022. Magnetic biochar for removal of perfluorooctane sulphonate (PFOS): interfacial interaction and adsorption mechanism. Environmental Technology & Innovation 28, 102593.

Hassan, M., Liu, Y., Naidu, R., Du, J., Qi, F., 2020. Adsorption of perfluorooctane sulfonate (PFOS) onto metal oxides modified biochar. Environmental Technology & Innovation 19, 100816.

Inyang, M., Dickenson, E.R., 2017. The use of carbon adsorbents for the removal of perfluoroalkyl acids from potable reuse systems. Chemosphere 184, 168–175.

Issabayeva, G., Hang, S.Y., Wong, M.C., Aroua, M.K., 2018. A review on the adsorption of phenols from wastewater onto diverse groups of adsorbents. Reviews in Chemical Engineering 34 (6), 855–873.

Kabiri, S., Navarro, D.A., Hamad, S.A., Grimison, C., Higgins, C.P., Mueller, J.F., et al., 2023. Physical and chemical properties of carbon-based sorbents that affect the removal of per-and polyfluoroalkyl substances from solution and soil. Science of the Total Environment 875, 162653.

Karoyo, A.H., Wilson, L.D., 2013. Tunable macromolecular-based materials for the adsorption of perfluorooctanoic and octanoic acid anions. Journal of Colloid and Interface Science 402, 196–203.

Krebsbach, S., He, J., Adhikari, S., Olshansky, Y., Feyzbar, F., Davis, L.C., et al., 2023. Mechanistic understanding of perfluorooctane sulfonate (PFOS) sorption by biochars. Chemosphere 330, 138661.

Kujawa, J., Cerneaux, S., Kujawski, W., 2014. Characterization of the surface modification process of Al2O3, TiO2 and ZrO2 powders by PFAS molecules. Colloids and Surfaces A: Physicochemical and Engineering Aspects 447, 14–22.

Kumar, A., Bhattacharya, T., Shaikh, W.A., Chakraborty, S., Owens, G., Naushad, M., 2022a. Valorization of fruit waste-based biochar for arsenic removal in soils. Environmental Research 213, 113710.

Kumar, A., Bhattacharya, T., Shaikh, W.A., Chakraborty, S., Sarkar, D., Biswas, J.K., 2022b. Biochar modification methods for augmenting sorption of contaminants. Current Pollution Reports 8 (4), 519–555.

Kumar, A., Bhattacharya, T., Shaikh, W.A., Roy, A., Chakraborty, S., Vithanage, M., et al., 2023a. Multifaceted applications of biochar in environmental management: a bibliometric profile. Biochar 5 (1), 11.

Kumar, A., Shaikh, W.A., Maqsood, H.M., Parikh, S.J., Biswas, J.K., 2025. Harnessing sustainable biochar-based composites for effective PFAS removal from wastewater. Current Opinion in Environmental Science & Health 100594.

Kumar, P., Singhania, R.R., Sumathi, Y., Kurrey, N.K., Chen, C.W., Patel, A.K., et al., 2024. Investigating innovative techniques for biochar modification to enhance the removal of heavy metals from aqueous environments: a comprehensive review. Clean Technologies and Environmental Policy 1–23.

Kumar, S., Yadav, S., Kataria, N., Chauhan, A.K., Joshi, S., Gupta, R., et al., 2023b. Recent advancement in nanotechnology for the treatment of pharmaceutical wastewater: sources, toxicity, and remediation technology. Current Pollution Reports 9 (2), 110–142.

Kurwadkar, S., Dane, J., Kanel, S.R., Nadagouda, M.N., Cawdrey, R.W., Ambade, B., et al., 2022. Per-and polyfluoroalkyl substances in water and wastewater: a critical review of their global occurrence and distribution. Science of the Total Environment 809, 151003.

Lenka, S.P., Kah, M., Padhye, L.P., 2021. A review of the occurrence, transformation, and removal of poly-and perfluoroalkyl substances (PFAS) in wastewater treatment plants. Water Research 199, 117187.

Liu, N., Wu, C., Lyu, G., Li, M., 2021. Efficient adsorptive removal of short-chain perfluoroalkyl acids using reed straw-derived biochar (RESCA). Science of the Total Environment 798, 149191.

Liu, Z., Xu, Z., Xu, L., Buyong, F., Chay, T.C., Li, Z., et al., 2022. Modified biochar: synthesis and mechanism for removal of environmental heavy metals. Carbon Research 1 (1), 8.

Li, L., Wang, J., Sun, W., Peng, X., Qi, X., 2025. Production of algae-derived biochar and its application in pollutants adsorption—a mini review. Separations 12 (2), 32.

Lyu, X., Xiao, F., Shen, C., Chen, J., Park, C.M., Sun, Y., et al., 2022. Per-and polyfluoroalkyl substances (PFAS) in subsurface environments: occurrence, fate, transport, and research prospect. Reviews of Geophysics 60 (3) e2021RG000765.

Murtaza, G., Ahmed, Z., Valipour, M., Ali, I., Usman, M., Iqbal, R., et al., 2024. Recent trends and economic significance of modified/functionalized biochars for remediation of environmental pollutants. Scientific Reports 14 (1), 217.

Parthasarathy, P., Al-Ansari, T., Mackey, H.R., Narayanan, K.S., McKay, G., 2022. A review on prominent animal and municipal wastes as potential feedstocks for solar pyrolysis for biochar production. Fuel 316, 123378.

Patel, S., Marzbali, M.H., Hakeem, I.G., Veluswamy, G., Rathnayake, N., Nahar, K., et al., 2023. Production of H2 and CNM from biogas decomposition using biosolids-derived biochar and the application of the CNM-coated biochar for PFAS adsorption. Waste Management 159, 146–153.

Rafiee, A., Faridi, S., Sly, P.D., Stone, L., Kennedy, L., Mahabee-Gittens, E.M., 2024. Asthma and decreased lung function in children exposed to perfluoroalkyl and polyfluoroalkyl substances (PFASs): an updated meta-analysis unveiling research gaps. Environmental Research, 119827. Available at https://doi.org/10.1016/j.envres.2024.119827.

Rahman, M.S., Saha, N., Kumar, S., Khan, M.D.H., Islam, A.R.M.T., Khan, M.N.I., 2022. Coupling of redundancy analysis with geochemistry and mineralogy to assess the behavior of dust arsenic as a base of risk estimation in Dhaka, Bangladesh. Chemosphere 287, 132048.

Rani, S., Sabharwal, H., Kumar, P., Chauhan, A.K., Khoo, K.S., Kataria, N., 2024. Comparative behavior of carbon-based materials for the removal of emerging bisphenol A from water: adsorption modelling and mechanism. Groundwater for Sustainable Development 25, 101121.

Rayne, S., Forest, K., 2009. Perfluoroalkyl sulfonic and carboxylic acids: a critical review of physicochemical properties, levels and patterns in waters and wastewaters, and treatment methods. Journal of Environmental Science and Health Part A 44 (12), 1145–1199.

Rodrigo, P.M., Navarathna, C., Pham, M.T., McClain, S.J., Stokes, S., Zhang, X., et al., 2022. Batch and fixed bed sorption of low to moderate concentrations of aqueous per-and poly-fluoroalkyl substances (PFAS) on Douglas fir biochar and its Fe3O4 hybrids. Chemosphere 308, 136155.

Ross, I., McDonough, J., Miles, J., Storch, P., Thelakkat Kochunarayanan, P., Kalve, E., et al., 2018. A review of emerging technologies for remediation of PFASs. Remediation Journal 28 (2), 101–126.

Salehi, M., 2025. Surface complexation at charged organic surfaces. Reviews in Mineralogy and Geochemistry 91 (1), 149–173.

San Miguel, G., Lambert, S.D., Graham, N.J.D., 2001. The regeneration of field-spent granular-activated carbons. Water Research 35 (11), 2740–2748.

Sheehan, N.P., Ponge, C.A., Pankratz, A., Hutchison, J.M., Laird, B.B., Nguyen, N.P., et al., 2025. Interference of PFAS sorption on zeolites from natural water characteristics. Chemosphere 378, 144414.

Shekoohiyan, S., Sajadi, A., Moussavi, G., Heidari, M., 2025. Hydrothermal carbonization of plastic wastes and effect of influential parameters on performance and challenges: a review. International Journal of Environmental Science and Technology 1–42.

Sühnholz, S., Kopinke, F.D., Weiner, B., 2018. Hydrothermal treatment for regeneration of activated carbon loaded with organic micropollutants. Science of the Total Environment 644, 854–861.

Sun, Y., Jia, J., Liu, Z., Liu, Z., Huo, L., Zhao, L., et al., 2024. Nitrogen-doped biochar derived from corn straw for CO2 adsorption: a new vision on nitrogen sources comparison. Carbon Research 3 (1), 61.

US Department of Health and Human Services, 1999. Agency for toxic substances and disease registry-ATSDR.

Vakili, M., Cagnetta, G., Deng, S., Wang, W., Gholami, Z., Gholami, F., et al., 2024. Regeneration of exhausted adsorbents after PFAS adsorption: a critical review. Journal of Hazardous Materials 134429.

Varkolu, M., Gundekari, S., Palla, V.C.S., Kumar, P., Bhattacharjee, S., Vinodkumar, T., 2025. Recent advances in Biochar production, characterization, and environmental applications. Catalysts 15 (3), 243.

Velusamy, K., Isabel, J.B., Periyasamy, S., Thiruvenkadam, A., Ravikumar, H., Gupta, S.K., et al., 2024. Role of biochar as a greener catalyst in biofuel production: production, activation, and potential utilization–a review. Journal of the Taiwan Institute of Chemical Engineers 105732.

Vo, H.N.P., Nguyen, T.M.H., Ngo, H.H., Guo, W., Shukla, P., 2022. Biochar sorption of perfluoroalkyl substances (PFASs) in aqueous film-forming foams-impacted groundwater: effects of PFASs properties and groundwater chemistry. Chemosphere 286, 131622.

Wang, W., Maimaiti, A., Shi, H., Wu, R., Wang, R., Li, Z., et al., 2019. Adsorption behavior and mechanism of emerging perfluoro-2-propoxypropanoic acid (GenX) on activated carbons and resins. Chemical Engineering Journal 364, 132–138.

Wang, F., Shih, K., Leckie, J.O., 2015. Effect of humic acid on the sorption of perfluorooctane sulfonate (PFOS) and perfluorobutane sulfonate (PFBS) on boehmite. Chemosphere 118, 213–218.

Wang, H., Zhang, H., He, L., Wang, J., Wang, S., Shi, X., et al., 2025. Environmental behavior of per-and polyfluoroalkyl substances (PFASs) and the potential role of biochar for its remediation: a review. Biochar 7 (1), 1–20.

Weber, K., Quicker, P., 2018. Properties of biochar. Fuel 217, 240–261.

Wu, Y., Liu, Z., Jiang, W., Wang, J., 2021. Research progress on removal of several common emerging pollutants by biochar. Chemical Industry and Engineering Progress 40 (5), 2839.

Xiao, X., Ulrich, B.A., Chen, B., Higgins, C.P., 2017. Sorption of poly-and perfluoroalkyl substances (PFASs) relevant to aqueous film-forming foam (AFFF)-impacted groundwater by biochars and activated carbon. Environmental Science & Technology 51 (11), 6342–6351.

Xing, Y., Zhou, Y., Liu, P., Lee, H.K., Huang, Z., 2025. Application of bovine serum albumin-decorated magnetic biochar for the extraction of per-and polyfluoroalkyl substances in food samples. Microchemical Journal 211.

Yaashikaa, P.R., Kumar, P.S., Varjani, S., Saravanan, A.J.B.R, 2020. A critical review on the biochar production techniques, characterization, stability and applications for circular bioeconomy. Biotechnology reports 28, e00570.

Yu, J., Nickerson, A., Li, Y., Fang, Y., Strathmann, T.J., 2020. Fate of per-and polyfluoroalkyl substances (PFAS) during hydrothermal liquefaction of municipal wastewater treatment sludge. Environmental Science: Water Research & Technology 6 (5), 1388–1399.

Zaggia, A., Conte, L., Falletti, L., Fant, M., Chiorboli, A., 2016. Use of strong anion exchange resins for the removal of perfluoroalkylated substances from contaminated drinking water in batch and continuous pilot plants. Water Research 91, 137–146.

Zhang, Y., Ren, M., Tang, Y., Cui, X., Cui, J., Xu, C., et al., 2022a. Immobilization on anionic metal (loid) s in soil by biochar: a meta-analysis assisted by machine learning. Journal of Hazardous Materials 438, 129442.

Zhang, Y., Tan, X., Lu, R., Tang, Y., Qie, H., Huang, Z., et al., 2023a. Enhanced removal of polyfluoroalkyl substances by simple modified biochar: adsorption performance and theoretical calculation. ACS ES&T Water 3 (3), 817–826.

Zhang, Y., Thomas, A., Apul, O., Venkatesan, A.K., 2023b. Coexisting ions and long-chain per-and polyfluoroalkyl substances (PFAS) inhibit the adsorption of short-chain PFAS by granular activated carbon. Journal of Hazardous Materials 460, 132378.

Zhang, D.Q., Zhang, W.L., Liang, Y.N., 2019. Adsorption of perfluoroalkyl and polyfluoroalkyl substances (PFASs) from aqueous solution-a review. Science of the Total Environment 694, 133606.

Zhang, X., Zhao, B., Liu, H., Zhao, Y., Li, L., 2022b. Effects of pyrolysis temperature on biochar's characteristics and speciation and environmental risks of heavy metals in sewage sludge biochars. Environmental Technology & Innovation 26, 102288.

CHAPTER

# 18

# Metal oxides: solid scaffolding for per- and polyfluoroalkyl substances removal

*Samar Dhiman[1,2,#], Hemant Veer Jain[1,2,#], and Nasreen Ghazi Ansari[1,2]*

[1]**Analytical Chemistry Laboratory, Analytical Sciences and Accredited Testing Services Group, Analytical Sciences & Services, Industrial Support through Technological Solutions (ASSIST) Division, CSIR-Indian Institute of Toxicology Research (CSIR-IITR), Lucknow, Uttar Pradesh, India [2]Academy of Scientific and Innovative Research (AcSIR), Ghaziabad, Uttar Pradesh, India**

## 18.1 Introduction

Per- and polyfluoroalkyl substances (PFAS) pollution is a global problem that has drawn plenty of scientific attention in recent years. The PFAS compounds, formerly regarded as miracle molecules because of their water-proof and grease-proof properties, are now a problem that needs to be removed quickly and effectively. PFAS are man-made compounds that resemble other hydrocarbons except for the functional groups, where F atoms take the place of hydrogen atoms. These compounds have a $-C_nF_{2n+1}$ or $-C_nF_{2n-1}$ ($n \geq 1$) moiety, according to the Organization for Economic Cooperation and Development. These substances also fall into two categories: polymeric and nonpolymeric PFAS. These PFAS molecules' aforementioned characteristics have led to their widespread use in numerous industrial applications (Sunderland et al., 2019). Proponents of the Anthropocene, including overindustrialization, construction, tourism, careless waste management, dumping, and incineration, are primarily responsible for the introduction of PFAS into the environment (Dhiman and Ansari, 2024). PFAS contamination has been discovered in snow that is in close proximity to the world, in addition to the soil, water, and air. Notably, soil, air, and water are the foundation of life; if any of these elements are contaminated, PFAS will eventually move into the others. The food chain is impacted by PFAS in environmental components,

[#] Equal Contribution.

*Per- and Polyfluoroalkyl Substances (PFAS) as Environmental Contaminants*
DOI: https://doi.org/10.1016/B978-0-443-43856-1.00002-2

which leads to their bio-accumulation and bio-magnification at various trophic levels. The 2019 Stockholm Convention also addressed PFAS contamination, and PFAS have been classified as a kind of persistent bioaccumulative toxins (PBT). It's interesting that short-chain PFAS (GenX), which are their equivalents, were introduced into industrial applications after the hazardous characteristics of PFAS with longer carbon chains than seven (legacy) were discovered.

However, many regulatory bodies around the globe, such as the European Chemicals Agency, UNEP, and the United States Environmental Protection Agency (US EPA), have also raised concerns about PFAS. Of note, the US EPA has revised the MCL of PFAS to 4 parts per trillion (ppt), which was 70 ppt earlier, only for 2 PFAS (perfluorooctanoic acid [PFOA] and perfluorooctane sulfonic acid [PFOS]), while for perfluorohexane sulfonic acid [PFHXs], PFAN, and hexafluoropropylene oxide dimer acid (HFPO-Da), the limit is 10 ppt. However, as the European Union's new directive on PFAS, the latest set limit for PFAS in drinking water is 500 ppt for total PFAS and 100 ppt for 20 PFAS. But studies have reported that more than 9000 PFAS compounds are used in industries, but there are no set acceptable limits. The contamination by PFAS has now become a vicious issue that requires immediate action. Many studies have been reported in the literature focused on eradicating PFAS from different matrices. The regular and traditional technologies used for water treatments, such as flocculation, sedimentation, reverse osmosis, and bioremediation by modified bacteria, are incapable of PFAS removal (Rahman et al., 2014). As per reports, the granular activated carbon (GAC) and powdered activated carbon (PAC) have also been explored to remove PFAS in water. But on the contrary, GAC and PAC cannot be considered as a practical solution, because the functional group and the number of carbon atoms attached to PFAS molecules affect the adsorption, thus affecting the removal efficiencies (Malovanyy et al., 2023), though these are cost-effective. Moreover, the challenge of regeneration is an issue that persists in the activated carbon-based formulation (Larasati et al., 2021).

Nonetheless, the fine amalgamation of metal oxides (MOs) with various technologies such as nanoparticles (NPs), composites, MIPs, metal–organic frameworks (MOFs), anion exchange resin, electrodes, hydrogels, nano films, and nanotubes has turned the tables in the remediation strategies. These MOs offer advantages in flexible structure, enhanced surface area, and selective molecule-based removal (Karbassiyazdi et al., 2023). As reported in the literature, minerals, especially mineral clays, have found wide-ranging and substantial applicability in remediation strategies (Mukhopadhyay et al., 2019). These clay minerals offer advantages in bio compatibility, cost effectiveness, and their ability to reduce the bioavailability of PFAS in soil, thus helping to prevent translocation of PFAS from soil to other components (Hearon et al., 2022). Of note, MOs have also been found to have applications as a major component in PFAS sensing technologies (Dhiman and Ansari, 2024). Different MOs are extensively used in the sample preconcentration and sample clean-up to investigate PFAS from different matrices. This chapter delves into the potential of MO and mineral incorporation within various formulations as a strategy to bridge the technological gap in PFAS remediation. Not only from the remediation point of view, but also the MOs have been a promising choice in sample clean-up and preconcentration. It critically evaluates the inherent advantages and disadvantages of employing MOs, and minerals, offering a balanced perspective on their application. Furthermore, the emphasizes on future research

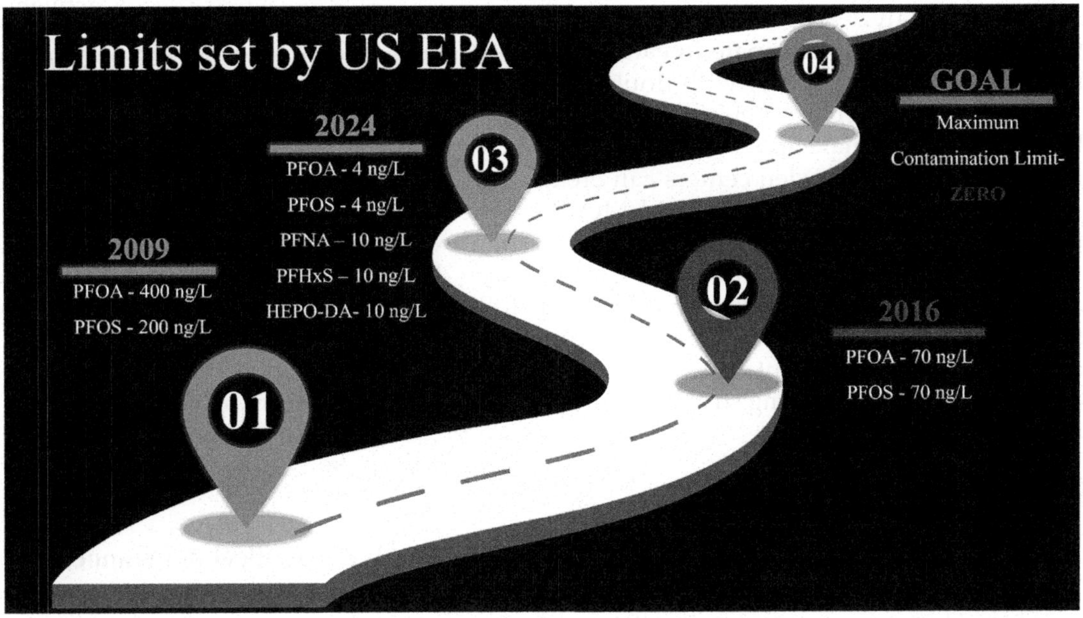


**FIGURE 18.1** Amendments to the limits set by United States Environmental Protection Agency (US EPA).

directions that prioritize the development of sustainable and comprehensive remediation technologies, aiming to mitigate PFAS contamination while concurrently preventing the generation or mobilization of other environmental pollutants during the treatment process. This forward-looking approach is crucial for establishing effective and environmentally responsible strategies for diminishing the complex challenge of PFAS contamination. The various amendments made by the US EPA in setting the PFAS limits in water are represented in Fig. 18.1.

Maybe a more detailed legend to this nice figure: progressive decrement of regulatory limits for PFAS in drinking water as set by the US EPA.

## 18.2 Metal oxides/mineral oxides

MOs are a family of compounds with a wide range of characteristics that span chemistry, physics, and that result from the interaction of metal ions with oxygen ions are referred to as MOs, and are also commonly known as ores and are primarily used to extract metals (He, 2020). Furthermore, MOs are a type of mineral frequently separated from other minerals containing oxygen, including carbonates, phosphates, silicates, sulfates, and borates. However, it has been reported that mineral oxides or minerals are a broader category that often includes hydroxyl minerals, which have a hydroxyl group $OH^-$ in place of $O^{2-}$, and oxyhydroxides, which have both a hydroxyl group and oxygen, and hydrated oxides of Si, Fe, Mn, Al, and Ti (Nadoll, 2017). In addition to their capacity to catalyze acid-base reactions, MOs attracted interest since the 1950s because of their propensity to participate in catalytic events (small areas where an acid and a base participate simultaneously) with double acid-base sites. Generally, unsaturated cationic metal centers, which are Lewis acid sites, and anionic oxygen centers, which are Lewis base sites ($O_2^-$), make up the catalytic surface of MOs. Due to their strong electronegativity, anionic oxygen centers can create ionic connections in their interactions with metals (Araújo et al., 2023). Several MOs have been included in the research for novel materials for sample pretreatment because of their exceptional mechanical, chemical, and thermal stabilities. Astonishingly, when MOs are doped in different formulations, the enhanced adsorption is observed due to the following factors: an increase in magnetic separation, metal-based functional groups enhance the adsorption capacity, stronger aggregation with pollutants due to the ligand and ion exchange (de Jesus et al., 2022; Lu and Astruc, 2020). Owing to these properties, MOs have been a part of different remediation strategies.

## 18.3 Metal oxides and minerals in per- and polyfluoroalkyl substances remediations

Different MOs used extensively in PFAS remediations, are explained here.

### 18.3.1 Metal oxide electrodes

There have been reports of the breakdown of PFAS molecules using metal electrodes through electro-oxidation. Electrochemical reactions and electron transfers occurring at the surface of working electrodes form the ground for PFAS breakdown. Direct and indirect oxidation can be used for the PFAS decomposition. An electron transport from the PFAS chemical to the metal anode is thus typically a part of the direct oxidation pathway (anodic oxidation). Under the influence of the electric field, an electron from the primary end of the PFAS chain is transported throughout the disintegration process, producing degradable PFAS radicals.

Though, these free radicals of the PFAS chain combine with OH free radicals, further undergoing degradation (Su et al., 2019). The degradation reactions include free radical reaction, hydrolysis, and hydroxylation (Pierpaoli et al., 2021). Additionally, electro-Fenton reaction and electrocoagulation have been reported inliterature when the degradation of the PFAS molecules is based on MO-based electrodes. Therefore, the intrinsic properties of working MO electrodes significantly impact the sensitivity and selectivity of electrochemical reactions occurring on the surface (Rezaei and Irannejad, 2022). Along with intermittent and straightforward operation under moderate circumstances, this method provides excellent energy economy, scalability, adaptability, and ease of setup. Additionally, the procedure may be applied at ambient temperature and atmospheric pressure without extra reagents. In contrast to other oxidative processes, where supporting chemicals are necessary to achieve greater efficiency, the electrolyte is required for the reaction to enhance the process in terms of cost-effectiveness. However, the electrode's nature directly affects the degradation's effectiveness since a larger anodic potential is necessary to achieve the rate-limiting direct electron transfer reaction step (Uwayezu et al., 2021). Furthermore, MO electrodes such as tin oxide ($SnO_2$), iridium dioxide ($IrO_2$), ruthenium dioxide ($RuO_2$), lead dioxide ($PbO_2$), and boron-doped diamond, known as "non-active" anodes have a great choice for the electrochemical destruction of these notorious PFAS molecules. Astonishingly, it has been reported that these MOs have the ability to generate $OH^{\cdot}$ without even having a strong interaction between the electrode and reactant (Sharma et al., 2022). Additionally, mixed MOs, when incorporated in the defluorination of PFAS in the aqueous medium, are more effective than the single MO. However, it has been reported that (Barisci and Suri, 2021) the number of carbon atoms attached to PFAS molecules greatly affects the electrochemical degradation under different conditions, such as interelectrode gap, current density, and concentration of the supporting analyte. Moreover, it has been reported that as the hydrophobicity of the PFAS chain increases with the increase in the chain length, the rate of degradation increases. However, apart from the electrochemical oxidation (EO), electrochemical reduction of the PFAS molecules has also been reported in the literature. Three steps/mechanisms have been reported for the electrochemical reductive breakdown of PFOA: (1) PFOA approach and adsorption on the cathode surface; (2) the production and supply of highly reductive species, such as $H^{\bullet}$; and (3) the hydro-defluorination reaction (Zhu et al., 2022; Li et al., 2021). Furthermore, in Table 18.1, different MOs that have been used in the degradation and mineralization of the PFAS are mentioned.

### 18.3.2 Limitations of metal oxide electrodes

The aspect that MO-based electrodes have found great applicability in the PFAS degradation has helped somewhat in the eradication of these forever chemicals. Every coin has two sides, the MOs such as $PbO_2$, $IrO_2$, and $Ti_4O_7$, have been an integral part of the EO process, leading to the generation of $ClO_4^-$ during the process. This shows that the generation of perchlorate is a crucial and difficult problem in the EO process. Effective scale-up and process parameter optimization need a field demonstration (Mukherjee et al., 2023). Although even after treatment, the concentration of short-chain PFAS may rise due to the conversion of precursors. Nonetheless, EO can be a promising "green" technology for PFAS remediation because it utilizes less energy than thermal incineration (>1100°C), but energy consumption is there, which cannot be neglected. However, as per the retrospective studies conducted, it has been reported that current density plays a crucial role. As per the studies reported (Maldonado et al., 2021), when the current density was treated at 100 to 200 mA $cm^{-2}$ for 2 h, the elimination effectiveness of PFAS was reported to be more than 90%. After 8 h, the same removal efficiency was attained when the current density was changed to 50 mA $cm^{-2}$.

Nonetheless, EO has the potential to convert hazardous by-products that can be more harmful. In addition, EO can efficiently disintegrate multisolute solutions containing a mixture of PFAS, which is not possible with single-solute PFAS systems (Nienhauser et al., 2022). There is a dearth of knowledge on possible PFAS by-products. The defluorination occurring during these degradation processes generates $F^-$ ions, which may lead to the fluoride ion and organic content in the environment. Astonishingly, the development of fluorinated products, which are more persistent than parent PFAS, is frequently observed

**TABLE 18.1** Different types of metal oxide electrodes have been reported in per- and polyfluoroalkyl substances (PFAS) remediation.

| Metal oxide | PFAS removed | Removal (%) | Removal time (hours) | References |
|---|---|---|---|---|
| Ti/$SnO_2$–Sb–Ce ($SnO_2$), Ti/$SnO_2$–Sb/Ce–$PbO_2$($PbO_2$), and Ti/BDD (BDD) | PFNA and PFDA | 90 | 3 | Lin et al. (2013) |
| Ceramic $Ti_4O_7$ | PFOA and PFOS | 91 and 97 | 24 | Lin et al. (2018) |
| $CeO_2$ | PFOA | 94 | 12 | Liu et al. (2025) |
| $PbO_2$-PTFE anode | PFOA | 98.9 | 5 | Lou et al. (2022) |
| Ti/$SnO_2$-$Sb_2O_5$/$PbO_2$-PTFE | PFOA | 90 | 3 | Zhuo et al. (2020) |
| Ti/$SnO_2$-$Sb_2O_5$/$PbO_2$ | PFOA | 94 | 2 | Singh et al. (2021) |
| F-doped Ti/$SnO_2$ | PFOA | 99 | 0.5 | Yang et al. (2015) |
| $Ti_4O_7$-based electrode loaded with amorphous Pd clusters | PFOA | 86.7 | 1 | Huang et al. (2020) |
| Zr-$PbO_2$ electrode | PFOA | 97 | 1.5 | Xu et al. (2017) |
| Ce-doped modified porous nanocrystalline film electrode | PFBA, PFPeA, PFHxA, PFHpA, and PFOA | 31.8, 41.4, 78.2,97.9, and 96.7. | 1.5 | Niu et al. (2012) |
| Ti/$IrO_2$ | PFCASs ($C_4$-$C_{14}$), 4:2 FTS, 6:2 FTS and 8:2 FTS, Perfluorooctane sulfonamidoacetic acids, perfluoroalkyl sulfonamides | 54.5–98.1 | 3 | Uwayezu et al. (2021) |

*PFNA*, Perfluorononanoic acid; *PFDA*, perfluorodecanoic acid; *PFOA*, perfluorooctanoic acid; *PFOS*, perfluorooctanesulfonic acid; *PFBA*, perfluorobutanoic acid; *PFPeA*, perfluoropentanoic acid; *PFHxA*, perfluorohexanoic acid; *PFHpA*, perfluoroheptanoic acid; *PFCAS*, perfluoroalkyl carboxylic acids.

when incomplete defluorination with PFAS degradation occurs. Identifying transition products and investigating their mechanisms are thus of particular importance when it comes to the remediation of PFAS pollution. The MO-based electrodes may leach after some time, and after a certain number of regeneration cycles, the life span of the electrodes is reduced. It is not surprising that the safe disposal of the MO and solvents that are used in these EO processes is still in question. The knowledge gap in the safe disposal that is part of the PFAS degradation process and studies needs to be addressed.

### 18.3.3 Metal oxide formulations

The different nano formulations synthesized from ferric oxides, cobalt oxides, copper oxides, zinc oxides, titanium oxides, magnesium oxides, cerium oxides, aluminum oxides, and other MOs have all demonstrated effective removal and degradation of pollutants like heavy metals and persistent organic pollutants (POPs). However, MO nano formulation has now become an integral part of the removal strategies. Large surface/volume area, a high number of active sites, nano- or/and macro-porous structure, stability, recovery, and low toxicity are some of the unique physical and chemical characteristics that allow them to be distinguished from their equivalent bulk material, which is impacted by their size (1–100 nm) (de Jesus et al., 2022). The synergistic combination of thermal, catalytic, and photochemical properties, high superficial area/volume ratio, and tunable optical, magnetic, electrical, mechanical properties, emanating from their bulk and surface properties, makes MO nanostructures unique in this context compared to other materials (Lu and Astruc, 2020). MO-based NPs, nanotubes, nanofibers, nanowires, and nanomembranes (Lu and Astruc, 2020; Dehghani et al., 2022) have found great applicability in PFAS removal from different matrices, as shown in Fig. 18.2. Of note, the different MOs in the nano form have been used to adsorb the PFAS from different environmental matrices. One of the few studies reported in the literature that explored the adsorption efficiency of nano MO in the removal of PFAS from aqueous media. It has been well reported in the literature that the adsorption phenomenon is influenced by the proximal environmental conditions, such as coexisting ions and pH (Gong et al., 2016). However, it has been reported that nano-MOs are positively charged when the pH is lower than their point of zero charge (PZC), their adsorption towards negatively charged anions is improved, and enhances the adsorption capacity of the PFAS in acidic environmental media. So, it can be stated that typically MOs, such as $Fe_2O_3$, have greater adsorption efficiencies/kinetics at low pH, which is associated with the PZC. In an acidic environment, for example, pH equal 3, the OH-groups on the surface of $Fe_2O_3$ are neutralized, exposing positively charged cations that increase affinity for PFASs (Gao and Chorover, 2012). $TiO_2$, $Fe_2O_3$, and $Al_2O_3$ have PZC values of 5.4, 7.6, and 7.3, respectively. Astonishingly, the PZC is the point at a certain pH where the charge of the adsorbent becomes zero. As it is clear that PFAS molecules exist as negative species because of the presence of a functional group that dissipates positive ions, the low pH values when these nano MOs behave as positive species have a greater affinity towards these dissociated negatively charged PFAS species (Vu and Wu, 2022). Apart from the pH, it has been observed that the presence of other divalent ions also influences PFAS adsorption, so when the divalent ions, such as Cu (II) and Pb (II), are combined with the nano-MOs such as $Fe_3O_4$, $TiO_2$, and $SiO_2$, better removal efficiency is observed. For this reason, they form

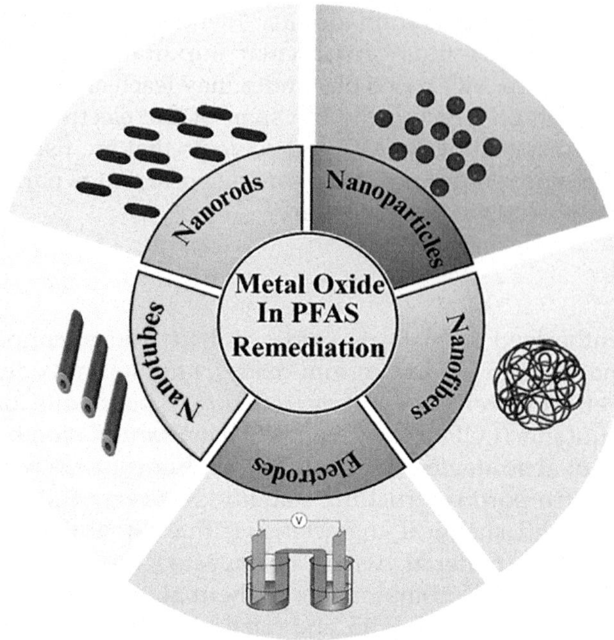


**FIGURE 18.2** Different metal oxides used in per- and polyfluoroalkyl substances (PFAS) remediation.

inner-sphere complexes that have enhanced adsorption capacity towards PFOA (Lu et al., 2016). Subsequently, the adsorption efficiency of PFAS has been reported to be enhanced when nano MO is amalgamated with carbon-based materials. Electrostatic and hydrophobic interactions enhance the adsorption effectiveness between the so-formed MOs doped nanocomposite adsorbents and PFAS chemicals. Inner-sphere compounds are formed between PFAS and the composites when nano MOs are doped into carbon-based formulations.

Metal hydroxides (M–OH) can be generated when metals or MOs combine with oxygen functional groups on the surface of carbon formulations (Sun et al., 2023; Xu et al., 2020a; Rhakho et al., 2024). The surface of the nanocomposite becomes positively charged upon hydrolysis of $OH^-$. However, apart from the adsorption as a removal mechanism, the photocatalytic degradation of the PFAS using MO-based nano formulation has been extensively explored and is well reported in the literature. Biosynthesized $TiO_2$ NPs show potential as photocatalytic materials for removing PFOS from water through defluorination and degradation (Saheed et al., 2023).

## 18.3.4 Photocatalytic remediation

These PFAS molecules have been reported to be degraded under exposure to shorter-wavelength ultraviolet radiation. The shorter wavelength of UV radiation possesses a high energy and breaks the C–F bond in the PFAS chains (Chen et al., 2015). Nano MO, titanium oxide ($TiO_2$), gallium oxide ($Ga_2O_3$), and indium oxide ($In_2O_3$) have been shown to remove PFCAs by photocatalysis (Zhao et al., 2015). Photoactive nanomaterials doped with MO or noble metals for enhancing the electron transfer capabilities of the photocatalysts with conductive nanomaterials can increase the photocatalytic activity and photodegradation of PFAS (Yin and Villagrán, 2022). Howbeit, stable and effective photocatalysis that effectively has removed PFOA at ppm concentrations (100 mg $L^{-1}$) at acidic conditions (pH 2–3) in 1 h was recorded on doping nano-$In_2O_3$ with minor amounts of nano $CeO_2$ Given that there is an increase in negative charges on the surface of the nanocomposite, raising the pH can decrease the effectiveness of PFOA degradation (Jiang et al., 2016). Doping nano-$TiO_2$ and $In_2O_3$ with noble metals (Ag, Pt, and Pd) can also accelerate the breakdown of PFOA in aqueous systems, as per the studies (Xu et al., 2017; Lin et al., 2012). As per the studies, the carbonaceous material also finds applicability in the PFAS removal strategies because of the large specific surface area and higher adsorption capacity, which is seen in carbon nanotubes (CNTs) as compared to activated carbon. Subsequently, it has been reported that when CNTs are impregnated with MOs, they enhance the adsorption behavior compared to the parent precursor (Yin et al., 2023). Additionally, graphene oxide in amalgamation with the MO has been reported in the literature and has shown excellent and enhanced photocatalytic degradation of the PFAS chemicals (Chowdhury and Choi, 2023). However, Table 18.2, mentioned below, lists different types of nano metal formulations that have been reported in literature.

### 18.3.5 Limitations of metal oxide-based formulations

MO-based formulations have drawbacks that cannot be neglected or overlooked, though these formulations have played an imperative role in the PFAS eradication. Nonetheless, as per the studies, it has been stated that the nano adsorbents cannot be easily regenerated and in some cases may be more expensive than bulk materials (Gagliano et al., 2020). Therefore nanoadsorbents with high capacity and low cost need to be developed (Lu et al., 2020). Subsequently, effects due to environmental changes, such as variance in pH or temperature, cannot be compromised, as in the real case scenarios, the minute changes in any of the abovementioned factors may cause PFAS desorption and result in recontamination. Subsequently, the use of MOe-based nano formulation is limited, and mostly $TiO_2$ and $Fe_2O_3$ are used. As it's well known that more than 5000 PFAS compounds have been reported. Moreover, adsorption studies focus on a single PFAS contaminant in the laboratory (often PFOA or PFOS). However, different PFAS are used in many industrial applications, whereas in the environment, a mixture of PFAS is found. Remediation regimes with grounds on adsorption are required to work on the mixture instead of a single compound (Gobelius et al., 2018). There are drawbacks to using nanotechnology and nanomaterials that should be taken into consideration as well. The safety and health effects of nanomaterials have long been a concern (Mauter et al., 2018; Teow et al., 2011). Although using nanomaterials has been considered a great choice, the disposal of nanomaterials is still in question after being used (Liu and Xia, 2020). So, it's moderately substantial to state that eradicating the PFAS from aqueous media using nanomaterials incorporating MOs requires long-term evaluation to use these nanotechnologies effectively (Yin and Villagrán, 2022).

### 18.3.6 Clay minerals

Apart from the MOs mentioned above, clay minerals (mainly, Fe, Al, and/or Si oxides) have drawn much attention in remediation methodologies due to their affordability,

**TABLE 18.2** Different types of metal oxide formulations have been reported for per- and polyfluoroalkyl substances (PFAS) remediation.

| Different metal oxide formulations used | PFAS targeted | Removal (%) | Synthesis method | References |
|---|---|---|---|---|
| *Nanoparticles* | | | | |
| $Al_2O_3$, $Fe_2O_3$, $SiO_2$, and $TiO_2$ | PFOS | 95 | Commercially procured | Lu et al. (2016) |
| $TiO_2$ | PFOA | 68 | Commercially procured | Duan et al. (2022) |
| Surface-functionalized $Fe_2O_3$ | PFOA and PFOS | 95.68, 96.99 | Sol-gel method | Chen et al. (2023) |
| $TiO_2$ | PFOA | 90 | Commercially procured | Gatto et al. (2015) |
| $MnO_2$polymorphs | PFOS | 90 | Hydrothermal method | Chang et al. (2020) |
| MMO-$TiO_2$ | PFOS | 95.99 ± 1.49 | Coprecipitation method | Yang et al. (2024) |
| $TiO_2$ | PFOS | 95.6 | Chemical precipitation | Saheed et al. (2023) |
| $WO_3/TiO_2$ | PFOA, PFHxS, PFBS, 6:2 FTS, GenX | 4 to 26 | Sol-gel method | Lashuk et al. (2022) |
| $(Ga_2O_3)$/ peroxymonosulphate | PFOA | 100 | Commercially procured | Xu et al. (2020b) |
| $Bi_4O_7$-$Ga_2O_3$ composite | PFOA | 59.6 | Hydrothermal method | Chen et al. (2024) |
| $Fe_2O_3$ | PFOA | 96 | Coprecipitation method | Gong et al. (2016) |
| $In_2O_3$ | PFOA | 80 | Commercially procured | Li et al. (2012) |
| $FeTiO_2$, $CuTiO_2$ | PFOA | 90 | Photo-deposition method | Chen et al. (2015) |
| *Nano fibers* | | | | |
| ZnO | PFOS | 99 | Chemical precipitation | Dehghani et al. (2022) |
| *Nano rods* | | | | |
| $Pt/In_2O_3$ | PFOA | 98 | Hydrothermal method | Xu et al. (2017) |
| *Carbonaceous materials* | | | | |
| $Fe_2O_3$ doped multiwalled carbon nanotubes | PFOA | 20–80 | Chemical precipitation | Liu et al. (2018) |
| $MgAl_2O_4$@CNTs | PFOS | 100 | Sol-gel method | Yin et al. (2023) |
| (MWCNT)-modified carbon-doped $TiO_2$ | PFOA | 90 | Chemical precipitation | Wang et al. (2023) |
| $TiO_2$ coated with reduced graphene oxide | PFOA | 98 | Hydrothermal method | Chowdhury and Choi, (2023) |

environmental friendliness. The minerals like kaolinite (KL), halloysite, montmorillonite (MMT), and sepiolite have gained broad applicability as adsorbents due to their strong chemical affinity for pollutants, ease of availability, and affordability. These clay minerals have varied charge densities because of their distinct surface topologies (Tsujimoto et al., 2013). Clay-based remediation media work on three main principal mechanisms/steps for PFAS removal, i.e., hydrophobic interaction between the PFAS perfluorocarbon chain and hydrophobic moiety on the adsorbent surface, electrostatic interactions between PFAS and positively charged surface of adsorbents, and ligand exchange between –OH–mineral groups of clay minerals and PFAS compounds. However, it has been observed that the number of carbons in the chain increases, and chain lengths containing C–F bonds increase the lipophilic nature of PFAS. Furthermore, lipophilic interaction may also occur between organoclay minerals and long-chained PFAS during adsorption, especially $C_8$–$C_{14}$. Nonetheless, hydrophobic interactions also come into play between negatively charged PFAS compounds and negatively charged mineral surfaces after the defluorination of PFAS molecules. Nevertheless, the hydrophobicity of PFAS also increases with the increase in carbon chain lengths. Interestingly, the electrostatic interaction between the carboxylate or sulfonate groups of PFAS and the cationic clay-intercalated species can also result in PFAS adsorption on modified clay adsorbents (Mukhopadhyay et al., 2021). Mineral adsorbents are surface-modified with organic and inorganic agents that contain different functional groups to improve surface functionality, efficacy, and selectivity, as natural clay mineral adsorbents are not very effective at removing bulk contaminants from aqueous systems (Sarkar et al., 2019). Many clay minerals that have been used for PFAS are mentioned in Table 18.3.

Furthermore, the literature has reported that most of these mineral clays have great potential for eradicating PFOS. However, it has been reported that PFOS primarily forms outer-sphere surface complexes because of the hydrophobic and electrostatic interactions crucial to its

**TABLE 18.3** Clay minerals reported for per- and polyfluoroalkyl substances (PFAS) remediation

| Mineral clay used | PFAS removed | Types of interaction | References |
|---|---|---|---|
| Montmorillonite | PFOS | Hydrophobic partition | Zhou et al. (2013) |
| | PFOA and PFOS | Photolysis | Tian et al. (2016) |
| | PFOA | | Chen et al. (2019) |
| Kaolinite (KL) and montmorillonite | PFOS | Hydrophobic and Electrostatic interaction | Zhang et al. (2014) |
| Montmorillonite (MMT), KL, and hematite | PFOS, PFHxS, PFOA, PFHxA | Electrostatic attractions | Zhao et al. (2014) |
| Organo-bentonite | PFOS | Electrostatic interaction | Khodabakhshloo et al. (2021) |
| Alumina | PFOA, PFOS | Electrostatic interaction | Wang and Shih, (2011) |
| Boehmite | PFOS and PFBS | Hydrophobic interaction | Wang et al. (2015) |
| Mg-amino clay | PFOA, PFNA, PFDA, PFOS | Adsorption and reduction | Arvaniti et al. (2015) |

adsorption on KL and montmorillonite. The sulfonate group of PFOS and the hydroxyl group on the surfaces of clay minerals exhibit selective chemisorption in addition to nonspecific physical interactions. Depending on the structure of the clay minerals, interlayer intercalation might affect the PFOS adsorption. For example, PFOS does not exhibit interlayer adsorption on KL, but on montmorillonite, interlayer insertion occurs (Zhang et al., 2014). Nonetheless, one significant mechanism that aids in the dispersion and movement of PFAS in water systems is their sorption onto minerals. The length of the carbon chain and the terminal functional groups significantly impacted the sorption. PFSAs had a substantially greater sorption on the minerals than PFCAs with the same carbon length, suggesting that PFCAs are more likely to leach from the sediment and be susceptible to long-distance groundwater transport. The PFAS exhibit decreased sorption on the minerals at low ionic strength and high pH.

### 18.3.7 Limitations of mineral clay-based formulations

Interestingly, mineral clays have been found to be applicable to removing and eradicating the PFAS from the water samples. These clay-based approaches have their limitations that cannot be overlooked, such as low adsorption capacity because of the hydrophilic properties of natural clay minerals and the oleophobic and hydrophobic properties of PFAS; poor particle-size sensitivity and selectivity; high desorption rate, high cost of synthesizing certain organoclays using particular surfactants; and potential ecotoxic effects of some surfactant-modified organoclays (Amari et al., 2021; Momina et al., 2018; Sarkar et al., 2013, 2010). It was discovered that long-chain PFAS outcompeted short-chain PFAS by hydrophobic interaction while adhering to organically modified KL. The carboxylic functional head groups result in lower adsorption in the short-chain compared to long-chain PFAS with sulfonic functional groups, causing the desorption rate of short-chain PFAS to be significantly higher (Kabiri and McLaughlin, 2021). Therefore while developing an efficient PFAS remediation strategy, it is essential to take into account the possibility that short-chain PFAS may re-pollute the groundwater more frequently than long-chain PFAS due to leaching and desorption from the adsorbents. Weak ion-dipole interactions, commonly known as electrostatic attraction, can arise between PFAS compounds and certain polar functionalities (dipoles), such as $-NH_2$ and -OH. Additionally, the minerals had negative charges at pH 5.0–7.0 due to the PZC of KL and MMT being 3.2 and 2.0, respectively, making the use of natural KL and MMT in PFOS removal inefficient (Zhang et al., 2014). Of note, the anionic character of PFOS ($pK_a = -3.27$ over a broad pH range) results in a stronger electrostatic repulsion between PFOS and MMT than with KL, which lowers the clay mineral's ability to absorb PFOS. These findings have significant environmental complications: if the background solution is basic and has a low ionic strength, the leaching of PFAAs from sediment would accelerate. Nonetheless, the inherent processes behind the sorption of PFAAs on minerals with varying compositions and structures require more investigation. Furthermore, it has been observed that the organic matter in the media competes with the PFAS, limiting its removal. This effect is more pronounced for shorter chain PFAS molecules (Zhao et al., 2011). The presence of organic matter also plays a crucial role in the rate of adsorption of PFAS molecules. Fulvic and humic acid coatings are commonly utilized in clay-based formulations to counterbalance the abovementioned issue. Furthermore, the safe and environmentally friendly disposal of these clay-based formulations is necessary in addition to the disadvantages above (Mukhopadhyay et al., 2021).

### 18.3.8 Miscellaneous

MOs, such as iron oxides, also play a key role in the biochemical reduction of contaminants. It has been well reported in previous studies that electron-accepting behavior of iron oxides helps some special bacteria known as iron- reducing bacteria (FeRB) in the decomposition of PFAS such as PFOA, PFOS (Huang and Jaffé, 2019). Beta proteobacteria, Alpha proteobacteria, Delta proteobacteria, and Actinobacteria are some of the FeRB that have been reported in the literature for the biodegradation of POPS, such as PFAS (Yi et al., 2019). The literature has reported that iron-based minerals offer favorable bioavailability, adsorption, and biological reduction. Due to these characteristics, microbial degradation of the pollutants comes into play. Moreover, the reduction of iron oxides accelerates the electron transfer (Bai et al., 2020; Fan et al., 2023). Furthermore, 50% of the PFOS and 63% of PFOA have been reported to be removed by the FeRB within 100 days (Huang and Jaffé, 2019). Recently, iron-oxidizing bacteria have also been reported for the biodegradation of PFAS (Ma et al., 2024). Nonetheless, iron oxide-based PFPE and composite polymers prepared by Ion exchange methods have shown great potential in PFAS removal (Park et al., 2020; Tan et al., 2022; Kundu and Radian, 2022), such as perfluoroheptanoic acid, PFOA, perfluorononanoic acid, perfluorodecanoic acid, PFOS, perfluorobutanoic acid, and perfluorohexanoic acid. Furthermore, MO-based polymer-derived formulations have also gained wide attention in PFAS removal (Elanchezhiyan et al., 2021). Nonetheless, the safe disposal of iron oxides is still in question, like previously mentioned formulations. However, more studies are required in the bioremediation of the PFAS molecules, as these bioremediation techniques take a long time to degrade these molecules. On the contrary, these bioremediations are not a fast and most rapid solution.

## 18.4 Role of metal oxides in sample preparation and preconcentration

Though the abovementioned aspects of the MOs, and mineral clays have been explored and are well reported in the retrospective studies. MOs from different matrices are used in sample cleanup and preconcentration in PFAS investigations. It is well known that the complexity of a matrix may cause issues like signal interference, matrix effects, and instrumental contamination. Nonetheless, sample preparation is paramount to enhance accuracy, sensitivity, and specificity, along with measuring analyte samples, preceding instrumental investigation. In PFAS analysis, sample pretreatment can be considered a preanalytical separation procedure that primarily entails the enrichment of analytes, the elimination of interfering species, and the selective isolation of analytes of interest from the sample matrix. However, the matrix effect plays a crucial role in the analysis (Dhiman and Ansari, 2024).

MOs can retain their particular morphology when the solvent systems change, and some MOs can tolerate high temperatures and extreme acidic or/and alkaline pH. The capacity to function as both ion and ligand exchangers is an attribute that is exhibited by all

MO phases. Furthermore, MOs varied composition and structure, enabling a broad range of surface reactivity, which introduces the potential of surface functionalization, for example, by using the reactivity of surface-bound hydroxyl groups as grounding sites to adhere the analyte. Thus, MOs' stability and diverse surface chemistry allow for greater flexibility in the development of analytical techniques. Xu et al. (2016); Song et al. (2023) have reported using a composite, referred to as multi walled carbon nano tubes@$Fe_3O_4$@C, in magnetic solid phase extraction (MSPE) adsorbent to enrich the six PFAS from the water samples. Nonetheless, perfluorinated compounds at trace levels in water samples have been enriched and analyzed using covalent triazine-based framework/$Fe_2O_3$ composites as MSPE (Ren et al., 2018). To extract and enrich PFAS from river water using the dispersive magnetic solid phase extraction (DMSPE) technique, an efficient and effective iron oxide/MOF composite, $Fe_3O_4$@MIL-101(Cr), was successfully synthesized and utilized as an adsorbent (Selahle et al., 2022). Furthermore, it has been observed that the dSPE sorbents that constitute a significant part in extraction processes mostly center around using the iron oxide-based formulations because of their inherent magnetic properties. As mentioned, due to the electrostatic and hydrophilic interactions between the PFAS and sorbents, these magnetic particles lead to enhanced entrapment of the PFAS from the matrix. However, after the entrapment of the PFAS in this extraction, it can be recovered by a suitable solvent and analyzed by chromatographic techniques. Though, the aspect that adsorbent dose, extraction time, salt concentration, elution time, and desorption time play an essential role in these solid phase extractions.

## 18.5 Conclusion

PFAS, "the forever chemical," that has made human life effortless by its unique properties, has now become an environmental load. The harmful nature of these chemicals is no longer hidden. The contamination loop of these chemicals is infinite. Irrespective of the fact that a large number of chemicals have made an entry point in the environment and food cycle, very few PFAS molecules are taken into consideration by stakeholders around the globe. On the contrary, approximately 9000 PFAS molecules have been reported to be used in everyday consumer products. As per global reports, most of the regulations center around the water, and no such regulation on soil has been updated until now. Although PFAS molecules have now become an irreplaceable part of daily consumer products, many efforts have been made by researchers around the globe to eradicate the molecules.

- In the play of removal and eradication strategies of the PFAS, MOs, and minerals are one of the protagonists. The different MO-based technologies have emerged as efficient in the PFAS removal, such as MO electrodes, MO formulation, and many more. Mostly, these technologies focus on the photodegradation or adsorption of these PFAS molecules onto their surface.
- But these technologies come with limitations, such as the generation of shorter-chain PFAS molecules from the disintegration of longer-chain PFAS molecules. In the process of the photodegradation, the $F^-$ ions are generated along with the organic content. Indirectly, in the process of photodegradation, more pollution is created.
- Besides, the MO-based electrodes require safe disposal after use. Nonetheless, the MO-based nano formulations offer a large surface area and greater flexibility. But again, the same question still arises in the context of the safe disposal of these formulations.

- Moreover, in context with the mineral clays, the removal is influenced by the number of carbon atoms attached to molecules. In removing longer chain PFAS molecules, shorter chain molecules get dropped back. Besides, the inextraction methodologies involve the process in which the analyte is first adsorbed and then desorbed, so the possibility of the loss of the analyte is always there.

In a nutshell, extraction and removal technologies lack in PFAS as most of the studies focus on a limited number of PFAS molecules, and there is a great need to minimize the use of PFAS to rein in the contamination. Furthermore, more than 9000 PFAS molecules, but a very limited number of reports have been published that provide an ultimate solution for PFAS. However, bioremediation of PFAS that uses bacteria can turn out to be one of the most promising tools. Nevertheless, mechanochemical processes/technologies, such as sonication, have gained attention in recent years for their great potential in PFAS disintegration. Nonetheless, understanding the current global situation, there is a great need to establish more strict regulations for the use of the PFAS in industrial applications. Nevertheless, more hybrid and sustainable technologies focusing more on bio remediation with proper disposal should be incorporated.

## Declaration of competing interest

The authors certify that they have no affiliations with or involvement in any organization or entity with any financial or nonfinancial interest in the subject matter or materials discussed in manuscripts.

## Author contributions

**Samar Dhiman and Hemant Veer Jain**: Data curation, writing and formal analysis, Proof Reading, Methodology, Validation.

**Nasreen G. Ansari**: Resource, Visualization, Conceptualization, Supervision, Writing - review & editing.

## Data availability

Data are compiled from the list of references.

## Acknowledgment

The authors would like to acknowledge the Director, CSIR-Indian Institute of Toxicology Research, Lucknow, India, for supporting and approving the work under the manuscript number IITR/SECC-PME/MSS/2025/039 provided by the Scientific ethics committee. Additionally, the images used in manuscript are designed by Microsoft

Office and BioRender. Also acknowledge the University Grants commission and Ministry of minority affairs have funded this work under the NTA Reference numbers 210510415660 and 210510390513.

## Abbreviations

| | |
|---|---|
| **ppt** | Parts per trillion |
| **ppm** | Parts per million |
| **ppb** | Parts per billion |
| **ng** | Nanogram |
| **L** | Liter |
| **US EPA** | United States Environmental Protection Agency |
| **mg** | Milligram |
| **ml** | Milliliter |
| **PFAS** | Per/poly fluoro alkyl substances |
| **PFOS** | Perfluoro octane sulfonate |
| **PFHxS** | Perfluoro hexane sulfonate |
| **PFOA** | perfluorooctanoic acid |
| **PFHxA** | Perfluoro hexanoic acid |
| **PFBA** | Perfluorobutanoic acid |
| **PFHpS** | Perfluoro heptane sulfonic acid |
| **MMT** | Montmorillonite |
| **PZC** | Point of zero charge |
| **DMSPE** | Dispersive magnetic solid phase extraction |
| **MSPE** | Magnetic solid phase extraction |
| **dSPE** | Dispersive solid phase extraction |
| **EU** | European Union |
| **MWCNTS** | Multiwalled carbon nanotubes |
| **FeRB** | Iron-reducing bacteria |
| **FeOB** | Iron oxidizing bacteria |

## References

Amari, A., et al., 2021. Clay-polymer nanocomposites: preparations and utilization for pollutants removal. Materials 14, 1365.

Araújo, E.S., et al., 2023. A review on the use of metal oxide-based nanocomposites for the remediation of organics-contaminated water via photocatalysis: fundamentals, bibliometric study and recent advances. Toxics 11.

Arvaniti, O.S., et al., 2015. Reductive degradation of perfluorinated compounds in water using Mg-aminoclay coated nanoscale zero valent iron. Chemical Engineering Journal 262, 133–139. Available at https://doi.org/10.1016/j.cej.2014.09.079.

Bai, Y., Sun, T., Angenent, L.T., Haderlein, S.B., Kappler, A., 2020. Electron hopping enables rapid electron transfer between quinone-/hydroquinone-containing organic molecules in microbial iron(III) mineral reduction. Environmental Science & Technology 54, 10646–10653. Available at https://doi.org/10.1021/acs.est.0c02521.

Barisci, S., Suri, R., 2021. Electrooxidation of short- and long-chain perfluoroalkyl substances (PFASs) under different process conditions. Journal of Environmental Chemical Engineering 9, 105323. Available at https://doi.org/10.1016/j.jece.2021.105323.

Chang, Y., Cho, Y.-C., Lin, Y.-P., 2020. Degradation of PFOS by a MnO2/H2O2 process. Environmental Science: Water Research & Technology 6, 3476–3487. Available at https://doi.org/10.1039/D0EW00739K.

Chen, J., Wang, S., Wang, S., Tang, R., Chen, H., 2024. Facilely prepared Bi4O7 modified commercial Ga2O3 with optimized photocatalytic efficiency for perfluorooctanoic acid degradation. Applied Surface Science 673, 160877. Available at https://doi.org/10.1016/j.apsusc.2024.160877.

Chen, M.-J., Lo, S.-L., Lee, Y.-C., Huang, C.-C., 2015. Photocatalytic decomposition of perfluorooctanoic acid by transition-metal modified titanium dioxide. Journal of Hazardous Materials 288, 168–175. Available at https://doi.org/10.1016/j.jhazmat.2015.02.004.

Chen, X., et al., 2023. Two-in-one platform based on conjugated polymer for ultrasensitive ratiometric detection and efficient removal of perfluoroalkyl substances from environmental water. Science of the Total Environment 860, 160467. Available at https://doi.org/10.1016/j.scitotenv.2022.160467.

Chen, Z., et al., 2019. Application of surfactant modified montmorillonite with different conformation for photo-treatment of perfluorooctanoic acid by hydrated electrons. Chemosphere 235, 1180–1188. Available at https://doi.org/10.1016/j.chemosphere.2019.07.032.

Chowdhury, N., Choi, H., 2023. Photocatalytic degradation of perfluorooctanoic acid on Pb-doped TiO2 coated with reduced graphene oxide. Water Environment Research 95, e10871. Available at https://doi.org/10.1002/wer.10871.

de Jesus, R.A., et al., 2022. Biodegradation and Biodeterioration at the Nanoscale. Elsevier, pp. 529–560 (eds Hafiz M.N., I., Muhammad Bilal, T.A. Nguyen, & G. Yasin).

Dehghani, M., Naseri, M., Nadeem, H., Banaszak Holl, M.M., Batchelor, W., 2022. Sunlight-driven photocatalytic per- and polyfluoroalkyl substances degradation over zinc oxide/cellulose nanofiber catalyst using a continuous flow reactor. Journal of Environmental Chemical Engineering 10, 108686. Available at https://doi.org/10.1016/j.jece.2022.108686.

Dhiman, S., Ansari, N.G., 2024. A review on extraction, analytical and rapid detection techniques of per /poly fluoro alkyl substances in different matrices. Microchemical Journal 196, 109667. Available at https://doi.org/10.1016/j.microc.2023.109667.

Duan, L., et al., 2022. Titanium oxide improves boron nitride photocatalytic degradation of perfluorooctanoic acid. Chemical Engineering Journal 448, 137735. Available at https://doi.org/10.1016/j.cej.2022.137735.

Elanchezhiyan, S.S., Preethi, J., Rathinam, K., Njaramba, L.K., Park, C.M., 2021. Synthesis of magnetic chitosan biopolymeric spheres and their adsorption performances for PFOA and PFOS from aqueous environment. Carbohydrate Polymers 267, 118165. Available at https://doi.org/10.1016/j.carbpol.2021.118165.

Fan, Q., et al., 2023. Iron redox cycling in layered clay minerals and its impact on contaminant dynamics: a review. Science of the Total Environment 855, 159003. Available at https://doi.org/10.1016/j.scitotenv.2022.159003.

Gagliano, E., Sgroi, M., Falciglia, P.P., Vagliasindi, F.G.A., Roccaro, P., 2020. Removal of poly- and perfluoroalkyl substances (PFAS) from water by adsorption: role of PFAS chain length, effect of organic matter and challenges in adsorbent regeneration. Water Research 171, 115381. Available at https://doi.org/10.1016/j.watres.2019.115381.

Gao, X., Chorover, J., 2012. Adsorption of perfluorooctanoic acid and perfluorooctanesulfonic acid to iron oxide surfaces as studied by flow-through ATR-FTIR spectroscopy. Environmental Chemistry 9, 148–157. Available at https://doi.org/10.1071/EN11119.

Gatto, S., et al., 2015. Surface fluorination on TiO2 catalyst induced by photodegradation of perfluorooctanoic acid. Catalysis Today 241, 8–14. Available at https://doi.org/10.1016/j.cattod.2014.04.031.

Gobelius, L., et al., 2018. Per- and polyfluoroalkyl substances in Swedish groundwater and surface water: implications for environmental quality standards and drinking water guidelines. Environmental Science & Technology 52, 4340–4349. Available at https://doi.org/10.1021/acs.est.7b05718.

Gong, Y., Wang, L., Liu, J., Tang, J., Zhao, D., 2016. Removal of aqueous perfluorooctanoic acid (PFOA) using starch-stabilized magnetite nanoparticles. Science of the Total Environment 562, 191–200. Available at https://doi.org/10.1016/j.scitotenv.2016.03.100.

He, H., 2020. Solution Processed Metal Oxide Thin Films for Electronic Applications. Elsevier, pp. 7–30 (eds Z. Cui & G. Korotcenkov).

Hearon, S.E., et al., 2022. Montmorillonite clay-based sorbents decrease the bioavailability of per- and polyfluoroalkyl substances (PFAS) from soil and their translocation to plants. Environmental Research 205, 112433. Available at https://doi.org/10.1016/j.envres.2021.112433.

Huang, D., et al., 2020. Amorphous Pd-loaded Ti4O7 electrode for direct anodic destruction of perfluorooctanoic acid. Environmental Science & Technology 54, 10954–10963. Available at https://doi.org/10.1021/acs.est.0c03800.

Huang, S., Jaffé, P.R., 2019. Defluorination of perfluorooctanoic acid (PFOA) and perfluorooctane sulfonate (PFOS) by acidimicrobium sp. strain A6. Environmental Science & Technology 53, 11410–11419. Available at https://doi.org/10.1021/acs.est.9b04047.

Jiang, F., Zhao, H., Chen, H., Xu, C., Chen, J., 2016. Enhancement of photocatalytic decomposition of perfluorooctanoic acid on CeO2/In2O3. RSC Advances 6, 72015–72021. Available at https://doi.org/10.1039/C6RA09856H.

Kabiri, S., McLaughlin, M.J., 2021. Durability of sorption of per- and polyfluorinated alkyl substances in soils immobilized using common adsorbents: 2. Effects of repeated leaching, temperature extremes, ionic strength and competing ions. Science of the Total Environment 766, 144718. Available at https://doi.org/10.1016/j.scitotenv.2020.144718.

Karbassiyazdi, E., et al., 2023. A juxtaposed review on adsorptive removal of PFAS by metal-organic frameworks (MOFs) with carbon-based materials, ion exchange resins, and polymer adsorbents. Chemosphere 311, 136933. Available at https://doi.org/10.1016/j.chemosphere.2022.136933.

Khodabakhshloo, N., Biswas, B., Moore, F., Du, J., Naidu, R., 2021. Organically functionalized bentonite for the removal of perfluorooctane sulfonate, phenanthrene and copper mixtures from wastewater. Applied Clay Science 200, 105883. Available at https://doi.org/10.1016/j.clay.2020.105883.

Kundu, S., Radian, A., 2022. Surface confinement of per-fluoroalkyl substances on an iron-decorated clay-cyclodextrin composite enables rapid oxidation by hydroxyl radicals. Chemical Engineering Journal 431, 134187. Available at https://doi.org/10.1016/j.cej.2021.134187.

Larasati, A., Fowler, G.D., Graham, N.J.D., 2021. Insights into chemical regeneration of activated carbon for water treatment. Journal of Environmental Chemical Engineering 9, 105555. Available at https://doi.org/10.1016/j.jece.2021.105555.

Lashuk, B., Pineda, M., AbuBakr, S., Boffito, D., Yargeau, V., 2022. Application of photocatalytic ozonation with a WO3/TiO2 catalyst for PFAS removal under UVA/visible light. Science of The Total Environment 843, 157006. Available at https://doi.org/10.1016/j.scitotenv.2022.157006.

Li, J., et al., 2021. Hydrodechlorination and deep hydrogenation on single-palladium-atom-based heterogeneous catalysts. Applied Catalysis B: Environmental 282, 119518. Available at https://doi.org/10.1016/j.apcatb.2020.119518.

Li, J., Yao, J., Li, Y., Shao, Y., 2012. Controlled release and retarded leaching of pesticides by encapsulating in carboxymethyl chitosan /bentonite composite gel. Journal of Environmental Science and Health. Part. B, Pesticides, Food Contaminants, and Agricultural Wastes 47, 795–803. Available at https://doi.org/10.1080/03601234.2012.676421.

Lin, H., et al., 2018. Development of macroporous Magnéli phase Ti4O7 ceramic materials: as an efficient anode for mineralization of poly- and perfluoroalkyl substances. Chemical Engineering Journal 354, 1058–1067. Available at https://doi.org/10.1016/j.cej.2018.07.210.

Lin, H., et al., 2013. Highly efficient and mild electrochemical mineralization of long-chain perfluorocarboxylic acids (C9–C10) by Ti/SnO2–Sb–Ce, Ti/SnO2–Sb/Ce–PbO2, and Ti/BDD electrodes. Environmental Science & Technology 47, 13039–13046. Available at https://doi.org/10.1021/es4034414.

Lin, H., Niu, J., Ding, S., Zhang, L., 2012. Electrochemical degradation of perfluorooctanoic acid (PFOA) by Ti/SnO2–Sb, Ti/SnO2–Sb/PbO2 and Ti/SnO2–Sb/MnO2 anodes. Water Research 46, 2281–2289. Available at https://doi.org/10.1016/j.watres.2012.01.053.

Liu, L., Liu, Y., Li, C., Ji, R., Tian, X., 2018. Improved sorption of perfluorooctanoic acid on carbon nanotubes hybridized by metal oxide nanoparticles. Environmental Science and Pollution Research 25, 15507–15517. Available at https://doi.org/10.1007/s11356-018-1728-5.

Liu, S., Xia, T., 2020. Continued efforts on nanomaterial-environmental health and safety is critical to maintain sustainable growth of nanoindustry. Small (Weinheim an der Bergstrasse, Germany) 16, 2000603. Available at https://doi.org/10.1002/smll.202000603.

Liu, X., et al., 2025. Electrochemical destruction of PFAS at low oxidation potential enabled by CeO2 electrodes utilizing adsorption and activation strategies. Journal of Hazardous Materials 486, 137043. Available at https://doi.org/10.1016/j.jhazmat.2024.137043.

Lou, Z., et al., 2022. Strong hydrophobic affinity and enhanced •OH generation boost energy-efficient electrochemical destruction of perfluorooctanoic acid on robust ceramic/PbO2-PTFE anode. Separation and Purification Technology 280, 119919. Available at https://doi.org/10.1016/j.seppur.2021.119919.

Lu, D., Sha, S., Luo, J., Huang, Z., Zhang Jackie, X., 2020. Treatment train approaches for the remediation of per- and polyfluoroalkyl substances (PFAS): a critical review. Journal of Hazardous Materials 386, 121963. Available at https://doi.org/10.1016/j.jhazmat.2019.121963.

Lu, F., Astruc, D., 2020. Nanocatalysts and other nanomaterials for water remediation from organic pollutants. Coordination Chemistry Reviews 408, 213180. Available at https://doi.org/10.1016/j.ccr.2020.213180.

Lu, X., et al., 2016. Adsorption behavior and mechanism of perfluorooctane sulfonate on nanosized inorganic oxides. Journal of Colloid and Interface Science 474, 199–205. Available at https://doi.org/10.1016/j.jcis.2016.04.032.

Ma, H., et al., 2024. Strengthening effect of different iron minerals for perfluorooctanoic acid and perfluorooctane sulphonic acid removal in constructed wetlands: mechanisms of electron transfer and microbial effect. Chemical Engineering Journal 486, 150199. Available at https://doi.org/10.1016/j.cej.2024.150199.

Maldonado, V.Y., et al., 2021. A flow-through cell for the electrochemical oxidation of perfluoroalkyl substances in landfill leachates. Journal of Water Process Engineering 43, 102210. Available at https://doi.org/10.1016/j.jwpe.2021.102210.

Malovanyy, A., et al., 2023. Comparative study of per- and polyfluoroalkyl substances (PFAS) removal from landfill leachate. Journal of Hazardous Materials 460, 132505. Available at https://doi.org/10.1016/j.jhazmat.2023.132505.

Mauter, M.S., et al., 2018. The role of nanotechnology in tackling global water challenges. Nature Sustainability 1, 166–175. Available at https://doi.org/10.1038/s41893-018-0046-8.

Momina, Shahadat, M., Isamil, S., 2018. Regeneration performance of clay-based adsorbents for the removal of industrial dyes: a review. RSC Advances 8, 24571–24587. Available at https://doi.org/10.1039/C8RA04290J.

Mukherjee, P., Sathiyan, K., Zidki, T., Nadagouda, M.N., Sharma, V.K., 2023. Electrochemical degradation of per- and poly-fluoroalkyl substances in the presence of natural organic matter. Separation and Purification Technology 325, 124639. Available at https://doi.org/10.1016/j.seppur.2023.124639.

Mukhopadhyay, R., et al., 2019. Fe-exchanged nano-bentonite outperforms Fe3O4 nanoparticles in removing nitrate and bicarbonate from wastewater. Journal of Hazardous Materials 376, 141–152. Available at https://doi.org/10.1016/j.jhazmat.2019.05.025.

Mukhopadhyay, R., et al., 2021. Natural and engineered clays and clay minerals for the removal of poly- and perfluoroalkyl substances from water: state-of-the-art and future perspectives. Advances in Colloid and Interface Science 297, 102537. Available at https://doi.org/10.1016/j.cis.2021.102537.

Nadoll, P., 2017. Encyclopedia of Geochemistry: A Comprehensive Reference Source on the Chemistry of the Earth (ed William M. White). Springer International Publishing, pp. 1–4.

Nienhauser, A.B., et al., 2022. Boron-doped diamond electrodes degrade short- and long-chain per- and polyfluorinated alkyl substances in real industrial wastewaters. Journal of Environmental Chemical Engineering 10, 107192. Available at https://doi.org/10.1016/j.jece.2022.107192.

Niu, J., Lin, H., Xu, J., Wu, H., Li, Y., 2012. Electrochemical mineralization of perfluorocarboxylic acids (PFCAs) by Ce-doped modified porous nanocrystalline PbO2 film electrode. Environmental Science & Technology 46, 10191–10198. Available at https://doi.org/10.1021/es302148z.

Park, M., Daniels, K.D., Wu, S., Ziska, A.D., Snyder, S.A., 2020. Magnetic ion-exchange (MIEX) resin for perfluorinated alkylsubstance (PFAS) removal in groundwater: roles of atomic charges for adsorption. Water Research 181, 115897. Available at https://doi.org/10.1016/j.watres.2020.115897.

Pierpaoli, M., et al., 2021. Electrochemical oxidation of PFOA and PFOS in landfill leachates at low and highly boron-doped diamond electrodes. Journal of Hazardous Materials 403, 123606. Available at https://doi.org/10.1016/j.jhazmat.2020.123606.

Rahman, M.F., Peldszus, S., Anderson, W.B., 2014. Behaviour and fate of perfluoroalkyl and polyfluoroalkyl substances (PFASs) in drinking water treatment: a review. Water Research 50, 318–340. Available at https://doi.org/10.1016/j.watres.2013.10.045.

Ren, J.-Y., et al., 2018. Magnetic covalent triazine-based frameworks as magnetic solid-phase extraction adsorbents for sensitive determination of perfluorinated compounds in environmental water samples. Analytical and Bioanalytical Chemistry 410, 1657–1665. Available at https://doi.org/10.1007/s00216-017-0845-1.

Rezaei, B., Irannejad, N., 2022. In: Maruccio, G., Narang), J. (Eds.), Electrochemical Sensors. Woodhead Publishing, pp. 51–78.

Rhakho, N., et al., 2024. Transformative dynamics: self-assembly of iron oxide hydroxide nanorods into iron oxide microcubes for enhanced perfluoroalkyl substance remediation. Langmuir: The ACS Journal of Surfaces and Colloids 40, 10184–10194. Available at https://doi.org/10.1021/acs.langmuir.4c00472.

Saheed, M., et al., 2023. Photocatalytic degradation and defluorination of per- and poly-fluoroalkyl substances (PFASs) using biosynthesized TiO2 nanoparticles under UV–visible light. Engineering Proceedings 37, 114.

Sarkar, B., Megharaj, M., Shanmuganathan, D., Naidu, R., 2013. Toxicity of organoclays to microbial processes and earthworm survival in soils. Journal of Hazardous Materials 261, 793–800. Available at https://doi.org/10.1016/j.jhazmat.2012.11.061.

Sarkar, B., Megharaj, M., Xi, Y., Krishnamurti, G.S.R., Naidu, R., 2010. Sorption of quaternary ammonium compounds in soils: implications to the soil microbial activities. Journal of Hazardous Materials 184, 448–456. Available at https://doi.org/10.1016/j.jhazmat.2010.08.055.

Sarkar, B., Rusmin, R., Ugochukwu, U.C., Mukhopadhyay, R., Manjaiah, K.M., 2019. Modified Clay and Zeolite Nanocomposite Materials. Elsevier, pp. 113–127 (eds M. Mercurio, B. Sarkar, & A. Langella).

Selahle, S.K., Mpupa, A., Nomngongo, P.N., 2022. Liquid chromatographic determination of per- and polyfluoroalkyl substances in environmental river water samples. Arabian Journal of Chemistry 15, 103960. Available at https://doi.org/10.1016/j.arabjc.2022.103960.

Sharma, S., Shetti, N.P., Basu, S., Nadagouda, M.N., Aminabhavi, T.M., 2022. Remediation of per- and polyfluoroalkyls (PFAS) via electrochemical methods. Chemical Engineering Journal 430, 132895. Available at https://doi.org/10.1016/j.cej.2021.132895.

Singh, S., Shah, K.J., Mehta, N., Srivastava, V.C., Lo, S.-L., 2021. Electrochemical oxidation of perfluorooctanoic acid (PFOA) from aqueous solution using non-active Ti/SnO2-Sb2O5/PbO2 anodes. Advances in Wastewater Treatment II 102, 48.

Song, X.-L., et al., 2023. Application of magnetic carbon nanotube composite nanospheres in magnetic solid-phase extraction of trace perfluoroalkyl substances from environmental water samples. Talanta 253, 123930. Available at https://doi.org/10.1016/j.talanta.2022.123930.

Su, Y., et al., 2019. Potential-driven electron transfer lowers the dissociation energy of the C–F bond and facilitates reductive defluorination of perfluorooctane sulfonate (PFOS). ACS Applied Materials & Interfaces 11, 33913–33922. Available at https://doi.org/10.1021/acsami.9b10449.

Sun, X.-L., et al., 2023. The role of surface functional groups of iron oxide, organic matter, and clay mineral complexes in sediments on the adsorption of copper ions. Sustainability 15, 6711. Available at https://doi.org/10.3390/su15086711.

Sunderland, E.M., et al., 2019. A review of the pathways of human exposure to poly- and perfluoroalkyl substances (PFASs) and present understanding of health effects. Journal of Exposure Science & Environmental Epidemiology 29, 131–147. Available at https://doi.org/10.1038/s41370-018-0094-1.

Tan, X., et al., 2022. Efficient removal of perfluorinated chemicals from contaminated water sources using magnetic fluorinated polymer sorbents. Angewandte Chemie International Edition 61, e202213071. Available at https://doi.org/10.1002/anie.202213071.

Teow, Y., Asharani, P.V., Hande, M.P., Valiyaveettil, S., 2011. Health impact and safety of engineered nanomaterials. Chemical Communications 47, 7025–7038. Available at https://doi.org/10.1039/C0CC05271J.

Tian, H., Gao, J., Li, H., Boyd, S.A., Gu, C., 2016. Complete defluorination of perfluorinated compounds by hydrated electrons generated from 3-indole-acetic-acid in organomodified montmorillonite. Scientific Reports 6, 32949. Available at https://doi.org/10.1038/srep32949.

Tsujimoto, Y., Chassagne, C., Adachi, Y., 2013. Comparison between the electrokinetic properties of kaolinite and montmorillonite suspensions at different volume fractions. Journal of Colloid and Interface Science 407, 109–115. Available at https://doi.org/10.1016/j.jcis.2013.05.080.

Uwayezu, J.N., et al., 2021. Electrochemical degradation of per- and poly-fluoroalkyl substances using boron-doped diamond electrodes. Journal of Environmental Management 290, 112573. Available at https://doi.org/10.1016/j.jenvman.2021.112573.

Vu, C.T., Wu, T., 2022. Recent progress in adsorptive removal of per- and poly-fluoroalkyl substances (PFAS) from water/wastewater. Critical Reviews in Environmental Science and Technology 52, 90–129. Available at https://doi.org/10.1080/10643389.2020.1816125.

Wang, F., Shih, K., 2011. Adsorption of perfluorooctanesulfonate (PFOS) and perfluorooctanoate (PFOA) on alumina: influence of solution pH and cations. Water Research 45, 2925–2930. Available at https://doi.org/10.1016/j.watres.2011.03.007.

Wang, F., Shih, K., Leckie, J.O., 2015. Effect of humic acid on the sorption of perfluorooctane sulfonate (PFOS) and perfluorobutane sulfonate (PFBS) on boehmite. Chemosphere 118, 213–218. Available at https://doi.org/10.1016/j.chemosphere.2014.08.080.

Wang, Y., Wang, C., Luo, P., Hu, Q., 2023. Removal of perfluorooctanoic acid by MWCNT-modified carbon-doped titanium dioxide in a peroxymonosulfate/simulated sunlight system. Applied Surface Science 614, 156251. Available at https://doi.org/10.1016/j.apsusc.2022.156251.

Xu, B., et al., 2020b. Improved photocatalysis of perfluorooctanoic acid in water and wastewater by Ga2O3/UV system assisted by peroxymonosulfate. Chemosphere 239, 124722. Available at https://doi.org/10.1016/j.chemosphere.2019.124722.

Xu, J., Wu, P., Ye, E.-C., Yuan, B.-F., Feng, Y.-Q., 2016. Metal oxides in sample pretreatment. TrAC Trends in Analytical Chemistry 80, 41–56. Available at https://doi.org/10.1016/j.trac.2016.02.027.

Xu, T., et al., 2020a. Enhanced adsorption and photocatalytic degradation of perfluorooctanoic acid in water using iron (hydr)oxides/carbon sphere composite. Chemical Engineering Journal 388, 124230. Available at https://doi.org/10.1016/j.cej.2020.124230.

Xu, Z., Yu, Y., Liu, H., Niu, J., 2017. Highly efficient and stable Zr-doped nanocrystalline PbO2 electrode for mineralization of perfluorooctanoic acid in a sequential treatment system. Science of the Total Environment 579, 1600–1607. Available at https://doi.org/10.1016/j.scitotenv.2016.11.180.

Yang, B., et al., 2015. Highly efficient electrochemical degradation of perfluorooctanoic acid (PFOA) by F-doped Ti/SnO2 electrode. Journal of Hazardous Materials 299, 417–424. Available at https://doi.org/10.1016/j.jhazmat.2015.06.033.

Yang, H., Park, S.-J., Lee, C.-G., 2024. Enhanced removal of perfluoroalkyl substances using MMO-TiO2 visible light photocatalyst. Alexandria Engineering Journal 87, 31–38. Available at https://doi.org/10.1016/j.aej.2023.12.008.

Yi, L., et al., 2019. Enhanced degradation of perfluorooctanoic acid by a genome shuffling-modified Pseudomonas parafulva YAB-1. Environmental Technology 40, 3153–3161. Available at https://doi.org/10.1080/09593330.2018.1466918.

Yin, S., Villagrán, D., 2022. Design of nanomaterials for the removal of per- and poly-fluoroalkyl substances (PFAS) in water: strategies, mechanisms, challenges, and opportunities. Science of the Total Environment 831, 154939. Available at https://doi.org/10.1016/j.scitotenv.2022.154939.

Yin, S., López, J.F., Solís, J.J.C., Wong, M.S., Villagrán, D., 2023. Enhanced adsorption of PFOA with nano MgAl2O4@CNTs: influence of pH and dosage, and environmental conditions. Journal of Hazardous Materials Advances 9, 100252. Available at https://doi.org/10.1016/j.hazadv.2023.100252.

Zhang, R., Yan, W., Jing, C., 2014. Mechanistic study of PFOS adsorption on kaolinite and montmorillonite. Colloids and Surfaces A: Physicochemical and Engineering Aspects 462, 252–258. Available at https://doi.org/10.1016/j.colsurfa.2014.09.019.

Zhao, B., et al., 2015. ß-Ga2O3 nanorod synthesis with a one-step microwave irradiation hydrothermal method and its efficient photocatalytic degradation for perfluorooctanoic acid. Photochemistry and Photobiology 91, 42–47. Available at https://doi.org/10.1111/php.12383.

Zhao, D., Cheng, J., Vecitis, C.D., Hoffmann, M.R., 2011. Sorption of perfluorochemicals to granular activated carbon in the presence of ultrasound. The Journal of Physical Chemistry. A 115, 2250–2257. Available at https://doi.org/10.1021/jp111784k.

Zhao, L., Bian, J., Zhang, Y., Zhu, L., Liu, Z., 2014. Comparison of the sorption behaviors and mechanisms of perfluorosulfonates and perfluorocarboxylic acids on three kinds of clay minerals. Chemosphere 114, 51–58. Available at https://doi.org/10.1016/j.chemosphere.2014.03.098.

Zhou, Q., Pan, G., Shen, W., 2013. Enhanced sorption of perfluorooctane sulfonate and Cr(VI) on organo montmorillonite: influence of solution pH and uptake mechanism. Adsorption 19, 709–715. Available at https://doi.org/10.1007/s10450-013-9496-5.

Zhu, J., et al., 2022. Feasibility study on the electrochemical reductive decomposition of PFOA by a Rh/Ni cathode. Journal of Hazardous Materials 422, 126953. Available at https://doi.org/10.1016/j.jhazmat.2021.126953.

Zhuo, Q., Han, J., Niu, J., Zhang, J., 2020. Degradation of a persistent organic pollutant perfluorooctane sulphonate with Ti/SnO2–Sb2O5/PbO2-PTFE anode. Emerging Contaminants 6, 44–52. Available at https://doi.org/10.1016/j.emcon.2019.11.002.

CHAPTER

# 19

# Minerals/metal oxide-based photocatalysts for degradation/removal of per- and polyfluoroalkyl substances from water

*Solomon Oluwaseun Akinnawo*

**Department of Chemical Sciences, Olusegun Agagu University of Science and Technology, Okitipupa, Ondo State, Nigeria**

## 19.1 Introduction

Per- and polyfluoroalkyl substances (PFAS) are obstinate environmental pollutants that have been extensively detected in water, soil, and plants around the world (Li et al., 2022; Akinnawo, 2024a). PFAS are partial or completely fluorinated compounds, identified for their highly water-soluble, chemical stability, along with their resistance to environmental degradation, cause significant perils to the ecosystem and human health attributable to their highly bio-accumulative nature (Li et al., 2022; Akinnawo, 2024a). As a consequence of the broad utilizations of PFAS in both consumer and industrial products, coupled with their improper discharge into the environment, there has been a global occurrence of PFAS, specifically perfluorooctane sulfonic acid (PFOS) as well as perfluorooctanoic acid (PFOA) in surface and groundwater (Ahrens, 2011; Dzierlenga et al., 2021; Sidnell et al., 2022). Yet, the prime setback that has been linked with the application versatility of PFAS is its effect on humans at part-per-billion concentration levels that results in hormonal imbalance from endocrine disruption, as well as organ dysfunction and the development of various forms of cancer (Ersan et al., 2023).

Given the adverse effects of PFAS on human health and the ecological system, it is necessary to effectively remove PFAS from drinking water as well as wastewater before discharge into receiving water bodies (Akinnawo, 2024a). Several techniques that include adsorption, coagulation-flocculation, precipitation, and membrane filtration have been investigated for water treatment (Jin and Peydayesh, 2012; Akinnawo et al., 2023;

*Per- and Polyfluoroalkyl Substances (PFAS) as Environmental Contaminants*
DOI: https://doi.org/10.1016/B978-0-443-43856-1.00015-0

Adegoke et al., 2022, 2023; Akinnawo, 2021, 2025a; undefined; undefined). The adsorption technique has been connected with a low percentage removal of short-chain PFAS and challenges in recycling PFAS-spent adsorbents (Akinnawo, 2024a). Comparatively, coagulation-flocculation techniques have been associated with a low percentage removal of PFAS from contaminated water and the generation of a large volume of sludge that is often costly to dewater and discard due to the contamination of soil and water, especially when it involves chemical-based coagulants such as aluminum that can leach into the environment (Akinnawo, 2023; Zoumpouli et al., 2023).

To this end, photocatalysis has been considered a suitable technique with significant success for the effective elimination of PFAS from water. This remarkable consideration is based on factors that include the ability of the photocatalyst to degrade PFAS and evade the problem of secondary pollution that has been associated with adsorption and coagulation techniques via the improper discarding of PFAS-spent adsorbents as well as sludge. Also, the adsorption selectively towards PFAS, resulting in poor removal of short-chain PFAS in comparison to their analogous long-chain PFAS, is another factor that spotlights the advantageous usage of photocatalysis for PFAS elimination from water (Akinnawo, 2024a; Rodrigo et al., 2022; Riegel et al., 2023). Among the array of photocatalysts that have been successfully deployed for degradation and effective removal of PFAS from water, mineral/metal-oxide-based photocatalysts have been recounted with exceptional performance. However, the capability of a photocatalyst to effectively degrade and remove PFAS from water hinges on certain factors that include the efficient separation and migration of photogenerated electron/hole pairs to the surface of the photocatalyst, where decarboxylation and cleavage of the carbon-fluorine (C–F) bond take place, leading to defluorination and degradation of PFAS (Verma et al., 2021, 2024; Liu et al., 2022). In addition, the light-harvesting potential, which hinges on the band gap of the photocatalyst, is another crucial factor that has been asserted to significantly sway the percentage degradation and removal of PFAS from water. Thus the modification of mineral/metal-oxide-based photocatalysts to improve light harvesting and efficient separation and transfer of photogenerated electron/hole pair, either by adjusting the band gap, introduction of defects, vacancies, sensitizers, or dopants, has been considered a way forward in improving photocatalytic degradation and effective removal of PFAS from water. In comparison to previous research findings, this study is aimed at investigating the role of modification strategies in improving the visible light absorption capacity of minerals/metal oxide-based photocatalysts for efficient degradation and defluorination of PFAS for their removal from water.

## 19.2 Modification of mineral/metal oxide-based photocatalyst for per- and polyfluoroalkyl substances degradation

For the purpose of enhancing the performance of mineral and metal oxide-based photocatalysts towards PFAS degradation, a series of modification strategies has been investigated by various researchers. These modification strategies often involve altering the structure and surface of the minerals/metal oxides to boost photocatalytic activity via improvement in light absorption and harvesting, efficient separation and migration of the photogenerated charge carriers, and reaction kinetics of the photocatalytic process. Some of the recent strategies are given below:

### 19.2.1 Doping

This is usually referred to as elemental doping due to the introduction of metals or nonmetal ions/elements as dopants into the structural configuration of metal oxide-based photocatalyst. The introduced dopant often alters the electronic properties and results in the synthesis of a doped metal oxide-based photocatalyst with a narrower band gap that promotes and boosts light absorption, charge separation, and migration. For instance, Chen et al. (2015) and Li et al. (2016) utilized elemental deposition of Ag, Pd, and Pt on $TiO_2$ to synthesize Ag-$TiO_2$, $TiO_2$-Pd, in addition to $TiO_2$-Pt, which displayed higher photocatalytic performance of 54%, 94.4%, and 100% for PFOA degradation and removal from water in comparison to pristine $TiO_2$ with lower photocatalytic efficiency. On the use of nonmetals for doping approach, Li et al. (2020a), synthesized $BiOI_{0.95}Br_{0.05}$ by elemental bromine doping of BiOI for enhanced photocatalytic performance. The synthesized $BiOI_{0.95}Br_{0.05}$ was used to achieve absolute (100%) degradation and 65% defluorination of PFOA at a rate constant (0.0205 $min^{-1}$), compared to the lower photocatalytic PFOA degradation outcome obtained via the undoped BiOI photocatalyst.

### 19.2.2 Surface defect engineering

This approach usually involves the introduction of defects that are not limited to oxygen vacancies (OVs) on the surface of a photocatalyst. The introduced vacancies create accessible active binding sites on the surface of the adapted photocatalyst to effectively bind with PFAS molecules. For instance, engineered OVs on the photocatalyst surface have been noted to enhance PFAS adsorption and promote the generation of reactive oxidation species (ROS) that play a crucial part in effective degradation of PFAS. Wu et al. (2019) deposited MnOx cocatalyst on the surface of an oxygen vacancy-engineered $In_2O_3$ photocatalyst to develop a MnOx-modified OVs engineered $In_2O_3$ photocatalyst with a higher photocatalytic outcome for the degradation of PFOA compared to the use of conventional photocatalyst materials. Also, Song et al. (2019) utilized different alkali to engineer OVs on BiOCl and obtained OVs-tunable BiOCl nanosheet for application in photocatalytic degradation of PFOA. A linear relationship describing the amount of OVs in BiOCl and photocatalytic degradation of PFOA was established by Song et al. (2019), with the attainment of 3.7 times (degradation) and 2.6 times (defluorination) of PFOA when the operational ratio of OVs in BiOCl nanosheet was increased from 0.573 to 0.981. Thus the introduction of OVs in BiOCl nanosheets was designated as the contributing factor for improving the hole-oxidation of PFOA via the trapping of photogenerated electrons by providing localized states and active sites for adsorbing PFOA. A study by Cao et al. (2024) delves into the degradation efficiency of BiOI with abundant iodine vacancies ($BiOI_{1-x}$) and recounts higher photocatalytic degradation of PFOA using $BiOI_{1-x}$ compared to pristine BiOI and attributes the holes and $^{\cdot}OH$ radicals in the PFOA degradation process.

### 19.2.3 Surface sensitization

This modification technique involves the introduction of sensitizers such as dyes, metal sulfides, or other sensitizing materials on the surface of the photocatalysts. The high ability of these sensitizers to absorb light enhances the performance of a modified metal oxide-based photocatalyst to effectively absorb light in the visible region and boost their photocatalytic activities (Diaz-Angulo et al., 2021). This process undergoes some significant stages that start with dye molecules absorption of light, which results in the excitation and transfer of the absorbed light energy to the metal oxide-based photocatalyst (Diaz-Angulo et al., 2021). Upon receiving the absorbed light energy, the photocatalyst produces photogenerated electron-hole pairs that react with adsorbed water or oxygen molecules to effectively yield ROS for the degradation of PFAS molecules (Akinnawo and Ediagbonya, 2025). Comparatively, dye sensitization of photocatalyst allows the use of visible light that is relatively more abundant and readily available, unlike the conventional photocatalyst that frequently relies on UV light irradiation (Akinnawo and Ediagbonya, 2025). An analysis by Liu et al. (2025), reportedly prepared a 2D hexagonal ZnIn2S4 nanosheets as a ternary metal sulfphide semiconductor for degradation of sodium-p-perfluorosnonenoxybenzenesulfonate (OBS), an emerging PFAS developed primarily to substitute PFOS. Compared to the conventional P25 $TiO_2$, the synthesized 2D hexagonal $ZnIn_2S_4$ nanosheets displayed an enhanced photocatalytic activity for the removal of OBS with percentage degradation of 98% and 99% after 15 and 90 mins under sunlight and visible light, respectively. The enhanced photocatalytic outcome of the synthesized 2D hexagonal $ZnIn_2S_4$ nanosheets was attributed to the ability of the metal sulfide to act as a sensitizer in improving light absorption and effective charge separation.

### 19.2.4 Surface functionalization

The introduction of functional groups on the surface of metal oxide-based photocatalysts has been described as efficient in enhancing photocatalytic degradation of PFAS from water. Compounds with sulphonated functional groups have been demonstrated in enhancing photocatalytic performance towards PFAS degradation. From the study outcome by Zhu et al. (2020), complete (100%) degradation of PFOA obtained using $TiO_2$ quantum dots loaded sulfonated graphene aerogel may be attributed to higher surface area that resulted from incorporating sulfonated graphene aerogel on the surface of $TiO_2$ quantum dots. Similarly, the carboxylic functional group of oxalic acid has been reportedly used in enhancing photocatalytic activity. For instance, pristine $TiO_2$ has been reportedly modified using oxalic acid to prepare an oxalic acid-modified $TiO_2$ photocatalyst that displayed 97% degradation of PFOA; this outcome may be attributed to the effect of oxalic acid modification (Wang and Zhang, 2011).

### 19.2.5 Morphological engineering

This technique considers the alteration of morphology in terms of controlling the size, shape, and structure of a photocatalyst. Thus photocatalysts are engineered to have anticipated sizes, shapes, and structures for specific photocatalytic properties such as surface area, light absorption and harvesting, effective separation and migration of photogenerated charge carriers (Akinnawo, 2025b). For instance, $TiO_2$ nanosheets have been noted to yield enhanced photocatalytic performance toward PFAS degradation in comparison to the bulk $TiO_2$ photocatalyst. Similarly, a study by Zhao and Zhang (2009) narrated the characteristic wide band gap exhibited by $\beta$-$Ga_2O_3$ photocatalyst as the reason for the low percentage (36%) photocatalytic degradation of PFOA. Providing a solution to this snag, Zhao et al. (2015) deployed a morphological engineering strategy to synthesize a nanostructured $\beta$-$Ga_2O_3$ photocatalyst with a higher percentage (98.8%) degradation and removal of PFOA from contaminated water. On this premise, Shao et al. (2013a,b) synthesized $Ga_2O_3$ needle-like and $Ga_2O_3$ sheaf-like photocatalysts that displayed complete (100%) photocatalytic degradation of PFOA. On a similar note, nanostructured $In_2O_3$-based photocatalysts such as $In_2O_3$ microsphere, $In_2O_3$ nanocubes, $In_2O_3$ nanoplates, $In_2O_3$ nanoporous nanosphere, and $In_2O_3$ porous nanoplates have been investigated to result in complete (100%) photocatalytic degradation of PFOA. In comparison to pristine $In_2O_3$, which displayed 75% degradation of PFOA from water at optimum conditions (Li et al., 2012, 2013, 2014; Liu et al., 2021).

## 19.2.6 Heterojunction photocatalyst formation

Heterojunction photocatalysts are prepared by combining two or more photocatalysts with different properties with the aim of creating a photocatalyst that possesses a synergistic effect of the individual photocatalysts. Thus heterojunction photocatalysts usually exhibit enhanced charge separation and photocatalytic outcome, in comparison to the individual photocatalyst. A comparative study by Yao et al. (2021) obtained a higher percentage (81.7%) photocatalytic outcome on PFOA degradation within 120 min duration using $Sb_2O_3/TiO_2$ heterojunctions photocatalyst. Moreover, the photocatalytic outcome on the degradation and removal of PFOA from water proceeded at about 4.2 times faster rate with the use of $Sb_2O_3/TiO_2$ heterojunctions, in comparison to the use of commercial P25 as well as the as-synthesized mesoporous $TiO_2$. In another study, flower-shaped $Bi_5O_7I/ZnO$ n-n heterojunction microspheres and p-n $BiOI@Bi_5O_7I$ heterojunction photocatalysts were used to obtain higher photocatalytic outcome on PFOA degradation from water compared to the use of pristine BiOI photocatalyst (Yang et al., 2021; Wang et al., 2020a,b). Also, Gates et al. (2025) harnessed reduced graphene oxide (r-GO) and $WO_3$ nanoflower-based to synthesized r-GO/$WO_3$ nanoflower heterostructure for photocatalytic degradation and elimination of perfluorononanoic acid (PFNA). A complete (100%) degradation and 14% defluorination of PFNA via carbon-carbon (C–C) bond cleavage and C–F bond breakage was obtained as the respective major and minor pathway under 365 nm light irradiation within 22 h duration in the presence of isopropyl alcohol and glucose as respective $^{\cdot}OH$ and hole quenchers. Comparatively, Gates et al. (2025) recounted a wavelength-dependent impact on the use of r-GO/$WO_3$ nanoflower heterostructure with the attainment of 22% degradation of PFNA under 532 nm light irradiation for 24 h, and attributed the degradation of PFNA molecules to C–F bond breakage as the sole mechanistic pathway.

## 19.3 Factors influencing photocatalytic degradation of per- and polyfluoroalkyl substances

### 19.3.1 Effect of light

Light irradiation, especially in the UV and visible region, is essential for photocatalytic generation of ROS that include hydroxyl ($^{\cdot}OH$) and superoxide ($^{\cdot}O_2^-$) radicals that are required for PFAS degradation (Park et al., 2024). Thus the intensity and type of light source crucially influence the degradation efficiency of a photocatalyst (Akinnawo, 2024b). Moreover, different light sources that include UV, visible, and solar can have varying effects on the photocatalytic PFAS degradation. For instance, UV light has been reported to generate higher degradation efficiency due to its higher energy potential to excite and generate more electrons in a photocatalyst, in comparison to visible and solar light, which are considered to be less expensive and more eco-friendly. Moreover, the intensity of light has been noted with direct proportionality to photocatalytic performance for PFAS degradation. Higher light intensity has been reportedly linked to fostering faster PFAS degradation. However, excessive light intensity may result in a heat-up that may potentially damage the photocatalyst and inhibit its efficiency in degrading PFAS from water. An experimental study on the effects of light intensity by Moreira et al. (2024), using a hybrid graphemic iron oxide catalyst, revealed a higher fraction of photocatalytic degradation of PFOA with the utilization of UV irradiated light at a wavelength region of 254 nm. Moreira et al. (2024), obtained the lowest residual fraction concentration of PFOA in water after irradiation with UV irradiated light, compared to the use of simulated solar light (AM 1.5 G) with a higher fraction concentration of PFOA in the treated water, while the highest residual fraction concentration of PFOA in the treated water was recorded in a dark experiment without the use of light. This outcome denotes the impact of irradiated light intensity in improving the photocatalytic PFAS degradation.

### 19.3.2 Effect of pH

The solution pH has a decisive effect on the PFAS degradation efficiency of a photocatalyst due to the role of pH in influencing the surface chemistry of the photocatalyst, PFAS degradation, and the interaction between the molecules of PFAS and the photocatalyst. Most notably, higher PFAS degradation has been associated with lower solution pH (in the acidic range), in comparison to higher solution pH (in the alkaline range). For instance, PFAS molecules are more likely to exist in the protonated form at lower pH, thus facilitating an electrostatic interaction between the protonated PFAS molecules and the photocatalyst. On the contrary, at higher solution pH, PFAS molecules have a higher tendency to exist in an anionic form that hinders reaction with the photocatalyst, due to the negative charges exhibited by both PFAS and photocatalyst at higher pH values. According to the research outcome on using biogenic $TiO_2$ nanoparticles by Mustapha et al. (2023), an increase in solution pH led to a consistent diminution in the degradation and removal of PFOS from water, as portrayed in Fig. 19.1. However, a higher photocatalytic degradation (95.62%) and defluorination (56.13%) of PFOS was obtained at a lower solution pH value of

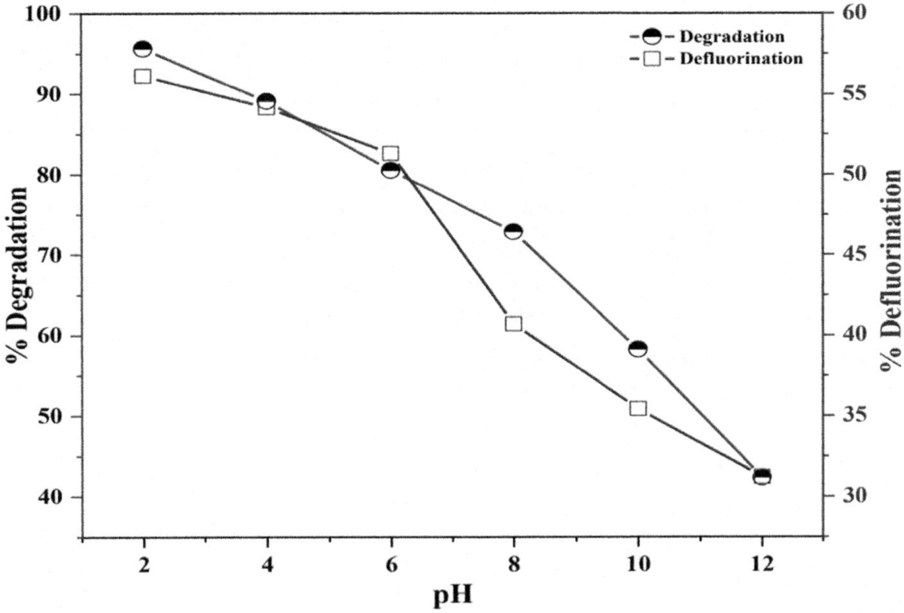


**FIGURE 19.1** The effect of solution pH on photocatalytic perfluorooctane sulfonic acid degradation and defluorination. Source: *Adapted from Mustapha, S., Tijani, J.O., Elabor, R., Etsuyankpa, M.B., Amigun, A.T., Shuaib, D.T., et al., 2023. Photocatalytic degradation and defluorination of per- and poly-fluoroalkyl substance (PFAS) using biosynthesized TiO2 nanoparticles under UV-visible light. Engineering Proceedings. 37 (x), https://doi.org/10.3390/xxxxx, under a Creative Commons license (CC BY 4.0).*

2, in comparison to a higher pH value of 12 that resulted in lower photocatalytic PFOS degradation and defluorination as depicted in Fig. 19.1.

The declining percentage of PFOS degradation and defluorination at higher solution pH, as described by Mustapha et al. (2023), may be attributed to the ensuing electrostatic repulsive forces between the negatively charged photocatalyst and anionic PFOS at higher solution pH. Similarly, under visible light irradiation, Xu et al. (2020c) reportedly obtained complete (100%) photocatalytic PFOA degradation at pH 3 using $TiO_2$ and peroxymonosulfate (PMS). In contrast, the percentage photocatalytic degradation of PFOA diminished to 90%, 45%, and 24% when the solution pH was increased to 5, 7, and 9, respectively. Moreover, Boonchata et al. (2024) reportedly obtained a percentage increase in the residual PFOS and PFOA concentration in treated water when the solution pH was increased from 3 to 10 by means of titanium dioxide/graphene oxide nanocomposite immobilized on polyvinyl alcohol film, as depicted in Fig. 19.2. Thus indicating that an increase in solution pH will give rise to a decrease in the photocatalytic degradation of PFOS and PFOA.

### 19.3.3 Effect of initial per- and polyfluoroalkyl substances concentration

The initial concentration of PFAS is a crucial factor in estimating the efficiency of a photocatalyst. Several studies have associated lower performance of a photocatalyst at higher initial PFAS concentration attributable to the high tendency of the surface of the

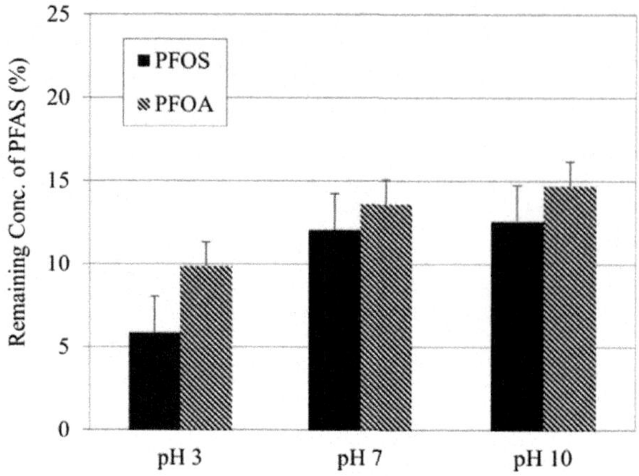


**FIGURE 19.2** The effects of solution pH on the remaining concentration of per- and polyfluoroalkyl substances (PFAS) in treated water. Source: *Reprinted from Boonchata, P., Boontanon, N., Jindal, V., Aung, H.K.Z., Visvanathan, C., Fujii, S., et al., 2024. Photocatalytic degradation of perfluorooctane sulfonic acid and perfluorooctanoic acid using titanium dioxide/graphene oxide nanocomposite immobilized on polyvinyl alcohol film. Case Studies in Chemical and Environmental Engineering. 10, 100862. https://doi.org/10.1016/j.cscee.2024.100862 with permission from Elsevier.*

photocatalyst to be saturated with the formation of multiple layers of PFAS at higher concentration. Thus an increase in the initial PFAS concentration will result in lower efficiency of the photocatalyst to degrade PFAS due to the effects of the multiple layers that limit the direct contact between irradiated light and the photocatalyst. Consequently, inhibiting the absorption of irradiated light and the generation of ROS is required for the photodegradation of PFAS (Akinnawo, 2024a).

According to the research outcome on the use of lead and reduced graphene oxide doped bismuth ferrite photocatalyst by Shang et al. (2018), an increase in the PFOA degradation was observed with increasing initial concentration of PFOA from 10 to 50 mg $L^{-1}$, as evident from the corresponding increase in the rate constant ranging from 0.012 to 0.081 $h^{-1}$. Nevertheless, Shang et al. (2018), noted that a further increase in the initial PFOA concentration to 100 mg $L^{-1}$, resulted in a decline in the value of the rate constant to 0.0333 $h^{-1}$ due to the formation of multiple layers of PFOA that led to an oversaturation of the active sites on the surface of the photocatalyst, which hinders the absorption of irradiated photons. Likewise, Wang et al. (2020a), narrated an inverse relationship occurring between the initial PFAS concentration and the photocatalytic degradation efficiency of p-n $BiOI@Bi_5O_7I$ heterojunction photocatalyst. On this premise, Wang et al. (2020a) recounted an increase in the residue fraction of PFOA in treated water when the initial concentration was increased in a range of 0.5 to 50 mg $L^{-1}$.

Further study on the effect of initial PFAS concentration has been delved into by Fernando et al. (2024), using ilmenite (iron titanium oxide [$FeTiO_3$]), an ore of titanium, as a cost-effective photocatalyst for UV-assisted degradation and mineralization of PFAS in a concentration range of 200 to 1000 ppb. As portrayed in Fig. 19.3C and D, Fernando et al. (2024) highlighted the ilmenite/UV-induced photocatalytic degradation kinetics of PFOA and PFOS to be best described using the pseudo-first-order kinetic model and obtained the highest rate constant of 0.8 and 1.2 $h^{-1}$ for PFOA and PFOS, respectively, at the least concentration of 200 ppb. This indicated that the highest degradation efficiency and removal rate of PFOA and PFOS is obtainable at the lowest initial concentration of PFOA and PFOS, while an increase in initial PFOA and PFOS concentration resulted in a consistent decrease

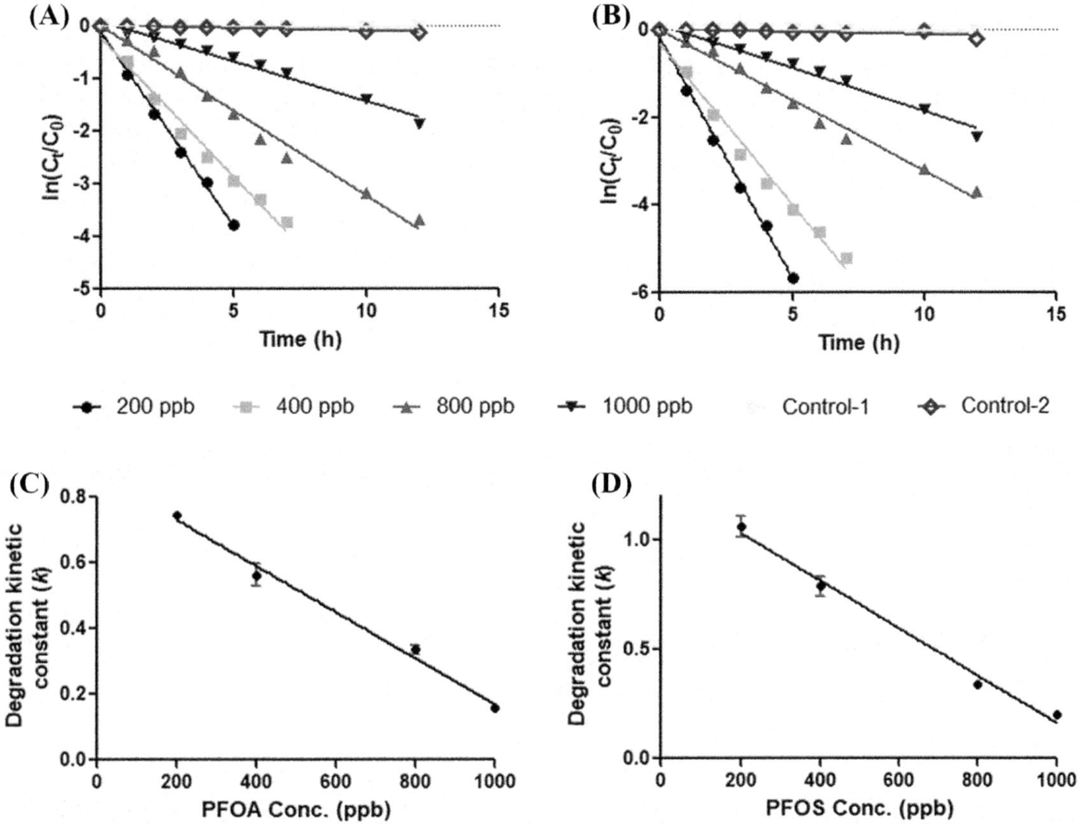


**FIGURE 19.3** The effects of initial per- and polyfluoroalkyl substances concentration on ilmenite/UV-induced photocatalytic residual concentration of (A) perfluorooctanoic acid (PFOA), and (B) perfluorooctane sulfonic acid (PFOS), degradation kinetics of (C) PFOA and (D) PFOS. Source: *Adapted from Fernando, E.Y., Sarkar, D., Rodwihok, C., Satpathy, A., Zhang, J., Rahmati, R., et al., 2024. Unrefined and milled ilmenite as a cost-effective photocatalyst for UV-assisted destruction and mineralization of PFAS. Materials. 17 (3801), https://doi.org/10.3390/ma17153801, under a Creative Commons license (CC BY 4.0).*

in the degradation efficiency and removal rate of PFAS from water due to the saturation of the catalytic reactive sites by high concentrations of PFAS molecules. In comparison to the maximum removal efficiencies of 99.7%, 99.1%, 98.6%, and 97.1%, and 99.49%, 99.2%, 98.8%, and 97.3% obtained for PFOA and PFOS at initial concentrations of 200, 400, 800, and 100 ppb, respectively, under ilmenite/UV irradiation. As depicted in Fig. 19.3A and B, Fernando et al. (2024) pointed out that the control-1 experiment performed in the absence of ilmenite as photocatalyst under UV irradiation yielded 0.09% PFOA degradation and a nondetectable removal of PFOS. Comparatively, the control-2 experiment performed in the presence of ilmenite but in the absence of UV-irradiation yielded higher PFOA and PFOS percentage removal of 7.72% and 8.91%, respectively. This outcome demonstrated that the role of surface adsorption of PFOA and PFOS on ilmenite is less than 10% of the overall removal efficiency of both PFAS from water.

## 19.3.4 Effect of contact time

The operational contact time of a photocatalytic process is a crucial factor that quantifies the duration at which a substance interacts with a photocatalyst under light irradiation (Akinnawo and Ediagbonya, 2025). Research studies have utilized the prospect of a higher chance of photocatalytic reaction occurring at extended contact time to generalize a direct proportional increase in photocatalytic PFAS degradation with an increase in contact time. This is based on the higher propensity of PFAS molecules to interact with the surface of the catalyst for further photocatalytic reaction due to the extended contact time. In addition, an extended contact time paves the way for the generation of more ROS that oxidize and break down the bonds in PFAS molecules, thus ensuing an increase in photocatalytic degradation of PFAS molecules (Park et al., 2024). However, at equilibrium, the reaction rate of PFAS degradation may not increase with an incessant increase in contact time, thus making any further increase in contact time cost-ineffective. Thereby, necessitating the need to optimize the contact time for PFAS degradation. For instance, an investigative outcome by Boonchata et al. (2024) revealed a direct proportionality between the percentage removal of PFAS and an increase in contact time utilizing titanium dioxide/graphene oxide nanocomposite immobilized on a polyvinyl alcohol film prepared at varied durations of 1, 3, and 5 h corresponding to photocatalyst H-1, H-3, and H-5, as shown in Fig. 19.4. Thus increasing the contact time was described to result in a consistent increase in the percentage removal of PFOS and PFOA, as shown in Fig. 19.4.

## 19.3.5 Effect of photocatalyst dosage

Photocatalyst dosage is highly significant in influencing an efficient PFAS degradation and removal from water (Akinnawo, 2024a). An increase in the photocatalyst dosage will initially result in a higher degradation effectiveness for PFAS removal from water owing to the increase in the number of available active sites required for the degradation of PFAS, as depicted in Fig. 19.5. However, at excessively high levels of photocatalyst dosage or beyond the optimal dosage, this effect will result in little or no impact, and

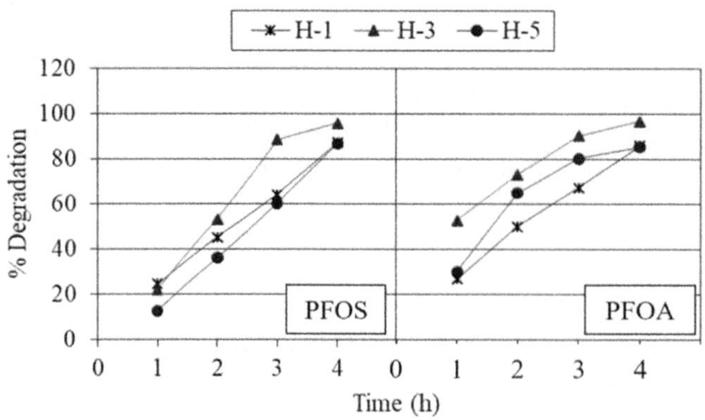


**FIGURE 19.4** The effects of contact time on the performance of metal oxide-based photocatalysts. Source: *Reprinted from Boonchata, P., Boontanon, N., Jindal, V., Aung, H.K.Z., Visvanathan, C., Fujii, S., et al., 2024. Photocatalytic degradation of perfluorooctane sulfonic acid and perfluorooctanoic acid using titanium dioxide/graphene oxide nanocomposite immobilized on polyvinyl alcohol film. Case Studies in Chemical and Environmental Engineering. 10, 100862. https://doi.org/10.1016/j.cscee.2024.100862 with permission from Elsevier.*

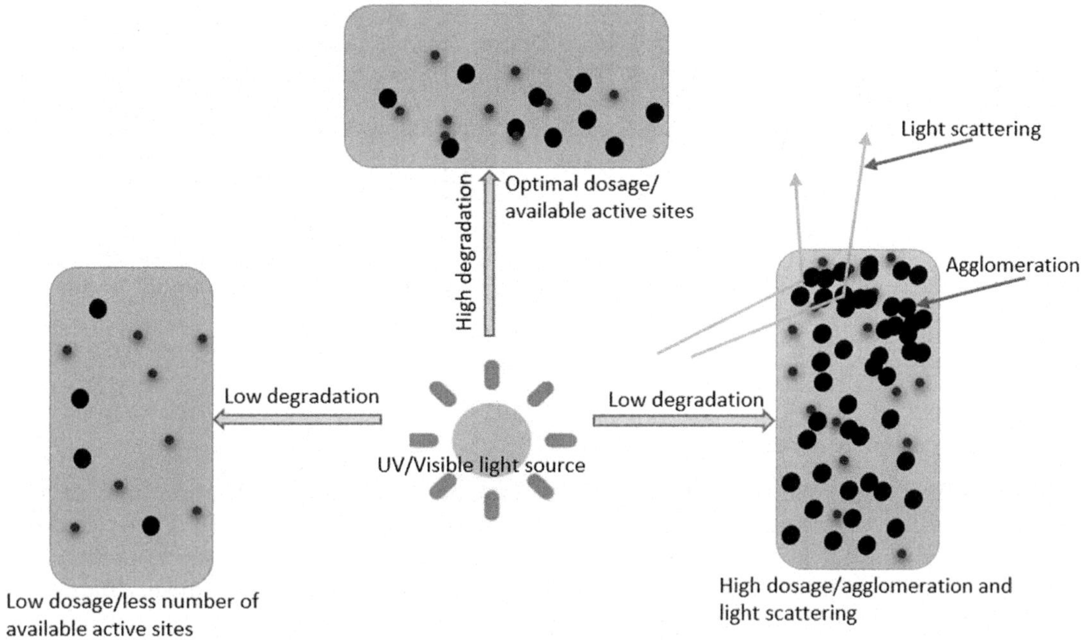


**FIGURE 19.5** Schematic diagram of the effect of photocatalyst dose on the performance of a photocatalyst.

may even result in a decrease in the photocatalytic degradation outcome of PFAS due to the aggregation of photocatalyst to form large particles with lower surface area that inhibit interaction with PFAS molecules. Thus as illustrated in Fig. 19.5, an overdose of a photocatalyst can increase the turbidity of a solution, which will result in the scattering of irradiated light and a consequential decline in the efficiency of a photocatalyst to degrade PFAS (Akinnawo, 2024a; Mabaso et al., 2024). Due to the effect of photocatalyst dosage, it is important to optimize the dosage of a photocatalyst for effective degradation of PFAS from water.

On this claim, Wang et al. (2020a) reported a directly proportional increase in the PFOA degradation ranging from 30% to 80% when the dosage of p-n $BiOI@Bi_5O_7I$ heterojunction was increased in a range of 0.1 to 0.7 $g\,L^{-1}$. Conversely, a decline in the outcome of photocatalytic PFOA degradation to 50% was reported as a result of a further increase in the quantity of photocatalyst to a dosage of 2.0 $g\,L^{-1}$. This outcome may be due to the resultant effect of the lower surface area of the photocatalyst and the light scattering effect from the aggregation of the photocatalyst at high dosage. Likewise, Yao et al. (2021) reported a decrease in the residual fraction of PFOA concentration remaining in a solution when the dosage of mesoporous $Sb_2O_3/TiO_2$ heterojunction photocatalyst was increased from a range of 0.1 to 0.75 $g\,L^{-1}$, as illustrated in Fig. 19.6. The increase in the photocatalytic outcome for PFOA degradation may be attributable to the increasing number of accessible active sites required for interaction with PFOA molecules.

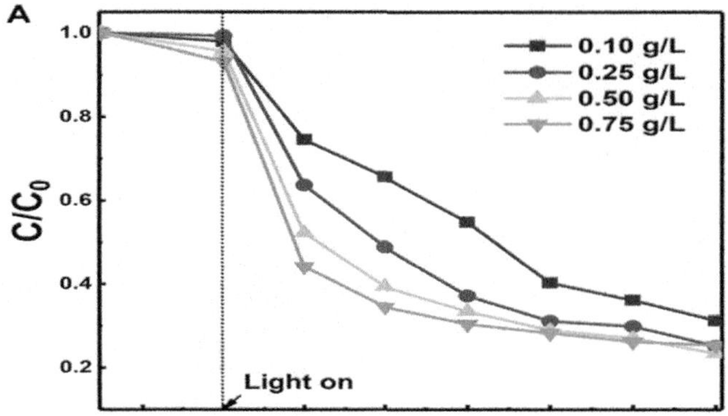


FIGURE 19.6 The effects of photocatalyst dosage on the performance of the $Sb_2O_3/TiO_2$ heterojunction photocatalyst to degrade PFAS. Source: *Adapted from Yao, X., Zuo, J., Wang, Y.J., Song, N.N., Li, H.H., Qiu, K., 2021. Enhanced photocatalytic degradation of perfluorooctanoic acid by mesoporous Sb2O3/TiO2 heterojunctions. Front ChemFrontiers in Chemistry. 9, 1–10. https://doi.org/10.3389/fchem.2021.690520, under a Creative Commons license (CC BY 4.0).*

## 19.3.6 Effect of temperature

Temperature has a significant impact on enhancing PFAS degradation rate at high temperature due to the occurrence of high interfacial mass transfer of PFAS molecules at high temperature that favors an increased kinetic energy for collision and interaction with the photocatalyst, with a resultant effect of PFAS dissociation. On this premise, Zhang et al. (2016) narrated an enhanced increase in the photodegradation of PFOA from 34% to 99% when the temperature was increased in a range of 25°C to 85°C for 60 min duration under UV light. On a similar note, an increase in the defluorination of PFOA from 8% to 50% was recounted when the temperature was increased in a range of 25°C to 85°C for 60 min duration under UV light. An overview by Tshangana et al. (2025) recounted an increase in solution temperature with a resultant increase in photocatalytic activity towards PFAS degradation. However, an extreme increase in temperature destroys the structural morphology of a photocatalyst, thus decreasing its performance for PFAS degradation.

## 19.3.7 Effects of photocatalyst properties and composition constituents

The characteristic properties of a catalyst play a crucial role in its capacity to perform photocatalytic activity (Akinnawo and Ediagbonya, 2025; Navidpour, 2023). The characteristic property of a photocatalyst, such as the band gap, is significant in the region of light absorption and harvesting. For instance, Fu et al. (2022) attributed the photocatalytic performance of a photocatalyst to be proportional to its characteristic band gap. As noted by Fu et al. (2022), the PFOA photocatalytic degradation performance of some investigated photocatalysts followed the sequence of $Ga_2O_3 > TiO_2 > CeO_2 > In_2O_3 >$ CdS, which was posited to directly match the energy band gap from largest to narrowest of the investigated photocatalysts as $Ga_2O_3 > TiO_2 > CeO_2 > In_2O_3 >$ CdS. The role of the composition constituent of a photocatalyst is crucial in its efficiency to degrade PFAS in water. According to the study outcome by Boonchata et al. (2024), an increase in the dosage of graphene oxide constituent (from 0 to 25 wt.%) integrated with $TiO_2$ to form titanium dioxide/graphene oxide nanocomposite has been noted to yield a decline in the residual fraction of PFOA and PFOS concentration remaining in water after photocatalytic degradation, as illustrated in Fig. 19.7. However, a further increase in the dosage of incorporated graphene oxide (beyond

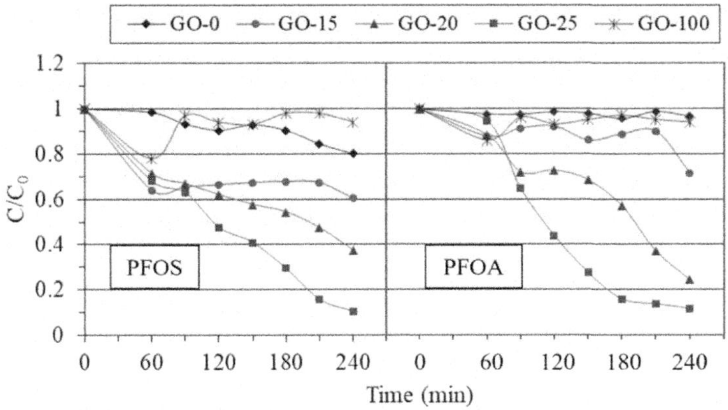


**FIGURE 19.7** The effect of graphemic oxide composition in titanium dioxide/graphene oxide nanocomposite photocatalyst for degradation of PFOS and PFOA from water. Source: *Reprinted from Boonchata, P., Boontanon, N., Jindal, V., Aung, H.K.Z., Visvanathan, C., Fujii, S., et al., 2024. Photocatalytic degradation of perfluorooctane sulfonic acid and perfluorooctanoic acid using titanium dioxide/graphene oxide nanocomposite immobilized on polyvinyl alcohol film. Case Studies in Chemical and Environmental Engineering. 10, 100862. https://doi.org/10.1016/j.cscee.2024.100862, with permission from Elsevier.*

25 wt.%) with $TiO_2$ to form titanium dioxide/graphene oxide nanocomposite increased the residual fraction of PFOA and PFOS concentration that remains in water after photocatalytic degradation. The increased residual fraction of PFOA and PFOS concentration that remains in water after photocatalytic degradation may be attributed to the decrease in the quantity of light absorption and harvesting due to the clogging of graphene oxide at higher constituent dosage. Thus leading to lower degradation efficiency of the photocatalyst prepared with 100 wt.% graphene oxide constituents. Similar findings have been detailed by Moreira et al. (2024), with an increase in wt.% Fe content in the hybrid graphemic and iron oxide photocatalyst, leading to an increase in photocatalytic PFOA degradation.

### 19.3.8 Effects of radicals as reductants and oxidants

Radicals play a crucial part in the photocatalytic degradation of PFAS, on account of their ability to directly attack and break the strong C–F bonds in PFAS molecules. These radicals often function as oxidants and reductants, and they can be activated using UV irradiation to generate secondary reactive species to enhance photocatalytic degradation and removal of PFAS from water. The commonly used oxidants include iodate ($NaIO_4$), hypochlorite (NaOCl), bromate ($NaBrO_3$), hydroxyl ($H_2O_2$), and sulfate ($Na_2S_2O_8$), while sulfite is among the commonly used reductants for photocatalytic PFAS degradation (Chowdhury et al., 2021). Radicals such as $^{\cdot}OH$ and $^{\cdot}O_2^-$ are ROS that are generated when electron-hole pairs react with oxygen or other molecules in the aqueous solution. However, the efficiency of radicals such as $^{\cdot}OH$ can be affected by quenching when they react with other molecules in the solution. Thus leading to a significant reduction in the quantity of $^{\cdot}OH$ required to induce the breakdown of PFAS molecules. Moreover, the ability to generate ROS is contingent on the nature of the photocatalyst. Thus different photocatalysts exhibit varying abilities to generate ROS (Boonchata et al., 2024; Shao et al., 2017).

According to Shao et al. (2017) and Sabri et al. (2023), sulfate radicals ($SO_4^{\bullet -}$) can be generated from persulfate (PS) and PMS oxidants using UV irradiation and have the ability to increase the efficiency of a photocatalyst to degrade organic pollutants. Xu et al. (2020c) recounted the successful activation of PMS using 0.25 g $L^{-1}$ of $TiO_2$ for photocatalytic PFOA degradation under visible light irradiation for a duration of 8 h. Likewise, Navidpour (2023) narrated an enhanced photocatalytic PFOA degradation in the presence of PMS (as an oxidant) under UV/visible irradiated light. The author indicated that a low dose of PMS as an oxidant for ZnO resulted in low photocatalytic PFOA degradation, while an increase in the dosage of PMS, along with the use of ZnO, increased the percentage photocatalytic degradation of PFOA. This outcome suggests a synergistic effect between PMS and ZnO for enhanced photocatalytic degradation and mineralization of PFAS.

Similarly, Park et al. (2024) attributed the generation of strong oxidant $^{\cdot}OH$ radical to the enhanced photocatalytic degradation of PFOS and PFNA at 97% and 93%, respectively, using a bimetallic catalyst of $Fe_2O_3/Mn_2O_3$ photocatalyst within a 10 min duration. Chowdhury et al. (2021) analyzed the effect of PS oxidant on photocatalytic PFOA and PFOS degradation and defluorination using $TiO_2$ under UV irradiation, as described in Fig. 19.8A and C. The authors obtained higher degradation and percentage defluorination of PFOA using PS as oxidant along with $TiO_2$ under UV irradiation, compared to the sole use of PS, $TiO_2$, and UV irradiation. According to the report by Chowdhury et al. (2021), PS/$TiO_2$/UV irradiation produces various radicals of sulfate radicals (SRs) and hydroxyl

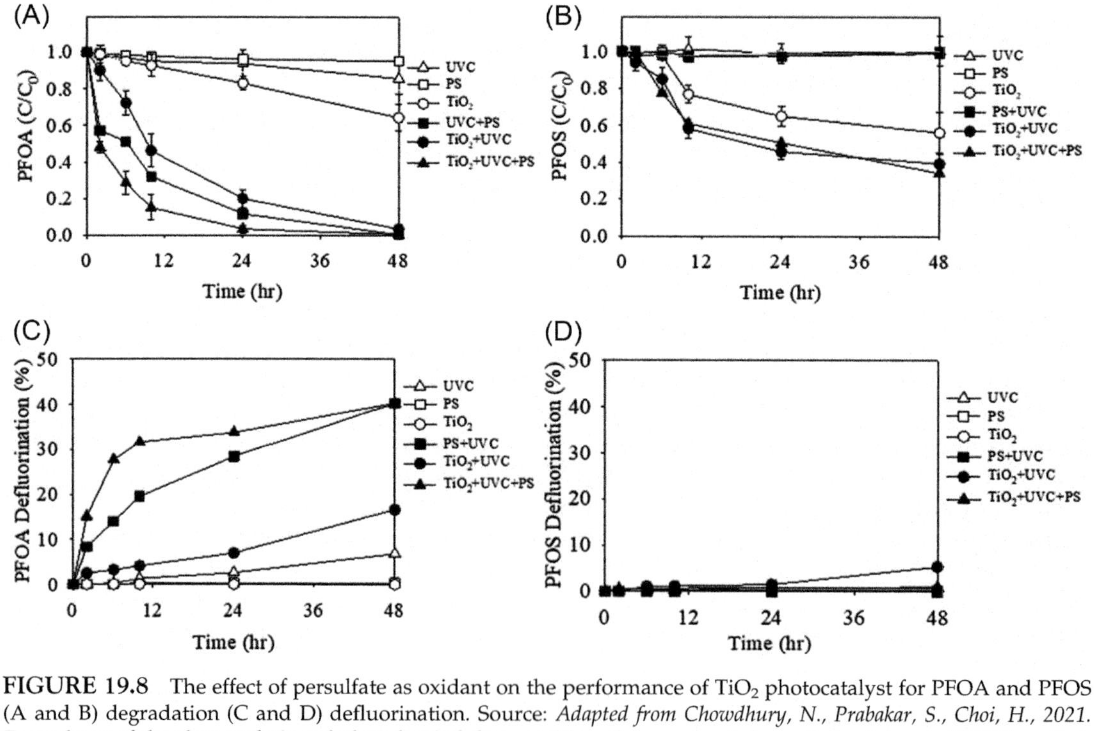


**FIGURE 19.8** The effect of persulfate as oxidant on the performance of $TiO_2$ photocatalyst for PFOA and PFOS (A and B) degradation (C and D) defluorination. Source: *Adapted from Chowdhury, N., Prabakar, S., Choi, H., 2021. Dependency of the photocatalytic and photochemical decomposition of per- and polyfluoroalkyl substances (PFAS) on their chain lengths, functional groups, and structural properties. Water Science & Technology. 84 (12), 3738–3754. https://doi.org/10.2166/wst.2021.458, under a Creative Commons license (CC BY 4.0).*

radicals (HRs), while PS/UV irradiation and $TiO_2$/UV irradiation respectively, produce SRs and HRs. A previous study outcome by Liang and Su (2009) pointed out that SRs are better species in comparison to HRs for direct electron transfer reaction to enhance the degradation of PFAS and their reaction by-products. Thus this denotes the role of PS as an oxidant in enhancing photocatalytic PFOA degradation and removal from water. However, PS/UV irradiation was not effective for photocatalytic degradation of PFOS as highlighted by Chowdhury et al. (2021), in Fig. 19.8B and D.

Comparatively, Chowdhury et al. (2021) demonstrated the effect of sulfite as a reductant for photocatalytic PFOA and PFOS degradation and defluorination from water using $TiO_2$ under UV irradiation. The study outcome revealed that the sole use of sulfite, $TiO_2$, and UV irradiation was not effective for PFAS degradation and defluorination as portrayed in Fig. 19.9. However, using sulfite as a reductant radical resulted in enhanced degradation and defluorination of PFOA and PFOS.

## 19.3.9 Effects of salinity (salt) and organic matter

The presence of salinity and organic matter has a significant impact on the photocatalytic degradation and mineralization of PFAS. For instance, the presence of salt may enhance the surface adsorption of PFAS molecules on the photocatalyst via electrostatic interaction. Thus leading to an improved degradation of PFAS (Akinnawo, 2024a). As a result, an increase in

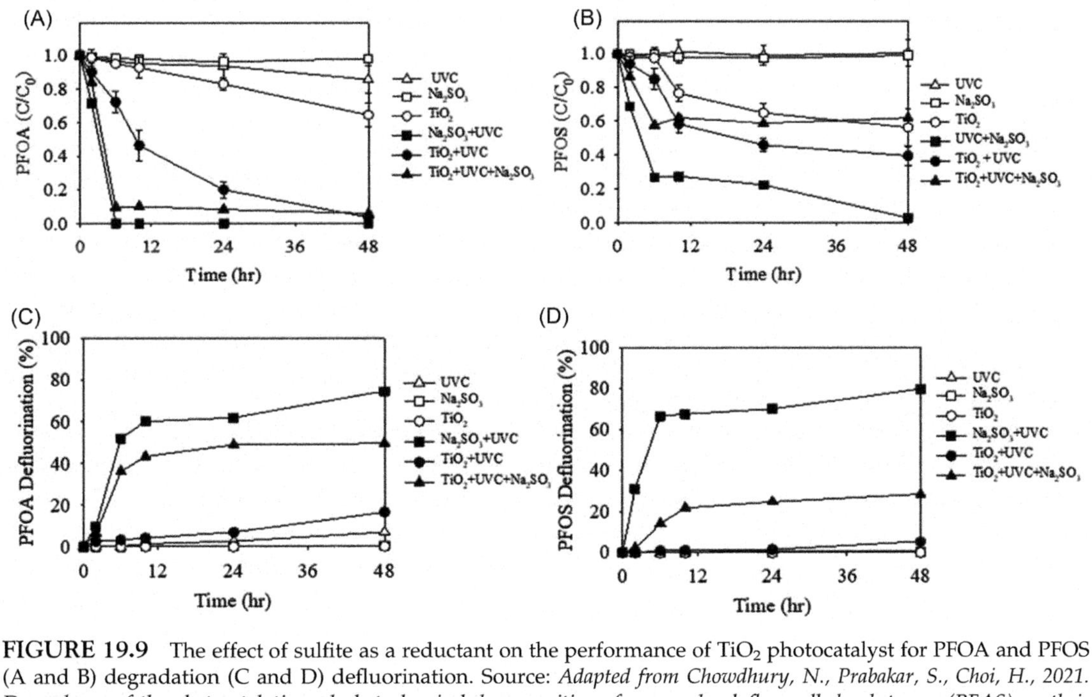


**FIGURE 19.9** The effect of sulfite as a reductant on the performance of $TiO_2$ photocatalyst for PFOA and PFOS (A and B) degradation (C and D) defluorination. Source: *Adapted from Chowdhury, N., Prabakar, S., Choi, H., 2021. Dependency of the photocatalytic and photochemical decomposition of per- and polyfluoroalkyl substances (PFAS) on their chain lengths, functional groups, and structural properties. Water Science & Technology. 84 (12), 3738–3754. https://doi.org/10.2166/wst.2021.458, under a Creative Commons license (CC BY 4.0).*

salinity will lead to a stronger interaction between the photocatalyst and PFAS molecule, thus improving the adsorption of PFAS on the photocatalyst and facilitating photocatalytic degradation of PFAS. Conversely, organic matter competes with PFAS molecules for surface adsorption on the active sites of the photocatalyst. Moreover, the adsorption of organic matter on the surface of a photocatalyst results in a hydrophobic interaction between the photocatalyst and molecules of PFAS (Akinnawo, 2024a). Thus repelling surface interaction with PFAS molecules and inhibiting the photocatalyst's efficiency to degrade PFAS molecules.

## 19.4 Mechanism for photocatalytic degradation of per- and polyfluoroalkyl substances from water

Catalysts are essential for the photocatalytic degradation of PFAS in the presence of light to generate the required reactive species for the breakdown of these pollutants. The mechanism for photocatalytic PFAS degradation typically consists of photo-oxidation and photo-reduction, which leads to C–C bond cleavage, which is the major (most important) pathway for PFAS degradation, while C–F bond breaking is considered to be a minor pathway for PFAS degradation. The photo-oxidation and photo-reduction are contrasting photocatalytic processes that are initiated by light absorption and characterized by electron transfer in opposite pathways to respectively obtain photogenerated holes ($h^+$) in the valence band as well as photogenerated electrons in the conduction band.

As portrayed in Fig. 19.10, the PFAS photocatalytic degradation mechanism is initiated by the absorption of irradiated light (specifically in the UV region) by the photocatalyst, for the excitation and migration of electrons from the valence band region to the conduction band region. This resulting photoexcitation step is preceded by the migration of photogenerated electron-hole pairs to the surface of the photocatalyst, where they react with adsorbed water and oxygen to produce $^{\cdot}OH$ and $^{\cdot}O_2^-$ as ROS. Then the ROS react with PFAS molecules, resulting in C–C bond cleavage and C–F bond breaking, either by photo-oxidation or photo-reduction as depicted in Fig. 19.10. The photo-oxidation proceeds by the direct reaction of holes ($h^+$) and oxidative radicals such as $^{\cdot}OH$ radicals with PFAS molecules or water to generate more $^{\cdot}OH$ radicals. Thus ensuing in the degradation of PFAS molecules via the cleavage of C–C bonds and the breaking of C–F bonds.

As described in Fig. 19.10, photo-oxidation mainly results in C–C bond cleavage, which is often followed by C–F bond cleavage, as ensuing electron transfer during photo-oxidation results in the formation of unstable intermediate perfluoroalkyl radicals ($C_nF_{2n+1}^{\bullet}$) via the Kolbe decarboxylation process that breaks down C–C bonds (Alam et al., 2023; Mazen et al., 2025). Furthermore, the formed strong oxidizing agents ($h^+$), and radicals of hydroxyl (OH), and sulfate ($\bullet SO_4^-$), obtained from photo-oxidation, react with water to produce intermediate fluorinated alcohols that can undergo defluorination via hydrogen fluoride elimination and as a result contribute to an indirect cleavage of the C–F bond (Esfahani et al., 2022; Kim et al., 2025). Thus shortening the carbon chain length of PFAS by a $CF_2$ unit in each degradation and defluorination cycle and a consequential mineralization of PFAS into $CO_2$ and fluoride ions as depicted in Fig. 19.10.

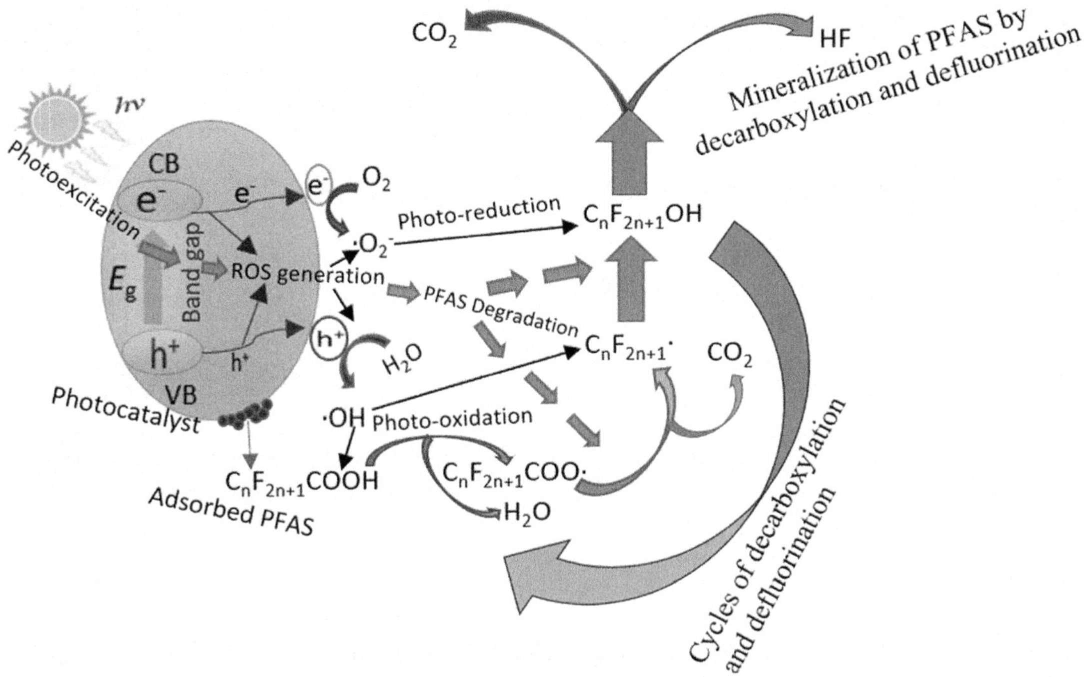


**FIGURE 19.10** Proposed photocatalytic mechanism pathway for per- and polyfluoroalkyl substances degradation.

Complementary, photo-reduction proceeds via the direct interaction between electrons or reductive species such as $\cdot O_2^-$ with PFAS molecules or water for the degradation of PFAS as described in Fig. 19.10. A direct cleavage of C–F bond takes place under the photo-reduction process due to the influence of a strong reducing agent such as hydrated electrons ($e_a q^-$) in a direct attack on the C–F bond that is specifically located at the alpha-position adjacent to the head of the PFAS molecules. This direct attack weakens the C–F bond, consequently facilitating hydro-defluorination, which is regarded as the process of cleavage and replacement of fluorine atoms with hydrogen atoms. However, the process of hydro-defluorination results in defluorination of PFAS via fluoride ions elimination and the formation of hydrofluorinated by-products (Verma et al., 2024; Esfahani et al., 2022, 2024).

## 19.5 Mineral/metal oxide-based photocatalyst

### 19.5.1 $TiO_2$-based photocatalyst

Extensive studies on using $TiO_2$-based photocatalysts have revealed higher photocatalytic PFAS removal from water using modified $TiO_2$, in comparison to pristine $TiO_2$, as portrayed in Table 19.1. As described by Panchangam et al. (2009), a lower percentage photocatalytic PFOA removal of 15% and 30% was obtained using $TiO_2$ and P25 $TiO_2$. Comparatively,

**TABLE 19.1** Photocatalytic degradation and elimination of per- and polyfluoroalkyl substances from water.

| Photocatalyst | Condition | Pollutants | Efficiency | References |
|---|---|---|---|---|
| $TiO_2$ | t (12 h) | PFOA | 15% | Panchangam et al. (2009) |
| P25 $TiO_2$ | t (3 h) | | 30% | Chen et al. (2011) |
| Titanate nanotubes | t (24 h) | | 59% | Zhu et al. (2020) |
| $TiO_2$ QD/graphene | t (10 h) | | 90% | |
| $TiO_2$-MWCNT | Dose (1.6 g $L^{-1}$), t (7 h), I.C (30 mg $L^{-1}$), pH (5), 300 W lamp source at 365 nm | PFOA | 100% | Song et al. (2012) |
| $TiO_2$-Oxalic acid | t = 3 h | PFOA | 86.7% | Wang and Zhang (2011) |
| | | $Fl^-$ | 16.5% | Panchangam et al. (2009) |
| $TiO_2$-$HClO_4$ | t = 7 h | PFOA | 97% | |
| | | $Fl^-$ | 38% | |
| $TiO_2$-MIP | t = 10 h | PFOA | 81% | Wu et al. (2016) |
| | | $Fl^-$ | 30.2% | |
| $TiO_2$-rGO | Dosage (0.1 g $L^{-1}$), pH (3.8), t (12 h), I.C (100 mg $L^{-1}$), 150 W lamp source at 200–600 nm | PFOA | 93% | Gomez-Ruiz et al. (2018) |
| | | $Fl^-$ | 20% | |
| $TiO_2$-Ag | | PFOA | 57.7% | Chen et al. (2015) |
| $TiO_2$-Pd | Dosage (0.5 g $L^{-1}$), t (7 h), I.C (144.9 μM), pH (3) | $Fl^-$ | 8.1% | Li et al. (2016) |
| | | PFOA | 94.2% | |
| | | $Fl^-$ | 25.9% | |
| $TiO_2$-Pt | | PFOA | 100% | |
| | | $Fl^-$ | 34.8% | |
| $TiO_2$-Fe | Dosage (0.5 g $L^{-1}$), t (7 h), I.C (121 μM), pH (5), 400 W lamp source at 254 nm | PFOA | 60% | Chen et al. (2015) |
| $TiO_2$-Cu | dosage (0.5 g $L^{-1}$), t (4.5 h), I.C (121 μM), pH (5) | PFOA | 91% | |
| $TiO_2$-Pb | | PFOA | 90% | Chen et al. (2016) |
| | | $Fl^-$ | 22.4% | |
| Pb-$BiFeO_3$/rGO | Dosage (100 mg $L^{-1}$), pH (2), I.C (50 ppm), t (8 h) | PFOA | 69.6% | Shang et al. (2018) |
| | | $Fl^-$ | 37.6% | |
| Fe/TNTs@AC | Dose (1 mg $L^{-1}$), t (4 h), I.C (0.1 mg $L^{-1}$), pH (7), 18 W lamp at 254 nm t (4 h), under exposed UV light irradiation | PFOA | 90% | Li et al. (2020a,b) |
| | | $Fl^-$ | 62% | |
| Ga/TNTs@AC | | PFOS | 75% | Zhu et al. (2021) |
| | | $Fl^-$ | 66.2% | |

*(Continued)*

**TABLE 19.1** (Continued)

| Photocatalyst | Condition | Pollutants | Efficiency | References |
|---|---|---|---|---|
| β-$Ga_2O_3$ | | PFOA | 36% | Zhao and Zhang (2009) |
| $Ga_2O_3$ needle-like | Dosage (0.5 g $L^{-1}$), t (3 h), I.C (40, 0.5 mg $L^{-1}$), 15 W light source (254 nm) | $Fl^-$ | 15% | Shao et al. (2013a,b) |
| $Ga_2O_3$ sheaf-like | | PFOA | 100% | Leonello et al. (2021) |
| | | $Fl^-$ | 58% | |
| | | PFOA | 100% | |
| | | $Fl^-$ | 60% | |
| β-$Ga_2O_3$ nanorods | Dosage (0.5 g $L^{-1}$), t (1 to 1.5 h), I.C (0.5 mg $L^{-1}$) | PFOA | 98.8% | Zhao et al. (2015) |
| In-doped $Ga_2O_3$ nanosheets | | PFOA | 100% | Tan et al. (2020) |
| $In_2O_3$ | Dosage (3.4 g $L^{-1}$), t (4 h), I.C (40 mg $L^{-1}$), 23 W light source (254 nm) | PFAS | 75% | Li et al. (2012a) |
| | | $Fl^-$ | 33% | |
| $In_2O_3$ nanocubes | Dosage (0.5 g $L^{-1}$), t (2, 17 and 42 min), I.C (30 mg $L^{-1}$) | PFOA | 100% | Li et al. (2012b) |
| $In_2O_3$ microsphere | | PFOA | 100% | Li et al. (2014) |
| $In_2O_3$ nanoplates | | PFOA | 100% | Liu et al. (2021) |
| $In_2O_3$ nanoporous nanosphere | Dosage (0.5 g $L^{-1}$), I.C (30 mg $L^{-1}$), t (30 min) | PFOA | 100% | Li et al. (2012b) |
| | | PFOA | 100% | Li et al. (2014) |
| $In_2O_3$ porous nanoplates | | | | Li et al. (2013a) |
| $In_2O_3$/graphene composite | Dosage (0.5 g $L^{-1}$), t (3 h), I.C (30 mg $L^{-1}$), 15 W UV lamp source at 254 nm | PFOA | 90%–100% | Li et al. (2013b) |
| MOF-derived $In_2O_3$ nanospheres | | PFAS | 100% | Leonello et al. (2021) |
| $In_2O_3/MnO_x$ | Dosage (0.5 g $L^{-1}$), t (3 h), I.C (50 mg $L^{-1}$), pH (3.8), 500 W UV lamp source | PFOA | 100% | Wu et al. (2019) |
| $In_2O_3/Ce_2O_3$ | Dosage (0.4 g $L^{-1}$), t (1 h), I.C (100 mg L-1), pH (2.83), 500 W UV lamp source | PFOA | 100% | Jiang et al. (2016) |
| InOOH | Dosage (0.25 g $L^{-1}$), t (3 h), I.C (20 mg $L^{-1}$) | PFAS | 83.4% | Xu et al. (2017a,b) |
| Pt/$In_2O_3$ nanorod | Dosage (0.4 g $L^{-1}$), t (1 h), I.C (200 ppm) | PFOA | 98% | Xu et al. (2017a,b) |
| BiOCl | Dosage (0.5 g $L^{-1}$), t (12 h), pH (4.8), I.C (20 μM) | PFAS | 100% | Song et al. (2017) |
| | | $Fl^-$ | 59.3% | |
| BiOCl with oxygen vacancies | Dosage (1 g $L^{-1}$), pH (2), I.C (50 μM), t (8.5 h) | PFAS | 68.8% | Song et al. (2019) |
| | | $Fl^-$ | 12.86% | |
| BiOF | t (6 h), I.C (15 mg $L^{-1}$) | PFOA | 100% | Wang et al. (2021a) |
| | | $Fl^-$ | 56.8% | |

*(Continued)*

**TABLE 19.1** (Continued)

| Photocatalyst | Condition | Pollutants | Efficiency | References |
|---|---|---|---|---|
| In-MOF/BiOF | Dosage (0.5 g $L^{-1}$), I.C (15 mg $L^{-1}$), t (3 h) | PFOA | 100% | Wang et al. (2022) |
| | | $Fl^-$ | 34% | |
| BiOI | t (3 h), I.C (100 mg $L^{-1}$) | PFOA | 35% | Khan et al. (2023) |
| | | $Fl^-$ | 20% | |
| BiOI/*Cunninghamella elegans* | t (3 h), I.C (100 mg $L^{-1}$) | PFOA | 90% | Khan et al. (2023) |
| | | $Fl^-$ | 60% | |
| $BiOI_{0.95}Br_{0.05}$ | Dosage (0.4 g), t (2 h), I.C (20 mg $L^{-1}$) | PFOA | 100% | Li et al. (2020a,b) |
| | | $Fl^-$ | 65% | |
| $Bi/BiOI_{1-x}F_x$ composite | Dosage (0.4 g), pH (3 to 5), I.C (40 mg $L^{-1}$), t (2 h) | PFOA | 100% | Wang et al. (2021b) |
| | | $Fl^-$ | 10% | |
| $BiOI@Bi_5O_7I$ | Dosage (0.5 g), pH (3), I.C (15 mg $L^{-1}$), t (6 h) | PFOA | 81% | Wang et al. (2020a) |
| | | $Fl^-$ | 60% | |
| $BiOI_{1-x}$ | t (5 h), 5 W LED irradiation | PFOA | 51.9% | Cao et al. (2024) |
| BiOI/ZnO composite | Dosage (0.5 g), I.C (15 mg $L^{-1}$), t (6 h), pH (4) | PFOA | 91% | Yang et al. (2021) |
| | | $Fl^-$ | 52.2% | |
| BiOCl/Zn-Al hydrotalcite | Dosage (0.5 g $L^{-1}$), pH (2), I.C (500 μg $L^{-1}$), t (6 h) | PFOA | 90% | Yang et al. (2020) |
| $Bi_3O(OH)(PO_4)_2$ Microparticle | Dosage (1.8 g $L^{-1}$), pH (4), I.C (54 mg $L^{-1}$), t (2 h) | PFOA | 100% | Sahu et al. (2018) |
| | | $Fl^-$ | 60% | |
| $Bi_3O(OH)(PO_4)_2$ Microparticle/carbon spheres composite | Dosage (1 g $L^{-1}$), pH (7), I.C (0.2 mg $L^{-1}$), t (4 h) | PFOA | 100% | Xu et al. (2020a) |
| | | $Fl^-$ | 32.5% | |
| $Bi_2O_2S$ | Dosage (0.25 g $L^{-1}$), pH (4), I.C (50 mg $L^{-1}$), t (10 h) | PFOA | 99% | Gu et al. (2022) |
| | | $Fl^-$ | 60.6% | |
| $Bi_2O_2Se$ | Dosage (0.25 g $L^{-1}$), pH (4), I.C (50 mg $L^{-1}$), t (10 h) | PFOA | 99% | Gu et al. (2022) |
| | | $Fl^-$ | 54.2% | |
| $Pt–Bi_2O_4$ | Dosage (200 mg $L^{-1}$), t (6 h) I.C (50 ppm) | PFOA | 45.0% | Wang et al. (2020b) |
| | | $Fl^-$ | 25% | |
| Pt-SiC | Dosage (0.5 g $L^{-1}$), pH (7), I.C (0.1 ppm), t (1.5 h) | PFOA | 99% | Huang et al. (2018) |
| | | $Fl^-$ | 50.6% | |
| $Ti3C_2/TiO_2$ | Dosage (0.2 g $L^{-1}$), I.C (8.82 mg $L^{-1}$), pH (3), t (16 h) | PFOA | 99% | Shang et al. (2018) |
| | | $Fl^-$ | 49% | |
| Boron nitride | Dosage (0.5 g $L^{-1}$), t (4 h), I.C (50 ppm), pH = 3.2, | PFOA | 100% | Duan et al. (2020) |
| | | $Fl^-$ | 52% | |

*(Continued)*

TABLE 19.1 (Continued)

| Photocatalyst | Condition | Pollutants | Efficiency | References |
|---|---|---|---|---|
| $TiO_2$-BN composite | Dosage (0.5 g $L^{-1}$), t (7 h), I.C (50 ppm), pH (3.2), | PFOA | 99% | Duan et al. (2020) |
| | | $Fl^-$ | 68% | |
| FeO/CS | Dosage (1 g $L^{-1}$), t (4 h), I.C (0.2 ppm), pH (7) | PFOA | 95.2% | Xu et al. (2020b) |
| | | $Fl^-$ | 57.2% | |
| $Sb_2O_3/TiO_2$ heterojunction | Dosage (0.25 g), t (120 min), I.C (10 mg $L^{-1}$) | PFOA | 81.7% | Yao et al. (2021) |
| ZnO | Dosage (1.06 g $L^{-1}$) under UV irradiation. | PFOA | 64% | Navidpour, (2023) |
| ZnO-rGO | Dosage (100 mg $L^{-1}$), t (1 h) I.C (100 ppm) | PFOA | 90.9% | Ong et al. (2017) |
| Zn@CA-modified GCN | Dosage (530 mg $L^{-1}$), I.C (53 ppm) 300 xenon lamp (254 nm UV light) | PFOA | 54% | Navidpour, (2023) |
| | | | 59% | |
| Zn@CA-modified GCN/PS | Dosage (530 mg $L^{-1}$), I.C = 53 ppm 300 xenon lamp (254 nm UV light) | PFOA | 74% | Navidpour, (2023) |
| | | | 79% | |
| $CeO_2$@Ni–Al LDH | Dosage (0.5 g $L^{-1}$), t (5 h), I.C (50 ppm) | PFOA | 90.2% | Tang et al. (2021) |
| | | $Fl^-$ | 87.3% | |
| $Fe_3O_4@SiO_2$-BiOBr | Dosage (100 g $L^{-1}$), t (1 h), I.C (20 ppm) | PFOA | 92.9% | Wang et al. (2017) |
| | | $Fl^-$ | 32.8% | |
| Iron oxide/graphenic carbon (Fe/g-C) | Dosage (1 g $L^{-1}$), pH (3.5), I.C (1 ppm), t (30 min) | PFOA | 85% | Moreira et al. (2024) |
| Bimetallic $Fe_2O_3/Mn_2O_3$ | Dosage (0.1 g $L^{-1}$), t (2–20 min), I.C (2 ppm), pH = 7.3 | PFOS | 97% | Park et al. (2024) |
| | | PFNA | 80%–93 % | |
| Fe-loaded zeolites (Fe-BEA35) | pH (5.5), t (24 h) | PFOA | 90% | Qian et al. (2021) |
| $Fe_3O_4$/ZnO@Ag | pH (3) | PFOS | 87.37% | Pouresmaeil et al. (2005) |
| SiC@Graphene | Dosage (0.5 g $L^{-1}$), pH (7), I.C (50 ppm), t (6 h) | PFOA | 60% | Huang et al. (2016) |

Chen et al. (2011) demonstrated the effect of nano-sized modification in attaining a higher percentage (59%) degradation of PFOA from water using titanate nanotubes. Similarly, higher photocatalytic PFOA degradation of 90% and 100% has been respectively obtained using $TiO_2$-multiwall carbon nanotube ($TiO_2$-MWCNT) composites and $TiO_2$ quantum dots loaded sulfonated graphene aerogel ($TiO_2$ QD/graphene). The boosted photocatalytic PFOA degradation may be attributed to the obtained higher surface area that resulted from incorporating QD/graphene and MWCNT into $TiO_2$ (Zhu et al., 2020; Song et al., 2012). Similar outcome on the effect of higher surface area in enhancing photocatalytic effectiveness of $TiO_2$ attributable to the incorporation of modifying agents such as activated carbon, reduce graphene oxide have been respectively recounted by Wu et al. (2016), Li et al. (2020b), and Gomez-Ruiz et al. (2018) in accomplishing 81%, 90%, 93% degradation and 62% defluorination of PFOA from water using $TiO_2$-MIP, Fe/TNTs@AC and $TiO_2$-rGO.

As posited by Akinnawo and Ediagbonya (2025), surface morphology modification via cocatalyst deposition or elemental doping is effective in boosting the photocatalytic degradation of organic pollutants from water. On this note, Chen et al. (2015), and Li et al. (2016), deployed cocatalyst deposition of Ag, Pd in addition to Pt on $TiO_2$ to develop $TiO_2$-Ag, $TiO_2$-Pd as well as $TiO_2$-Pt with respectively percentage degradation of 57%, 94.4% and 100% for PFOA removal from water under operational condition of initial concentration (144.9 μM), time (7 h), dose (0.5 g $L^{-1}$) and pH (3) using 135 W lamp source at 365 nm. Further studies on transition-metal modified $TiO_2$, as investigated by Chen et al. (2015), reportedly obtained 60% and 91% photocatalytic degradation of PFOA by utilizing $TiO_2$-Fe and $TiO_2$-Cu, respectively. However, using $TiO_2$-Pb as described by Chen et al. (63), was effective in obtaining 90% photocatalytic degradation of PFOA from water. A study by Wang and Zhang (2011) revealed an enhanced photocatalytic performance of $TiO_2$ in obtaining 97% degradation of PFOA due to oxalic acid modification, compared to pristine $TiO_2$. This photocatalytic efficiency may be an outcome of the generation of higher quantity of ROS due to oxalic acid modification.

Heterojunction is another modification strategy that has been demonstrated to enhance the photocatalytic effectiveness of a $TiO_2$-based photocatalyst. For instance, the development of mesoporous $Sb_2O_3/TiO_2$ heterojunctions for photocatalytic PFOA degradation has been described by Yao et al. (2021), to yield 81.7% degradation of PFOA in operational time of 120 min. Moreover, $Sb_2O_3/TiO_2$ heterojunctions exhibited 4.2 times faster PFOA degradation rate in comparison to using commercial P25 and unmodified mesoporous $TiO_2$ for PFOA removal from water.

### 19.5.2 $In_2O_3$-based photocatalyst

$In_2O_3$ is categorized as a metal oxide with significant photocatalytic activity towards the degradation and elimination of organic pollutants from water. However, the wide band gap exhibited by $In_2O_3$ has been noted to restrict its photocatalytic activities due to poor light harvesting and recombination of obtainable photogenerated electron/hole pairs. Thus pristine $In_2O_3$ exhibits lower photocatalytic activity in comparison to the modified $In_2O_3$-based photocatalyst. According to a study by Li et al. (2012a), at an initial PFAS concentration (40 mg $L^{-1}$), in addition to pH (5), dose (3.4 g $L^{-1}$), time (4 h) under 23 W light source at a wavelength of 254 nm, 75% degradation and 33% defluorination of PFAS was obtained using pristine $In_2O_3$. A similar outcome on using a wide band gap p-block metal oxyhydroxide, InOOH, has been recounted in obtaining 83% degradation of PFOA at an initial concentration (20 mg $L^{-1}$), in addition to dose (0.25 g $L^{-1}$) and time (3 h) under an 18 W light source at 254 nm (Xu et al., 2017a,b). However, modification strategies that generated nanostructured $In_2O_3$-based photocatalysts have been described in obtaining higher photocatalytic PFAS degradation from water, as portrayed in Table 19.1. For instance, complete (100%) degradation and elimination of PFOA from water have been reported with the use of $In_2O_3$ microspheres, $In_2O_3$ nanocubes, $In_2O_3$ nanoplates, $In_2O_3$ nanoporous nanospheres, and $In_2O_3$ porous nanoplates (Li et al., 2012b; 2013a, 2014; Liu et al., 2021). A similar outcome on obtaining complete degradation and elimination of PFOA from water has been recounted using metal-organic frameworks-derived $In_2O_3$ nanospheres (Leonello et al., 2021). Further, $In_2O_3$-based photocatalytic composites have been noted to exhibit enhanced degradation of PFOA. This outcome has been ascribed to the synergy between the composite materials. Research outcomes on photocatalytic composite materials such as $In_2O_3$/graphene, $In_2O_3/Ce_2O_3$, and $In_2O_3/MnO_x$ have been noted to produce complete photocatalytic PFAS degradation from water, as shown in Table 19.1 (Li et al., 2013b; Jiang et al., 2016; Wu et al., 2019). Similar findings have been narrated by Xu et al. (2017a,b), in obtaining nearly complete degradation of PFOA using platinum modified indium oxide nanorods ($Pt/In_2O_3$ nanorod) at a photocatalyst dosage of 0.4 g $L^{-1}$, initial PFOA concentration of 200 ppm in 1 h contact time.

### 19.5.3 $Ga_2O_3$-based photocatalyst

The photocatalytic performance of $Ga_2O_3$ is contingent on its band gap and surface morphological structure. For instance, Zhao and Zhang (2009), attributed the low percentage degradation (36%) and defluorination (15%) of PFOA to the characteristic wide band gap exhibited by $\beta$-$Ga_2O_3$ photocatalyst. Comparatively, the morphological nanostructured $\beta$-$Ga_2O_3$ photocatalyst has been used to obtain a higher percentage (98.8%) degradation and removal of PFOA from water (Zhao and Zhang, 2009; Shao et al., 2013a,b). Thus modifying the morphological structure of $Ga_2O_3$-based photocatalysts is a significant methodology to enhancing photocatalytic activity for eliminating PFAS from water, as described in Table 19.1. According to the findings by Shao et al. (2013a,b), complete (100%) degradation of PFOA along with 58% and 60% defluorination can be obtained using $Ga_2O_3$ needle-like and $Ga_2O_3$ sheaf-like photocatalysts, respectively. Another significant methodology to improve the photocatalytic activity of $Ga_2O_3$ is via doping with semiconductors such as indium to produce Indium-modified $Ga_2O_3$. According to Tan et al. (2020), the complete degradation (100%) and removal of PFOA from water using Indium-modified $Ga_2O_3$ hierarchical nanosheets may be attributed to the synergistic interaction between Indium and $Ga_2O_3$ nanosheets. Conversely, lower percentage degradation (75%) and defluorination (66.2%) of PFOS have been recounted utilizing gallium-doped carbon-modified titanate nanotubes (Ga/TNTs@AC) (Zhu et al., 2021). This outcome may be ascribed to the adsorption-desorption interaction between the sulfonate functional group of PFOS and the carbon-modified titanate nanotubes, which may result in the reintroduction of PFOS via desorption into the treated water.

### 19.5.4 Bismuth-based photocatalysts

Bismuth-based catalysts (Bi-based NPs) are emerging catalysts that exhibit high physical and chemical stability along with exceptional catalytic performance in the degradation and elimination of organic pollutants from wastewater. Yet, their performance can be influenced by experimental conditions, and enhanced by modification strategies such as incorporating other elements to enhance absorption of irradiated light in the visible region (Akinnawo and Ediagbonya, 2025). A study by Yang et al. (2021) reportedly developed flower-shaped

$Bi_5O_7I/ZnO$ n-n heterojunction microspheres with enhanced visible-light absorption potential for respective 91% degradation and 52.2% defluorination of PFOA in a contact duration of 6 at 0.0002 $min^{-1}$ rate constant under optimized working conditions of pH 4, dosage (0.5 g), and initial PFOA concentration (1 mg $L^{-1}$).

Similarly, Wang et al. (2020a,b) developed an effective p-n $BiOI@Bi_5O_7I$ heterojunction photocatalyst for enhanced PFOA degradation under operational conditions of simulated solar light irradiation, and accomplished 60% defluorination and a high percentage degradation (81%) of PFOA under experimental conditions of solution pH 3, 0.5 g catalyst dosage, and 15 mg $L^{-1}$ of initial PFOA concentration. The use of elemental bromine to dope BiOI, as demonstrated by Li et al. (2020a), yielded $BiOI_{0.95}Br_{0.05}$ with a total (100%) degradation and 65% defluorination of PFOA at a faster rate constant (0.0205 $min^{-1}$). In comparison to the singular use of BiOI and BiOF photocatalysts, which yielded 66% and 100% degradation of PFOA from water at lower rate constant. Wang et al. (2021a,b) prepared a $Bi/BiOI_{1-x}F_x$ composite with a characteristic lower band gap to obtain complete (100%) degradation of PFOA from water at a higher activity of 0.0375 μmol$g^{-1}$ $h^{-1}$. This outcome is a result of the enhanced light-harvesting capacity of $Bi/BiOI_{1-x}F_x$ composite, which is a function of the loaded Bi nanoparticles into a blend of integrated bismuth oxyhalides that contains BiOI with BiOF.

Furthermore, photocatalytic reaction using 1 g $L^{-1}$ catalyst dosage of BiOCl with OVs at a higher initial PFOA concentration (50 μM), solution pH of 2 and 8.5 h reaction time, have been confirmed by Song et al. (2019) to yield 68.8% degradation and 12.86% defluorination of PFOA. In comparison, Song et al. (2019), employed BiOCl nanosheets in achieving complete (100%) degradation and considerably high defluorination (59.3%) of PFOA from water under optimized working condition of initial PFOA concentration (20 μM), catalyst dosage (0.5 g $L^{-1}$), solution pH of 4.8 and 12 h contact time. This result indicated the influence of the operation condition and the morphological structure of the photocatalyst in the degradation and elimination of PFOA from water.

Yang et al. (2020), synthesized a spherical-shaped BiOCl/Zn-Al hydrotalcite (B-BHZA) by integrating Zn-Al hydrotalcite with BiOCl. The efficacy of the B-BHZA was notably influenced by the operating condition, with 90% photocatalytic degradation of PFOA achieved under the working condition of dosage (0.5 g $L^{-1}$), initial PFOA concentration (500 μg $L^{-1}$), and solution pH of 2, within a 6 h contact duration under UV irradiation. The research outcomes by Sahu et al. (2018), and Xu et al. (2020c), respectively, utilized $Bi_3O(OH)(PO_4)_2$ microparticles and $Bi_3O(OH)(PO_4)_2$ microparticles/carbon spheres composite to obtain complete degradation of PFOA. Yet, a higher percentage defluorination (60%) of PFOA was obtained using $Bi_3O(OH)(PO_4)_2$ microparticles, compared to using $Bi_3O(OH)(PO_4)_2$ microparticles/carbon spheres that yielded 32.5% defluorination of PFOA. The role of fungal biocatalysts has been investigated in boosting the photocatalytic degradation of PFAS from contaminated water. According to the analysis outcome by Khan et al. (2023), fungal treatment of BiOI Photocatalysts, such as both $Bi_5O_7I$ and $Bi_4O_5I_2$ phases, yielded a composite photocatalyst of BiOI/*Cunninghamella elegans* that displayed 90% degradation and 60% defluorination of PFOA, compared to 35% to 40% degradation and 20% to 30% defluorination of PFOA obtained using $Bi_5O_7I$ and $Bi_4O_5I_2$.

### 19.5.5 Iron-based photocatalysts

Moreira et al. (2024) proposed a less complex and cost-effective approach to obtain a synthesized iron oxide/graphenic carbon (Fe/g-C) hybrid-based photocatalyst that displayed over 85% degradation of PFOA from water within a 3 h duration under irradiated ultraviolet light, as described in Table 19.1. Moreover, the utilized Fe/g-C reportedly demonstrated high chemical stability that accounted for the sustained high PFOA degradation rate over an extended period of reuse. Thus denoting a long-term potential reuse of Fe/g-C over an extended period. In comparison, Xu et al. (2020c), reportedly achieved 95.2% (photocatalytic degradation) and 57.2% (defluorination) of PFOA by using a hybridized FeO CS photocatalyst.

Investigative study on using bimetallic $Fe_2O_3/Mn_2O_3$ photocatalyst aided with $H_2O_2$ as oxidant by Park et al. (2024), attributed the rapid photocatalytic removal of PFAS within a 10 min duration to the spontaneous generation of HR by photogenerated pairs of electrons and holes. Moreover, Park et al. obtained 97% removal of PFOS and 80% to 93% removal of PFNA, using a bimetallic $Fe_2O_3/Mn_2O_3$ photocatalyst aided with $H_2O_2$ as oxidant and recounted the role of $SiO_2$ as a support material in providing a higher surface area for the development of bimetallic $Fe_2O_3/Mn_2O_3$ photocatalyst. However, despite the increase in the surface area provided by the use of $SiO_2$ as a support material. There was a reported decline in the photocatalytic degradation efficiency of bimetallic $Fe_2O_3/Mn_2O_3$ for PFNA removal from water. This photocatalytic outcome may be ascribed to the impeded generation of $^{\cdot}OH$ radicals due to the effect of surface adsorption in deterring the effective contact between the active reaction sites of the bimetallic $Fe_2O_3/Mn_2O_3$ photocatalyst and $H_2O_2$.

Similarly, Qian et al. (2021) utilized zeolite as a support to develop Fe-loaded zeolites (Fe-BEA35) photocatalytic adsorbent for PFOA removal, with 90% PFOA removal at a solution pH of approximately 5.5 in a 24 h reaction time. The high percentage degradation of PFOA may be ascribed to the constructive formation of complexes between PFOA and ferric ions on Zeolite that aided PFOA access to the active sites on the photocatalytic adsorbent Fe-BEA3. A study on PFOS removal from water using $Fe_3O_4$/ZnO@Ag nano-photocatalyst under exposure to visible light, as described by Pouresmaeil et al. (2005), resulted in obtaining 87.37% degradation of PFOS from water at a solution pH of 3. Also, the functionalization of $Fe_3O_4$ with $SiO_2$ and BiOBr to obtain $Fe_3O_4@SiO_2$-BiOBr has been narrated with 92.9% degradation and 32.8% defluorination of PFOA from water (Wang et al., 2017).

### 19.5.6 Boron-based photocatalyst

Boron nitride (BN) has been delved into as an efficient photocatalyst for the degradation of PFAS such as PFOA and Gen X in water. According to Duan et al. (2020), BN exhibits a wide band gap that inhibits its ability to absorb light, thus requiring modification for photocatalytic reactions. On this premise, Duan et al. (2020) considered the ability of a mechanical modification approach, such as ball milling, to induce defects on BN material and obtained approximated 20% to 40% enhanced photocatalytic efficiency, resulting in 99% degradation and 52% defluorination of PFOA using BN with induced defects compared to pristine BN. The outcome of the boosted photocatalytic degradation of PFOA was ascribed to the improved light absorption due to the vacancy on the defect-induced BN. Moreover, Duan et al. (2020), recounted no decline in the photocatalytic PFOA degradation efficiency of defect-induced BN after 3 consecutive cycles of usage. Further study by Duan et al. (2020) revealed that coupling $TiO_2$ to BN resulted in the production of a $TiO_2$-BN heterojunction photocatalyst that degraded PFOA under sunlight conditions to obtain 99% degradation and 68% defluorination of PFOA. This outcome demonstrated the synergistic effect between $TiO_2$ to BN in improving photocatalytic degradation of PFOA.

## 19.5.7 Zinc oxide-based photocatalysts

ZnO has been posited to exhibit a higher proportion of visible light absorption capacity along with cost-effectiveness, appropriate electron abundance, and mobility in comparison to $TiO_2$, thus making ZnO a suitable alternative photocatalyst (Navidpour et al., 2017). An experimental finding by Navidpour (2023) utilized ZnO for water treatment operation and obtained an enhanced photocatalytic PFOA degradation of 46%, 61%, and 64% with an increase in the dosage of ZnO photocatalyst corresponding to 0.27, 0.53, and 1.06 $g\,L^{-1}$, respectively, under UV irradiation. Additionally, Navidpour (2023), employed nanocomposites of ZnO@citric acid-modified carbon nitride in obtaining 54% and 59% photocatalytic degradation of PFOA under exposure to visible light irradiation (300 W xenon lamp) and irradiated light from UV at 254 nm, respectively, for a 2 h duration at optimal working conditions. Relatively, Navidpour (2023) reportedly used PMS as an aid to enhance the efficiency of ZnO@citric acid-modified carbon nitride nanocomposite in achieving 74% and 79% photocatalytic degradation of PFOA, respectively, under visible (300 W xenon lamp) and UV (254 nm) for a 2 h duration at optimal working conditions. This outcome signifies

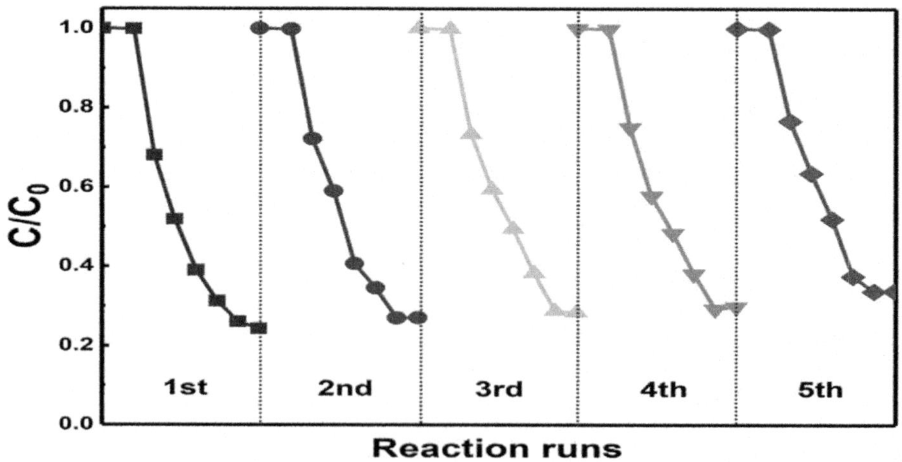


FIGURE 19.11 The reusability of perfluorooctanoic acid-spent mesoporous $Sb_2O_3/TiO_2$ after five consecutive cycles. Source: *Adapted from Yao, X., Zuo, J., Wang, Y.J., Song, N.N., Li, H.H., Qiu, K., 2021. Enhanced photocatalytic degradation of perfluorooctanoic acid by mesoporous Sb2O3/TiO2 heterojunctions. Front ChemFrontiers in Chemistry. 9, 1–10. https://doi.org/10.3389/fchem.2021.690520, under a Creative Commons license (CC BY 4.0).*

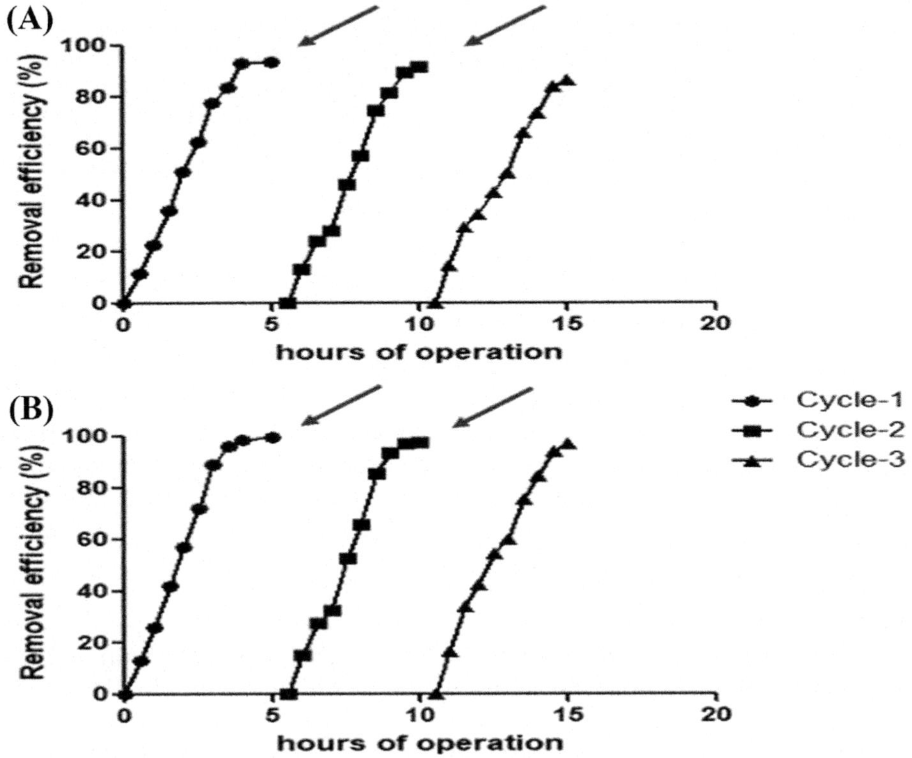


FIGURE 19.12 The reusability of ilmenite as a photocatalyst for degradation (A) perfluorooctanoic acid and (B) perfluorooctane sulfonic acid. Source: *Adapted from Fernando, E.Y., Sarkar, D., Rodwihok, C., Satpathy, A., Zhang, J., Rahmati, R., et al., 2024. Unrefined and milled ilmenite as a cost-effective photocatalyst for UV-assisted destruction and mineralization of PFAS. Materials. 17 (3801), https://doi.org/10.3390/ma17153801, under a Creative Commons license (CC BY 4.0).*

the effect of oxidant in enhancing photocatalytic degradation and elimination of PFOA in water.

An assessment by Ong et al. (2017) integrated reduced graphene oxide on prepared ZnO to prepare ZnO-rGO composite that displayed 90.9% photocatalytic degradation of PFOA at optimal conditions of dosage (100 mg), and 100 ppm initial PFOA concentration in a reaction period of 1 h under solar irradiation. The high percentage degradation of PFOA exhibited by ZnO-rGO composite may be accounted for by the synergistic effect of rGO and ZnO in enhancing the adsorption of light in the visible region for producing the required photogenerated pairs of electrons and holes to improve photocatalytic reactions.

### 19.5.8 Cerium oxide-based photocatalysts

$CeO_2$ photocatalyst has been noted with significant photocatalytic activity. From the experimental outcome by Fu et al. (2022), the PFOA photocatalytic degradation performance of the $CeO_2$ photocatalyst has been posited to match its characteristic energy band gap as compared to other photocatalysts for the degradation and elimination of PFOA in water. Moreover, Tang et al. (2021) demonstrated that an increase in the catalytic dosage of $CeO_2$ from 0.2 to 0.6 mg $L^{-1}$ correspondingly increased the percentage photocatalytic degradation of PFOA by means of using $CeO_2$ and $CeO_2$@NiAl-LDHs catalyst (where LDHs denotes layered double hydroxides).

## 19.6 Regeneration and reusability of per- and polyfluoroalkyl substances-spent metal oxide-based photocatalyst

The reusability of minerals/metal oxide-based photocatalysts for effective degradation and elimination of PFAS is a crucial aspect of economic value. On this premise, there have been studies on the regeneration of PFAS-spent metal oxide photocatalyst. For instance, Yao et al. (2021), utilized deionized water and ethanol to wash PFOA-spent mesoporous $Sb_2O_3$/$TiO_2$ that was later oven-dried at 60°C and reused under the same condition for five consecutive cycles, with about 88% of its initial photocatalytic PFOA degradation efficiency retained as portrayed in Fig. 19.11. A similar outcome as presented in Fig. 19.12, has been highlighted by Fernando et al. (2024), in retaining over 96% and 88% removal efficiency for PFOS and PFOA, respectively, at the end of three consecutive cycles of using ilmenite photocatalyst material under UV irradiation. This denotes a potential long-term usage of mineral/metal oxide-based photocatalyst for the degradation and elimination of PFAS from water.

Despite the successful use of chemicals for regenerating PFAS-spent metal oxide-based photocatalysts for consecutive degradation and elimination of PFAS. It is essential to know that using harsh chemicals for regenerating PFAS-spent photocatalyst may damage the structural morphology, with potential leaching of some elemental constituents of these photocatalysts. Consequently, affecting the efficiency of these photocatalysts to degrade PFAS for consecutive cycles of usage (Akinnawo, 2024a).

Another significant approach that has gained wide acceptance for regenerating PFAS-spent metal oxide-based photocatalysts is the thermal regeneration method. However, lower photocatalytic performance has been reported for thermally regenerated metal oxide-based photocatalysts, in comparison to fresh or chemically regenerated photocatalysts. This is due to the possible alteration in the structural framework of metal oxide-based photocatalysts as a result of heating at high temperature (Akinnawo, 2024a). Moreover, the use of electrochemical techniques may be considered a milestone in the regeneration of spent metal oxide-based photocatalysts for the degradation and elimination of PFAS from water. However, the setback associated with the electrochemical regeneration approach includes the formation of salt and other residual products at the cathode as a contributing factor that alters the solution pH and consequently impacts the efficiency of the regenerated photocatalyst. Another major step to overcome this setback is to utilize high electric current; paradoxically, this will contribute to the high cost of operation, which makes the electrochemical regeneration approach less feasible in countries of the global south with poor generation of electricity (Akinnawo, 2024a).

## 19.7 Challenges and future perspective

The incapability of PFAS molecules to absorb light in the visible region is a crucial challenge that inhibits the degradation of PFAS from water using photocatalysts that can not absorb irradiated light in the visible region which constituent a large quota of sunlight. Thus contributing to the high cost of using UV light sources for photocatalytic degradation and elimination of PFAS from water. Additionally, the formation of intermediate products due to incomplete mineralization of PFAS is another significant challenge because some of these intermediate products may be more toxic than the target PFAS. This has necessitated the need to use various modification strategies to modify the photocatalyst to absorb exposed light in the visible region, in addition to enhancing the effectiveness of the photocatalyst for complete degradation and mineralization of PFAS. As a result, future studies should be focused on cost-effective modification strategies that enhance photocatalytic performance for mineralization of PFAS as well as absorption of exposed light in the visible region in order to evade the high cost of using UV light sources.

Moreover, the energy consumption and high cost of synthesizing or purchasing photocatalysts for large-scale photocatalytic PFAS treatment are of economic significance that necessitate the need for regeneration and reusability of spent photocatalysts. However, some regeneration methods, such as heating and chemical treatment, have been associated with altering the morphological structure and elemental constituents of the photocatalyst. Thus inhibiting the efficiency of regenerated photocatalysts for degradation of PFAS from water. On this premise, future studies should focus on cost-effective regeneration methods that preserve the elemental constituent and morphological structure of the photocatalyst in order to retain the physical and chemical characteristics that enhance the performance of the photocatalyst for PFAS degradation.

## 19.8 Concluding remarks

PFAS contamination of water resources is a significant ecological concern owing to its adverse aftermaths on humans and the ecosystem. Despite the vast array of techniques for water treatment, challenges such as the reintroduction of PFAS into the environment via the improper disposal of PFAS-spent adsorbents and coagulants, the cost of electrochemical treatment methods, and membrane fouling have necessitated the need for a photocatalytic approach for PFAS removal from water. Minerals/metal oxide-based photocatalysts have proved effective in this regard, but their efficiency hinges on operating parameters that are not limited to photocatalyst dosage, initial PFAS concentration, in addition to solution pH, contact time, and light intensity, as well as the characteristic nature and elemental composition of the photocatalyst. Moreover, incomplete PFAS degradation and the formation of intermediate compounds that may be more toxic have been associated with photocatalytic PFAS degradation. To this end, various modification strategies that include doping, in addition to surface defect engineering, morphological engineering, surface sensitization, surface functionalization, and heterojunction construction, have been successfully used in enhancing PFAS degradation performance of various photocatalysts. Modified photocatalysts have been observed to yield a higher percentage degradation and elimination of PFAS from water, in comparison to the pristine photocatalyst. Hence, insight and further investigation into cost-effective modification strategies for enhancing photocatalytic performance to yield complete degradation, mineralization, and elimination of PFAS in water are needed.

## References


Adegoke, K.A., Akinnawo, S.O., Ajala, O.A., Adebusuyi, T.A., Maxakato, N.W., Bello, O.O., 2022. Progress and chellenges in batch and optimization sudies on the adsorptive removal of heavy metals using modified biomass-based adsorbents. Bioresource Technology Reports 19, 101115. Available at https://doi.org/10.1016/j.biteb.2022.101115.

Adegoke, K.A., Akinnawo, S.O., Adebusuyi, T.A., Ajala, O.A., Adegoke, R.O., Maxakato, N.W., et al., 2023. Modified biomass adsorbents for removal of organic pollutants: a review of bathc and optimization studies. International Journal of Environmental Science and Technology 20, 11615–11644. Available at https://doi.org/10.1007/s13762-023-04872-2.

Ahrens, L., 2011. Polyfluoroalkyl compounds in the aquatic environment: a review of their occurrence and fate. Journal of Environmental Monitoring: JEM 13 (1), 20–31. Available at https://doi.org/10.1039/c0em00373e.

Akinnawo, S.O., 2021. Chemical precipitation and reduction methods for the restoration of water from aquaculture operation. Coastal Journal of School Science 3 (1), 517–533.

Akinnawo, S.O., 2023. Eutrophication: causes, consequences, physical, chemical and biological techniques for mitigation strategies. Environmental Challenges 12, 100733. Available at https://doi.org/10.1016/j.envc.2023.100733.

Akinnawo, S.O., 2024a. Per- and polyfluoralkyl substances sequestration by adsorption and photocatalysis: technical challenges and pragmatic solution. Desalination and Water Treatment 319, 100437. Available at https://doi.org/10.1016/j.dwt.2024.100437.

Akinnawo, S.O., 2024b. Covalent organic frameworks: design, synthesis, characterization, and applications. ChemPhysMater 3, 36–63. Available at https://doi.org/10.1016/j.chphma.2023.08.003.

Akinnawo, S.O., 2025a. Adsorptive removal of pesticides from wastewater using conventional and framework materials. Desalination and Water Treatment 321, 100905. Available at https://doi.org/10.1016/j.dwt.2024.100905.

Akinnawo, S.O., 2025b. A review on the green synthesis of nanoparticles for energy conversion and storage. Kuwait Journal of Science 52, 100434. Available at https://doi.org/10.1016/j.kjs.2025.100434.

Akinnawo, S.O., Ayadi, P.O., Oluwalope, M.T., 2023. Chemical coagulation and biological techniques for wastewater treatment. Ovidius University Annals of Chemistry 34 (1), 14–21.

Akinnawo, S.O., Ediagbonya, T.F., 2025. Advances on modification of photocatalyst for degradation/removal of organic pollutants from water. Cleaner Chemical Engineering 11, 100176. Available at https://doi.org/10.1016/j.clce.2025.100176.

Alam, D., Lee, S., Hong, J., Fletcher, D.F., McClure, D., Cook, D., et al., 2023. Experimental investigations of per-and poly-fluoroalkyl substances (PFAS) degradation by non-themal plasma in aqueous solutions. Journal of Environmental Chemical Engineering 11, 111588. Available at https://doi.org/10.1016/j.jece.2023.111588.

Boonchata, P., Boontanon, N., Jindal, V., Aung, H.K.Z., Visvanathan, C., Fujii, S., et al., 2024. Photocatalytic degradation of perfluorooctane sulfonic acid and perfluorooctanoic acid using titanium dioxide/graphene oxide nanocomposite immobilized on polyvinyl alcohol film. Case Studies in Chemical and Environmental Engineering 10, 100862. Available at https://doi.org/10.1016/j.cscee.2024.100862.

Cao, X., Ding, L., Peng, J., Wang, W., Zhang, Y., Chang, Y., et al., 2024. Efficient photocatalytic decomposition of PFOA over BiOI1-x with low power LED light. The Science of the Total Environment 951, 175492. Available at https://doi.org/10.1016/j.scititenv,2024.175492.

Chen, Y.-C., Lo, S.-L., Kuo, J., 2011. Effects of titanate nanotubes synthesized by a microwave hydrothermal method on photocatalytic decomposition of perfluorooctanoic acid. Water Research 45 (14), 4131–4140.

Chen, M.-J., Lo, S.-L., Lee, Y.-C., Huang, C.-C., 2015. Photocatalytic decomposition of perfluorooctanoic acid by transition-metal modified titanium dioxide. Journal of Hazardous Materials 288, 168–175. Available at https://doi.org/10.1016/j.jhazm.

Chen, M.-J., Lo, S.-L., Lee, Y.-C., Kuo, J., Wu, C.-H., 2016. Decomposition of perfluorooctanoic acid by ultraviolet light irradiation with Pb-modified titanium dioxide. Journal of Hazardous Materials 303, 111–118 https://doi.org/10.1.

Chowdhury, N., Prabakar, S., Choi, H., 2021. Dependency of the photocatalytic and photochemical decomposition of per- and polyfluoroalkyl substances (PFAS) on their chain lengths, functional groups, and structural properties. Water Science & Technology 84 (12), 3738–3754. Available at https://doi.org/10.2166/wst.2021.458.

Diaz-Angulo, J., Lara-Ramos, J.A., Machuca-Martinez, F., 2021. Dye photosensitization on heterogenous photocatalysis process, fundamentals and application. Nanostructured Photocatalysts 331–362.

Duan, L., Wang, B., Heck, K., Guo, S., Clark, C.A., Arredondo, J., et al., 2020. Efficient photocatalytic PFOA degradation over boron nitride. *Environmental Science & Technology Letters* 7, 613–619. Available at https://doi.org/10.1021/acs.estlett.0c00434.

Dzierlenga, M.W., Keast, D.R., Longnecker, M.P., 2021. The concentration of several perfluoroalkyl acids in serum appears to be reduced by dietary fiber. Environment International 146, 6292. Available at https://doi.org/10.1016/j.envint.2020.106292.

Ersan, G., Ersan, M.S., Perrreault, F., Garcia-Segura, S., 2023. Enabling in situ electro-regeneration systems for PFOA-laden spent activated carbon adsorbents reuse. Journal of Environmental Chemical Engineering 11, 111369. Available at https://doi.org/10.1016/j.jece.2023.111369.

Esfahani, E.B., Asadi, F., Zhang, S., Mohseni, M., 2022. Photo-chemical/catalytic oxidative/reductive decomposition of per-and polyfluoroalkyl substances (PFAS), decomposition mechanisms and effects of key factors: a review. Environmental Science 8, 5. Available at https://doi.org/10.1039/D1EW00774B.

Esfahani, E.B., Asadi, F., Zhang, S., Mohseni, M., 2024. Photo-chemical/catalytic oxidative/reductive decomposition of polyfluorooctanoic acid (PFOA) and its common alternatives: mechanism and kinetic modelling. Journal of Water Process 61, 105332. Available at https://doi.org/10.1016/j.jwpe.2024.105332.

Fernando, E.Y., Sarkar, D., Rodwihok, C., Satpathy, A., Zhang, J., Rahmati, R., et al., 2024. Unrefined and milled ilmenite as a cost-effective photocatalyst for UV-assisted destruction and mineralization of PFAS. Materials 17 (3801). Available at https://doi.org/10.3390/ma17153801.

Fu, C., Xu, X., Zheng, C., 2022. Photocatalysis of aqueous PFOA by common catalysts of In2O3, Ga2O3, TiO2, CeO2 and CdS: influence factors and mechnistic insights. Environmental Geochemistry and Health 44, 2943–2953. Available at https://doi.org/10.1007/s10653-021-01127-2.

Gates, K., RAi, S., Pramanik, A., Kolawole, O.P., Kundu, S., Ucak-Astarlioglu, M., et al., 2025. Insight into the photocatalytic degradation mechanism for "Forever Chemicals" PFNA by reduced graphene oxide/WO3 nanoflower heterostructures. ACS Omega 10 (10), 10675–10684. Available at https://doi.org/10.1021/acsomega.5c0054.

Gomez-Ruiz, B., Ribao, P., Diban, N., Rivero, M.J., Ortiz, I., Urtiaga, A., 2018. Photocatalytic degradation and mineralization of perfluorooctanoic acid (PFOA) using a composite TiO2 –rGO catalyst. Journal of Hazardous Materials 344, 950–957. Available at https://doi.org/10.1016/j.jhazmat.2017.11.048.

Gu, M., Li, S., Fan, X., Huang, J., Yu, G., 2022. Effective breaking of the fluorocarbon chain by the interface Bi2Ox——PFOA complex strategy via coordinate Se on construction of the internal photogenerated carrier pathway. ACS Applied Materials & Interfaces 14, 654–667. Available at https://doi.org/10.1021/acsami.1c17406.

Huang, D., Yin, L., Niu, J., 2016. Photoinduced hydrodefluorination mechanisms of perfluorooctanoic acid by the SiC/Graphene catalyst. Environmental Science & Technology 50, 5857–5863. Available at https://doi.org/10.1021/acs.est.6b00652.

Huang, D., De Vera, G.A., Chu, C., Zhu, Q., Stavitski, E., Mao, J., et al., 2018. Single-atom Pt catalyst for effective C-F bond activation via hydrodefluorination. ACS Catalysis 8, 9353–9358. Available at https://doi.org/10.1021/acscatal.8b02660.

Jiang, F., Zhao, H., Chen, H., Xu, C., Chen, J., 2016. Enhancement of photocatalytic decomposition of perfluorooctanoic acid on CeO2/In2O3. RSC Advances 6 (76), 72015–72021. Available at https://doi.org/10.1039/C6RA09856H.

Jin, T., Peydayesh, M., Mezzenga, R., 2012. Membrane-based technologies for per-and poly-fluoroalkyl substances (PFAS) removal from water: removal mechanisms, applications, challenges and perspectives. Environment Internation 157, 106876.

Khan, M.F., Paul Guin, J., Thampi, R.K., Sullivan, J.A., Murphy, C.D., 2023. Enhanced removal of perfluorooctanoic acid with sequential photocatalysis and fungal treatment. Environmental Science and Pollution Research 30 (39), 91478–91486. Available at https://doi.org/10.1007/s11356-023-28588-5.

Kim, H.-H., Choi, J.S., Jung, B., An, S., Yang, N., Her, N., et al., 2025. A review of photocatalytic decomposition of per- and polyfluoroalkyl substances in aqueous solution. Journal of Environmental Management 389, 126142. Available at https://doi.org/10.1016/j.jenvman.2025.126142.

Leonello, D., Fendrich, M.A., Parrino, F., Patel, N., Orlandi, M., Miotello, A., 2021. Light-induced advanced oxidation processes as pfas remediation methods: a review. Applied Sciences 11 (18). Available at https://doi.org/10.3390/app11188458.

Li, X., Zhang, P., Jin, L., Shao, T., Li, Z., Cao, J., 2012a. Efficient photocatalytic decomposition of perfluorooctanoic acid by indium oxide and its mechanism. Environmental Science & Technology 46 (10), 5528–5534. Available at https://doi.org/10.1021/es204.

Li, Z., Zhang, P., Shao, T., Li, X., 2012b. In2O3 nanoporous nanosphere: a highly efficient photocatalyst for decomposition of perfluorooctanoic acid. Applied Catalysis B: Environmental 125, 350–357. Available at https://doi.org/10.1016/j.apcatb.2012.06.017.

Li, Z., Zhang, P., Shao, T., Wang, J., Jin, L., Li, X., 2013a. Different nanostructured In2O3 for photocatalytic decomposition of perfluorooctanoic acid (PFOA). Journal of Hazardous Materials 260, 40–46. Available at https://doi.org/10.1016/j.jhazmat.2013.04.042.

Li, Z., Zhang, P., Li, J., Shao, T., Jin, L., 2013b. Synthesis of In2O3-graphene composites and their photocatalytic performance towards perfluorooctanoic acid decomposition. Journal of Photochemistry and Photobiology A: Chemistry 271, 111–116. Available at https://doi.org/10.1016/j.jphotochem.2013.08.012.

Li, Z., Zhang, P., Li, J., Shao, T., Wang, J., Jin, L., 2014. Synthesis of In2O3 porous nanoplates for photocatalytic decomposition of perfluorooctanoic acid (PFOA). Catalysis Communications 43, 42–46. Available at https://doi.org/10.1016/j.catcom.2013.09.004.

Li, M., Yu, Z., Liu, Q., Sun, L., Huang, W., 2016. Photocatalytic decomposition of perfluorooctanoic acid by noble metallic nanoparticles modified TiO2. Chemical Engineering Journal 286, 232–238. Available at https://doi.org/10.1016/j.cej.2015.10.

Li, T., Wang, C., Wang, T., Zhu, L., 2020a. Highly efficient photocatalytic degradation toward perfluorooctanoic acid by bromine doped BiOI with high exposure of (001) facet. Applied Catalysis B: Environmental 268, 18442.

Li, F., Wei, Z., He, K., Blaney, L., Cheng, X., Xu, T., et al., 2020b. A concentrate-and-destroy technique for degradation of perfluorooctanoic acid in water using a new adsorptive photocatalyst. Water Research 185, 116219. Available at https://doi.org/10.1016/j.watres.2020.116219.

Li, J., Sun, J., Li, P., 2022. Exposure routes, bioaccumulation and toxic effects of per- and polyfluoroalkyl substances (PFAS) on plants: a critical review. Environmental International 158, 106891. Available at https://doi.org/10.1016/j.envint.2021.106891.

Liang, C., Su, H.W., 2009. Identification of sulfate and hydroxyl radicals in thermally activated persulfate. Industrial & Engineering Chemistry Research 48 (11), 5558–5562. Available at https://doi.org/10.1021/ie9002848.

Liu, X., Xu, B., Duan, X., Hao, Q., Wei, W., Wang, S., et al., 2021. Facile preparation of hydrophilic In2O3 nanospheres and rods with improved performances for photocatalytic degradation of PFOA. Environmental Science: Nano 8 (4), 1010–1018. Available at https://doi.org/10.1039/D0EN01216E.

Liu, F., Guan, X., Xiao, F., 2022. Photodegradation of per- and polyfluoroalkyl substances in water: a review of fundamentals and applications. Journal of Hazardous Materials 439, 129580. Available at https://doi.org/10.1016/j.jhazmat.2022.129580.

Liu, F., Xue, H., Kang, T., Lei, Q., Chen, J., Zuo, Z., et al., 2025. Efficient photodegradation of perfluoroalkyl substances under visible light by ZnIn2S4 nanosheets. Journal of Environmental Sciences 148, 116–125. Available at https://doi.org/10.1016/j.jes.2025.01.014.

Mabaso, N.S.N., Tshangana, C.S., Muleja, A.A., 2024. Efficient removal of PFASs using photocatalysis, membrane separation and photocatalytic membrane reactors. Membranes 14 (217). Available at https://doi.org/10.3390/membranes14100217.

Mazen, K., Venkatesh, G., Samara, F., Kanan, S., 2025. ZnO/BiOI heterojuction catalyst: efficient photocatalytic decomposition of perfluorooctanoic acid (PFOA) and mixed dye pollutants in wastewater. Separation and Purification Technology 376 (2). Available at https://doi.org/10.1016/seppur.2025.133953 *133953* 14.

Moreira, R., Esfahani, E.B., Zeidabadi, F.A., Rostami, P., Thuo, M., Mohseni, M., et al., 2024. Hybrid graphenic and iron oxide photocatalysts for the decomposition of synthetic chemicals. Communications Engineering 114. Available at https://doi.org/10.1038/s44172-024-00267-4.

Mustapha, S., Tijani, J.O., Elabor, R., Etsuyankpa, M.B., Amigun, A.T., Shuaib, D.T., et al., 2023. Photocatalytic degradation and defluorination of per- and poly-fluoroalkyl substance (PFAS) using biosynthesized TiO2 nanoparticles under UV-visible light. Engineering Proceedings 37 (x). https://doi.org/10.3390/xxxxx.

Navidpour, H., 2023. Photocatalytic degradation of perfluorooctanoic acid in water by free and immobilized ZnO, g-C3N4 and their composite under UV and visible light irradiation. PhD Thesis. Sydney: University of Technology Sydney.

Navidpour, A.H., et al., 2017. Plasma-sprayed photocatalytic zinc oxide coatings. Journal of Thermal Spray Technology 26 (4), 717–727. Available at https://doi.org/10.1007/s11666-017-0541-x.

Ong, C.B., Mohammad, A.W., NgL, Y., Mahmoudi, E., Azizkhani, S., Hayati, H., 2017. Solar photocatalytic and surface enhancement of ZnO/rGO nanocomposite: degradation of perfluorooctanoic acid and dye. Process Safety and Environmental Protection 112, 298–307. Available at https://doi.org/10.1016/j.psep.2017.04.031.

Panchangam, S.C., Lin, A.Y.-C., Shaik, K.L., Lin, C.F., 2009. Decomposition of perfluorocarboxylic acids (PFCAs) by heterogenous photocatalysis in acidic aqueous medium. Chemosphere 77 (2), 242–248.

Park, J., Choe, J.K., Bae, J., Baek, S., 2024. Rapid degradation of perfluorooctane sulfonic acid (PFOS) and perfluorononanoic acid (PFNA) through bimetallic catalyst of Fe2O3/Mn2O3 and unravelling the effect of support SiO2. Heliyon 10, 34199. Available at https://doi.org/10.1016/j.heliyon.2024.e34199.

Pouresmaeil, H., Nabi, B.G.R., Bagjdado, M., 2005. The removal of PFAS by Fe3O4/ZnO@Ag nano-photocatalyst under visible light irradiation. International Journal of Environmental Science and Technology. Available at https://doi.org/10.1007/s13762-025-06520-3.

Qian, L., Kopinke, F.-D., Georgi, A., 2021. Photodegradation of perfluorooctanesulfonic acid on Fe-Zeolites in water. Environmental Science and Technology 55 (1), 614–622. Available at https://doi.org/10.1021/acs.est.0c04558.

Riegel, M., Haist-Gulde, B., Sacher, F., 2023. Sorptive removal of short-chain perfluoroalkyl substances (PFAS) during drinking water treatment using activated carbon and anion exchanger. Environmental Sciences Europe 35 (1). Available at https://doi.org/10.1186/s12302-023-00716-5.

Rodrigo, P.M., Navarathna, C., Pham, M.T.H., McClain, S.J., Stokes, S., Zhang, X., et al., 2022. Batch and fixed bed sorption of low to moderate concentrations of aqueous per- and poly-fluoroalkyl substances (PFAS) on Douglas fir biochar and its Fe2O3 hybrids. Chemosphere 308, 136155. Available at https://doi.org/10.1016/j.chemosphere.2022.136155.

Sabri, M., Habibi-Yangjeh, A., Pouran, S.R., Wang, C., 2023. Titania-activated persulfate for environmental remediation: the-state-of-the-art. Catalysis Reviews 65 (1), 118–173. Available at https://doi.org/10.1080/01614940.2021.1996776.

Sahu, S.P., Qanbarzadeh, M., Ateia, M., Torkzadeh, H., Maroli, A.S., Cates, E.L., 2018. Rapid degradation and mineralization of perfluorooctanoic acid by a new petitjeanite Bi3O(OH)(PO4)2 microparticle ultraviolet photocatalyst. Environmental Science & Technology Letters 5 (8), 533–538. Available at https://doi.org/10.1021/acs.estlett.8b00395.

Shang, E., Li, Y., Niu, J., Li, S., Zhang, G., Wang, X., 2018. Photocatalytic degradation of perfluorooctanoic acid over Pb-BiFeO3/rGO catalyst: kinetics and mechanism. Chemosphere 211, 34–43. Available at https://doi.org/10.1016/j.chemosphere.2018.07.130.

Shao, T., Zhang, P., Jin, L., Li, Z., 2013a. Photocatalytic decomposition of perfluorooctanoic acid in pure water and sewage water by nanostructured gallium oxide. Applied Catalysis B: Environmental 142–143, 654–661. Available at https://doi.org/10.1016/j.apcatb.2013.05.074.

Shao, T., Zhang, P., Li, Z., Jin, L., 2013b. Photocatalytic decomposition of perfluorooctanoic acid in pure water and wastewater by needle-like nanostructured gallium oxide. Chinese Journal of Catalysis 34 (8), 1551–1559. Available at https://doi.org/10.1016/S1872-2067(12)60612-3.

Shao, H., Zhao, X., Wang, Y., Mao, R., Wang, Y., Qiao, M., et al., 2017. Synergetic activation of peroxymonosulfate by Co3O4 modified g-C3N4 for enhanced degradation of diclofenac sodium under visible light irradiation. Applied Catalysis B: Environmental 218, 810–818. Available at https://doi.org/10.1016/j.apcatb.2017.07.016.

Sidnell, T., Wood, R.J., Hurst, J., Lee, J., Bussmaker, J., 2022. Sonolysis of per- and polyfluoroalkyl substances (PFAS): a meta-analysis. Ultrasonics Sonochemistry 87, 105944. Available at https://doi.org/10.1016/j.ultsonch.2022.105944.

Song, C., Chen, P., Wang, C., Zhu, L., 2012. Photodegradation of perfluorooctanoic acid by synthesized TiO2–MWCNT composites under 365nm UV irradiation. Chemosphere 86 (8), 853–859. Available at https://doi.org/10.1016/j.chemosphere.2011.11.034.

Song, Z., Dong, X., Wang, N., Zhu, L., Luo, Z., Fang, J., et al., 2017. Efficient photocatalytic defluorination of perfluorooctanoic acid over BiOCl nanosheets via a hole direct oxidation mechanism. Chemical Engineering Journal 317, 925–934. Available at https://doi.org/10.1016/j.cej.2017.02.126.

Song, Z., Dong, X., Fang, J., Xiong, C., Wang, N., Tang, X., 2019. Improved photocatalytic degradation of perfluorooctanoic acid on oxygen vacancies-tunable bismuth oxychloride nanosheets prepared by a facile hydrolysis. Journal of Hazardous Materials 377, 371–380. Available at https://doi.org/10.1016/j.jhazmat.2019.05.084.

Tan, X., Chen, G., Xing, D., Ding, W., Liu, H., Li, T., et al., 2020. Indium-modified Ga2O3 hierarchical nanosheets as efficient photocatalysts for the degradation of perfluorooctanoic acid. Environmental Science: Nano 7 (8), 2229–2239. Available at https://doi.org/10.1039/D0EN00259C.

Tang, H., Wuijian, Z., Yue, M., Shengjie, X., 2021. A direct Z-scheme heterojunction with boosted transportation of photogenerated charge carriers for highly efficient photodegradation of of photogenerated charge carriers for highly efficient photodegradation of PFOA: reaction kinetics and mechanism. Applied Catalysis B: Environmental 285, 119851. Available at https://doi.org/10.1016/j.apcatb.2020.119851.

Tshangana, C.S., Nhlengethwa, S.T., Glass, S., Denison, S., kuvarega, A.T., Nkambule, T.T.I., et al., 2025. Technology status to treat PFAS-contaminated water and limiting factors for their effective full-scale application. npj Clean Water 8, 41. Available at https://doi.org/10.1038/s41545-025-00457-3.

Verma, S., Varma, R.S., Nadagouda, M.N., 2021. Remediation and mineralization processes for per- and polyfluoroalkyl substances (PFAS) in water: a review. The Science of the Total Environment 794, 148987. Available at https://doi.org/10.1016/j.scitotenv.2021.148987.

Verma, S., Mezgebe, B., Hejase, C.A., Sahle-Demessie, E., Nadagouda, M.N., 2024. Photodegradation and photocatalysis of per- and polyfluoroalkyl substances (PFAS): a review of recent progress. Mater 2, 100077. Available at https://doi.org/10.1016/j.nxmate.2023.100077.

Wang, Y., Zhang, P., 2011. Photocatalytic decomposition of perfluorooctanoic acid (PFOA) by TiO2 in the presence of oxalic acid. Journal of Hazardous Materials 192 (3), 1869–1875. Available at https://doi.org/10.1016/j.jhazmat.2011.07.026.

Wang, L., Sun, L., Yu, Z., Hou, Y., Peng, Z., Yang, F., et al., 2017. Synergetic decomposition performance and mechanism of perfluorooctanoic acid in dielectric barrier discharge plasma system with Fe3O4@SiO2-BiOB rmagnetic photocatalyst. Molecular Catalysis 441, 179–189. Available at https://doi.org/10.1016/j.mcat.2017.08.014.

Wang, J., Cao, C., Wang, Y., Wang, Y., Sun, B., Zhu, L., 2020a. In situ preparation of p-n BiOI@Bi5O7I heterojunction for enhanced PFOA photocatalytic degradation under simulated solar light irradiation. Chemical Engineering Journal 391, 123530. Available at https://doi.org/10.1016/j.cej.2019.123530.

Wang, W., Chen, Y., Li, G., Gu, W., An, T., 2020b. Photocatalytic reductive defluorination of perfluorooctanoic acid in water under visible light irradiation: the role of electron donor. Environmental Science: Water Research & Technology 6 (6), 1638–1648. Available at https://doi.org/10.1039/D0EW00205D.

Wang, J., Cao, C., Zhang, Y., Zhang, Y., Zhu, L., 2021a. Underneath mechanisms into the super effective degradation of PFOA by BiOF nanosheets with tunable oxygen vacancies on exposed (101) facets. Applied Catalysis B: Environmental 286, 119911. Available at https://doi.org/10.1016/j.apcatb.2021.119911.

Wang, J., Wang, Y., Cao, C., Zhang, Y., Zhang, Y., Zhu, L., 2021b. Decomposition of highly persistent perfluorooctanoic acid by hollow Bi/BiOI1-xFx: synergistic effects of surface plasmon resonance and modified band structures. Journal of Hazardo Journal of Hazardous Materials 402, 123459. Available at https://doi.org/10.1016/j.jhazmat.2020.123459.

Wang, J., Cao, C., Wang, J., Zhang, Y., Zhu, L., 2022. Insights into highly efficient photodegradation of poly/perfluoroalkyl substances by In-MOF/BiOF heterojunctions: built-in electric field and strong surface adsorption. Applied Catalysis B: Environmental 304, 121013. Available at https://doi.org/10.1016/j.apcatb.2021.121013.

Wu, Y., Li, Y., Tian, A., Mao, K., Liu, J., 2016. Selective removal of perfluorooctanoic acid using molecularly imprinted polymer-modified TiO2 nanotube arrays. International Journal of Photoenergy 7368795. Available at https://doi.org/10.1155/2016/736879.

Wu, Y., Li, Y., Fang, C., Li, C., 2019. Highly efficient degradation of perfluorooctanoic acid over a MnOx-modified oxygen-vacancy-rich In2O3 photocatalyst. ChemCatChem 11 (9), 2297–2303. Available at https://doi.org/10.1002/cctc.201900273.

Xu, C., Qiu, P., Chen, H., Jiang, F., 2017a. Platinum modified indium oxide nanorods with enhanced photocatalytic activity on degradation of perfluorooctanoic acid (PFOA). Journal of the Taiwan Institute of Chemical Engineers 80. Available at https://doi.org/10.1016/j.jtice.2017.09.018.

Xu, J., Wu, M., Yang, J., Wang, Z., Chen, M., Teng, F., 2017b. Efficient photocatalytic degradation of perfluorooctanoic acid by a wide band gap p-block metal oxyhydroxide InOOH. Applied Surface Science 416, 587–592. Available at https://doi.org/10.1016/j.apsusc.2017.04.040.

Xu, T., Zhu, Y., Duan, J., Xia, Y., Tong, T., Zhang, L., et al., 2020a. Enhanced photocatalytic degradation of perfluorooctanoic acid using carbon-modified bismuth phosphate composite: effectiveness, material synergy and roles of carbon. Chemical Engineering Journal 395, 124991. Available at https://doi.org/10.1016/j.cej.2020.124991.

Xu, T., Ji, H., Gu, Y., Tong, T., Xia, Y., Zhang, L., et al., 2020b. Enhanced adsorption and photocatalytic degradation of perfluorooctanoic acid in water using iron (hydr)oxides/carbon sphere composite. Chemical Engineering Journal 388, 124230. Available at https://doi.org/10.1016/j.cej.2020.124230.

Xu, B., Ahmed, M.B., Zhou, J., Altaee, A., 2020c. Visible and UV photocatalysis of aqueous perfluorooctanoic acid by TiO2 and peroxymonosulfate: process kinetics and mechanistic insights. Chemosphere 243, 125366. Available at https://doi.org/10.1016/j.chemosphere.2019.125366.

Yang, Y., Zheng, Z., Yang, M., Chen, J., Li, C., Zhang, C., et al., 2020. In-situ fabrication of a spherical-shaped Zn-Al hydrotalcite with BiOCl and study on its enhanced photocatalytic mechanism for perfluorooctanoic acid removal performed with a response surface methodology. Journal of Hazardous Materials 399, 123070. Available at https://doi.org/10.1016/j.jhazmat.2020.123070.

Yang, Y., Ji, W., Li, X., Zheng, Z., Bi, F., Yang, M., et al., 2021. Insights into the degradation mechanism of perfluorooctanoic acid under visible-light irradiation through fabricating flower-shaped Bi5O7I/ZnO n-n heterojunction microspheres. Chemical Engineering Journal 420, 129934. Available at https://doi.org/10.1016/j.cej.2021.129934.

Yao, X., Zuo, J., Wang, Y.J., Song, N.N., Li, H.H., Qiu, K., 2021. Enhanced photocatalytic degradation of perfluorooctanoic acid by mesoporous Sb2O3/TiO2 heterojunctions. Frontiers in Chemistry 9, 1–10. Available at https://doi.org/10.3389/fchem.2021.690520.

Zhang, T., Pan, G., Zhou, Q., 2016. Temperature effect on photolysis decomposing perfluorooctanoic acid. Journal of Environmental Science 42, 126–133. Available at https://doi.org/10.1016/j.jes.2015.05.008.

Zhao, B., Zhang, P., 2009. Photocatalytic decomposition of perfluorooctanoic acid with β-Ga2O3 wide bandgap photocatalyst. Catalysis Communications 10 (8), 1184–1187. Available at https://doi.org/10.1016/j.catcom.2009.01.017.

Zhao, B., Li, X., Yang, L., Wang, F., Li, J., Xia, W., et al., 2015. ß-Ga2O3 nanorod synthesis with a one-step microwave irradiation hydrothermal method and its efficient photocatalytic degradation for perfluorooctanoic acid. Photochemistry and Photobiology 91 (1), 42–47. Available at https://doi.org/10.1111/php.12383.

Zhu, C., Xu, J., Song, S., Wang, J., Li, Y., Liu, R., et al., 2020. TiO2 quantum dots loaded sulfonated graphene aerogel for effective adsorption-photocatalysis of PFOA. Science of the Total Environment 698, 134275. Available at https://doi.org/10.1016/j.scitotenv.2019.134275.

Zhu, Y., Xu, T., Zhao, D., Li, F., Liu, W., Wang, B., et al., 2021. Adsorption and solid-phase photocatalytic degradation of perfluorooctane sulfonate in water using gallium-doped carbon-modified titanate nanotubes. Chemical Engineering Journal 421, 129676. Available at https://doi.org/10.1016/j.cej.2021.129676.

Zoumpouli, G.A., Herron, D., Thornton, A., Jefferson, B., Campo, 2023. The role of coagulation on the fate of PFAS, brominated flame retardants and other trace contaminants in tertiary wastewater treatment for phosphorus control. Sceince of the Total Environment 887, 163982. Available at https://doi.org/10.1016/j.scitotenv.2023.163982.

CHAPTER

# 20

# Membrane processes and materials for remediation of per- and polyfluoroalkyl substances from water and wastewater

*Ayah Jabr*[1], *Laith Jaber*[1], *Amani Al-Othman*[1,2], *Muhammad Tawalbeh*[3,4], *and Mohammad Mahdi A. Shirazi*[5]

[1]Department of Chemical and Biological Engineering, American University of Sharjah, Sharjah, United Arab Emirates [2]Energy, Water and Sustainable Environment Research Center, American University of Sharjah, Sharjah, United Arab Emirates [3]Sustainable and Renewable Energy Engineering Department, University of Sharjah, Sharjah, United Arab Emirates [4]Water and Environmental Conservation Center, RISE, University of Sharjah, Sharjah, United Arab Emirates [5]Department of Civil, Geological & Environmental Engineering, University of Saskatchewan, Saskatoon, SK, Canada

## 20.1 Introduction

The PFAS compounds containing functional groups are widely adopted due to their unique chemical and physical properties, including resistance to heat, chemical stress, and corrosion, leading to their use as surfactants, process aids, stain repellents, foam stabilizers, and components in coatings, firefighting foams, electronics, and lubricants, among other high-performance applications (Dickman and Aga, 2022).

These PFAS substances are identified by their unusual persistence of PFAS in the environment and the human body. Investigations revealed that a majority of the population had detectable levels of PFAS in their blood, and exposure has been linked to numerous health risks, including cancer, liver damage, thyroid disorders, and developmental issues (Bonefeld-Jørgensen et al., 2014). These findings urged regulatory

*Per- and Polyfluoroalkyl Substances (PFAS) as Environmental Contaminants*
DOI: https://doi.org/10.1016/B978-0-443-43856-1.00012-5

responses, particularly in the United States, with several governmental agencies restricting and limiting the use of PFAS in specific applications (Panieri et al., 2022). It should be noted in this context that PFAS usually exists in the form of negatively charged anions, perfluorooctane sulfonate (PFOS) or perfluorooctanoic acid (PFOA). It should be noted that Perfluorooctanesulfonic acid and PFOS both form a conjugate acid–base pair, and they are forms of PFSA.

For example, the U.S Department of Defense firefighting updated their foam specifications in 2017 to limit PFOA and perfluorooctanesulfonic acid (Cousins et al., 2020). Despite these increased regulations, the use of PFAS continues to be widespread in many modern applications, although steps are being taken to replace some compounds gradually with safer alternatives.

The PFAS are a group of synthetic chemicals characterized by a carbon backbone chain that is fully or partially saturated with fluorine atoms, along with at least one functional group that varies across different types of PFAS. These functional groups include carboxylic acid, sulfonic acid, or amine, with each functional group determining the charge and slightly altering the behavior of each PFAS. The combination of a highly fluorinated hydrophobic carbon "tail" and a polar anionic headgroup gives many PFAS surfactant-like properties that strongly influence their molecular size, charge distribution, and mobility in water and soils (Su and Rajan, 2021). The high level of fluorine saturation in these compounds results in their high stability, due to strong carbon—fluorine bonds. This makes PFAS resistant to chemical degradation, leading to their classification as "forever chemicals" (Panieri et al., 2022; Cousins et al., 2020). The PFAS vary in carbon chain length, functional group type, and molecular structure. These compounds range from short-chain low molecular-weight compounds to long-chain and polymeric forms, such as polytetrafluoroethylene (PTFE). These structural differences affect membrane rejection mechanisms, as longer, highly charged chain PFAS generally exhibit higher rejection due to their increased size exclusion, electrostatic repulsion, and hydrophobic interactions, while short-chain PFAS remain more difficult to remove (Su and Rajan, 2021). This chapter aims to discuss the environmental effects of PFAS materials as well as their removal techniques. The focus will be on membrane-based technologies and their removal efficiency.

## 20.2 Conventional methods for per- and polyfluoroalkyl substances remediation

The physical and chemical properties of PFAS make these chemicals have serious environmental and health risks (Su and Rajan, 2021). The chemicals' high mobility, due to their properties of resistance to degradation, enables the long-range pervasiveness of these chemicals into numerous regions around the world. Their presence has also been detected in remote regions like the Arctic and Antarctic. In a study where Antarctic and subantarctic seabirds were investigated, total PFAS concentrations ranged from 0.53 to 53 $ng\ g^{-1}$ wet weight, despite their absence in direct nearby sources (Roscales et al., 2019). Additionally, their chemical stability causes environmental persistence, as PFAS have the tendency to bioaccumulate, causing an increase in the PFA concentration up the food chain and posing a high risk of accumulating harmful levels of these chemicals in living organisms (Cousins et al., 2020; MacInnis et al., 2019). Humans can also be exposed to PFAs through inhalation, ingestion of contaminated food and water, and dermal absorption, leading to serious health concerns. There have been reports of PFAs having developmental effects in children, including behavioral changes, accelerated puberty, and in newborns, decreased birth weight (Panieri et al., 2022).

The conventional techniques for PFAS remediation from water and wastewater can be categorized into physical, biological, and chemical techniques. Physical techniques for PFAS removal include adsorption using biochar, activated carbon, membrane filtration, and ion exchange (Li et al., 2022; Abuwatfa et al., 2021). Activated carbon-based adsorbents, both granular and powdered activated carbon, have been shown to have a high degree of PFAS removal property (Sharma et al., 2024). However, these physical techniques exhibit slow adsorption kinetics in comparison to other methods (Li et al., 2022). Membrane filtration methods have gained commercial popularity for the removal of PFAS from water, as the removal efficiency relies entirely on the membrane's properties. The most widely used membrane filtration technologies are reverse osmosis (RO) and nanofiltration (NF) (Bentel et al., 2019).

Chemical methods, on the other hand, include defluorination and electrochemical oxidation. These methods aim at breaking down the PFAS molecule instead of separating it from water. The efficiency of PFAS defluorination using UV-generated hydrated electrons was found to depend significantly on molecular structure. Specifically, factors such as carbon chain length, the nature of the head functional group, and the position and number of carbon–fluorine bonds influenced the degradation process (Meegoda et al., 2020). However, since this method is dependent on the headgroup and the fluoroalkyl chain length, the rate of the PFAS degradation will be variable depending on the size of these components within the molecule. In some cases, the decay of the parent structures was faster than the intermediates, causing the PFAS to degrade at a slower rate (Bentel et al., 2019). Regarding electrochemical oxidation, a technique that uses an electric current and reactive electrodes to generate strong oxidants that break down the stable carbon–fluorine bonds in the PFAS molecule, when boron-doped diamond (BDD) anodes were investigated for removing PFAS from water, the efficiency reached above 95% (Sharma et al., 2024). Despite its efficiency, this method must subject the PFAS to a defluorination pretreatment that eventually reaches a state of little to no degradation of PFAS. Additionally, long-term efficiency will be hindered as the anode will be subjected to degradation, and the changing surface chemistry will impact the system's reliability across different cycles (Yanagida et al., 2022). Biological methods include microbial bioremediation, which involves certain bacteria. While some microbes can transform PFAS into less toxic forms, microbial biotransformation can unintentionally increase PFAS levels by converting fluorinated precursors into terminal PFAS through microbial activity (Bentel et al., 2019). The PFAS remediation methods are summarized in Table 20.1 (Li et al., 2022; Sharma et al., 2024; Bentel et al., 2019; Yanagida et al., 2022).

**TABLE 20.1** Summary of the conventional methods for per- and polyfluoroalkyl substances (PFAS) remediation (Li et al., 2022; Sharma et al., 2024; Bentel et al., 2019; Yanagida et al., 2022).

| Method | Technique type | Mechanism | Advantages | Disadvantages |
|---|---|---|---|---|
| Adsorption (activated carbon) | Physical | Adsorbs PFAS onto granular activated carbon or powdered activated carbon | High PFAS removal capacity | Slow adsorption kinetics; less effective for short-chain PFAS |
| Membrane filtration | Physical | Physically separates PFAS based on membrane pore size and charge interactions | High efficiency and commercially available | Depends heavily on membrane properties; costly; requires maintenance |
| Defluorination | Chemical | Uses hydrated electrons to break C–F bonds | Degrades PFAS instead of separating it | Efficiency depends on chain length and functional groups; incomplete breakdown of intermediates |
| Electrochemical oxidation | Chemical | Applies electric current with reactive electrodes to oxidize PFAS | High removal (>95%) of PFAS | node degradation over time; requires pretreatment; reduced long-term reliability |
| Microbial remediation | Biological | Uses bacteria to degrade PFAS | Reduces PFAS in natural/engineered environments | Biotransformation may generate terminal PFAS; slower rates; limited to certain strains/environments |

## 20.3 Membrane processes for per- and polyfluoroalkyl substances remediation from water and wastewater

Membrane processes have emerged among the most effective technologies for the remediation of various contaminants, oily water, and PFAS from water and wastewater (Chaudhary et al., 2023; Al Sharabati et al., 2021; Tawalbeh et al., 2018). These techniques rely on the selective permeability of engineered membranes to either separate or transform PFAS compounds, offering distinct advantages such as operational simplicity, scalability, and compatibility with existing treatment systems (Saawarn et al., 2022). The membrane filtration process is illustrated in Fig. 20.1 (Suhalim et al., 2022), where a feed mixture is passed through a membrane by a driving force, leaving behind the solute particles and allowing the smaller solvent particles to pass. This section discusses the different membrane processes' efficiency and percent developments in PFAS removal, including the commercial, synthesized, and hybrid membrane systems.

### 20.3.1 Nanofiltration and reverse osmosis

The NF and RO are among the most widely used membrane filtration technologies for treating PFAS-contaminated water. There has been extensive research conducted to enhance the performance and efficiency of these systems. This section explores the separation mechanisms of both NF and RO membranes, as well as the development of commercial, hybrid, and synthesized membranes specifically designed for PFAS removal. There are two main separation processes for water purification that use membranes, and these include NF and RO (Saawarn et al., 2022).

The NF is a membrane filtration mechanism developed in the 1980s that is known for producing high-quality water with a small footprint and lower energy requirements (Chaudhary et al., 2023). The NF membranes possess a thin-film composite (TFC) structure comprising a thin rejection layer (most commonly polyamide), a porous substrate layer, and a non-woven fabric support. This structure enables the independent optimization of the rejection and support layers for an optimized higher performance of the NF membrane. These membranes feature a pore size of approximately 0.5–2 nm with a molecular weight cut-off between 100 and 1000 Da (Guo et al., 2022). The main separation mechanism in NF membranes occurs through size exclusion and electrostatic repulsion forces. Solutes larger

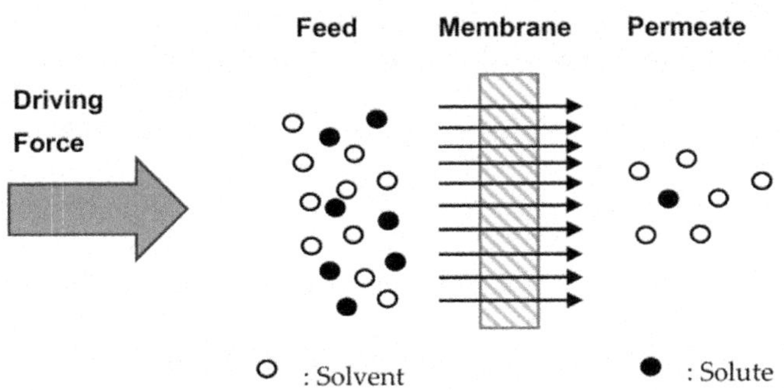


FIGURE 20.1 Schematic representation of the membrane filtration process (Suhalim et al., 2022).

than the pore size get excluded, and charged solutes in the feed get rejected depending on the membrane's surface charge (Suhalim et al., 2022; Guo et al., 2022). The NF membrane's properties are unique since the surface charge is dependent on the functional groups of the membrane (Suhalim et al., 2022). Due to their unique structure, NF membranes effectively remove suspended solids, colloidal particles, bacteria, and organics, along with partial removal of dissolved ions. The NF membranes can be reused through physical and chemical cleaning processes that remove accumulated foulants from the membrane surface (Wang et al., 2015). These cleaning procedures help restore permeate flux and maintain PFAS rejection efficiency during operation. However, overextended use, membrane fouling, scaling, and organic accumulation may gradually reduce membrane performance and require periodic maintenance or replacement (Baudequin et al., 2014). The NF membranes offer several advantages in comparison to conventional water treatment methods, including a small footprint and ease of automation, removal of a wide variety of water contaminants, and the ability to adapt to different feed water conditions. However, membrane fouling remains a critical issue (Tawalbeh et al., 2023; Alyazouri et al., 2025). Additionally, the cost and long-term stability should be taken into account when assessing NF as a water treatment method (Guo et al., 2022).

In contrast, RO operates under a different mechanism. The RO process is a pressure-driven membrane-based process that uses a semipermeable membrane, typically made of polymers, to allow water molecules to pass while blocking solutes like salt (Ali et al., 2025). The separation characteristics of RO depend on the membrane, which relies on the chemical properties of the membrane material (Banks et al., 2020). In RO, separation occurs because water molecules can enter and move through the dense polymer structure of the membrane, whereas dissolved solutes have much lower mobility and permeability within the membrane material. This results in the rejection of the solutes under applied pressure (Suhalim et al., 2022; Banks et al., 2020). The RO membranes don't have distinct pores; they rely on a highly dense, web-like structure that forces water to take a tortuous path, effectively rejecting the smallest particles, including monovalent ions, as opposed to other membranes, including NF, which are designed to remove particles of increasing size (Yang et al., 2019; Greenlee et al., 2009). The RO membranes can be cleaned and reused during operation, although their lifespan is typically limited to about 2–5 years due to fouling and operational wear (Baudequin et al., 2014). However, studies indicate that even chemically cleaned or partially degraded RO membranes can still remove more than 85% of PFAS, suggesting potential for membrane reuse and reduced waste generation in PFAS treatment systems (Wang et al., 2015). This mechanism allows for producing highly purified water, making RO the simplest and most efficient method for water desalination (Matin et al., 2011).

Commercial NF and RO membranes are highly effective in removing PFAS from water, with NF membranes showing a higher rejection rate. The RO membranes remove 96%–100% of PFAS, including both long- and short-chain compounds. Studies on PFOS-contaminated industrial wastewater reported RO rejection rates of >99% across a concentration range of 0.5–1500 ppm (Hara-Yamamura et al., 2022). Furthermore, NF membranes have a higher rejection rate, particularly for long-chain PFAS, such as PFOA and PFOS. The reported removal efficiencies ranged from 90%–99%. For instance, NF90 membranes, with a lower molecular weight cut-off, achieved 94%–99% rejection. This outperformed the NF270 membrane, which removed 95% or more of most PFAS long-chain compounds; however, only 72% of the smaller Perfluoropentanoic acid (PFPeA) compound, which is a short-chain PFAS, with a lower molecular weight. Other commercial NF membranes, such as DK, NF200, and NTR7450, also exhibited >95% rejection rates for various PFAS types (Liu et al., 2022). However, short-chain PFAS generally show lower rejection rates with NF, highlighting RO's superior performance for comprehensive PFAS treatment. The efficiency of PFAS removal by NF/RO membranes based on molecular weight and rejection rates can be seen in Fig. 20.2 (Liu et al., 2022).

Additionally, the performance and PFAS rejection efficiency of commercial NF and RO membranes are influenced by membrane properties, operating conditions, and the chemical composition of the feed water, including the presence of coexisting solutes (Deng, 2025). The key operating parameters for PFAS removal are transmembrane pressure, the pH of the feed solution, electrostatic interactions, ionic strength due to salinity of the feed, and the natural organic matter present. High transmembrane pressure enhances permeate recovery and PFAS rejection, particularly for RO membranes, by increasing the driving force for water transport through the membrane while limiting PFAS passage (Ma et al., 2024). Furthermore, pH adjustments can alter membrane surface charge and pore size, as lower pH levels can compress membrane pores and reduce electrostatic repulsion, which can lead to the enhancement of PFAS rejection by shrinking effective pore diameters (Das and Ronen, 2022). However, electrostatic interactions also play a role when PFAS molecules are smaller than the pore size, especially in NF membranes. In such cases, a negatively charged membrane surface repels the anionic PFAS molecules, improving rejection performance (Bentel et al., 2019; Nguyen et al., 2017).

The effect of coexisting ions plays an important part in the rejection efficiency. Divalent and trivalent cations (e.g., $Ca^{2+}$, $Fe^{3+}$) can bind with PFAS to form larger complexes, which are more easily removed by size exclusion, thereby increasing overall PFAS rejection rates

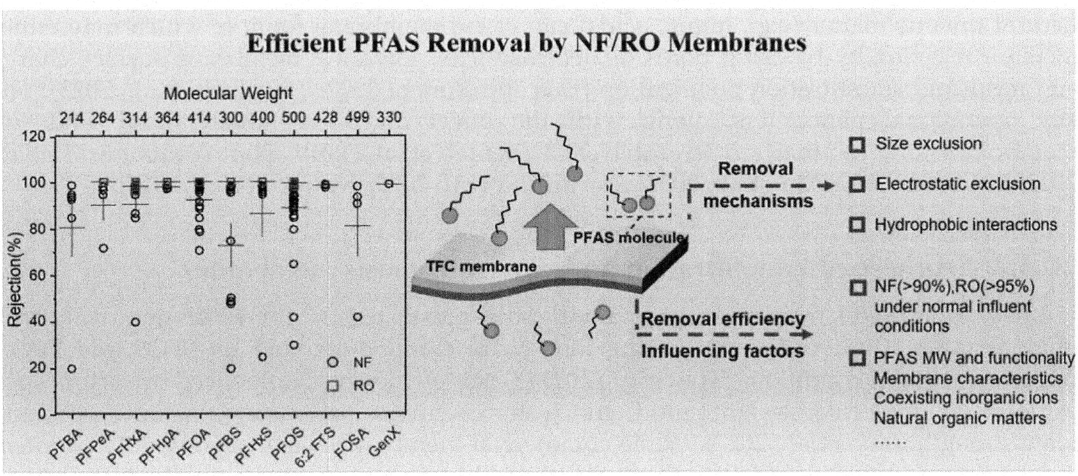


**FIGURE 20.2** Efficiency of per- and polyfluoroalkyl substances (PFAS) removal by nanofiltration (NF)/reverse osmosis (RO) membranes based on molecular weight and rejection rates (Liu et al., 2022).

**TABLE 20.2** Key operational parameters and effects on NF and RO per- and polyfluoroalkyl substances (PFAS) rejection (Bentel et al., 2019; Hara-Yamamura et al., 2022; Liu et al., 2022; Ma et al., 2024; Das and Ronen, 2022; Nguyen et al., 2017).

| Parameter | Operating range | Adjustment | Effect on rejection |
|---|---|---|---|
| Transmembrane pressure | 5–15 bar (NF)<br>10–60 bar (RO) | High pressure | High permeate recovery and PFAS rejection |
| pH | pH 3–10 | Lower pH | Compresses pores, leading to increased rejection as it alters surface charge and reduces electrostatic repulsion |
| Electrostatic interactions | Negative charge at pH > 5–6 | Repulsion between negatively charged PFAS and the membrane surface | Improves rejection |
| Ionic strength (e.g., NaCl) | 1–100 mM NaCl | Salinity reduces surface charge (charge shielding), decreases electrostatic repulsion | Decreases rejection (especially short-chain PFAS) |
| Divalent/trivalent ions (e.g., $Ca^{2+}$, $Fe^{3+}$) | 1–10 mM | Form larger PFAS complexes | Leads to a higher size exclusion, increasing rejection |
| Natural organic matter (e.g., humic acids) | 2–20 mg $L^{-1}$ DOC | May cause fouling | Higher rejection (pore blocking) or lower rejection (reduced surface charge) |

(Das and Ronen, 2022). On the other hand, increased ionic strength (e.g., from NaCl) can reduce membrane surface charge via charge shielding, weakening electrostatic repulsion, and decreasing rejection, particularly for short-chain PFAS (Ma et al., 2024). Similarly, natural organic matter (e.g., humic acids) can cause membrane fouling, which may either increase rejection by blocking pores or decrease it by lowering membrane surface charge and inducing concentration polarization (Das and Ronen, 2022; Tawalbeh et al., 2023). The key operational parameters, along with the effect on the commercial NF and RO membranes, are summarized in Table 20.2 (Bentel et al., 2019; Hara-Yamamura et al., 2022; Liu et al., 2022; Ma et al., 2024; Das and Ronen, 2022; Nguyen et al., 2017).

## 20.3.2 Synthesized nanofiltration and reverse osmosis membranes

The NF and RO membranes are among the most studied for PFAS removal, with demonstrated effectiveness in rejecting long-chain compounds such as PFOS and PFOA under a variety of conditions (Tow et al., 2021). Their efficiency is influenced by factors such as pore size, molecular weight cutoff, and water chemistry parameters, including pH and ionic strength (Guo et al., 2022). While commercial membranes have been widely adopted, there exists a drawback in which they are limited in terms of short-chained PFAS rejection, generic membrane pore size (i.e., not specifically tailored for PFAS removal), and weak electrostatic repulsion forces of the membrane under high ionic feed, all of which lead to an inconsistent rejection performance. Therefore, researchers developed synthesized NF and RO membranes to treat PFAS while addressing these limitations. A recent study reported the fabrication of a hollow fiber NF membrane from poly(m-phenylene isophthalamide) via dry-jet wet spinning, representing a synthesized approach for PFAS treatment using modified NF membranes (Wang et al., 2015). The membrane exhibited a dense selective layer with an average pore size of 0.404 nm and a MWCO of 904 Da, tailored for PFOS rejection (MW 538 Da). Under optimized conditions, including increased pH (3.2–9.5) and higher $Ca^{2+}$ concentration (0.1–2 mM), PFOS rejection improved from ~91% to >99%, with overall efficiencies above 97% for feed concentrations of 50–500 μg $L^{-1}$, as demonstrated in Fig. 20.3 (Wang et al., 2015).

Regarding synthesized RO membranes, one study investigated the efficiency of removing PFAS from contaminated water using a fabricated polyamide thin-film membrane. The researchers used synthesized polyamide composite membranes, specifically XLE and FT-30, which are thin-film polyamide membranes, as well as cellulose acetate membranes, which are made from cellulose diacetate and cellulose triacetate or cellulose monoacetate blends, to evaluate their effectiveness in removing PFAS from water without incorporating any hybrid or nanoparticle-based modifications (Zahmatkesh et al., 2024). On a laboratory and industrial scale, the membranes achieved rejection rates of 99.4% and 99.9% for PFOA in both commercial and domestic water sources (Zahmatkesh et al., 2024). Furthermore, another study focused on treating firefighting waters, particularly from fluorinated surfactants, using a non-hybrid polyamide RO membrane. The membrane used was an RO SG membrane, which was a flat-sheet thin-film polyamide. The membrane showed consistent removal rates of 99.94%–99.97% for fluorinated surfactants at varying concentrations, with permeate concentrations reduced to as low as 10.47 μg $L^{-1}$, meeting the target of <0.1 mg $L^{-1}$ (Baudequin et al., 2014).

The PFAS usually exist in the form of negatively charged anions, PFOS or PFOA. The application of coagulation using aluminum chloride ($AlCl_3$) received attention recently (Al-Qodah et al., 2020). In a study that investigated the use of a hybrid coagulation NF

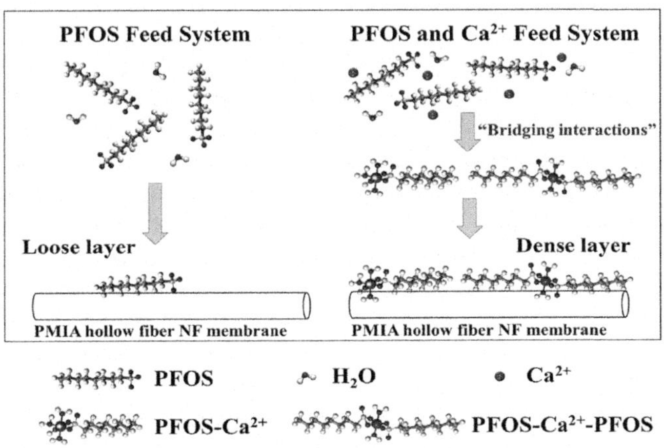


**FIGURE 20.3** Schematic diagram of $Ca^{2+}$ concentration improving per- and polyfluoroalkyl substances (PFOS) rejection efficiency (Wang et al., 2015).

system to remove PFOS from water (Yu et al., 2016), it was found that coagulation (using $AlCl_3$) as a pretreatment improved the removal efficiency of two different NF membranes (HYDRACORE and NF270). Coagulation pretreatment in hybrid NF systems forms loose, permeable flocs that trap PFOS and alleviate fouling pressure on the membrane surface. It was found that PFOS removal was significantly enhanced along with a reduction in membrane fouling. In the system, the coagulation pretreatment removed up to 60% PFOS, while the addition of NF further improved removal efficiency, as the NF270 membrane achieved over 97% rejection, and HYDRACORE improved from 55% to 86% after pretreatment. This hybrid approach offers a cost-effective and efficient strategy for treating trace levels of PFAS in water, especially in natural waters containing humic substances (Yu et al., 2016). The results showed that PFOS removal efficiency was 86% with Coagulation + NF and the HYDRACORE membrane, while >97% efficiency was achieved with the NF270 membrane.

In contrast, the literature surrounding hybrid RO systems for PFAS treatment is not as abundant as that of NF membranes (Liu et al., 2022). However, recent studies have begun to explore the potential of RO-based hybrid systems by incorporating advanced nanomaterials into thin-film nanocomposites (TFN) membrane structures for enhanced PFAS removal. Incorporating materials such as MXene nanosheets into RO-type TFN membranes has been shown to significantly improve PFAS removal efficiency and water permeability, owing to increased hydrophilicity, negative surface charge, and combined electrostatic and size-exclusion mechanisms (Le et al., 2022). For instance, MXene-based TFN membranes achieved over 97% rejection of perfluorohexane sulfonic acid (PFHxS) and perfluorohexanoic acid (PFHxA) by leveraging enhanced hydrophilicity and strong hydrogen bonding interactions between functional membrane surfaces and PFAS moieties (Yogarathinam et al., 2025). Similarly, MOF-incorporated TFN membranes fabricated by capillary-assisted interfacial polymerization achieved PFAS rejection rates ranging from 58% to 95%, with performance influenced by PFAS chain length (Yogarathinam et al., 2025). These nanomaterial-improved membranes also showed improved antifouling behavior, allowing for compatibility with other treatment processes like advanced oxidation processes (Yogarathinam et al., 2025).

### 20.3.3 Ion-exchange membranes

Ion-exchange membranes are an alternative solution to the remediation of PFAS from water. They operate based on electrostatic interactions rather than pressure-driven separation, unlike NF and RO. Ion-exchange membranes selectively target charged PFAS species under the influence of an applied electric field (Amen et al., 2023). In recent years, ion-exchange membranes have gained popularity as a treatment method due to their high selectivity, low energy requirements compared to pressure-driven systems, and compatibility with electrodialysis and hybrid treatment configurations. This section covers ion-exchange membrane systems, including commercial, synthesized, and hybrid membranes.

Ion-exchange membranes operate by using an external electric potential to drive charged species of PFAS through a membrane containing fixed ionic functional groups (Taher et al., 2024). Those functional groups interact with the PFAS molecules through electrostatic forces. During the water treatment, and as mentioned before in this context, PFAS usually exist in the form of negatively charged anions, PFOS or PFOA. This is due to their carboxylate (–$COO^-$) and sulfonate (–$SO_3^-$) headgroups. When these membranes are integrated into electrodialysis systems, an applied electric field induces the migration of these anionic PFAS toward oppositely charged membranes, where they are selectively transported and concentrated into a separate brine stream. The separation efficiency of this system is primarily determined by the magnitude and orientation of the applied electric potential, as well as the membrane charge properties. Additionally, highly charged or larger PFAS molecules tend to be separated more efficiently due to their strong electrostatic interactions with the charged sites on the membrane (Zeidabadi et al., 2023). Ion-exchange resins also offer an advantage in terms of reusability because the adsorption process is reversible and can be restored, allowing the resins to be regenerated and reused after saturation during PFAS removal (Baudequin et al., 2014).

Commercial anion exchange membranes, particularly Type I and Type II, are effective in removing PFAS, such as PFOS and PFOA, from water using electrodialysis (Yadav et al., 2022). These membranes showed high removal efficiencies reaching up to ~99%. These efficiencies were achieved at standard environmental PFAS concentrations typically in the ng $L^{-1}$ to low μg $L^{-1}$ range. The ranges were specific to real wastewater concentrations rather than idealized lab solutions (Yadav et al., 2022). The ion-exchange membrane's performance was highly dependent on charge density and ion exchange capacity. These parameters ensured that the PFAS remained fully ionized, promoting strong electrostatic interactions (Taher et al., 2024).

However, the presence of other competing ions (including chlorides, sulfates, and nitrates) at high concentrations competed with PFAS for availability at the membrane's exchange site, which compromised the efficiency of these membranes (Yadav et al., 2022). This effect is magnified at low PFAS concentrations as studies report that short-chain PFAS (such as PFBA and PFPeA) exhibit noticeably lower removal than long-chain compounds under comparable operating conditions, particularly in the presence of background electrolytes (Taher et al., 2024). Overall, commercial ion-exchange membranes have been reported to achieve high removal rates for long-chain PFAS due to their strong electrostatic interactions. In contrast, there is limited selectivity when it comes to short-chained PFAS (such as PFBA and PFHxA), and they require higher charge densities or enhanced selectivity to achieve a similar removal rate (Amen et al., 2023).

Synthesized ion-exchange membranes have been developed to overcome the limited selectivity and performance variability of commercial membranes, in particular regarding short-chained PFAS and non-standard water PFAS concentrations (Amen et al., 2023). This was achieved by tailoring the membrane chemistry and charge densities of the membrane to enable stronger and more selective electrostatic interactions with anionic PFAS species. For example, membranes composed of quaternized poly(arylene ether ketone) have demonstrated >95% removal of PFOA and PFOS in electrodialysis systems. This result is attributed to their permanent cationic charge and controlled surface chemistry, which reduces competition from background ions (Yadav et al., 2022). One key innovation in these membranes is the fabrication of ultrathin selective layers, often under 100 nm, via methods such as interfacial polymerization. These thin selective layers increase PFAS selectivity without reducing high water permeability. This process is shown in Fig. 20.4 (Park et al., 2020).

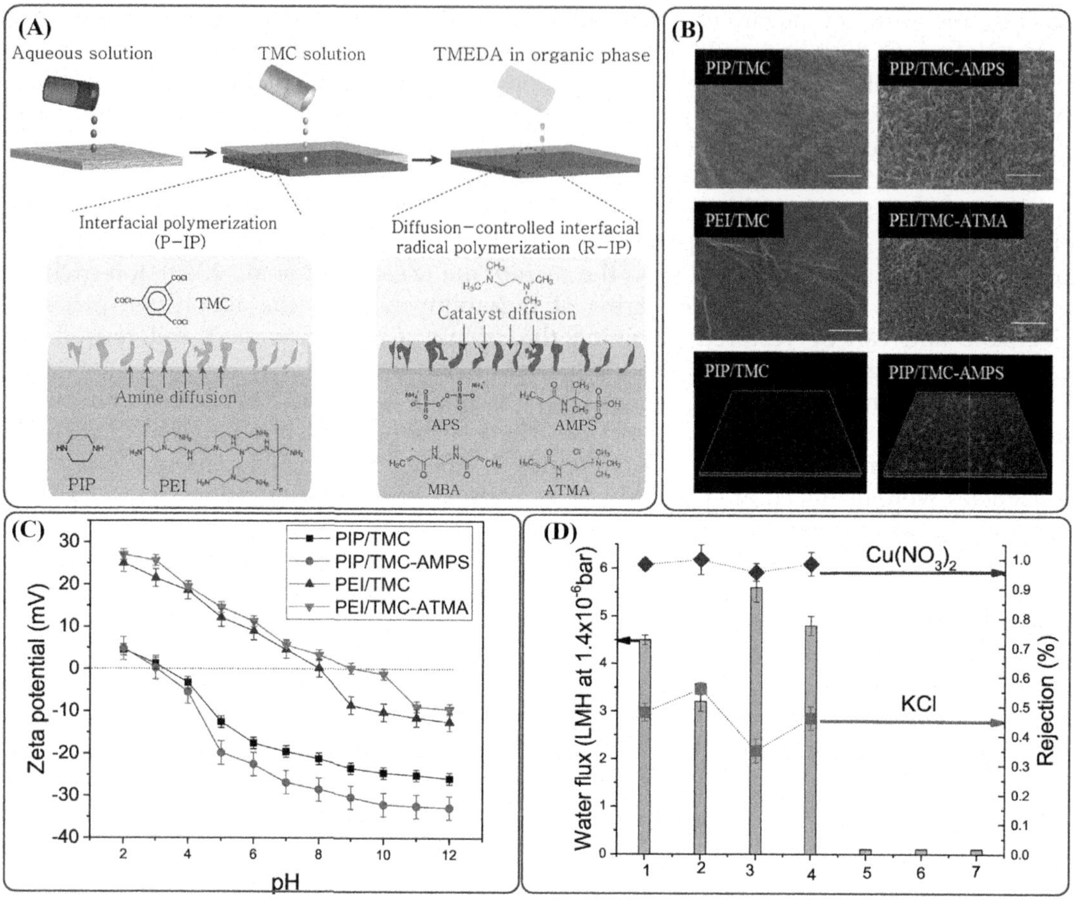


FIGURE 20.4 Schematic diagram of: (A) interfacial polymerization process (B) Scanning electron microscope images of the thin-film membranes (C) shows the zeta potentials of the membranes as a function of pH, and (D) shows water flux and salt rejection for these membranes (Park et al., 2020).

Additionally, some synthesized membranes make use of blended polymers, including poly(ethyleneimine) combined with sulfonated compounds, to optimize charge density and hydrophilicity (Park et al., 2018). These modifications enhance the number of available binding sites, optimizing the adsorption of long- and short-chain PFAS (Murray et al., 2023). Cross-linking strategies are also utilized to enhance mechanical resilience and tolerance to extreme water conditions, such as pH and salinity changes (Zeidabadi et al., 2023). In addition, applying an electric field to synthesized membranes can further enhance performance. Through an electric field, PFAS are driven towards the membrane surface, improving efficiency in removal and making it possible to achieve low-energy usage compared to pressure-driven systems (Amen et al., 2023). The majority of synthesized membranes remain at a pilot or lab scale, and further evaluation under continuous operation and realistic water compositions is required before large-scale implementation.

Hybrid ion-exchange membrane systems, on the other hand, have been developed to solve the limitations of the traditional ion-exchange membrane techniques. These systems combine the selective separation capabilities of ion-exchange membranes with processes that either destroy PFAS or improve operational performance. Depending on the combination, hybrid systems can be designed to either separate and retain PFAS for later treatment or actively degrade it within the same process. One example is the electrochemical electrodialysis-electrooxidation system, which combines anion exchange membranes and electrochemical oxidation. In this setup, electrodialysis is used to concentrate PFAS into an intermediate stream by electrodialysis. This is then degraded using BDD electrodes with an electric current (Sharma et al., 2022). This integrated technique has accomplished over 99% PFAS removal with near-complete destruction of persistent chemicals such as PFOA, thereby lowering the generation of secondary waste as opposed to techniques that work by separating the PFAS only (Yadav et al., 2022).

Moreover, other hybrid configurations have been designed to focus solely on PFAS separation rather than destruction. One system is the flow-electrode capacitive deionization and modular electrodialysis, which utilizes ion-exchange membranes to selectively isolate PFAS from complex or saline water solutions. This allows for a continuous, energy-efficient treatment at low PFAS concentrations, although downstream treatment is required for PFAS disposal (Amen et al., 2023; Park et al., 2020). An alternative hybrid strategy combines ion exchange with NF sequentially. In this approach, NF serves as a pretreatment step to remove the majority of long-chain PFAS, while ion-exchange membranes polish residual short-chain species that pass through the NF membrane. Reducing the PFAS load before ion exchange limits material exhaustion, leading to longer operating periods between regeneration cycles without compromising overall removal performance (Park et al., 2020).

### 20.3.4 Membrane distillation

Membrane distillation (MD) is a thermally driven membrane process that has received increasing attention for the treatment of PFAS-contaminated water. The process relies on hydrophobic microporous membranes and a temperature gradient to separate water vapor from non-volatile compounds (Bahutair et al., 2025). The following section discusses the fundamental separation mechanisms of MD, commercial, synthesized, and hybrid MD systems developed to improve overall PFAS removal.

The MD membranes operate by isolating PFAS from the water through a hydrophobic, microporous membrane that permits only water vapor to permeate in response to a vapor pressure gradient formed between a heated feed and a cooler permeate stream (Yousefi et al., 2024). The PFAS compounds (having very low volatility) are retained effectively in the feed since they cannot evaporate along with the water vapor (Su and Rajan, 2021). The membrane is impermeable to liquid and dissolved components; thus, PFAS are rejected primarily due to their low volatility and the tendency to attach to the hydrophobic membrane surface. Some long-chained PFAS, such as PFOS and PFHxS, adsorb strongly to the membrane surface and exhibit over 99% rejection (Taher et al., 2024). Surfactant-like behavior is observed where hydrophobic PFAS tails attach to the membrane surface and resist transport in the vapor phase. In the case of direct contact membrane distillation (DCMD), adsorption and diffusion resistance are the key to the separation mechanism.

Although PFAS are not vaporizable, surface diffusion is present, notably in short-chained PFAS, such as PFPeA, moving along membrane pore walls without phase change (Chen et al., 2020). This mechanism is demonstrated in Figs. 20.5 and 20.6. Increasing the feed temperature enhances this effect, leading to decreased PFAS rejection due to higher molecular mobility. At low temperatures (e.g., 50°C), PFAS rejection is relatively high; however, under high temperatures (e.g., 70°C), it declines noticeably (Chen et al., 2020). Eventually, PFAS, which accumulate over time on the membrane surface, lead to fouling, which changes the physical properties of the membrane, such as roughness, contact angle, and vapor flux (Yadav et al., 2022). The reusability of MD membranes can be affected by fouling, pore wetting, and chemical interactions with PFAS, which may degrade the membrane structure over time and reduce long-term performance (Taher et al., 2024).

Commercial MD membranes are mostly composed of PTFE and polyvinylidene fluoride (PVDF). They have demonstrated high, consistent performance in the remediation of PFAS, removing more than 99% (Amen et al., 2023). These membranes are designed to be highly hydrophobic and thermally stable in order to withstand elevated temperatures, which are common in MD systems. Commercial MD models are available in flat-sheet and hollow fiber configurations, with each possessing their own operating advantages. Flat sheets offer a more even temperature profile and easier cleaning, while hollow fibers have a lower footprint but are more prone to fouling, as illustrated in Fig. 20.7. In practical applications, commercial MD systems based on these membranes have successfully treated both

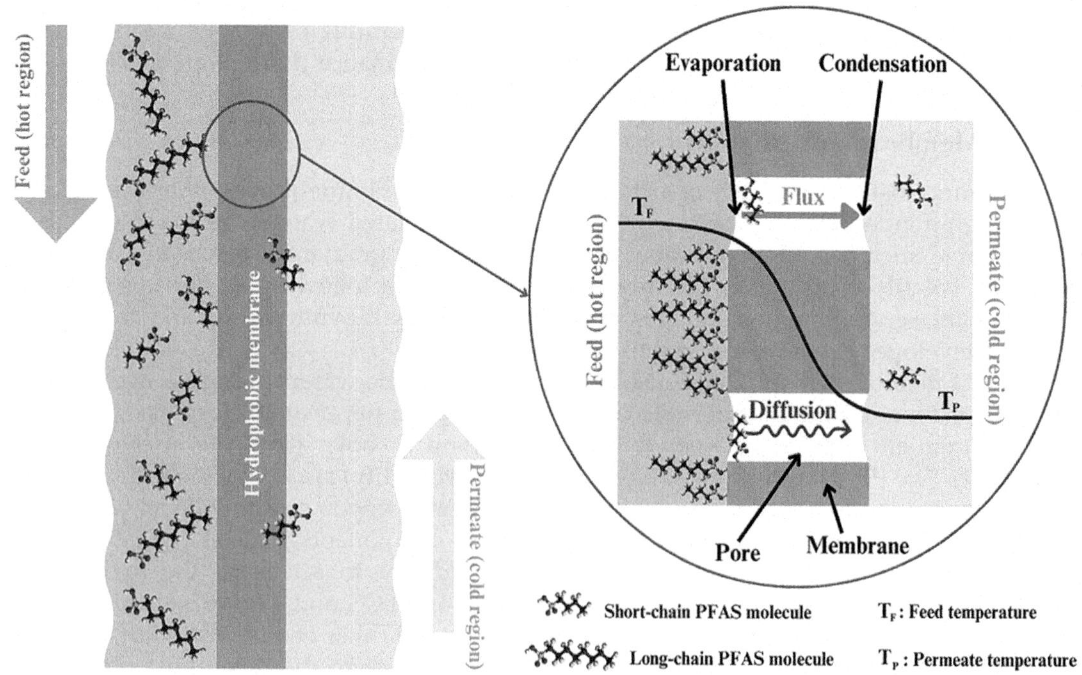


**FIGURE 20.5** Per- and polyfluoroalkyl substances (PFAS) separation mechanism in the membrane distillation (MD) process (Taher et al., 2024).

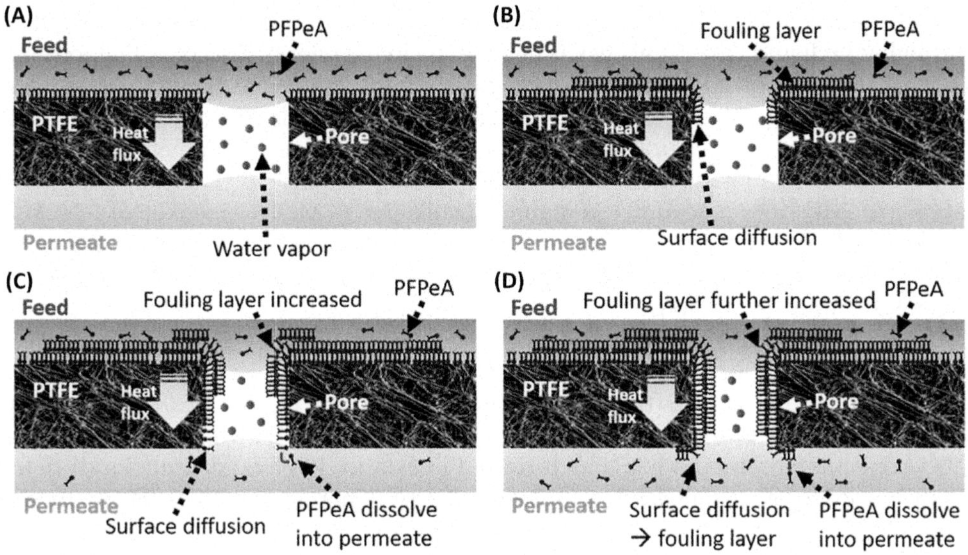


FIGURE 20.6 Direct contact membrane distillation (DCMD) mechanism for perfluoropentanoic acid (PFPeA) removal: (A) initial surface deposition, (B) adsorption forming aggregates, (C) surface diffusion within pores, (D) pore saturation due to PFAS accumulation (Chen et al., 2020).

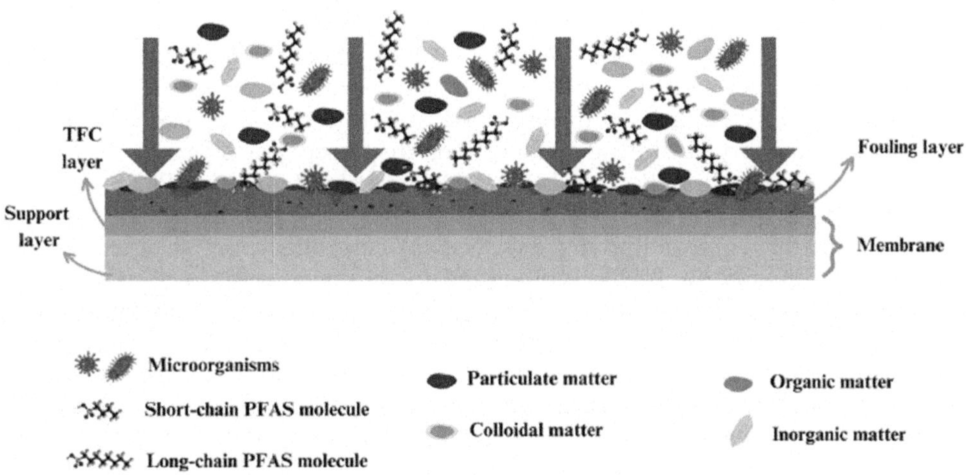


FIGURE 20.7 Overview of fouling in membrane-based technology for per- and polyfluoroalkyl substances (PFAS) removal from water and wastewater (Taher et al., 2024).

municipal and industrial wastewater streams, including mixed PFAS profiles (Nguyen et al., 2017).

The performance of commercial MD membranes is strongly influenced by operating conditions and feed water chemistry. Parameters such as feedwater temperature, flow rate, and vapor pressure gradient directly affect permeate flux and PFAS rejection (Yousefi et al., 2024).

Additionally, maintaining a dry membrane interface is critical. If the membrane wets, PFAS may pass through with liquid water. This risk is heightened in the presence of coexisting ions like salts, bicarbonates, and dissolved organic matter, which can destabilize membrane surfaces and promote pore wetting (Yadav et al., 2022). To avoid this, membrane hydrophobicity and hydraulics must be well controlled by the operating manager. Commercial systems have integrated such a membrane into sophisticated configurations such as air gap membrane distillation (AGMD) and vacuum membrane distillation (VMD) (Amen et al., 2023), which extend energy efficiency and ensure high rejection performance. When properly maintained, these membranes offer long-term durability, minimal degradation, and reliable PFAS removal even under harsh treatment conditions.

Synthesized MD membranes were designed to avoid the disadvantage of current commercial configurations by enhancing wetting, fouling, and structure degradation resistance under harsh PFAS treatment conditions (Chen et al., 2020). In one reported study, a synthesized dual-layer PTFE membrane operated in DCMD and resulted in more than 99% rejection of total PFAS, including PFOS and PFHxS, during landfill leachate treatment without permanent loss of hydrophobicity (Chen et al., 2020). These membranes are generally designed through dual-layer configurations, which comprise the top surface designed to be superhydrophobic and the bottom surface for mechanical stability and thermal support (Zhang et al., 2022). Surface modifications using fluorinated silanes, silica nanoparticles, or inorganic additives, including titanium dioxide ($TiO_2$) and silicon dioxide ($SiO_2$), have tremendously promoted membrane stability and functionality. These additives not only raised the contact angle, indicating superior repellency to water, but also provided stability to membranes under long-term exposure to high temperatures. As such, most synthesized MD membranes have reported PFAS rejection efficiency above 99.8%, even under mixed PFAS species or high contaminant concentrations (Amen et al., 2023).

Apart from surface coatings, controlling pore structure has also seen advancements to provide higher reliability. Uniformly small, well-dispersed pores ensure that the liquid phase is just vapor, excluding dissolved PFAS. Documented failures in PFAS rejection were highly dependent on non-uniform or oversized pores that promoted early wetting and contaminant breakthrough, reinforcing the need for controlled pore structure in synthetic MD membranes (Chen et al., 2020). Synthesized membranes have more control over pore structure, which can be adjusted to achieve maximum vapor flux without compromising on selectivity. Moreover, synthesized membranes have high chemical and thermal stability, which provides higher durability to MD in practical applications such as organically rich wastewater or seawater (Yadav et al., 2022). Having the ability to precisely modify surface, energy, and pore architecture, as well as material composition, synthesized MD membranes are yet another highly emerging path to ensure long-term, high-efficiency PFAS treatment.

Hybrid MD processes complement PFAS treatment by integrating thermal separation and posttreatments to achieve more effective removal and destruction. An important strategy is based on utilizing BDD electrodes, as in hybrid ion exchange systems. In hybrid MD electrochemical oxidation setups, MD effectively concentrates PFAS into a smaller volume stream, after which BDD electrodes achieve near-complete electrochemical destruction of the concentrated PFAS, demonstrating combined volume reduction and contaminant destruction (Chen et al., 2020). Other hybrid MD setups pair the system with adsorption media, nanomaterial-enhanced heat exchangers, or pretreatment filters. For example, after

MD reduces water volume and concentrates PFAS, activated carbon or ion exchange media can capture residual PFAS in the permeate, providing a secondary barrier (Zhang and Xian, 2024). These combinations improve water recovery, minimize membrane fouling, and reduce energy consumption, while extending membrane life (Amen et al., 2023). By coupling MD's thermal-driven concentration with complementary destruction or polishing steps, hybrid systems offer a versatile and scalable solution for PFAS treatment in both drinking water and industrial wastewater contexts.

### 20.3.5 Amyloid fibril-based membranes

Amyloid fibril-based membranes are a sustainable approach for the removal of PFAS from water through the dual mechanisms of electrostatic attraction and hydrophobic interaction. These membranes, derived from β-lactoglobulin protein fibrils, present a strong affinity for both long- and short-chain PFAS, with potential further enhancement through hybridization with activated carbon (Jin et al., 2021). To date, however, only one study has reported the development of such fibril-based membranes, along with a corresponding patent, which highlights both the novelty of the approach and the need for further investigation (Jin et al., 2021). This section explores the separation mechanism of amyloid fibril membranes, the potential for commercial application, new advancements in synthesized membranes, and the integration of hybrid systems across a broad molecular range.

Amyloid fibril-based membranes use a combination of electrostatic attraction and hydrophobic interaction to remove PFAS from water. Under acidic conditions ($pH < 5$), β-lactoglobulin fibrils carry a net positive charge (above their isoelectric point), which enhances the capture of negatively charged PFAS head groups via strong Coulombic interactions (Jin et al., 2021). Simultaneously, the hydrophobic domains of the fibril network allow the highly fluorinated carbon tails of PFAS molecules to partition within them, further enhancing adsorption. Because these protein nanofibrils self-assemble into a dense, interconnected mat with high surface area, even long-chain PFAS (e.g., PFOS, PFHxS) are retained with high efficiency, while short-chain species benefit from the dual binding modes that compensate for their lower hydrophobicity. The membrane material can be separated and replaced once the adsorption sites become saturated, enabling practical use in filtration-based treatment systems. In this process, PFAS molecules are removed primarily through adsorption onto the amyloid fibrils, which provide active binding sites for contaminants. Because the membrane is structured as a solid filtration medium rather than dispersed adsorbent particles, it can be easily handled, removed, and replaced during operation. This design simplifies the management of spent membrane material and supports its application in continuous water treatment systems (Park et al., 2020).

These hybrid amyloid–activated carbon membranes have been further engineered to boost uptake capacity and hydraulic performance by integrating granular activated carbon into the fibril matrix. This hybridization increases available adsorption sites and porosity, yielding over 96% removal of challenging short-chain PFAS, such as PFBA, in passive filtration tests (Dharpure et al., 2023). Most notably, this approach is reliant purely on physicochemical adsorption; no applied voltage or redox reactions are involved, making this method a completely passive and chemical-free process that is suitable for both centralized facilities and decentralized point-of-use systems (Dharpure et al., 2023). Over time, accumulated PFAS can lead to fouling of the composite membrane; however, regeneration protocols (e.g., low-pH backwash) restore performance by disrupting electrostatic bonds and desorbing bound contaminants without damaging the fibrillar architecture.

The application of amyloid fibril-based membranes for PFAS removal in water or wastewater treatment is a relatively new concept and thus lacks extensive literature about the topic. However, commercial applications for the amyloid fibril-based membranes have been studied regarding other pollutants, including heavy metals, arsenic, and radioactive waste, demonstrating their versatility and scalability (Dharpure et al., 2023). In the study referenced earlier, the synthesized membranes that were composed purely of β-lactoglobulin amyloid fibrils were able to efficiently remove over 99% of long- and medium-chain PFASs (with molecular weights ranging from 214–714 Da) (Jin et al., 2021). The pH environment significantly influenced the membrane's performance. At pH ~2, the amyloid fibrils became more positively charged, enhancing electrostatic attraction to the negatively charged PFAS molecules. In addition, hydrophobic interactions were identified as a dominant mechanism for long-chain PFAS removal, as these molecules readily adsorbed to the hydrophobic surface of the amyloid fibrils.

To combat the limitations of the commercial amyloid fibril-based membranes, which have difficulty in removing short-chain PFAS, a hybrid membrane was proposed that combined the porous properties and the adsorption capabilities of activated carbon. The amyloid–-carbon hybrid membrane possesses higher permeability and is able to remove PFAS at a trace level of concentration (Jin et al., 2021). The study tested the hybrid membrane against a cellulose acetate membrane and a pure amyloid fibril membrane, made from β-lactoglobulin protein fibrils. The results showed that the permeability of the hybrid membrane exceeded that of the cellulose acetate and amyloid fibril membranes. Additionally, the PFBA removal efficiency was >96% for the hybrid membrane, noting that PFBA has the lowest molecular weight among the PFAS and is therefore the most difficult to remove. The results from the experiments of this study are summarized in Table 20.3.

The amyloid fibrils in the hybrid membrane offer strong electrostatic and hydrophobic interactions, while the activated carbon offers high surface area and physical adsorption, making the hybrid membrane uniquely powerful, especially for removing the difficult short-chain PFASs that neither component alone could fully address. In addition, the hybrid membrane ranks higher in environmental friendliness, capital expenditures, and operating expenses, energy requirements, long and short-chain PFAS removal efficiency, and water recovery. These criteria are demonstrated in Fig. 20.8 (Jin et al., 2021).

**TABLE 20.3** A summary of the experimental study of the hybrid amyloid–carbon membrane.

| Membrane type | Permeability ($L\ m^{-2}\ h^{-1}\ bar^{-1}$) | PFBA removal | Remarks |
|---|---|---|---|
| Cellulose acetate membrane | 7114.0 ± 280.0 | Poor | Control membrane |
| Amyloid fibril membrane | 301.8 ± 32.7 | Low | Strong for long-chain PFASs only |
| Amyloid–carbon hybrid | 1738.9 ± 176.5 | >96% | Strong for both long- and short-chain PFASs |

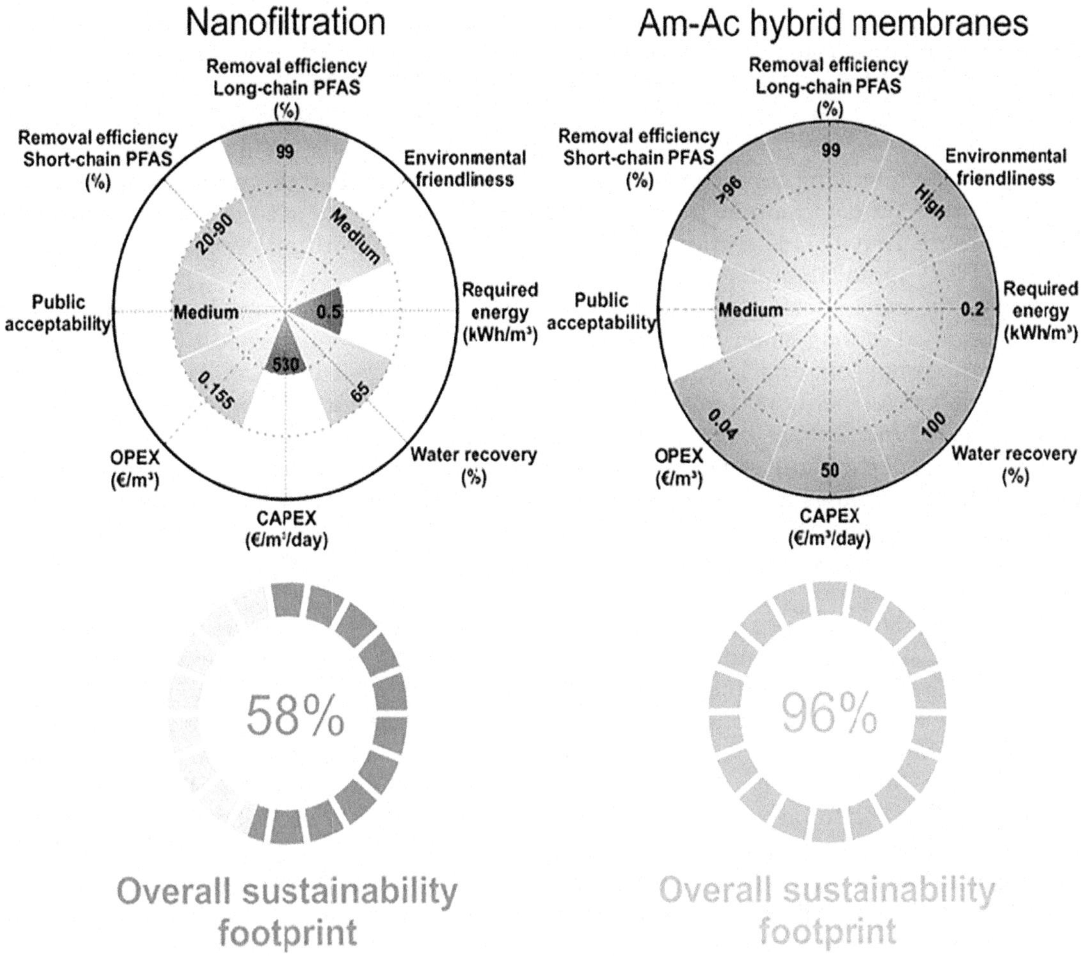


FIGURE 20.8 Comparative analysis of nanofiltration and amyloid–carbon hybrid membranes under different parameters for per- and polyfluoroalkyl substances (PFAS) removal (Jin et al., 2021).

## 20.4 Opportunities of membrane technologies in the remediation of per- and polyfluoroalkyl substances from water and wastewater

Membrane technologies have been considered an effective option for the removal of PFAS from water and wastewater because of their ability to combine physical separation methods with electrostatic and chemical interactions that target these persistent compounds. This section discusses the feasibility of using different membrane methods for PFAS removal, including NF and RO. By describing the operational principles, performance levels, and potential applications, the discussion explores both the opportunities these methods provide for achieving high removal rates and the challenges that can limit their independent use. Overall, this section highlights how membrane technologies can play a major role in advancing PFAS removal strategies, whether applied individually or as part of a hybrid treatment system.

Opportunities for improving PFAS removal across different water bodies depend on innovative technologies whose objectives are to achieve higher efficiency, improved selectivity, and greater reliability for worldwide applications. While existing methods work well for removing long-chain PFAS, new research is creating opportunities to better target short-chain compounds, lower energy use, and make the systems more suitable for large-scale treatment. This section highlights how these opportunities shape the development of PFAS removal strategies.

Among the most effective methods that involve the use of membrane-based technologies for the removal of long-chain PFAS are NF and RO, which are governed by size exclusion, electrostatic repulsion, and hydrophobic interactions between PFAS molecules and membrane surfaces. As mentioned previously, RO membranes have a PFAS rejection rate above 90%, while NF membranes have a removal efficiency between 90% and 99%, depending on membrane properties and water chemistry. Most commercial NF and RO systems use TFC polyamide membranes with ultra-thin active layers. This provides an opportunity to optimize NF and RO membrane performance for long-chain PFAS by adjusting active layer density and operating pressure to maximize efficiency without great flux loss (Yang et al., 2019).

The water matrix composition is the main driver for the removal efficiency in NF and RO. Divalent ions like calcium and magnesium can enhance the size-based rejection property of NF and RO by increasing the molecular size of PFAS, particularly for long-chain compounds. This offers an opportunity to control feed-water chemistry to be able to further improve PFAS retention without membrane modification. Likewise, Hydrophobic organic matter can also enhance the removal of PFAS by interacting with the PFAS molecules, increasing their retention at the membrane surface through adsorption and fouling-layer formation. However, these interactions may also be the reason for accelerated membrane fouling, which leads to higher maintenance and reduced membrane lifespan (Liu et al., 2022). Regardless of the high removal efficiency for long-chain PFAS, NF and RO remain unreliable methods at retaining short-chain PFAS due to their smaller molecular size and weaker hydrophobic interaction with the membranes. Furthermore, both methods produce a PFAS-concentrated brine stream that requires further treatment and handling (Sharma et al., 2022).

Ion-exchange membranes, on the other hand, provide a clear opportunity for PFAS treatment by allowing selective separation of charged PFAS molecules, especially in streams with high PFAS contamination. Electrodialysis is one main opportunity within this approach; it is able to concentrate PFAS into a smaller brine volume, which makes it useful as a pretreatment or preconcentration step before a handling procedure, as it can reduce the overall treatment load (Park et al., 2018).

More opportunities arise from combining ion-exchange membranes with electrically driven systems that include capacitive deionization and electrochemical treatment. These systems allow PFAS to be selectively separated with lower energy demand compared to traditional pressure-driven techniques, and they also enable in situ oxidation reactions to occur at reactive electrodes. Studies report that membrane charge density and thickness directly control the trade-off between PFAS permselectivity and electrical resistance, making membrane design a key lever for improving system efficiency. This separation–degradation approach is particularly promising for the treatment of persistent PFAS molecules. Although ion-exchange membranes are most effective as part of a multistage treatment system, their high selectivity and versatility make them a valuable method in a complete PFAS treatment process (Greenlee et al., 2009).

The MD provides a strong opportunity for PFAS treatment due to its ability to achieve a rejection efficiency of both long and short-chain PFAS through a vapor phase transport across a hydrophobic membrane. As mentioned previously, MD achieves a removal efficiency up to 99% since the separation technique is mainly driven by volatility rather than molecular size or charge. Experimental studies have shown that PFAS transport in MD is governed by their negligible vapor pressure, resulting in near-complete rejection even when molecular size-based mechanisms fail. The DCMD has been the most widely studied approach due to its ability to be easily integrated into existing treatment systems. Other configurations, such as VMD and AGMD, are used in situations where reducing heat loss or improving water recovery is important. Sweeping gas membrane distillation, although less commonly applied, offers greater flexibility in controlling the separation process and is suitable for PFAS-contaminated streams that also contain other cocontaminants.

Opportunities in MD development for PFAS treatment are highly dependent on refinements in membrane materials and systems design. Hydrophobic polymers such as PTFE, PVDF, and polypropylene provide a strong foundation because of their chemical resistance and low surface energy, which limit pore wetting and support stable operation in PFAS-saturated environments (Zeidabadi et al., 2023). Reported liquid entry pressure (LEP) values for these membranes exceed operating pressures in MD systems, creating a design margin that enables long-term operation without pore flooding. Ongoing material research, such as nanofiber surface layers, carbon nanotube-enhanced membranes, surface functionalization, and composite structures, provides multiple opportunities that can further increase vapor flux and operational durability without reducing effectiveness. Additionally, in settings where thermal energy is normally available, MD poses a significant opportunity for deployment in industrial and decentralized treatment, mainly due to its ability to utilize low-grade or waste heat sources (Taher et al., 2024).

Amyloid fibril-based membranes can be used as a standalone method or incorporated into hybrid composites such as amyloid–activated carbon membranes, which improve surface area, selectivity, and removal efficiency. The amyloid fibril network forms nanoscale channels with high binding affinity for fluorinated chains, enabling adsorption-driven PFAS capture even at low concentrations. The hybrid membranes have shown improved efficiency of 96% PFAS removal, even in short-chain PFAS like PFBA, making them suitable for settings that require high performance, like municipal or industrial settings. In terms of fabrication, the use of β-lactoglobulin, a byproduct of dairy processing, makes the membrane production cost-effective and environmentally sustainable (Chen et al., 2020). The membranes can be customized via pH adjustments, particularly at $pH < 5$, to optimize their charge profile and enhance selectivity for negatively charged PFAS. Adding porous carbon scaffolds improves the membrane's strength and flow performance, especially in real-world water conditions. These membranes have already been used in both lab and pilot-scale setups, ranging from batch filtration systems to crossflow setups and modular units, showing they can work in everything from small projects to larger treatment systems.

The use of bio-based nanofibrils also aligns with emerging green chemistry approaches, reducing reliance on synthetic polymers and enhancing biodegradability (Dharpure et al., 2023).

## 20.5 Challenges in per- and polyfluoroalkyl substances removal approaches

Challenges in PFAS removal approaches are due to the technical, operational, and cost-related limitations that limit their wide-scale application. Membrane technologies, ion exchange processes, biochar adsorption, and electrochemical methods all face obstacles, such as fouling, flux decline, membrane wetting, and limited selectivity. In addition, many of these methods face drawbacks such as slow treatment rates, high energy requirements, or limited stability, which further reduce their efficiency and reliability. This section reviews the key challenges associated with different technologies, highlighting the obstacles that must be addressed to improve the effectiveness and long-term reliability of PFAS removal.

The main disadvantages of the NF and RO membranes when used in PFAS water treatment are the fouling, flux decline, and the compromise in permeability. Fouling of the high-pressure membranes can be attributed to the organic matter that forms a layer on the membrane, reducing permeate flux and possibly altering rejection performance. Fouling can either decrease or increase PFAS retention depending on interactions between membrane surface properties, foulants, and PFAS type. Due to fouling, membranes will also start to experience flux decline. The cake-enhanced concentration polarization (CECP) effect may increase PFAS concentration near the membrane surface, decreasing rejection (Yang et al., 2019). In the study that explored synthesized RO membranes for the removal of fluorinated surfactants, fouling was linked to adsorption. This was modeled successfully using the resistance-in-series and Langmuir adsorption models, reinforcing the idea that surfactant-membrane interactions (especially from free monomeric surfactants) are a serious operational limitation. Even though the flux decline in the study stabilized over time, the initial drop was steep, and cleaning (with ethanol/salt solution) did not fully restore the original permeability (Baudequin et al., 2014). These disadvantages hinder the membrane's performance, which impacts its long-term operational efficiency and cost. Additionally, the adsorption of PFAS compounds onto membrane surfaces can lead to incomplete recovery of permeability even after membrane cleaning (Baudequin et al., 2014). Moreover, even though RO and NF membranes are widely used and effective, common materials like cellulose are still sensitive to operating conditions, such as chlorine, which can damage the membrane and reduce its performance over time (Yang et al., 2019). This shows the need for more durable materials that can maintain performance during long-term PFAS treatment.

Like other membranes, ion-exchange membranes are susceptible to fouling and scaling as well. Fouling can occur from organic matter, silica, hardness, and suspended solids, which lowers the ion-exchange efficiency and shortens the membrane's lifespan (Greenlee et al., 2009). Ion-exchange membranes are also selective in filtering out the type of PFAS, as shorter chain compounds slip through the membrane. In addition, the ion-exchange membranes have limited adsorption capacities. Once the shared binding sites are saturated, the membranes can no longer filter out the PFAS; therefore, ion-exchange membranes can't handle the prolonged exposure to PFAS-concentrated feeds (Léniz-Pizarro et al., 2023). Another drawback is the multistep process needed to treat contaminated water using ion-exchange membranes. Since the membranes can separate the PFAS anions through electrostatic interactions, but not degrade them, additional treatment is needed to break down these compounds, such as electrochemical oxidation. Electrochemical oxidation stages are normally limited by high energy demand and electrode material degradation, which in turn hinders their ability to be in an ion exchange multistep process. This reliance on multistep processes adds complexity and increases both operational and capital costs (Araújo et al., 2022).

Membrane wetting is one of the biggest challenges for the MD technique in treating PFAS-contaminated waters. This is due to the PFAS reducing the membrane's natural hydrophobic properties. Once that barrier is weakened, water and contaminants can leak through, making the treatment less effective. Short-chain PFAS pose a challenge, especially for membrane wetting, since the molecules are smaller and more mobile, which tends to cause membrane wetting more quickly (Yadav et al., 2022). To overcome this issue, researchers modify membranes with special coatings or additives; however, those fixes can increase costs and may shorten how long the membrane lasts. Additionally, fouling is a common problem in MD. Over time, the accumulation of salts, natural organic matter, and microbial growth on the membrane surface can obstruct flow pathways and degrade performance, ultimately requiring frequent cleaning and maintenance (Zeidabadi et al., 2023). The PFAS is able to lower LEP by accumulating at the vapor-liquid interface within the membrane pores, which directly speeds up pore formation at MD operating conditions (Yadav et al., 2022). In comparison to pressure-driven processes such as RO, MD typically exhibits lower water flux, so it processes a smaller volume of water over the same period. As a result, achieving higher treatment capacities will require the use of larger membrane surface areas.

From an energy standpoint, MD systems rely on thermal input to operate. Although they can utilize low-grade or waste heat sources, this requirement limits applicability in settings where such heat sources are unavailable or inconsistent. Moreover, MD is a separation process rather than a destructive one; while it effectively separates PFAS, it does not break them down. Consequently, the resulting concentrated waste still demands further treatment to ensure safe disposal (Taher et al., 2024).

Similar to ion-exchange membranes, amyloid fibril-based membranes' biggest setback is the multistep process required to treat the PFAS-contaminated wastewater. Since the membranes work through adsorption and not degradation, extra steps are taken to destroy the complex molecules, otherwise creating a secondary waste burden. Additionally, water sensitivity is an important problem, as the presence of competing contaminants can block the active site of the membrane and interfere with selective PFAS binding (Chen et al., 2020). Another issue is the cost and complexity of the materials needed to build the amyloid fibril-based membranes. The assembly of the functional fibrils requires controlled heating and specific acidic conditions, which affect the system's consistent reproducibility across large-scale batches. Furthermore, amyloid-fibril membranes have reduced efficiency in the removal of short-chained PFAS compounds due to their weaker hydrophobic interactions (Dharpure et al., 2023).

## 20.6 Future directions and challenges

To address the limitations noted above, future research should focus on improving the reliability and lifespan of the membranes under real-life water matrices. In practical applications, membrane fouling remains one of the main factors limiting long-term performance, as adsorption, pore blockage, and cake layer formation can significantly reduce efficiency and shorten membrane lifespan. Addressing this issue requires greater attention to fouling control strategies, some of which include pretreatment processing, chemical and physical cleaning methods, and the development of more fouling-resistant membrane materials (Najid et al., 2025). Operational cost and energy demand also continue to limit widespread implementation. In order to reduce operational costs and energy consumption without compromising the membrane performance, a solution proposed was to develop membranes with higher specific fluxes while maintaining the PFAS rejection rate (Liu et al., 2025). Advancing these aspects will be necessary before membrane-based PFAS treatment can be reliably applied on a large scale. The thermal degradation process, on the other hand, can be used to degrade or break PFAS compounds. Extremely high temperatures (980°C–1200°C) and long residence time are needed to overcome PFAS's thermal stability. The degradation results in converting PFAS into smaller inorganic molecules, such as carbon monoxide, carbon dioxide, water, and hydrogen fluoride, among others (Meegoda et al., 2022). Furthermore, a comparative overview of current PFAS treatment technologies based on their practicality and stage of development is presented in Fig. 20.9. Established systems such as RO, NF, and ion-exchange resins are positioned closer to mature implementation, while MD and hybrid membrane processes remain largely experimental (Taher et al., 2024). This comparison highlights that beyond removal efficiency, parameters such as operational stability, scalability, and energy demand must be considered during membrane design and application.

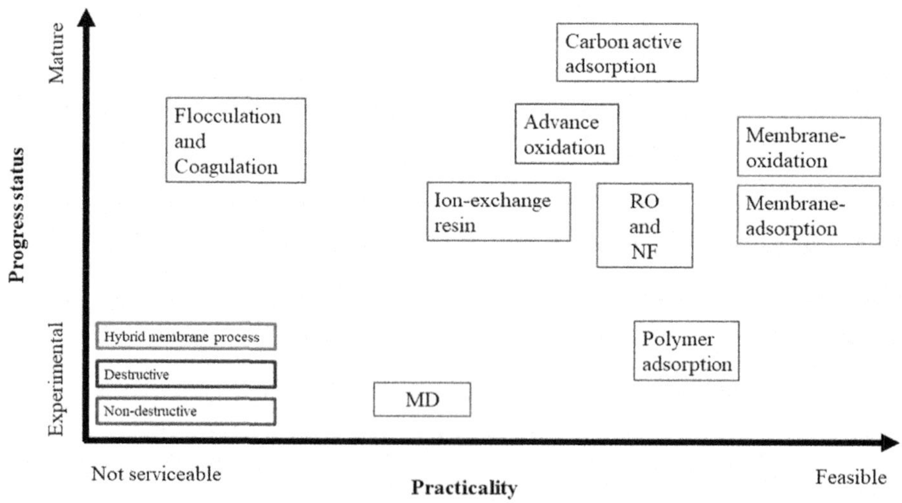


FIGURE 20.9 Investigated technologies for remediation of per- and polyfluoroalkyl substances (PFAS) from the aqueous environment, along with their maturity and feasibility status (Taher et al., 2024).

## 20.7 Conclusions

Despite the continuous development, PFAS treatment remains limited by the inability of existing methods to successfully capture and destroy all PFAS types, highlighting the continuous need for enhanced development of hybrid and integrated systems. This chapter discussed membrane technologies, such as NF, RO, MD, and ion-exchange membranes, which are all flexible and reliable options for treating PFAS in water and wastewater. These methods operate on effective mechanisms that include size exclusion, electrostatic repulsion, adsorption, and vapor transport, which makes them appropriately fitting for multiple scales of water treatment applications; however, high-volume scaling of the treatment method remains a challenge. Additionally, an important consideration after separating PFAS from water is its management or disposal. One way where PFAS can be destroyed is through thermal degradation, where extremely high temperatures (980°C–1200°C) and long residence time are needed to overcome PFAS's thermal stability. The degradation results in converting PFAS into smaller inorganic molecules, such as carbon monoxide, carbon dioxide, water, and hydrogen fluoride, among others (Meegoda et al., 2022). Another method is supercritical water oxidation, where PFAS, which don't normally oxidize at normal temperatures and pressures, can be broken down at water's supercritical phase (Krause et al., 2022). While PFAS destruction methods are abundant, they still need to be further studied as their large-scale implementation remains limited by the potential formation of hazardous byproducts, high energy demands, and insufficient full-scale validation. Moreover, combining multiple removal strategies in order to enhance progress, stability, and selectivity shows great potential in hybridized and synthesized membrane technologies, as demonstrated in hybrid amyloid–carbon membranes and ion exchange resins. However, the continuous development and optimization of these materials, as well as cumulative techno-economic assessments and practical stability evaluations, remain a priority. Achieving large-scale use and meaningful reductions in PFAS contamination depend on continued research, improved membrane designs, and supportive regulations that encourage further development of these treatment methods. This chapter also concludes that more research into the stability, cost, and sustainable production of these integrated systems is needed before they can be practically applied.

## References

Abuwatfa, W.H., Al-Muqbel, D., Al-Othman, A., Halalsheh, N., Tawalbeh, M., 2021. Insights into the removal of microplastics from water using biochar in the era of COVID-19: a mini review. Case Studies in Chemical and Environmental Engineering 4, 100151. Available at https://doi.org/10.1016/j.cscee.2021.100151.

Al Sharabati, M., et al., 2021. Biodegradable polymers and their nano-composites for the removal of endocrine-disrupting chemicals (EDCs) from wastewater: a review. Environmental Research 202, 111694. Available at https://doi.org/10.1016/j.envres.2021.111694.

Al-Qodah, Z., Tawalbeh, M., Al-Shannag, M., Al-Anber, Z., Bani-Melhem, K., 2020. Combined electrocoagulation processes as a novel approach for enhanced pollutants removal: a state-of-the-art review. The Science of the Total Environment 744, 140806. Available at https://doi.org/10.1016/j.scitotenv.2020.140806.

Ali, A.A., Al-Othman, A., Tawalbeh, M., Ali, A., Quist-Jensen, C.A., Shirazi, M.M.A., 2025. Membrane technologies for sustainable development goals: a critical review of bright horizons. Journal of Environmental Chemical Engineering 13 (1), 114998. Available at https://doi.org/10.1016/j.jece.2024.114998.

Alyazouri, S., Tawalbeh, M., Al-Othman, A., 2025. Mitigation techniques of membranes' biofouling in bioelectrochemical cells (BEC cells): recent advances. Membranes (Basel) 15 (11), 332. Available at https://doi.org/10.3390/membranes15110332.

Amen, R., Ibrahim, A., Shafqat, W., Hassan, E.B., 2023. A critical review on PFAS removal from water: removal mechanism and future challenges. Sustainability (Switzerland) 15 (23). Available at https://doi.org/10.3390/su152316173.

Araújo, R.G., et al., 2022. Detection and tertiary treatment technologies of poly-and perfluoroalkyl substances in wastewater treatment plants. Frontiers in Environmental Science 10. Available at https://doi.org/10.3389/fenvs.2022.864894.

Bahutair, W., Darra, R., Al-Othman, A., Tawalbeh, M., McPhedran, K., Shirazi, M.M.A., 2025. Purifying the future: membrane technologies for ultrapure water supply in hydrogen production. Desalination 614, 119174. Available at https://doi.org/10.1016/J.DESAL.2025.119174.

Banks, D., Jun, B.M., Heo, J., Her, N., Park, C.M., Yoon, Y., 2020. Selected advanced water treatment technologies for perfluoroalkyl and polyfluoroalkyl substances: a review. Separation and Purification Technology 231. Available at https://doi.org/10.1016/j.seppur.2019.115929.

Baudequin, C., et al., 2014. Removal of fluorinated surfactants by reverse osmosis - role of surfactants in membrane fouling. Journal of Membrane Science 458. Available at https://doi.org/10.1016/j.memsci.2014.01.063.

Bentel, M.J., et al., 2019. Defluorination of per- and polyfluoroalkyl substances (PFASs) with hydrated electrons: structural dependence and implications to PFAS remediation and management. Environmental Science & Technology 53 (7). Available at https://doi.org/10.1021/acs.est.8b06648.

Bonefeld-Jørgensen, E.C., Long, M., Fredslund, S.O., Bossi, R., Olsen, J., 2014. Breast cancer risk after exposure to perfluorinated compounds in Danish women: a case–control study nested in the Danish National Birth Cohort. Cancer Causes & Control: CCC 25 (11). Available at https://doi.org/10.1007/s10552-014-0446-7.

Chaudhary, M., Sela-Adler, M., Ronen, A., Nir, O., 2023. Efficient PFOA removal from drinking water by a dual-functional mixed-matrix-composite nanofiltration membrane. npj Clean Water 6 (1). Available at https://doi.org/10.1038/s41545-023-00286-2.

Chen, X., et al., 2020. Direct contact membrane distillation for effective concentration of perfluoroalkyl substances – impact of surface fouling and material stability. Water Research 182. Available at https://doi.org/10.1016/j.watres.2020.116010.

Cousins, I.T., et al., 2020. Strategies for grouping per-and polyfluoroalkyl substances (PFAS) to protect human and environmental health. Environmental Science: Processes and Impacts 22 (7). Available at https://doi.org/10.1039/d0em00147c.

Das, S., Ronen, A., 2022. A review on removal and destruction of per-and polyfluoroalkyl substances (PFAS) by novel membranes. Membranes 12 (7). Available at https://doi.org/10.3390/membranes12070662.

Deng, Y., 2025. Membrane processes for the removal of PFAS from water, In: Per- and Polyfluoroalkyl Substance Treatment Technologies. doi: 10.1039/9781837671779-00062.

Dharpure, R., Pramanik, S., Pradhan, A., 2023. In silico analysis decodes transthyretin (TTR) binding and thyroid disrupting effects of per- and polyfluoroalkyl substances (PFAS). Archives of Toxicology 97 (3). Available at https://doi.org/10.1007/s00204-022-03434-8.

Dickman, R.A., Aga, D.S., 2022. A review of recent studies on toxicity, sequestration, and degradation of per- and polyfluoroalkyl substances (PFAS). Journal of Hazardous Materials 436. Available at https://doi.org/10.1016/j.jhazmat.2022.129120.

Greenlee, L.F., Lawler, D.F., Freeman, B.D., Marrot, B., Moulin, P., 2009. Reverse osmosis desalination: water sources, technology, and today's challenges. Water Research 43 (9), 2317–2348. Available at https://doi.org/10.1016/j.watres.2009.03.010.

Guo, H., et al., 2022. Nanofiltration for drinking water treatment: a review. Frontiers of Chemical Science and Engineering 16 (5), 681–698. Available at https://doi.org/10.1007/s11705-021-2103-5.

Hara-Yamamura, H., Inoue, K., Matsumoto, T., Honda, R., Ninomiya, K., Yamamura, H., 2022. Rejection of perfluorooctanoic acid (PFOA) and perfluorooctane sulfonate (PFOS) by severely chlorine damaged RO membranes with different salt rejection ratios. Chemical Engineering Journal 446. Available at https://doi.org/10.1016/j.cej.2022.137398.

Jin, T., Peydayesh, M., Joerss, H., Zhou, J., Bolisetty, S., Mezzenga, R., 2021. Amyloid fibril-based membranes for PFAS removal from water. Environmental Science: Water Research & Technology 7 (10). Available at https://doi.org/10.1039/d1ew00373a.

Krause, M.J., et al., 2022. Supercritical water oxidation as an innovative technology for PFAS destruction. Journal of Environmental Engineering 148 (2). Available at https://doi.org/10.1061/(asce)ee.1943-7870.0001957.

Le, T., Jamshidi, E., Beidaghi, M., Esfahani, M.R., 2022. Functionalized-MXene thin-film nanocomposite hollow fiber membranes for enhanced PFAS removal from water. ACS Applied Materials & Interfaces 14 (22). Available at https://doi.org/10.1021/acsami.2c03796.

Liu, C., et al., 2022. Evaluating the efficiency of nanofiltration and reverse osmosis membrane processes for the removal of per- and polyfluoroalkyl substances from water: a critical review. Separation and Purification Technology 302. Available at https://doi.org/10.1016/j.seppur.2022.122161.

Liu, C., Ho, H.J., Iizuka, A., 2025. Filtration and electrical membrane-based treatment methods for PFAS-contaminated water and preparation methods for the membranes employed. Applied Water Science 15 (11). Available at https://doi.org/10.1007/s13201-025-02644-6.

Li, R., Adarsh, N.N., Lu, H., Wriedt, M., 2022. Metal-organic frameworks as platforms for the removal of per- and polyfluoroalkyl substances from contaminated waters. Matter 5 (10). Available at https://doi.org/10.1016/j.matt.2022.07.028.

Léniz-Pizarro, F., Rudel, H.E., Briot, N.J., Zimmerman, J.B., Bhattacharyya, D., 2023. Membrane functionalization approaches toward per- and polyfluoroalkyl substances and selected metal ion separations. ACS Applied Materials & Interfaces 15 (37). Available at https://doi.org/10.1021/acsami.3c08478.

MacInnis, J.J., et al., 2019. Fate and transport of perfluoroalkyl substances from snowpacks into a lake in the High Arctic of Canada. Environmental Science & Technology 53 (18). Available at https://doi.org/10.1021/acs.est.9b03372.

Matin, A., Khan, Z., Zaidi, S.M.J., Boyce, M.C., 2011. Biofouling in reverse osmosis membranes for seawater desalination: phenomena and prevention. Desalination 281 (1), 1–16. Available at https://doi.org/10.1016/j.desal.2011.06.063.

Ma, Q., Lei, Q., Liu, F., Song, Z., Khusid, B., Zhang, W., 2024. Evaluation of commercial nanofiltration and reverse osmosis membrane filtration to remove per-and polyfluoroalkyl substances (PFAS): effects of transmembrane pressures and water matrices. Water Environment Research: A Research Publication of the Water Environment Federation 96 (2). Available at https://doi.org/10.1002/wer.10983.

Meegoda, J.N., Bezerra de Souza, B., Casarini, M.M., Kewalramani, J.A., 2022. A review of PFAS destruction technologies. International Journal of Environmental Research and Public Health 19 (24), 25. Available at https://doi.org/10.3390/ijerph192416397.

Meegoda, J.N., Kewalramani, J.A., Li, B., Marsh, R.W., 2020. A review of the applications, environmental release, and remediation technologies of per-and polyfluoroalkyl substances. International Journal of Environmental Research and Public Health 17 (21). Available at https://doi.org/10.3390/ijerph17218117.

Murray, C.C., et al., 2023. PFAS adsorbent selection: the role of adsorbent use rate, water quality, and cost. Journal of Hazardous Materials 454. Available at https://doi.org/10.1016/j.jhazmat.2023.131481.

Najid, N., Gourich, B., Vial, C., Necibi, M.C., 2025. Fouling dynamics in water purification membranes: understanding, modeling, and mitigation strategies, In: Membrane Assisted Water Purificatio. Available at https://doi.org/10.1016/B978-0-443-32968-5.00013-9.

Nguyen, T.T., Bui, X.T., Luu, V.P., Nguyen, P.D., Guo, W., Ngo, H.H., 2017. Removal of antibiotics in sponge membrane bioreactors treating hospital wastewater: comparison between hollow fiber and flat sheet membrane systems. Bioresource Technology 240, 42–49. Available at https://doi.org/10.1016/j.biortech.2017.02.118.

Panieri, E., Baralic, K., Djukic-Cosic, D., Djordjevic, A.B., Saso, L., 2022. PFAS molecules: a major concern for the human health and the environment. Toxics 10 (2). Available at https://doi.org/10.3390/toxics10020044.

Park, C.H., et al., 2020. Foldable multiple-energies harvester consisting of a thin ion-exchange membrane prepared by a two-step interfacial polymerization. Desalination 476. Available at https://doi.org/10.1016/j.desal.2019.114242.

Park, C.H., et al., 2018. Thin film composite membrane prepared by interfacial polymerization as an ion exchange membrane for salinity gradient power. Journal of Industrial and Engineering Chemistry 59. Available at https://doi.org/10.1016/j.jiec.2017.10.044.

Roscales, J.L., Vicente, A., Ryan, P.G., González-Solís, J., Jiménez, B., 2019. Spatial and interspecies heterogeneity in concentrations of perfluoroalkyl substances (PFASs) in seabirds of the southern ocean. Environmental Science & Technology 53 (16). Available at https://doi.org/10.1021/acs.est.9b02677.

Saawarn, B., Mahanty, B., Hait, S., Hussain, S., 2022. Sources, occurrence, and treatment techniques of per- and polyfluoroalkyl substances in aqueous matrices: a comprehensive review. Environmental Research 214. Available at https://doi.org/10.1016/j.envres.2022.114004.

Sharma, N., et al., 2024. A comprehensive review on the need for integrated strategies and process modifications for per- and polyfluoroalkyl substances (PFAS) removal: current insights and future prospects. Case Studies in Chemical and Environmental Engineering 9. Available at https://doi.org/10.1016/j.cscee.2024.100623.

Sharma, S., Shetti, N.P., Basu, S., Nadagouda, M.N., Aminabhavi, T.M., 2022. Remediation of per- and polyfluoroalkyls (PFAS) via electrochemical methods. Chemical Engineering Journal 430. Available at https://doi.org/10.1016/j.cej.2021.132895.

Suhalim, N.S., et al., 2022. Rejection mechanism of ionic solute removal by nanofiltration membranes: an overview. Nanomaterials 12 (3), 437. Available at https://doi.org/10.3390/NANO12030437/S1.

Su, A., Rajan, K., 2021. A database framework for rapid screening of structure-function relationships in PFAS chemistry. Scientific Data 8 (1). Available at https://doi.org/10.1038/s41597-021-00798-x.

Taher, M.N., Al-Mutwalli, S.A., Barisci, S., Koseoglu-Imer, D.Y., Dumée, L.F., Shirazi, M.M.A., 2024. Progress on remediation of per- and polyfluoroalkyl substances (PFAS) from water and wastewater using membrane technologies: a review. Journal of Water Process Engineering 59. Available at https://doi.org/10.1016/j.jwpe.2024.104858.

Tawalbeh, M., Al Mojjly, A., Al-Othman, A., Hilal, N., 2018. Membrane separation as a pre-treatment process for oily saline water. Desalination 447, 182–202. Available at https://doi.org/10.1016/J.DESAL.2018.07.029.

Tawalbeh, M., Aljaghoub, H., Qasim, M., Al-Othman, A., 2023. Surface modification techniques of membranes to improve their antifouling characteristics: recent advancements and developments. Frontiers of Chemical Science and Engineering 17 (12), 1837–1865. Available at https://doi.org/10.1007/s11705-023-2347-3.

Tawalbeh, M., Qalyoubi, L., Al-Othman, A., Qasim, M., Shirazi, M., 2023. Insights on the development of enhanced antifouling reverse osmosis membranes: industrial applications and challenges. Desalination 553, 116460. Available at https://doi.org/10.1016/j.desal.2023.116460.

Tow, E.W., et al., 2021. Managing and treating per-and polyfluoroalkyl substances (Pfas) in membrane concentrates. AWWA Water Science 3 (5). Available at https://doi.org/10.1002/aws2.1233.

Wang, T., Zhao, C., Li, P., Li, Y., Wang, J., 2015. Fabrication of novel poly(m-phenylene isophthalamide) hollow fiber nanofiltration membrane for effective removal of trace amount perfluorooctane sulfonate from water. Journal of Membrane Science 477. Available at https://doi.org/10.1016/j.memsci.2014.12.038.

Yadav, S., et al., 2022. Updated review on emerging technologies for PFAS contaminated water treatment. Chemical Engineering Research and Design 182. Available at https://doi.org/10.1016/j.cherd.2022.04.009.

Yanagida, A., Webb, E., Harris, C.E., Christenson, M., Comfort, S., 2022. Using electrochemical oxidation to remove PFAS in simulated investigation-derived waste (IDW): laboratory and pilot-scale experiments. Water (Switzerland) 14 (17). Available at https://doi.org/10.3390/w14172708.

Yang, Z., Zhou, Y., Feng, Z., Rui, X., Zhang, T., Zhang, Z., 2019. A review on reverse osmosis and nanofiltration membranes for water purification. Polymers 11 (8). Available at https://doi.org/10.3390/polym11081252.

Yogarathinam, L.T., Baig, U., Waheed, A., Aljundi, I.H., 2025. Recent progress and challenges in nanocomposite membranes for efficient removal of hazardous per-/polyfluoroalkyl substances from water sources. Journal of Environmental Chemical Engineering 13 (2).

Yousefi, A., et al., 2024. Evaluating the efficiency of modified hydrophobic PVDF membrane for the removal of PFOA substances from water by direct contact membrane distillation. Desalination 579. Available at https://doi.org/10.1016/j.desal.2024.117509.

Yu, Y., Zhao, C., Yu, L., Li, P., Wang, T., Xu, Y., 2016. Removal of perfluorooctane sulfonates from water by a hybrid coagulation-nanofiltration process. Chemical Engineering Journal 289. Available at https://doi.org/10.1016/j.cej.2015.12.048.

Zahmatkesh, S., Chen, Z., Khan, N.A., Ni, B.J., 2024. Removing polyfluoroalkyl substances (PFAS) from wastewater with mixed matrix membranes. Science of the Total Environment 912. Available at https://doi.org/10.1016/j.scitotenv.2023.168881.

Zeidabadi, F.A., Esfahani, E.B., Mohseni, M., 2023. Effects of water matrix on per- and poly-fluoroalkyl substances (PFAS) treatment: physical-separation and degradation processes – a review. Journal of Hazardous Materials Advances 10. Available at https://doi.org/10.1016/j.hazadv.2023.100322.

Zhang, J., Huang, Z., Gao, L., Gray, S., Xie, Z., 2022. Study of MOF incorporated dual layer membrane with enhanced removal of ammonia and per-/poly-fluoroalkyl substances (PFAS) in landfill leachate treatment. The Science of the Total Environment 806. Available at https://doi.org/10.1016/j.scitotenv.2021.151207.

Zhang, H., Xian, H., 2024. Review of hybrid membrane distillation systems. Membranes 14 (1). Available at https://doi.org/10.3390/membranes14010025.

PART 4

# Policy and regulation

CHAPTER

# 21

# Regulatory approaches to per- and polyfluoroalkyl substance contamination in the United States and Canada

*Sai Sharanya Pulimamidi*[1], *Rushikesh Bandal*[1], *Srikanth Ponneganti*[2], *Saurabh Bodake*[1], *Dhanavath Dattu Naik*[1], *and Roshan M. Borkar*[1]

**[1]Department of Pharmaceutical Analysis, National Institute of Pharmaceutical Education and Research, Guwahati, Changsari, Assam, India [2]Department of Food and Science and Technology, The Ohio State University, Columbus, OH, United States**

## 21.1 Introduction

Per- and polyfluoroalkyl substances (PFAS) are a group of synthetic fluorinated organic compounds characterized by diverse functional groups and chain lengths. Due to their exceptional chemical and thermal stability, PFAS have been widely used in the industrial production of consumer goods across more than 200 categories, including surfactants, waterproof fabrics, firefighting foams, textiles, leather, oil-repellent containers, semiconductors, nonstick cookware, food packaging, shampoos, photographic films, cleaning agents, and masks. Their persistence has led to extensive use but has also raised significant global concern about health hazards, including disruptions in metabolic functions, immune response, reproductive health, and an increased risk of cancer (Rekik et al., 2024; Zhang et al., 2022; Perfluoroalkyl and Polyfluoroalkyl Substances PFAS, National).

PFAS were developed in the 1930s and came into widespread use during the 1940s and 1950s. They have since been used in a wide range of industrial and consumer products, including firefighting foams, aerospace applications, and nonstick or water-repellent materials. These compounds contain highly stable carbon-fluorine bonds, among the

*Per- and Polyfluoroalkyl Substances (PFAS) as Environmental Contaminants*
DOI: https://doi.org/10.1016/B978-0-443-43856-1.00001-0

strongest in organic chemistry, making PFAS exceptionally resistant to heat, water, oil, and degradation (O'Hagan, 2008; Brendel et al., 2018).

Federal regulation of PFAS in the United States is conducted under several laws, including the Toxic Substances Control Act (TSCA), the Safe Drinking Water Act (SDWA), and the Comprehensive Environmental Response, Compensation, and Liability Act (CERCLA) (Per- and Polyfluoroalkyl Substances PFAS, 2025). In recent years, the US Environmental Protection Agency (EPA) has implemented a series of actions to address PFAS, beginning with the release of its PFAS action plan, which outlines strategies to reduce public health risks associated with PFAS exposure (Fact Sheet, 2010; Federal Register: Revisions to the Unregulated Contaminant Monitoring Regulation UCMR 3 for Public Water Systems, 2025; EPA's Per- and Polyfluoroalkyl Substances PFAS, Action; Fourth national report on human exposure to environmental chemicals: updated tables, 2019). In the United States, awareness of PFAS risks increased after biomonitoring studies detected legacy compounds such as perfluorooctane sulfonic acid (PFOS) in nearly the entire population, prompting regulatory and industry responses (Calafat et al., 2007; Fourth national report on human exposure to environmental chemicals: updated tables, 2019; Lewis et al., 2015). Meanwhile, state-level initiatives and actions by other federal agencies have created a patchwork of advisories, enforceable limits, and remediation programs across different jurisdictions (States Step Up as Federal PFAS Regulations Retreat, Baker).

The Canadian government recognizes the need for precautionary and comprehensive guidelines to address the potential impacts of PFAS on human health and the environment. Although guidelines exist for specific compounds such as PFOS and perfluorooctanoic acid (PFOA), significant uncertainty remains regarding other PFAS. This has led to recommendations for a cautious and inclusive approach to managing these substances. A report by the ministers of the environment and health highlights the importance of evaluating the entire class of PFAS due to their potential environmental and health risks, emphasizing the need for stronger regulations and informed decision-making in Canada (State of per- and polyfluoroalkyl substances PFAS report, 2025; Canadian Environmental Protection Act, 1999). Canadian PFAS regulation falls under a number of regulations, such as the Canadian Environmental Protection Act (CEPA), 1999, the new substances notification regulations (chemicals and polymers) (NSNR), and the Prohibition of Certain Toxic Substances Regulations (PCTSR), 2012. Key agencies involved include the Canadian Council of Ministers of the Environment (CCME), Environment and Climate Change Canada (ECCC), and Health Canada (HC) (State of per- and polyfluoroalkyl substances PFAS report, 2025; Ccme, Canadian; Ecological screening assessment report on perfluorooctane sulfonate, Salts; Risk Management Strategy for Perfluorooctane Sulfonate and Its Salts and Precursors, Canada). Several provinces, such as the Atlantic provinces, Alberta, British Columbia, Ontario, and Québec, have developed or published provisional screening values, soil and groundwater guidance, sampling standard operating procedures, and contaminated site management frameworks to address PFAS in soil and water. Many provinces have also issued interim groundwater or soil guideline values for various PFAS and provided operational guidance for utilities and site managers (State of per- and polyfluoroalkyl substances PFAS report, 2025; Guidance for the Assessment and Remediation of Per- and Polyfluoroalkyl Substances in British Columbia, 2019; Alberta Tier 1 soil and groundwater remediation guidelines open, Government; Atlantic RBCA - Human Health-Based

Tier I Environmental Quality Standards EQS for Soil - All Land Uses; Non-Potable Groundwater Condition, mg/kg; Health Ontario, P).

The US EPA has implemented a range of actions to address PFAS, while the Canadian government has emphasized the necessity for precautionary and comprehensive guidelines to mitigate their potential effects on human health and the environment. Building on these national initiatives, this chapter examines PFAS contamination and management in the USA and Canada, focusing on regulatory frameworks, detection and monitoring techniques, remediation strategies, legal challenges, international comparisons, and future directions for effective governance and risk reduction.

## 21.2 USA: per- and polyfluoroalkyl substances regulatory guidelines

### 21.2.1 Overview of environmental issues in the USA

Environmental policies are primarily aimed at safeguarding human health and the environment. Achieving this objective is difficult if the regulations are not comprehensive and inclusive. Although the regulations may cover nine environmental threats, the remaining tenth threat can still be harmful, highlighting the need for truly comprehensive protection. It is due to such complexity that environmental regulations must be detailed and extensive to offer protection to a meaningful extent (Rogers, 2022).

The US EPA regulated only six pollutants between 1970 and 1990, which were limited to individual chemicals with a small impact. The issue of environmental pollution of air, water, and the soil prompted changes in regulations, most notably the Clean Air Act (CAA) amendments of 1990, which introduced a new regulatory regime on a larger scale. The CAA was created as an act of the Air Quality Act of 1967 and was meant to conserve the environment and protect public health; support in funding national air pollution studies; offer financial and technical assistance to states; and promote air pollution control on a regional basis. CAA has since become one of the most robust environmental policies in the world, and it still raises contentious policy during its discussions.

Along with laws concerning air quality, significant US regulations address water pollution. Human health and conservation of the environment are dependent on clean water. By the 1960s, the anthropogenic pollution of US surface waters had been so high that federal government intervention was necessary. The Clean Water Act, which was passed in the early 1970s, is a law aimed at restoring and preserving the integrity of the water of the country (Summary of the Clean Water Act, US).

To this end, in 1974, the SDWA came into force to safeguard the drinking of both surface and ground water sources in the country (Summary of the Safe Drinking Water Act, US). Within the past 50 years, the CWA and SDWA have enhanced water quality in the US; nonetheless, it is not enough to introduce regulatory measures. The success of the future will be determined by the organization's sustainability efforts beyond the existing environmental regulations, primarily to address the scarcity of water and the pollution due to unrestricted sources.

Regarding land pollution, the Resource Conservation and Recovery Act (RCRA) of 1976 emerged as the main law in the US regarding the disposal of solid and hazardous waste. It came about after the increase in industrialization and population-caused waste. Along with the CAA, CWA, and SDWA, RCRA became the basis of the US protection legislation, except for those on legacy site remediation 334. RCRA was influential in modifying the collective behavior, especially in promoting the recycling of materials. Wastes that have become common resources are tyres, plastics, and metals. Nonetheless, RCRA has not successfully controlled municipal waste, though in most cases, the latter contains hazardous components.

To correct the pollution of earlier operations, CERCLA or Superfund was passed in 1980. CERCLA not only gives the US EPA the authority to clean up hazardous and toxic waste sites and compensate damages to the environment and ecology, but it also created another US government agency, the Agency for Toxic Substances and Disease Registry (ATSDR) (Superfund: CERCLA Overview, US EPA, https).

Together, RCRA and CERCLA form a robust framework to address land-based pollution; the former addresses past issues, while the latter focuses on preventing future problems. The laws are currently part of environmental policy in most parts of the world and have been adopted by many countries as examples of sustained regulations.

Nevertheless, PFAS present emerging and developing regulatory challenges, which require modifications to regulatory frameworks, practical monitoring, and innovative, well-funded cleanup options. The chapter will explore the evolving paradigm of PFAS regulation in the United States, including new federal and state laws, enforcement mechanisms, risk assessment, and the development of sustainable solutions that aim to preserve the environment and human health.

### 21.2.2 Regulatory framework for per- and polyfluoroalkyl substances

PFAS were introduced after their development in the 1930s and have been extensively used since the 1940s and 1950s. They have been used in numerous industrial and consumer products, including firefighting foams, aerospace technologies, and nonstick or water-repellent consumer items. These substances are characterized by highly stable carbon-fluorine compounds, some of the most tenacious in organic chemistry, so PFAS is highly resistant to high temperatures, water, oil, or degradation (O'Hagan, 2008; Brendel et al., 2018). The environmental persistence and bioaccumulation properties of PFAS have raised global concerns regarding their long-term impact on human health. A US study published by the Centers for Disease Control and Prevention (CDC), which revealed the presence of PFOS in the blood serum of about 98% of the US population in 2003, sparked early alarm in the United States (Lewis et al., 2015; M et al., 2007). Recent statistics from 2013–14 confirm the widespread exposure of Americans to PFOS, PFOA, perfluorononanoic acid (PFNA), and perfluorohexane sulfonic acid (PFHxS) (Fourth national report on human exposure to environmental chemicals: updated tables, 2019).

Though the levels of PFOS and PFOA have decreased in the blood since manufacturers such as 3 M and DuPont began voluntarily phasing out these compounds in the early 2000s, they are still regularly detected in the blood of a large proportion of Americans (Fourth National Report on Human Exposure to Environmental Chemicals, 2021; Regulations.Gov.Perfluorooctanoic Acid PFOA Site Related Environmental Assessment Program, 2005). Animal studies and human epidemiologic research have found an association of PFOS and PFOA exposure with high serum cholesterol, thyroid disease, gestational hypertension, ulcerative colitis, and cancers (Vaughn et al., 2013; Maria-Jose et al., 2012). Children appear particularly vulnerable to PFAS-related health effects; a systematic review links early-life and prenatal exposure to mainly increased dyslipidemia, immune response impairment, including decreased vaccine efficacy, reduced kidney performance, and an earlier age of menarche (Rappazzo et al., 2017). Attempts to mitigate PFAS contaminants at the global, national, and local levels have included regulating the production and importation of PFAS, establishing a maximum acceptable level of PFAS in food and drinking water, and regulating the disposal of PFAS in products. Nonetheless, there is still inconsistency in regulatory standards and regulations, as uniform standards for this type of chemical are not in place. The question of whether to regulate PFAS individually or as a group of chemicals is one of the most controversial. Although some are also in support of regulating individual compounds, it appears that regulating thousands of PFAS is impractical. As a result, their structural and chemical stability is being regulated as a class, which is becoming increasingly popular (Andreea et al., 2025; Henry et al., 2018; Kwiatkowski et al., 2020).

Regulatory activity across various jurisdictions has increased, driven by growing scientific validation of health risks associated with PFAS. Though an emerging literature describes current and suggested PFAS policies at the US state and federal levels (Our Current Understanding of the Human Health and Environmental Risks of PFAS, US). Effective regulation necessitates a systems-level comprehension that includes historical context, an understanding of feedback loops that influence the policymaking process (Pierson, 1993; Daugbjerg and Kay, 2020). International, federal, and state-level initiatives focus on a diverse range of PFAS policies being discussed, presented, and/implemented.

#### *21.2.2.1 Per- and polyfluoroalkyl substances regulations at the US national level*

The US EPA regulates PFAS under a variety of legislative authorities, including but not limited to the TSCA, the SDWA, and the CERCLA. The PFAS regulatory work done by the EPA centered on compounds like PFOS and PFOA, which were voluntarily discontinued by large manufacturers (Per- and Polyfluoroalkyl Substances PFAS, US, EPA). The agency's monitoring commenced in 2002 when it issued two significant new use rules. These rules require manufacturers to notify the EPA before synthesizing or importing 13 PFAS compounds. This was later expanded to include an additional 75 PFAS compounds that had been discontinued by the chemical giant 3 M between 2000 and 2002 (Federal Register: Perfluoroalkyl Sulfonates, Proposed). In 2003, the EPA again collaborated with PFAS manufacturers, including 3 M and DuPont, to negotiate the term invited to join the agreement to reduce PFOA emissions and contents in products by 95% by 2010 (Fact Sheet, 2010).

In 2009, the US EPA began regulating PFAS under the SDWA by issuing provisional health advisories for PFOS and PFOA. From 2013 to 2015, PFOS, PFOA, and PFNA were monitored in the EPA unregulated contaminant monitoring rule 3 (UCMR 3), which is a program designed to keep track of contaminants not formally regulated in the SDWA (Fact Sheet, 2010; Federal Register: Revisions to the Unregulated Contaminant Monitoring Regulation UCMR 3 for Public Water, Systems). In 2016, the EPA established a lifetime health advisory value of 70 parts per trillion (ppt) for PFOS and PFOA in drinking water (Federal Register : Lifetime Health Advisories and Health Effects Support Documents for Perfluorooctanoic Acid and Perfluorooctane, Sulfonate), which is significantly higher than that included in the ATSDR 2018 draft toxicological profile on perfluoroalkyls (Toxicological Profile for Perfluoroalkyls).

It should be noted that these health advisory guidelines are not enforceable and are intended only to help federal, state, tribal, and local governments safeguard human health if PFAS are detected in drinking water. In February 2019, the EPA published its PFAS action plan, outlining strategies to address public health risks associated with PFAS exposure (Epa, U). The proposal initiated the regulatory procedure to impose the maximum contaminant levels (MCLs) on PFOA and PFOS and provided notice that the agency is planning to designate both hazardous substances under the CERCLA, which is otherwise known as Superfund. Such a designation would enable greater responsibility in cleanup efforts. Yet, so far, the EPA has still not adopted the MCL and CERCLA hazardous substance designation of PFOA and PFOS.

Federal lawmakers and environmental advocates have criticized the actions of the EPA regarding widespread PFAS contamination. In response to concerns about PFAS, several legislative proposals were introduced during the $116^{th}$ Congress aimed at improving the regulatory process. Senate Bill 1473, known as the Protect Drinking Water from PFAS Act of 2019, sought to require the EPA to establish a MCL and a national drinking water standard for specific PFAS compounds (S.1473 - 116th Congress 2019-2020: Protect Drinking Water from PFAS Act of 2019, Congress.Gov, Library). Similarly, House Resolution 2605, referred to as the PROTECT Act of 2019, mandated that the EPA add PFAS to its list of hazardous air pollutants within 180 days. Additionally, House Resolution 2596, titled the Protecting Communities from New PFAS Act, called for comprehensive reporting on all manufacturing activities related to PFAS (H.R.2596 - 116th Congress 2019-2020, Protecting).

Most of these initiatives were incorporated into H.R. 535, the PFAS Action Act of 2019, and it was passed by the House on January 1, 2020. A comprehensive bill of regulation on regulating the PFAS contamination and safeguarding the residents' health was outlined in this bill's provisions (H.R.535). Other proposed bills in the Senate included S.1507, which would have set out to add some of the PFAS to the Toxics Release Inventory (S.1507 - 116th Congress 2019-2020: PFAS Release Disclosure and Protection Act of 2019, Congress). and S.1372, which would have set out to get the federal agencies to sign agreements to carry out remedies on the PFAS (S.137 - 116th Congress 2019-2020: Right Start Child Care and Education Act of 2019, Congress.Gov, Library). The Senate, however, did not introduce an encompassing PFAS bill similar to the PFAS Action Act. Instead, it tried to attach PFAS provisions to the regularly occurring statute authorizing Army Corps of Engineers projects, the Water Resources Development Act (H.R.7575 - 116th Congress, Water). Such provisions were struck out in the final version of the House.

To reduce the exposure of PFAS among communities living in and around operational and decommissioned military sites, the US Department of Defence has taken steps to minimize the risks of exposure to these communities. Since the 1970s, military bases have been using aqueous film-forming foam (AFFF), which consists of AFFF containing PFOS and sometimes PFOA, to extinguish fires involving petroleum-based materials. A Government Accountability Office review reported that the DOD spent about USD 200 million on PFAS assessment and cleanup, identified 401 locations where PFAS releases likely occurred, and confirmed completed cleanup at 32 sites (Drinking water: DOD has acted on some emerging contaminants but should improve internal reporting on regulatory compliance, U.S.). The newer pollution evaluations depict that PFAS contamination might occur in up to 600 DOD installations, with groundwater levels of over 1 million ppt in 14 installations (PFAS Could Contaminate More Than 600 Military Installations, DOD). Federal officials have criticized the DOD for its delayed and insufficient response.

The National Defense Authorization Act (NDAA) of 2020 (Text - H.R.2500 - 116th Congress, 2020). Introduced major PFAS-related provisions, including the phase-out of PFAS-based firefighting foams, offering blood testing services to military firefighters, providing greater attention to PFAS-contaminated water sources associated with the operations of the military, mandating EPA action under TSCA, and a ban on specific PFAS uses. The 2021 NDAA, signed in July 2020, allocated even more funding, including USD 1.5 billion for PFAS cleanup, USD 150 million for research on remediation, and USD 15 million for a CDC study on PFAS exposure and health impacts at military bases (Text - S.4049 - 116th Congress, 2021).

In another related move, the US FDA announced in July 2020 a voluntary agreement with three manufacturers to discontinue the use of 6:2 fluorotelomer alcohol in the grease-proof coating of paper-based food packaging. The phase-out commenced in January 2021 and is expected to be completed within 3 years. According to the FDA, the method balances the health of the people with preventing disturbances in the supply chain (FDA Announces the Voluntary Phase-Out by Industry of Certain PFAS Used in Food Packaging, FDA). In summary, the regulation of PFAS in the United States at the national level is a sign of increased awareness of the environmental and health hazards. Although initial responses focused mainly on voluntary industry efforts and temporary health advisories, the past several years have been characterized by concerted efforts at the legislative and regulatory front, including proposed mandatory drinking water standards, the designation of hazardous substances under CERCLA, and full-fledged cleanup requirements at military bases. However, serious challenges remain, including regulatory loopholes, inadequate control over new PFAS, and the debate over whether to regulate compounds specifically or in a class-based manner. Federal leadership, scientific research, and interagency coordination will be crucial in establishing an integrated, enforceable, and health-protective national policy for PFAS.

#### 21.2.2.2 *Per- and polyfluoroalkyl substances regulations at the US state-level*

As of May 1, 2020, in the absence of a federal drinking water standard to regulate PFAS, at least nine US states had established enforceable limits, or guidance values, of PFOS and PFOA, sometimes more stringent than those provided by the EPA (Post, 2021). Other PFAS compounds, including PFNA and PFHxS, are also subject to advisory levels in several states, based on assessments by the ATSDR and European Food Safety Authority (Post, 2021). This patchwork of regulations highlights a disjointed policy landscape that addresses various PFAS exposure routes, including water, firefighting foams, and food packaging. Interestingly, in 2019, the State of California enacted Assembly Bill 756, which gives the State Water Board the capacity to order PFAS testing in publicly owned water systems and publicly announce or source removal when PFAS concentration exceeds the state levels (Bill Text - AB-756 Public Water Systems, Perfluoroalkyl). Vermont also passed Act 21 (Senate Bill 49) the same year to mandate water sources that supply the communities in the state and create a PFAS monitoring program. The systems were mandated to offer a do-not-drink advisory upon the discovery of high levels above the 20 ppt health advisory level of Vermont until the system can be adequately treated (PFAS in Surface Waters, Department; Agency Of Natural Resources Department Of Environmental Conservation Drinking Water And Groundwater Protection Division Advance Notice On The Regulation Of Perfluoroalkyl, Polyfluoroalkyl Substances Pfas As A Class, 2020).

In September 2020, California strengthened its PFAS rules by prohibiting the production and sale of firefighting foams with PFAS. The legislation also mandates the manufacturers of firefighter protective gear to disclose any PFAS levels in their products. The initial violation may lead to a fine of up to USD 10,000 (Bill Text - SB-1044 Firefighting Equipment and Foam, PFAS). Moreover, the state enacted four other bills related to PFAS to expand its regulatory system.

Likewise, in May 2020, Maryland moved forward with its own PFAS restriction, sending a bill to the governor that clamped down on using class B firefighting foam containing intentionally added PFAS, at least for testing and training scenarios. The restriction took effect on October 1, 2021. Offenders could face fines ranging from $500 to $1000 per violation. The bill makes a clear exception for real emergencies, allowing it to be deployed during actual firefighting or fire prevention operations (Environment-Use of Fire-Fighting Foam Containing Perfluoroalkyl and Polyfluoroalkyl Substances-Prohibition and PFAS, Chemicals).

In 2018, Washington adopted a bill called "Concerning the use of perfluorinated chemicals in food packaging," which banned the use of PFAS in food packaging if a safer alternative was available. The prohibition on the application of PFAS in certain types of food packaging will take effect 2 years after the publication of a report on a safer alternative (RCW 70A.222.070, Prohibition). Besides the necessary legislation directly minimizing human exposure to PFAS, some states have also established measures that provide funding to testing locations where the presence of PFAS is suspected. They have also gone to enforce actions against companies that have released PFAS into the environment and to cover the associated cleanup expenses. Indicatively, the Tennessee Department of Environment and Conservation is undertaking a statewide project to sample the drinking water systems of the population for the presence of PFAS (TDEC's Strategy to Assess the Impact of Per-and Polyfluoroalkyl Compounds on Public Drinking Water Sources inTennessee, https). When the contamination of the water system is detected, the state seeks to establish the sources of PFAS.

The other strategies have included forming coalitions and suing PFAS suppliers in court. In Michigan, a multiagency working group called the PFAS Action Response Team (MPART) was established as part of executive directive 2017-4, which aims to identify and investigate the sources of PFAS contamination in the state and protect the state's water supply against PFAS. Instead of passing laws, the state of Vermont filed a lawsuit against Saint-Gobain performance plastics after contamination of PFOA was detected in wells near a disused plant that manufactures plastics (Saint-Gobain Dropping Lawsuit Against the State of Vermont, Community).

The state eventually reached a settlement with Saint-Gobain, whereby the company agreed to fund the water line extensions to approximately 200 households that had been impacted by contaminated water (State Reaches Agreement with Saint-Gobain, Office). With more than half of the states actively involved in PFAS activities, it is essential to understand the structure of the legislation and the issues that may influence policy implementation.

New Jersey establishes PFAS standards in drinking water, demonstrating a strong commitment to effective PFAS regulation and public health. In 2020, the MCLs for PFOA and PFOS were set at 14 ppt, and 13 ppt, respectively, and the MCL for PFNA has been 13 ppt since 2018 (New Jersey Sets Stringent Limits for PFOA, PFOS). The New Jersey public water system needs to be monitored for PFOA, PFOS, and PFNA, and consumers should be notified if the levels exceed the MCLs, including private wells, under the Private Well Testing Act. Additionally, the groundwater quality standard is aligned with the same limit of MCLs and formally classifies PFOA and PFOS as hazardous substances.

The Massachusetts Department of Environmental Protection has set MCLs for PFOS, PFOA, PFHxS, PFNA, perfluoroheptanoic acid (PFHpA), and perfluorodecanoic acid (PFDA) at 20 ppt for all public water systems, except for nontransient noncommunity water systems. All six PFAS compounds must be monitored in public water systems using the EPA methods 537 or 537.1. The frequency of the six PFAS monitoring systems will increase monthly if the contamination level exceeds 10 ppt. The quarterly average PFAS contamination level rose above 20 ppt, prompting the public to take notice and implement preventive measures to reduce PFAS levels. Thus Massachusetts surpasses the federal EPA by establishing an MCL for PFAS and serves as a model for other states.

In 2018, New Hampshire implemented a regulatory program for PFAS. The New Hampshire Department of Environmental Services, in collaboration with the New Hampshire Department of Health and Human Services, established MCLs for several PFAS compounds, including PFOA, PFOS, PFHxS, and PFNA, due to their widespread presence in the state (Update on New Hampshire PFAS Drinking Water Standards MCLs, DES). This initiative was formalized in House Bill 1264, which was signed into law in July 2020. The established MCLs are set at 12 ppt for PFOA, 15 ppt for PFOS, 18 ppt for PFHxS, and 11 ppt for PFNA. This adoption of MCLs is a significant step toward protecting public health from PFAS exposure through drinking water.

In 2020, New York State established stringent regulations to address PFOA and PFOS contamination in drinking water standards and consumer products, thereby protecting public health. The MCL for PFOA and PFOS is 10 ppt in drinking water standards for emerging contaminants. New York State is also working to pass 23 more PFAS drinking standards. All water systems are required to test for contamination and monitor PFOA and PFOS by periodically monitoring. If the MCL exceeds 10 ppt, notify the public about confirmed exceedances and implement preventive measures to reduce PFOA and PFOS levels. The new federal MCL limit for PFOA and PFOS is 4 ppt, and New York State must meet the new limit in its drinking water system by 2029 (Public Water Systems and Drinking Water Standards for PFAS and Other Emerging, Contaminants). In 2023, New York State restricted the use of food packaging materials that intentionally contain PFAS, and the New York State Department of Environmental Conservation may conduct inspections and compliance reviews (PFAS In Food Packaging, NYSDEC).

## 21.2.3 Limits on per- and polyfluoroalkyl substances in the USA

### 21.2.3.1 *per- and polyfluoroalkyl substances in water*

The EPA has finalized a National Primary Drinking Water Regulation that establishes MCLs for six PFAS compounds. For PFOA and PFOS, the enforceable MCL is set at 4.0 ppt for each compound. The MCL Goal, which is a nonenforceable health objective, is set at zero. This reflects the determination that no level of exposure to these substances is without risk. For PFNA, PFHxS, and GenX Chemicals, the enforceable MCL is established at 10 ppt for each compound, with the MCL Goal also set at 10 ppt (EPA Finalizes PFAS Drinking Water Regulation, nsights).

### 21.2.3.2 *Per- and polyfluoroalkyl substances in soil*

The United States EPA has established Regional Screening Levels (RSLs) for various PFAS substances. Below is a list of RSLs for some of the most commonly regulated PFAS compounds. It is important to note that RSLs are advisory benchmarks rather than enforceable standards. However, both the EPA and state regulatory agencies often use RSLs when assessing and characterizing potential remediation sites (PFAS in Soil: State Regulations, BCLP). Table 21.1 provides information regarding the limits of PFAS in soil.

### 21.2.3.3 *Per- and polyfluoroalkyl substances in air*

Although there are currently no federal regulations specifically governing PFAS emissions in the air, seven states have established regulations, rules, or health-based standards addressing this issue. This trend is expected to continue unless there is federal intervention. Furthermore, some states, such as Connecticut, have indicated plans to create their own PFAS air emission standards once reliable methods for measuring PFAS emissions in the air are developed (PFAS Air Emissions Restrictions, BCLP). Details about the airborne PFAS limitations are given in Table 21.2.

**TABLE 21.1** Limits of per- and polyfluoroalkyl substances in soil according to US EPA.

| PFAS substance | Residential soil level (mg kg$^{-1}$) | Industrial soil level (mg kg$^{-1}$) |
|---|---|---|
| PFOS | 0.0063 | 0.058 |
| PFOA | 0.000019 | 0.000078 |
| PFNA | 0.19 | 2.5 |
| PFHO/DA | 0.23 | 3.5 |
| PFHxS | 1.3 | 16 |
| PFBS | 19 | 250 |
| PFHxA | 32 | 410 |
| PFBA | 78 | 1200 |

**TABLE 21.2** Limits of per- and polyfluoroalkyl substances (PFAS) in air according to US EPA.

| | Types of PFAS ($\mu g\ m^{-3}$) | | | | | | |
|---|---|---|---|---|---|---|---|
| **State** | **PFOA** | **PFOS** | **PFHxS** | **PFBS** | **PFHxA** | **PFBA** | **PFIB** |
| Michigan | 0.0001 | 0.0004 | NA | NA | NA | NA | NA |
| Minnesota | 0.063 | 0.011 | 0.034 | 0.3 | 1 | 10 | NA |
| New Hampshire | 0.024 | NA | NA | NA | NA | NA | NA |
| New Jersey | 0.07 | 0.07 | NA | NA | NA | NA | NA |
| New York | 0.0053 | NA | NA | NA | NA | NA | NA |
| Texas | 0.005 | 0.01 | NA | NA | NA | NA | NA |
| Wisconsin | 0.24 | NA | NA | NA | NA | NA | 8.18 |

Concentration: $\mu g\ m^{-3}$, NA: not available.

#### 21.2.3.4 *Per- and polyfluoroalkyl substances limits on food products*

The EPA has not yet established federal regulations for PFAS in food. However, specific states, such as California and New York, have prohibited distributing and selling food packaging containing intentionally added PFAS or exceeding specific concentration limits (PFAS Food Packaging Regulations State by State, https). Additionally, a lawsuit has been initiated against the FDA, advocating for establishing enforceable limits for PFAS in food (FDA Must Set Limits on PFAS in Food, Lawsuit Says, The).

### 21.2.4 Per- and polyfluoroalkyl substances detection and monitoring

In the UCMR, the water systems of the population are obliged to measure the contaminants that are listed in the UCMR. The latest list, the UCMR 5, contains PFAS. In efforts to help with the monitoring of these contaminants, the EPA has proposed Method 537.1. This is a more advanced form of Method 537 capable of identifying 18 kinds of PFAS as opposed to the 14 obtainable by the old method.

The entire EPA methods, as identified in Table 21.3 have the general approach of sample processing. The samples are then labeled with isotopically labeled analogs and then subjected to solid-phase extraction (SPE). Samples are then eluted and concentrated, and the concentrated extracts are measured using liquid chromatography tandem mass spectrometry (LC-MS) in order to measure PFAS levels. The draft Method 1633 of the EPA is developed to analyze PFAS in a range of different samples, including both aqueous (wastewater) and solid (biosolids and fish tissue) ones. The method of testing is done by spiking the samples with PFAS analog isotopes prior to the extraction process, which involves many other steps (US EPA PFAS, Method).

EPA's Method 45 other test method (OTM-45) is a draft methodology currently under review and will be updated as more data from stakeholders becomes available. OTM-45 analyses semivolatile compounds (with a boiling point greater than 100°C) and particulate-bound PFAS in air emissions from stationary sources. Additionally, this method can collect

**TABLE 21.3** Summary of Environmental Protection Agency (EPA) methods for the detection of per- and polyfluoroalkyl substances (PFAS) in various matrices.

| EPA method | Types of PFAS detected | Matrix tested in | Detection range | References |
|---|---|---|---|---|
| 533 | The total consists of 25 PFAS, each with 12 or fewer carbon atoms, encompassing PFAS acids, sulfonates, telomers, and carboxylic acids. | Drinking water | LCMRL of 1.6–9.1 ppt | Epa (2018) |
| 537.1 | 18 PFAS acids | Drinking water | LCMRL of 0.53–6.3 ppt | PFAS Method 537.1 - List Details - SRS, US; Method 537.1 Determination of Selected Per- and Polyflourinated Alkyl Substances in Drinking Water by Solid Phase Extraction and Liquid Chromatography/Tandem Mass Spectrometry LC/MS/MS, Science |
| 8327 | 24 PFAS | Nonpotable water for soil and solid waste under the RCRA | Not applicable. Uses LLOQ | SW-846 Test Method (8327): Per- and Polyfluoroalkyl Substances PFAS by Liquid Chromatography/Tandem Mass Spectrometry LC/MS/MS, US |
| Draft method 1633 (in partnership with the DoD) | 40 PFAS | Wastewater, surface water, groundwater, biosolids, landfill leachates, fish tissue | Not applicable. Typical reporting limits for aqueous: 1.6–40 ppt, for solids: 0.2–5 $\mu g\ kg^{-1}$, and for tissue: 0.5–12.5 $\mu g\ kg^{-1}$ | Epa (2021b) |
| Other test method 45 | 50 PFAS | PFAS in air emissions from stationary sources | Not applicable. Uses MDL, range: 0.08–2.08 $ng\ m^{-3}$ | Other Test Method 45 OTM-45, Measurement |

and recover ionic and covalent PFAS for nontargeted analysisof PFAS compounds. The study is performed using liquid chromatography tandem mass spectrometry (LC-MS/MS) in multiple reaction monitoring mode (Other Test Method 45 (OTM-45)).

The lowest concentration minimum reporting limit defines the detection range of EPA Methods 533 and 537.1, that is, the smallest spiking concentration that can be reliably identified and measured (Technical Basis for the Lowest Concentration Minimum Reporting Level LCMRL Calculator, 2010; Epa, 2018). They provide high analytical sensitivity, and the analysis has a detection level well below the EPA health advisory level of 70 ppt for PFAS.

These methods offer high analytical sensitivity, with detection limits significantly below the EPA's health advisory level of 70 ppt for PFAS. In contrast, EPA Method 8327 defines detection capability using the lower limit of quantitation, which refers to the lowest concentration of a PFAS compound that can be consistently quantified with acceptable accuracy and precision (Epa, 2021a). One limitation of these techniques is that they can only measure PFAS compounds specifically listed in the protocol, as they require isotopically labeled analogs to measure them quantitatively. PFAS lacking an isotope-labeled counterpart may go undetected. These methods employ analytical-grade chemicals and implement specific sample handling precautions to minimize the risk of sample cross-contamination.

### 21.2.4.1 *Other per- and polyfluoroalkyl substances analytical methods*

The analytical approaches mentioned above can only identify a limited number of the thousands of PFAS compounds that enter the environment through various processes, such as waste handling and incineration. Consequently, to gain a more comprehensive understanding of PFAS presence in the environment, it is essential to employ additional analytical techniques to detect a broader spectrum of PFAS.

Recognizing this gap, the EPA developed total organic fluorine, which estimates the overall concentration of organic fluorine in a sample. The technique involves adsorbing PFAS and other organofluorine chemicals to activated carbon or other materials before analysis (Epa, 2021a). Nevertheless, due to its ability to capture the complexity of all organofluorine forms, the method is not specific to PFAS. Moreover, it is hard to detect volatile and nonpolar PFAS in high-abundance air samples.

The total oxidizable precursor (TOP) assay was developed to address some of these limitations. The TOP method is aimed at PFAS precursors that can break down to yield terminal perfluoroalkyl acids such as PFOA and PFOS, potentially providing a mechanism to estimate the burden of unquantified or precursor PFAS in environmental matrices (PFAS Analysis Toolkit: LC/MS/MS, Total).

The problem of determining volatile nonpolar precursors is solved by the total oxidizable precursor (TOP) method. Such a method identifies PFAS precursors that can be transformed into other PFAS, such as the persistently occurring perfluorinated compounds, such as the environmentally persistent PFOA and PFOS (PFAS Analytical Methods Development and Sampling Research, US).

#### 21.2.4.1.1 SW-846 Method 0010

The EPA's SW-846 is a compilation of test methods designed to measure chemicals in solid waste. These methods are categorized into chapters and series, with the chapters offering guidance on utilizing the techniques found in the series (Basic Information about How to Use SW-846 U S, EPA). Among the methods mentioned in this article, only Method 0010 is part of SW-846. This particular method is used to analyze semivolatile compounds.

#### 21.2.4.1.2 Modified Method toxic organics-15

It is an analytical method that has been developed to analyze toxic organics, although it was originally developed to analyze volatile organic compounds, it is also undergoing a modification to measure PFAS. It is a test method that requires the collection of samples into a canister, and its analytes are analyzed by gas chromatography-mass spectrometry (Watson, PFAS).

#### 21.2.4.1.3 Other US federal per- and polyfluoroalkyl substances analytical methods

Some federal agencies have developed specific procedures to identify and quantify PFAS in various matrices, in addition to those established by the US EPA. US FDA uses liquid chromatography tandem mass spectrometry (LC-MS/MS) to measure 16 PFAS in processed food products, supporting its efforts to assess dietary exposure (FDA) (Testing Food for PFAS and Assessing Dietary Exposure, FDA).

Similarly, the US Geological Survey has used LC-MS/MS to screen 17 PFAS in source waters and finished drinking water across public water systems, which adds to a wider observation of PFAS prevalence in the hydrological environments (Per- and Polyfluoroalkyl Substances PFASs detected in Source Waters and Treated Public Water Supplies, U.S.).

The DOD employs this technique to measure PFAS in both aqueous and solid samples, including soils and sludges. Besides that, the DOD incorporates the draft version of the EPA Method 1633, specifically to measure PFAS in complex matrices like AFFF, a known source of contamination at military bases (Wingard, 2009).

US CDC utilizes a very sensitive analysis system, online SPE coupled with high-performance liquid chromatography and turbo ionspray tandem mass spectrometry (SPE-High-Performance Liquid Chromatography [HPLC]-TIS-MS/MS). Through this technique, the CDC can measure the 17 PFAS in human serum and other biological fluids, which aids in determining the exposure level that exists in a population, as part of programs like the National Health and Nutrition Examination Survey (NHANES) program (Botelho and Pirkle, Laboratory).

Case studies of different PFAS compounds in the US are included in Table 21.4, together with information on their detection, concentration, and impacts on human health and the environment. Using a range of environmental matrices and biological samples. They provide comprehensive information on the prevalence, distribution, and risks of PFAS.

## 21.2.5 Remediation efforts of per- and polyfluoroalkyl substances

The chemical stability and high mobility of PFAS remain major challenges in addressing PFAS contamination. The EPA's Water Treatability Database (TDB) offers a centralized resource for understanding how effective and ineffective treatment technologies perform against PFAS (Drinking Water Treatability, Database). Several water treatment techniques, including aeration, air stripping, biological filtration, chlorination, and UV irradiation, are generally ineffective in removing PFAS from drinking water, according to the TDB. In

**TABLE 21.4** Case studies reported on per- and polyfluoroalkyl substances (PFAS) in the USA.

| Type of PFAS | Study (year) | Study participants/ location | Sample type | Detection frequency | Concentration reported | Outcomes/ inference | References |
|---|---|---|---|---|---|---|---|
| 18 compounds, including PFOA, PFOS, HFPO-DA (GenX), PFMOAA | September 2017 and October 2019 | Cape Fear River | Riverine and coastal sediments | NA | Total PFAS: below LOQ to 7.47 ng $g^{-1}$ dry weight | PFAS, both legacy and emerging, are stored in sediments, which can be washed back into the water in response to disturbances. | Shimizu et al. (2022) |
| 51 PFAS | 2019 (Sampling before, during, and after Hurricane Dorian) | St. Augustine area | Surface water samples | PFOS, PFBS, PFHxA, PFHpA and PFOA (100%) PFNA (92%) PFBA (71%) | nd | Hurricane Dorian led to a 177% increase in PFOS, demonstrating that storms can redistribute PFAS and highlighting the need for monitoring during storm events. | Martinez et al. (2022) |
| 45 PFAS analytes | 2023 | The Cape Fear River | Drinking water samples | NA | PFOA: 39.8 ng $L^{-1}$ PFOS: 205.3 ng $L^{-1}$ | Known and emergent PFAS, such as PFOA and PFOS, were detected in excess of the EPA limits in the analysis. | Weed et al. (2024) |
| 40 PFAS total; specific congeners include PFOS, 7:3 FTCA (fluorotelomer carboxylic acid), PFHpS, PFNS, MeFOSAA, EtFOSAA. | 2021 | Detroit and St. Clair Rivers | Muscle biopsy and serum samples of fish | PFOS (100%) PFNA (11.1%) PFDA (77.8%) PFDoA (44.4%) | Muscle PFOS concentration in Detroit River walleye (WAE): 13–19 ng | PFOS are bioaccumulating and biomagnifying in fish from the Great Lakes' Detroit and St. | George et al. (2023) |

(*Continued*)

**TABLE 21.4** (Continued)

| Type of PFAS | Study (year) | Study participants/ location | Sample type | Detection frequency | Concentration reported | Outcomes/ inference | References |
|---|---|---|---|---|---|---|---|
| | | | | | | Clair Rivers. ecological and health risks, and support nonlethal monitoring as a valuable tool. | |
| 32 PFAS and 18 hydrocarbon surfactants | 2021 | In-Vitro and In-silico study | Human serum albumin (HSA) | NA | PFOS, PFOA, PFBS, and other PFAS compounds strongly bind to human serum albumin protein with log Ka (noncovalent binding constant) values like $4.1 \pm 0.2\ M^{-1}$ for PFPS. | Firefighting foam PFAS bind firmly to serum albumin, pointing to a high level of bioaccumulation; an experimental study provides a feasible risk measurement approach. | Li et al. (2021) |
| 57 analytes including L-PFOS, PFOA, PFNA, L-PFHxS, and other per- and polyfluoroalkyl substances | 2023 | | Milk powder, milk-based infant formula, meat-based baby food puree, fish and fish oil, fresh egg, and soluble coffee | NA | nd | Validated LC-MS/MS method detects 57 PFAS in food at $ng\ kg^{-1}$ levels, meeting EU standards for future monitoring. | Theurillat et al. (2023) |
| Me-PFOSA-AcOH, PFDeA, PFNA, PFOSA, PFOS, PFOA | 2015 | LIFE study cohort | Serum samples and semen samples | NA | nd | Supports growing concerns about PFAS exposure and its potential impact on human fertility. | Louis et al. (2015) |

| | | | | | | | |
|---|---|---|---|---|---|---|---|
| PFOS, EtFOSAA | 2019 | Child Health and Development Studies (CHDS) cohort | Maternal perinatal serum | NA | nd | Due to PFAS exposure, there exists a disruption of maternal metabolism, which influences inflammation, lipids, and oxidative stress. | Hu et al. (2019) |
| PFOS, PFOA, GenX (HFPO-DA), ADONA, 9Cl-PF3ONS, PFBA (perfluorobutanoate) | 2016 | | Urine and serum | Urine PFBA (56%) Serum PFHpA (2%), PFOA (98%), Sb-PFOA (2%), PFNA (100%) PFA (40%), PFUnDA (8%), PFHxS (92%), PFHpS (96%), n-PFOS (98%), SmPFOS (86%), MeFOSAA (46%) | LOD- 0.1 ng $mL^{-1}$ | Sensitive approaches measured PFAS in US serum but at a low level in urine, suggesting little short-chain PFAS exposure. | Kato et al. (2018) |
| PFOS, EtFOSAA, EtFOSAA-to-PFOS ratio | 2020 | CHDS cohort | Maternal perinatal serum samples | NA | PFOS (median level 33.9 ng $mL^{-1}$) (PFHxS) (median 2.29 ng $mL^{-1}$)\ PFOA (median 0.40 ng $mL^{-1}$) and EtFOSAA (median 0.28 ng $mL^{-1}$) | Exposure in pregnancy has shown that PFAS modify maternal metabolism, especially amino acids, lipids, and energy metabolism. | Hu et al. (2020) |
| PFOS, PFCAs, PFBA, PFPeA, PFUdA, | 2015 | The Northwestern | Liver, muscle, and adipose | NA | 22.91–70.48 ng $g^{-1}$ wet weight (ww) | The results indicate an | Chu et al. (2015) |

*(Continued)*

**TABLE 21.4** (Continued)

| Type of PFAS | Study (year) | Study participants/ location | Sample type | Detection frequency | Concentration reported | Outcomes/ inference | References |
|---|---|---|---|---|---|---|---|
| PFTrDA, PFNA, PFDA, PFDoA | | Hawaiian Islands | tissues of dead black-footed albatrosses | | (PFOS) in liver; 3.01–6.59 ng $g^{-1}$ ww (PFOS) in muscle; 0.53–8.35 ng $g^{-1}$ ww (PFOS) in adipose. PFUdA: 8.04–18.70 ng $g^{-1}$ ww in liver, higher in muscle than PFOS. | increase in PFAS concentrations at the top level of marine predators since the 1990s, thus showing long-term ecological hazards of contamination. | |
| PFOS, PFHxS | 2010 | | Maternal serum, cord serum | NA | nd | A LC-MS/MS method was developed to detect PFOS and PFHxS in maternal and cord serum with higher sensitivity (LODs: 0.100.20 ng/mL) and reduced analysis time to improve prenatal biomonitoring. | Kato et al. (2011) |
| PFOA, PFOS | 2006 | American Red Cross North Central Blood Services | Plasma, Serum | NA | PFOS - 2000: 33.1 ng $mL^{-1}$ (geometric mean), 2005: 15.1 ng $mL^{-1}$ (geometric mean). PFOA - 2000: 4.5 ng $mL^{-1}$ (geometric mean), | A pilot study discovered acute decreases in PFOS and PFOA among blood donors in Minneapolis-St. Paul (2000–05), which demonstrates the | Olsen et al. (2007) |

| | | | | | | | |
|---|---|---|---|---|---|---|---|
| | | | | | 2005: 2.2 ng $mL^{-1}$ (geometric mean) | utility of expanded biomonitoring. | |
| PFOA, PFOS, PFHS | 2007 | Retirees from the 3 M Company, Decatur, Alabama | Serum | NA | PFOS: 799 ng $mL^{-1}$ (mean, range: 145-3490 ng $mL^{-1}$) PFHS: 290 ng $mL^{-1}$ (mean, range: 16-1295 ng $mL^{-1}$) PFOA: 691 ng $mL^{-1}$ (mean, range: 72-5100 ng $mL^{-1}$) | The study reported that humans have long half-lives of PFOS, PFHxS, and PFOA, with PFHxS having the longest half-life, suggesting slow removal, possibly due to processes such as renal resorption. | W et al. (2007) |
| 17 PFAS | 2023 | US | Drinkingwater samples from private wells (269 locations) and public supply tap water (447 locations) across the US | NA | Range from 0.348 to 346 ng $L^{-1}$ for cumulative concentrations (sum of detected PFAS) | Approximately 45 per cent of analyzed drinking water in the US contained at least one PFAS, so there is a necessity to evaluate the cumulative health impact, especially in nonmonitored wells on private property. | Smalling et al. (2023) |
| PFOA, PFOS | 2023 | US | Serum samples from 427 pregnant women | NA | nd | This US study of 427 pregnant women revealed that exposure to PFAS differs according to location, demographics, housing, and | DeLuca et al. (2023) |

(*Continued*)

TABLE 21.4 (Continued)

| Type of PFAS | Study (year) | Study participants/ location | Sample type | Detection frequency | Concentration reported | Outcomes/ inference | References |
|---|---|---|---|---|---|---|---|
| | | | | | | activities, with local points of origin as key contributors, in addition to raising environmental justice concerns and informing exposure mitigation measures. | |
| 29 PFAS, including perfluoroalkyl acids and fluoroethers (e.g., PMPA (perfluoro-2-(perfluoromethoxy) propanoic acid) and GenX (hexafluoropropylene oxide dimer acid)) | 2016–21 | CFR basin (Piedmont and coastal plain regions of NC) | Groundwater and stream water samples | na | Total quantified PFAS (ΣPFAS) in groundwater ranged from 20 to 4773 ng $L^{-1}$ (mean = 1863 ng $L^{-1}$); stream water concentrations ranged from 426 to 3617 ng $L^{-1}$ (mean = 1717 ng $L^{-1}$) | The study demonstrates that groundwater contamination near PFAS sources can sustain long-term PFAS loading into surface waters (dozens of kg/ year), posing ongoing risks to drinking water and ecosystems downstream of point sources. | Pétré et al. (2021) |
| 17 PFAS | 2020 | Las Vegas Wash, Lake Mead, Truckee River, Lake Tahoe, and Pyramid Lake in Nevada | Surface water and sediment samples | PFPeA and PFHxA (100%), PFOA (90%), PFBS (80%), PFHpA and PFHxS (60%), | Truckee River water: Total PFAS = 441.7 ng $L^{-1}$. Las Vegas Wash water: Total PFAS = 2234.3 ng $L^{-1}$. Sediments: Total | PFAS levels were significantly higher in Las Vegas Wash, with short-chain PFAS dominating water samples and long- | Bai and Son (2021) |

| | | | | | | | |
|---|---|---|---|---|---|---|---|
| | | | | PFOS (50%), PFDS (40%) PFBA (20%) and PFUA (10%) | PFAS in Truckee River sediments $= 272.9\ \mu g\ kg^{-1}$ (dry weight). Sediments in Las Vegas Wash: Total PFAS $= 345.7\ \mu g\ kg^{-1}$ (dry weight) | chain PFAS accumulating in sediments. Seasonal variations showed lower PFAS in summer. | |
| 14 of 24 PFAS, including PFOA, PFOS | 2021 | 5 aquifer systems | Groundwater samples | groundwater samples At least one PFAS (54%) ≥2 PFAS (47%) | nd | PFAS were present in 60% of public water wells and 20% of private wells. Higher PFAS levels were linked to urban land use, proximity to fire-training sites, and other chemical indicators of human activity. | McMahon et al. (2022) |
| PFOS, PFOA, PFDA | 2021 | Agricultural feedstock in the western US | Surface soil (0–30 cm), deeper soil (up to 90 cm), and underlying groundwater (up to 17 m deep) | NA | nd | Their mobility and the lack of updated assessments of their risks are reflected by the study that showed many PFAS in soils and groundwater at a biosolid-irrigated site. | Johnson (2022) |
| PFOA | 2021 | Green Mountain National Forest | Soil and groundwater samples | Downwind area PFHpA (74%) PFHxA (53%) other areas | Up to 100 ppt (parts per trillion) PFOA in groundwater springs and seeps | The findings highlight that industrial airborne emissions can | Schroeder et al. (2021) |

(*Continued*)

**TABLE 21.4** (Continued)

| Type of PFAS | Study (year) | Study participants/ location | Sample type | Detection frequency | Concentration reported | Outcomes/ inference | References |
|---|---|---|---|---|---|---|---|
| | | | | PFHpA (29%) PFHxA (10%) | | lead to widespread PFAS contamination in areas not typically considered downgradient of a source. | |
| 33 target PFAS | 2023 | Pennsylvania Water Quality Network | Surface water samples | PFCA (75%), PFSA (56%), FTSA (5%) PFOA (70%), PFHxA (60%), PFOS (47%), PFBS (45%), and PFPeA (42%) | nd | In this 161-Pennsylvania-stream assessment, the researchers identified elevated PFAS primarily in streams with urban development, wastewater treatment, and electronics-related industries; this finding could assist in formulating regulatory and mitigation measures. | Breitmeyer et al. (2023) |

| | | | | | | | |
|---|---|---|---|---|---|---|---|
| PFOA, PFOS | 2021 | US FDA datasets | Fish fillets | NA | Median PFOS in fish fillets: 9500 ng $kg^{-1}$ (rivers and streams), 11,800 ng $kg^{-1}$ (Great Lakes). Total PFAS: Median 9500 ng $kg^{-1}$ in rivers and streams; 11,800 ng $kg^{-1}$ in the Great Lakes. | This research identified elevated PFAS levels, particularly PFOS, in US freshwater fish caught locally, which have higher exposure risks compared to fish obtained from grocery stores. It sheds light on the ecological unfairness of subsistence fishers and highlights the necessity of limiting PFAS contamination. | Barbo et al. (2023) |
| PFOA, PFOS, PFNA, PFHxS | 2021 | NHANES study | Red blood cells and Serum | NA | nd | In this extensive NHANES study, an increased blood folate level was associated with reduced PFAS levels in both the adolescent and adult groups, suggesting that folate potential could be used to mitigate PFAS accumulation in adults and adolescents, and potentially inform | Zhang et al. (2023) |

(*Continued*)

**TABLE 21.4** (Continued)

| Type of PFAS | Study (year) | Study participants/ location | Sample type | Detection frequency | Concentration reported | Outcomes/ inference | References |
|---|---|---|---|---|---|---|---|
| | | | | | | future mitigation efforts. | |
| PFOA, PFOS, PFHxA, PFBA, PFPeA, PFBS | 2021 | Contaminated sites in Michigan | Leachates | NA | concentrations in leachates and effluents ranging from 5 –2000 g $yr^{-1}$ for leachates and up to 19 times greater PFAS concentrations in effluent compared to influent in WWTPs. | The study found that landfills, wastewater treatment plants (WWTPs), and contaminated sites in Michigan are major sources of PFAS pollution, contributing to environmental cycling of both legacy and short-chain PFAS. Landfills had the highest PFAS mass flux, especially from older sites. | Helmer et al. (2022) |
| PFOA, PFOS, PFHxS, PFNA | 2021 | Birth cohort in the Greater Cincinnati area | Maternal serum | PFOA, PFOS, PFNA (100%) PFHxS (99.6%) | 5.3 ng $mL^{-1}$ (PFOA), 12.8 ng $mL^{-1}$ (PFOS), 1.5 ng $mL^{-1}$ (PFHxS), and 0.9 ng $mL^{-1}$ (PFNA) | This analysis revealed that greater prenatal exposure to PFOS, PFNA, and PFHxS was correlated with greater ADHD-related traits in kids, with PFOA not demonstrating a relevant effect, indicating a compound-specific neurobehavioral hazard. | Vuong et al. (2021). |

contrast, activated carbon adsorption, ion exchange, and high-pressure membrane filtration technologies have demonstrated high removal efficiencies, often exceeding 90% for certain PFAS compounds. However, challenges in regeneration, concerns over disposal, and variable performance across different PFAS types are the limitations of such technologies. These highlight the challenges of developing a comprehensive and sustainable treatment strategy for PFAS-contaminated water sources.

#### 21.2.5.1 *Activated carbon*

Activated carbon treatment is one of the most thoroughly studied and widely used methods for removing PFAS from drinking water systems. It is used to adsorb contaminants, such as PFAS, onto the surface of activated carbon particles via physical and chemical interactions within the liquid-solid interface. Activated carbon is beneficial in removing a broad array of organic compounds, including taste and odor compounds, synthetic chemicals, and natural organic matter, due to its large surface area and high porosity. The most effective method of PFAS mitigation involves granular activated carbon (GAC), a product made from carbon-rich substances, such as coal, wood, or lignite. GAC has demonstrated particular success in treating PFAS in treated water when implemented as a flow-through filter following particulate removal. Nevertheless, it depends greatly on various factors, such as the nature and depth of carbon media, the water flow rate, temperature, and the individual targeted PFAS compound (Reducing PFAS in Drinking Water with Treatment Technologies, US).

#### 21.2.5.2 *Ion exchange resins*

Ion exchange resins can be used to decontaminate PFAS using anion exchange. The resins are prepared using long-lasting, porous beads of hydrocarbon-like polymers that are capable of resisting water, acids, or bases. An ideal way to remove negatively charged PFAS ions is by using anion exchange resins (AER) with a positive charge. Particularly effective are those single-use AERs that are then incinerated after usage, avoiding the creation of secondary waste. In contrast to GAC, AERs have a greater binding ability to a wider variety of PFAS at a higher cost. They can remove close to 100% of the targeted PFAS under favorable conditions, and their efficiency varies according to the resin type, bed design, flow rate, and other water components (Reducing PFAS in Drinking Water with Treatment Technologies, US).

#### 21.2.5.3 *Membrane separation technology*

Nanofiltration (NF) and reverse osmosis (RO) membranes can be highly efficient at removing PFAS, with efficiencies exceeding 90%, even when used to remove short-chain PFAS. Compared to RO membranes, NF membranes are more selective and reject almost all salts. Still, they permit the limited passage of low-molecular-weight salts, such as sodium chloride, and do not remove hardness minerals. Approximately 84% of the feed water is converted into treated effluent, with 20% of it serving as concentrate waste. This PFAS-contaminated waste stream is difficult to manage, so these technologies are more comfortably applied in a household point-of-use system where the smaller volumes and easier waste management mean disposal is more feasible and less environmentally impactful (Reducing PFAS in Drinking Water with Treatment Technologies, US).

#### 21.2.5.3.1 Electrochemical treatment

Electrochemical treatment breaks down PFAS into unusable residue, unlike GAC, which creates spent waste. It acts by oxidizing pollutants with an electrical current, which breaks the carbon-fluorine bond and releases the fluoride ions. It operates at ambient conditions and is both expensive and effective in centralized and decentralized systems. However, chloride can also form perchlorate in wastewater, which is hazardous (Table 21.3). EPA has further intentions of researching to mitigate the presence of perchlorate in drinking water (EPA's Plan to Address Perchlorate Contamination, 2022; Berg et al., 2022).

#### 21.2.5.3.2 Plasma technology

Plasma is a partially or fully ionized gas produced through electrical discharges that contains electrons, ions, and free radicals capable of breaking strong covalent bonds such as the carbon-fluorine bond in PFAS. The use of catalysts, such as ferrous iron, can enhance this process. Plasma can be generated in thermal or nonthermal ways, with nonthermal (cold plasma) being more energy efficient. Pulsed or gliding arc discharges tend to produce nonthermal plasma, currently deemed to be preferred in PFAS treatment (Jiang et al., 2014). Based on the recommendations of the EPA on drinking water TDB and other studies, cold plasma has been reported to be effective in removing PFAS from groundwater through free electron activation, which breaks up the components Table 21.3 (Stratton et al., 2017). Another study revealed total elimination of PFOS, major acquisition of PFOA (83%) (Palma et al., 2021), and PFHxS (63%). This nonthermal plasma process employs a rotating electromagnetic field to scrub more than 90% of PFAS out of water using a system developed by AquaTech (Sustainable PFAS Destruction Works at Low Light Levels, Aquatech). Nevertheless, it underperforms in the presence of natural organic matter in the water, underscoring one of the main weaknesses of this otherwise promising technology (Palma et al., 2022).

#### 21.2.5.3.3 Ozone, including the use of hydrogen peroxide

In the EPA drinking water TDB, ozonation is demonstrated as a viable technique to remove PFAS, with over 98% PFAS reduction (Daugbjerg and Kay, 2020). In this process, ozone gas is added to the impaired water to create bubbles that react with PFAS because of their surfactant characteristics. The bubble surfaces incorporate PFAS molecules with hydrophobic tails and hydrophilic heads, creating a foam that can be retrieved and treated. Ozone was bubbled into raw water flowing through three ozofractionator columns, with foam containing PFAS rising to the top and solids at the bottom. The test eliminated over 95% of PFOA, PFOS, and PFHxS. The treated water, such as that resulting from adsorption, can then be further processed to remove any remaining contaminants. Another study focused on alkaline ozonation, showing up to 97% PFOS degradation and complete PFOA, PFNA, and 6:2 fluorotelomer sulfonate reduction, making ozonation a promising solution to remove PFAS-contaminated groundwater (Thomas et al., 2020).

## 21.2.6 Legal and policy changes

DuPont and Chemours have been involved in several major legal disputes related to PFAS. They also settled a significant class-action lawsuit in West Virginia in 2017 for $671 million, over allegations of exposure to PFOS/PFOA related to emissions from their chemical plant. Most recently, in 2023, 3 M settled with US public water suppliers in a record-breaking agreement of 10.3 billion dollars to address widespread PFAS contamination cases (Exposure: Poisoned Water, Corporate).

Under SDWA, the EPA established national drinking water standards in April 2024 to regulate the presence of six PFAS. These guidelines are referred to as MCLs and specify rigorous guidelines with limits of PFOA and PFOS as low as 4 ppt (Per- and Polyfluoroalkyl Substances PFAS, ECHA).

In 2023, the EPA proposed a rule under Section 8(a)(7) of the TSCA that would compel manufacturers of PFAS to provide detailed reporting per the intent of the act on the use and release of PFAS since 2011 (Interim Guidance on the Destruction and Disposal of PFAS and Materials Containing PFAS, US).

In 2024, the EPA advanced the proposal to list PFOA and PFOS as hazardous substances under the CERCLA, or the Superfund law. This new classification has the potential to help the agency force cleanup and make the polluters liable for the PFAS contamination (Per- and Polyfluoroalkyl Substances (PFAS), US).

Nevertheless, considerable gaps remain, despite these regulatory trends. The lack of a cohesive federal approach to the entire PFAS is a main concern. Currently, a tiny percentage of the more than 12,000 different PFAS compounds are being regulated at the federal level. Very short-chain PFAS and novel alternatives, such as GenX and perfluorobutane sulfonic acid (PFBS), are widely avoided due to extensive regulation.

Currently, no federal law disposal standard exists for waste containing PFAS. Thermal treatment procedures, such as incineration, are also controversial because they raise concerns that PFAS compounds may not be destroyed and that harmful emissions may be released into the atmosphere.

Moreover, some states, such as Michigan, New Jersey, and California, have adopted their PFAS laws, creating a patchwork system that complicates nationwide compliance (Per- and Polyfluoroalkyl Substances PFAS, State; Federal Role in Responding to Potential Risks of Per-and Polyfluoroalkyl Substances PFAS, 2022). Industries delayed PFOS regulation by withholding data (Gaber et al., 2023), lobbying against standards (Industry Lobbying and Eye-Watering Costs of PFAS Pollution, Revealed). Funding favorable research and settling PFAS contamination claims with a $10.3 billion agreement in 2023 (3M Gets Court Approval of $10 Billion PFAS, Settlement).

### *21.2.6.1 Public awareness*

The presence of PFAS in US drinking water has garnered increased attention from the media, federal agencies, and public health researchers in recent years. Nonetheless, PFAS is unknown to the general public. The research involved a national survey to assess awareness of PFAS, the discovery of PFAS-containing products, and concerns about drinking water pollution. The findings indicated that 45.1% of the respondents were unaware of PFAS, and 31.6% were aware but lacked a clear understanding of it. The most effective predictor of

PFAS knowledge was awareness of community exposure to PFAS. The results underscore the necessity of a well-organized, clear public communication from government bodies, utilities, and researchers to improve awareness and informed decision-making (Berthold et al., 2023).

### 21.2.7 Comparison with international regulations

The regulation of PFAS differs widely across countries. Certain countries have even decided to prohibit certain PFAS chemicals, while others have limited their use or established upper permitted levels of PFAS in drinking water. The United States EPA has declared a nonenforceable lifetime health advisory on PFAS in drinking water. The European Union (EU), on the other hand, has been more aggressive in this regard by limiting or prohibiting the usage of particular PFAS. Canada has set guidelines for drinking water by HC. In the meantime, Australia has developed the voluntary phase-out of certain PFAS and limited their possible application in firefighting foam.

### 21.2.8 Future directions and recommendations

Several actions and policy initiatives have been proposed to enhance the regulation of PFAS and protect public health. To begin with, the decision to declare PFOS and PFOA as hazardous substances under the CERCLA section must be completed and implemented. This action would strengthen the framework of liability, where the polluters should carry a financial burden of cleanup, particularly in military and industrial locations (Designation of Perfluorooctanoic Acid PFOA and Perfluorooctanesulfonic Acid PFOS as CERCLA Hazardous Substances, US EPA, http). Furthermore, national drinking water standards developed in April 2024 are expected to extend the list of PFAS substances to include PFBS, perfluorohexanoic acid (PFHxA), and perfluorobutanoic Acid (PFBA). An alternative recommended regulation method is the utilization of a class-based control system that regulates the whole PFAS family instead of individual chemicals (Per- and Polyfluoroalkyl Substances PFAS, ECHA).

Monitoring and risk assessment are also essential. A countrywide program should track PFOS in water, air, soil, and food, with reporting required under TSCA Section 8(a)(7). The funding of long-term health studies involving birth cohort and cancer risk studies should be carried out to provide information about risk thresholds (TSCA Section 8a7 Reporting and Recordkeeping Requirements for Perfluoroalkyl and Polyfluoroalkyl Substances, US). Federal disposal and destruction guidelines, which are apparent, should be established to dispose of the waste PFOS in the form of landfilling, incineration, and deep well injection, as well as control of the harmful byproducts and emissions. The use of advanced technologies, such as the use of Plasma treatment, electrochemical oxidation, and bioremediation, is advised (Interim Guidance on the Destruction and Disposal of PFAS and Materials Containing PFAS, US).

There is also a pressing need to prioritize vulnerable populations and address environmental justice issues. These risk assessments must focus on low-income and indigenous groups, especially those residing near military bases or relying on fish and groundwater resources. The practices of PFOS mitigation should be incorporated into the screenings, education practices, and access to clean water in the sphere of public health (National PFAS Testing Strategy, US).

Finally, legislative action is vital. The act, such as the PFAS Action Act, should be passed by Congress to require cleanups, prohibit nonessential uses, and increase funding to support research and remediation efforts. Designation of a specific PFAS research center or task force under the NIH or CDC could facilitate coordinating toxicological research and enable the central collection of information (H.R.2467 - 117th Congress 2021-2022, PFAS).

## 21.3 Canada: per- and polyfluoroalkyl substances regulatory guidelines

### 21.3.1 Overview of environmental issues in Canada

Canada ranks among the leading producers of fourteen mined minerals, including potash, uranium, gemstones, platinum, indium, and gold, in the global market. Along with benefiting the economy, there are great environmental costs that accompany mining, including deforestation, the contamination of water, and the forced displacement of communities. Canada also produces the fourth-largest amount of oil and exports most of its oil to the US as crude oil to oil refineries, which are mostly located in Alberta. Alberta has almost a quarter of its economy based on the mining and energy industries, which are associated with water pollution. Moreover, freshwater is polluted as a consequence of the use of salt in the winter, which has long periods in Canada. Many organisms are susceptible to sodium chloride, which can inhibit the growth of plants in different soils, since it is used to melt ice. The environment is facing increasingly harsh realities caused by climate change, such as increased wildfires and increasing air and water pollution due to industry and mining (The Biggest Environmental Issues In Canada, WorldAtlas; 5 Pressing Environmental Issues in Canada, 2024). Another pollutant in Canadian waters, albeit increasing, is pollution by endocrine-disrupting chemicals (EDCs). The EDCs, which are present in pesticides, plastics, industrial chemicals, and personal care products, have been found in water, soil, air, and human biomonitoring studies. These chemicals may interfere with the functions of hormones, thereby affecting reproduction, development, and overall health. In accordance with the CEPA, the EDCs are regarded as chemicals that hamper the natural functioning of hormones in organisms (SC 1999, c 33, 1999; Marlatt et al., 2022; Barton-Maclaren et al., 2022). The Canadian system for managing the risks of EDCs is the Chemicals Management Plan (CMP). Some of the actions taken are the prohibition of bisphenol A in baby bottles (Committee Report, 2024; Consideration of Endocrine-Related Effects in Risk Assessment - Canada.Ca, 2024; Gewurtz et al., 2021).

PFAS have been found persistently in a wide range of environmental samples across all regions in Canada, such as in air, surface waters, landfill leachate, treated wastewater, biosolids, and aquatic and terrestrial wildlife. Initially, few PFAS compounds could be monitored in human biomonitoring studies; now, the range of analyzed PFAS compounds is growing, and present studies report the highest PFAS detection rates in different biological matrices (State of Per- and Polyfluoroalkyl Substances PFAS Report - Canada.Ca, 2024). Recent Canadian studies have investigated PFAS exposure and its health outcomes

(Borghese et al., 2022). Studied changes in liver biomarkers of liver functionality due to single and combined PFAS exposure (Borghese et al., 2022). Caron-Beaudoin et al. (2019) studied the relevance of PFAS relationships with thyroid functions on the First Nations children/youth population in Quebec (Caron-Beaudoin et al., 2019). Aker et al. (2023) presented a study on Inuit cardiometabolic health in terms of PFAS (Aker et al., 2023).

In Europe and the US, the biomonitoring data have demonstrated that the legacy PFAS (such as PFOA and PFOS) production levels have declined because of the regulatory response, whereas the levels of newer PFAS alternatives have been on the increase. This pattern is reflected by Canadian PFAS trends, but there is a regional difference, predominantly driven by industrial activities and rules. The level of PFAS in the Canadian blood and serum is comparable to that in the US, although it is higher in some groups of the population, namely, Indigenous and firefighters, because their exposure routes are different (Per- and Polyfluoroalkyl Substances PFAS in Canadians, Canada).

The presence of PFAS in the environment and human populations raises significant health and environmental concerns, particularly when contamination levels become excessive. To tackle these issues, monitoring and regulation of PFAS levels across different samples are essential. In response, the Government of Canada has established guidelines aimed at effectively managing and regulating PFAS, which will be detailed in the following discussion, and Table 21.5 provides an overview of the guidelines.

### 21.3.2 Regulatory framework for per- and polyfluoroalkyl substances

The Canadian government has created guidelines aimed at safeguarding human health and the environment, which include federal environmental quality guidelines (FEQGs) for PFOS in surface water and for different wildlife diets and tissues (ECCC, 2018). Guidelines for Canadian soil and groundwater quality regarding PFOS have also been developed (Canadian Council of Ministers of the Environment, CCME 2021b). HC published soil screening values based on human contact that apply to other PFAS because no Canadian soil quality guidelines (SQGs) for these PFAS exist at this time, although they have not received the same level of review by the CCME. Due to uncertainties surrounding PFAS evaluations, a precautionary approach is suggested. Additionally, federal guidelines regarding PFOA and its impact on wildlife and bird eggs are also available. Some provinces also developed guidelines tailored to their specific needs within their jurisdictions to guide the management of provincial lands, industrial properties, and private properties. The Minister of the Environment and the Minister of Health have issued a report under Section 68 of the CEPA, 1999 (Canadian Environmental Protection Act, 1999) on the classification of PFAS. This report summarizes the PFAS origin, behavior, presence, and related risks for people and ecosystems. This group of substances is a priority due to scientific evidence indicating potential environmental and human health effects. The other PFAS groups, which are being used as substitutes for the regulated PFAS (such as PFOS, PFOA, and long-chain perfluorocarboxylic acids [LC-PFCAs] and their salts and precursors), may also pose similar risks. This conclusion, along with the findings of this report, should guide decision-making regarding PFAS as a class in Canada (State of Per- and Polyfluoroalkyl Substances PFAS Report - Canada.Ca, 2024).

**TABLE 21.5** Overview of regulatory guidelines in Canada.

| Authority | Scope | Key PFAS | Guideline values/ criterion |
|---|---|---|---|
| Environment and Climate Change Canada—federal environmental quality guidelines (2018) | Water | PFOS | 6.8 $\mu g\ L^{-1}$ |
| | Fish tissues | | 9.4 $mg\ kg^{-1}$ ww |
| | Wildlife diet (mammalian) | | 4.6 $\mu g\ kg^{-1}$ ww food |
| | Wildlife diet (avian) | | 8.2 $\mu g\ kg^{-1}$ ww food |
| | Bird egg | | 1.9 $\mu g\ g^{-1}$ ww |
| Health Canada | Drinking water | PFOA | 0.2 $\mu g\ L^{-1}$ |
| | | PFOS | 0.6 $\mu g\ L^{-1}$ |
| | | Sum of 25 individual PFAS | 30 $ng\ L^{-1}$ |
| | Human health protection—direct contact residential soil screening value | PFOA | 0.70 $mg\ kg^{-1}$ |
| Canadian Council of Ministers of the Environment (CCME) | Soil quality guideline for environmental health in agricultural, residential/parkland, commercial, and industrial soil | PFOS | 0.01($mg\ kg^{-1}$ dry weight [dw]) |
| CCME<br>Ontario Ministry of the Environment, Conservation and Parks | Soil quality guideline for human health ($SoQG_{HH}$) in agriculture and residential soil | PFOS<br>key 11 PFAS | 0.01 ($mg\ kg^{-1}$ dry weight [dw]) |
| | $SoQG_{HH}$ in commercial and industrial soil | | 0.2 ($mg\ kg^{-1}$ dry weight [dw]) for the course soil |
| | $SoQG_{HH}$ in commercial and industrial soil<br>groundwater quality guideline | | 0.1 ($mg\ kg^{-1}$ dry weight [dw]) for the fine soil |
| | | | 0.0006 $mg\ L^{-1}$ for course and fine types of soil |
| | Drinking water | | 70 $ng\ L^{-1}$ |
| Atlantic Provinces (Nova Scotia, Prince Edward Island, New Brunswick, and Newfoundland and Labrador) government | Potable groundwater standards for agricultural, residential/parkland, commercial, and industrial sites | PFDS<br>OA | 0.2 $\mu g\ L^{-1}$ |
| Atlantic Provinces (Nova Scotia, Prince Edward Island, New Brunswick, and Newfoundland and Labrador) government | Potable groundwater standards for agricultural, residential/parkland, commercial, and industrial sites | PFOS | 0.6 $\mu g\ L^{-1}$ |

(*Continued*)

TABLE 21.5 (Continued)

| Authority | Scope | Key PFAS | Guideline values/ criterion |
|---|---|---|---|
| | Soil standards for agriculture and residential/parkland sites | | |
| | | PFBA | 30 μg $L^{-1}$ |
| | | PFBS | 15 μg $L^{-1}$ |
| | | PFHxS | 0.6 μg $L^{-1}$ |
| | | PFPeA | 0.2 μg $L^{-1}$ |
| | | PFHxA | 0.2 μg $L^{-1}$ |
| | | PFHpA | 0.2 μg $L^{-1}$ |
| | | PFNA | 0.2 μg $L^{-1}$ |
| | | PFOA | 0.7 mg $kg^{-1}$ |
| | Soil standards for agriculture and residential/parkland sites<br>Soil standards for commercial sites | PFBA | 114 mg $kg^{-1}$ |
| | | PFBS | 61 mg $kg^{-1}$ |
| | | PFHxS | 2.3 mg $kg^{-1}$ |
| | | PFPeA | 0.8 mg $kg^{-1}$ |
| | | PFHxA | 0.8 mg $kg^{-1}$ |
| | | PFHpA | 0.8 mg $kg^{-1}$ |
| | | PFNA | 0.08 mg $kg^{-1}$ |
| | | PFOA | 1.05 mg $kg^{-1}$ |
| | Soil standards for commercial sites<br>Soil standards for industrial sites | PFBA | 173 mg $kg^{-1}$ |
| | | PFBS | 92 mg kg-1 |
| | | PFHxS | 3.5 mg $kg^{-1}$ |
| | | PFPeA | 1.21 mg $kg^{-1}$ |
| | | PFHxA | 1.21 mg $kg^{-1}$ |
| | | PFHpA | 1.21 mg $kg^{-1}$ |
| | | PFNA | 0.13 mg $kg^{-1}$ |
| | | PFOA | 9.94 mg $kg^{-1}$ |
| | Soil standards for industrial sites<br>Soil standards (with potable groundwater) for all sites | PFBA | 1630 mg $kg^{-1}$ |
| | | PFBS | 872 mg $kg^{-1}$ |
| | | PFHxS | 33 mg $kg^{-1}$ |
| | | PFPeA | 11.41 mg $kg^{-1}$ |

*(Continued)*

TABLE 21.5 (Continued)

| Authority | Scope | Key PFAS | Guideline values/ criterion |
|---|---|---|---|
| | | PFHxA | 11.41 mg $kg^{-1}$ |
| | | PFHpA | 11.41 mg $kg^{-1}$ |
| | | PFNA | 1.2 mg $kg^{-1}$ |
| | | PFOS | 0.35 mg $kg^{-1}$ |
| | Soil standards (with nonpotable groundwater) for agriculture and residential/parkland sites | PFOS | 2.1 mg $kg^{-1}$ |
| | Soil standards (with nonpotable groundwater) for commercial sites | | 3.2 mg $kg^{-1}$ |
| | Soil standards (with nonpotable groundwater) for Industrial sites | | 30.5 mg $kg^{-1}$ |

### *21.3.2.1 Government agencies and regulations Involved for per- and polyfluoroalkyl substances*

#### 21.3.2.1.1 Canadian Environmental Protection Act, 1999

There are three different subgroups of PFAS that have been evaluated under CEPA in Canada. They have been identified as an environmental issue and are therefore included in Schedule 1 of CEPA: PFOS and its salts and precursors (Carreau and Buccini, 1999). PFOA and salts and precursors of PFOA and LC-PFCAs and salts and precursors of LC-PFCAs (Ecological Screening Assessment Report on Perfluorooctane Sulfonate, Salts).

#### 21.3.2.1.2 Prohibition of Certain Toxic Substances Regulations 2012

In Canada, the three primary groups of PFAS (PFOS, PFOA, and LC-PFCAs, and their salts and precursors) are added in Schedule 1 of CEPA due to risk to the environment and their continuous bioaccumulation; they are regulated under PCTSR (2012). This risk control measure deals with 94 PFAS on the Domestic Substances List (DSL).

Since 2016, PCTSR has banned the manufacture, use, sale, offer for sale, and import of PFOA, LC-PFCAs, their salts, and precursors, as well as products containing them, with a limited number of exemptions. To illustrate this, the exemptions to the use of PFOA and LC-PFCAs in AFFF were allowed as limited, and all manufactured products were provided (State of Per- and Polyfluoroalkyl Substances PFAS Report - Canada.Ca, 2024).

#### 21.3.2.1.3 New substances notification regulations (chemicals and polymers)

It is estimated that the established NSNR resulted in the simultaneous regulation and/or prohibition, on ministerial conditions, or under the CEPA, in significant new activity (SNAc) notification of approximately 100 or higher of the more than 290 PFAS so notified. Some sections of the NSNR have been superseded by the introduction of other legislation that prohibits the same substances, and which will not result in any threat to human health and the environment, such as the Ozone-Depleting Substances and Halocarbon Alternatives Regulations (ODSHAR).

New substance evaluation takes into account potential risks to the notified activities and other potential activities that may use the substance. SNAc with respect to a new substance (SNAc) may be effected under Section 85 of CEPA through the publication of SNAc notice in the Canada Gazette, Part I, in cases where there is a suspicion that the SNAc may render the substance toxic (State of Per- and Polyfluoroalkyl Substances PFAS Report - Canada.Ca, 2024).

#### 21.3.2.1.4 Canadian Council of Ministers of the Environment

Under CCME, the SQGs and GWQGs guidelines exist for PFOS to ensure human health and the environment. For soil quality guideline for environmental health the PFOS limit is 0.01 (mg $kg^{-1}$ dry weight [dw]) in the agricultural, residential/parkland, commercial and industrial soil and for soil quality guideline for human health the limit is 0.01 in the agriculture and residential soil and for commercial and industrial soil the limit is 0.2 for the course soil and for fine soil it is 0.1 (mg $kg^{-1}$ dry weight [dw]).

The lowest final-ground water quality guideline of pathway-specific guidelines based on ecological receptors and humans, and other management factors such as substance solubility, detection limits, and background levels, is 0.0006 mg $L^{-1}$, considering the coarse and fine types of soil (Canadian Environmental Protection Act, 1999).

#### 21.3.2.1.5 Environment and Climate Change Canada

Under the ECCC 2018 guidelines, federal environmental quality standards have been established for PFOS levels in surface water to protect various aspects of aquatic life. The guidelines specify the following values: the water value is 6.8 g $L^{-1}$, fish tissue is 9.4 g $kg^{-1}$ (ww), wildlife diet, mammalian is 4.6, and bird egg is 1.9 μg $g^{-1}$ (ww) (Ccme, Canadian).

#### 21.3.2.1.6 Health Canada

In December 2018, HC published guidelines for Canadian drinking water quality for PFOS and PFOA (HC, 2018a; HC, 2018b). It has established a maximum acceptable concentration (MAC) of 0.2 mcg $L^{-1}$ for PFOA and a MAC of 0.6 mcg $L^{-1}$ against PFOS. In addition, HC has also provided a statement according to which, in cases where PFOS and PFOA are both present in the drinking water, the ratio between the PFOA concentration and its MAC, as well as the ratio between the PFOS concentration and its MAC, should not exceed one, in case they are combined. In case it is equal to or less than one, then the water is assessed to be safe to drink by HC (Guidance for the Assessment and Remediation of Per- and Polyfluoroalkyl Substances in British Columbia, 2019).

In 2024, the objective for Canadian drinking water quality was released by HC: per-polyfluoroalkyl compounds, which implies a single value based on the value of PFAS, in this case, a group of PFAS (HC 2024). The 30 ng $L^{-1}$ standard includes the sum of 25 individual PFAS and ways of minimizing probable exposure to PFAS via drinking water, while the official rules are being updated (State of Per- and Polyfluoroalkyl Substances PFAS Report - Canada.Ca, 2024).

#### 21.3.2.2 Local/state government bodies or provincial regulations and standards

##### 21.3.2.2.1 British Columbia

British Columbia, under the Contaminated Sites Regulation (BC CSR), gives a standard for PFBS, PFOS, and PFOA. For PFOS, BC CSR developed standards for soil and groundwater(aquatic and drinking water), for PFBS, standards are developed for soil and groundwater (drinking water), and for PFOA, standards for groundwater are developed by BC CSR (Guidance for the Assessment and Remediation of Per-and Polyfluoroalkyl Substances in British Columbia, 2019). BC has also established guidelines for water quality regarding PFOS (protection of freshwater aquatic life and drinking water) and PFOA (protection of drinking water) (B.C. Ministry of Environment and Climate Change Strategy, 2020, 2021) (State of Per- and Polyfluoroalkyl Substances PFAS Report - Canada.Ca, 2024).

##### 21.3.2.2.2 Alberta

Alberta establishes Tier 1 limits for PFOS in soil across various land uses, as well as separate Tier 1 limits for PFOS and PFOA in groundwater. Both substances are also included in Alberta's Tier 2 soil and groundwater guideline (Alberta Tier 1 Soil and Groundwater Remediation Guidelines - Open, Government).

##### 21.3.2.2.3 Quebec

The Michigan Department of Environment, Great Lakes, and Energy's surface water PFOS and PFOA standards for drinking water and aquatic food consumption have been accepted by Quebec (State of Per- and Polyfluoroalkyl Substances PFAS Report - Canada.Ca, 2024).

##### 21.3.2.2.4 Ontario

When a "widespread PFAS release" from firefighting foam was discovered to have affected groundwater in 2017, the Ontario MOE, Conservation and Parks advised private drinking water well owners to think about treatment if the total of 11 distinct PFAS was greater than 70 ng $L^{-1}$ (Health Ontario, P). In May 2021, the province published human health toxicity reference values for PFOS and PFOA to be used at contaminated sites in Ontario (State of Per- and Polyfluoroalkyl Substances PFAS Report - Canada.Ca, 2024).

##### 21.3.2.2.5 Atlantic Provinces

The Atlantic governments of the Four Atlantic Provinces (Nova Scotia, Prince Edward Island, New Brunswick, and Newfoundland and Labrador) obtained Atlantic risk-based corrective action environmental quality standards that stimulate various PFAS in water and soil. Agricultural, residential/parkland, commercial, and industrial groundwater potable standards have been established as follows (in µg $L^{-1}$): PFOA-0.2, PFOS-0.6, PFBA-30, PFBS-15, PFHxS-0.6, PFPeA-0.2, PFHxA-0.2, PFHpA-0.2, and PFNA-0.02 (Atlantic RBCA-Human Health-Based Tier I Environmental Quality Standards EQS for Groundwater-All Land Uses; Potable Groundwater Condition, mg/L). It also depends on land use in soil standards. In agricultural and residential/parkland sites (mg $kg^{-1}$), PFOA-0.7, PFBA-114, PFBS-61, PFHxS-2.3, PFPeA-0.8, PFHxA-0.8, PFHpA-0.8, PFNA-0.08. On commercial sites (in mg $kg^{-1}$): PFOA-1.05, PFBA-173, PFBS-92, PFHxS-3.5, PFPeA-1.21, PFHxA-1.21, PFHpA-1.21, PFNA-

0.13. For Industrial sites (mg $kg^{-1}$) PFOA-9.94, PFBA-1630, PFBS-872, PFHxS-33, PFPeA-11.41, PFHxA-11.41, PFHpA-11.41, PFNA-1.2 (Atlantic RBCA-Human Health-Based Tier I Environmental Quality Standards EQS for Groundwater-All Land Uses; Potable Groundwater Condition, mg/L). Also sets soil standards (with potable groundwater) for all sites at 0.35 mg $kg^{-1}$. Sets soil standards (with nonpotable groundwater) for: agriculture and residential/parkland sites: 2.1 mg $kg^{-1}$, commercial sites: 3.2 mg $kg^{-1}$, industrial sites: 30.5 mg $kg^{-1}$ for PFOS (Atlantic RBCA-Human Health-Based Tier I Environmental Quality Standards EQS for Groundwater-All Land Uses; Potable Groundwater Condition, mg/L).

## 21.3.3 PFAS detection and monitoring

### *21.3.3.1 Environmental monitoring*

Canada began monitoring PFAS in passive air samples in 2004, covering C4–C14, C16, and C18 PFCAs and C4, C6, C8, and C10 perfluorosulfonic acids (PFSAs) and their precursors. Sampling took place at 13 sites across the country through the Global Atmospheric Passive Sampling network. During the same period, the levels of PFSA, such as PFBS, PFHxS, and PFOS, have increased drastically at some Canadian sites. Additionally, the levels of PFCA compounds, including PFHxa, PFHpA, PFOA, PFNA, and PFDA, increased in 2017. PFAS emissions associated with the waste management sector, comprising wastewater treatment plants (WWTPs) and landfills, have accumulated. In Ontario, air samples were taken at and near one WWTP and two solid waste landfills in 2009 (PFAS). These samples were tested across five categories of PFAS. The total concentration of PFAS in air was observed to be 3 to 15 times greater at the WWTP and 5 to 30 times greater at landfill sites (State of Per- and Polyfluoroalkyl Substances PFAS Report - Canada.Ca, 2024). The Government of Canada conducts monitoring of freshwater at various sites across the country. PFAS in 29 sites were analyzed between 2013 and 2020 to assess the levels and patterns of surface waters. The sampling covered all provinces apart from Alberta and Prince Edward Island. PFAS were detected in the surface water of all the tested provinces. In Canada, 13 types of PFAS were detected in most of the freshwater samples; the highest concentration value was 138 ng $L^{-1}$ (PFBS), and the lowest concentration was at the laboratory detection limit (between 0.4 and 1.6 ng $L^{-1}$). Even though the level of concentration of PFOS and PFOA has declined within this period, other compounds have increased for example PFBA and PFPeA. Moreover, PFOS levels in surface water samples (2007–17) from different drainage areas in Canada were consistently lower than the FEQG for PFOS. PFAS are also measured in whole-body homogenates of fish collected from Canadian waters. Such continuous monitoring is essential for knowing the existence and concentration of PFAS in the aquatic ecosystem. Furthermore, the concentration of PFAS has also been measured in fish and birds across different regions of the country.

Anthropogenic sources of PFAS in an active fire-training area on the territory of an international civilian airport in Ontario, 70 AFFF-derived PFAS were detected in fish, eggs, and blood of nestling peregrine falcons- a species of predator birds that feeds on other birds in the south of Ontario and along the north shore of Lake Superior (State of Per- and Polyfluoroalkyl Substances PFAS Report - Canada.Ca, 2024). Between 2008 and 2014, landfill leachate was collected as part of CMP environmental monitoring and surveillance and was collected in 13 large municipal solid waste (MSW) landfills across Canada, each of which was permitted to receive 40,000 tonnes of waste annually. Between 2009 and 2011, leachate samples from 12 landfills were analyzed for PFAS (PFCAs, C4, C6, C8 PFSAs, and perfluorooctanesulfonamide). The total PFAS concentration in the leachate was between 320 and 9400 ng $L^{-1}$, untreated, with a median of 3227 ng $L^{-1}$. Following on-site leachate treatment, the overall level of PFAS ranged from 800 to 14,201 ng $L^{-1}$. It is interesting to note that the total concentration of PFAS tended to improve following on-site treatment. The Government of Canada conducted a study between 2019 and 2022 that involved sampling leachate from 10 operational MSW landfills. The purpose of this study was to assess the presence and concentration of various substances, including PFAS, in the landfill leachate. This effort was part of the initiative aimed at protecting and recovering endangered whale populations. PFBs were more commonly found in the raw leachate samples, with the most common being PFNA (52% of samples recorded), 6:2 FTSA (66%), PFOS (74%), PFBS (84%), PFHxS (90%), 96:2 FTSA (94%), PFPeA (98%), PFOA (98%), and PFHxa (99%) (State of Per- and Polyfluoroalkyl Substances PFAS Report - Canada.Ca, 2024).

The Government of Canada has also implemented partnerships with local governments around the country to determine the highly different types of WWTP in Canada (primary, secondary, advanced, and lagoon treatment plants), as well as different geographical regions (mountainous, prairie, Great Lakes/St Lawrence, and coastal regions). In Canada, the authors have studied the destiny and behavior of 13 PFAS in influent, effluent, and solid media at 15 Canadian WWTPs. Of the identified PFAAs, PFOA was the most common, ranging from 2.2 to 150 ng $L^{-1}$ in the influent and from 1.9 to 140 ng $L^{-1}$ in the effluent. PFOS were found in primary sludge, waste biological sludge, and treated biosolids in 6.4–2900 ng $g^{-1}$; 9.7–8200 ng $g^{-1}$ and 2.1–17,000 ng–$g^{-1}$ as dry weight, respectively. In more recent studies, the sources and retention of 42 PFAS were tracked in 27 Canadian WWTPs. These investigations revealed a general downward trend in concentrations of certain PFAS, which can be attributed to the termination of industrial production and regulatory activities. Nevertheless, in wastewater, PFOS concentrations have not been shown to decrease over time, indicating that the effects of regulatory action and the PFOS phase-out are not being reflected in wastewater in a timely manner. Short-chain PFCAs and PFSAs in wastewater influents and effluents showed an increase in concentration over time (2009–21), indicating a persistence of short-chain PFAS usage as second-generation reactions to longer-chained PFAS, which were regulated and phased out (State of Per- and Polyfluoroalkyl Substances PFAS Report - Canada.Ca, 2024).

#### *21.3.3.2 Human biomonitoring*

Nine PFAS were measured in Canada as part of the Canadian Health Measures Survey (CHMS), which has been conducted since 2007 to estimate the presence of environmental chemicals in the urine or blood of Canadians. CHMS is measured every 2 years and is indicative of the noninstitutionalized population and the general Canadian population, with the exception of residents on reserve, full-time Canadian Forces, residents in institutions, and inhabitants of some sparsely settled areas, comprising less than 4% of the population. Excluding that, published biomonitoring results for selected groups not included in the CHMS, such as Indigenous groups in specific regions, are also measured. Biomonitoring results for PFAS from CHMS are available from four cycles from 2007 to 2019. Cycle 1 (2007–09) measured PFOA, PFHxS, and PFOS. In subsequent cycles-Cycle 2 (2009 to 2011), Cycle 5 (2016 to 2017), and Cycle 6 (2018 to 2019), nine PFAS were measured: PFBA, PFHxA , PFOA, PFNA, PFDA, perfluoroundecanoic acid (PFUnDA) , PFBS, PFHxS, and PFOS.

CHMS results show a strong decline in PFOA, PFNA, PFDA, PFHxS, and PFOS levels in Canadians aged 12 or 20–79 years ($P<.001$) (Per- and Polyfluoroalkyl Substances PFAS in Canadians, Canada). Compared with 2007 and 2019, the geometric means (GMs) of the CHMS data on individuals aged 20–79 years showed a 52% and 67% reduction in plasma concentrations of PFOA and PFOS, respectively. Despite such reductions, PFOA and PFOS remain in most of the population. Similar to other national and regional biomonitoring surveys, CHMS cycle 6 results of the general population aged 3 to 79 years in Canada have shown that PFOS at 3 to 79 years (GM = 2.5 ug $L^{-1}$) has the highest amount of the chemical (PFOS), followed by PFOA (GM = 1.2 ug $L^{-1}$) (State of Per- and Polyfluoroalkyl Substances PFAS Report - Canada.Ca, 2024).

PFHxS was analyzed in the four instances of the CHMS, and it was found that the plasma concentration geometrically declined, showing a 64% decrease between 2007 and 2019 in the adult population of Canada aged 20–79 years. PFHxS was reported in over 99% of the 3–79-year-old population in Cycle 6, with a GM plasma concentration of 0.76 mg $L^{-1}$ (limit of detection [LOD] = 0.063 mg $L^{-1}$). The concentrations of PFOA, PFHxS, and PFOS were higher in male participants than in female participants over the four cycles of the CHMS. Moreover, PFAS levels were generally higher in the Canadian adult population compared to children. PFNA, PFDA, and PFUnDA were monitored in Cycles 2, 5, and 6. The PFNA GM plasma concentration of 0.44 μg $L^{-1}$ was detected in more than 98% of the population between the age groups of 3 to 79 years in cycle 6. It was the fourth-highest measured PFAS concentration of CHMS participants after PFOS, PFOA, and PFHxS. PFDA concentration in Cycle 6 was found to have a GM of 0.12 μg $L^{-1}$, with lower concentrations observed in individuals aged 3 to 79 years. PFUnDA was detected in 36%of the samples (LOD: 0.12 μg $L^{-1}$) in cycle 6; therefore the GM was calculated, as a large proportion of samples were below the LOD threshold. PFNA and PFDA concentrations in Canadians (12–79) fell by 47 and 36, respectively, between 2009 and 2019 and did not differ between sexes, as was the case with other PFAS compounds (Aker et al., 2023). PFBA, PFHxa, and PFBS had a low rate of detection during Cycles 2, 5, and 6 of the CHMS. When the samples are 40% or less below LOD, GMs are not calculated, as was the case in PFBA, PFHxa, and PFBS. PFBA, PFHxa, and PFBS have low biological half-lives, which may be the reason why they are not detected with as high a frequency as these PFAS. Nonetheless, other studies that have employed more sensitive detection methods have shown that a greater percentage of samples exceed the detection level. A small study in Oslo measured PFBS in adult plasma and serum. Detection rates were above 100% in plasma and 51% in serum, based on MDLs of 0.018 and 0.009 μg $L^{-1}$ (Poothong et al., 2017). The plasma and serum sensitive cutoffs of plasma and serum are lower than 0.066 ug $L^{-1}$, which is the PFBS LOD in Cycle 6 of the CHMS.

There are data on the levels of PFAS in the plasma or serum of Inuit societies and other indigenous or northern societies in Canada. In a single study, PFAS and variants of the chemical (PFBA, PFHxA, PFOA, PFNA, PFDA, PFUnDA, PFBS, PFHxS, and PFOS) were analyzed in Inuit serum in Nunavik communities among pregnant women aged 16 to 40 years. They were members of the Nutaratsaliit Qanuingisiangingit Niqituinnanut (NQN) Pregnancy Wellness with Country Food Project, which was carried out from 2016 to 2017.

Data on PFAS in plasma or serum are available for First Nations communities, Inuit groups, and other indigenous or northern populations in Canada. The article examined PFBA, PFHxS, PFNA, PFDA, PFUnDA, PFBS, PFHxS, and PFOS in the serum of pregnant Inuit women aged 16–40 years in Nunavik who participated in the NQN Pregnancy Wellness with Country Food Project (2016–17). A third study also found a decrease in the levels of maternal serum PFOA, PFHxS, and PFOS between 2004 and 2007. These trends aligned with national CHMS statistics. PFHxS and PFOA were identified to be lower in the NQN cohort than in cycle 5 of the CHMS (Caron-Beaudoin et al., 2020). PFAS can also be detected in the plasma and cord blood plasma of expectant mothers, and findings of these studies show that infants in Canada can be exposed to several PFAS by way of consumption of human milk (Caron-Beaudoin et al., 2020). Table 21.6 below provides an overview of research studies conducted on PFAS in Canada.

## 21.3.4 Remediation efforts of per- and polyfluoroalkyl substances

The strategy implemented in 2006 in Canada aimed to reduce the concentration of PFOS in the environment. In 2008, regulations were introduced that banned the use of PFOS and provided some exemptions. In 2010, a voluntary agreement was reached with the companies to eliminate residual PFOA, long-chain PFCAs, and their precursors in products sold in Canada within a 5-year period.

PFOA, LC-PFCAs, their salts, precursors, and products with these substances have been banned under the PCTSR since 2016, with few exemptions. These exceptions involved the exemption of PFOA and an LC-PFCA in AFFF, used in limited and specified applications, as well as in all manufactured products. The current legislation of the PCTSR controls 94 PFAS detected in the Canadian trade through the DSL, along with some other PFAS whose existence in Canada is uncertain.

In 2018, a consultation document proposed changes to the PCTSR with the goal of gradually eliminating toxic substances that are already regulated. Some early exemptions for PFOS, PFOA, and LC-PFCAs were granted to allow certain sectors time to transition to alternative options. The proposed regulations were based on feedback from this consultation and issued in the Canada Gazette on May 14, 2022. Meanwhile, an Order was suggested to add PFOA and LC-PFCAs into the Export Control List, in accordance with Canada's obligations to the Rotterdam Convention. New substances not on the DSL that are found to have a specified trigger threshold should be reported to the government to assess the risk before their importation or production in Canada. Drugs are evaluated individually under the NSNR; new drugs are reported to the government at diverse occasions, and they are individually evaluated as to whether they may cause harm to the environment and human health through industrial and other purposes (consumer goods, cosmetics, and pharmaceuticals). More than 290 new PFAS have been reported to the New Substances program, and regulators are applying risk reduction (State of Per- and Polyfluoroalkyl Substances PFAS Report - Canada.Ca, 2024).

The most recent risk management improvements regarding the use of PFAS resulted in a mandatory notice issued under Section 71 of the CEPA on July 27, 2024, which requires a

**TABLE 21.6** Case Studies reported on PFAS in Canada.

| Type of PFAS | Study (year) | Participants/ location | Sample type | Detection frequency | Concentration reported | Outcomes/ inference | References |
|---|---|---|---|---|---|---|---|
| PFOS, PFOA, PFHxS, PFNA, PFDA | 2021 | Adults from Old Crow and Dehcho, Northern Canada | Blood serum | >50% for all listed PFAS | PFNA: 0.94 μg $L^{-1}$ (Old Crow), 1.42 μg $L^{-1}$ (Dehcho) | Similar to or lower than national; elevated PFNA suggests local sources | Garcia-Barrios et al. (2021) |
| Legacy and alternative PFAS | 2024 | Adult females (MIREC-ENDO) | Serum | 17 PFAS >50%; 7 PFAS >97% | ND | Wide exposure; elimination through pregnancy/ lactation | Borghese et al. (2024) |
| PFOS, PFOA, PFHxS | 2016–17 | General Canadian population (CHMS) | Blood serum | >98% for PFOS, PFOA, PFHxS | PFOS: 3.0 μg $L^{-1}$, PFOA: 1.3 μg $L^{-1}$, PFHxS: 0.9 μg $L^{-1}$ | ~50% decline since 2007–09; still widespread exposure | Cakmak et al. (2022) |
| PFOA, PFOS, others in milk | 2022 | 664 mothers (MIREC study), Canada | Human milk | PFOA and L-PFOS in all samples | ΣPFAS: 3.1–603 ng $L^{-1}$ (median 106 ng $L^{-1}$) | Higher in younger and first-time mothers; national baseline data | Rawn et al. (2022) |
| PFOSA, GenX, others | 2025 | Adults in Toronto (clinical trial) | Serum | PFOSA: 100%, GenX: 71% | ND | Some PFAS are linked to liver biomarkers; dietary fiber may reduce PFAS burden | Schlezinger et al. (2025) |
| PFOA, PFOS, PFHxS | 2021 | Inuit children (11 mo–4.5 y), Nunavik, Arctic Quebec (n = 86) | Plasma | 75% for major PFAS | PFOA: 1.62 μg $L^{-1}$, PFHxS: 0.33 μg $L^{-1}$, PFOS: 3.37 μg $L^{-1}$ | Evidence of early-life exposure in Arctic population | Lalonde and Garron (2022) |
| PFOA, PFOS, PFHxS, PFNA, PFDA, others | 2019 | First Nation youth and children, Nunavik (Quebec) | Plasma | Majority detected | ND | Linked to country foods and thyroid hormone changes | Caron-Beaudoin et al. (2019) |

| | | | | | | | |
|---|---|---|---|---|---|---|---|
| PFOS, PFOA, LC-PFCAs | 2013–20 | National (13 PFAS at 29 sites, all provinces except AB, PEI) | Surface and drinking water | PFOS: Detected in every province; PFOA high frequency | PFOS max: 13 ng $L^{-1}$; PFOA max: 8.1 ng $L^{-1}$; Range: <DL–138 ng $L^{-1}$ | Replacement PFAS (PFBA, PFPeA, PFHxA, PFHpA, PFBS) are increasingly frequent; concentrations are decreasing for PFOS/PFOA, increasing for some short-chain PFAS. | Lalonde and Garron (2022) |
| Multiple PFAS homolog classes | 2023 | Quebec (376 municipalities, 17 regions) | Tap water | 99.3% samples ≥1 PFAS; PFOA: 88%; PFOS: 80% | PFOS median: 0.15 ng $L^{-1}$, max: 13 ng $L^{-1}$; PFOA median: 0.27 ng $L^{-1}$; max: 8.1 ng $L^{-1}$; ΣPFAS: <DL–108 ng $L^{-1}$ (median 2 ng $L^{-1}$) | Higher PFAS in tap water from surface sources; 10 samples from 5 sites >30 ng $L^{-1}$ (Health Canada threshold); novel/emerging PFAS detected. | Objective for Canadian Drinking Water Quality Per- and Polyfluoroalkyl Substances, Canada |
| PFAS (PFBA, PFHxA, PFOA, PFOS, PFHpA, PFPeA) | 2022 | National freshwater sites, various locations | Surface water | PFBA highest, also PFHxA, PFOA; PFOS in 43% of samples | Range: <DL–138 ng $L^{-1}$ (PFBS) | Dominance of PFBA noted in high Arctic/urban sites; PFOS not exceeding Canadian guideline (6800 ng $L^{-1}$). | State of per- and polyfluoroalkyl substances PFAS report, Canada |
| Multiple PFAS (19 targeted compounds) | 2009–21 | National, 15 municipal WWTPs in six provinces | Wastewater, biosolids | All 15 WWTPs: 100% positive for ≥4 PFAS compounds | ΣPFAS in biosolids: 17.2–142 ng $g^{-1}$ dry wt; effluent: up to 540 ng $L^{-1}$; influent: up to 1050 ng $L^{-1}$ | Short-chain PFAS increasing over time; highest at WWTPs processing landfill leachate; widespread municipal contamination. | Gewurtz et al. (2024) |

(*Continued*)

TABLE 21.6 (Continued)

| Type of PFAS | Study (year) | Participants/ location | Sample type | Detection frequency | Concentration reported | Outcomes/ inference | References |
|---|---|---|---|---|---|---|---|
| PFOS (focus on biosolids regulation) | 2024 | Biosolids (commercial fertilizers, farm fields), nationwide | Biosolids | All domestic/ imported biosolids are regulated; data from regulatory monitoring | <50 ng $g^{-1}$ PFOS required by interim standard (2024, Health Canada and CFIA) | New biosolid products must not exceed 50 ng $g^{-1}$ PFOS; part of Canada's class-based and product-centered regulatory approach under CEPA revisions. | Government of Canada publishes State of Per- and Polyfluoroalkyl Substances PFAS Report and proposed Risk Management Approach, Canada |
| PFAS (hotspot drinking water) | 2023–24 | National Defense-impacted wells (e.g., Saguenay, QC) | Private wells, groundwater | 4 of 24 wells above 30 ng $L^{-1}$ | Above 30 ng $L^{-1}$ (Health Canada's objective); precise value variable | Hotspots near military/fire training sites can exceed safe thresholds; well water at risk for nearby residents; site-specific monitoring and mitigation needed. | Mapping Canada's PFAS Hotspots, CBC |

reporting date of January 29, 2025. This notification seeks to gather information on individual and/or mixture/products/component PFAS chemicals in Canadian businesses in the year 2023. In September 2024, a consultation document was issued proposing the addition of 131 separate PFAS to the National Pollutant Release Inventory (NPRI). The reporting obligations will be applicable to the organizations fulfilling the particular definitions and thresholds presented in the notice. An ultimate ruling on the reporting requirements is anticipated to be released in the Canada Gazette in 2025, and the reporting of subsequent PFAS releases, which occurred in 2025, will soon follow.

To advance knowledge on PFAS, the United States, Singapore, the University of Ottawa, and the University of Birmingham are among the jurisdictions and academic institutions leading a joint case study through the international governmental effort, accelerating the pace of chemical risk assessment (APCRA). The transcriptomic points of departure were determined through microtissues of the human liver that were subjected to PFAS in this case study, to evaluate potency and additivity. Furthermore, the analyses are associated with the chemical potency of subcategorized PFAS, specifically the length of the carbon chain or the composition of the PFAS mixture (State of Per- and Polyfluoroalkyl Substances PFAS Report - Canada.Ca, 2024).

The Canadian Total Diet Study and the Canadian Drinking Water Survey are used to monitor environmental exposures to PFAS. Additional examination of PFAS in dust samples collected during the Canadian House Dust Study (Canada Gazette, Government). It is also underway. In a similar manner, studies supported by the Northern Contaminants Program are shedding light on the exposure to PFAS in Northern First Nations, Métis, and Inuit communities (State of Per- and Polyfluoroalkyl Substances PFAS Report - Canada.Ca, 2024). Canada is rolling out an action plan to protect firefighters from harmful chemicals. Early studies have measured PFAS in turnout gear and in dust from fire stations.

Innovation, Science, and Economic Development Canada launched a challenge in October 2024 to develop safe, scalable, and low-cost solutions for the destruction of PFAS at contaminated sites. This initiative aims to develop innovative solutions to mitigate the impact of PFAS compounds in contaminated solid or aqueous media.

### 21.3.5 Comparison with international regulations

Other regions, such as the EU and certain states within the US, are either undertaking or planning to collectively address PFAS. The Canadian government is working together with other countries in a number of programs, such as the Stockholm Convention on persistent organic pollutants (POPs), Organisation for Economic Cooperation and Development (OECD), and trilateral cooperation with the US EPA and European Chemicals Agency on the APCRA program to communicate and respond to scientific and regulatory needs (State of Per- and Polyfluoroalkyl Substances PFAS Report - Canada.Ca, 2024).

The Stockholm Convention on POPs aims to safeguard human health and the environment from harmful substances. One of the features of the POPs included in the Convention is their persistence, bioaccumulation, long-range transport, and their capability to have severe adverse effects on human health and/or the environment. Under the Stockholm Convention on POPs, Canada has been actively involved in initiatives to manage and phase out PFAS substances. For example, in 2021, Canada submitted a nomination for LC-PFCAs, along with their salts and related compounds, for inclusion in the convention, with the goal of promoting international efforts to eliminate these substances (State of Per- and Polyfluoroalkyl Substances PFAS Report - Canada.Ca, 2024; Risk Management Approach for Per- and Polyfluoroalkyl Substances PFAS, Excluding Fluoropolymers, Canada).

The Montreal Protocol regulates chlorofluorocarbons, hydrochlorofluorocarbons, and hydrofluorocarbons, which align with the OECD definition of PFAS. The ODSHAR framework enforces Canada's international commitments as described in the Montreal Protocol (State of Per- and Polyfluoroalkyl Substances PFAS Report - Canada.Ca, 2024). Different countries and regions have implemented various measures to manage and mitigate the risks associated with PFASs. Among all the reviewed regulatory databases, the only regulator that contains environmental requirements and categories of listed PFAS is the Canadian DSL (Yu et al., 2025).

## 21.3.6 Future directions and recommendations

In accordance with the stipulations of Section 77(6)(c)(ii) of CEPA, the government released the risk management approach alongside the final report to address the risks associated with the specified PFAS class. The primary objectives of the risk management approach are: (1) to diminish the release of PFAS into the Canadian environment to prevent harmful effects while balancing environmental stewardship with economic viability; and (2) to lower human exposure to PFAS, particularly for populations disproportionately affected. The government aims to accomplish these goals through the following phased strategy:

Phase 1: The prohibition on using PFAS (without fluoropolymers) in firefighting foams (currently not subject to regulation) poses a significant risk of environmental and human exposure.

Phase 2: Prohibition of the use of PFAS (no fluoropolymers) in consumer goods in cases where there are alternative products. Such uses of prohibition will be prioritized based on costs and benefits, availability of suitable alternatives, and other socio-economic factors. Some of the important products to be considered are cosmetics, natural health products, and over-the-counter drugs, food packaging materials and food additives, paints and coatings, adhesives and sealants, and other building materials available to the consumer, and consumer blends such as cleaning products, waxes, and polishes, textiles (including firefighting turnout gear), and ski waxes.

Phase 3: A ban on the use of PFAS (except the fluoropolymers) that ought to be given further thought by use of stakeholder involvement and other assessments, especially the sectors that do not have viable alternatives, such as: fluorinated gas use, prescription drugs (both human and animal), medical devices, industrial food contact substances, sectors like mining and petroleum, and transport and military use (Risk Management Approach for Per- and Polyfluoroalkyl Substances PFAS, Excluding Fluoropolymers - Canada.Ca, 2025; Canada Moves to Add PFAS to List of Toxic Substances, Proposes; Canada Releases Final State of PFAS Report and Proposed Risk Management Approach, Bergeson).

### *21.3.6.1 Trends in per- and polyfluoroalkyl substances regulation*

Canada is implementing a class-based strategy to regulate PFAS under the CCEPA. This approach focuses on the entire PFAS chemical group, rather than specific compounds. The government is highlighting precautionary actions to address the risks associated with PFAS, which include recommending their inclusion on the list of toxic substances. Fluoropolymers are excluded from the PFAS category because of their distinct hazard and exposure characteristics. The monitoring of existing databases is being conducted to check for PFAS presence in various products, including cosmetics, natural health products, and nonprescription medications (State of Per- and Polyfluoroalkyl Substances PFAS Report - Canada.Ca, 2024).

## 21.4 Conclusion

The US regulatory landscape for PFAS has evolved rapidly from voluntary industry phase-outs and advisory guidance to more prescriptive and enforceable measures. Recent federal actions, including the establishment of MCLs for selected PFAS, proposals to list PFOA and PFOS under CERCLA, and expanded reporting requirements under TSCA, mark significant progress. Yet major challenges remain: disposal and destruction standards are still undefined, many PFAS compounds and emerging alternatives lack comprehensive regulation, and authority remains fragmented across federal and state levels. While the US has made notable advances, a harmonized, class-based, and enforceable approach supported by broader monitoring, research funding, and remediation capacity is needed to protect public health and promote equitable outcomes.

The United States has mounted an extensive regulatory and technical response to PFAS, applying multiple statutory authorities and agency programs to protect human health and the environment. Under key federal laws, the TSCA, the SDWA, and the CERCLA, agencies such as the US EPA, the Department of Defense, and public health bodies like the CDC and ATSDR have undertaken monitoring, reporting, and remediation efforts. EPA initiatives include nationwide monitoring programs such as the UCMR, advances in analytical methods and laboratory capacity, rulemaking to establish enforceable drinking water limits for select PFAS, and proposals to strengthen reporting and site-cleanup requirements. In parallel, many state governments have adopted their own advisories and enforceable standards, creating a diverse landscape of state-level limits and remediation programs.

Environmental surveillance and research have documented PFAS contamination in drinking water, surface water, groundwater, landfill leachate, wastewater effluents, biosolids, and biota. National biomonitoring programs such as NHANES, along with targeted investigations of communities affected by AFFF use and industrial discharges, have confirmed widespread human exposure to both legacy and emerging PFAS.

Risk management in the US combines federal regulation, targeted cleanup actions (notably at DoD sites), expanded reporting and testing requirements, and collaboration with states, academia, and industry. Persistent challenges include coordination across overlapping legal authorities, addressing the vast and evolving array of PFAS compounds, and developing validated, scalable technologies for the safe destruction and disposal of PFAS-contaminated materials.

Canada's PFAS regulatory framework reflects a proactive, class-based approach supported by robust monitoring programs, including national surveys and targeted assessments of vulnerable communities. Still, Canada continues to refine its reporting requirements, substance listings, and risk management strategies to address emerging PFAS chemistries. Sustained international collaboration, expanded monitoring of short-chain PFAS, and investment in scalable destruction technologies will be crucial to reducing exposure and supporting remediation, particularly for Indigenous and northern communities that face disproportionate risks.

The Government of Canada has adopted a class-based approach to controlling PFAS, with the goal of establishing guidelines to protect human health and the environment. PFOS and its salts and precursors, PFOA and its salts and precursors, and LC-PFCAs and their salts and precursors, are some of the PFAS subsets, which have been assessed and controlled under the CEPA. Some of the instruments and agencies involved in regulatory oversight include the New Substances Notification Regulations, the PCTSR, ECCC, HC, and the CCME, as well as complementary provincial guidelines.

Monitoring and research have identified PFAS in various environmental media, including air, drinking and surface water, municipal landfills, and WWTPs. Human biomonitoring programs, including the CHMS and the Maternal-Infant Research on Environmental Chemicals study, as well as targeted research on Indigenous and northern communities, have further confirmed widespread exposure to PFAS.

Risk management in Canada employs multiple measures, including adding specific PFAS to the NPRI, strengthening collaboration with other jurisdictions and academic institutions, and advancing innovation solutions initiatives aimed at developing and applying destruction technologies for PFAS in contaminated media.

In general, the United States and Canada have made significant progress in identifying and addressing the threat of PFAS. The US has progressed with enforceable federal standards and extensive surveillance, whereas Canada has insisted on a class-based regulatory framework with a national biomonitoring system and a precautionary approach to management. Despite these efforts, critical challenges remain, including regulating the entire range of PFAS, developing effective destruction technology, and maintaining regular risk communication. Further collaboration between federal and provincial or state authorities, as well as enhanced international cooperation, will become critical to achieving long-term reductions in PFAS contamination and securing both human health and the environment.

## List of abbreviations

| | |
|---|---|
| **ADONA** | Ammonium 4,8-dioxa-3H perfluorononanoate |
| **AER** | Anion exchange resins |
| **AFFF** | Aqueous film-forming foams |
| **APCRA** | Accelerating the pace of chemical risk assessment |
| **ATSDR** | Agency For Toxic Substances And Disease Registry |
| **BC** | British Columbia |
| **BPA** | Bisphenol A |
| **CAA** | Clean Air Act |
| **CCME** | Canadian Council Of Ministers Of The Environment |

| | |
|---|---|
| **CDC** | Centers For Disease Control And Prevention |
| **CEPA** | Canadian Environmental Protection Act |
| **CERCLA** | Comprehensive Environmental Response, Compensation, And Liability Act |
| **CFC** | Chlorofluorocarbons |
| **CHDS** | Child Health And Development Studies |
| **CHMS** | Canadian Health Measures Survey |
| **CMP** | Chemicals Management Plan |
| **CSR** | Contaminated Sites Regulation |
| **CWA** | Clean Water Act |
| **DHHS** | Department Of Health And Human Services |
| **DSL** | Domestic Substances List |
| **EC** | Environment Canada Ec |
| **ECCC** | Environment And Climate Change Canada |
| **ECHA** | European Chemicals Agency |
| **EDC** | Endocrine-disrupting chemicals |
| **EPA** | Environmental Protection Agency |
| **EU** | European Union |
| **FDA** | Food And Drug Administration |
| **FEQG** | Federal environmental quality guideline |
| **FOSAA** | Perfluorooctane sulfonamidoacetic acid |
| **FTCA** | Fluorotelomer carboxylic acid |
| **FTSA** | Fluorotelomer sulfonic acid |
| **GAC** | Granular activated carbon |
| **GAO** | Government Accountability Office |
| **GAPS** | Global Atmospheric Passive Sampling |
| **GM** | Geometric mean |
| **GWQG** | Groundwater quality guideline |
| **GWQGF** | Lowest final-ground water quality guideline |
| **HC** | Health Canada |
| **HCFC** | Hydrochlorofluorocarbon |
| **HFC** | Hydrofluorocarbons |
| **HFPO-DA** | Hexafluoropropylene oxide dimer acid |
| **HPLC** | High-performance liquid chromatography |
| **HSA** | Human serum albumin |
| **ISED** | Innovation Science And Economic Development Canada |
| **LCMRL** | Lowest concentration minimum reporting limit |
| **LC-MS** | Liquid chromatography tandem mass spectrometry |
| **LC-PFCA** | Long-chain perfluorocarboxylic acids |
| **LLOQ** | Lower limit of quantitation |
| **LOD** | Limit of detection |
| **LOQ** | Limit of quantitation |
| **MAC** | Maximum acceptable concentration |
| **MassDEP** | Massachusetts Department Of Environmental Protection |
| **MCLs** | Maximum contaminant levels |
| **MDL** | Method detection limit |
| **MECP** | Ontario Ministry Of The Environment, Conservation And Park's |
| **MIREC** | Maternal-Infant Research On Environmental Chemicals |
| **MPART** | Michigan Pfas Action Response Team |
| **MSW** | Municipal solid waste |
| **NA** | Not available |
| **ND** | Not detected |
| **NDAA** | National Defense Authorization Act |
| **NF** | Nanofiltration |

| | |
|---|---|
| **NHANES** | National Health And Nutrition Examination Survey |
| **NHDES** | New Hampshire Department Of Environmental Services |
| **NPRI** | National Pollutant Release Inventory |
| **NQN** | Nutaratsaliit Qanuingisiarningit Niqituinnanut |
| **NSNR** | New substances notification regulations (chemicals and polymers) |
| **NTA** | Nontargeted analysis |
| **ODS** | Ozone-Depleting Substances |
| **ODSHAR** | Ozone-Depleting Substances and Halocarbon Alternatives Regulations |
| **OECD** | Organisation for Economic Cooperation and Development |
| **OTM** | Other test method |
| **PCTSR** | Prohibition of Certain Toxic Substances Regulations |
| **PEI** | Polyethyleneimine |
| **PFAS** | Per- and polyfluoroalkyl substances |
| **PFBA** | Perfluorobutanoic acid |
| **PFBS** | Perfluorobutane sulfonic acid/Perfluorobutane sulfonate |
| **PFCA** | Perfluorocarboxylic acids |
| **PFDA** | Perfluorodecanoic acid |
| **PFDS** | Perfluorodecanesulfonic acid |
| **PFHO** | Perfluorohexyloctane |
| **PFHpA** | Perfluoroheptanoate |
| **PFHS** | Perfluorohexane sulfonate |
| **PFHxA** | Perfluorohexanoic acid |
| **PFHxS** | Perfluorohexane sulfonic acid |
| **PFIB** | Perfluoroisobutylene |
| **PFMOAA** | Perfluoro-2-methoxyacetic acid |
| **PFNA** | Perfluorononanoic acid |
| **PFNS** | Perfluorononanesulfonic acid |
| **PFOA** | Perfluorobutanoic acid/Perfluorobutanoate |
| **PFOS** | Perfluorooctane sulfonate |
| **PFOSA** | Perfluorooctanesulfonamide |
| **PFPeA** | Perfluoropentanoate |
| **PFPS** | Perfluoropentanesulfonic acid |
| **PFSA** | Perfluorosulfonic acids |
| **PFTrDA** | Perfluorotridecanoic acid |
| **PFUnDA** | Perfluoroundecanoic acid |
| **PMPA** | Perfluoro-2-methoxypropanoic acid |
| **POPs** | Persistent organic pollutants |
| **PWTA** | Private Well Testing Act |
| **RBCA** | Atlantic Risk-Based Corrective Action |
| **RCRA** | Resource Conservation And Recovery Act |
| **RO** | Reverse osmosis |
| **RSL** | Regional Screening Levels |
| **SDWA** | Safe Drinking Water Act |
| **SNAc** | Significant activity |
| **SNURs** | Significant new use rules |
| **$SoQG_E$** | Soil quality guideline for environmental health |
| **$SoQG_{HH}$** | Soil quality guideline for human health |
| **SPE** | Solid-phase extraction |
| **SQG** | Soil quality guidelines |
| **TDB** | Treatability Database |
| **TDS** | Total Diet Study |
| **TO** | Toxic organics |
| **TOF** | Total organic fluorine |

| | |
|---|---|
| **TOP** | Total oxidizable precursor |
| **TRV** | Toxicity reference values |
| **TRVs** | Toxicity reference values |
| **TSCA** | Toxic Substances Control Act |
| **UCMR** | Unregulated contaminant monitoring regulations |
| **USA** | United States of America |
| **USEPA** | United States Environmental Protection Agency |
| **USGS** | U.S Geological Survey |
| **VOC** | Volatile organic compounds |
| **WWTPs** | Wastewater treatment plants |

# References

3M Gets Court Approval of $10 Billion PFAS Settlement. https://www.nacwa.org/news-publications/news-detail/2024/04/01/3m-gets-court-approval-of-10-billion-pfas-settlement. (accessed 14.07.25).

5 Pressing Environmental Issues in Canada in 2024, Earth.Org. https://earth.org/environmental-issues-in-canada/. (accessed 14.07.25).

Agency Of Natural Resources Department Of Environmental Conservation Drinking Water And Groundwater Protection Division Advance Notice On The Regulation Of Perfluoroalkyl, Polyfluoroalkyl Substances (Pfas) As A Class. 2020. https://dec.vermont.gov/sites/dec/files/PFAS/20180814-PFAS-as-a-Class.pdf. (accessed 15.07.25).

Aker, A., Ayotte, P., Caron-Beaudoin, É., Ricard, S., Gaudreau, É., Lemire, M., 2023. Cardiometabolic health and per and polyfluoroalkyl substances in an inuit population. Environment International 181, 108283. Available at https://doi.org/10.1016/j.envint.2023.108283.

Alberta Tier 1 soil and groundwater remediation guidelines - Open Government. https://open.alberta.ca/publications/1926-6243. (accessed 16.07.25).

Alberta Tier 1 Soil and Groundwater Remediation Guidelines - Open Government. https://open.alberta.ca/publications/1926-6243. (accessed 14.07.25).

Andreea, B.S., Chander, M.V., Fengmao, G.D., Maurice, A.A., 2025. Regulating PFAS as a chemical class under the california safer consumer products program. Environmental Health Perspectives 129 (2), 25001. Available at https://doi.org/10.1289/EHP7431.

Atlantic RBCA - human health-based tier I Environmental quality standards (EQS) for Soil - All Land Uses; Non-potable Groundwater Condition (mg/kg). https://atlanticrbca.com/wp-content/uploads/2021/07/Atlantic_RBCA_-_Human_Health-Based_Tier_I_Environmental_Quality_Standards_EQS_for_Soil_-_All_Land_Uses_Non-Potable_Groundwater_Condition.pdf. (accessed 21.10.25).

Atlantic RBCA-human health-based tier I Environmental quality standards (EQS) for groundwater-all land uses; potable groundwater condition (mg/L). https://atlanticrbca.com/wp-content/uploads/2021/07/Atlantic_RBCA_-_Human_Health-Based_Tier_I_Environmental_Quality_Standards_EQS_for_Soil_-_All_Land_Uses_Non-Potable_Groundwater_Condition.pdf. (accessed 21.10.25).

Bai, X., Son, Y., 2021. Perfluoroalkyl substances (PFAS) in surface water and sediments from two urban watersheds in Nevada, USA. The Science of the Total Environment 751, 141622. Available at https://doi.org/10.1016/j.scitotenv.2020.141622.

Barbo, N., Stoiber, T., Naidenko, O.V., Andrews, D.Q., 2023. Locally caught freshwater fish across the United States are likely a significant source of exposure to PFOS and other perfluorinated compounds. Environmental Research 220, 115165. Available at https://doi.org/10.1016/j.envres.2022.115165.

Barton-Maclaren, T.S., Wade, M., Basu, N., Bayen, S., Grundy, J., Marlatt, V., et al., 2022. Innovation in regulatory approaches for endocrine disrupting chemicals: the journey to risk assessment modernization in Canada. Environmental Research 204, 112225. Available at https://doi.org/10.1016/j.envres.2021.112225.

Basic Information about How to Use SW-846 | US EPA. https://www.epa.gov/hw-sw846/basic-information-about-how-use-sw-846. (accessed 17.07.25).

Berg, C., Crone, B., Gullett, B., Higuchi, M., Krause, M.J., Lemieux, P.M., et al., 2022. Developing innovative treatment technologies for PFAS-containing wastes. Journal of the Air & Waste Management Association 72 (6), 540–555. Available at https://doi.org/10.1080/10962247.2021.2000903.

Berthold, T.A., McCrary, A., deVilleneuve, S., Schramm, M., 2023. Let's talk about PFAS: inconsistent public awareness about PFAS and its sources in the United States. PLoS One 18 (11), e0294134. Available at https://doi.org/10.1371/journal.pone.0294134.

Bill text - AB-756 public water systems: perfluoroalkyl substances and polyfluoroalkyl substances. https://leginfo.legislature.ca.gov/faces/billNavClient.xhtml?bill_id=201920200AB756. (accessed 15.07.25).

Bill text - SB-1044 firefighting equipment and foam: PFAS chemicals. https://leginfo.legislature.ca.gov/faces/billTextClient.xhtml?bill_id=201920200SB1044. (accessed 15.07.25).

Borghese, M.M., Liang, C.L., Owen, J., Fisher, M., 2022. Individual and mixture associations of perfluoroalkyl substances on liver function biomarkers in the Canadian Health Measures Survey. Environ. Heal. 21 (1), 85. Available at https://doi.org/10.1186/s12940-022-00892-6.

Borghese, M.M., Ward, A., MacPherson, S., Manz, K.E., Atlas, E., Fisher, M., et al., 2024. Serum concentrations of legacy, alternative, and precursor per- and polyfluoroalkyl substances: a descriptive analysis of adult female participants in the MIREC-ENDO study. Environmental Health: A Global Access Science Source 23 (1), 55. Available at https://doi.org/10.1186/s12940-024-01085-z.

Botelho, J., Pirkle, J.L., Laboratory procedure manual analytes: perfluoroalkyl and polyfluoroalkyl substances matrix: serum method: online solid phase extraction-high performance liquid chromatography-turbo ion spray-tandem mass spectrometry (Online SPE-HPLC-TIS-MS/MS). https://wwwn.cdc.gov/nchs/data/nhanes/2017-2018/labmethods/PFAS-J-MET-508.pdf. (accessed 14.07.25).

Breitmeyer, S.E., Williams, A.M., Duris, J.W., Eicholtz, L.W., Shull, D.R., Wertz, T.A., et al., 2023. Per- and polyfluorinated alkyl substances (PFAS) in Pennsylvania surface waters: a statewide assessment, associated sources, and land-use relations. The Science of the Total Environment 888, 164161. Available at https://doi.org/10.1016/j.scitotenv.2023.164161.

Brendel, S., Fetter, É., Staude, C., Vierke, L., Biegel-Engler, A., 2018. Short-chain perfluoroalkyl acids: environmental concerns and a regulatory strategy under REACH. Environmental Sciences Europe 30 (1), 9. Available at https://doi.org/10.1186/s12302-018-0134-4.

Cakmak, S., Lukina, A., Karthikeyan, S., Atlas, E., Dales, R., 2022. The association between blood PFAS concentrations and clinical biochemical measures of organ function and metabolism in participants of the Canadian Health Measures Survey (CHMS). The Science of the Total Environment 827, 153900. Available at https://doi.org/10.1016/j.scitotenv.2022.153900.

Calafat, A.M., Wong, L.-Y., Kuklenyik, Z., Reidy, J.A., Needham, L.L., 2007. Polyfluoroalkyl chemicals in the U.S. population: data from the national health and nutrition examination survey (NHANES) 2003-2004 and comparisons with NHANES 1999-2000. Environmental Health Perspectives 115 (11), 1596–1602. Available at https://doi.org/10.1289/ehp.10598.

Canada Gazette – Government Notices. https://gazette.gc.ca/rp-pr/p1/2015/2015-07-25/html/notice-avis-eng.html. (accessed 14.07.25).

Canada Moves to Add PFAS to List of Toxic Substances; Proposes Risk Management Strategy, Insights, Torys LLP. https://www.torys.com/our-latest-thinking/publications/2025/03/canada-moves-to-add-pfas-to-list-of-toxic-substances. (accessed 14.07.25).

Canada Releases Final State of PFAS Report and Proposed Risk Management Approach - Bergeson & Campbell, P.C. https://www.lawbc.com/canada-releases-final-state-of-pfas-report-and-proposed-risk-management-approach/. (accessed 14.07.25).

Canadian Environmental Protection Act, 1999. https://laws-lois.justice.gc.ca/eng/acts/c-15.31/. (accessed 21.10.25).

Caron-Beaudoin, É., Ayotte, P., Laouan Sidi, E.A., Gros-Louis McHugh, N., Lemire, M., 2019. Exposure to perfluoroalkyl substances (PFAS) and associations with thyroid parameters in first nation children and youth from Quebec. Environment International 128, 13–23. Available at https://doi.org/10.1016/j.envint.2019.04.029.

Caron-Beaudoin, É., Ayotte, P., Blanchette, C., Muckle, G., Avard, E., Ricard, S., et al., 2020. Perfluoroalkyl acids in pregnant women from Nunavik (Quebec, Canada): trends in exposure and associations with country foods consumption. Environment International 145, 106169. Available at https://doi.org/10.1016/j.envint.2020.106169.

Carreau, G., Buccini, J. CEPA Environmental Registry - Proposed Risk Management Strategy for Perfluorooctane Sulfonate (PFOS), Its Salts and Its Precursors. 1999. https://www.canada.ca/en/environment-climate-change/services/canadian-environmental-protection-act-registry/publications/risk-management-strategy-perfluorooctane-sulfonate.html. (accessed 14.07.25).

Ccme. Canadian Soil and Groundwater Quality Guidelines for the Protection of Environmental and Human Health: Perfluorooctane Sulfonate (PFOS). https://ccme.ca/en/res/pfosfactsheeten.pdf. (accessed 21.10.25).

Chu, S., Wang, J., Leong, G., Woodward, L.A., Letcher, R.J., Li, Q.X., 2015. Perfluoroalkyl sulfonates and carboxylic acids in liver, muscle and adipose tissues of black-footed albatross (Phoebastria Nigripes) from Midway Island, North Pacific Ocean. Chemosphere 138, 60–66. Available at https://doi.org/10.1016/j.chemosphere.2015.05.043.

Committee Report – July 18-19 2018 - Canada.Ca. https://www.canada.ca/en/health-canada/services/chemical-substances/chemicals-management-plan/science-committee/meeting-records-reports/committee-report-july-18-19-2018.html.5 Pressing Environmental Issues in Canada in 2024. (accessed 14.07.25).

Consideration of Endocrine-Related Effects in Risk Assessment - Canada.Ca. https://www.canada.ca/en/health-canada/services/chemical-substances/fact-sheets/consideration-endocrine-related-effects-risk-assessment.html.5 Pressing Environmental Issues in Canada in 2024. (accessed 14.07.25).

Daugbjerg, C., Kay, A., 2020. Policy feedback and pathways: when change leads to endurance and continuity to change. Policy Sciences 53 (2), 253–268. Available at https://doi.org/10.1007/s11077-019-09366-y.

DeLuca, N.M., Thomas, K., Mullikin, A., Slover, R., Stanek, L.W., Pilant, A.N., et al., 2023. Geographic and demographic variability in serum PFAS concentrations for pregnant women in the United States. Journal of Exposure Science & Environmental Epidemiology 33 (5), 710–724. Available at https://doi.org/10.1038/s41370-023-00520-6.

Designation of Perfluorooctanoic Acid (PFOA) and Perfluorooctanesulfonic Acid (PFOS) as CERCLA Hazardous Substances, US EPA. https://www.epa.gov/superfund/designation-perfluorooctanoic-acid-pfoa-and-perfluorooctanesulfonic-acid-pfos-cercla. (accessed 14.07.25).

Drinking Water Treatability Database. https://tdb.epa.gov/tdb/contaminant?id=11020. (accessed 14.07.25).

Drinking Water: DOD Has Acted on Some Emerging Contaminants but Should Improve Internal Reporting on Regulatory Compliance, U.S. GAO. https://www.gao.gov/products/gao-18-78. (accessed 15.07.25).

ECCC. Perfluorooctane Sulfonate in the Canadian Environment: Environmental Monitoring and Surveillance in Support of the Chemical Management Plan. 2018, 22. https://publications.gc.ca/collections/collection_2013/ec/En14-96-2013-eng.pdf. (accessed 14.07.25).

Ecological screening assessment report on perfluorooctane sulfonate, salts and precursors - Canada.ca. https://www.canada.ca/en/environment-climate-change/services/canadian-environmental-protection-act-registry/publications/ecological-screening-report-sulfonate.html. (accessed 16.07.25).

Ecological Screening Assessment Report on Perfluorooctane Sulfonate, Salts and Precursors - Canada.Ca. https://www.canada.ca/en/environment-climate-change/services/canadian-environmental-protection-act-registry/publications/ecological-screening-report-sulfonate.html. (accessed 14.07.25).

Environment-Use of Fire-Fighting Foam Containing Perfluoroalkyl and Polyfluoroalkyl Substances-Prohibition and PFAS Chemicals. https://mgaleg.maryland.gov/2020RS/Chapters_noln/CH_276_hb0619t.pdf. (accessed 15.07.25).

EPA Finalizes PFAS Drinking Water Regulation, nsights, Holland & Knight. https://www.hklaw.com/en/insights/publications/2024/04/epa-finalizes-pfas-drinking-water-regulation. (accessed 15.07.25).

Epa, U. EPA's Per- and Polyfluoroalkyl Substances (PFAS) Action Plan. https://www.epa.gov/sites/default/files/201902/documents/pfas_action_plan_021319_508compliant_1.pdf. (accessed 15.07.25).

Epa, U. S. Method 533: Determination Of Per-And Polyfluoroalkyl Substances In Drinking Water By Isotope Dilution Anion Exchange Solid Phase Extraction And Liquid Chromatography/Tandem Mass Spectrometry 533-I. 2018. https://www.epa.gov/sites/default/files/2019-12/documents/method-533-815b19020.pdf. (accessed 15.07.25).

Epa, U.; of Science, O. Draft Method 1633 Analysis of Per- and Polyfluoroalkyl Substances (PFAS) in Aqueous, Solid, Biosolids, and Tissue Samples by LC-MS/MS; DRAFT - August 2021. 2021a. https://www.epa.gov/system/files/documents/2021-09/method_1633_draft_aug-2021.pdf. (accessed 15.07.25).

Epa, U.; of Science, O. Draft Method 1633 Analysis of Per- and Polyfluoroalkyl Substances (PFAS) in Aqueous, Solid, Biosolids, and Tissue Samples by LC-MS/MS; DRAFT - August 2021. 2021b. https://www.epa.gov/system/files/documents/2021-09/method_1633_draft_aug-2021.pdf. (accessed 17.07.25).

EPA's Per- and Polyfluoroalkyl Substances (PFAS) Action Plan. https://www.epa.gov/sites/default/files/201902/documents/pfas_action_plan_021319_508compliant_1.pdf (accessed 15.07.25).
EPA's Plan to Address Perchlorate Contamination, 2022. https://www.epa.gov/system/files/documents/2022-03/epa-plan-to-address-perchlorate.final_.pdf. (accessed 14.07.25).
Exposure: Poisoned Water, Corporate Greed, and One Lawyer's Twenty-Year Battle Against DuPont - The Legal Writer's Toolkit – Syracuse University. https://legalwriterstoolkit.law.syr.edu/alternative-reading-list/2020-alternative-reading-list/exposure-poisoned-water-corporate-greed-and-one-lawyers-twenty-year-battle-against-dupont/ (accessed 14.07.25).
Fact Sheet: 2010/2015 PFOA Stewardship Program, US EPA. https://www.epa.gov/assessing-and-managing-chemicals-under-tsca/fact-sheet-20102015-pfoa-stewardship-program (accessed 14.07.25).
FDA Announces the Voluntary Phase-Out by Industry of Certain PFAS Used in Food Packaging, FDA. https://www.fda.gov/food/hfp-constituent-updates/fda-announces-voluntary-phase-out-industry-certain-pfas-used-food-packaging. (accessed 15.07.25).
FDA Must Set Limits on PFAS in Food, Lawsuit Says - The New Lede. https://www.thenewlede.org/2025/01/fda-must-set-limits-on-pfas-in-food-lawsuit-says/. (accessed 15.07.25).
Federal Register: Lifetime Health Advisories and Health Effects Support Documents for Perfluorooctanoic Acid and Perfluorooctane Sulfonate. https://www.federalregister.gov/documents/2016/05/25/2016-12361/lifetime-health-advisories-and-health-effects-support-documents-for-perfluorooctanoic-acid-and. (accessed 14.07.25).
Federal Register: Perfluoroalkyl Sulfonates; Proposed Significant New Use Rule. https://www.federalregister.gov/documents/2002/03/11/02-5747/perfluoroalkyl-sulfonates-proposed-significant-new-use-rule. (accessed 14.07.25).
Federal Register: Revisions to the Unregulated Contaminant Monitoring Regulation (UCMR 3) for Public Water Systems. https://www.federalregister.gov/documents/2012/07/02/C1-2012-9978/revisions-to-the-unregulated-contaminant-monitoring-regulation-ucmr-3-for-public-water-systems. (accessed 14.07.25).
Federal Role in Responding to Potential Risks of Per-and Polyfluoroalkyl Substances (PFAS). 2022. https://www.congress.gov/crs-product/R45986. (accessed 14.07.25).
Fourth national report on human exposure to environmental chemicals: updated tables, January 2019, Volume one. https://stacks.cdc.gov/view/cdc/75822 (accessed 21.10.25).
Fourth National Report on Human Exposure to Environmental Chemicals, Updated tables, March 2021: volume three: analysis of pooled serum samples for select chemicals, NHANES 2005–2016. Atlanta, GA. https://stacks.cdc.gov/view/cdc/105344. (accessed 21.10.25).
Gaber, N., Bero, L., Woodruff, T.J., 2023. The devil they knew: chemical documents analysis of industry influence on PFAS science. Annals of Global Health 89 (1), 37. Available at https://doi.org/10.5334/aogh.4013.
Garcia-Barrios, J., Drysdale, M., Ratelle, M., Gaudreau, É., LeBlanc, A., Gamberg, M., et al., 2021. Biomarkers of poly- and perfluoroalkyl substances (PFAS) in sub-arctic and arctic communities in Canada. International Journal of Hygiene and Environmental Health 235, 113754. Available at https://doi.org/10.1016/j.ijheh.2021.113754.
George, S.E., Baker, T.R., Baker, B.B., 2023. Nonlethal detection of PFAS bioaccumulation and biomagnification within fishes in an urban- and wastewater-dominant great lakes watershed. Environmental Pollution (Barking, Essex 1987) 321, 121123. Available at https://doi.org/10.1016/j.envpol.2023.121123.
Gewurtz, S.B., Auyeung, A.S., De Silva, A.O., Teslic, S., Smyth, S.A., 2024. Per- and polyfluoroalkyl substances (PFAS) in Canadian municipal wastewater and biosolids: recent patterns and time trends 2009 to 2021. The Science of the Total Environment 912, 168638. Available at https://doi.org/10.1016/j.scitotenv.2023.168638.
Gewurtz, S.B., Tardif, G., Power, M., Backus, S.M., Dove, A., Dubé-Roberge, K., et al., 2021. Bisphenol A in the Canadian environment: a multimedia analysis. The Science of the Total Environment 755, 142472. Available at https://doi.org/10.1016/j.scitotenv.2020.142472.
Government of Canada publishes state of per- and polyfluoroalkyl substances (PFAS) report and proposed risk management approach - Canada.ca. https://www.canada.ca/en/environment-climate-change/news/2025/03/government-of-canada-publishes-state-of-per--and-polyfluoroalkyl-substances-pfas-report-and-proposed-risk-management-approach.html. (accessed 16.07.25).
Guidance for the assessment and remediation of per-and polyfluoroalkyl substances in british columbia, 2019. https://csapsociety.bc.ca/wp-content/uploads/Final-PFAS-Guidance-Document-dated-2019-05-28.pdf. (accessed 21.07.25).

H.R.2467 - 117th Congress (2021-2022): PFAS Action Act of 2021 I Congress.Gov I Library of Congress. https://www.congress.gov/bill/117th-congress/house-bill/2467. (accessed 14.07.25).

H.R.2596 - 116th Congress (2019-2020) Protecting Communities from New PFAS Act, Congress.Gov, Library of Congress. https://www.congress.gov/bill/116th-congress/house-bill/2596. (accessed 15.07.25).

H.R.7575 - 116th Congress (2019-2020): Water Resources Development Act of 2020, Congress.Gov, Library of Congress. https://www.congress.gov/bill/116th-congress/house-bill/7575.(accessed 15.07.25).

Health Ontario, P., Per-and Poly-Fluoroalkyl Substances (PFAS) 1 focus on Per-and Poly-Fluoroalkyl Substances (PFAS). https://www.publichealthontario.ca/-/media/Documents/P/2023/pfas-per-poly-fluoroalkyl-substances.pdf?rev=270b2dc8401c4653903d3a082579a108&sc_lang=en. (accessed 21.10.25).

Helmer, R.W., Reeves, D.M., Cassidy, D.P., 2022. Per- and polyfluorinated alkyl substances (PFAS) cycling within michigan: contaminated sites, landfills and wastewater treatment plants. Water Research 210, 117983. Available at https://doi.org/10.1016/j.watres.2021.117983.

Henry, B.J., Carlin, J.P., Hammerschmidt, J.A., Buck, R.C., Buxton, L.W., Fiedler, H., et al., 2018. A critical review of the application of polymer of low concern and regulatory criteria to fluoropolymers. Integrated Environmental Assessment and Management 14 (3), 316–334. Available at https://doi.org/10.1002/ieam.4035.

Hu, X., Li, S., Cirillo, P.M., Krigbaum, N.Y., Tran, V., Jones, D.P., et al., 2019. Metabolome wide association study of serum poly and perfluoroalkyl substances (PFASs) in pregnancy and early postpartum. Reproductive Toxicology (Elmsford, N.Y.) 87, 70–78. Available at https://doi.org/10.1016/j.reprotox.2019.05.058.

Hu, X., Li, S., Cirillo, P.M., Krigbaum, N.Y., Tran, V., Jones, D.P., et al., 2020. Reprint of "metabolome wide association study of serum poly and perfluoroalkyl substances (PFASs) in pregnancy and early postpartum" Reproductive Toxicology (Elmsford, N.Y.) 92, 120–128. Available at https://doi.org/10.1016/j.reprotox.2020.01.001.

Industry Lobbying and Eye-Watering Costs of PFAS Pollution Revealed. https://chemtrust.org/news/industry-lobbying-costs-pfas-pollution/. (accessed 14.07.25).

Interim Guidance on the Destruction and Disposal of PFAS and Materials Containing PFAS, US EPA. https://www.epa.gov/pfas/interim-guidance-destruction-and-disposal-pfas-and-materials-containing-pfas. (accessed 14.07.25).

Jiang, B., Zheng, J., Qiu, S., Wu, M., Zhang, Q., Yan, Z., et al., 2014. Review on electrical discharge plasma technology for wastewater remediation. Chemical Engineering Journal 236, 348–368. Available at https://doi.org/10.1016/j.cej.2013.09.090.

Johnson, G.R., 2022. PFAS in soil and groundwater following historical land application of biosolids. Water Research 211, 118035. Available at https://doi.org/10.1016/j.watres.2021.118035.

Kato, K., Basden, B.J., Needham, L.L., Calafat, A.M., 2011. Improved selectivity for the analysis of maternal serum and cord serum for polyfluoroalkyl chemicals. Journal of Chromatography. A 1218 (15), 2133–2137. Available at https://doi.org/10.1016/j.chroma.2010.10.051.

Kato, K., Kalathil, A.A., Patel, A.M., Ye, X., Calafat, A.M., 2018. Per- and polyfluoroalkyl substances and fluorinated alternatives in urine and serum by on-line solid phase extraction–liquid chromatography–tandem mass spectrometry. Chemosphere 209, 338–345. Available at https://doi.org/10.1016/j.chemosphere.2018.06.085.

Kwiatkowski, C.F., Andrews, D.Q., Birnbaum, L.S., Bruton, T.A., DeWitt, J.C., Knappe, D.R.U., et al., 2020. Scientific basis for managing PFAS as a chemical class. Environmental Science & Technology Letters 7 (8), 532–543. Available at https://doi.org/10.1021/acs.estlett.0c00255.

Lalonde, B., Garron, C., 2022. Perfluoroalkyl substances (PFASs) in the Canadian freshwater environment. Archives of Environmental Contamination and Toxicology 82 (4), 581–591. Available at https://doi.org/10.1007/s00244-022-00922-x.

Lewis, R.C., Johns, L.E., Meeker, J.D., 2015. Serum biomarkers of exposure to perfluoroalkyl substances in relation to serum testosterone and measures of thyroid function among adults and adolescents from NHANES 2011–2012. International Journal of Environmental Research and Public Health 6098–6114. Available at https://doi.org/10.3390/ijerph120606098.

Li, W., Hu, Y., Bischel, H.N., 2021. In-vitro and in-silico assessment of per- and polyfluoroalkyl substances (PFAS) in Aqueous film-forming foam (AFFF) binding to human serum albumin. Toxics. Available at https://doi.org/10.3390/toxics9030063.

Louis, G.M.B., Chen, Z., Schisterman, E.F., Kim, S., Sweeney, A.M., Sundaram, R., et al., 2015. Perfluorochemicals and human semen quality: the LIFE study. Environmental Health Perspectives 123 (1), 57–63. Available at https://doi.org/10.1289/ehp.1307621.

M, C.A., Lee-Yang, W., Zsuzsanna, K., A, R.J., L, N.L., 2007. Polyfluoroalkyl chemicals in the U.S. population: data from the National Health and Nutrition Examination Survey (NHANES) 2003–2004 and comparisons with NHANES 1999–2000. Environmental Health Perspectives 115 (11), 1596–1602. Available at https://doi.org/10.1289/ehp.10598.

Mapping Canada's PFAS Hotspots, CBC News. https://newsinteractives.cbc.ca/features/2025/pfas-canada-map/. (accessed 14 july 2025).

Maria-Jose, L.-E., Debapriya, M., Ben, A., S, B.M., Tony, F., 2012. Thyroid function and perfluoroalkyl acids in children living near a chemical plant. Environmental Health Perspectives 120 (7), 1036–1041. Available at https://doi.org/10.1289/ehp.1104370.

Marlatt, V.L., Bayen, S., Castaneda-Cortès, D., Delbès, G., Grigorova, P., Langlois, V.S., et al., 2022. Impacts of endocrine disrupting chemicals on reproduction in wildlife and humans. Environmental Research 208, 112584. Available at https://doi.org/10.1016/j.envres.2021.112584.

Martinez, B., Da Silva, B.F., Aristizabal-Henao, J.J., Denslow, N.D., Osborne, T.Z., Morrison, E.S., et al., 2022. Increased levels of perfluorooctanesulfonic acid (PFOS) during hurricane dorian on the East Coast of Florida. Environmental Research 208, 112635. Available at https://doi.org/10.1016/j.envres.2021.112635.

McMahon, P.B., Tokranov, A.K., Bexfield, L.M., Lindsey, B.D., Johnson, T.D., Lombard, M.A., et al., 2022. Perfluoroalkyl and polyfluoroalkyl substances in groundwater used as a source of drinking water in the eastern United States. Environmental Science & Technology 56 (4), 2279–2288. Available at https://doi.org/10.1021/acs.est.1c04795.

Method 537.1 Determination of Selected Per- and Polyflourinated Alkyl Substances in Drinking Water by Solid Phase Extraction and Liquid Chromatography/Tandem Mass Spectrometry (LC/MS/MS) | Science Inventory | US EPA. https://cfpub.epa.gov/si/si_public_record_report.cfm?dirEntryId=348508&Lab=CESER&simpleSearch=0&showCriteria=2&searchAll=537.1&TIMSType=&dateBeginPublishedPresented=03%2F24%2F2018. (accessed 17.07.25).

National PFAS Testing Strategy, US EPA. https://www.epa.gov/assessing-and-managing-chemicals-under-tsca/national-pfas-testing-strategy. (accessed 14.07.25).

New Jersey Sets Stringent Limits for PFOA, PFOS in Drinking Water - The Environmental Council of the States (ECOS). https://www.ecos.org/news-and-updates/new-jersey-enacts-stringent-limits-for-pfoa-pfos-in-drinking-water/. (accessed 15.07.25).

O'Hagan, D., 2008. Understanding Organofluorine Chemistry. An Introduction to the C–F Bond. Chemical Society Reviews 37 (2), 308–319. Available at https://doi.org/10.1039/B711844A.

Objective for Canadian Drinking Water Quality Per- and Polyfluoroalkyl Substances - Canada.Ca. https://www.canada.ca/en/health-canada/services/publications/healthy-living/objective-drinking-water-quality-per-polyfluoroalkyl-substances.html. (accessed 14.07.25).

Olsen, G.W., Mair, D.C., Reagen, W.K., Ellefson, M.E., Ehresman, D.J., Butenhoff, J.L., et al., 2007. Preliminary evidence of a decline in perfluorooctanesulfonate (PFOS) and Perfluorooctanoate (PFOA) concentrations in american red cross blood donors. Chemosphere 68 (1), 105–111. Available at https://doi.org/10.1016/j.chemosphere.2006.12.031.

Other Test Method 45 (OTM-45) Measurement of Selected Per-and Polyfluorinated Alkyl Substances from Stationary Sources Background on OTM-45. https://www.epa.gov/sites/default/files/2021-01/documents/otm_45_semivolatile_pfas_1-13-21.pdf. (accessed 15.07.25).

Our Current Understanding of the Human Health and Environmental Risks of PFAS, US EPA. https://www.epa.gov/pfas/our-current-understanding-human-health-and-environmental-risks-pfas (accessed 17.07.25).

Palma, D., Papagiannaki, D., Lai, M., Binetti, R., Sleiman, M., Minella, M., et al., 2021. PFAS degradation in ultrapure and groundwater using non-thermal plasma. Molecules (Basel, Switzerland). Available at https://doi.org/10.3390/molecules26040924.

Palma, D., Richard, C., Minella, M., 2022. State of the art and perspectives about non-thermal plasma applications for the removal of PFAS in water. Chemical Engineering Journal Advances 10, 100253. Available at https://doi.org/10.1016/j.ceja.2022.100253.

Per- and Polyfluoroalkyl Substances (PFAS) - ECHA. https://echa.europa.eu/hot-topics/perfluoroalkyl-chemicals-pfas. (accessed 14.07.25).

Per- and Polyfluoroalkyl Substances (PFAS) | State Legislation and Federal Action. https://www.ncsl.org/environment-and-natural-resources/per-and-polyfluoroalkyl-substances#packaging. (accessed 14.07.25).

Per- and Polyfluoroalkyl Substances (PFAS) in Canadians - Canada.Ca. https://www.canada.ca/en/health-canada/services/environmental-workplace-health/reports-publications/environmental-contaminants/human-biomonitoring-resources/per-polyfluoroalkyl-substances-canadians.html. (accessed 14.07.25).

Per- and Polyfluoroalkyl Substances (PFAS), US EPA. https://www.epa.gov/pfas. (accessed 16.07.25).

Per- and Polyfluoroalkyl Substances (PFAS), US EPA. https://www.epa.gov/sdwa/and-polyfluoroalkyl-substances-pfas (accessed 15.07.25).

Per- and Polyfluoroalkyl Substances (PFAS), US EPA. https://www.epa.gov/sdwa/and-polyfluoroalkyl-substances-pfas (accessed 14.07.25).

Per- and Polyfluoroalkyl Substances (PFASs) detected in Source Waters and Treated Public Water Supplies, U.S. Geological Survey. https://www.usgs.gov/programs/environmental-health-program/science/and-polyfluoroalkyl-substances-pfass-detected-source?qt-science_center_objects=0#qt-science_center_objects. (accessed 14.07.25).

Perfluoroalkyl and Polyfluoroalkyl Substances (PFAS), National Institute of Environmental Health Sciences. https://www.niehs.nih.gov/health/topics/agents/pfc. (accessed 24.07.25).

Pétré, M.-A., Genereux, D.P., Koropeckyj-Cox, L., Knappe, D.R.U., Duboscq, S., Gilmore, T.E., et al., 2021. Per- and Polyfluoroalkyl Substance (PFAS) transport from groundwater to streams near a PFAS manufacturing facility in North Carolina, USA. Environmental Science & Technology 55 (9), 5848–5856. Available at https://doi.org/10.1021/acs.est.0c07978.

PFAS Air Emissions Restrictions, BCLP - Bryan Cave Leighton Paisner. https://www.bclplaw.com/en-US/events-insights-news/pfas-air-emissions-restrictions.html. (accessed 15.07.25).

PFAS Analysis Toolkit: LC/MS/MS, Total Oxidizable Precursors (TOPs) and Total Organic Fluorine (TOF). https://www.bvna.com/insight/pfas-analysis-toolkit-lcmsms-total-oxidizable-precursors-tops-and-total-organic-fluorine. (accessed 17.07.25).

PFAS Analytical Methods Development and Sampling Research, US EPA. https://www.epa.gov/water-research/pfas-analytical-methods-development-and-sampling-research. (accessed 17.07.25).

PFAS Could Contaminate More Than 600 Military Installations, DOD Says, Environmental Working Group. https://www.ewg.org/news-insights/news-release/pfas-could-contaminate-more-600-military-installations-dod-says. (accessed 16.07.25).

PFAS Food Packaging Regulations State by State. https://www.packagingdigest.com/food-safety/unpacking-pfas-food-packaging-regulations-in-the-us. (accessed 15.07.25).

PFAS In Food Packaging - NYSDEC. https://dec.ny.gov/environmental-protection/recycling-composting/businesses/pfas-in-food-packaging. (accessed 15.07.25).

PFAS in Soil: State Regulations, BCLP - Bryan Cave Leighton Paisner. https://www.bclplaw.com/en-US/events-insights-news/pfas-in-soil-state-regulations.html. (accessed 15.07.25).

PFAS in Surface Waters, Department of Environmental Conservation. https://dec.vermont.gov/watershed/tasc/pfas-surface-waters. (accessed 15.07.25).

PFAS Method 537.1 - List Details - SRS, US EPA. https://cdxapps.epa.gov/oms-substance-registry-services/substance-list-details/475. (accessed 15.07.25).

Pierson, P., 1993. When effect becomes cause: policy feedback and political change. World Politics 45 (4), 595–628. Available at https://doi.org/10.2307/2950710.

Poothong, S., Thomsen, C., Padilla-Sanchez, J.A., Papadopoulou, E., Haug, L.S., 2017. Distribution of novel and well-known poly- and perfluoroalkyl substances (PFASs) in human serum, plasma, and whole blood. Environmental Science & Technology 51 (22), 13388–13396. Available at https://doi.org/10.1021/acs.est.7b03299.

Post, G.B., 2021. Recent US state and federal drinking water guidelines for per- and polyfluoroalkyl substances. Environmental Toxicology and Chemistry / SETAC 40 (3), 550–563. Available at https://doi.org/10.1002/etc.4863.

Public Water Systems and Drinking Water Standards for PFAS and Other Emerging Contaminants. https://www.health.ny.gov/environmental/water/drinking/emerging_pfas_publicwater.html. (accessed 15.07.25).

Rappazzo, K.M., Coffman, E., Hines, E.P., 2017. Exposure to perfluorinated alkyl substances and health outcomes in children: a systematic review of the epidemiologic literature. International Journal of Environmental Research and Public Health. Available at https://doi.org/10.3390/ijerph14070691.

Rawn, D.F.K., Dufresne, G., Clément, G., Fraser, W.D., Arbuckle, T.E., 2022. Perfluorinated alkyl substances in Canadian human milk as part of the Maternal-infant research on environmental chemicals (MIREC) study. The Science of the Total Environment 831, 154888. Available at https://doi.org/10.1016/j.scitotenv.2022.154888.

RCW 70A.222.070: prohibition on the manufacture, sale, or distribution of certain food packaging—safer alternatives assessment by department of ecology—publication of findings—report to legislature—prohibition effective date contingent on findings. https://app.leg.wa.gov/rcw/default.aspx?cite=70A.222.070. (accessed 15.07.25).

Reducing PFAS in Drinking Water with Treatment Technologies | US EPA. https://www.epa.gov/sciencematters/reducing-pfas-drinking-water-treatment-technologies (accessed 14.07.25).

Regulations.Gov.Perfluorooctanoic Acid (PFOA) Site Related Environmental Assessment Program, 3M Decatur, Alabama Site, Quarterly Status Report, January 25, 2005 to April 25, 2005 (Part 10 of 14). https://www.regulations.gov/document/EPA-HQ-OPPT-2004-0112-0031. (accessed 21.10.25).

Rekik, H., Arab, H., Pichon, L., El Khakani, M.A., Drogui, P., 2024. Per-and Polyfluoroalkyl (PFAS) eternal pollutants: sources, environmental impacts and treatment processes. Chemosphere 358, 142044. Available at https://doi.org/10.1016/j.chemosphere.2024.142044.

Risk Management Approach for Per- and Polyfluoroalkyl Substances (PFAS), Excluding Fluoropolymers - Canada.Ca. https://www.canada.ca/en/environment-climate-change/services/evaluating-existing-substances/risk-management-approach-per-polyfluoroalkyl-substances.html. (accessed 14.07.25).

Risk Management Strategy for Perfluorooctane Sulfonate and Its Salts and Precursors - Canada.Ca. https://www.canada.ca/en/environment-climate-change/services/canadian-environmental-protection-act-registry/publications/risk-management-strategy-perfluorooctane-sulfonate.html. (accessed 21.10.25).

Rogers, D.T., 2022. Environmental compliance handbook: sustainability and future environmental regulations. Environmental Compliance Handbook for Sustainable Future Environmental Regulations 4, 1–429. Available at https://doi.org/10.1201/9781003150121/environmental-compliance-handbook-volume-4-daniel-rogers/rights-and-permissions.

S.137 - 116th Congress (2019-2020): Right Start Child Care and Education Act of 2019, Congress.Gov, Library of Congress. https://www.congress.gov/bill/116th-congress/senate-bill/137. (accessed 15.07.25).

S.1473 - 116th Congress (2019-2020): Protect Drinking Water from PFAS Act of 2019, Congress.Gov, Library of Congress. https://www.congress.gov/bill/116th-congress/senate-bill/1473. (accessed 15.07.25).

S.1507 - 116th Congress (2019-2020): PFAS Release Disclosure and Protection Act of 2019, Congress.Gov Library of Congress. https://www.congress.gov/bill/116th-congress/senate-bill/1507. (accessed 15.07.25).

Saint-Gobain Dropping Lawsuit Against the State of Vermont. - Community Action Works. https://communityactionworks.org/saint-gobain-dropping-lawsuit-against-the-state-of-vermont/. (accessed 15.07.25).

SC 1999, c 33, Canadian Environmental Protection Act, 1999, CanLII. https://www.canlii.org/en/ca/laws/stat/sc-1999-c-33/latest/sc-1999-c-33.html.5 Pressing Environmental Issues in Canada in 2024. (accessed 14.07.25).

Schlezinger, J.J., Bello, A., Mangano, K.M., Biswas, K., Patel, P.P., Pennoyer, E.H., et al., 2025. Per- and polyfluoroalkyl substances (PFAS) in circulation in a canadian population: their association with serum-liver enzyme biomarkers and piloting a novel method to reduce serum-PFAS. Environmental Health: A Global Access Science Source 24 (1), 10. Available at https://doi.org/10.1186/s12940-025-01165-8.

Schroeder, T., Bond, D., Foley, J., 2021. PFAS soil and groundwater contamination via industrial airborne emission and land deposition in SW vermont and Eastern New York State, USA. Environmental Science: Processes & Impacts 23 (2), 291–301. Available at https://doi.org/10.1039/D0EM00427H.

Shimizu, M.S., Garcia, R.S., Avery, G.B., Kieber, R.J., Skrabal, S.A., Mead, R.N., 2022. Distribution of legacy and emerging per- and polyfluoroalkyl substances in riverine and coastal sediments of southeastern North Carolina, USA. Environmental Science: Processes & Impacts 24 (11), 2119–2128. Available at https://doi.org/10.1039/D2EM00246A.

Smalling, K.L., Romanok, K.M., Bradley, P.M., Morriss, M.C., Gray, J.L., Kanagy, L.K., et al., 2023. Per- and polyfluoroalkyl substances (PFAS) in United States tapwater: comparison of underserved private-well and public-supply exposures and associated health implications. Environment International 178, 108033. Available at https://doi.org/10.1016/j.envint.2023.108033.

State of per- and polyfluoroalkyl substances (PFAS) report - Canada.ca. https://www.canada.ca/en/environment-climate-change/services/evaluating-existing-substances/state-per-polyfluoroalkyl-substances-report.html. (accessed 23.07.25).

State of Per- and Polyfluoroalkyl Substances (PFAS) Report - Canada.Ca. https://www.canada.ca/en/environment-climate-change/services/evaluating-existing-substances/state-per-polyfluoroalkyl-substances-report.html.5 Pressing Environmental Issues in Canada in 2024. (accessed 14.07.25).

State of per- and polyfluoroalkyl substances (PFAS) report - Canada.ca. https://www.canada.ca/en/environment-climate-change/services/evaluating-existing-substances/state-per-polyfluoroalkyl-substances-report.html. (accessed 16.07.25).

State Reaches Agreement with Saint-Gobain I Office of Governor Phil Scott. https://governor.vermont.gov/press-release/state-reaches-agreement-saint-gobain. (accessed 15.07.25).

States Step Up as Federal PFAS Regulations Retreat I Baker Donelson. https://www.bakerdonelson.com/states-step-up-as-federal-pfas-regulations-retreat. (accessed 21.10.25).

Stratton, G.R., Dai, F., Bellona, C.L., Holsen, T.M., Dickenson, E.R.V., Mededovic Thagard, S., 2017. Plasma-based water treatment: efficient transformation of perfluoroalkyl substances in prepared solutions and contaminated groundwater. Environmental Science & Technology 51 (3), 1643–1648. Available at https://doi.org/10.1021/acs.est.6b04215.

Summary of the Clean Water Act, US EPA. https://www.epa.gov/laws-regulations/summary-clean-water-act. (accessed 14.07.25).

Summary of the Safe Drinking Water Act, US EPA. https://www.epa.gov/laws-regulations/summary-safe-drinking-water-act. (accessed 14.07.25).

Superfund: CERCLA Overview, US EPA. https://www.epa.gov/superfund/superfund-cercla-overview (accessed 14.07.25).

Sustainable PFAS Destruction Works at Low Light Levels, Aquatech. https://www.aquatechtrade.com/news/water-treatment/pfas-catch-destroy-sustainable-fast-solution-low-uv-level. (accessed 14.07.25).

SW-846 Test Method 8327: Per-and Polyfluoroalkyl Substances (PFAS) by Liquid Chromatography/Tandem Mass Spectrometry (LC/MS/MS), US EPA. https://www.epa.gov/hw-sw846/sw-846-test-method-8327-and-polyfluoroalkyl-substances-pfas-liquid-chromatographytandem. (accessed 17.07.25).

TDEC's Strategy to Assess the Impact of Per-and Polyfluoroalkyl Compounds on Public Drinking Water Sources in Tennessee. https://www.tn.gov/content/dam/tn/environment/policy-planning/documents/pfas/opp_pfas-sampling-strategy.pdf. (accessed 15.07.25).

Technical Basis for the Lowest Concentration Minimum Reporting Level (LCMRL) Calculator. 2010. https://nepis.epa.gov/Exe/ZyPDF.cgi?Dockey=P100J7CA.txt. (accessed 15.07.25).

Testing Food for PFAS and Assessing Dietary Exposure, FDA. https://www.fda.gov/food/process-contaminants-food/testing-food-pfas-and-assessing-dietary-exposure. (accessed 17.07.25).

Text - H.R.2500 - 116th Congress (2019-2020): National Defense Authorization Act for Fiscal Year 2020, Congress.Gov, Library of Congress. https://www.congress.gov/bill/116th-congress/house-bill/2500/text. (accessed 15.07.25).

Text - S.4049 - 116th Congress (2019-2020): National Defense Authorization Act for Fiscal Year 2021, Congress.Gov, Library of Congress. https://www.congress.gov/bill/116th-congress/senate-bill/4049/text. (accessed 15.07.25).

The Biggest Environmental Issues In Canada - WorldAtlas. https://www.worldatlas.com/articles/the-biggest-environmental-issues-in-canada.html. (accessed 14.07.25).

Theurillat, X., Mujahid, C., Eriksen, B., Griffin, A., Savage, A., Delatour, T., et al., 2023. An LC-MS/MS method for the quantitative determination of 57 per- and polyfluoroalkyl substances at Ng/Kg levels in different food matrices. Food Additives & Contaminants: Part A 40 (7), 862–877. Available at https://doi.org/10.1080/19440049.2023.2226771.

Thomas, R., Jenkins, K., Landale, B., Trigger, G., Holsen, T.M., Dore, S., et al., 2020. Evaluation of PFAS treatment technology: alkaline ozonation. Remediation Journal 30 (3), 27–37. Available at https://doi.org/10.1002/rem.21654.

Toxicological Profile for Perfluoroalkyls. https://www.atsdr.cdc.gov/toxprofiles/tp200.pdf. (accessed 14Food Additives & Contaminants: Part A25).

TSCA Section 8(a)(7) Reporting and Recordkeeping Requirements for Perfluoroalkyl and Polyfluoroalkyl Substances I US EPA. https://www.epa.gov/assessing-and-managing-chemicals-under-tsca/tsca-section-8a7-reporting-and-recordkeeping. (accessed 14.07.25).

Update on New Hampshire PFAS Drinking Water Standards (MCLs), DES - PFAS Blog. https://www.pfas.des.nh.gov/news-and-media/blog/update-new-hampshire-pfas-drinking-water-standards-mcls. (accessed 15.07.25).

US EPA PFAS Method 1633. https://www.bvna.com/insight/us-epa-pfas-method-1633. (accessed 15.07.25).

Vaughn, B., Andrea, W., Kyle, S., 2013. Perfluorooctanoic acid (PFOA) exposures and incident cancers among adults living near a chemical plant. Environmental Health Perspectives 121 (11–12), 1313–1318. Available at https://doi.org/10.1289/ehp.1306615.

Vuong, A.M., Webster, G.M., Yolton, K., Calafat, A.M., Muckle, G., Lanphear, B.P., et al., 2021. Prenatal exposure to per- and polyfluoroalkyl substances (PFAS) and neurobehavior in US children through 8 years of age: the home study. Environmental Research 195, 110825. Available at https://doi.org/10.1016/j.envres.2021.110825.

W, O.G., M, B.J., J, E.D., W, F.J., M, S.A., L, B.J., et al., 2007. Half-life of serum elimination of perfluorooctanesulfonate, perfluorohexanesulfonate, and perfluorooctanoate in retired fluorochemical production workers. Environmental Health Perspectives 115 (9), 1298–1305. Available at https://doi.org/10.1289/ehp.10009.

Watson, N. PFAS in Air: Sampling and Analytical Techniques for the Identification of Persistent Contaminants. https://apps.nelac-institute.org/nemc/2020/docs/presentations/pdf/8-4-20-Air%20Methods,%20Monitoring,%20and%20Technology-3.02-Watson.pdf. (accessed 17.07.25).

Weed, R.A., Campbell, G., Brown, L., May, K., Sargent, D., Sutton, E., et al., 2024. Non-targeted PFAS suspect screening and quantification of drinking water samples collected through community engaged research in North Carolina's Cape Fear River Basin. Toxics 12 (6). Available at https://doi.org/10.3390/toxics12060403.

Wingard, A. Department of Defense (DoD) Department of Energy (DOE) Consolidated Quality Systems Manual (QSM) for Environmental Laboratories. 2009. Department of Defense. https://www.denix.osd.mil/edqw/documents/manuals/qsm-version-5-4-final/QSM Version 5.4 FINAL.pdf. (accessed 14.07.25).

Yu, R.-S., Yu, H.-C., Yang, Y.-F., Singh, S., 2025. A global overview of per- and polyfluoroalkyl substance regulatory strategies and their environmental impact. Toxics. Available at https://doi.org/10.3390/toxics13040251.

Zhang, Y., Mustieles, V., Wang, Y.-X., Sun, Y., Agudelo, J., Bibi, Z., et al., 2023. Folate concentrations and serum perfluoroalkyl and polyfluoroalkyl substance concentrations in adolescents and adults in the USA (National Health and Nutrition Examination Study 2003-16): an observational study. The Lancet Planetary Health 7 (6), e449–e458. Available at https://doi.org/10.1016/S2542-5196(23)00088-8.

Zhang, Z., Sarkar, D., Biswas, J.K., Datta, R., 2022. Biodegradation of per- and polyfluoroalkyl substances (PFAS): a review. Bioresource Technology 344, 126223. Available at https://doi.org/10.1016/j.biortech.2021.126223.

CHAPTER

# 22

# Regulatory frameworks for per- and polyfluoroalkyl substances in Australia and the European Union

*Sai Sharanya Pulimamidi*[1], *Soumi Saha*[1], *Srikanth Ponneganti*[2], *Dhanavath Dattu Naik*[1], and *Roshan M. Borkar*[1]

[1]**Department of Pharmaceutical Analysis, National Institute of Pharmaceutical Education and Research, Guwahati, Changsari, Assam, India** [2]**Department of Food and Science and Technology, The Ohio State University, Columbus, OH, United States**

## 22.1 Introduction

Per- and polyfluoroalkyl substances (PFASs) are dangerous, synthetic, hazardous compounds that have negative impacts on the environment and human health. They form a large family of fluorinated compounds with partially or fully fluorinated carbon chains of varying lengths (Per- and Polyfluoroalkyl Substances PFASs, UNEP, 2025). PFAS have been widely used because of their unique positive properties. They are also stable during high temperatures. Many materials also serve as surfactants and can be used, for example, to repel water and grease. The most common industry types that use PFAS include aerospace and defense, automotive, aviation, food contact materials, textiles, leather and apparel, construction and household products, electronics, firefighting, food processing, and medical articles (Per- and polyfluoroalkyl substances PFAS - ECHA, 2025a). Their extensive presence in various environmental matrices and their resistance to biodegradation raised health concerns about their potential negative impacts on human health. These qualities, accumulated both in the environment and in the human body, lead to negative health outcomes (Hossini et al., 2025).

The combination of environmental stability, global mobility, and bioaccumulative potential of PFAS is truly unique, leading not only to its commercial utility but also to extensive environmental pollution, worldwide spread, and ubiquitous human exposure. There is mounting scientific evidence that PFAS are associated with varied health effects such as hormonal imbalance, immunosuppression, liver disease, lipid imbalance, kidney

*Per- and Polyfluoroalkyl Substances (PFAS) as Environmental Contaminants*
**DOI: https://doi.org/10.1016/B978-0-443-43856-1.00003-4**

disease, reproductive toxicity, and developmental defects (Fenton et al., 2021). Furthermore, the world feels an urgent need to regulate the use of such chemicals to protect human lives and the environment. It is important to note that Australia and Europe have prioritized PFAS as contaminants of high priority and made effective efforts in regulating them through regulatory frameworks, environmental management, and risk assessment methods (Abunada et al., 2020; Sonne et al., 2023).

In Australia, the Department of Health, Food Standards Australia New Zealand (FSANZ), and the National Health and Medical Research Council (NHMRC) have established health-based recommended levels for the key PFAS compound to protect human health through exposure to drinking water, food, and the environment (Perfluorinated Compounds, Food Standards Australia New Zealand, 2025; NHMRC, 2025).

Complementing these efforts, the Australian Industrial Chemicals Introduction Scheme (AICIS) oversees the introduction of PFAS-related substances through premarket risk assessment and registration, ensuring chemical safety and minimizing environmental release (Industrial Chemicals Law in Australia, Australian Industrial Chemicals Introduction Scheme AICIS, 2025). The PFAS National Environmental Management Plan (NEMP), developed collaboratively by the Australian government, provides nationally consistent guidance on PFAS investigation, assessment, remediation, and reuse of contaminated materials in Australia (PFAS National Environmental Management Plan, 2018). Additionally, several Australian jurisdictions, including Queensland, Victoria, and the Australian Capital Territory (ACT), have introduced region-specific PFAS management and phase-out strategies. Together, these collaborative efforts underscore Australia's adaptive approach to regulating PFAS (Firefighting Foam Management Policy, 2025; PFAS and Livestock, 2025; Per- and polyfluoroalkyl substances PFAS, 2025).

By comparison, the European Union has also developed a wholly comprehensive and legally enforceable set of rules and regulations to govern PFAS, in part due to its similarity to international treaties such as the Stockholm Convention on Persistent Organic Pollutants (POPs). The key elements of this framework are the Registration, Evaluation, Authorisation and Restriction of Chemicals (REACH) and the European Chemicals Agency (ECHA), which together regulate the registration, evaluation, and restriction of PFAS (Per- and poly-fluoroalkyl substances PFAS, 2025; REACH Regulation - Environment - European Commission, n.d.). The European Union has prohibited a number of long-chain PFAS, including perfluorooctane sulfonate (PFOS), perfluorooctanoic acid (PFOA), and perfluorohexane sulfonate (PFHxS), and has also proposed the prohibition of 10,000 more compounds (Per- and polyfluoroalkyl substances PFAS - ECHA, 2025b; All news - ECHA, 2025). Additionally, national agencies, such as the Danish Environmental Protection Agency (EPA), the Swedish Chemicals Agency (KEMI), and the Norwegian Environment Agency, play a role in monitoring, enforcement, and regulating PFAS. Comprehensively, the group approach to handling PFAS as a category and preventing them is adopted by Europe, ensuring wider environmental protection.

Australia and Europe have different approaches to regulatory measures for addressing the challenges posed by PFAS pollutants. Furthermore, despite inconsistencies in enforcement as part of their regulatory frameworks, both regions aim to achieve the common goal of mitigating the effects of PFAS contamination. This chapter analyzes the management and pollution of PFAS in Australia and European countries, examining their regulatory requirements, detection methods, monitoring methods, remediation activities, litigation, international analysis, and future opportunities.

## 22.2 Australia: per- and polyfluoroalkyl substances regulatory guidelines

### 22.2.1 Overview of environmental issues in Australia

Australia is now experiencing a gradual deterioration in the state of its environment due to the mounting threats posed by climate change, habitat destruction, invasive species, pollutant emissions, and excessive resource use. Some of the most pressing issues are the widespread pollution and contamination of air, water, soil, and groundwater. It is estimated that 16,000 to 80,000 sites in Australia are polluted, with some 7500 different types of pollution resulting from industrial and governmental actions (Khudur et al., 2018). Many exclusion areas have been established due to the high levels of contamination. Plastics, specifically microplastics, are major pollutants that impact 85% of waterways (Reisser et al., 2013), with an estimated 130,000 tonnes entering the environment every year. More than 16,000 sites are impacted by hydrocarbon pollution, and over 3500 are affected by herbicides and pesticides (Huang et al., 2023). More than 100 sites are contaminated with toxic metals, such as lead, mercury, arsenic, and radioactive materials (Molino et al., 2019). Moreover, an average of 4.2–4.5 tonnes of bisphenols and 5.9–47 tonnes of phthalates are released into the environment annually as a result of sewage treatment facility operations (Moore et al., 2023). Although manufactured elsewhere, PFASs have become one of the most significant environmental issues in Australia due to their widespread use in firefighting foams, textiles, and cookware. Research has established concentrations of up to 25,000 $ng\ L^{-1}$ of PFAS in landfill leachate and has found them to be ubiquitous in wastewater, as well as in households, offices, and transportation. PFAS have also been detected in human biological samples (e.g., blood and follicular fluids) and are known to persist in the long term, with potential health implications. Australia is regarded as a global hotspot of PFAS pollution, despite not having a domestic production of the chemical, primarily due to its use in past firefighting operations (Ackerman Grunfeld et al., 2024).

Considering the ubiquity and potential health consequences of PFAS and other pollutants, the need for robust regulatory systems has become increasingly concerning. This chapter analyzes the regulatory measures, enforcement, and remediation measures that Australia has implemented to combat environmental contamination, particularly PFAS. This section will examine the effectiveness of environmental governance in Australia and identify areas for improvement to safeguard the health and ecological integrity of the population through a review of monitoring activities, risk assessment, and current remediation programs.

#### 22.2.1.1 *Overview of the regulatory framework of Australia*

The chemical management system in Australia is regulated by several national regulatory bodies that oversee the importation, production, and use of chemicals, as well as the safety assessment of chemicals used in various sectors. These include:

1. **Australian Pesticides and Veterinary Medicines Authority (APVMA):** The Agricultural and Veterinary Chemicals Code Act 1994 is the name of the act that regulates the use of

chemicals in agriculture, forestry, fisheries, veterinary work, and household pest management in Australia, commonly known as the Agvet Code (Federal Register of Legislation, 2025). The framework ensures the regulation of agricultural and veterinary (Agvet) chemicals, ensuring they are safe, effective, and have a low environmental impact. Under the control of the APVMA and the Department of Agriculture, it ensures that products are correctly labeled, safe, effective, and that there are no risks to humans, the environment, or nontarget species (Australian Pesticides and Veterinary Medicines Authority, 2025).

2. **Australian Industrial Chemicals Introduction Scheme (AICIS):** In 2020 following the enactment of the Industrial Chemicals Act 2019, the National Industrial Chemicals Notification and Assessment Scheme was replaced by AICIS. AICIS controls chemicals in products such as cosmetics and paints, evaluates and assesses risks, maintains a chemical inventory, and complies with and cooperates both domestically and internationally to ensure chemical safety and fulfill international responsibilities (Industrial Chemicals Law in Australia, Australian Industrial Chemicals Introduction Scheme AICIS, 2025).
3. **FSANZ:** FSANZ makes and implements the Food Standards Code to control additives, processing aids, food composition, and labeling. It provides safety in food innovations, such as genetic modification, and requires allergy warnings, which apply to packaged and unpackaged food throughout Australia and New Zealand (Food Standards Australia New Zealand, 2025a).
4. **Therapeutic Goods Administration (TGA):** In Australia, medicines, medical devices (including AI-based devices), biologicals, blood and tissues, disinfectants, and other therapeutic products are regulated by the TGA. It also regulates unapproved products that are not registered in the Australian Register of Therapeutic Goods, ensuring that safety, efficacy, and quality standards are met (Therapeutic Goods Administration TGA, 2025).
5. **The Stockholm Convention:** The Stockholm Convention is the international agreement that was created to safeguard human health and the environment against the POPs chemicals that were found to last long in the environment, cross-region, settle in the fatty tissue of both humans and wildlife, and have detrimental impacts. These encompass cancers, birth anomalies, impairments in the immune and reproductive systems, and brain damage. POPs are cross-border pollutants; therefore they cannot be addressed by individual countries. Adopted in 2001 and coming into force in 2004, the convention obliges signatories to reduce or eliminate the emission of these pollutants. Australia signed the convention in 2001 and ratified it on August 18, 2004, thereby assuming an obligation to take measures to reduce POP emissions (Listing of POPs in the Stockholm Convention, 2025). Nevertheless, Australia has yet to ratify amendments that enumerate PFOS and PFOA as POPs on the basis of intricate management issues. In this regard, the Australian government is developing a new national standard for the environmental risk management of industrial chemicals to enhance the coordination of national government activities related to the control of chemicals such as PFAS (Ibrahim et al., 2025; Stockholm Convention on Persistent Organic Pollutants POPs - DCCEEW, 2025). These regulatory bodies collaborate based on the broader policies of environmental and health protection in Australia, utilizing risk-based methods of assessment and control of chemical hazards, including newly identified contaminants such as PFAS.

### 22.2.2 Regulatory framework of per- and polyfluoroalkyl substances

The Department of Health, FSANZ, and the NHMRC have established health-based guidance values for key PFAS, including PFOS, PFOA, and PFHxS. These values serve as precautionary tools during site investigations and help guide public health advice to reduce PFAS exposure in affected communities. The aim is to safeguard the population from PFAS exposure through food, drinking water, and recreational water sources. The guidance values are detailed in FSANZ's hazard assessment report for PFOS, PFOA, and PFHxS, as well as in NHMRC's Australian Drinking Water Guidelines (2011) and Guidelines for Managing Risks in Recreational Water (2008).

Tolerable daily intake levels are 20 ng $Kg^{-1}$ of body weight for PFOS (including PFHxS) and 160 ng $kg^{-1}$ of PFOA. The guideline limits for PFAS in drinking water are 70 ng $L^{-1}$ for PFOS + PFHxS and 560 ng $L^{-1}$ for PFOA. For recreational water, the respective values are 2 and 10 μg $L^{-1}$. These thresholds are applied during environmental assessments across Australia (Health Based Guidance Values for PFAS for Use in Site Investigations in Australia, 2025). An overview of regulatory guidelines in Australia is mentioned in Table 22.1.

i. **Australian Drinking Water Guidelines for per- and polyfluoroalkyl substances**
The NHMRC has been regularly updating the Australian Drinking Water Guidelines since 2011 to reflect the latest scientific evidence on water safety. As part of its recent review, the NHMRC evaluated international health advice and scientific data, particularly recent findings assessed by the US EPA, to determine whether revisions to PFAS-related recommendations were necessary (Australian Drinking Water Guidelines - Public Consultation on Draft Guidance for Per- and Polyfluoroalkyl Substances PFAS - NHMRC Consultation Hub - Citizen Space, 2025). Based on this evaluation, updated health guideline values for PFAS in drinking water have been proposed. For PFOA and PFOS, the guideline values were significantly lowered to 200 and 4 ng $L^{-1}$, respectively, due to concerns about cancer risk and effects on bone marrow. For PFHxS and perfluorobutane sulfonic acid (PFBS), the updated values are 30 and 1 μg $L^{-1}$, respectively, due to their potential thyroid-related effects (NHMRC Review of PFAS in Australian Drinking Water, 2025).

ii. **Australian and New Zealand Guidelines for Fresh and Marine Water Quality**
Australia and New Zealand provide comprehensive Water Quality Guidelines to help manage the quality of natural and seminatural water resources. The Australian and New Zealand Guidelines for Fresh and Marine Water Quality (2023) outline interim default guideline values for PFOS and PFOA to protect aquatic ecosystems. These ecological guideline values are categorized based on species protection levels: 99%, 95%, 90%, and 80%. For freshwater, the PFOS values are: 99% protection: 23 ng $L^{-1}$; 95% protection: 130 ng $L^{-1}$; 90% protection: 2 μg $L^{-1}$; 80% protection: 31 μg $L^{-1}$. For PFOA in freshwater, the values are: 99% protection: 0.019 ng $L^{-1}$; 95% protection: 0.22 ng $L^{-1}$; 90% protection: 0.63 ng $L^{-1}$; 80% protection: 1.82 ng $L^{-1}$. The same values apply to marine water as interim guidance (Fresh and Marine Water Australia and New Zealand, 2025; Default Guideline Values, 2025).

iii. **PFAS NEMP**
Due to growing concerns about the environmental and health risks posed by high levels of PFAS exposure, Australia prioritized a coordinated national response.

**TABLE 22.1** Overview of regulatory guidelines in Australia.

| Authority | Scope | Key PFAS | Guideline values | Actions |
|---|---|---|---|---|
| The Department of Health, Food Standards Australia New Zealand, and the National Medical Research Council (NHMRC) | TDI (Food) | Total PFOS + PFHxS | 20 ng $kg^{-1}$ bw $day^{-1}$ | Developed health-based guidance values to minimize exposure to PFAS. |
| | | PFOA | 160 ng $kg^{-1}$ bw $day^{-1}$ | |
| | Drinking water | Total PFOS + PFHxS | 70 ng $L^{-1}$ | |
| | | PFOA | 560 ng $L^{-1}$ | |
| | Recreational water | Total PFOS + PFHxS | 2000 ng $L^{-1}$ | |
| | | PFOA | 10,000 ng $L^{-1}$ | |
| NHMRC proposed review | Drinking water | PFOA | 200 ng $L^{-1}$ | Ensure that they remain current with the most recent scientific findings about safe drinking water. |
| | | PFOS | 4 ng $L^{-1}$ | |
| | | PFHxS | 30 ng $L^{-1}$ | |
| | | PFBS | 1000 ng $L^{-1}$ | |
| Australian and New Zealand Guidelines for Fresh and Marine Water Quality (2023)—interim default guideline values | Freshwater and marine water | PFOS | 0.00023 ng $L^{-1}$ (99% species protection) | |
| | | | 0.13 ng $L^{-1}$ (95% species protection) | |
| | | | 2 ng $L^{-1}$ (90% species protection) | |
| | | | 31 ng $L^{-1}$ (80% species protection) | |
| | | PFOA | 19 ng $L^{-1}$ (99% species protection) | |
| | | | 220 ng $L^{-1}$ (95% species protection) | |
| | | | 632 ng $L^{-1}$ (90% species protection) | |
| | | | 1824 ng $L^{-1}$ (80% species protection) | |

| | | | | |
|---|---|---|---|---|
| NEMP biota guideline values for wildlife diet values (based on Environment and Climate Change Canada (2018) methods and adjusted using Australian mammal dietary data) | Ecological direct exposure for the wildlife diet | Sum of PFOS and PFHxS | 3100 ng $L^{-1}$: mammalian diet—consumption of biota as wet weight food<br>8200 ng $L^{-1}$: avian diet—consumption of biota as wet weight food | Assess and manage the risk of PFAS exposure to wildlife through their diet, helping to protect animal health and ecosystems in the Australian environment. |
| | | PFOA | 2800 ng $L^{-1}$: mammalian diet—consumption of biota as wet weight food<br>2400 ng $L^{-1}$: avian diet—consumption of biota as wet weight food | |
| | Ecological exposure protective of birds | Sum of PFOS and PFHxS | 200 ng $L^{-1}$ whole bird egg as wet weight | |
| | | PFOA | 15 ng $L^{-1}$ whole bird egg as wet weight | |
| The Australian Industrial Chemicals Introduction Scheme (AICIS) | | PFAS (including those that may break down to form perfluorinated substances such as PFOS and PFOA) | | Banned import and export of PFOS and PFOA, unless approved by AICIS under the Rotterdam Convention.<br>Made compulsory for importers and manufacturers to register and categorize their chemical introductions before legal entry into Australia. |

To address this, the Australian Environment Ministers asked the heads of EPAs (the HEPA) in Australia and New Zealand to partner with the Australian Government Department of the Environment and Energy. This paved the way to the PFAS NEMP. The NEMP is a consistent, transparent, and usable national mechanism for managing PFAS-affected materials and locations in Australia. It provides guidelines about environmental regulation, site investigation, sampling of PFAS, treatment, and disposal of the product. The strategy aims to support the government, industry players, and regulators in working effectively to mitigate PFAS risks and ensure environmental and public health safety (PFAS National Environmental Management Plan 1.0 - DCCEEW, 2025).

iv. **PFAS NEMP 1.0**

In January 2018, the Environment Ministers of Australia endorsed the first version of the PFAS NEMP 1.0. The national plan provided guidelines for the management of PFOS, PFOA, PFHxS, and their precursors, which are the most significant indicators of overall PFAS contamination. It recognized how PFAS are persistent, how difficult it is to break them down, and how landfills and wastewater treatment plants contribute to environmental pollution.

NEMP 1.0 emphasized the precautionary method involving the minimization of human exposure, allowance of site- and catchment-specific dealing with risk. It also emphasized the need to change the management of PFAS through the implementation of evolving scientific evidence. Though it did not govern continued use of products containing PFAS, it created guidelines to enable environmental regulators to act accordingly, following the applicable legislation falling within their jurisdictions.

v. **PFAS NEMP 2.0**

The 2018 version was replaced in some Australian jurisdictions by the PFAS NEMP 2.0, published on February 28, 2019, in the ACT, Tasmania, the Northern Territory (NT), and Queensland jurisdictions. It was developed through the comments of EPA heads, approved by environment ministers, and officially adopted in October 2019.

NEMP 2.0 complemented and elaborated on the advice in NEMP 1.0. It provided updates on the four priority areas outlined above, including environmental guideline values, soil reuse, wastewater management, and on-site containment. Besides such revisions, it is also necessary to note that NEMP 2.0 provides a more detailed description and enhancement of certain aspects of the initial plan, making it easier to implement both nationally and regionally.

The updated document aims to bring more effective, transparent, and consistent national-level management of PFAS contamination in Australia (PFAS National Environmental Management Plan 1.0 - DCCEEW, 2025; PFAS National Environmental Management Plan Version 2.0, 2025).

vi. **PFAS NEMP 3.0**

Developed since 2020, the PFAS NEMP 3.0 was formally endorsed by Australian Environment Ministers and is scheduled for publication on December 10, 2024. It has replaced NEMP 2.0 and includes major changes and additional guidance on the most relevant areas.

NEMP 3.0 presents updated guideline levels for site investigation and risk assessment, frameworks for correcting contaminated sites, and elaborate standards for the reuse of PFAS-contaminated resources and other product recovery areas. It further focuses on risk-based decision-making styles to facilitate the steady management of environmental practices between jurisdictions.

The new guidance was prepared by the National Chemical Working Group and endorsed by the HEPA in July 2024. This version indicates changing scientific information and regulatory requirements with a more coordinated approach to deal with PFAS risks on a national level in Australia (PFAS National Environmental Management Plan 1.0 - DCCEEW, 2025; Department of climate change, 2025).

**vii. The AICIS**

AICIS is involved in ensuring the safe use of industrial chemicals in Australia. It evaluates the risks posed by imported chemicals and, more importantly, manufactures them to safeguard the health of people and the environment. As a way forward, AICIS has pressed upon industries to address PFAS and associated toxicants by aggressively seeking safer options. One way that companies can make sure that AICIS evaluates a new perfluorinated chemical, such as those added to firefighting foams, is to declare it to AICIS (Australian Government PFAS Taskforce, 2025). AICIS also keeps the Australian Inventory of Industrial Chemicals, which contains all the details of those registered PFAS, including the chemical names, formulas, and registration numbers, to be transparent to the industry and the general public (Department of climate change, 2025).

**viii. Intergovernmental agreement on a national framework for responding to PFAS contamination**

In 2018 the Australian government, in conjunction with state and territory governments, except for Tasmania, signed an intergovernmental agreement to develop a whole-of-nation approach to handling PFAS contamination. The agreement also guarantees a uniform response on the international level, with references to ecology and the reduction of risks to human health. It shall contain among its essential instruments the plan on NEMP on PFAS and other guidelines regarding roles, responsibilities, and unified procedures to be used in managing PFAS pollution (Australian Government Department Of Health, 2025; International Cooperation, 2025; Per- and Poly-Fluoroalkyl Substances PFASs - DCCEEW, 2025; Intergovernmental Agreement on a National Framework for Responding to PFAS Contamination, 2025).

#### 22.2.2.1 *State per- and polyfluoroalkyl substances regulations*

**1. Queensland's regulations**

Queensland has adopted a proactive approach that has facilitated its management process regarding PFAS. The Queensland Government introduced the Environmental Management of Firefighting Foam—Operational Policy (Firefighting Foam Policy) to regulate suppliers and users who may be guided in 2016, according to the Environmental Protection Act 1994 (Firefighting Foam Management Policy, 2025).

This policy aimed to eliminate the use of fluorinated firefighting foams (such as aqueous film-forming foam (AFFF), fluoropolymers, and forming fluoroprotein foams), which contain the long-lived and undesirable PFAS, by 2019. The policy would facilitate the transition to nonpersistent, PFAS-free solutions, in addition to covering the proper handling, use, and disposal of firefighting foams, in support of public health and the environment. As of July 7, 2019, with amendments made in October 2021, all entities involved in the use, storage, transport, or disposal of firefighting foams must comply with this policy. Compliance includes using containment systems (e.g., bunding) to prevent foam releases and submitting disposal plans for PFAS-based foams. Nonpersistent foams must follow standard waste disposal procedures, while PFAS-containing foams require stricter controls. For the temporary use of PFAS-based foams, government approval is mandatory under the Queensland Environmental Protection Act 1994 (Firefighting Foam Management Policy, 2025).

2. **Victoria**

Five major sites of PFAS management are managed by the Department of Defense (DOD) in Victoria, including: Royal Australian Air Force (RAAF) Base Williams, Laverton, RAAF Base Point Cook, HMAS Cerberus, and the Bandiana Military Area (Per- and Poly-Fluoroalkyl Substances PFASs - DCCEEW, 2025). These sites are part of current initiatives to monitor, control, and remediate the presence of PFAS pollution resulting from the past application of firefighting foams and other sources.

Furthermore, Agriculture Victoria plays a significant role in facilitating PFAS management in the agricultural industry. The agency collaborates with local authorities and communities to conduct human health risk assessments and raise awareness among livestock producers about potential PFAS effects. Their endeavors are geared towards ensuring that the livestock-related industries are well-informed and adequately protected against the risks encountered by exposure to PFAS (PFAS and Livestock, 2025).

3. **South Australia**

The Environment Protection Authority (EPA) of South Australia has provided elaborate requirements on how to handle contamination relating to PFA. The guidelines align with the PFAS NEMP, which provides a uniform framework for managing PFAS-contaminated substances and locations. To address the issue of PFAS-contaminated waste, the EPA has established specific disposal requirements and environmental management actions applicable to landfills. Furthermore, provisional advice on how PFAS can be managed in waste soils has also been published, pending any future changes to regulations (Guidance for Managing PFAS in SA, 2025).

It is interesting to note that South Australia was also the first state in Australia to pass an act prohibiting the use of PFAS-containing fluorinated firefighting foams. It is an amendment to the Environment Protection (Water Quality) Policy 2015, and is officially considered in force as of January 30, 2018, following a 2 year transition period that allowed industries to transition to fluorine-free options. Throughout this period, the EPA collaborated with stakeholders, providing technical assistance and compliance advice (Per- and Poly-Fluoroalkyl Substances PFAS, 2025). South Australia Health also informs the public about the routes of exposure and the potential health effects of PFAS, supporting the broader effort to mitigate environmental and human health risks.

4. **ACT**

The Environment Protection Act 1997 is the primary legislation governing environmental protection and the management of chemical contamination, including PFAS, in the

ACT. The ACT EPA has the role of informing the process of identifying, evaluating, and remediating contaminated land, which includes PFAS-impacted sites. Moreover, the Health Protection Service provides information about possible health risks associated with exposure to PFAS in the territory (PFAS Advice, Australian Government PFAS Taskforce, 2025).

In August 2019, the ACT government published the PFAS Action Plan ACT, outlining specific strategies, responsibilities, and activities to address PFAS contamination. The EPA is the spearhead agency in the execution of this plan. PFAS Action Plan complements the larger goals of the intergovernmental agreement on a National Framework to Respond to PFAS Contamination to provide a coherent and effective response across all levels of government (Report on Australian Capital Territory ACT, PFAS ACTION PLAN, 2025).

The ACT EPA portal is a source of additional information and regulatory details (Environment Protection Authority - Access Canberra, 2025).

5. **New South Wales (NSW)**

Several organizations exist in NSW to manage and investigate PFAS contamination. The NSW Department of Primary Industries is a source of information on agriculture, fisheries, biosecurity, and food safety. It also provides dietary and fishing advisories in regions where PFAS contamination poses a threat, as well as supporting communities affected by it (PFAS Response, 2025).

The NSW EPA is undertaking a detailed PFAS Investigation Program to establish the existence, distribution, and potential effects of legacy PFAS in a wide area of the state (The NSW Government PFAS Investigation Program, 2025).

Moreover, Fire and Rescue NSW has introduced an environmental assessment to examine and control PFAS contamination at its operational sites and in the surrounding areas. It is part of a broader campaign to understand the ecological persistence of PFAS in firefighting operations (PFAS Environmental Investigation - Fire and Rescue NSW, 2025).

These collaborative actions demonstrate the NSW government's commitment to reducing the risk of PFAS to human health and the environment by adopting a coordinated approach to monitoring, risk assessment, and community communication.

6. **Tasmania**

In Tasmania, the Tasmanian EPA provides region-specific advice on the issues associated with PFAS contamination. The EPA is overseeing and managing the area-wide contaminated land issues, that is, the PFAS, to safeguard the environment and the health of the people (PFAS Contamination, 2025).

7. **NT**

The NT EPA provides guidance on local issues related to PFAS in environmental matters. NT Power and Water Corporation has been undertaking tests on PFAS in drinking water at different locations, and the findings have been compared with national health-based guidelines values to determine any possible risk (PFAS, 2025; PFAS Monitoring, 2025).

8. **Western Australia**

In Western Australia, investigations of PFAS are conducted by the Department of Water and Environmental Regulation, which coordinates the management of contaminated sites. Moreover, West Australia Health provides detailed data on the pathways of exposure and health risks of PFAS, which can assist in raising awareness and precautionary measures at the community level (Western Australian Government, 2025; Per- and Poly-Fluoroalkyl Substances, 2025).

## 22.2.3 Per- and polyfluoroalkyl substances detection and monitoring

The National Measurement Institute, Australia's foremost measurement entity, provides exceptional environmental measurement services and solutions. Contemporary laboratories can aid in meeting quality and regulatory standards. The PFAS testing protocol adheres to United States Environmental Protection Agency (USEPA) Draft Method 821-R-11-007, employing solid phase extraction (SPE) for water samples and liquid chromatography and mass spectrometry (LC-MS)/MS detection, alongside USEPA 537, which outlines the determination of selected perfluorinated alkyl acids in drinking water via SPE and LC-MS/MS. This approach possesses NATA accreditation for biosolids, soil, and water samples (Department of Industry Science and Resources, 2025). The combination of LC-MS/MS, is being executed at several laboratories in Australia for the purpose of conducting analysis. Within a wide range of sample matrices, such as water, soil, blood/serum, and biota, the LC-MS/MS technology offers a method that is both quick and accurate for detecting PFAS. Additionally, the total oxidizable precursor assay is utilized in order to evaluate "unknown" PFAS materials that have been utilized in the past as AFFF in Australia (Nahar et al., 2023; NATA Accredited Envirolab Services, 2025). Studies conducted in Australia on various environmental matrices using different analytical techniques to detect PFAS are presented in Table 22.2.

### 22.2.3.1 *Overview of monitoring programs*

The latest version of the PFAS NEMP outlines two main categories of monitoring programs to track the presence and behavior of PFAS in the environment.

#### 22.2.3.1.1 Ambient monitoring programs

Ambient monitoring programs operate at broader geographic scales compared to site-specific programs. Their purpose is to assess the distribution, concentration, and types of PFAS present across various regions, including catchments, urban areas, and entire jurisdictions. These programs also include reference sites located distant from known PFAS sources with minimal human activity, to establish background levels. In addition to characterizing regional PFAS presence, ambient monitoring helps identify previously unrecognized sources, transport pathways, or accumulation zones through preliminary screening data.

***Queensland ambient per- and polyfluoroalkyl substances monitoring program report*** Between May 2019 and March 2020, the Queensland Government implemented the Queensland Ambient PFAS Monitoring Program to evaluate environmental PFAS levels. Water samples were collected bi-monthly from 55 coastal sites, spanning from the Wet Tropics to Southeast Queensland. Findings from the study indicated that PFAS concentrations were generally low or undetectable near conservation

**TABLE 22.2** Case studies reported on per- and polyfluoroalkyl substances (PFAS) in Australia.

| Type of PFAS | Study (year) | Study participants/ location | Sample type | Detection frequency | Concentration reported | Outcomes/inference | References |
|---|---|---|---|---|---|---|---|
| 47 PFAS | 2025 | Australian states and territories, Victoria, Queensland, Northern Territory, Western Australia, South Australia, Australian Capital Territory, and Tasmania | Serum | PFBA, PFOA, PFHxS, PFOS (100%), PFNA (99%), PFDA (73%), PFHpS (43%), 9Cl-F53B (18%), 5:3 FTCA (15%), F53B, 5:3 FTCA, PFUnDA, PFDoDA, PFHxDA, 6:2 FTS (<10%) | PFBA (0.06–0.19 ng $mL^{-1}$), PFOA (0.79–2.8 ng $mL^{-1}$), PFHxS (0.49–3.2 ng $mL^{-1}$), PFOS (1.4 to 16 ng $mL^{-1}$), PFNA (<0.11–0.87 ng $mL^{-1}$), PFDA (<0.09–1.34 ng $mL^{-1}$), PFHpS (<0.06–0.29 ng $mL^{-1}$), 9Cl-F53B (<0.09–1.0 ng $mL^{-1}$), 5:3 FTCA (<0.05–1.82 ng $mL^{-1}$), F53B, 5:3 FTCA, PFUnDA, PFDoDA, PFHxDA, and 6:2 FTS (<0.50 ng $mL^{-1}$). | Based on 8000 individuals, this study found consistent serum concentrations across age, sex, and regions in Australia, which supports Queensland's data. This data can serve as the national reference for background PFAS serum concentrations. | Nilsson et al. (2025) |
| PFHxS, PFOA, PFOS, PFAS. | 2025 | 5 sewage treatment plants (STPs) in the Warragamba catchment rivers in Australia | River/stream water samples | PFHxS (15%), PFOA (0%), PFOS (5%), PFAS (20%) | PFHxS (1.5 ng $L^{-1}$), PFOA (<1 ng $L^{-1}$), PFOS (0.5 V), and PFAS (2 ng $L^{-1}$). | This study is evidence that STPs are a significant source of PFAS contamination in the Warragamba catchment rivers in Australia. | Warwick et al. (2025) |
| | | | STP effluent | PFHxS (50%), PFOA (35%), PFOS (65%), PFAS (65%) | PFHxS (10.5 ng $L^{-1}$), PFOA (4 ng $L^{-1}$), PFOS (16 ng $L^{-1}$), and PFAS (29 ng $L^{-1}$). | | |
| 33 PFAS congeners from eight classes (PFCA, PFSA, FTUCA, FASA, FASAA, FASE, FTS, diPAP). | 2023 | 65 sites comprising constructed urban stormwater treatment wetlands (45) and streams (20) with various land-use types in the Greater Melbourne area, Southeastern Australia | Surface water | PFBA (89%), PFPeA (91%), PFHxA (95%), PFHpA (92%), PFOA (95%), PFNA (77%), PFDA (78%), PFUdA (16%), PFDoA (11%), PFTeDA (2%), PFBS (84%), PFPeS (31%), PFHxS (98%), PFHpS (14%), PFOS (98%), PFOSA (9%), N-MeFOSAA (3%), N-EtFOSAA (2%), 6:2 FTS (17%), 8:2 FTS (3%), total 33 PFAS (98%) | $\Sigma_{33}$ PFAS (63.5 ng $L^{-1}$), PFBA (11.3 ng $L^{-1}$), PFHxA (9.2 ng $L^{-1}$), PFOA (8.3 ng $L^{-1}$), PFOS (8.0 ng $L^{-1}$), PFPeA (7.5 ng $L^{-1}$), PFHpA (3.2 ng $L^{-1}$), and PFHxS (2.9 ng $L^{-1}$). | This study highlights the presence of PFBA even at background levels, emphasizing the need for continuous monitoring of PFAS to better understand their environmental impact and sources. | Paige et al. (2024) |

(*Continued*)

TABLE 22.2 (Continued)

| Type of PFAS | Study (year) | Study participants/ location | Sample type | Detection frequency | Concentration reported | Outcomes/inference | References |
|---|---|---|---|---|---|---|---|
| | | | Sediments | Total 33 PFAS (8.1%), PFOS (3.2%), N-MeFOSAA (3.2%), N-EtFOSAA (3.2%), N-EtFOSE (1.6%) | | | |
| 11 types of PFA, including PFBS, PFHxS, PFHpS, PFOS, 6:2 FTS, PFHxA, PFHpA, PFOA, PFNA, PFDA, PFUnDA. | 2022–24 | People aged 12 years and over who voluntarily provided urine and/or blood samples for the National Health Measures Survey 2022–24, in which biomarkers of chronic disease and nutrition were measured | Urine and/or blood samples | PFOS (98.6%), PFOA (96.1%), PFHxS (88.1%), PFNA (40.8%), PFHpS (19.6%), PFDA (9.7%), PFUnDA (4.0%), PFHpA (1.5%), PFBS (0.2%), PFHxA (0.1%), 6:2 FTS (0.1%) | PFOS (3.62 ng $mL^{-1}$), PFHxS (0.70 ng $mL^{-1}$), and PFOA (0.83 ng $mL^{-1}$) etc. | PFOS, PFHxS, and PFOA were predominantly detected in the population; however, the results revealed that overall PFAS levels remain lower compared to those internationally reported. | Kizhakket-hil et al. (2025) |
| 12 PFAS, including PFPeA, PFHxA, PFHpA, PFOA, PFNA, PFDA, PFBS, PFHxS, PFOS, PFDS, 8:2 FTS, 6:2 FTS | 2022 | Wastewater treatment plants (WWTP) in each Australian state and territory, except for Western Australia | 76 wastewater influent samples | PFPeA (91%), PFHxA (100%), PFHpA (96%), PFOA (99%), PFNA (14%), PFDA (35%), PFBS (99%), PFHxS (91%), PFOS (92%), PFDS (95%), 8:2 FTS (35%), 6:2 FTS (43%), total PFCAs (100%), total PFSAs (100%), total PFASs (100%) | PFPeA (5.7 ng $L^{-1}$), PFHxA (11 ng $L^{-1}$), PFHpA (3.8 ng $L^{-1}$), PFOA (7.4 ng $L^{-1}$), PFNA (2.1 ng $L^{-1}$), PFDA (3.4 ng $L^{-1}$), PFBS (8.4 ng $L^{-1}$), PFHxS (16 ng $L^{-1}$), PFOS (16 ng $L^{-1}$), PFDS (6.2 ng $L^{-1}$), 8:2 FTS (15 ng $L^{-1}$), 6:2 FTS (34 ng $L^{-1}$), total PFCAs (30 ng $L^{-1}$), total PFSAs (43 ng $L^{-1}$), and total PFASs (93 ng $L^{-1}$) | Approximately 1 kg of PFAS (12) enters the Australian WWTP in a day, of which 59% and 41% are background and specific point sources, including landfills and industry. These highlight their significant contribution to PFAS contamination in WWTPs. | Nguyen et al. (2022) |

| | | | | | | | |
|---|---|---|---|---|---|---|---|
| 21 PFAS including PFBA, PFPeA, PFHxA, PFHpA, PFOA, PFNA, PFDA, PFUdA, PFDoA, PFTrA, PFTeA, PFBS, PFPeS, PFHxS, PFHpS, PFOS, PFDS, 6:2 FTS, 8:2 FTS, 6:2 Cl-PFESA, 8:2 Cl-PFESA. | 2019 | From Australian WWTP | Aqueous samples (influent, primary effluent, secondary effluent, final effluent, and recycled water) collected from WWTPs | PFBA (100%), PFPeA (96%), PFHxA (100%), PFHpA (100%), PFOA (100%), PFNA (97%), PFDA (98%), PFUdA (12%), PFDoA (49%), PFTrA (19%), PFTeA (25%), PFBS (98%), PFPeS (77%), PFHxS (95%), PFHpS (76%), PFOS (99%), PFDS (23%), 6:2 FTS (99%), 8:2 FTS (82%), 6:2 Cl-PFESA (4%), 8:2 Cl-PFESA (0%) | PFBA (13 ng L$^{-1}$), PFPeA (8.3 ng L$^{-1}$), PFHxA (21 ng L$^{-1}$), PFHpA (6.1 ng L$^{-1}$), PFOA (19 ng L$^{-1}$), PFNA (0.92 ng L$^{-1}$), PFDA (2.3 ng L$^{-1}$), PFUdA (<0.6 ng L$^{-1}$), PFDoA (0.47 ng L$^{-1}$), PFTrA (<0.6 ng L$^{-1}$), PFTeA (0.27 ng L$^{-1}$), PFBS (4 ng L$^{-1}$), PFPeS (1.9 ng L$^{-1}$), PFHxS (13 ng L$^{-1}$), PFHpS (0.86 ng L$^{-1}$), PFOS (15 ng L$^{-1}$), PFDS (0.21 ng L$^{-1}$), 6:2 FTS (7.3 ng L$^{-1}$), 8:2 FTS (0.53 ng L$^{-1}$), 6:2 Cl-PFESA (<0.6 ng L$^{-1}$), and 8:2 Cl-PFESA (<0.6 ng L$^{-1}$). | Many PFAS, such as FPeA to PFDA and 6:2 FTS, were found to be increasing through the WWTPs in Australia, where 6:2 FTS was the primary contributor. These highlighted the need for an investigation into the origin and environmental fate of PFAS. | Coggan et al. (2019) |
| | | | Solid samples (primary sludge, secondary sludge, lagoon sludge, and one lagoon sludge dredge pile) | PFBA (29%), PFPeA (20%), PFHxA (82%), PFHpA (54%), PFOA (84%), PFNA (50%), PFDA (84%), PFUdA (54%), PFDoA (94%), PFTrA (70%), PFTeA (90%), PFBS (44%), PFPeS (14%), PFHxS (50%), PFHpS (26%), PFOS (94%), PFDS (42%), 6:2 FTS (26%), 8:2 FTS (42%), 6:2 Cl-PFESA (16%), 8:2 Cl-PFESA (8%) | PFBA (0.45 ng g$^{-1}$), PFPeA (<0.6 ng g$^{-1}$), PFHxA (1.9 ng g$^{-1}$), PFHpA (0.30 ng g$^{-1}$), PFOA (2.6 ng g$^{-1}$), PFNA (0.20 ng g$^{-1}$), PFDA (5.1 ng g$^{-1}$), PFUdA (<0.6 ng g$^{-1}$), PFDoA (3.8 ng g$^{-1}$), PFTrA (0.32 ng g$^{-1}$), PFTeA (0.69 ng g$^{-1}$), PFBS (0.83 ng g$^{-1}$), PFPeS (<0.6 ng g$^{-1}$), PFHxS (1.1 ng g$^{-1}$), PFHpS (0.29 ng g$^{-1}$), PFOS (14 ng g$^{-1}$), PFDS (0.78 ng g$^{-1}$), 6:2 FTS (0.26 ng g$^{-1}$), 8:2 FTS (<0.73 ng g$^{-1}$), 6:2 Cl-PFESA (<0.6 ng g$^{-1}$), and 8:2 Cl-PFESA (<0.6 ng g$^{-1}$). | | |
| 16 PFAS including PFOA, PFNA, PFDA, PFUdA, PFHxS, PFOS, PFOSA, PFDoA, PFTrDA, PFTeDA, PFPeS, PFHpS, | 2017–20 | Australian pinnipeds from multiple colonies | Liver samples | - | **Australian fur seal**: PFOA (0.32–52.2 ng g$^{-1}$), PFNA (0.91–44.2 ng g$^{-1}$), PFDA (0.50–38.0 ng g$^{-1}$), PFUdA (0.50–11.0 ng g$^{-1}$), PFHxS (0.52–37.4 ng g$^{-1}$), PFOS (10.52–2119 ng g$^{-1}$), PFOSA (0.50–6.43 ng g$^{-1}$), PFDoA (0.50–13.2 ng g$^{-1}$), PFTrDA (1.00–3.95 ng g$^{-1}$), PFTeDA | This study revealed maternal transfer and species-specific accumulation of PFAS in pinnipeds, emphasizing the need for health impact assessments and urgent monitoring to | Taylor et al. (2021) |

(*Continued*)

TABLE 22.2 (Continued)

| Type of PFAS | Study (year) | Study participants/ location | Sample type | Detection frequency | Concentration reported | Outcomes/inference | References |
|---|---|---|---|---|---|---|---|
| PFNS, PFDS, PFBS. | | | | | (1.00–3.28 ng $g^{-1}$), PFPeS (0.50–0.78 ng $g^{-1}$), PFHpS (0.50–70.2 ng $g^{-1}$), PFNS (1.00–8.52 ng $g^{-1}$), PFDS (1.00–23.1 ng $g^{-1}$), and PFBS (0.50–1.12 ng $g^{-1}$); **Australian sea lion**: PFOA (0.32–11.2 ng $g^{-1}$), PFNA (0.61–8.22 ng $g^{-1}$), PFDA (0.50–0.90 ng $g^{-1}$), PFUdA (0.50–1.05 ng $g^{-1}$), PFHxS (0.50–3.87 ng $g^{-1}$), PFOS (1.00–16.9 ng $g^{-1}$), PFOSA (0.50–0.66 ng $g^{-1}$), PFDoA (<0.50 ng $g^{-1}$), PFTrDA (<1.00 ng $g^{-1}$), PFTeDA (<1.00 ng $g^{-1}$), PFPeS (<0.50 ng g), PFHpS (<0.50 ng $g^{-1}$), PFNS (<1.00 ng g), PFDS (<1.00 ng $g^{-1}$), and PFBS (<0.50 ng $g^{-1}$). | protect marine species and ecosystems. | |
| 20 PFAS, including PFBA, PFPeA, PFHxA, PFHpA, PFOA, PFNA, PFDA, PFBS, PFPeS, PFHxS, PFOS, PFDS, 6:2 FTS, etc. | 2017 – 18 | 20 monitoring wells, all within the irrigation district in Melbourne, Australia, across two sampling campaigns in August 2017 and May 2018 | Groundwater samples | **In Campaign 1:** PFBA (56%), PFPeA (67%), PFHxA (78%), PFHpA (44%), PFOA (67%), PFNA (22%), PFDA (22%), PFBS (78%), PFPeS (33%), PFHxS (33%), PFOS (100%), PFDS (11%), 6:2 FTS (33%); **in Campaign 2:** PFBA (79%), PFPeA (43%), PFHxA (64%), PFHpA (46%), PFOA (82%), PFNA (50%), PFDA (57%), PFBS (86%), PFPeS (64%), PFHxS (68%), PFOS (96%), PFDS (11%), 6:2 FTS (57%) | **In Campaign 1**: PFBA (4.9 ng $L^{-1}$), PFPeA (2.1 ng $L^{-1}$), PFHxA (3 ng $L^{-1}$), PFHpA (1.3 ng $L^{-1}$), PFOA (2.4 ng $L^{-1}$), PFNA (<0.26 ng $L^{-1}$), PFDA (<0.25 ng $L^{-1}$), PFBS (4.8 ng $L^{-1}$), PFPeS (0.8 ng $L^{-1}$), PFHxS (3.7 ng $L^{-1}$), PFOS (16 ng $L^{-1}$), PFDS (<0.22 ng $L^{-1}$), and 6:2 FTS (2.2 ng $L^{-1}$); **in Campaign 2:** PFBA (6.1 ng $L^{-1}$), PFPeA (1 ng $L^{-1}$), PFHxA (3.7 ng $L^{-1}$), PFHpA (0.8 ng $L^{-1}$), PFOA (2.2 ng $L^{-1}$), PFNA (0.4 ng $L^{-1}$), PFDA (0.31 ng $L^{-1}$), PFBS (4.4 ng $L^{-1}$), PFPeS (1 ng $L^{-1}$), PFHxS (5.9 ng $L^{-1}$), PFOS (11 ng $L^{-1}$), PFDS (<0.22 ng $L^{-1}$), and 6:2 FTS (1.3 ng $L^{-1}$). | This study is evidence that recycled water use can lead to PFAS contamination in groundwater, because it has higher concentrations than the background levels, which are similar to those found in wastewater effluent. | Szabo et al. (2018) |

| | | | | | | | |
|---|---|---|---|---|---|---|---|
| PFPeA, PFHxA, PFHpA, PFOA, PFNA, PFDA, PFUnA, PFDoA, PFTrDA, PFTeDA. | 2016–24 | 75 WWTPs across Australia during the 2016 census period | Effluent of the wastewater treatment plants (WWTPs) | PFPeA (100%), PFHxA (100%), PFHpA (100%), PFOA (100%), PFNA (94%), PFDA (91%), PFBS (98.5%), PFHxS (100%), PFOS (99.3%), PFDS (9%), 6:2 FTS (71.6%), 8:2 FTS (8.2%) | PFPeA (14 $ngL^{-1}$), PFHxA (22 $ng\,L^{-1}$), PFHpA (4.7 $ng\,L^{-1}$), PFOA (23 $ng\,L^{-1}$), PFNA (1.7 $ng\,L^{-1}$), PFDA (4.5 $ng\,L^{-1}$), PFBS (6.9 $ng\,L^{-1}$), PFHxS (14 $ng\,L^{-1}$), PFOS (12 $ng\,L^{-1}$), PFDS (1.7 $ng\,L^{-1}$), 6:2 FTS (16 $ng\,L^{-1}$), and 8:2 FTS (2.2 $ng\,L^{-1}$). | Approximately 250 kg of total PFAS are released through the effluent and biosolids, with PFCA increasing to 90% after treatment. It indicates that PFCAs are not effectively removed. | Nguyen et al. (2024) |
| | | | Biosolids | PFPeA (58%), PFHxA (93%), PFHpA (58%), PFOA (96%), PFNA (83%), PFDA (96%), PFUnA (87%), PFDoA (96%), PFTrDA (51%), PFTeDA (85%) | PFPeA (1.2 $ng\,g^{-1}$), PFHxA (1.9 $ng\,g^{-1}$), PFHpA (0.61 $ng\,g^{-1}$), PFOA (7.5 $ng\,g^{-1}$), PFNA (1.3 $ng\,g^{-1}$), PFDA (25 $ng\,g^{-1}$), PFUnA (1.6 $ng\,g^{-1}$), PFDoA (9.4 $ng\,g^{-1}$), PFTrDA (0.7 $ng\,g^{-1}$), and PFTeDA (3.7 $ng\,g^{-1}$). | | |

zones, agricultural lands, and forested/grazing regions. In contrast, the highest concentrations were recorded at urban and industrial sites in Southeast Queensland, aligning with trends observed in other Australian and international regions. Of the 55 locations sampled, 8 had no detectable PFAS, and 21 exhibited very low PFOS concentrations. Biota sampling, which included aquatic organisms such as fish, mangrove crabs, and worms, also confirmed the presence of PFAS in many samples from Southeast Queensland. These results align with those observed in other urban and industrial regions across the country.

*Victoria ambient per- and polyfluoroalkyl substances monitoring program* EPA Victoria conducted a comprehensive PFAS assessment at 87 freshwater locations, including rivers, lakes, and streams, categorized by four land-use types: remote, agricultural, mixed, and urban. The study found PFAS in 91% of water samples, 34% of sediment samples, and 28% of riparian soil samples. Among the remote sites, only one water sample contained PFOS at a low concentration. In agricultural and mixed-use areas, short-chain perfluoroalkyl carboxylic acids (PFCAs) were commonly found, with perfluorobutanoic acid (PFBA) showing the highest median concentration of approximately 0.01 $\mu g\ L^{-1}$. This compound significantly contributed to total PFAS and PFCA levels in those premises. PFAS profiles were more diverse in urban locations. PFBA, perfluoropentanoic acid, undecafluorohexanoic acid (PFHxA), PFOA, PFHxS, and PFOS had median concentrations on the order of 0.010 $g\ L^{-1}$, with each contributing an order of 10%–20% to the total PFAS load. These results indicate the influence of land use on the distribution of PFAS and underline the necessity of specific monitoring in urban environments (Sardiña et al., 2024; PFAS National Environmental Management Plan Version 3.0, 2025).

*Site-specific monitoring programs* The site-specific PFAS monitoring program is designed to assess the nature, quantity, and location of PFAS associated with a known source at a specific location. The programs are designed to assess the on-site and off-site effects, including risks to the environment and potential dangers to human life, as well as the extent of pollution that will be released into the nearby vicinity The National Environment Protection (Assessment of Site Contamination) Measure (ASC measure) in Australia establishes a uniform framework for conducting such measurements. On the same note, in New Zealand, it is facilitated by the contaminated land management guidelines (guidelines 1, 2, and 5). The creation of a powerful conceptual site model based on the physical nature of the site, land use, and possible or known PFAS sources is a major element of site-specific monitoring. The sampling techniques and risk opinions are guided by this model, hence ensuring effective evaluation and proper planning of remediation (PFAS National Environmental Management Plan Version 3.0, 2025).

#### 22.2.3.1.2 Human biomonitoring program

*Australian National University per- and polyfluoroalkyl substances health study* The study, conducted by the Australian National University's National Centre for Epidemiology and Population Health, was commissioned by the Australian government to examine the potential health consequences of exposure to PFAS in three localities: Williamtown (NSW), Katherine (NT), and Oakey (QLD). This was referred to as the PFAS health study and involved the assessment of blood serum concentrations of PFAS and a comprehensive list of biochemical health markers, completion of self-reporting health and exposure history surveys, and linking of information to previous and follow-up studies, with respect to perinatal outcomes, development, and mortality in childhood (Queensland Government, Department of Environment, 2019; Lazarevic et al., 2021).

A total of 2587 and 702 blood samples in the PFAS-affected communities and the control ones were analyzed. The findings in this regard depicted a remarkable elevation in the level of PFOS and PFHxS within the exposed groups. Long-term habitation in the affected areas, drinking bore water, and working with firefighting foam were among the most significant risk factors. PFAS were associated with high levels of cholesterol and uric acid (Todd et al., 2025).

***Human biomonitoring study*** To decipher the extent to which the populace in the nation is exposed to chemical compounds, the Australian Academy of Health and Medical Sciences has advised the federal government to initiate a countrywide HBM program that will carry out a systematic examination of the chemical pollutants built up in the population, especially PFAS. The proposed program would be instrumental in tracking the long-term exposure to harmful substances and the health consequences of their exposure, particularly among vulnerable groups like pregnant women (Academy Calls for a National Human Biomonitoring Program to Track the Health Impacts of Forever Chemicals - AAHMS, 2025).

Australia has been conducting HBM efforts since 2002, generating valuable data on serum concentrations of PFAS in the general public. A recent 2024 study further expanded this effort by evaluating regional differences in PFAS levels across all states and territories. The findings underscore the need for a coordinated national approach to better understand and manage the public health impacts of PFAS exposure (Nilsson et al., 2025).

***Queensland Firefighting Foam Survey (2017)*** As part of the Queensland Operational Policy on Environmental Management of Firefighting Foam, a voluntary survey was completed in 2017 to collect information on the foam stockpiles, practices used to contain foams, and usage and compliance. The questionnaire covered high-risk industries, namely mining activities, chemical industries, and fuel storage facilities. Of the 468 responses received, some 425,000 kg was reported to be firefighting foam, which is mostly stored at chemical and fuel facilities. In a disturbing development, 98% of such sites had foam products containing PFAS, and only 5% of the sites met the policy's requirements in total.

***EPA Victoria PFAS Pilot Study (2017)*** In 2017, EPA Victoria conducted a pilot study to assess the levels of PFAS in various environmental media, including soil, water, marine organisms, wastewater treatment facilities, and landfills. PFASs were identified in specific samples; however, they were not identified in all test sites. The report also suggested continued environmental monitoring to track

PFAS trends with time, based on these results (PFAS National Environmental Management Plan Version 3.0, 2025).

***Australian Total Diet Study*** The most well-rounded program that tracks chemicals, nutrients, and other contaminants in foods that Australians are likely to consume is the Australian Total Diet Study (ATDS). Its major aim is to determine whether the concentrations of these substances are detrimental to the health of the populace and to maintain food safety nationwide (Food Standards Australia New Zealand, 2025b).

The ATDS (27th edition) specifically measured the occurrence of PFAS in the Australian food supply. It was found that the concentrations of PFAS in the sampled foods were very low, indicating that the general population was not highly exposed to them through their diet. On the basis of these findings, FSANZ declared that no immediate food safety issues are present, and no additional regulatory measures, including maximum residue levels in PFAS, are required at this time, according to the Australia New Zealand Food Standards Code (FSANZ 27th Australian Total Diet Study, 2025).

## 22.2.4 Remediation efforts of per- and polyfluoroalkyl substances

### *22.2.4.1 National per- and polyfluoroalkyl substances management program by Airservices Australia*

The Australian government has allocated $130.5 million to the National PFAS Management Program, overseen by Airservices Australia, to investigate and manage PFAS contamination at civilian airports where firefighting foams containing PFAS were historically used. This program focuses on airports where the Commonwealth previously provided firefighting services.

The initiative includes comprehensive site assessments, along with independent audits, to determine the extent and nature of PFAS contamination. Based on these assessments, the program develops and implements robust site-specific management and mitigation strategies to address any identified environmental and health risks. Key components of the program include whole-of-site investigations, regular environmental monitoring, and management interventions.

This proactive strategy ensures that PFAS-related risks at affected airports are appropriately managed in alignment with national environmental protection goals (PFAS—Airports Investigation Program, 2025; National PFAS Management Program - Airservices, 2025).

### *22.2.4.2 National per- and polyfluoroalkyl substances investigation and management program Department of Defence*

The Australian DOD has initiated a nationwide program to assess, manage, and reduce the impacts of PFAS contamination across several of its facilities throughout the country. The objective of this effort is to identify scenes that have resulted in environmental interference due to historical applications of PFAS-based firefighting foams. With major scientific experts, industry partners, and every layer of government, Defence conducts a thorough analysis to survey the extent of PFAS dispersion in the environment.

Such observations inform remediation and risk management efforts of a specific nature. These are some of the major operations: the Installation of 11 water treatment plants, more than 11.5 billion litres of contaminated water treated, and over 160,000 tonnes of PFAS-affected soil removed or treated. The program focuses on transparency and validated test results, and site reports are published online frequently. This will make sure that the communities and other stakeholders are aware of the disasters being undertaken and the results (PFAS Investigation and Management Program, 2025; PFAS Site Investigations, 2025).

#### *22.2.4.3 SourceZone® Technology—Soil Treatment Plant 2018 (Ventia)*

The Australian DOD has collaborated with Ventia at RAAF Base Edinburgh, in South Australia, to complete the world-first full-scale trial of SourceZone® technology to remediate soil contaminated with PFAS. This is a developed, sustainable method that has consistently performed excellently in eliminating PFAS contamination. Among the limited number of globally available soil-treatment tools to address PFAS, SourceZone® is a technology that has been successfully used to recover a considerable amount of PFAS from contaminated soils. The volume of the extracted one during this trial is approximately equal to the pollution caused during 25 years of working in the area of firefighting. The project marks a milestone in the active cooperation between the Defense and industry to find viable solutions to the worldwide PFAS problem. The fact that the project accounts for a significant portion of the reduction in PFAS mass in the soil can help mitigate the risk of environmental pollution and, ultimately, reduce the impacts of possible environmental contamination on workers and nearby communities. Such a project not only indicates the efficacy of the new remediation technology but also provides a precedent for further work on PFAS control in Australia or even globally (Ventia. PFAS Remediation, 2025; Ventia. PFAS Remediation at RAAF Base Edinburgh, 2025).

#### *22.2.4.4 Australian Research Council per- and polyfluoroalkyl substances remediation research program*

Through its special research initiative on PFAS Remediation, the Australian Research Council has been funding scientific projects that have taken sophisticated steps towards addressing PFAS contamination. This program, since 2017, has been actively funded by the Australian government to promote research on more advanced technologies and treatment techniques for removing PFAS contaminants from contaminated areas and products. The program promotes networked, high-impact research that is collaborative between leading academic institutions, industry specialists, and government entities. Its main aim is to develop realistic and scaled-up solutions that can be applied in real-life environmental cleanup projects within Australia. The program is nationwide, which demonstrates the government's stance on science-based practices of managing POPs such as PFAS, and adds to the worldwide knowledge base in this area (Class Action Settled but PFAS Issues Far from Resolved - Corrs Chambers Westgarth, 2025; Research. Australian Government PFAS Taskforce, 2025).

### 22.2.5 Legal and policy challenges

PFAS regulation in Australia has experienced big legal and administrative barriers. The major problem has been the ambiguity in the roles of the agency and councils; in most cases, the two overlapped, or there have been situations where no authority took the responsibility of regulation. Such confusion has kept any response and enforcement ineffective (Class Action Settled but PFAS Issues Far from Resolved - Corrs Chambers Westgarth, 2025).

Firefighting training sites and Defense facilities have been at the center of PFAS contamination due to the prolonged use of PFAS-based firefighting foams. The issue has sparked a wave of legal action. In May 2023, the Australian Federal Government agreed to a $132.7 million class action settlement covering over 30,000 affected individuals, acknowledging harm caused by PFAS contamination around Defense properties. Overall, the DOD has spent over $366 million to resolve PFAS-related claims and began active remediation efforts in 2021 (Chemicals, Forever: How Do You Fix a Problem like PFAS?, 2025; PFAS Litigation Risks Mount as EU Claims Rise - Kennedys, 2025). In a legal precedent set in December 2024, the Supreme Court of Victoria heard the first consumer-led class action against 3M Australia Pty Ltd, presented by consumers in Australia. The lawsuit claims that 3M did not provide sufficient warnings to consumers about the dangers of its products, which may contain PFAS, thereby causing environmental pollution and health concerns among the public. It further charges the company with negligence, misleading action, and nuisance to the people. The case has become a landmark case involving PFAS litigation in Australia, extending the litigation beyond governmental liability into corporate responsibility (PFAS Litigation: A Deep Dive into the Rise of Consumer-Led Claims and Its Implications - Wotton Kearney, 2025).

### 22.2.6 Comparison with international regulations

Although Australia has formulated national and state-level frameworks for the management of PFAS, a significant portion of its policies is based on international principles established by international bodies. These are the Organisation for Economic Co-operation and Development (in monitoring the environment and risk assessment), the EU REACH regulation (for control of industrial chemicals), the US EPA, and the World Health Organization in establishing guideline values of the drinking water safety. The Australian strategy is primarily guidance-focused and will provide only recommendations, rather than enforcement. Conversely, regulatory bodies such as the US EPA and the EU REACH system have implemented more stringent, legally enforceable regulations, particularly regarding restrictions on products and acceptable levels of PFAS in the environment. Moreover, the current Australian limit on both water and consumer products of PFAS is usually relaxed compared to the rules applied by other countries. The Australian system of regulations permits PFAS in most applications, unlike the European Union and the United States, which are proceeding towards outright bans and precautionary regulations, with few restrictions on consumer products.

### 22.2.7 Future directions and recommendations

The PFAS NEMP identifies six priority areas for regulating and controlling PFAS pollution in Australia. The following directions are intended to enhance data quality, regulatory clarity, and environmental management practices:

**PFAS chemical family:** There are also attempts to refine the analysis methods (e.g., the TOP assay), expand the analysis beyond the limits of PFOS, PFOA, and PFHxS, and assess the necessity of additional guidance on the precursors of perfluoroalkyl acids (PFAA).

**Environmental data and monitoring:** A national system is proposed to share data, collate available monitoring data, and develop a digital platform (currently underway) to provide readily available PFAS information.

**Water:** It is advisable to incorporate bioaccumulation into risk assessment to enhance guidance on PFAS handling in wastewater, biosolids, and groundwater, and fund research to develop fine-tuned ecological guideline values.

**Soil:** Future efforts will involve setting direct and indirect ecological values as guidelines, understanding the behavior of PFAA in soil chemistry, and establishing standards for managing PFAS in soil and reuse activities.

**Resource recovery and waste management:** This includes developing guidance for PFAS in both organic and nonorganic waste streams, particularly in less common waste types such as construction materials.

**Site-specific application:** A focus on refining protocols for PFAS containment, site investigation, sampling, and remediation. Future initiatives also aim to establish sediment quality criteria for estuarine and marine ecosystems (PFAS National Environmental Management Plan Version 3.0, 2025).

**Upcoming ban on PFAS under Industrial Chemicals Environmental Management Standard (IChEMS):** In an attempt to reduce the level of environmental hazard, Australia has presented the IChEMS, which is a nationally consistent system of governance over the use, aftermath, and utilization of industrial chemicals, and their admission into the country. IChEMS is a joint development of the Commonwealth, states, and territories, which offers national coordination of chemical management. With this system, more than 500 PFAS-related chemical substances, including PFOS, PFOA, PFHxS, and related compounds, have been subject to regulation. As a piece of Australia that aims to reduce harmful chemical exposure and enhance environmental safety, the import, use, and production of some of the PFAS chemicals will be banned starting July 1, 2025, under IChEMS (Overview of Risk Reduction Approaches for PFASs, 2025).

## 22.3 Europe: per- and polyfluoroalkyl substances regulatory guidelines

### 22.3.1 Overview of environmental issues in the European Union

The European Union is grappling with growing environmental issues stemming from industrial growth, intensive farming, urbanization, and the unsustainable exploitation of natural resources. Among the most pressing issues is the widespread pollution of air, water, land, and groundwater by long-term and harmful chemicals. Among them, PFAS, also known as "forever chemicals," are particularly problematic due to their high environmental stability and ability to bioaccumulate in living organisms, also referred to as biota. According to the latest estimates, it has been established that more than 17,000 sites in the

European Union are contaminated with PFAS, and an additional 21,000 are considered probable hotspots. Decades of industrialization, military use, and the application of PFAS-based firefighting foams, as well as simple consumer products like nonstick cookware and waterproof fabrics, have contaminated these areas (Exposed: 17,000 European Sites Contaminated by Forever Chemicals, 2025; Revealed: The Massive Contamination of Europe by PFAS Forever Chemicals, 2025). These pollutants are infiltrating the human population. The median blood levels of PFOS, PFOA, PFHxS, and perfluorononanoic acid (PFNA) were 2.13, 0.97, and 0.41 $\mu g\ L^{-1}$, respectively, in a European-wide large-scale biomonitoring study of teenagers. Worryingly, in certain areas, adolescents' levels exceeded the recommended values by up to 24% above the health-based guidance established by the European Food Safety Authority (EFSA) (Richterová et al., 2023; How Pesticides Impact Human Health and Ecosystems in Europe — European Environment Agency, 2025). In addition to PFAS, other dangerous pollutants are present in Europe. Lead, cadmium, copper, and zinc are examples of heavy metals that remain in soil due to historical emissions from industrial sources and improper waste disposal. Approximately one-tenth of the agricultural land in the European Union is assumed to exceed at least one regulatory limit for heavy metal contamination (Environmental Statement 2023 and European Environment Agency's Home Page, 2025). Pesticide usage is also a major contributor to environmental and health hazards. In the European Union, approximately 350,000 tonnes of pesticides are sold annually, with residues found in more than 22% of supervised surface waters. Recent studies have demonstrated that 84% of Europeans have been found to have detectable amounts of pesticide chemicals in their bodies (Environmental Statement 2023 and European Environment Agency's Home Page, 2025). Microplastics are another significant emerging pollutant. Pollution from sources such as tire wear, synthetic textiles, paints, and agricultural products emits between 0.7 and 1.8 million tonnes per year into the atmosphere. EU farmland alone can be supplied with up to 42,000 tonnes of microplastics each year, and related endocrine-disrupting chemicals such as phthalates and brominated flame retardants (Microplastic Releases to the Environment Indicator, 2025).

To address such complex issues, the European Union has introduced a range of innovative environmental measures. The European Green Deal, the Zero Pollution Action Plan, and the EU strategy on climate adaptation all aim to reduce pollution, limit hazardous chemicals, and make the environment more resilient in the long run. Policy advancements encompass the revision of the Drinking Water Directive, which currently establishes binding maximums for PFAS, and an unprecedented proposal to eliminate over 10,000 PFAS substances (A Ban on Forever Chemicals Would Be Best, 2025). Nevertheless, the issue of chemical contamination is quite complex and in-depth. The chapter reviews EU environmental governance initiatives with respect to PFAS and other chemical contaminants. It critically evaluates existing policies and identifies gaps and opportunities that could be used to enhance the EU's ability to safeguard the health and ecological systems of future generations.

#### *22.3.1.1 Overview of environmental and chemical regulations in the European Union*

The European union has established a robust legal framework to address environmental and chemical pollution, encompassing issues related to air, water, and soil pollution.

The Water Framework Directive aims to achieve good ecological and chemical conditions in all water bodies by promoting the sustainable use of water and reducing pollution (Water Framework Directive - European Commission, 2025). In conformity with this, the Urban Wastewater Treatment Directive provides for the quality treatment of municipal wastewater, and the Drinking Water Directive provides strict quality standards for the water that can be used by humans (Water - European Commission).

To control air quality, the European Union implements laws such as the National Emission Ceilings Directive, which sets limits on significant air pollutants (European Environment Agency's, 2025). The proposed Soil Monitoring Law will restore and defend soil health by combating degradation problems, such as soil erosion and the disappearance of organic matter (Soil Health - European Commission, 2025). Zero Pollution Action Plan envisions the minimization of pollutant levels to the level of harmlessness in soil, water, and air by 2050 (Zero Pollution Action Plan - European Commission, 2025).

The REACH regulation is the staple of chemical management in the European Union. It is tailored with the aim of safeguarding human health and the environment against chemical hazards and includes chemicals that are present in industrial operations and consumer products, such as paints, detergents, and furniture, among others (ECHA, 2025). REACH also promotes the replacement of hazardous materials, innovation, and alternative testing methods to minimize the excessive use of animal tests (REACH Regulation - European Commission, 2025).

The primary regulatory body responsible for implementing REACH is the ECHA. It is the source of scientific and technical advice to the European Commission and the member states of the European Union, ensuring that chemical safety regulations are effectively enforced throughout the Union (ECHA, 2025).

### 22.3.2 Regulatory framework for per- and polyfluoroalkyl substances

The European Union has enacted a variety of legislative tools to affect PFAS, partly due to the existing international agreements like the Stockholm Convention on POPs that the European Union is part of, having 186 signatories and 186 countries of signatories. The idea of the global convention is to protect human health and the environment from persistent and dangerous chemicals. This convention prohibited the use of PFOS and its derivatives since 2009. The European Union also implements this through the POPs regulation, which reflects its strict stance on the elimination of long-term environmental pollutants (Lallas, 2001). At the regional level, the chemicals regulated by the European Union are primarily those covered by the REACH regulation (introduced in 2007), which requires companies to evaluate and manage the risks presented by chemicals they produce or import. Materials must be registered with the ECHA, and data on their hazards and safe use must be provided. If certain substances are unacceptable, they can be limited or prohibited. REACH has also played a key role in the elimination of hazardous chemicals and the promotion of less harmful alternatives (REACH Regulation - European Commission, 2025; Per- and Polyfluoroalkyl Substances PFAS - ECHA, 2025).

Within the context of REACH, chemicals that pose severe health or environmental risks, which may manifest over a prolonged period, are designated as substances of very great concern. They are also included in the list of candidates maintained by ECHA. Between 2019 and 2023, three major groups of PFAS were added to the SVHC list: 2,3,3,3-tetrafluoro-2-(heptafluoropropoxy)propionic acid (HFPO-DA), including salts and acyl halides; PFBS and its salts; perfluoroheptanoic acid (PFHpA) and its salts. These PFAs have been classified as persistent, mobile, and toxic, posing a hazard due to their long-term environmental impact. Their properties fall within the requirements of substances being persistent, bioaccumulative, and toxic, or very persistent and very bioaccumulative. As a result of these properties, they are classified as posing threats similar to those of carcinogens, mutagens, and reproductive toxicants.

Since February 2023, the European Commission has imposed an EU-wide ban on PFCAs with carbon chain lengths of $C_9$–$C_{14}$, their salts, and compounds of this type. The other family of restrictions is that of PFHxA and other associated substances, and it is set to come into effect in April 2026 throughout the European Union/European Environment Agency (EEA). The source of this regulatory measure is a 2019 proposal by German authorities, which was supported by the scientific committees of the ECHA in 2021. At the same time, Germany, Denmark, the Netherlands, Norway, and Sweden submitted a proposal to further restrict PFAS through a wide range of applications. This initiative is supported by the policy discussions at the EU Environment Council meeting of December 2019, which is now under consideration by the scientific committee of ECHA. In January 2022, ECHA separately suggested restrictions specifically on PFAS in firefighting foams, and its scientific committees gave this proposal unanimous support by June 2023. A final decision will be made by the European Commission together with the other European Union member states. Due to their hazardous properties, several PFAS, including PFOA, ammonium pentadecafluorooctanoate, PFNA, perfluorodecanoic acid, and PFHpA, have been assigned to harmonized classifications and labeling requirements as stipulated in the classification, labelling and packaging regulation.

In January 2023, the ECHA received the Universal PFAS Restriction Proposal, which was jointly prepared by the authorities of Denmark, Germany, the Netherlands, Norway, and Sweden. The proposed regulations constitute a comprehensive regulatory initiative aimed at reducing the emissions of over 10,000 PFAS into the environment, thereby making products and industrial processes safer. The proposal is a pillar of the EU Chemicals Strategy for Sustainability and supports the objectives outlined in the Zero Pollution Action Plan (Per- and Polyfluoroalkyl Substances PFAS - ECHA, 2025; All News - ECHA, 2025a). In addition, supporting these objectives, the regulation (EU) 2024/573 on fluorinated greenhouse gases contains provisions addressing certain PFAS, in line with the objectives of the European Green Deal, particularly those related to climate neutrality and environmental protection (Regulation - EU - 2024/573 - EN - EUR-Lex, 2025). To ensure the safety of the population, the EFSA is a crucial body that evaluates the health risks associated with PFAS and determines the maximum levels of intake (Per- and Polyfluoroalkyl Substances PFAS, 2025). At EFSA, PFAS persistence and toxicity are collaboratively reviewed with ECHA to aid in the formulation of substance-specific rules. In the future, further PFAS will be discussed by the Community Rolling Action Plan. These tests aim to identify preliminary concerns regarding potential risks to human health and the environment, particularly in terms of production volume and exposure risks (Per- and Polyfluoroalkyl Substances PFAS - ECHA, 2025).

## 22.3.3 Limits on per- and polyfluoroalkyl substances in the European Union

### *22.3.3.1 Permissible levels of per- and polyfluoroalkyl substances in drinking water, soil, air, and food*

PFAS are introduced into the environment by industrial effluents, firefighting foams, and consumer products, polluting soil, water, and food. They pose a significant concern to the environment and public health due to their persistence and potential for bioaccumulation. To counter this, safety limits have been established by various regulatory bodies across different forms of media. In January 2021, the EU Drinking Water Directive was revised to establish a common limit of $0.5\,\mu g\,L^{-1}$ for all PFAS, utilizing a grouping approach to more effectively manage the overall effects of these substances (EFSA. Science, Safe Food, Sustainability, 2025). According to the EFSA, the main sources of PFAS dietary exposure are fish, meat, eggs, and fruits. PFOS and PFOA, among other Long chain PFAS are particularly likely to bioaccumulate. In September 2020, EFSA proposed a group tolerable weekly intake (TWI) of $4.4\,ng\,kg^{-1}$ of body weight of four key LCPs, specifically, PFOA, PFOS, PFNA, and PFHxS, to manage risk in food safety. The action will help avoid approximately 13,200 tonnes of PFAS emissions over 30 years, particularly in the oil and gas industries and seveso-regulated locations (All News - ECHA, 2025b).

### *22.3.3.2 Government agencies involved in per- and polyfluoroalkyl substances regulation*

The PFAS regulation is a multigovernmental initiative in the European Union, with multiple governmental institutions having different tasks to mitigate environmental and human health risks. The European Commission is at the center of strategic direction, presenting the EU Chemicals Strategy for Sustainability, which proposes eliminating the use of PFAS, except in instances where the substances are vital and have no alternatives. The EEA monitors the levels of environmental PFAS and provides information-based recommendations to inform regulatory measures. Its activity is aimed at monitoring PFAS in the air, water, and soil, and assisting in the development of measures to reduce emissions and contact. The ECHA plays a key role in regulating PFAS by enforcing various laws governing chemicals, including REACH. ECHA evaluates the hazards of PFAS, proposes restrictions, and has significantly contributed to the effort to prohibit PFAS in firefighting foams (Per- and Polyfluoroalkyl Substances PFAS, 2025).

The EFSA assesses PFAS in the food chain and suggests scientific advice to lead policy. To limit dietary exposure, EFSA has set a TWI of key PFAS compounds, such as PFOA, PFOS, PFNA, and PFHxS (Per- and polyfluoroalkyl substances PFAS and EFSA, 2025). Besides these EU-wide institutions, each EU member country has national regulatory and health institutions responsible for implementing EU PFAS policies and conducting country-specific monitoring and risk assessments. The national bodies collaborate with ECHA, EFSA, and the European Commission to establish consistent regulations and enforcement of PFAS on a regional scale. Table 22.3 provides a summary of national agencies involved in regulating PFAS in EU countries.

**TABLE 22.3** National agencies involved in per- and polyfluoroalkyl substances (PFAS) regulation across EU countries.

| Country | National authority | Role in PFAS regulation |
|---|---|---|
| Germany | Federal Environment Agency (UBA) | Conducts research, monitors pollution, and provides regulatory recommendations. |
| Denmark | Danish Environmental Protection Agency | Banned PFAS in food packaging and actively contributes to EU-level regulatory discussions. |
| Sweden | Swedish Chemicals Agency (KEMI) | Advocates stricter regulations and collaborates with other EU countries on proposals. |
| The Netherlands | National Institute for Public Health and the Environment | Performs risk assessments, monitors PFAS levels, and provides guidance to policymakers. |
| Norway | Norwegian Environment Agency | Monitors contamination and supports EU initiatives despite being outside the European Union. |

## 22.3.4 Per- and polyfluoroalkyl substances detection and monitoring

The European Union has been leading efforts to finance research aimed at improving the monitoring, detection, and control of PFAS, thereby mitigating adverse environmental and health outcomes (EU research on PFAS). A number of large-scale projects financed by the European Union are currently addressing these challenges (Research And Innovation - Per- And Polyfluorinated Substances Pfas, 2025). The PROMISCES project aims to investigate the behavior and distribution of PFAS at riverbank filtration locations, and how they can be better analyzed and sources identified. The PANORAMIX project assesses the exposure to PFAS among the European population by using the biomonitoring data to determine the level of health risks. In the meantime, ZeroPM develops preventative measures to restrict PFAS emissions and decrease the amount in the environment. The SCENARIOS project is a flagship initiative within the EU Horizon 2020 program, addressing the issue of PFAS pollution through innovative detection and remediation technologies. The project focuses on the development of intelligent detection technology, such as Surface- Enhanced Raman Scattering and electrochemical microsensors, which are sensitive to the detection of parts-per-trillion of PFAS in both environmental matrices (soil, water, air) and biological fluids. Besides the 34 PFAS congeners, such as PFOA and PFOS, which are given a higher priority, SCENARIOS is used to conduct a toxicological evaluation of the 34 congeners using complex cell models to determine the health impact of these congeners. Additionally, SCENARIOS is used to characterize other less-studied emerging compounds. Another area explored by the project is quasi zero-energetic remediation, where the problem of cleaning up PFAS-contaminated areas will be achieved by minimizing the reliance on chemicals through scrubbed interventions, thereby promoting the idea of sustainable cleanup operations. Besides scientific research, SCENARIOS directly influences the EU policy, where the required critical information is presented, and regulatory policies are informed. Its inputs meet the standards of the EU Zero Pollution Action Plan, as it suggests long-term environmental and population health benefits through innovation and sustainability.

The PROMISCES project plays a crucial role in the EU's efforts to combat PFAS pollution. The goal of this project is to develop better approaches that can analyze and quantify PFAS in complex matrices such as soil, sediment, and water systems. These methods are tested in real-life conditions to determine their effectiveness and precision. The project will combine environmental monitoring, open-source technologies, and artificial intelligence to create predictive models that simulate the behavior, transportation, and fate of PFAS in the environment. Modeling can be applied to predict the risk of contamination and direct response strategies. Additionally, the PROMISCES initiative showcases innovative remediation methods that ensure the effective treatment of PFAS in groundwater, wastewater, and drinking water, thereby restoring the environment and protecting human health (PROMISCES, 2025).

Another project funded by the European Union and operating under Horizon 2020 is ZeroPM, which commenced in October 2021 and is scheduled to run until October 2026. It brings together leading researchers, regulators, and specialists in green chemistry with the aim of encouraging the use of alternatives to PFAS that are nontoxic in the production and consumption of goods, thereby helping to avoid future pollution. ZeroPM also investigates high-efficiency technologies to eliminate PFAS in water, soil, and other affected environments. The project encourages partnerships between policymakers, scientists, and industry stakeholders with the intention of facilitating the adoption of sustainable and effective solutions in accordance with the EU Zero Pollution and Circular Economy Action Plans.

The PANORAMIX project, under Horizon 2020, aims to develop the scientific and regulatory knowledge of chemical mixtures, including PFAS. It examines the routes of PFAS exposure, particularly the pathways by which the toxin is transferred through the environment to humans, including maternal-to-fetal routes of exposure. It evaluates the risks involved in their health, such as the possibility that it may impact neurodevelopment and the reproductive health of children. PANORAMIX is producing scientific evidence to inform EU policies on the European Green Deal and Zero Pollution objectives and is also creating user-friendly tools to assist regulators and stakeholders in assessing risks and making science-informed decisions (Research And Innovation - Per- And Polyfluorinated Substances Pfas, 2025; Bil et al., 2023).

Case studies of PFAS in the European Union are presented in Table 22.4 below.

### 22.3.5 Remediation efforts of per- and polyfluoroalkyl substances

European countries are utilizing various new technologies to combat PFAS contamination in soil, water, and air. The most common ones are granular activated carbon (GAC) and ion exchange resin, which are effective in removing PFAS in contaminated water. Nevertheless, these options need proper management and disposal of PFAS-contaminated waste substances. PFAS compounds can be thermally destroyed by high temperature burning, and electrochemical oxidation is a possible solution to this problem, utilizing electric currents to break down PFAS molecules into less toxic products in both aqueous and soil conditions. Simultaneously, a viable alternative is being investigated: bioremediation, which employs the assistance of certain microorganisms to break down PFAS in an environmentally friendly and sustainable manner. This aligns with the overall EU commitment to green innovation. It is worth noting that the SCENARIOS project, as part of Horizon 2020, is an example of how the

**TABLE 22.4** Case studies reported on per- and polyfluoroalkyl substances (PFAS) in the European Union.

| S. no. | Type of PFAS | Study year | Study participants/ location | Sample type | Detection frequency | Concentration reported | Outcomes | Inferences | References |
|---|---|---|---|---|---|---|---|---|---|
| 1. | PFBA, PFBS, PFHxS, PFOA, PFOS, PFNA, PFHpA, PFDA | 2025 | UK; wastewater treatment plants (WWTP) emissions during aeration | $PM_{10}$ air particles | NA | Autumn: 15.49 pg $m^{-3}$; spring: 4.25 pg $m^{-3}$; PFBA most abundant | Aerosolised PFAS detected in ambient air | WWTPs are recognized as emerging airborne PFAS sources | Kizhakkethil et al., 2025 |
| 2. | Legacy and novel PFAS | 2025 | Veneto, Italy; 30 municipa-lities | Groundwater, drinking water | NA | Multiyear trend from 2013–23; legacy and novel PFAS tracked | Persistent contamination from industrial discharge | Long-term PFAS risk remains despite mitigation | Abstract EGU25-12081 (2025) |
| 3. | Various PFAS | 2025 | EU populations across various nations | Hair, nails, urine | Consistent detection across matrices | Consistent PFAS presence; long-chain PFAS are lower in urine | Hair and nails are viable for monitoring | Encourages harmonized PFAS biomoni-toring | Nadal and Domingo (2025) |
| 4. | Various PFAS | 2025 | EU-wide | Blood, milk, soil, air, urine, water | NA | PFAS are widespread across all sample types | Health and exposure risk remain high | Advocates for unified EU PFAS regulation | Dehghani et al. (2025) |
| 5. | Trifluoroacetic acid (TFA) and short-chain PFAS | 2024 | 23 rivers and 11 EU countries | Surface, tap, bottled water | High (79% had TFA > 500 ng $L^{-1}$) | TFA >500 ng $L^{-1}$ in 79% of samples; tap water max 4100 ng $L^{-1}$ | TFA dominates PFAS profile; nearly universal in EU water | Current water treatment fails to remove TFA; urgent ban on fluorinated pesticides/ F-gases needed | TFA in Water: Dirty PFAS Legacy Under the Radar (2025) |

| | | | | | | | | | |
|---|---|---|---|---|---|---|---|---|---|
| 6. | Ultrashort and short-chain PFAS (56 types) | 2023 | Netherlands, 18 locations | Raw and treated drinking water | Detected at all sites | Ultrashort PFAS: 300–1100 ng $L^{-1}$; C4+ PFCAs: 0.4–95 ng $L^{-1}$ | Several locations exceeded EU health benchmarks | Short-chain PFAS are widespread; they require regulatory reassessment | Sadia et al. (2023) |
| 7. | PFOS, PFOA, PFHxS, PFNA | 2022 | 1957 adolescents (12–18 year) from 9 EU countries | Blood serum | NA | PFOS GM 2.13 μg $L^{-1}$; PFOA GM 0.97 μg $L^{-1}$ | Higher levels in Northern and Western European Union; seafood-linked | ~9% exceeded HBM guidance values; exposure linked to diet | Richterová et al. (2023) |
| 8. | PFOS, PFOA, others | 2020 | Teenagers in Veneto, Italy (near PFAS source) | Serum | NA | Elevated PFAS serum; associated lipid profile disruption | Local drinking water caused measurable health effects | Strong evidence of exposure risk linkage | Pitter et al. (2020) |

European Union is integrating its PFAS management approach, which aims at smart detection, risk assessment, and sustainable remediation. These attempts are indicative of the European desire to achieve long-term pollution reduction through science-oriented, environmentally friendly technologies (Strategies for Health Protection, 2025; PFAS Contamination and Soil Remediation Signal. European Zero Pollution Dashboards, 2025).

PFAS contamination poses significant barriers to remediation by the European Union due to the extreme persistence of the chemicals, their mobility in groundwater, and their affinity to soils. These properties make it harder to remove or contain them. Traditional systems, such as pump-and-treat systems, remain in high demand; however, their high energy needs and carbon emissions raise concerns about their long-term sustainability. New methods, including in situ sequestration and enhanced attenuation, provide a more sustainable alternative but are often constrained by expense and scalability. PFAS are also challenging to monitor, as they are complex chemically and their concentrations require detection at extremely low levels. The innovation of remediation technologies has proven effective in several case studies. Indicatively, research conducted at an exfire training site, where AFFF had been applied, revealed tendencies in the movement and natural attenuation of PFAS (McGuire et al., 2014). Another experimental design was of silica-based granular media impregnated with titanium dioxide, which was introduced into a column reactor under UV light. This system produced reactive species that facilitated PFAS degradation through photocatalysis and nucleophilic reactions, leading to major defluorination (McIntyre et al., 2022). Other interesting remediation technologies include PlumeStop colloidal activated carbon, which was implemented at a commercial airport to mitigate PFAS leaching and mobility without requiring active pumping or energy input. An array of sophisticated methods that focus on various matrices (soil, groundwater, and air), have been investigated in various studies (McIntyre and Hart, 2021; McIntyre et al., 2021; Hannah et al., 2023; McKenzie et al., 2016; McNamara et al., 2018). These case studies highlight the EU's desire to incorporate innovative, cost-effective, and environmentally friendly technologies in PFAS remediation, thereby promoting environmental sustainability and human health.

## 22.3.6 Legal and policy challenges

The regulation of PFAS in the European Union is fraught with legal and policy challenges, primarily shaped by the sheer number of more than 10,000 distinct chemically different PFAS compounds and their persistence. Although the EU regulation on the use of PFAS (REACH) has led to proposals to limit its use, implementing such limits across various industries is a daunting challenge. Numerous industries still lack viable, scalable alternatives, making compliance both technically and economically demanding.

The principle of essential use is a central issue in current policy debates and therefore relates to confining the applications of PFAS to those that are exceptionally necessary, such as in specific medical equipment, and gradually removing nonessential applications, including textiles and cosmetics. It is a precautionary-driven strategy, and this practice has raised eyebrows among industries that are highly dependent on PFAS, including the firefighting, electronics, and apparel industries. The current state of these industries is that switching is very expensive and presents numerous logistical challenges in substituting PFAS-based materials. The proposed restriction of PFAS levels, as outlined in Annex XV, has sparked further concern about supply chain disruption and the challenge of meeting compliance timelines. Moreover, litigation risks are also increasing due to growing community awareness. PFAS lawsuits could reach the European Union, just as they have in the United States, but the European Union is demonstrating a growing interest in collective redress mechanisms. Lastly, monitoring and enforcement are issues because no standardized systems exist to detect PFAS at ultra-trace concentrations, and therefore regulation is both technically difficult and resource intensive.

#### *22.3.6.1 Legal framework for per- and polyfluoroalkyl substances*

PFAS, or forever chemicals, have been at the center of legal and regulatory activities in the European region, where these chemicals persist in the environment and raise serious health concerns. Although the European Union has made significant efforts to control PFAS through the REACH regulation, critics suggest that the strategy has been disjointed and focused on individual substances, rather than the entire group of PFAS (PFAS Regulation and Development at the European Level with Focus on Germany and France, 2025). Criminal prosecution is on the rise in the EU member states. Companies like 3M, Chemours, Miteni, Arkema, and even military and municipal entities (e.g., the water authority of Ronneby and the US/NATO Air Base Spangdahlem) have been sued in connection with the contamination of PFAS and the associated risks to the population (Just the Start': The Growing Legal Battle over PFAS in Europe, 2025). As the European Union has begun to expand its regulatory prism, its previous approach has been criticized as aiming too narrowly at individual substances (such as PFOA or PFOS) instead of regulating PFAS as a family of chemicals. But the tide is turning: in 2023 the EU-wide class action law was introduced, which presented a legal option of collective litigation, which is reminiscent of similar efforts to do so in the United States (From Contamination to Accountability: Europe's Drive to Regulate PFAS and Protect Water Resources, 2025). The opposition of the industry is a significant problem. Industries with a high dependency on PFAS include aerospace, electronics, and textiles, claiming that chemicals are essential in maintaining the performance of products, especially where waterproofing, wearability, and heat resistance are required. These industries have been lobbying for a transition period and research on safer alternatives, rather than outright bans (The PFAS Crackdown: How 2025 Regulations Are Reshaping Industries, 2025).

PFAS is a contaminant that has caught the attention of the public due to investigative initiatives, such as the Forever Pollution Project, which has geo-mapped over 23,000 contaminated sites in Europe. Pollution with PFAS and health risks related to its use are not well understood by the population, despite the growing awareness. There are still educational campaigns and consumer-oriented programs to bridge this gap and ensure informed choices (From Contamination to Accountability: Europe's Drive to Regulate PFAS and Protect Water Resources, 2025).

### 22.3.7 Comparison with international per- and polyfluoroalkyl substances regulations

The European Union has adopted one of the most comprehensive and forward-looking approaches to regulating PFAS, positioning itself as a global leader in chemical safety.

The European Union, in contrast to most other nations that have been doing silico testing on individual compounds, strives to regulate PFAS not as individual compounds, but as a chemical family, with more than 10,000 substances under a multiverse restriction regime. A revised Drinking Water Directive by the European Union has required stringent PFAS levels, which must be met by every member state by 2026. Conversely, the US EPA has adopted a substance-specific approach by recently establishing maximum contaminant levels for six separate PFAS chemicals, including PFOA and PFOS. Although this is significant progress, it is limited compared to the group-based system of the European Union. Following Brexit, the United Kingdom began to diverge from the EU's comprehensive framework on PFAS. The current control definition in the United Kingdon encompasses thousands of chemically diverse PFAS, which would be subject to EU regulation, thereby limiting the scope of its actions. Similarly, the Chemicals Management Plan in Canada targets certain PFAS chemicals, for example, PFOA and PFOS, and the rest are yet to be regulated.

The other regions of the world also respond differently. Australia has implemented restrictions on some of the PFAS substances but has yet to adopt a group-wide ban. Japan has primarily regulated the use of PFAS in industrial applications, and some controls have been implemented in China, particularly in manufacturing and export sectors. The European Union is distinguished by an intended group-wide ban on PFAS in firefighting foams that is more extensive than those currently in place in the United Kingdom and other nations. It also emphasizes restricting PFAS in consumer goods and ensuring environmental and human health protection. To conclude, it is fair to state that the EU's group-based regulation strategy, combined with high thresholds and broad sectoral coverage, makes it the global model in PFAS regulation.

## 22.3.8 Future directions and recommendations

In an effort to mitigate the environmental and health impacts of PFAS, scholars within the European Union have proposed a set of progressive recommendations. One of the most important is the encouragement of the transition to PFAS-free alternatives in industries such as textiles, food packaging, and firefighting, which are highly impacted by the chemicals due to their ability to resist heat and water. Researchers and regulatory bodies encourage the faster development of safer, nonpersistent alternatives that do not contribute to long-term contamination (Understanding New PFAS Regulations in the U.S. and EU and Roland Berger, 2025). Research and development of improved PFAS treatment technologies is also necessary. PFAS have been shown to be mostly resistant to traditional water treatment systems. Another area of focus is increased openness, which experts recommend through high levels of remediation, such as GAC, ion exchange, high-pressure membrane, and emerging degradation technologies that reduce PFAS into harmless products (EU Advances PFAS Restriction and Chemical Reclassification - Foresight, 2025). To ensure responsibility and adequate decision-making, policymakers demand that PFAS be disclosed as required in the product labeling and industrial processes. By defining tracking, labeling, and reporting standards, consumers and companies would be able to determine and verify the presence of PFAS much more effectively (Scientific Evaluation of the Proposal to Restrict Per-and Polyfluoroalkyl Substances PFAS-Current Status, 2025). Lastly, it is important to invest more in the research of PFAS. Current initiatives involve creating biodegradable substitutes and developing more sensitive and cost-efficient detectors. It is believed that cross-sectoral cooperation and long-term, consistent public investment are necessary to develop sustainable, pragmatic solutions that strike a balance between environmental protection and product performance.

## 22.4 Conclusion

The regulatory trends of Australia and Europe are distinct yet harmonized approaches that manage the persistent challenges associated with PFAS contamination. Europe's regulatory framework is grounded in REACH, the Stockholm Convention on POPs, and the integrated environmental bodies, which follow a precautionary and proactive model aimed at reducing and ultimately phasing out the harmful PFAS wherever possible. By establishing a group-based regulation of PFAS and setting a goal of a PFAS-free economy, Europe demonstrates a coherent and ambitious vision that positions it as a global leader in chemical safety and sustainable governance. On the other hand, Australia follows a collaborative approach, which is characterized by intergovernmental coordination and responsiveness to emerging evidence. Their risk-based PFAS NEMP, chemical management under the AICIS, and state-level bans on PFAS-containing firefighting foams enable Australia to create a balanced system that simultaneously protects public health and the environment, while also facilitating industrial adoption.

Both jurisdictions reflect a global shift toward sustainable chemical management, rooted in innovation and accountability. In addition, advances in analytical platforms, toxicological understanding, and innovative remediation technologies are continually enhancing the effectiveness of these regulations in both regions. However, there is a need for more sustained international cooperation and policy harmonization to combat the persistent and transboundary nature of PFAS. Lastly, although Australia and Europe's regulatory architectures differ in terms of enforcement, they still converge on the shared vision of minimizing PFAS contamination, protecting the ecosystem as well as public health, and ultimately pioneering a PFAS-free future.

## List of abbreviations

**AAHMS** Australian Academy of Health and Medical Sciences
**ACT** Australian Capital Territory
**AFFF** aqueous film-forming foam
**AICIS** Australian Industrial Chemicals Introduction Scheme
**ANU** Australian National University
**ANZG** Australian And New Zealand Guidelines
**APFO** Ammonium pentadecafluorooctanoate
**APVMA** Australian Pesticides and Veterinary Medicines Authority
**ARC** Australian Research Council
**ARTG** Australian Register of Therapeutic Goods
**ASC** Assessment of Site Contamination
**ATDS** Australian Total Diet Study
**CLP** classification, labelling, and packaging
**CSM** conceptual site model

| | |
|---|---|
| **diPAP** | Polyfluoroalkyl Phosphate Diesters |
| **DOD** | Department of Defense |
| **DoEE** | Australian Government Department of the Environment And Energy |
| **DPI** | Department of Primary Industries |
| **DWER** | Department of Water And Environmental Regulation |
| **ECCC** | Environment and Climate Change Canada |
| **ECHA** | European Chemicals Agency |
| **EDC** | endocrine-disrupting chemicals |
| **EEA** | European Environment Agency |
| **EFSA** | European Food Safety Authority |
| **EPA** | Environmental Protection Agency |
| **EU** | European Union |
| **FASA** | perfluoroalkane sulfonamides |
| **FASAA** | perfluoroalkyl sulfonamide |
| **FASE** | Fluoroalkanesulfonyl compounds |
| **FFFP** | forming fluoroprotein foams |
| **FOSAA** | Perfluorooctane Sulfonamidoacetic acid |
| **FOSE** | Perfluorooctanesulfonamido-Ethanol |
| **FP** | fluoropolymers |
| **FSANZ** | Food Standards Australia New Zealand |
| **FTCA** | Fluorotelomer Carboxylic acid |
| **FTS** | Fluorotelomer Sulfonate |
| **FTUCA** | Fluorotelomer Unsaturated Carboxylic acid |
| **GAC** | granular activated carbon |
| **HBM** | human biomonitoring |
| **HEPA** | heads of EPAs |
| **HFPO** | heptafluoropropoxy propionic acid |
| **IGA** | intergovernmental agreement |
| **KEMI** | Swedish Chemicals Agency |
| **LC-MS** | liquid chromatography and mass spectrometry |
| **MCL** | maximum contaminant levels |
| **NA** | Not Available |
| **NCWG** | National Chemical Working Group |
| **NEMP** | National Environmental Management Plan |
| **NEPM** | National Environment Protection Measure |
| **NHMRC** | National Health and Medical Research Council |
| **NHMS** | National Health Measures Survey |
| **NICNAS** | National Industrial Chemicals Notification and Assessment Scheme |
| **NMI** | National Measurement Institute |
| **NSW** | New South Wales |
| **OECD** | Organisation for Economic Co-operation and Development |
| **PFAS** | Per- and polyfluoroalkyl substances |
| **PFBA** | perfluorobutanoic acid |
| **PFBS** | perfluorobutane sulfonate |
| **PFCA** | perfluorocarboxylic acids |
| **PFDA** | Perfluorodecanoic acid |
| **PFDoA** | Perfluorododecanoic acid |
| **PFDS** | Perfluorodecanesulfonic acid |
| **PFHxS** | Perfluorohexane Sulfonate |
| **PFNA** | perfluorononanoic acid |
| **PFNS** | Perfluorononanesulfonic acid |
| **PFOA** | Perfluorooctanoic acid |
| **PFOS** | Perfluorooctane sulfonate |

| | |
|---|---|
| **PFOSA** | Perfluorooctanesulfonamide |
| **PFPeA** | Perfluoropentanoic acid |
| **PFPeS** | Perfluoropentanesulfonic acid |
| **PFSA** | Perfluorosulfonic acids |
| **PFT** | Perfluorinated Tensides, |
| **PFUnDA** | Perfluoroundecanoic acid |
| **POP** | Persistent Organic Pollutants |
| **RAAF** | Royal Australian Air Force |
| **REACH** | Registration, Evaluation, Authorisation And Restriction of Chemicals |
| **RIVM** | National Institute for Public Health and The Environment |
| **SEAC** | Socio-Economic Analysis |
| **SERS** | Surface-Enhanced Raman Scattering |
| **SGM** | silica-based granular media |
| **SPE** | solid phase extraction |
| **TFA** | trifluoroacetic acid |
| **TGA** | Therapeutic Goods Administration |
| **TOP** | total oxidizable precursor |
| **TWI** | tolerable weekly intake |
| **UBA** | Federal Environment Agency |
| **USEPA** | United States Environmental Protection Agency |
| **WFD** | Water Framework Directive |
| **WHO** | World Health Organisation |
| **WWTP** | Wastewater treatment plants |

# References


A Ban on Forever Chemicals Would Be Best, Pfas. The Guardian. Available at https://www.theguardian.com/environment/2025/jun/25/a-ban-on-forever-chemicals-would-be-best. (accessed 15.07.25).

Abstract EGU25-12081. Available at https://meetingorganizer.copernicus.org/EGU25/EGU25-12081.html. (accessed 15.07.25.).

Abunada, Z., Alazaiza, M.Y.D., Bashir, M.J.K., 2020. An overview of per- and polyfluoroalkyl substances (PFAS) in the environment: source, fate, risk and regulations. Water. Available at https://doi.org/10.3390/w12123590.

Academy Calls for a National Human Biomonitoring Program to Track the Health Impacts of Forever Chemicals - AAHMS. Available at https://aahms.org/news/academy-calls-for-a-national-human-biomonitoring-program-to-track-the-health-impacts-of-forever-chemicals/. (accessed 15.07.25.).

Ackerman Grunfeld, D., Gilbert, D., Hou, J., Jones, A.M., Lee, M.J., Kibbey, T.C.G., et al., 2024. Underestimated burden of per- and polyfluoroalkyl substances in global surface waters and groundwaters. Nature Geoscience 17 (4), 340–346. Available at https://doi.org/10.1038/s41561-024-01402-8.

All news - ECHA. Available at https://echa.europa.eu/-/echa-s-committees-eu-wide-pfas-ban-in-firefighting-foams-warranted. (accessed 15.07.25.).

All News - ECHA. Available at https://echa.europa.eu/-/highlights-from-june-2025-rac-and-seac-meetings. (accessed 15.07.25a.).

All News - ECHA. Available at https://echa.europa.eu/-/echa-s-committees-eu-wide-pfas-ban-in-firefighting-foams-warranted. (accessed 15.07.25b.).

Australian Drinking Water Guidelines - Public Consultation on Draft Guidance for Per- and Polyfluoroalkyl Substances (PFAS) - NHMRC Consultation Hub - Citizen Space. Available at https://consultations.nhmrc.gov.au/environmental-health/australian-drinking-water-guidelines-2024-pfas/. (accessed July 2025).

Australian Government Department Of Health, And Aged Care, Enhealth Guidance Statement, Per- and Polyfluoroalkyl Substances (PFAS). Available at https://www.health.gov.au/sites/default/files/2024-02/enhealth-guidance-statement-per–and-polyfluoroalkyl-substances-pfas.pdf. (accessed 16.07.25.).

Australian Government PFAS Taskforce. Regulatory Guidance. Available at https://www.pfas.gov.au/government-action/regulatory-guidance (accessed 15.07.25.).

Australian Pesticides and Veterinary Medicines Authority. APVMA Basics. Available at https://www.apvma.gov.au/about/about-us/apvma-basics (accessed July 2025).

Bil, W., Govarts, E., Zeilmaker, M.J., Woutersen, M., Bessems, J., Ma, Y., et al., 2023. Approaches to mixture risk assessment of PFASs in the European population based on human hazard and biomonitoring data. International Journal of Hygiene and Environmental Health 247, 114071. Available at https://doi.org/10.1016/j.ijheh.2022.114071.

Chemicals, Forever: How Do You Fix a Problem like PFAS? Available at https://www.uts.edu.au/news/2024/04/chemicals-forever-how-do-you-fix-problem-pfas. (accessed 15.07.25.).

Class Action Settled but PFAS Issues Far from Resolved - Corrs Chambers Westgarth. Available at https://www.corrs.com.au/insights/class-action-settled-but-pfas-issues-far-from-resolved. (accessed 15.07.25.).

Coggan, T.L., Moodie, D., Kolobaric, A., Szabo, D., Shimeta, J., Crosbie, N.D., et al., 2019. An investigation into per- and polyfluoroalkyl substances (PFAS) in nineteen australian wastewater treatment plants (WWTPs). Heliyon 5 (8), e02316. Available at https://doi.org/10.1016/j.heliyon.2019.e02316.

Default Guideline Values. Available at https://www.waterquality.gov.au/anz-guidelines/guideline-values/default (accessed 15.07.25.).

Dehghani, M.H., Aghaei, M., Bashardoust, P., Rezvani Ghalhari, M., Nayeri, D., Malekpoor, M., et al., 2025. An insight into the environmental and human health impacts of per- and polyfluoroalkyl substances (PFAS): exploring exposure pathways and their implications. Environ. Sci. Eur. 37 (1), 81. Available at https://doi.org/10.1186/s12302-025-01122-9.

Department of climate change, Energy, the environmet and water. Available at https://www.dcceew.gov.au/sites/default/files/documents/pfas-nemp-3.pdf. (accessed 15.07.25.).

Department of Industry Science and Resources. Per- and Poly-Fluoroalkyl Substances. Available at https://www.industry.gov.au/national-measurement-institute/nmi-services/testing-and-analytical-services/environmental-testing-and-analytical-services/and-poly-fluoroalkyl-substances. (accessed 15.07.25.).

ECHA. Available at https://echa.europa.eu/. (accessed 15.07.25.).

EFSA. Science, Safe Food, Sustainability. Available at https://www.efsa.europa.eu/en. (accessed 15.07.25.).

Environment Protection Authority - Access Canberra. Available at https://www.accesscanberra.act.gov.au/city-services/environment-protection-authority. (accessed 15.07.25.).

Environmental Statement 2023 | European Environment Agency's Home Page. Available at https://www.eea.europa.eu/en/analysis/publications/environmental-statement-2023. (accessed 15.07.25.).

EU Advances PFAS Restriction and Chemical Reclassification - Foresight. Available at https://www.useforesight.io/news/eu-advances-pfas-restriction-and-chemical-reclassification. (accessed 15.07.25.).

European Environment Agency's. Available at https://www.eea.europa.eu/en. (accessed 15.07.25.).

Exposed: 17,000 European Sites Contaminated by Forever Chemicals. Where Are the Worst Spots? Euronews. Available at https://www.euronews.com/green/2023/02/24/exposed-17000-european-sites-contaminated-by-forever-chemicals-where-are-the-worst-spots. (accessed 15.07.25.).

Federal Register of Legislation. Agricultural and Veterinary Chemicals Code Act 1994. Available at https://www.legislation.gov.au/C2004A04723/2016-10-21/text. (accessed 15.07.25.).

Fenton, S.E., Ducatman, A., Boobis, A., DeWitt, J.C., Lau, C., Ng, C., et al., 2021. Per- and polyfluoroalkyl substance toxicity and human health review: current state of knowledge and strategies for informing future research. Environmental Toxicology and Chemistry / SETAC 40 (3), 606–630. Available at https://doi.org/10.1002/etc.4890.

Firefighting Foam Management Policy, 2025. Environment, Land and Water. Queensland Government. Available at https://www.qld.gov.au/environment/management/environmental/pfas/firefighting-foam (accessed 23 October).

Food Standards Australia New Zealand. Available at https://www.foodstandards.gov.au/ (accessed July 2025a).

Food Standards Australia New Zealand. Australian Total Diet Study. Available at https://www.foodstandards.gov.au/science-data/monitoring-safety/australian-total-diet-study. (accessed 15.07.25.b.).

Fresh and Marine Water (Australia and New Zealand). Available at https://www.waterquality.gov.au/guidelines/anz-fresh-marine. (accessed 15.07.25.).

From Contamination to Accountability: Europe's Drive to Regulate PFAS and Protect Water Resources. 2025. Available at https://smartwatermagazine.com/news/smart-water-magazine/contamination-accountability-europes-drive-regulate-pfas-and-protect-water.

FSANZ 27th Australian Total Diet Study, Australian Government PFAS Taskforce. Available at https://www.pfas.gov.au/news/fsanz-27th-australian-total-diet-study. (accessed 15.07.25.).

Guidance for Managing PFAS in SA, Engage EPA. Available at https://engage.epa.sa.gov.au/guidance-for-managing-pfas-in-sa. (accessed 15.07.25.).

Hannah, M., Vidit, M., Elisabeth, H., Rula, D., Megan, H., 2023. Development and laboratory scalability of ultraviolet-activated silica-based granular media as an engineered system for the degradation of per- and polyfluoroalkyl substances in concentrated liquid waste. Journal of Environmental Engineering 149 (9), 4023049. Available at https://doi.org/10.1061/JOEEDU.EEENG-7228.

Health Based Guidance Values for PFAS for Use in Site Investigations in Australia. Available at https://www.health.gov.au/sites/default/files/documents/2022/07/health-based-guidance-values-for-pfas-for-use-in-site-investigations-in-australia_0.pdf. (accessed 15.07.25.).

Hossini, H., Massahi, T., Parnoon, K., Nouri, M., 2025. Per-and polyfluoroalkyl substances (PFAS) in milk and dairy products: a literature review of the occurrence, contamination sources, and health risks. Food Additives & Contaminants: Part A 42 (9), 1284–1296. Available at https://doi.org/10.1080/19440049.2025.2538224.

How Pesticides Impact Human Health and Ecosystems in Europe — European Environment Agency. Available at https://www.eea.europa.eu/publications/how-pesticides-impact-human-health. (accessed 15.07.25.).

Huang, L., Yao, S.-M., Jin, Y., Xue, W., Yu, F.-H., 2023. Co-contamination by heavy metal and organic pollutant alters impacts of genotypic richness on soil nutrients. Frontiers in Plant Science Volume 14.

Ibrahim, M., Li, Y., Danjaji, H.A., Wang, Y., 2025. Per- and polyfluoroalkyl substances and global water resources: an in-depth review of existing regulatory frameworks worldwide. International Journal of Environmental Science and Technology 22 (9), 8259–8280. Available at https://doi.org/10.1007/s13762-024-06256-6.

Industrial Chemicals Law in Australia, Australian Industrial Chemicals Introduction Scheme (AICIS). Available at https://www.industrialchemicals.gov.au/about-us/industrial-chemicals-law-australia.(accessed 23.10.25.).

Intergovernmental Agreement on a National Framework for Responding to PFAS Contamination, Federation. Available at https://federation.gov.au/about/agreements/intergovernmental-agreement-national-framework-responding-pfas-contamination (accessed 15.07.25.).

International Cooperation, Australian Government PFAS Taskforce. Available at https://www.pfas.gov.au/government-action/international-cooperation. (accessed 16.07.25.).

Just the Start': The Growing Legal Battle over PFAS in Europe. Available at https://chemsec.org/just-the-start-the-growing-legal-battle-over-pfas-in-europe/. (accessed 15.07.25.).

Khudur, L.S., Gleeson, D.B., Ryan, M.H., Shahsavari, E., Haleyur, N., Nugegoda, D., et al., 2018. Implications of co-contamination with aged heavy metals and total petroleum hydrocarbons on natural attenuation and ecotoxicity in Australian soils. Environmental Pollution (Barking, Essex: 1987) 243, 94–102.

Kizhakkethil, J.P., Shi, Z., Bogush, A., Kourtchev, I., 2025. Measurement report: per-and polyfluoroalkyl substances (PFAS) in particulate matter (PM10) from activated sludge aeration. Atmospheric Chemistry and Physics 25 (11), 5947–5958.

Lallas, P.L., 2001. The stockholm convention on persistent organic pollutants. The American Journal of International Law 95 (3), 692–708. Available at https://doi.org/10.2307/2668517.

Lazarevic, N., Smurthwaite, K., Trevenar, S., Este, C.D., Batterham, P., Lane, J. et al., 2021. PFAS Health Study. December, pp.1–6.

Listing of POPs in the Stockholm Convention. Available at https://www.pops.int/TheConvention/ThePOPs/AllPOPs/tabid/2509/Default.aspx. (accessed 15.07.25.).

McGuire, M.E., Schaefer, C., Richards, T., Backe, W.J., Field, J.A., Houtz, E., et al., 2014. Evidence of remediation-induced alteration of subsurface poly- and perfluoroalkyl substance distribution at a former firefighter training area. Environmental Science & Technology 48 (12), 6644–6652. Available at https://doi.org/10.1021/es5006187.

McIntyre, H.M., Hart, M.L., 2021. Photocatalytic porous silica-based granular media for organic pollutant degradation in industrial waste-streams. Catalysts. Available at https://doi.org/10.3390/catal11020258.

McIntyre, H., Minda, V., Gutheil, W., Hart, M., 2021. Degradation and defluorination of aqueous perfluorooctane sulfonate by silica-based granular media using batch reactors. Journal of Environmental Engineering 147. Available at https://doi.org/10.1061/(ASCE)EE.1943-7870.0001922.

McIntyre, H., Minda, V., Hawley, E., Deeb, R., Hart, M., 2022. Coupled photocatalytic alkaline media as a destructive technology for per- and polyfluoroalkyl substances in aqueous film-forming foam impacted stormwater. Chemosphere 291, 132790. Available at https://doi.org/10.1016/j.chemosphere.2021.132790.

McKenzie, E.R., Siegrist, R.L., McCray, J.E., Higgins, C.P., 2016. The influence of a non-aqueous phase liquid (NAPL) and chemical oxidant application on perfluoroalkyl acid (PFAA) fate and transport. Water Research 92, 199–207. Available at https://doi.org/10.1016/j.watres.2016.01.025.

McNamara, J., Franco, R., Mimna, R., Zappa, L., 2018. Comparison of activated carbons for removal of perfluorinated compounds from drinking water (In Press). Journal - American Water Works Association 110. Available at https://doi.org/10.5942/jawwa.2018.110.0003.

Microplastic Releases to the Environment (Indicator). Euro.pean Zero Pollution Dashboards. Available at https://www.eea.europa.eu/en/european-zero-pollution-dashboards/indicators/microplastic-releases-to-the-environment. (accessed 15.07.25.).

Molino, P.J., Bentham, R., Higgins, M.J., Hinds, J., Whiley, H., 2019. Public health risks associated with heavy metal and microbial contamination of drinking water in Australia. International Journal of Environmental Research and Public Health. Available at https://doi.org/10.3390/ijerph16203982.

Moore, B., He, C., Knight, E., Mueller, J.F., Tscharke, B., 2023. Bisphenols and phthalates in Australian wastewater: a statistical approach for estimating contributions from diffuse and point sources. Water Research 246, 120680. Available at https://doi.org/10.1016/j.watres.2023.120680.

Nadal, M., Domingo, J.L., 2025. Non-invasive matrices for the human biomonitoring of PFAS: an updated review of the scientific literature. Toxics. Available at https://doi.org/10.3390/toxics13020134.

Nahar, K., Zulkarnain, N.A., Niven, R.K., 2023. A review of analytical methods and technologies for monitoring per- and polyfluoroalkyl substances (PFAS) in water. Water. Available at https://doi.org/10.3390/w15203577.

NATA Accredited Envirolab Services. PFAS Testing. Available at https://www.envirolab.com.au/Capabilities/PFAS. (accessed 15.07.25.).

National PFAS Management Program - Airservices. Available at https://www.airservicesaustralia.com/community/environment/pfas/. (accessed 15.07.25.).

Nguyen, H.T., McLachlan, M.S., Tscharke, B., Thai, P., Braeunig, J., Kaserzon, S., et al., 2022. Background release and potential point sources of per- and polyfluoroalkyl substances to municipal wastewater treatment plants across Australia. Chemosphere 293, 133657. Available at https://doi.org/10.1016/j.chemosphere.2022.133657.

Nguyen, H.T., Thai, P.K., Kaserzon, S.L., O'Brien, J.W., Mueller, J.F., 2024. Nationwide occurrence and discharge mass load of per- and polyfluoroalkyl substances in effluent and biosolids: a snapshot from 75 wastewater treatment plants across Australia. Journal of Hazardous Materials 470, 134203. Available at https://doi.org/10.1016/j.jhazmat.2024.134203.

NHMRC. NHMRC Review of PFAS in Australian drinking water. Available at https://www.nhmrc.gov.au/health-advice/environmental-health/water/PFAS-review (accessed 16.07.25.).

NHMRC Review of PFAS in Australian Drinking Water. Available at https://www.nhmrc.gov.au/health-advice/environmental-health/water/PFAS-review. (accessed July 2025).

Nilsson, S., Kucharski, N., Orr, J., Bräunig, J., Thompson, K., Jolliet, O., et al., 2025. Serum concentrations of PFAS across Australian States and territories. International Journal of Hygiene and Environmental Health 265, 114542. Available at https://doi.org/10.1016/j.ijheh.2025.114542.

Overview of Risk Reduction Approaches for PFASs. Available at https://www.oecd.org/content/dam/oecd/en/topics/policy-sub-issues/risk-management-risk-reduction-and-sustainable-chemistry2/pfas-country-information/Australia.pdf. (accessed 15.07.25.).

Paige, T., De Silva, T., Buddhadasa, S., Prasad, S., Nugegoda, D., Pettigrove, V., 2024. Background concentrations and spatial distribution of PFAS in surface waters and sediments of the greater melbourne area, Australia. Chemosphere 349, 140791. Available at https://doi.org/10.1016/j.chemosphere.2023.140791.

Per- and Poly-Fluoroalkyl Substances. Available at https://www.health.wa.gov.au/Articles/F_I/Guidance-Statements-on-Perfluorinated-Chemicals. (accessed 15.07.25.).

Per- and Poly-Fluoroalkyl Substances (PFAS), SA Health. Available at https://www.sahealth.sa.gov.au/wps/wcm/connect/public+content/sa+health+internet/conditions/chemicals+and+contaminants/pfas/per-+and+poly-fluoroalkyl+substances+pfas. (accessed 15.07.25.).

Per- and poly-fluoroalkyl substances (PFAS), Australian information portal. Available at https://www.dcceew.gov.au/about/initiatives/pfas. (accessed 23.10.25.).

Per- and Poly-Fluoroalkyl Substances (PFASs) - DCCEEW. Available at https://www.dcceew.gov.au/environment/protection/chemicals-management/pfas. (accessed 15.07.25.).

Per- and Polyfluoroalkyl Substances (PFAS). EFSA. Available at https://www.efsa.europa.eu/en/topics/per-and-polyfluoroalkyl-substances-pfas. (accessed 15.07.25.).

Per- and polyfluoroalkyl substances (PFAS) - ECHA. Available at https://echa.europa.eu/hot-topics/perfluoroalkyl-chemicals-pfas (accessed 23.10.25a.).

Per- and polyfluoroalkyl substances (PFAS) - ECHA. Available at https://echa.europa.eu/hot-topics/perfluoroalkyl-chemicals-pfas (accessed 16.07.25b.).

Per- and Polyfluoroalkyl Substances (PFAS) - ECHA. Available at https://echa.europa.eu/hot-topics/perfluoroalkyl-chemicals-pfas. ( accessed 15.07.25c.).

Per- and polyfluoroalkyl substances (PFAS) | EFSA. Available at https://www.efsa.europa.eu/en/topics/per-and-polyfluoroalkyl-substances-pfas (accessed 16.07.25.).

Per- and Polyfluoroalkyl Substances (PFASs), UNEP - UN Environment Programme. Available at https://www.unep.org/topics/chemicals-and-pollution-action/pollution-and-health/persistent-organic-pollutants-pops/and (accessed 23.10.25.).

Perfluorinated Compounds, Food Standards Australia New Zealand. Available at https://www.foodstandards.gov.au/consumer/chemicals/Perfluorinated-compounds.(accessed 23.10.25.).

PFAS, NTEPA. Available at https://ntepa.nt.gov.au/your-environment/pfas. (accessed 15.07.25.).

PFAS Advice, Australian Government PFAS Taskforce. Available at https://www.pfas.gov.au/advice. (accessed 15.07.25.).

PFAS and Livestock, Agriculture Victoria. Available at https://agriculture.vic.gov.au/livestock-and-animals/livestock-health-and-welfare/pfas-and-livestock. (accessed 23.10.25.).

PFAS Contamination, EPA Tasmania. https://epa.tas.gov.au/environment/land/area-wide-contaminated-land-issues/pfas-contamination. (accessed 15.07.25.).

PFAS Contamination and Soil Remediation (Signal). European Zero Pollution Dashboards. Available at https://www.eea.europa.eu/en/european-zero-pollution-dashboards/indicators/pfas-contamination-and-soil-remediation-signal. Scientific Evaluation of the Proposal to Restrict Per-and Polyfluoroalkyl Substances (PFAS)-Current Status. 2025. https://echa.europa.eu/documents/10162/67348133/upfas_evaluation_state_of_play_en.pdf/d1ad6892-e726-84a7-d2dd-74bbf8fa09af?t=1727262011016 (accessed July 2025).

PFAS Environmental Investigation - Fire and Rescue NSW. Available at https://www.fire.nsw.gov.au/page.php?id=9170. (accessed 15.07.25.).

PFAS Investigation and Management Program. Available at https://www.defence.gov.au/about/locations-property/pfas. (accessed 15.07.25.).

PFAS Litigation Risks Mount as EU Claims Rise - Kennedys. Available at https://kennedyslaw.com/en/thought-leadership/article/2023/pfas-litigation-risks-mount-as-eu-claims-rise/. (accessed 15.07.25.).

PFAS Litigation: A Deep Dive into the Rise of Consumer-Led Claims and Its Implications - Wotton Kearney. Available at https://www.wottonkearney.com/pfas-litigation-a-deep-dive-into-the-rise-of-consumer-led-claims-and-its-implications/. (accessed 15.07.25.).

PFAS Monitoring, Power and Water Corporation. Available at https://www.powerwater.com.au/about/what-we-do/water-supply/drinking-water-quality/pfas. (accessed 15.07.25.).

PFAS National Environmental Management Plan. 2018. Available at https://www.dcceew.gov.au/environment/protection/publications/pfas-nemp.(accessed 23.10.25.).

PFAS National Environmental Management Plan 1.0 - DCCEEW. Available at https://www.dcceew.gov.au/environment/protection/publications/pfas-nemp. (accessed 15.07.25.).

PFAS National Environmental Management Plan Version 2.0. Available at https://www.dcceew.gov.au/sites/default/files/documents/pfas-nemp-2.pdf (accessed 15.07.25.).

PFAS National Environmental Management Plan Version 3.0. Available at https://www.dcceew.gov.au/environment/protection/publications/pfas-nemp-3. (accessed 15.07.25.).

PFAS Regulation and Development at the European Level with Focus on Germany and France. Available at https://www.genre.com/us/knowledge/publications/2025/march/pfas-regulation-and-development-at-the-european-level-en. (accessed 15.07.25.).

PFAS Response. Available at https://www.dpi.nsw.gov.au/biosecurity/pfas-response. (accessed July 2025).

PFAS Site Investigations, Environment, Land and Water. Queensland Government. Available at https://www.qld.gov.au/environment/management/environmental/pfas/sites?utm_source. (accessed 15.07.25.).

PFAS—Airports Investigation Program, Department of Infrastructure, Transport, Regional Development, Communications, Sport and the Arts. Available at https://www.infrastructure.gov.au/infrastructure-transport-vehicles/aviation/pfas-airports-investigation-program. (accessed 15.07.25.).

Pitter, G., Da Re, F., Canova, C., Barbieri, G., Zare Jeddi, M., Daprà, F., et al., 2020. Serum levels of perfluoroalkyl substances (PFAS) in adolescents and young adults exposed to contaminated drinking water in the veneto region, Italy: a cross-sectional study based on a health surveillance program. Environmental Health Perspectives 128 (2), 27007. Available at https://doi.org/10.1289/EHP5337.

PROMISCES. Research and Innovation. Available at https://projects.research-and-innovation.ec.europa.eu/en/strategy/strategy-2020-2024/environment-and-climate/european-green-deal/green-deal-projects-support/green-deal-projects-interactive-map/projects/promisces (accessed 25.07.25.).

Queensland Government, Department of Environment, Q. 2019. Queensland Ambient PFAS Monitoring Program Report 2019-2020. Available at https://www.qld.gov.au/environment/management/environmental/pfas/monitoring-program-report (accessed 15.07.25.).

REACH Regulation - Environment - European Commission, n.d. Available at https://environment.ec.europa.eu/topics/chemicals/reach-regulation_en.

REACH Regulation - European Commission. Available at https://environment.ec.europa.eu/topics/chemicals/reach-regulation_en (accessed 13.07.25.).

Regulation - EU - 2024/573 - EN - EUR-Lex. Available at https://eur-lex.europa.eu/eli/reg/2024/573/oj. (accessed 15.07.25.).

Reisser, J., Shaw, J., Wilcox, C., Hardesty, B.D., Proietti, M., Thums, M., et al., 2013. Marine plastic pollution in waters around australia: characteristics, concentrations, and pathways. PLoS One 8 (11), e80466. Available at https://doi.org/10.1371/journal.pone.0080466.

Report on Australian Capital Territory (ACT), PFAS ACTION PLAN. Available at https://www.accesscanberra.act.gov.au/__data/assets/pdf_file/0006/2867784/Report-on-2019-ACT-PFAS-Action-Plan-Implementation.pdf. (accessed 15.07.25.).

Research And Innovation - Per- And Polyfluorinated Substances (Pfas). Available at https://Ec.Europa.Eu/Newsroom/Rtd/Items/821889/En. (accessed 15.07.25.).

Research. Australian Government PFAS Taskforce. Available at https://www.pfas.gov.au/government-action/research. (accessed 15.07.25.).

Revealed: The Massive Contamination of Europe by PFAS Forever Chemicals. Available at https://www.lemonde.fr/en/les-decodeurs/article/2023/02/23/revealed-the-massive-contamination-of-europe-by-pfas-forever-chemicals_6016906_8.html. (accessed 15.07.25.).

Richterová, D., Govarts, E., Fábelová, L., Rausová, K., Rodriguez Martin, L., Gilles, L., et al., 2023. PFAS Levels and determinants of variability in exposure in european teenagers - results from the HBM4EU aligned studies (2014-2021). International Journal of Hygiene and Environmental Health 247, 114057. Available at https://doi.org/10.1016/j.ijheh.2022.114057.

Sadia, M., Nollen, I., Helmus, R., Ter Laak, T.L., Béen, F., Praetorius, A., et al., 2023. Occurrence, fate, and related health risks of PFAS in raw and produced drinking water. Environmental Science & Technology 57 (8), 3062–3074. Available at https://doi.org/10.1021/acs.est.2c06015.

Sardiña, P., Sharp, S., Saaristo, M., Coggan, T., Hoak, M., Leahy, P., 2024. A quantitative classification method of land uses and assessment of per-and poly-fluoroalkyl substances (PFAS) occurrence in freshwater environments. Environmental Pollution (Barking, Essex: 1987) 363, 125272. Available at https://doi.org/10.1016/j.envpol.2024.125272.

Scientific Evaluation of the Proposal to Restrict Per-and Polyfluoroalkyl Substances (PFAS)-Current Status, 2025. Available at https://echa.europa.eu/documents/10162/67348133/upfas_evaluation_state_of_play_en.pdf/d1ad6892-e726-84a7-d2dd-74bbf8fa09af?t=1727262011016 (accessed 15.07.25.).

Soil Health - European Commission. Available at https://environment.ec.europa.eu/topics/soil-health/soil-health_en. (accessed 15.07.25.).

Sonne, C., Jenssen, B.M., Rinklebe, J., Lam, S.S., Hansen, M., Bossi, R., et al., 2023. EU need to protect its environment from toxic per- and polyfluoroalkyl substances. The Science of the Total Environment 876, 162770. Available at https://doi.org/10.1016/j.scitotenv.2023.162770.

Stockholm Convention on Persistent Organic Pollutants (POPs) - DCCEEW. Available at https://www.dcceew.gov.au/environment/protection/chemicals-management/international-agreements/stockholm-convention. (accessed 15.07.25.).

Strategies for Health Protection, Pollution Control and Elimination of Next Generation Refractive Organic Chemicals from the Soil, Vadose Zone and Water. European Commission. https://cordis.europa.eu/project/id/101037509/reporting. Scientific Evaluation of the Proposal to Restrict Per-and Polyfluoroalkyl Substances (PFAS)-Current Status. 2025. Available at https://echa.europa.eu/documents/10162/67348133/upfas_evaluation_state_of_play_en.pdf/d1ad6892-e726-84a7-d2dd-74bbf8fa09af?t=1727262011016. (accessed July 2025).

Szabo, D., Coggan, T.L., Robson, T.C., Currell, M., Clarke, B.O., 2018. Investigating recycled water use as a diffuse source of per- and polyfluoroalkyl substances (PFASs) to groundwater in melbourne, Australia. The Science of the Total Environment 644, 1409–1417. Available at https://doi.org/10.1016/j.scitotenv.2018.07.048.

Taylor, S., Terkildsen, M., Stevenson, G., de Araujo, J., Yu, C., Yates, A., et al., 2021. Per and polyfluoroalkyl substances (PFAS) at high concentrations in neonatal Australian pinnipeds. The Science of the Total Environment 786, 147446. Available at https://doi.org/10.1016/j.scitotenv.2021.147446.

TFA in Water: Dirty PFAS Legacy Under the Radar | PAN Europe. Available at https://www.pan-europe.info/resources/reports/2024/05/tfa-water-dirty-pfas-legacy-under-radar. (accessed 15.07.25.).

The NSW Government PFAS Investigation Program, EPA. Available at https://www.epa.nsw.gov.au/Your-environment/Chemicals/PFAS-in-NSW/nsw-government-pfas-investigation-program. (accessed 15.07.25.).

The PFAS Crackdown: How 2025 Regulations Are Reshaping Industries. Available at https://www.deloachindustries.com/blog/the-pfas-crackdown-how-2025-regulations-are-reshaping-industries. (accessed 15.07.25.).

Therapeutic Goods Administration (TGA), Australian Government Department of Health, Disability and Aging. https://www.tga.gov.au/. (accessed 15.07.25.).

Todd, K., Walker, L., De Bellis Ayers, S., Harris, M., English, K., Rafferty, A. et al., 2025. Acknowledgements: This research is a project of the National Centre for Epidemiology and Population Health.

Understanding New PFAS Regulations in the U.S. and EU | Roland Berger. Available at https://www.rolandberger.com/en/Insights/Publications/Opportunities-challenges-in-new-regulations-of-forever-chemicals.html. (accessed 15.07.25.).

Ventia. PFAS Remediation. Available at https://www.ventia.com/what-we-do/capabilities/pfas-remediation. (accessed 15.07.25.).

Ventia. PFAS Remediation at RAAF Base Edinburgh. Available at https://www.ventia.com/what-we-do/projects/pfas-remediation-at-raaf-base-edinburgh. (accessed 15.07.25.).

Water - European Commission. Available at https://environment.ec.europa.eu/topics/water_en. (accessed 15.07.25.).

Warwick, K.G., Ryan, M.M., Nice, H.E., Wright, I.A., 2025. Contribution of treated sewage to nutrients and pfas in rivers within australia's most important drinking water catchment. Urban Science. Available at https://doi.org/10.3390/urbansci9060182.

Water Framework Directive - European Commission. Available at https://environment.ec.europa.eu/topics/water/water-framework-directive_en. (accessed 15.07.25.).

Western Australian Government. PFAS Investigations in Western Australia. Available at https://www.wa.gov.au/service/environment/environment-information-services/pfas-investigations-western-australia. (accessed 15.07.25.).

Zero Pollution Action Plan - European Commission. Available at https://environment.ec.europa.eu/strategy/zero-pollution-action-plan_en. (accessed 15.07.25.).

CHAPTER

# 23

# Regulatory approaches to per- and polyfluoroalkyl substances in Asia: implications for remediation, enforcement, and public health

*Sai Sharanya Pulimamidi*[1], *Srikanth Ponneganti*[2], *Saurabh Bodake*[1], *Suriyaprakash M.*[1], *Prakash Kumar Panda*[1], *and Roshan M. Borkar*[1]

**[1]Department of Pharmaceutical Analysis, National Institute of Pharmaceutical Education and Research, Guwahati, Changsari, Assam, India [2]Department of Food and Science and Technology, The Ohio State University, Columbus, OH, United States**

## 23.1 Introduction

Per- and polyfluoroalkyl substances (PFAS) are a diverse group of artificial synthetic chemicals with a carbon–fluorine bond, used in products that resist heat, oil, stains, and water. PFAS are often called "forever chemicals"; they persist in biological systems and the environment due to their stability. Their persistence has made them widely used, but also a growing global concern associated with health risks, including disruptions in metabolism, immunity, reproductive health, and increased cancer risk (Kurwadkar et al., 2022; Baluyot et al., 2021; National Institute of Environmental Health Sciences, Perfluoroalkyl).

PFAS resistance to degradation, bioaccumulation, and adverse health effects has caused them to be categorized as persistent organic pollutants (POPs). PFASs are typically used in fluoropolymer coatings and essential products, such as nonstick cookware, firefighting foams, and waterproof textiles. Numerous studies have shown that PFAS exposure causes ecological damage and human health issues, particularly among vulnerable populations (National Institute of Environmental Health Sciences, Perfluoroalkyl; Perera and Meegoda, 2024).

The mismanagement of waste (plastics, landfills, leachate, and wastewater) has massively worsened the PFAS pollution, leaving environmental damage and widespread contamination.

*Per- and Polyfluoroalkyl Substances (PFAS) as Environmental Contaminants*
DOI: https://doi.org/10.1016/B978-0-443-43856-1.00017-4

To address these issues, more stringent regulatory measures, improved waste management techniques, and eco-friendly alternatives to PFAS products are necessary. Advanced treatment methods like activated carbon filtration and the oxidation processes may not be adopted extensively because of their high cost and energy consumption (Mahinroosta and Senevirathna, 2020). Researchers and industry groups are testing alternative options that aim for similar performance without the toxicity or persistence associated with PFAS.

Most countries have their own regulatory standards on PFAS. At the same time, they have also been signatories to the Stockholm Convention, an international treaty seeking to safeguard human health and the environment against the impact of POPs (Stockholm Convention on Persistent Organic Pollutants, POPs).

In response to environmental and health concerns, Japan, China, India, and South Korea have implemented policies to manage PFAS across the Asia–Pacific region. Effective January 10, 2025, Japan bans the production, utilization, and importation of 138 PFAS compounds classified as Class I Specified Chemical Substances under the Chemical Substances Control Law (CSCL) (Cabinet Decision on the Cabinet, Order). The Ministry of Ecology and Environment (MEE) designates specific PFASs for strict control in China. On November 2, 2023, China's Solid Waste and Chemicals Management Center under the MEE published the reference list of perfluorooctanoic acid (PFOA), its salts, and PFOA-related compounds, comprising 363 chemicals along with their CAS numbers, HS codes, and regulatory criteria (China Releases, Reference; Yu et al., 2025). In South Korea, the Korean Registration, Evaluation, Authorization and Restriction of Chemicals (K-REACH) governs the registration, evaluation, and utilization of PFAS (Korean REACH, The). Although India lacks a comprehensive legislative framework for PFAS, it is a signatory to the Stockholm Convention, which regulates PFOS. However, it has not ratified the amendment, highlighting a significant policy gap compared to global efforts.

Due to their persistence, toxicity, and global distribution, PFAS have become a major environmental and human health issue. Although some advances have been made in certain regions of Asia regarding regulation, notable differences remain in how nations manage the risks associated with PFAS. This chapter analyzes PFAS contamination and management in key Asian countries, including Japan, China, India, and South Korea, addressing regulatory frameworks, detection and monitoring methods, remediation efforts, legal challenges, international comparisons, and future directions.

## 23.2 Asian countries

### 23.2.1 Overview of environmental issues in Asian countries

Towards the end of the 20th century, the concept of economic sustainability began to emerge as a guiding principle for a nation's long-term growth among economists. It is common knowledge that sustainable ecpersistent organic pollutants economic growth should not be confined to technological progress, capital investment, and labor force expansion, but should be achieved by maintaining natural ecosystems. In Asia, and especially in developing Asian countries, an emerging series of environmental challenges is likely to jeopardize economic stability, food security, and harmony within the region in the future. Again, Asia's most pressing ecological problems are air pollution, the effects of climate change, deforestation, land degradation, and poor water resource management. Other effects, such as the improper disposal of waste materials, biodiversity loss, and the disruption of marine ecosystems, are also critical. Moreover, the area is exposed to new pollutants, including PFAS, microplastics, personal care products, and drug residues. With rapid urbanization and industrialization that the Southeast Asia region faces nowadays, these nontraditional pollutants in the soil, water, and biota are increasingly becoming a health and environmental issue (Howes and Wyrwoll, 2012; Economic and Social Commission for Asia and the Pacific Committee on Environment and Development Fifth Session, 2018).

## 23.3 Japan: per- and polyfluoroalkyl substances regulatory guidelines

### 23.3.1 Overview of environmental issues in Japan

Renowned for its technological innovation and global influence, Japan faces notable internal challenges that complicate its environmental efforts. These include limited land availability, an ageing population, and a rising national debt, which strain its ability to implement effective large-scale ecological reforms (Environmental Issues in Japan and Solutions, https).

Decades of rapid industrialization, urban expansion, and economic growth have contributed to significant environmental pressures, including air, land, and marine pollution. Air pollution remains one of the most pressing concerns, mainly due to Japan's continued dependence on fossil fuels. Under Japan's Basic Environmental Law, the cornerstone of national environmental policy, air pollution is classified as the top priority among the seven key pollution categories, reflecting its substantial impact on public health and ecosystems (Environmental Pollution in Japan, Statidtics; The Basic Environment Law and Basic Environment Plan, Ministry; Arimura, 2024). Additionally, Japan faces complex challenges from environmental contamination caused by natural disasters. One of the most significant examples is the aftermath of the 2011 Fukushima nuclear disaster. As part of the long-term recovery strategy, Japanese law requires that all radioactive soil and debris generated through decontamination be removed from Fukushima Prefecture by 2045, a massive undertaking reflecting the scale of environmental restoration needed (BfS, Fukushima; Takada et al., 2024).

Environmental chemical monitoring in Japan has revealed the widespread presence of endocrine-disrupting chemicals (EDCs) in river water, sediments, wildlife, and human umbilical cord samples. Recognizing these substances' potential health and ecological threats, Japan has researched EDCs for over a decade. In 1998 the Ministry of the Environment (MOE) launched the Strategic Programs on Environmental Endocrine Disruptors (SPEED), focusing on screening suspected environmental endocrine disruptors. Initially identifying 67 chemicals for further study, this list was narrowed down to 65 chemicals by 2000 (Schug et al., 2016; Iguchi et al., 2002). To build on SPEED, Japan introduced the Extended Tasks on Endocrine

Disruption (EXTEND) 2010 program, which aimed to advance risk assessment methodologies for EDCs and implement regulatory actions as necessary. Under Japan's CSCL, chemicals found to have endocrine-disrupting effects through EXTEND may face restrictions or bans (Strategic Programs on Environmental Endocrine Disruptors '98 SPEED '98, Health; Coady et al., 2017; Endocrine Disruptors Regulations and Lists in Japan, https).

Ongoing research has continued to provide valuable insights. For example, the National Institute for Environmental Studies (NIES) reported in 2024 that bisphenol F may aggravate allergic asthma, and bisphenol S may induce anxiety-like behaviors in senescence-accelerated mouse models (NIES Annual Report, National). Another study explored using Japanese medaka (*Oryzias latipes*) as a model organism to identify EDCs that could impact human health. The findings suggested significant potential for applying laboratory-based data to support human health risk assessments (Dasmahapatra et al., 2023).

Human exposure studies have also been conducted. A pilot study involving 42 men from a Tokyo fertility clinic found that semen quality was significantly influenced by dietary habits (soy and coffee intake) and exposure to pyrethroid insecticides (Toshima et al., 2012). The Hokkaido Study on Environment and Children's Health continues to monitor long-term effects of chemical exposure. Urinary biomarkers collected from children at age seven revealed associations between prenatal and childhood exposure to substances such as phthalates, PFAS, polychlorinated biphenyls (PCBs), and dioxins, with increased risks of allergies, asthma, and neurodevelopmental disorders (Atsuko et al., 2023). These findings underscore the importance of continuous biomonitoring and comprehensive risk assessment strategies in gaining a better understanding of the cumulative and interactive effects of EDCs. They also underline the urgent need for stronger regulatory measures and policies to mitigate human and environmental exposure to these harmful substances.

POP pollution is a persistent problem in Japan. The nation is a signatory to the Stockholm Convention, an international treaty that safeguards individuals and the environment against persistent toxic chemicals (Stockholm Convention on Persistent Organic Pollutants, POPs). Following its inclusion of new substances in Annex B by the Fourth Conference of the Parties (COP), Japan designated PFOS, its salts, and perfluorooctane sulfonyl fluoride (PFOSF) as Class I Specified Chemical Substances in the CSCL. The rule took effect on April 1, 2010 (Substances under the Chemical Substances Control Law, Class).

PFAS are a large group of synthetic organic compounds that have been manufactured since the 1950s. While they have offered valuable properties such as heat, water, and oil resistance for commercial and industrial applications, their environmental persistence and tendency to bioaccumulate have raised serious concerns about their long-term impacts on ecosystems and human health (Lin et al., 2020; Megson et al., 2025). Once released into the environment, PFAS can contaminate soil, air, and water, and are now recognized as emerging pollutants due to their resistance to natural degradation. Their widespread and prolonged use has made environmental monitoring and regulatory intervention increasingly critical. A significant challenge in PFAS research is their environmental persistence, which allows them to accumulate in various ecological compartments to detectable levels (Xu et al., 2021; Peritore et al., 2023).

Several studies in Japan have documented the health risks associated with PFAS exposure. Data from the Japan Environment and Children's Study showed a possible association between maternal PFOS exposure in early pregnancy and chromosomal abnormalities in offspring (Hasegawa et al., 2024). Another study involving control subjects from a Japanese breast cancer case-control group found a correlation between PFAS levels in blood and increased global DNA methylation in leukocytes, suggesting possible epigenetic effects of PFAS exposure in women (Itoh et al., 2023). The Hokkaido Study on Environment and Children's Health also focused on prenatal PFAS exposure, reporting that higher exposure levels were associated with smaller birth size, hormonal disruptions, altered immune function, and changes in neurobehavioral development. Furthermore, prenatal PFAS exposure was found to significantly affect DNA methylation patterns at birth, indicating potential long-term developmental impacts (Atsuko et al., 2023).

Recognizing the environmental and public health threats posed by PFAS, Japan has been actively engaged in both domestic regulation and international efforts to address POPs. As a signatory to the Stockholm Convention since 2002, Japan designated PFOS, its salts, and PFOSF as Class I Specified Chemical Substances under its CSCL following their inclusion in Annex B of the Convention in 2010. The country has been a leading contributor to the Convention's work, sending experts to the POPs Review Committee (POPRC) since 2005 and to the Technical Working Group on Global POPs Monitoring since 2006 (MOE Japan) (Chemicals and Hazardous Waste, Stockholm).

The high risk of PFAS and other environmental contaminants, due to the prevalence and potential health effects of these chemicals, has increased the importance of a thorough and effective regulatory framework in Japan. This chapter discusses regulatory tools, enforcement practices, and remedial actions in Japan aimed at addressing chemical contamination, particularly PFAS. Evaluating Japan's progress in environmental monitoring, risk assessment, and remediation, the presented section demonstrates the effectiveness of the country's environmental governance system. It identifies the areas where policy and implementation can be strengthened to protect the health of the country's population and the ecosystem.

### 23.3.2 Regulatory framework for per- and polyfluoroalkyl substances

Japan has taken a proactive approach in dealing with environmental contamination due to PFAS, which pose significant environmental risks due to their implications for land, water, air, and human health. To control and reduce PFAS contamination, Japan has a well-developed chemical and environmental regulatory system that permits a variety of laws and international engagements. These include:

- CSCL (Act on the Evaluation of Chemical Substances and Regulation of Their Manufacture, etc.)
- Poisonous and Deleterious Substances Control Act
- Air Pollution Control Act
- Water Pollution Prevention Act (WPPA)
- Soil Contamination Countermeasures Act
- Stockholm Convention on POPs
- PRTR (Pollutant Release and Transfer Register) system.

Together, these mechanisms form the foundation of Japan's national strategy for regulating PFAS and other hazardous substances, aimed at protecting human health and the environment.

#### 23.3.2.1 *Chemical Substances Control Law (1973)*

The CSCL, formally titled the Act on the Evaluation of Chemical Substances and Regulation of Their Manufacture, etc., was enacted in 1973 to prevent environmental pollution from industrial chemicals harmful to human health and ecosystems. The law regulates both domestically manufactured and imported chemicals in Japan. Oversight is shared by three key ministries: the MOE, the Ministry of Economy, Trade and Industry (METI), and the Ministry of Health, Labor and Welfare (MHLW) (Cabinet Decision on the Cabinet, Order; Action Taken by Three Ministries against Violations of Chemical Substances Control, Law).

To align with global standards and fulfill Japan's obligations under the Stockholm Convention on POPs, the CSCL was significantly amended in 2009, with full implementation beginning on April 1, 2011. Before the amendment, CSCL could only enforce controls listed under Annex A of the Convention (complete bans). Postamendment, the law accommodates Annex B (restrictions) and Annex C (unintentional releases). Under the Cabinet Order for enforcement, Class I Specified Chemical Substances are defined as those with environmental persistence, high bioaccumulation potential, and long-term risks to human health or predators at higher trophic levels (Fukushima, 2021).

Under Japan's CSCL, key PFAS compounds are regulated by designating them as Class I Specified Chemical Substances, a category reserved for highly persistent, bioaccumulative, and toxic chemicals. In 2009 Japan classified perfluorooctane sulfonate (PFOS) and its salts, PFOSF, PFOA, and specific branched perfluoroalkanoic acids with eight carbon atoms, along with their salts, under this category. Once a compound is designated Class I, its manufacture, import, and use are effectively prohibited, including the import of products containing these substances. However, because PFOS and its derivatives were listed under Annex B (Restriction) rather than Annex A (Elimination) of the Stockholm Convention, some essential uses (e.g., in semiconductor manufacturing) are still permitted in semiconductors. This prompted Japan to revise the CSCL framework to incorporate the "essential use" concept for regulating PFAS (Listing of POPs in the Stockholm Convention, https; Summary of the Guideline on the Treatment of Wastes Containing Perfluorooctane Sulfonic Acid PFOS, and Its Salts in Japan, 2013).

In a recent development, Japan initiated a public consultation in September and October 2024 regarding the listing of 138 PFOA-related substances as Class I Specified Chemical Substances. This culminated in the issuance of Ministerial Ordinance No. 4 by the MHLW and other authorities in November 2024, with implementation scheduled for January 2025. These substances are now covered under entry 35(c) of the CSCL list (Japan Designates List of PFOA-Related Compounds as Class I Specified Substances, SGS; Substances under the Chemical Substances Control Law, Class).

#### 23.3.2.2 *Stockholm Convention on Persistent Organic Pollutants*

The Stockholm Convention is a global treaty aimed at protecting human health and the environment from POPs, which persist in the environment, accumulate in living organisms, and pose significant health risks. Adopted in 2001 and enforced since 2004, the Convention requires its member Parties to take specific regulatory actions to eliminate, restrict, or reduce the production, use, and release of POPs. Since its adoption, the COP has expanded the original list of 12 chemicals to 30 by early 2020, through nine decision-making sessions held between 2005 and 2019 (UNEP 2001–21) (Fiedler et al., 2008; Malisch et al., 2023). Japan became a Party to the Convention on August 30, 2002, and has played an active role in shaping and supporting the development of an enforceable international framework for the regulation of POPs. It has integrated the Convention's provisions into its domestic laws, notably through the CSCL (The National Implementation Plan of the Republic of Korea, https).

Key PFAS-related regulatory milestones under the Stockholm Convention and Japan's actions:

- 2009: In response to the Convention, Japan designated PFOS and its salts as Class I Specified Chemical Substances under the CSCL, marking the country's first regulatory action on PFAS (Cabinet Decision on the Cabinet, Order; Summary of the Guideline on the Treatment of Wastes Containing Perfluorooctane Sulfonic Acid PFOS, and Its Salts in Japan, 2013).
- 2019: Following the amendment to Part I of Annexe A, Article 4 of the Stockholm Convention (decision SC-9/12), member countries, including Japan, were required to eliminate the production and use of PFOA, its salts, and related compounds (UNEP, 2019; All POPs listed in the Stockholm Convention, Annex).
- 2022: At the Tenth Meeting of the COP, the Convention Parties agreed to include perfluorohexane sulfonic acid (PFHxS), its salts, and related compounds in Annex A, mandating their elimination (Tenth Meeting of the Conference of the Parties to the Stockholm Convention, https).
- 2022: During its 18th meeting, the POPRC evaluated long-chain perfluorocarboxylic acids (PFCAs) with 9–21 carbon atoms, their salts, and related substances. These were proposed for regulation under decision POPRC-18/5, in accordance with Article 8, paragraph 9 of the Convention (Report of the Persistent Organic Pollutants Review Committee on the work of its eighteenth, meeting).

#### *23.3.2.3 PRTR system*

Japan's PRTR system plays a vital role in environmental monitoring by identifying chemical emission sources, tracking the types and quantities of chemicals released annually, and ensuring public access to this information. The system was established under the 1999 Act on Confirmation of Release Amounts of Specific Chemical Substances in the Environment and Promotion of Improvements to the Management Thereof (PRTR Substances, PRTR; Chemical Management, Act).

The government developed Chemical Substance Management Guidelines to support effective implementation, which industries must follow. These guidelines promote better chemical handling practices and transparency by encouraging businesses to share information related to chemical management with the public (Japan's "PRTR Act", Eviliance).

Japan has added PFOS to the PRTR system, making annual reporting mandatory as part of its commitment to monitoring emerging contaminants. PFOS is classified as a Class I Designated Chemical Substance, one of 462 chemicals identified based on risk and exposure considerations, including their potential impacts on the ozone layer, ecosystems, biodiversity, and human health (PRTR Substances, PRTR).

In 2021 Japan extended this classification to include PFOA and its salts, restricting their manufacture and import. Under the PRTR Act, industries handling these substances are required to report their annual release and transfer volumes. This regulatory action is part of Japan's broader efforts to address PFAS pollution, which also includes setting temporary limits for PFAS in drinking water and enhancing public health protections (Japan's "PRTR Act", Eviliance).

#### 23.3.2.4 *Water Pollution Prevention Act*

The WPPA is Japan's primary legislation for controlling water pollution in groundwater and public water bodies. It regulates effluent discharges from manufacturing facilities, promotes wastewater treatment, and holds industries accountable for pollution-related damage to protect public health and the environment (Water Pollution Prevention Act, English).

Under the WPPA, workplaces with specified facilities must ensure that hazardous substances in their discharged water remain below the designated effluent standards. However, PFAS compounds are not currently classified as hazardous substances under this Act, and therefore, these requirements do not yet apply to facilities discharging PFAS-containing effluents.

Recognizing the emerging threat posed by PFAS, the MOE introduced a provisional guideline value in May 2020, setting a concentration limit of $50\,ng\,L^{-1}$ ($0.00005\,mg\,L^{-1}$) for PFOS and PFOA in groundwater and public drinking water. These substances were designated as "items requiring monitoring" (Japan's 5th Expert Meeting Proposed 50 ng/L Combined Limit for PFOS/PFOA in Water, Enviliance).

Further strengthening regulatory oversight, the MOE officially listed PFOA, PFOS, and their salts as designated substances in February 2023. As a result, operators of designated workplaces are now legally obligated to take emergency response measures to prevent the release or infiltration of these substances into the environment in the event of an accidental discharge.

In terms of liability, under the Japanese Civil Code, dischargers may be held legally responsible if a court establishes a causal link between PFOS/PFOA contamination in drinking or discharged water and human health impacts (Civil Code, Japanese).

#### 23.3.2.5 *Soil Contamination Countermeasures Act*

This Act enhances public health by implementing effective countermeasures against soil contamination. It achieves this by setting up the procedures to determine the occurrence of the specified harmful substances in soil and taking precautions to eliminate possible risks to human health. Currently, PFAS compounds are not classified as designated hazardous substances under this Act, nor are materials containing PFAS directly regulated. Nevertheless, constant research and evaluation by the MOE may contribute to the development of soil contamination standards or regulatory rules of PFAS in the future (Soil Contamination Countermeasures Act, English).

### 23.3.3 Limits on per- and polyfluoroalkyl substances in Japan

On April 1, 2020, the MHLW in Japan designated PFOS and PFOA substances as water quality management goals. A provisional target value of $50\,ng\,L^{-1}$ (combined for PFOS and

PFOA) was set. This limit is based on the fact that an individual weighing 50 kg will use two litres of drinking water per day throughout their life without incurring any health problems. The government has also indicated that it will soon convert this provisional guideline into a mandatory regulatory standard (Japan's 5th Expert Meeting Proposed 50 ng/L Combined Limit for PFOS/PFOA in Water, Enviliance; Regarding the "Review of Environmental Standards for the Protection of Human Health Related to Water Pollution", 5th).

#### *23.3.3.1 Government agencies participating in per- and polyfluoroalkyl substances regulation in Japan*

In Japan, multiple government agencies are involved in the regulation of PFAS, including:

- MHLW
- METI
- MOE

PFOA and PFOS have been identified as monitored substances by the MHLW, which set provisional guideline values for the presence of PFOA and PFOS in drinking water and river systems (Regarding the "Review of Environmental Standards for the Protection of Human Health Related to Water Pollution", 5th). The METI, in collaboration with other ministries, is enforcing a ban on 138 PFAS compounds, which are classified under Class I Specified Chemical Substances. This prohibition on manufacturing, importing, and using the same will take effect on January 10, 2025. METI also requires the reporting of the quantities of PFAS produced or imported under the responsibility of the CSCL (Japan Will Ban, 2025).

Meanwhile, the MOE has also published technical guidelines on how to treat PFOS waste. The document, a set of Technical Documents on the Treatment of Wastes Containing PFOS (revised March 2011), provides detailed interpretations of the Waste Management Law regarding solid and liquid wastes produced during the manufacturing and utilization of PFOS and Salts. It offers uniform orientation on how to handle waste, which involves the way of collection, transportation, storage, and disposal (Summary of the Guideline on the Treatment of Wastes Containing Perfluorooctane Sulfonic Acid PFOS, and Its Salts in Japan, 2013).

### 23.3.4 Per- and polyfluoroalkyl substances detection methods

To address industrial water and wastewater issues, Japan established a national analytical standard based on the first version of ISO 25101 (2009), incorporating technical adaptations specific to the country. Detection Procedures were provided in the Japanese Industrial Standard (JIS) K 0450-70-10, which outlines methods for quantifying perfluorinated compounds, such as the linear-chain isomers of PFOS and PFOA. This method is performed through solid-phase extraction and validated using high-performance liquid chromatography–tandem mass spectrometry (HPLC-MS/MS). It can be used in the industrial water and wastewater sectors, allowing for accurate monitoring of trace levels of PFAS (JISC, JIS K 0450-70-10).

In a similar study, the suspect in PFAS water samples collected in regions affected by firefighting foam and the drinking water treatment plant (DWTP) in Okinawa, Japan, was screened using liquid chromatography in conjunction with a quadrupole time-of-flight mass spectrometer. The results revealed a need to have sophisticated detection technologies to track the developing PFAS compounds in water systems within environments and industries (Yukioka et al., 2020).

### 23.3.5 Per- and polyfluoroalkyl substances monitoring techniques

To address PFAS pollution, Japan has emphasized the importance of robust environmental monitoring and management strategies. These efforts involve identifying pollution sources and collecting comprehensive data on PFAS concentrations in soil, water, air, and biota to understand the environmental distribution and persistence of these compounds (Sharma et al., 2024a). Management strategies include transitioning to PFAS-free alternatives, implementing best practices in industrial processes, and establishing risk-based standards for managing contaminated sites. As part of regulatory action, Japan has banned 138 PFAS compounds, further reinforcing the need for long-term surveillance.

Japan initiated continuous environmental monitoring in 1978 for wildlife and in 1986 for water and bottom sediments, with the goal of assessing long-term trends in the persistence of chemical pollutants (Chemicals and Hazardous Waste, Stockholm).

Specific monitoring activities included:

- Sediment and wildlife surveys in 2003 and 2005
- Water monitoring in 2002 and 2005
- Air quality assessments in FY2004.

PFOS has been monitored on a regular basis since fiscal year 2009. In the 2014 financial year, linear n-perfluorooctane sulfonic acids (n-PFOS) were detected and analyzed in wildlife samples, as well as in air, water, and sediment samples from various locations (see Chapter 3, 2016).

Although PFOS and PFOA were designated as "monitoring substances" for drinking water standards in 2009, the government did not enforce specific regulatory limits. Japan has no binding standards for PFOS, PFOA, or other PFAS compounds in tap water. Japan's Waste Management Law applies to materials containing fluorine and its compounds for the purpose of waste management. The 2011 technical guidelines for PFOS waste treatment outline proper disposal procedures, including handling solid and liquid waste generated during PFOS production and use (Japan PFAS Situation Report, 2019).

### 23.3.6 Remediation efforts of per- and polyfluoroalkyl substances

Ongoing monitoring and regular reporting play a critical role in guiding remedial actions to reduce the environmental and human health impacts of PFAS. These practices not only track the extent of contamination but also help ensure compliance with regulatory standards and assess the effectiveness of mitigation strategies (Yu et al., 2025). As part of its proactive regulatory approach, Japan is set to implement a ban on PFOA-related compounds effective January 10, 2025, to curb further environmental pollution across soil, water, air, and other environmental media. Extensive scientific research has documented PFAS contamination in key environmental matrices throughout Japan. For instance, elevated PFAS concentrations have been frequently observed in soils surrounding facilities that manufacture fluorochemicals and in areas used for firefighting training. Additionally, groundwater contamination has been widely reported in regions where aqueous film-forming foams were historically used in industrial processes or during firefighting drills. These studies underscore the persistence, mobility, and bioaccumulative nature of PFAS, which pose significant challenges to remediation efforts (Yu et al., 2025; Japan PFAS Situation Report, 2019).

#### 23.3.6.1 *Foam fractionation*

In anticipation of stricter future regulations, companies such as Shimizu are actively researching and testing solutions for remediating PFAS-contaminated soil and groundwater in the fiscal year 2021. Foam fractionation is one such technique, which includes pollutants adhering to the foam's surface and then being separated, and has been used to address groundwater contamination (Shimizu Commences, PFAS; Effectiveness of Shimizu PFAS Soil Remediation Technology Confirmed in the U.S., Company).

#### 23.3.6.2 *Phytoremediation and adsorbent-based approaches*

Adsorbent materials such as activated carbon are commonly used for PFAS remediation and have shown greater efficacy in removing long-chain PFAS from water than short-chain compounds. A novel approach in Japan utilizes phytoremediation through rice cultivation, where rice plants absorb PFAS from contaminated soil and irrigation water, primarily sourced from groundwater and rivers. The PFAS accumulated in the plant biomass can then be converted into biochar (activated carbon) through thermal treatment, effectively eliminating the contaminants. This integrated method helps remediate paddy fields and groundwater, contributing to a sustainable water purification cycle through rice production. Rice-based biochar represents a promising strategy for PFAS removal in agricultural settings (Yamashita et al., 2024).

#### 23.3.6.3 *Soil washing with water–organic solvent systems*

Soil washing using optimized water–organic solvent mixtures is emerging as a promising remediation strategy for PFAS-contaminated soils, particularly those with clay content. The study involved a systematic analysis of batch and column tests to evaluate the effectiveness of various solvents in extracting PFAS from clay soils. A total of 22 solvents were assessed, and it was found that short-chain PFCAs could be removed effectively using small concentrations of solvents (0%–5%). Conversely, long-chain PFCAs (perfluorosulfonic acids [PFSAs]) were resistant to solvent extraction, requiring 80% solvent to extract such compounds effectively. Of all the tested solvents, the values of the removal efficiency occurred in the sequence EtOH (Ethanol) = MeOH (Methanol) < ACN (Acetonitrile). These measurements emphasize the vital importance of optimizing solvent systems dependent on PFAS structure and chain length to maximize the results of soil remediation (Li et al., 2024a).

#### 23.3.6.4 *Per- and polyfluoroalkyl substances decomposition by visible light-emitting diode at ambient temperature*

Scientists at Ritsumeikan University have developed a new technique to degrade PFASs in an ambient environment using visible light-emitting diodes. Doing this will ensure that

PFOS, one of the most commonly found PFAS, is completely broken down within eight hours, even at room temperature. It is worth mentioning that the decomposition of PFAS was efficient, and the technique also enables fluorine recovery, thus supporting sustainable fluorine recycling. This development is a bright and energy-efficient substitute to the existing high-temperature processes, and also helps in lessening reliance on new production supplies of fluorine compounds (Arima et al., 2024).

#### 23.3.6.5 *Case studies reported on per- and polyfluoroalkyl substances in Japan*

Table 23.1 presents a series of case studies on other PFAS compounds in various areas of Japan, focusing on the detection, concentration, and potential health and environmental outcomes of these compounds. The studies employ a diverse range of sample types, including air, water, soil, consumer products, and biological samples from humans, and document both local incidental contamination situations and national exposure trends. Comprehensively, they can provide substantial information on the presence, distribution, and risks of PFAS in the Japanese arena.

### 23.3.7 Legal and policy challenges

#### 23.3.7.1 *Legal framework for per- and polyfluoroalkyl substances*

Japan has implemented several laws targeting PFAS, but its legal framework remains limited in scope. Under the CSCL, the manufacture and import of PFOS (since 2010), PFOA (since 2021), and PFHxS (since 2024) have been prohibited (Per and Poly-Fluorinated Chemicals PFAS, OECD). A 2024 report noted a decline in PFOS usage, particularly in firefighting foams, and emphasized the promotion of alternatives and ongoing monitoring. In 2020 Japan introduced a provisional target value of 50 ng $L^{-1}$ for the PFOS and PFOA in water. A revision to establish enforceable drinking water standards is planned for April 2026. If levels exceed the limit, remedial measures outlined by the MOE and the MHLW must be followed.

Waste with fluorinated compounds is managed under the Waste Management and Public Cleansing Law. The MOE provides technical guidance for handling 15 types of PFOS-contaminated waste, covering collection, transport, storage, and disposal (Waste Management And Public Cleansing, Law).

The Food Safety Commission of Japan (FSCJ) has also reviewed the health effects of synthetic organofluorine compounds, including PFAS, in food and beverages. Based on FSCJ evaluations, Japan plans to set quality standards for PFOS and PFOA in sterilized and filter-sterilized mineral water under the Water Supply Act (Japan Establishes New Standards for PFOS and PFOA in Mineral Water, RRMA; Food Sanitation Act, English).

#### 23.3.7.2 *Role of industries in resisting per- and polyfluoroalkyl substances regulations in Japan*

Several industries in Japan, including electronics, automotive manufacturing, and industrial coatings, have expressed concerns over the challenges of substituting PFAS compounds. PFAS are widely used in insulation, lubricants, seals, and specialized coatings due to their unique chemical stability and resistance to heat, oil, and water. Industry

**TABLE 23.1** Case studies reported on per- and polyfluoroalkyl substances (PFAS) in Japan.

| Type of PFAS | Study (year) | Study participants/ location | Sample type | Detection frequency | Concentration reported | Outcomes/inference | References |
|---|---|---|---|---|---|---|---|
| 18 PFAAs (PFSAs and PFCAs), HFPO–DA, DONA, 6:2 Cl-PFESA, 26 pre-PFAS | 2025 | Tokyo, Japan | Air samples | PFBS (100%)<br>PFBA (100%)<br>HFPO–DA and DONA (95% and 79%) | PFBS (117 pg $m^{-3}$)<br>PFBA (78.0 pg $m^{-3}$)<br>HFPO–DA and DONA (1.99 and 1.08 pg $m^{-3}$) | The study enhances understanding of PFAS atmospheric behavior and supports tailored mitigation strategies. | Asakawa et al. (2025) |
| 20 neutral PFAS<br>30 ionic PFAS | 2024 | Waste recycling facility in Japan | Air samples | | **Neutral PFAS**: 6:2 FTMAC (38 ng $m^{-3}$; 88 pmol $m^{-3}$)<br>6:2 FTOH, 4.9 to 17 ng $m^{-3}$ (13–47 pmol $m^{-3}$)<br>**Ionics PFAS:**<br>PFHxA (41 ng $m^{-3}$, 133 pmol $m^{-3}$) | This study will aid in developing PFAS emission control strategies for RPF production and waste management. | Kuribara et al. (2024) |
| PFSAs, PFCAs, FTUCAs, FOSAAs, FTSAs, FTOHs, FTIs, FDIAs, FOSEs, FOSAs, FIAs, 8:2 diPAP, HFPO–DA, DONA, 6:2 Cl-PFESA | 2021 | AIST West campus in Tsukuba City, Japan | Air samples | **Neutral PFASs**<br>6:2 and 8:2 FTOHs (40% and 32%)<br>**Ionizable PFASs**<br>PFCAs (78%), FTUCAs (11%), PFSAs (5%), FTSAs (5%), and other PFASs (1%) | **Neutral PFASs**<br>4:2, 6:2, 8:2, and 10:2 FTOH were 4.0, 190, 160, and 36 pg $m^{-3}$<br>**Ionizable PFASs**<br>PFASs: -11 pg $m^{-3}$ | Air samples showed an industrial shift in PFAS production from longer to shorter chains. Study findings indicate that the current approach enables a better understanding of PFASs' fates in the air. | Wu et al. (2021) |
| PFOS, PFOA, monoPAPs, diPAP | 2003 | Japan | House dust | | PFOS (15–2500 ng $g^{-1}$)<br>PFOA (69–3700 ng $g^{-1}$)<br>monoPAPs (3.7 to 1023 ng $g^{-1}$)<br>diPAPs (3.6 to 692 ng $g^{-1}$)<br>PFCAs (230 ng $g^{-1}$) | This investigation revealed that all household dust samples had high amounts of PFOS and PFOA. | Moriwaki et al. (2003) |

(*Continued*)

TABLE 23.1 (Continued)

| Type of PFAS | Study (year) | Study participants/ location | Sample type | Detection frequency | Concentration reported | Outcomes/inference | References |
|---|---|---|---|---|---|---|---|
| Suspect screening of 277 PFASs | 2020 | Okinawa | Water samples | | 116 candidate PFAS (105 negative and 22 positive compound groups) | This study used profile analysis and suspect screening to identify 116 potential PFAS contaminants in water affected by firefighting training. | Yukioka et al. (2020) |
| PFOS, PFHpA, PFOA, PFNA | | Tokyo | Groundwater samples | | PFOS (0.28–133 ng $L^{-1}$) PFHpA (<0.1–20 ng $L^{-1}$) PFOA (0.47–60 ng $L^{-1}$) PFNA (0.1–94 ng $L^{-1}$) | All groundwater samples from Tokyo contain PFAS compounds, which have resulted in groundwater contamination. | Murakami et al. (2009) |
| PFOS | 2014 | | Seawater (surface, intermediate, and bottom) and sediment samples | PFOS (100%) | Surface seawater (2.0–7.3 ng $L^{-1}$) Middle seawater (2.2–5.7ng $L^{-1}$) Bottom seawater (1.5–3.4 ng $L^{-1}$) Sediment levels (0.3–0.9 ng $g^{-1}$) | PFOS contamination in Tokyo Bay sediment and seawater was analyzed using a fate model based on river inflows. | Miyake et al. (2014) |
| PFAS | 2008 | Water purification plants in Osaka | Raw and treated tap water | PFOA (100%) | PFOA: 5.2–92 ng $L^{-1}$ (ppt) PFOS: 0.20–22 ng $L^{-1}$ (ppt) | PFAS clearance rates in many water treatment facilities were below 50% but improved with frequent activated carbon changes. In several factories, effluent PFAS levels were found to be higher than the influent levels. | Guruge et al. (2008) |

| | | | | | | | |
|---|---|---|---|---|---|---|---|
| PFOA, PFNA, PFDA, PFUnDA, PFDODA, and PFTrDA | | women | | PFDA, PFUnDA, and PFOS (100%) PFHxA (20%) PFHpA (50%) PFDoDA (97%) PFTrDA (97%) PFTeDA (13%) PFHxS (77%). | PFOA (24.88 ppb) PFNA (13.19 ppb) | infectious diseases in children is associated with prenatal exposure. | et al. (2018) |
| PFOA, PFHxS, PFNA, PFHpA | 2008 | Women | Breast milk | PFOA and PFHxS (92%) PFNA (13%) PFHpA (25%) | Highest level of PFOS (523 ppt) | Japanese breast milk at over 26 times the level allowed in drinking water. | Tao et al. (2008) |
| EtFOSE, | 2015 | | Consumer products | PFNA (16%) | EtFOSE (25,000 ng $g^{-1}$) PFNA (220 ng $g^{-1}$) | PFAS appear in textile sprays, car wash products, coating agents, and rust inhibitors. The authors suggest PFNA replace PFOA in Japan. | Ye et al. (2015) |
| PFNA, PFUnDA, PFDoDA, PFTrDA, and PFTeDA | 2018 | Dall's porpoises and harbor porpoises in Hokkaido | Liver, blood, and blubber | PFOA (94%) PFOS (98%) PFHxS (54%) PFUnDA and PFTrDA (70%) | PFUnDA (226 ng $g^{-1}$-wet) PFTrDA (102 ng $g^{-1}$ -wet) PFOA (5.2 ng $g^{-1}$ -wet) | High PFCA levels in Dall's porpoises show that PFAS reach deep seawater. The buildup of long-chain PFCAs indicates that FTOHs are a major source. | Fujii et al. (2018) |
| PFHxS, PFOS, PFOA, PFNA, PFDA, PFUnDA, PFDoDA, and PFTrDA | | Pregnant women | Maternal plasma | | PFOS (4.92 ng $mL^{-1}$) PFOA (2.01 ng $mL^{-1}$) PFUnDA (1.43 ng $mL^{-1}$) PFNA (1.18 ng $mL^{-1}$) | Prenatal PFHxS/PFOS exposure may impair children's immunity, increasing infectious disease risk. | Goudarzi et al. (2017a) |
| PFOS and PFOA | 2017 | Women and their children | Cord blood | | PFOS (5.1 ng $mL^{-1}$) and PFOA (1.4 ng $mL^{-1}$) | Prenatal exposure to PFAS may alter cord blood adiponectin levels and reduce birth size, while PFOS may disrupt fetal metabolism. | Minatoya et al. (2017) |

*(Continued)*

TABLE 23.1 (Continued)

| Type of PFAS | Study (year) | Study participants/ location | Sample type | Detection frequency | Concentration reported | Outcomes/inference | References |
|---|---|---|---|---|---|---|---|
| PFOS and PFOA | 2017 | Women and their children | Cord blood | | PFOS (5.1 ng $mL^{-1}$) and PFOA (1.4 ng $mL^{-1}$) | The authors suggest that disrupted glucocorticoid and DHEA balance at birth linked to PFAS exposure may harm the HPA axis and hormone balance later, indicating a public health concern. | Goudarzi et al. (2017b) |
| PFOS and PFOA | 2016 | Mother-infant pairs | Cord blood | | PFOS (5.1 ng $mL^{-1}$) and PFOA (1.4 ng $mL^{-1}$) | The authors indicate that low levels of PFOS and PFOA can disrupt fetal reproductive hormones and adversely affect steroid hormone synthesis and Leydig and Sertoli cell function. | Itoh et al. (2016) |
| PFAS | | Wild bird species from Miyagi and Tochigi prefectures | Livers | PFOS (72%), 7:3 FTCA (66%), and PFNA (61%) | PFOS (<0.25–210 ng $g^{-1}$) | The study found PFAS levels below toxic thresholds, but did not consider mixing effects, exposure timing, or the unclear age of the standards. | Guruge et al. (2011) |
| PFOS and PFOA | 2009 | Pregnant women | Serum | PFOS (100%) PFOA (92.8%) | PFOS (1.2–16.2 ng $mL^{-1}$) PFOA (ND–5.3 ng $mL^{-1}$) | The research found a negative correlation between birth weight and exposure to low levels of PFOS, despite lower blood PFOS levels than those affecting development in animal studies. | Washino et al. (2009) |

representatives have emphasized that replacing these substances could take 5 to 10 years, particularly given the complexities of global supply chains and the lack of viable alternatives in certain sectors.

In response to these concerns, the Japanese government has allowed limited exemptions for essential uses where suitable alternatives are not yet available, and the environmental risk is considered low. However, products designed for direct environmental release remain subject to strict import controls, with significant penalties for noncompliance. The METI clarified that such exemptions are closely monitored, and compliance is enforced through Japan's CSCL. Under CSCL, PFOA, perfluorohexanoic acid (PFHxA), and PFHxS are classified as "General Chemical Substances." Since 2011, manufacturers and importers producing over one tonne of these substances annually have been required to report quantities to METI, facilitating tracking and regulatory oversight (Acceptable Purposes, PFOS; Japan to Tighten Regulations on Dechlorane Plus to Comply with Stockholm Convention, Foresight; Act on the Regulation of Manufacture and Evaluation of Chemical Substances, English).

Japan's three major PFAS manufacturers, Daikin, Asahi Glass Company (AGC) (formerly Asahi Glass Co.), and Mitsui-DuPont, play a key role in this landscape:

Daikin, Japan's largest producer of PFAS, operates manufacturing facilities globally, including in China, France, and the United States. In August 2014, Daikin joined the US Environmental Protection Agency (EPA)'s 2010/2015 PFOA Stewardship Program. By the end of 2013, the company had phased out PFOA from its fluoropolymer products (Letter Inviting Participation in the PFOA Stewardship Program, US) and fully ceased PFOA production by the end of 2015. As replacements, Daikin developed shorter-chain fluorinated telomers such as PFHxA, which are believed to be less bioaccumulative (Daikin's Approach to PFAS, Fluorochemicals).

AGC also aligned with the Stewardship Program by launching fluorinated products with perfluoroalkyl chains of six carbon atoms or fewer. These were specifically designed to avoid forming PFOA or long-chain PFCAs during degradation (Regarding the PFAS, concern).

Despite these efforts, exemptions have been sought. Notably, Daikin submitted a request under the Stockholm Convention to use PFOI as an intermediate in manufacturing PFOB, which is used in pharmaceutical production. This exemption reflects the extension between industrial innovation and environmental safety within Japan's PFAS regulatory strategy (Requests for Extension of an Exemption in the Register, https).

#### 23.3.7.3 *Public awareness*

Widespread contamination by PFAS came to light following a series of comprehensive nationwide investigations initiated by the Japanese government in 2020. These findings drew significant media attention, which in turn heightened public awareness of the issue (Wang, 2025).

### 23.3.8 Comparison with international regulations

PFAS regulations vary widely across countries. Some nations have implemented complete bans on specific PFAS compounds, while others have opted for restricted usage based on assessed risks. Japan has proactively addressed PFAS contamination in its water, air, soil, and broader environment. Beginning January 10, 2025, a ban will take effect on 138 PFAS substances related to PFOA. In comparison, the European Union (EU) has been one of the most active regions in regulating PFAS. The European Chemicals Agency has advocated for a comprehensive ban, with a particular focus on PFAS in firefighting foams. The REACH regulation may restrict or prohibit the use of any chemical that creates an unacceptable risk to humans or the environment. In January 2023, five EU member states (Denmark, Germany, the Netherlands, Norway, and Sweden) submitted a proposal together to restrict C9–C14 PFCAs, including their salts and precursors, with a few exceptions provided in cases of essential application. Japanese regulations require the CSCL, which identified PFOS and PFOS salts as Class I Specified Chemical Substances in 2009. Japan also prohibits the export of PFOS under the Foreign Exchange and Foreign Trade Law in accordance with the Stockholm Convention. In Japan, the production and use of PFOS, PFOA, and related salts have been banned since 2020, and a provisional limit of 50 ng $L^{-1}$ of PFOS and PFOA has been set for drinking water. In the United States, the regulation of national PFAS is led by the EPA. It has even offered guidelines to limit the reintroduction of almost 300 PFAS compounds to the market and demand that the inclusion of PFAS in products be disclosed at any concentration level. Australia has also aligned its regulatory efforts with international standards. Under the Australian Industrial Chemicals Introduction Scheme, importers and exporters must obtain official clearance to trade PFOA-related compounds. Australia's regulatory approach is strongly influenced by developments in the US and the EU, reflecting a broader move toward global harmonization in PFAS management (Per- and Polyfluoroalkyl Substances; REACH Regulation - European Commission).

### 23.3.9 Future directions and recommendations

**Nationwide PFAS surveillance**: Japan should conduct comprehensive national surveys to assess the extent of PFAS contamination. Current studies are often limited in scope, either geographically or by chemical type, hindering a clear understanding of the overall situation. Greater transparency is essential, including regular public disclosure of data and findings to build public trust and awareness.

**Development of safer alternatives**: There is a pressing need to accelerate research and innovation for safer substitutes to PFAS. Leveraging its technological leadership, Japan has the potential to pioneer environmentally friendly materials and industrial processes that can replace PFAS across various sectors.

**Expand human biomonitoring (HBM) efforts**: A notable lack exists in nationwide HBM studies in Japan. To better evaluate long-term health risks, it is crucial to identify sources of exposure and reduce PFAS intake. Regular monitoring of biological samples (e.g., blood, breast milk, cord blood) will help in understanding population-level exposure trends and their health impacts.

**Strengthen international collaboration**: PFAS pollution is a global issue that transcends national borders. Japan should take a more active role in international cooperation, aligning regulations, sharing data, and participating in joint research initiatives to develop coordinated, science-based solutions to this environmental challenge.

# 23.4 China: per- and polyfluoroalkyl substances regulatory guidelines

## 23.4.1 Overview of environmental issues in china

China, as one of the most rapidly industrializing nations of the world, is experiencing increasingly sharp environmental challenges because of several decades of its rapid economic development, rapid urbanization process, and high energy consumption rate. The developments have contributed to the deterioration of air, water, and land quality, which is placing a significant burden on ecosystems and the healthcare system of the population. The emergence of PFAS and other persistent and bioaccumulative pollutants is an especially urgent issue, as those are now being discovered in various environmental matrices throughout the country.

Air pollution in China is a significant and alarming issue that has worsened over time, particularly in densely populated urban and industrial areas. The levels of fine particulate material (PM2.5) consistently exceed the health requirements developed internationally and are closely linked with elevated chances of developing heart disease or ailments related to breathing (Ambient Outdoor, Air). At the same time, the problem of water pollution poses a significant threat to the environment's integrity and the health of the population. According to government figures, nearly 90% of the city's groundwater supplies do not meet national quality standards. Not only PFAS, but also various industrial pollutants end up in waterways and increase the chemical burden (Hu and Cheng, 2013). Soil contamination further complicates the situation. According to the data published by the Ministry of Agriculture, about 16% of arable land has higher pollutant levels than safe above long-term safety thresholds, which adds to concerns regarding the sustainability of agricultural systems and food safety (Wen et al., 2022). Moreover, China's status as the world's largest emitter of greenhouse gases magnifies the urgency for integrated and sustainable pollution management strategies.

With this chapter discussing the regulatory governance relating to PFAS in China, it is critical to place the discussion within the context of China's larger environmental pressures and industrial history. PFAS are also challenging environmental management due to their persistence, long-range mobility potential, and toxicological profiles. Extensive use in industrial products, firefighting foams, textiles, and consumer products has resulted in measurable levels of PFAS in the air, water, soil, and in living organisms. There is an ever-increasing need for robust regulatory mechanisms, synergistic surveillance practices, and emerging remediation technologies in China's environmental policy landscape. The given chapter assesses the evolving strategy of PFAS management in China in relation to laws, regulatory norms, enforcement policies, and the development of new solutions to protect public health and environmental integrity.

## 23.4.2 Overview of environmental and chemical regulations

To address the urgent necessity of controlling pollution from both old and emerging contaminants, China has established a dynamic and extensively constructed environmental regulatory regime, with a focus on gaining control over PFAS, which is a growing concern. The country has major laws that govern its chemical and environmental policies. The Water

Pollution Prevention and Control Law mandates stringent controls on industrial wastewater discharge and emphasizes monitoring POPs, including PFAS. The fact that PFAS may also be included in this legal framework is another indication that the PFAS persistence in water bodies and possible threats that they may pose to citizens are becoming more acknowledged (Hu and Cheng, 2013).

Likewise, the Soil Pollution Prevention and Control Law provides a legal basis for identifying and remediating contaminated sites, including PFAS "hotspots" caused by industrial discharges or improper waste disposal. The Air Pollution Prevention and Control Law, while originally focused on conventional air pollutants, is now being adapted to address volatile PFAS compounds and thermally decomposed by-products generated during manufacturing or incineration processes (Jin et al., 2016). These laws form an increasingly integrated legislative framework through which PFAS and other emerging contaminants are under regulatory oversight.

## 23.4.3 Regulatory framework for per- and polyfluoroalkyl substances

Over the past decade, China has made significant efforts to introduce specific regulations targeting PFAS compounds. In 2019 the MEE published Announcement No. 10, which prohibited the production and use of PFOS and its salts except in minimal cases of industrial applications. This was a significant policy change compared to the previous policy of laxity toward PFAS-containing chemicals. Additional support took place in 2023 when the new List of Key Controlled New Pollutants was published, which officially incorporated other high-concern PFAS, including PFOA and PFHxS, and is moving toward a class-based framework of PFAS regulation (China Releases, Reference).

China is also beginning to establish numerical limits to acceptable levels of PFAS in different environmental media. The new national drinking water standard (Guobiao [GB] 5749–2022) establishes permissible levels of 80 and 40 $ng\,L^{-1}$ for PFOA and PFOS, respectively (China Implements First Chemical Park PFOA and PFOS, Control). These limitations are more stringent than certain global standards; however, they currently apply only to drinking water. The Environmental Quality Standard for Soils (GB 15618–1995) emphasizes the prevention of soil pollution and the protection of public health in areas such as farmland, vegetable fields, and nature reserves. Nevertheless, it may not specifically be a list of PFAS. MEE is directly addressing PFAS pollution in the soil by establishing regulations and limitations, as well as continually conducting risk evaluations. Precise and detailed protocols for PFAS in soil may still be under development or refinement.

Despite these advances, significant gaps remain in coverage regarding national regulatory limits for PFAS in air, food, and biosolids. Reports of PFAS levels above the upper limit in groundwater in the industrial provinces of Shandong serve as a reminder to come up with wider and more enforceable standards (Wang et al., 2025; Li et al., 2022).

### *23.4.3.1 Government agencies involved in per- and polyfluoroalkyl substances regulation*

The regulation of PFAS in China involves a network of governmental agencies with overlapping responsibilities. The MEE is the primary body responsible for implementing chemical and environmental policy at the national level. The MEE is involved in the registration of chemicals and the assessment of their safety levels through China REACH (Order No. 12), where it controls the hazards and their control measures of substances, such as PFAS. The Solid Waste and Chemicals Management Technology Center supports the work of MEE by keeping technical documentation containing a list of PFOA-related substances and their allowable use, etc., to adhere to the regulations (China Releases, Reference).

Concerning PFAS, public health surveillance is currently not focused on this substance; however, health-related governmental agencies are conducting surveillance of pollutants in water supplies and food. However, a national PFAS biomonitoring program is lacking, translating to a significant exposure assessment deficit. It is the mandate of local and provincial governments to enforce and implement national policies; however, the varying levels of technical competencies and resources result in inconsistent enforcement across different regions.

Companies in highly risk sectors, such as fluoropolymer manufacturers and textiles, are rigorously obliged to report and comply with requirements. Although compliance results are never uniform, companies are being held accountable for obligations in high-risk industries. All of these combined are constructed into a decentralized yet converging system of governance in the management of PFAS risks in China.

## 23.4.4 Per- and polyfluoroalkyl substances detection and monitoring

Over the past few years, there have been significant developments in PFAS detection and monitoring programs in China, as concerns have been raised about the issue of its ubiquitous environmental distribution. In a study of 526 drinking water samples collected in 66 urban areas throughout the country, more than 20% of the samples had a concentration of PFAS higher than recommended international health guidelines, including those of the US EPA and European agencies (Liu et al., 2021). This research, which covered a population of more than 450 million citizens, found hotspots in some cities, including Zigong, Jiujiang, and Lianyungang, whose PFAS were high. The authors translated PFAS levels into estimated daily intake values of PFAS. They concluded that the exposure was higher, especially among infants and children, surpassing international standards of PFOA and PFOS. The two above findings are an eye-opener about the need to adopt a standardized national surveillance system in China. Currently, PFAS is not monitored in water quality assessments regularly, and there is a lack of a centralized surveillance program to monitor their distribution. The absence of standardized detection procedures and limited data from rural and remote regions introduces essential vulnerabilities in exposure measurement. The research suggests the creation of a countrywide, systematic PFAS monitoring network that integrates advanced analytical techniques and region-specific risk thresholds, thereby aligning the context of the Chinese work on preventing environmental health hazards with the best efforts of the global community.

Taihu Lake study has shown that known PFAS represented only one-third of the total extractable organofluoride, implying that the contaminants represented discovery or new compounds. Furthermore, longitudinal surveys have shown that PFOA levels in surface waters have increased by 86% since 2011, even as PFOS levels decline. Based on the results of the review done by Jiao et al. (2022), it can be concluded that the transmission routes depend on the source of infection in both primary and secondary infections.

Further case-study reports on PFAS have been reported in China and are included in Table 23.2.

### 23.4.5 Remediation efforts of per- and polyfluoroalkyl substances

The China strategy for PFAS remediation incorporates both known and the latest technologies to mitigate environmental and human exposure risks. Conventional approaches to treating PFAS-contaminated groundwater and industrial wastewater include granular activated carbon (GAC) filtration and ion-exchange resins. Effectively applied to long-chain PFAS, these methods commonly fail to remove shorter-chain forms due to their greater solubility and mobility.

New technologies are being tested and applied to eliminate these limitations. In situ treatment methods, such as liquid-activated carbon, are gaining popularity due to their cost-effectiveness and minimal environmental impact. The most recent studies also allow consideration of electrochemical oxidation, nanofiltration, and degradation using plasma techniques as next-generation solutions, and some pilot projects have demonstrated high removal rates under controlled conditions. Nevertheless, there are hurdles to implementing these technologies on a large scale, especially regarding cost, operational stability, and the treatment of mixed PFAS waste streams (Alazaiza et al., 2025).

### 23.4.6 Legal and policy challenges

Although the efforts have been significant, the regulatory system concerning PFAS in China has several legal and policy challenges. One of the most significant issues is the lack of comprehensive, coordinated national standards for PFAS in all environmental matrices. The regulations governing drinking water exist, but the air, food, and biosolid matrices are not regulated, which does not bring clarity to enforcement and limits the scope of pollution control efforts. In addition, the province-to-province differences in implementing existing laws result in inconsistent oversight and monitoring, thereby undermining the effectiveness of the entire regulatory system. Regulatory development is also complicated by industrial resistance. The tough industries, including the textile manufacturing industry, semiconductor manufacturers, and fire-fighting foam manufacturers, have been reluctant to adopt PFAS-free alternatives, commonly citing high costs, technology-related concerns, and a loss of competitiveness. In certain instances, industrial stakeholder lobbying and a lack of visibility in the supply chain have hindered or slowed down the regulatory process.

The population's perception of PFAS and their health effects is increasing, although it is not yet comparable to the level of awareness in Western countries. Although nongovernmental organizations and research institutions are beginning to raise concerns, the advocacy at the grassroots level is being limited due to inaccessibility to environmental data, as well as the lack of widespread education on PFAS-related risks (Liu et al., 2021).

**TABLE 23.2** Case studies reported on per- and polyfluoroalkyl substances (PFAS) in China.

| S. No. | Study aim | Study year | Location and participants | Sample type | PFAS (conc.) measured | Key findings | Inference | References |
|---|---|---|---|---|---|---|---|---|
| 1. | Assessed PFAS biomigration in the aquatic food chain in the Fu River | 2025 | Wuhan, Fu River Basin; aquatic biota and residents | Water, sediment, fish, eggs | PFBS: 168 ng $L^{-1}$ (water), PFOS: 1510 ng $kg^{-1}$ (sediment) | PFOS is dominant in fish/eggs; short-chain PFAS are emerging | High dietary exposure risk due to the replacement of legacy compounds by short-chain PFAS. | Shi et al. (2025) |
| 2. | Evaluated PFAS in sewage sludge and transformation dynamics | 2025 | Multiple WWTPs across China | Sewage sludge | PFOS and PFOA: up to 1000 ng $g^{-1}$ | PFAS persist in sludge; transformation is influenced by sludge composition | Sludge is a PFAS reservoir, and heavy metals may affect its behavior and fate. | Guo et al. (2025) |
| 3. | Investigated the effect of PFAS exposure on immune health among white-collar workers in industrial areas | 2025 | Industrial region, white-collar employees | Human serum, seven common PFAS (including PFCAs, PFSAs) | Total PFAS =65.5 ng $mL^{-1}$; PFOA, PFOS dominant | Oxidative stress markers increased (MDA); IgM immune antibody levels decreased | Immune suppression can develop among nonindustrial workers due to environmental exposure to PFAS. | Gao et al. (2025) |
| 4. | Measured PFAS pollution in groundwater and surface water near a closed fluorochemical facility | 2025 | Industrial zone near the river basin | Groundwater and river water; 17 PFAS types | PFBS peaked at 11,463 ng $L^{-1}$; PFBA and PFHxS were also elevated | High contamination persists, especially in the rainy season | PFAS pollution persists long after industrial activities have ceased. | Xu et al. (2025) |

*(Continued)*

TABLE 23.2 (Continued)

| S. No. | Study aim | Study year | Location and participants | Sample type | PFAS (conc.) measured | Key findings | Inference | References |
|---|---|---|---|---|---|---|---|---|
| 5. | Compiled PFAS levels from over 350 studies across China to map national exposure | 2024 | Nationwide: food, blood, and water studies | Aggregated samples; focus on PFOS, PFOA, emerging PFAS | PFOA up to 53.9 ng $mL^{-1}$; PFOS 43.9 ng $mL^{-1}$ | PFOS and PFOA remain prevalent; limited data on soil and air | Short-chain and novel PFAS are increasing; stricter regulation is essential. | Li et al. (2024b) |
| 6. | Examined the relationship between maternal PFAS exposure and miscarriage risk | 2024 | Shanghai hospital cohort, 393 pregnant women | Maternal plasma tested for 30 PFAS | 15 PFAS detected in >90% of participants; PFOA linked with OR =2.99 | 13 PFAS showed a significant association with miscarriage | Prenatal PFAS exposure may increase miscarriage risk | Wang et al. (2024) |
| 7. | Monitored changes in PFAS blood levels over a decade in Hubei Province | 2023–24 | Tongji cohort, 216 individuals | Human serum, 30 PFAS measured | PFOS declined; PFOA increased; Cl-PFESA stable | Legacy PFAS are decreasing, while replacements like Cl-PFESA accumulate | New PFAS are becoming prominent in human exposure profiles | Cheng et al. (2025) |
| 8. | Performed risk assessment of PFAS in drinking water across 66 cities | 2021 | Nationwide (452 million population) | Drinking water | >20% cities exceeded Vermont MCL (20 ng $L^{-1}$) | East and SW China at higher risk; Yangtze basin hotspots | Urgent need for PFAS elimination and strict monitoring | Liu et al. (2021) |
| 9. | Measured PFAS in seafood imported from China | 2022 | Coastal regions; FDA sampling | Canned clams | PFOA exceeded safe exposure limits | Potential health risks from imported seafood | Highlights global exposure via trade | Further Research on PFAS Pollution in China, PFAS |

| | | | | | | | | |
|---|---|---|---|---|---|---|---|---|
| 10. | Evaluated human exposure near the fluorochemical industrial park | 2017 | Fuxin, Liaoning; 100 residents | Serum, drinking water | Serum PFOA: up to 160 ng $mL^{-1}$; Water: 8.5 ng $L^{-1}$ | PFOA levels 3× higher than in 2009; 35% exposure from water | Industrial emissions are a major contributor and an urgent need for source control | Bao et al. (2017) |
| 11. | Quantified PFAS in surface water and precipitation across China | 2010–20 | Nationwide, rivers, lakes, Tibetan Plateau | Surface water, rain | PFAS: $0.775–1.06 \times 10^6$ ng $L^{-1}$ (water); 4.2–191 ng $L^{-1}$ (rain) | PFOS/PFOA is dominant in urban areas; short-chain PFAS rising | Urbanization and FMFs drive PFAS load; shift to short-chain PFAS | Wang et al. (2022) |
| 12. | Historical PFOS production and regulatory response | 2002–09 | National production data | Industry records | PFOS production peaked in 2006; declined post-2009 | Stockholm Convention inclusion led to a production drop | Global treaties influenced domestic policy shifts | Further Research on PFAS Pollution in China, PFAS |

### 23.4.7 Comparison with international regulations

The Chinese advancements in regulating PFAS are admirable when compared to international standards, particularly in drinking water regulations. The PFOA maximum allowable concentration limit, 80 ng $L^{-1}$, and PFOS 40 ng $L^{-1}$, as per GB 5749–2022 standard, are lower than the interim health proposal of the US EPA, 70 ng L. Such values are similarly consistent with the cumulative criterion of the EU, which is 100 ng $L^{-1}$ for 20 PFAS chemicals (China Implements First Chemical Park PFOA and PFOS, Control). Nevertheless, unlike in some other countries that are implementing a class-based or grouping strategy, China's policies continue to be compound-specific, and regulation of a broader spectrum, particularly short-chain and emerging PFAS, is still under development.

### 23.4.8 Future directions and recommendations

China is likely to expand its regulatory framework on PFASs to newer compounds and enact a more systematic basis of enforcement in the future. A layer of technological advancement in PFAS remediation is also on course. Various new treatment methods, including electrochemical oxidation, nanofiltration, and plasma-based degradation, have demonstrated high efficacy in both laboratory and pilot-scale studies. Technologies that combine adsorptive, oxidative, and waste material filtration are also being tested for large-scale implementation. The innovations and partnering with the government are expected to contribute to the move towards safer and affordable mitigation of PFAS. To enhance control of PFAS, China needs to set up complete national standards, a harmonized enforcement system, and motivate cleaner technology in each sector. The impacts of such efforts will not only bring the country in line with international regulatory practice but also protect the health of the population and the sustainability of the environment.

## 23.5 India: per- and polyfluoroalkyl substances regulatory guidelines

### 23.5.1 Overview of environmental issues in India

India faces numerous environmental challenges. The country is currently experiencing environmental conditions that are both harmful and unsustainable. India's rapidly increasing population and economic development are key contributing factors to environmental pollution. The main issues are air pollution occurring through vehicle exhaust and industry, water pollution by untreated waste and municipal malpractices, depletion of water resources, land contamination, deforestation, land agriculture loss, habitat loss and depletion, loss of biodiversity, and general depletion of resources (Kumar et al., 2022; Pawar, Present).

EDCs, a rapidly expanding type of pollutant, are increasingly being discharged into the environment as a result of industrialization, agricultural development, and population growth (Chaturvedi et al., 2023). The chemical regulations in India are scattered across a dozen Acts and Rules, with general categorizations including consumer safety, occupational health, environmental protection, and trade. Although the future Chemicals (Management and Safety) Rules take account of endocrine disruption, there are no clear definitions or prioritization of EDCs (Sharma et al., 2022). In 2016 the MoEFCC prohibited lead in paint with concentrations higher than 90 ppm, in recognition of its endocrine-disrupting characteristics. However, investigations have found that, even now, lead is still present in some paints at levels thousands of times above the limit. In parallel, the banned bisphenol A (BPA) was found in falsely labeled BPA-free products in 2015, which had been prohibited in baby feeding bottles. Such results highlight the poor implementation and the necessity to regulate and monitor EDC more closely in India (Study on Lead in Paints in India, 2019; Toxics Link Report, 2018; Endocrine Disrupting Chemicals, https).

Studies have shown that the environment or human populations in developing states like India are exposed to toxic levels of endocrine disruptors, which are responsible for increasing cases of health problems like cancer and diabetes (Sharma et al., 2022). Nevertheless, EDC pollution in India is less studied than in the developed world. Recent research demonstrates contamination of drinking water with EDCs, particularly in rural regions of the country, with considerable gender differences in serum concentrations (Sharma et al., 2025; Hadibarata et al., 2020). There are also seasonal variations in the levels of phthalates and BPA contained in drinking water, which are not only harmful but also pose health risks, especially di(2-ethylhexyl) phthalate (Kumawat et al., 2022).

PFASs are the most commonly identified EDCs in India, appearing in food, drinking water, dust, breast milk, hair, and blood (Hadibarata et al., 2020; Gautam et al., 2025). The health risks of PFAS remain pervasive and persistent, with some studies confirming their continued use in India despite being included in the Stockholm Convention as a substance subject to restricted monitoring and controls (G et al., 2023; Mukherjee Das and Janardhanan, 2022). PFAS have even been found at distressing levels in environmental sampling throughout India, such as 136.27 ng $L^{-1}$ of PFAS present in Chennai's surface and groundwater, and 12,958 ng $L^{-1}$ of PFOS in the Periyar estuary, both of which are attributed to discharges from industrial sources (Binu et al., 2022; Koulini and Nambi, 2024). Contamination of road dust was observed in multiple cities, presumably due to emissions of automobiles (Yamazaki et al., 2023). There are more threats posed by bioaccumulation. Fish, shrimp, and dolphins living in the Ganges and Sundarbans contain PFAS at concentrations as high as 83.9 ng $g^{-1}$ wet weight (Jala et al., 2023; Corsolini et al., 2012; Yeung et al., 2009), and PFAS were found in 90% of teas sold in the Indian market (Jala et al., 2023). PFASs were found in breast milk samples of women in Chidambaram, Kolkata, and Chennai. PFOS levels (46 ppt) were higher than the 20 ppt advisory limit, and PFOA levels were at least 16 times higher than the advisory level (Tao et al., 2008). Although India signed the Stockholm Convention in 2006, it has not yet regulated PFAS in the country (Brennan et al., 2021).

The high prevalence of PFAS in the Indian environment and human bodies poses a significant public health and environmental threat, particularly in regions with high contamination rates. To address PFAS contamination, regular monitoring and implementing efficient policies and regulations are necessary. India does not yet have a comprehensive framework in place to address PFAS risks. The following section will discuss current data availability, areas of regulatory gaps, and the necessity for policy action to address PFAS contamination nationwide.

## 23.5.2 Regulatory framework for per- and polyfluoroalkyl substances in India

Despite increasing global interest in "forever chemicals," there is little regulation of PFAS in India. As industrialization has expanded and the widespread consumer product application of PFAS has grown to include textiles, food packaging, and even fire department foam operations, the sense of urgency of PFAS contamination is mounting. Despite emerging research and recognition of PFAS-associated health and environmental risks, India has yet to establish a comprehensive regulatory framework.

India was a signatory to the Stockholm Convention in 2006. Although PFOS was considered a restricted substance under the Convention and included in the list in 2009, India has not yet ratified this amendment. The Convention is designed to protect human health and the environment from POPs, to which PFAS compounds also belong. International guidance on alternatives to PFOS has been published; however, no specific implementation has been undertaken in India (India PFAS Situation Report, 2019). India has taken the first step toward regulating PFOA and PFOS by implementing standard sampling and analysis techniques. The Bureau of Indian Standards (BIS) has introduced three standardized methods for the sampling and analysis of PFOA and PFOS:

- ISO 3696:1987: Water for analytical laboratory use – Specifications and test methods.
- ISO 5667-1: Water quality sampling – Part 1: Guidance on the design of sampling programs and sampling techniques.
- ISO 8466-1:1990: Water quality, Calibration and evaluation of analytical methods – Part 1: Statistical evaluation of the linear calibration function.

Although these are initial steps towards monitoring, in India, there are currently no binding regulations, phase-out programs, or detailed PFAS control guidelines in place. Strong legal and environmental policies are still in development (Reply On Behalf Of Respondent No. 1 Central Pollution Control Board CPCB, https).

### *23.5.2.1 Per- and polyfluoroalkyl substances government agencies in the regulation*

A number of regulators in India have started the process of dealing with PFAS contamination, though explicit and comprehensive regulations are still lacking:

1. **BIS:** BIS has established standards on products like textiles and food packaging, which can indirectly tackle the issue of PFAS contamination. Despite the adoption of the International Standards Organization's criteria for sampling and testing PFOA and PFOS in 2020, additional initiatives are urgently required to address the unregulated and indiscriminate use of pervasive toxic chemicals such as PFAS, which are a primary source of exposure with significant detrimental effects on human health and the environment (Mukherjee Das and Janardhanan, 2022; Forever Chemicals, An).
2. **National Green Tribunal (NGT):** The NGT took suo motu cognizance over a study published by IIT Madras indicating the presence of PFAS in the lakes and drinking water of Chennai. The research associated contamination with raw sewage, industrial effluents, and dumping. The Tribunal advised the Central Pollution Control Board (CPCB) to examine the results and suggest parameters and rehabilitative measures under the Environment (Protection) Act, 1986, and the Water (Prevention and Control of Pollution) Act, 1974 (IIT Madras Study, 2024).

3. **Council of Scientific and Industrial Research (CSIR)–National Environmental Engineering Research Institute (NEERI):** CSIR–NEERI, under the Ministry of Environment, Forest and Climate Change, is undertaking a nationwide survey of the POPs, including PFAS, to contribute to the revision of the National Implementation Plan of India under the Stockholm Convention. The study also evaluates the manufacturing industries of PFAS. Nevertheless, timely regulatory action is required to limit contamination and protect public health (CSIR News, https).
4. **CPCB:** Although CPCB has started some initial research on PFAS contamination in water bodies, there are no formal directives on PFAS uses, disposal, and monitoring in India yet (Guidelines for Water Quality Monitoring Central Pollution Control Board Parivesh Bhawan East Arjun Nagar, Delhi).

These efforts represent an initial response to mounting environmental and public health concerns, although a robust, comprehensive regulatory framework for PFAS is still necessary.

### 23.5.3 Limits on per- and polyfluoroalkyl substances in India

Regulatory bodies worldwide have established guidelines or recommendations for permissible levels of PFAS in drinking water. However, India currently lacks such guidelines or awareness regarding PFAS.

### 23.5.4 Per- and polyfluoroalkyl substances detection and monitoring

In India, there is an urgent need to identify and track the presence of PFAS. Studies of PFAS gained momentum after 2015, driven by advancements in detection technologies and growing awareness of the environmental and health effects of PFAS.

The CSIR and the Indian Institute of Toxicology Research lead the capacity for PFAS detection in India. These efforts include developing advanced LC-MS/MS methods and working with the Food Safety and Standards Authority of India (FSSAI) and the United Nations Environment Program to study PFAS in food packaging, consumer products, and water (CSIR-IITR, Leads, CSIR-IITR Leads, India).

A comprehensive study was conducted across the nation utilizing an HPLC-MS/MS to examine 25 PFAS in 39 hair samples collected from 14 Indian cities, with 9 quantifiable PFAS, including PFHxS, PFOS, and PFOA. It was also the first to report PFAA precursors in human hair, suggesting the potential use of hair as a biomonitoring tool, although additional studies are required (Ruan et al., 2019).

According to a study, 12 PFAS compounds were identified in the groundwater of an Assam region called Kamrup, highlighting the importance of monitoring and assessing drinking water sources for PFAS, particularly in areas with significant manufacturing and industrial activities (Sharma et al., 2024b).

Researchers at IIT Kharagpur developed a new electrochemical sensor utilizing two-dimensional (2D) aluminum-based multicomponent quasicrystals (2D-Al QC) for the highly sensitive detection of PFOA at subpicomolar concentrations (LoD: $0.59 \pm 0.05$ pM), offering a promising tool for rapid and sensitive environmental monitoring (Sharma et al., 2024b).

A comprehensive overview of PFAS levels in India is provided in Table 23.3, which includes information on concentrations in various regions and environmental matrices.

## 23.5.5 Remediation efforts of per- and polyfluoroalkyl substances

The increasing presence of PFAS in the environment requires a comprehensive approach that combines preventive controls with solutions to address current contamination. Multiple remediation methods are being investigated to address PFAS in soil, water, sediments, and other matrices, with a focus on removal efficiency, stability, and the reduction of long-term exposure risks.

### *23.5.5.1 Remediation technologies for water treatment to remove per- and polyfluoroalkyl substances*

#### 23.5.5.1.1 Adsorption

PFAS are efficiently eliminated from water and wastewater by adsorption and ion exchange. A popular adsorbent, GAC, has a 99.9% removal effectiveness for PFAS at concentrations ranging from 0.6 to 35,500 $\mu g\ L^{-1}$. For more than a decade, GAC has been used to treat particular PFAS chemicals. Point-of-use -based GAC treatment requires a 4' x 4' room, but it has limitations due to frequent GAC regeneration and replacement issues. For eliminating PFAS from water, GAC remains the most effective technique.

#### 23.5.5.1.2 Coagulation/flocculation

By incorporating a coagulant, such as iron or aluminum chloride, into wastewater, suspended particles can be removed by 85%–90% using standard sedimentation and filtering methods. However, this kind of traditional coagulation technique doesn't effectively remove PFAS (Review of Water Treatment Systems for PFAS Removal, Concave).

#### 23.5.5.1.3 Foam fractionation

Foam fractionation can use multistage methods to reduce PFAS levels below 6000 $\mu g\ L^{-1}$, particularly at low concentrations. Both foam fractionation and removal of short-chain PFAS have trouble and necessitate an extra polishing unit. Adsorption and nanofiltration are necessary to enhance treatment quality, which increases the operational cost. Since adsorption using GAC produces a comparable outcome, foam fractionation is not generally advised (Review of Water Treatment Systems for PFAS Removal, Concave).

### *23.5.5.2 Approaches based on destruction*

#### 23.5.5.2.1 Electrochemical oxidation

Electrochemical oxidation utilizes electrons from electrodes to break down contaminants in liquids. The method relies on the applied potential; to remove PFAS from groundwater and industrial wastewater, boron-doped diamond electrodes are frequently used. However, this process can be costly, hazardous, and requires specific setup, which limits its effectiveness in treating PFAS (Gomez-Ruiz et al., 2017).

TABLE 23.3 Case studies reported on per- and polyfluoroalkyl substances (PFAS) in India.

| Type of PFAS | Study (year) | Study participants/ location | Sample type | Detection frequency | Concentration reported | Outcomes/inference | References |
|---|---|---|---|---|---|---|---|
| Perfluoroalkylsulfonates, perfluorocarboxylates, perfluorooctanesulfonamide, N-ethyl perfluorooctane sulfonamide, N-ethyl perfluorooctane sulfonamidoacetate | 2009 | Ganges River | Ganges River water and biota | Water samples PFOA (55%) PFDA, PFNA, PFPeA, and PFHxA (10%, 40%, 38%, and 43%) Fish samples PFUnDA, PFDA, PFNA, and PFOA (5%, 92%, 33%, 75%) Ganges River Dolphins PFUnDA, PFDA, and PFNA (80%, 87%, and 53%) | Water: <0.04–3.91 ng $L^{-1}$; biota: 0.248–27.9 ng $g^{-1}$ ww | India's production of fluoropolymers is not well documented. These results can be used as a starting point for further research, and tracking PFCs in the Ganges River may help to understand how they flow into the Bay of Bengal. | Yeung et al. (2009) |
| 34 PFAS (7 PFSAs, 13 PFCAs, 11 precursors, and 3 emerging PFAS) | 2023 | Road dust samples collected across India | Road dust | 21 PFAS (>30%) | 21 quantified PFAS (23.3–861 pg $g^{-1}$) | This is the first study to reveal legacy and rising PFAS and precursors in road dust nationally in India. The data can establish a PFAS cleanup road map for developing countries. | Yamazaki et al. (2023) |
| 12 PFAS (PFBS, PFHXS, PFOS, PFDCS, PFHXA, PFHPA, PFOA, PFNA, PFDA, PFUnA, PFDOA, PFOSA) | 2023 | Regions of Assam | Tea bag samples | ΣPFAS (26.58%) | ΣPFAS-22.36 ng $g^{-1}$ PFOS and PFHx 8.08 ng $g^{-1}$ and 5.90 ng $g^{-1}$ | This study highlights the health risks of long-term exposure to PFOS, PFHxS, and PFuNA in India, making it significant. The study emphasizes ongoing monitoring | Jala et al. (2023) |

(Continued)

TABLE 23.3 (Continued)

| Type of PFAS | Study (year) | Study participants/ location | Sample type | Detection frequency | Concentration reported | Outcomes/inference | References |
|---|---|---|---|---|---|---|---|
| | | | | | | and measurement of PFAS levels in regularly utilized items to maintain public safety. | |
| FOA and PFOS | 2016 | Chennai | Surface water | | PFOA (4 to 93 $ng\ L^{-1}$) and PFOS (3 to 29 $ng\ L^{-1}$) | The current analysis may be beneficial for evaluating future monitoring initiatives for PFOA and PFOS in surface water. | Sunantha and Vasudevan (2016) |
| PFCAs, PFSAs | 2016 | Ganges | River water and groundwater | | <MQL – 10.2 $ng\ L^{-1}$ | Ganges River basin PFAS releases and drinking water exposure are assessed. Residential and industrial waste pollutes groundwater, so people face exposure even when drinking water shows low PFAS levels. | Sharma et al. (2016) |
| PFOA, PFOS | 2012 | Ganges (Hugli) River | Surface sediment | | PFOA: <0.5 to 14.09 $ng^{-1}$ dry wt PFOS: <0.5 $ng\ g^{-1}$ dry wt | Untreated effluents and wastewater are likely the main sources of PFCs buildup in sediments. The current study can serve as a benchmark for future ecosystem monitoring projects. | Corsolini et al. (2012) |

| | | | | | | | |
|---|---|---|---|---|---|---|---|
| FTOHs, FOSAs/FOSEs and FTOs | 2024 | Howrah, Kolkata, Tripura, Assam, West Bengal, Chennai, Bangalore, Mumbai | Ambient air | | 54 to 820 pg $m^{-3}$ | The irregular geographical distributions of these PFCs highlight the importance of point sources in the dispersion of these pollutants throughout the Indian troposphere. | Li et al. (2011) |
| PFCAs, PFSAs, FOSA | | Delhi, Gujarat | Particulate matter | | | The data will help you understand how PFAS levels in air particles vary by region and particle size | Lin et al. (2020) |
| PFOS, PFHxS, PFOA, PFOSA | 2004 | Coimbatore | Human sera | | <3 ng $mL^{-1}$ | Blood contains several perfluorochemicals, according to the results. Hazard evaluation should encompass all chemicals and their cooccurrence, while ongoing research should examine human exposure sources and pathways. | Kannan et al. (2004) |
| PFOA, PFNA, PFDA, PFUnDA, and PFDoDA | 2008 | Chidambaram, Chennai, and Kolkata | Breast milk | PFOS (85%) | <1.66–335 pg $mL^{-1}$ | Breastfed Asian infants had 7–12 times greater daily PFOS exposure than previously reported studies. Assessing PFCs' impact in early childhood requires more research. | Tao et al. (2008) |

*(Continued)*

TABLE 23.3 (Continued)

| Type of PFAS | Study (year) | Study participants/ location | Sample type | Detection frequency | Concentration reported | Outcomes/inference | References |
|---|---|---|---|---|---|---|---|
| PFSAs, PFCAs, PFAA precursors | 2019 | States include Jammu and Kashmir, Rajasthan, Gujarat, Maharashtra, Kerala, Tamil Nadu, Andhra Pradesh, Odisha, West Bengal, Bihar, and Uttar Pradesh | Human hair | PFBS (21%), PFHxS (51%), PFOS (49%), PFDS (31%), N-EtFOSA (10%), PFBA (15%), and PFOA (59%) | ΣPFAS: <0.02 to 3.78 ng $g^{-1}$ | Further research on the accumulation process and sampling procedure is needed to properly interpret hair as a biomonitoring matrix for detecting human PFAS exposure. | Ruan et al. (2019) |
| PFNA, PFDA, PFUnDA, PFDoDA, PFTrDA, PFTeDA | 2011 | Lakshmikantapur, Diamond Harbor, Chingrighata, Sundarban | Carp, snakehead, catfish | | <0.05 to 0.2 ng $g^{-1}$ ww | Freshwater fish in tropical Asia exhibited lower PFS levels compared to those in developed countries. The exposure risk through fish remains a concern, so developing countries in the region need routine monitoring. | Murakami et al. (2011) |
| 12 PFA | | Kamrup region of Assam | Groundwater samples | PHxA (96.92%), PFOA (2.30%), PFDS (92.30%), PFDA (76.92%), PFHpA (75.38%), PFBS (60.00%), PFHxS (47.69%), and PFuD (43.07%) | PhxA(6), PFOA (11.90), PFDS (2.64), PFDA (6.80), PFHpA (13.33), PFBS (77.6, PFHxS) (4.22) and PfuD (0.76) in ng $L^{-1}$ | These findings underscore the importance of monitoring PFAS contamination in drinking water sources, particularly in areas such as Kamrup, where groundwater is the primary source. | Sharma et al. (2016) |

#### 23.5.5.2.2 Vacuum distillation with plasma destruction

PFAS are first concentrated using vacuum distillation, followed by complete breakdown through plasma treatment at extremely high temperatures (~10,000°C) and with the use of a high-voltage pulse generator. This method uses 4500 kWh $m^{-3}$ of energy, which makes it difficult to use because of its high cost and lack of reactor design knowledge.

### *23.5.5.3 Reduction technologies*

For in situ treatment, nanoscale zerovalent iron is a frequently utilized PFAS reducing agent. The zerovalent iron releases electrons that help break down PFAS into smaller, nontoxic compounds, offering a promising remediation strategy (Crane and Scott, 2012).

#### 23.5.5.3.1 Chemical oxidation

Chemical oxidation processes, including ozone, hydrogen peroxide, O3/UV, persulfate, and Fenton reagent, are effective in removing PFOA; however, they cannot remove PFOS. Highly oxidative reagents, including potassium dichromate and permanganate, are also not very effective against PFAS. In lab studies, high PFOA was removed under acidic pH, but obstacles such as inefficient defluorination and short-chain PFAS emerged, limiting further implementation (Thennakoon et al., 2010).

#### 23.5.5.3.2 Biodegradation

The majority of PFAS exhibit resistance to biological degradation, while certain polyfluorinated compounds are recognized to biotransform into their precursors. No information is provided regarding the degradation mechanism and by-product formation. Consequently, there is a paucity of research regarding the ineffectiveness of biodegradation in addressing PFAS contamination.

In summary, compared to destruction technologies, adsorption utilizing an activated carbon-based adsorbent is a technically viable and economically practical method for treating PFAS. Nonetheless, the unused adsorbent must be managed with caution throughout storage and incinerator treatment. Insufficient design for the disposal of unused adsorbents may lead to subsequent contamination by PFAS.

#### 23.5.5.3.3 Remediation of soil contaminated by per- and polyfluoroalkyl substances

Restoring PFAS-contaminated soil through excavation and disposal in landfills is the conventional method. Nonetheless, because PFAS is persistent, it is advisable to pretreat such waste via on-site incineration. The widely accepted field-scale technique is high-temperature incineration, typically applied when the soil concentrations of PFAS exceed 50 mg $g^{-1}$ (ATP3 1342/2014). Landfill boundaries are often treated with engineered caps to minimize leaching. In severely contaminated sites where the soil washing technique is employed, such as firefighting training sites, the PFAS is concentrated into a liquid phase that must be treated subsequently through degradation or adsorption. The use of agents such as activated carbon in soil stabilization can be helpful for PFAS immobilization. In general, excavation, thermal treatment, and stabilization are the most viable and field-proven methods (Reply On Behalf Of Respondent No. 1 Central Pollution Control Board CPCB, https; Ross et al., 2018).

Indian scientists have conducted several studies on PFAS treatment technologies, yielding positive results. As an example, adsorptive photocatalysis using activated carbon with $TiO_2$-P25 achieved 93% degradation of PFBA (perfluorobutanoic Acid) and full defluorination. Activated carbon with cetyltrimethylammonium bromide (CTAB) removed 90% of PFOA and reached an adsorption capacity of 455.8 mg $g^{-1}$. CTAB-treated sodium alginate hydrogels removed 95% and reached 382.1 mg $g^{-1}$ (Koulini et al., 2024).

Although these results are positive, they still require a more comprehensive study to determine their ultimate efficiency, expand the coverage area of PFAs, and explore mixed methods such as adsorption and biodegradation to create more lasting and practical solutions.

## 23.5.6 Legal and policy challenges

India experiences growing PFAS contamination, and its legal and regulatory response remains insufficient compared to global standards. There is no specific and enforceable framework that specifically focuses on PFAS, and the existing environmental legislation does not fully reflect the complexity of these chemicals. Although the limit of exposure to PFAS in the United States and EU has been rigorously determined, India has yet to formulate comprehensive PFAS exposure guidelines in water, soil, or food, despite setting up certain monitoring programs (Gautam et al., 2025).

Nevertheless, there are already some laws that can be amended to regulate PFAS:

1. PFAS may be regulated as hazardous compounds under the Environment Protection Act of 1986, which offers the general framework for controlling pollutants in India.
2. The Water (Prevention and Control of Pollution) Act of 1974 gives regulatory organizations the authority to manage PFAS contamination and other forms of pollution in water bodies.
3. Rules for Emergency Planning, Preparedness, and Response in Chemical Accidents, 1996: These regulations might be expanded to cover chemical contaminants and unintentional exposures of PFAS, even if they are not specifically related to them (The Chemical Accidents Emergency Planning, Preparedness and Response Rules, 1996; Forever Chemicals, -An).

To date, no significant PFAS-related lawsuits have been recorded in India. The lack of explicit norms and standards has constrained the basis for legal recourse against polluters or manufacturers.

### *23.5.6.1 Role of industries in resisting regulation*

The growing industrial use of PFAS in India is a rising concern. Speciality chemicals, textiles, automotive industries, and semiconductors rely on the properties of PFAS, such as chemical resistance, stability at high temperatures, and low friction. Nevertheless, growing international attention and governance are compelling Indian industries to reconcile performance demands with health and environmental safety. To meet these new standards, manufacturers are seeking sustainable options and are investing in innovation. Even minor origins, such as PFAS leaked during tire wear, underline the necessity of a powerful testing and monitoring network (India PFAS Situation Report, 2019; Forever Chemicals in India: An Emerging Environmental, Challenge).

#### 23.5.6.2 *Public awareness*

Epidemiological studies are essential in India to evaluate the impacts and underlying mechanisms of PFAS involved in early chronic diseases, particularly those affecting women and children. Additionally, there is a need for establishing ecological and HBM of PFAS in both children and adults to generate significant evidence for advocacy at the population level. Scalable interventions aimed at raising public and political awareness, as well as fostering behavioral change at both individual and population levels, to limit the availability and usage of PFAS-containing products, could mitigate persistent exposure (Mukherjee Das and Janardhanan, 2022).

### 23.5.7 Comparison with international regulations

PFAS are of significant environmental and health concern, and most countries have imposed strict regulations (Cousins et al., 2020). The US EPA established initial national drinking water limits for PFOS, PFOA, and PFHxS in April 2024 at 4 and 10 ppt. PFAS are also limited by the EU Drinking Water Directive, with a total PFAS limit of 0.5 $\mu g\ L^{-1}$ and the sum of 20 compounds in the series restricted to 0.1 $\mu g\ L^{-1}$ (PFAS National Primary Drinking Water Regulation, https). Other countries, such as Japan, Australia, and Canada, have also implemented restrictions on PFAS in drinking water and industrial effluents. However, India continues to respond to PFAS pollution without utilizing formal legislation. Instead, it depends on public attention and research studies. India's PFAS research is currently limited to discrete studies of specific industrial areas or environmental matrices, underscoring the need for increased funding, interdisciplinary cooperation, and more comprehensive investigations in contrast to the sophisticated regulatory systems of the United States and EU (Gautam et al., 2025). Although India does not yet have strict regulations on PFAS, there is an immediate need to initiate systematic discussions on establishing limits for the presence of these substances in drinking water, industrial effluents, and consumer products.

### 23.5.8 Future directions and recommendations

In order to effectively tackle the problem of PFAS pollution in India, a number of strategic moves need to be prioritized. Firstly, regular biomonitoring schemes should be established to capture both environmental and human exposure, as well as generate population-specific data, which will inform local policy. Awareness and political campaigns must also be implemented to promote behavioral change, both personally and socially, especially with the use of products containing PFAS. The ongoing exposure can be restricted by lowering demands through informed consumer decision-making. Based on available monitoring data, India should also establish regulatory limits for PFAS in drinking water, industrial effluents, and consumer products. All these measures will play a significant role in mitigating risks to health and preventing environmental damage. Lastly, the government must take urgent and synchronized steps towards alleviating the ever-increasing PFAS pollution, leading to the safeguarding of public health and the environment.

## 23.6 South Korea: per- and polyfluoroalkyl substances regulatory guidelines

### 23.6.1 Overview of environmental issues in South Korea

South Korea's environment, located on the southern part of the Korean Peninsula, faces several pressing challenges caused by rapid industrialization. The key environmental issues include air and water pollution, acid rain, drift net aquaculture, and contamination from sewage and industrial disposal. Industrialization has had a major impact on public health, quality of life, and socioeconomic systems (Park et al., 2025a).

According to a 2019 Ipsos survey, air pollution accounted for 35% of environmental concerns, followed by waste management at 34% (Climate change, Increases). The primary causes of air pollution in South Korea are vehicle emissions and the combustion of fossil fuels. Additionally, A substantial proportion of pollution, estimated to be between 30% and 50% of PM2.5, is attributed to transboundary pollution from China, which heavily depends on coal (Air quality in South Korea, http). South Korea passed the "Special Act on Fine Dust Reduction and Management" to combat deteriorating air quality, starting in February 2019. The law will serve as the foundation for the government's stringent dust management policies to reduce dust particles and improve air quality (Special Act, On).

EDCs, such as phthalates, bisphenols, parabens, polycyclic aromatic hydrocarbons, and volatile organic compounds, are commonly absorbed through ingestion, skin contact, and inhalation. Environmental studies in South Korea have revealed the widespread presence of EDCs, posing risks to humans and animals. In a Korean cohort study, 24 EDCs exposures in maternal urine were linked to perinatal depression and dietary–lifestyle factors (Moon et al., 2025). To address these risks, EDCs are specially regulated under the Act on K-REACH, which came into effect from January 1, 2015. Regulates the registration and evaluation of chemicals in the Korean market. Under Article 25(1) of the Act, the EDCs are encompassed based on the agreement with the relevant authorities. To manufacture, import, or use the EDCs, K-REACH requires prior approval from the Ministry of Environment of South Korea (Singh et al., 2023).

Alongside EDCs, residents of South Korea face increased health risks due to the presence of POPs in the environment and are identified as a hotspot for POPs. Moreover, the substantial seafood consumption among South Koreans is regarded as an additional source of POPs (Rokni et al., 2023). A study analyzed 257 food items from South Korea for dioxins present in the diet. Human blood levels remained stable, even though air POP levels declined (Shin et al., 2022). This shows continued exposure. In South Korea, the average PFOA level in breast milk increased by approximately 278% over a 12-year period (Kim et al., 2023).

To manage and regulate POPs, South Korea ratified the Stockholm Convention in 2007. The POPs Control Act was enacted by South Korea in the same year to comply with the provisions of this international convention. The Stockholm Convention categorizes PFOA, PFOS, PFHxS, and their salts as POPs because of their persistence, bioaccumulation, and environmental toxicity (Stockholm Convention National Implementation Plans, NIPs; National Implementation Plan for the Stockholm Convention Republic of Korea Updated in 2019 Contents, 2019).

The presence of PFAS in South Korea's environment and people poses a significant public health risk and environmental concern, particularly in areas with high pollution levels.

Addressing this issue requires regular monitoring and strict regulations. South Korea has made progress in its environmental policies by adhering to international agreements. Still, challenges remain to fully control PFAS exposure. The following section analyzes our existing knowledge, existing regulations, the gaps in their implementation, and the necessity of more effective approaches to effectively manage PFAS contamination across the nation.

## 23.6.2 Regulatory framework for per- and polyfluoroalkyl substances

PFAS are widely used in various industrial sectors and consumer products in South Korea due to their unique properties, including resistance to heat, oil, and water. Even with the implementation of PFAS laws, South Korea's existing standards remain less comprehensive compared to those of the United States, the EU, and other significant East Asian nations. Their environmental endurance and possible health hazards, however, are causing alarm and prompting more regulatory action. PFAS in South Korea are governed by the following regulatory frameworks:

### 23.6.2.1 *Regulation of per- and polyfluoroalkyl substances in compliance with the Stockholm Convention*

South Korea has associated itself with regulations pertaining to PFAS under the Stockholm Convention through the POPs Management Act, implemented in 2007. PFOA, PFOS, and PFHxS are among the perfluorinated substances identified as hazardous and designated as water pollutants; they are prohibited for production and use. Since 2007, a continuous survey on the status of perfluorinated compounds has been carried out at about 70 water purification facilities across the country (Kim et al., 2025a).

Further strengthening its commitment, the South Korean Ministry of Environment announced the most recent amended version of the Regulations on Types of POPs and Specific Exemptions (MOE Notice 2023-129), which will take effect on June 11, 2023. The ministry clarified that the 10th COP to the Stockholm Convention (June 2022), PFHxS, its salts and related substances were added to Annex A (Tenth Meeting of the Conference of the Parties to the Stockholm Convention, https; South Korea's Ministry of Environment, adds). This signaled the commitment to phasing out PFAS use entirely.

### 23.6.2.2 *The Act on Registration and Evaluation, etc. of Chemical Substances (K-REACH)*

In South Korea, the Ministry of Environment implemented K-REACH on May 22, 2013, which governs the registration and assessment of hazardous chemical compounds, including PFAS. The **Chemical Control Act** (CCA) governs the monitoring and control of accident-prone substances as defined by Article 39 of the CCA and hazardous chemicals identified by K-REACH (Korean REACH, TheAct). The regulation's primary amendment went into effect on January 1, 2019. On January 1, 2025, an act change took effect that altered the tonnage bar for the registration of new substances and increased compliance obligations for manufacturers. Manufacturers, importers, users, and distributors of chemicals are subject to strict K-REACH regulations. Violation may lead to penalties, including fines of up to 5% of annual sales or imprisonment for up to five years. Under this, the key regulatory requirements are: (1) any new substance manufactured or imported in quantities of 1 ton/year or more is subject to registration, (2) new substance prenotifications are required for manufacturing in amounts <1 ton/year (3) priority control substances are reported to the Ministry of Environment if they are produced or imported at a rate of more than one 1 ton/year and found in consumer goods at a rate of more than 0.1%, and (4) restricts the use of designated hazardous substances in consumer goods and articles (K-REACH, The; South Korea MOE, Released).

#### 23.6.2.3 *Per- and polyfluoroalkyl substances industry-specific regulations in South Korea*

In several cases, PFAS regulations are being implemented for specific industries. For instance, in 2022, the South Korean government proposed banning eight PFAS from the cosmetics sector to reduce exposure through personal care products (New Regulations Targeting Forever Chemicals, IDTechEx).

In a regulatory meeting on the current state of international PFAS regulations, South Korean authorities reviewed the distribution of PFAS for firefighting agents and discussed the procedure for selecting possible PFAS chemical substitutes for alternative development. The goal is to find viable substitutes for PFAS that will minimize environmental and health risks (REACH24H, Korea).

#### 23.6.2.4 *Persistent Organic Pollutants Control Act*

In South Korea, PFOA and PFOS have been regulated since 2019 and 2009, respectively, under the POPs Control Act, which was passed in 2008. While PFOA and PFOS are largely restricted, a special exemption for PFOS used in plating agents and PFOA used in semiconductor coatings was given until alternative methods are developed (Persistent Pollutants, Control). On September 4, 2024, the Ministry of Environment of South Korea announced Notification No. 2024-571, which covers major revisions to the regulation of persistent pollutants. In order to improve public health and environmental protection, this modification clarifies the exemption timeline and provides exact chemical names and CAS numbers in compliance with the POPs Control Act (New Regulations Targeting Forever Chemicals, IDTechEx).

### 23.6.3 Limits on per- and polyfluoroalkyl substances in South Korea

Since 2019, the drinking water quality monitoring standard in South Korea for perfluorinated chemicals has been set at 0.07 μg L-1 for PFOA, 0.48 μg L-1 for PFOS, and 0.07 μg L-1 for PFHxS (Special Act on Fine Dust Reduction and Management; Hong et al., 2024). Research is presently being conducted by the Ministry of Environment and the National Institute of Environmental Research to determine whether soil fluorine remediation guidelines need to be modified and to resolve any discrepancies (Ji et al., 2024). Similarly to soil, there are no PFAS limits in food products or the air, which suggests future regulations are expected. The Ministry of Food and Drug Safety has assessed the level of PFAS exposure from cookware and food packaging by measuring the quantity of PFAS transferred under typical cooking conditions, estimating daily intake at just $3.2 \times 10^{-5}$ $\mu g\ kg^{-1}$ bw $day^{-1}$ or 0.003% of the health-based advisory value for PFOA (Welcome to the Ministry of Food and Drug Safety, Ministry). Additionally, waste management strategies are under review, including establishing a limit for PFAS stockpiles, particularly for firefighting foams, and assessing appropriate disposal technologies.

#### 23.6.3.1 *Government agencies involved in per- and polyfluoroalkyl substances regulation*

Here is a key regulatory body responsible for environmental oversight in South Korea:

##### 23.6.3.1.1 Ministry of environment

All activities pertaining to the Stockholm Convention and the POPs Control Act are overseen by the Ministry of Environment. It establishes national standards for POPs, prohibits or limits their manufacture, import/export, usage, and manages the national POPs monitoring network. Additionally, it looks into the sources of emissions, establishes discharge limits, creates an emission inventory, and oversees handling, disposal, and management of wastes and stockpiles that contain or are contaminated with POPs (Welcome to the Ministry of Food and Drug Safety, Ministry).

##### 23.6.3.1.2 National Institute of Environmental Research

Scientific research and surveys aimed at preventing environmental contamination are carried out by the NIER. The K-REACH assigns the NIER the responsibility of assessing the toxicity and potential hazards of chemical compounds in South Korea (South Korea's, Chemical).

##### 23.6.3.1.3 Korea Environment Corporation

To manage hazardous compounds like PFAS, the Korea Environment Corporation (KECO), under the guidance of the Ministry of Environment, handles processing applications for preregistration and verifying exemptions from registration under the KREACH framework (Exemption From Registration of Chemical Substances, Korea). KECO is crucial for ensuring regulatory compliance and facilitating the effective implementation of chemical standards in South Korea.

### 23.6.4 Per- and polyfluoroalkyl substances detection and monitoring

PFAS research in South Korea primarily focuses on assessing environmental concentrations, particularly in rivers, drinking water systems, and wastewater treatment facilities. Monitoring efforts for PFAS in drinking water systems and wastewater treatment plants have increased, with national-scale systems established to monitor PFAS levels and develop technologies to lower the amount of PFAS entering the water supply (UNEP-POPS-NIP-Korea, Rep). Using the HPLC-MS/MS, 42 PFASs were detected in raw and treated water from DWTP in the Nakdong River basin. Gen-X, N-MeFOSE, and 6:2 fluorotelomer carboxylic acid were discovered in raw water for the first time in South Korea. Potentially as a result of industrial processes or precursor breakdown, short-chain PFAS were more prevalent and found in larger amounts in both raw and treated water (Kang et al., 2025).

In a similar investigation, the 13 PFASs were detected in sediment, river water, and crucian carp muscle tissues by extracting the river water samples using a modified solid phase extraction method, followed by LC-MS/MS. Despite the low usage of these chemicals due to worldwide prohibition, PFASs in water samples from Korea's three major rivers indicate that these pollutants continue to be a significant cause of contamination (Hung et al., 2025). Sediment and fish tissue samples were extracted using ion-pairing extraction to reveal the presence of PFASs in all fish species' blood, muscle, and liver samples. Long-chain PFASs, such as PFOA (C8), PFTeDA (C14), PFBS (C4), and PFHxS (C5), were predominant in the sediment samples. The level of PFASs in fish samples poses bioaccumulation risk in aquatic ecosystems (Hung et al., 2020). Furthermore, a study employing liquid chromatography–ion mobility spectrometry–quadrupole time Of flight–mass spectrometry identified 20 PFASs in wastewater, revealing that they are widely present in industrial and residential effluents (Yun et al., 2024). The Korean National Environmental Health Survey tracks exposure to environmental chemicals in a representative Korean population. The Ministry of Environment started the program in 2009 under the Environmental Health Act. The survey is conducted every three years and monitors the links between chemical exposure and health. In its fourth KoNEHS cycle, persistent pollutants, including PFOA, PFOS, PFHxS, PFNA, and PFDeA, have been studied in serum samples using a triple quadrupole HPLC-MS/MS. Serum levels of five perfluorinated compounds were found to be more than 99.7% in both adults and adolescents (Hong et al., 2024; Park et al., 2024). Additional case studies on PFAS examined in South Korea are listed in Table 23.4.

Recent developments in AI-assisted methods for PFAS contamination detection and management show great promise with a high accuracy (area under the curve of 0.90). AI models have achieved over 96% accuracy in identifying the sources of PFAS contamination and forecasting groundwater concentrations, whereas traditional methods are often ineffective and expensive. Furthermore, the removal methods in these models are optimized, with $R^2$ values higher than 0.93. But to effectively utilize AI's potential in PFAS management, especially for real-time monitoring, improved remediation techniques, and automated decision-making tools, issues including data availability, model selection, and large-scale deployment must be resolved (Park et al., 2025b).

## 23.6.5 Per- and polyfluoroalkyl substances remediation

### *23.6.5.1 Per- and polyfluoroalkyl substances elimination using drinking water treatment procedures*

Conventional water treatment techniques, including coagulation and chlorination, are ineffective for treating PFAS. Reverse osmosis, ion exchange systems, and GAC are examples of current PFAS treatment methods to meet stricter standards. Early water testing and monitoring of PFAS in utilities are crucial for improving their infrastructure and meeting future PFAS limitations. Key treatment methods include:

**(a)** GAC: The capacity of GAC to adsorb PFAS from water and efficiently capture these chemicals on the carbon material's surface makes this PFAS treatment technology popular.
**(b)** Ion exchange resin: This popular PFAS treatment method eliminates PFAS using ion exchange resin for industrial discharge and drinking water systems.

**TABLE 23.4** Case studies reported on per- and polyfluoroalkyl substances (PFAS) in South Korea.

| Type of PFAS | Study (year) | Study participants/ location | Sample type | Detection frequency | Concentration reported | Outcomes/ inference | References |
|---|---|---|---|---|---|---|---|
| Perfluorooctanoic acid (PFOA), perfluorooctane sulfonic acid (PFOS), perfluorohexane sulfonic acid (PFHxS) | 2025 | US military camps located in the Republic of Korea | Groundwater | | PFOA (125 ng $L^{-1}$), PFOS (103.7 ng $L^{-1}$), and PFHxS (127.8 ng $L^{-1}$) | This research demonstrates abundant PFAS pollution in groundwater around US Army installations in South Korea, highlighting the urgent need for remediation. | Kim et al. (2025b) |
| 11 perfluoroalkyl carboxylic acids (PFCAs, C4–C14), 9 perfluoroalkyl sulfonic acids (PFSAs, C4–C10), 5 precursors, and 3 emerging PFASs. | 2025 | Regional seas of South Korea | Surface and core sediments | PFUnDA (97%), PFOA (94%), and PFNA (79%) | <LOQ to 820 ng $g^{-1}$ OC | PFAS were found in 28 Korean water samples; additional studies on the environmental and human risks and sources are needed. | Yang et al. (2025) |
| PFOA, PFOS, perfluorononanoic acid (PFNA), PFHxS, and perfluorodecanoic acid (PFDeA) | 2025 | 828 participants from the fourth cycle of the KoNEHS (2018–20) | Human blood samples | | | PFAS disrupts lipids' metabolism, and children's long-term cardiovascular health should be investigated. | Shin et al. (2025) |
| 11 PFCAs (C4–C14), 6 PFSAs (C4–C8, C10), and 25 precursors/ alternatives | 2025 | Nakdong River basin | Raw and treated water samples | Detected in raw water: N-MeFOSE (42.2%), N-EtFOSE (8.9%), 6:2 FTUA (4.4%), 6:2 FTCA (77.8%), 10:2 FTCA (2.2%), 3:3 FTCA (4.4%), 5:3 FTCA (13.3%), ADONA (8.9%), Gen-X | 13.3 ng $L^{-1}$ to 189 ng $L^{-1}$ | The study tracks the PFAS in raw and treated drinking water, noting the patterns of legacy, precursor, and alternative compounds. | Kang et al. (2025) |

(*Continued*)

**TABLE 23.4** (Continued)

| Type of PFAS | Study (year) | Study participants/ location | Sample type | Detection frequency | Concentration reported | Outcomes/ inference | References |
|---|---|---|---|---|---|---|---|
| | | | | (66.7%), and F53B (major) (2.2%). In treated water: 6:2 FTCA (35.6%), N-MeFOSE (15.6%), and Gen-X (66.7%). | | | |
| PFCAs | 2018 | Cities in South Korea | Tap water samples | 10 PFAS were detected in tap water samples PFAS (63%–98%) and 2 PFAS (<50%) | 1.44 to 224 ng $L^{-1}$ human exposure via tap water consumption (46.8 ng $person^{-1}$ $day^{-1}$) | PFASs were detected in Korean tap water, probably due to nearby industrial pollution sources. | Park et al. (2018) |
| Perfluorinated carboxylic acids, perfluorinated sulfonates, and precursor compounds | 2024 | Changwon and Masan areas | Local streams | PFHxS (31%), PFHpA (17%), and PFOA (23%) | 0.00 ng $L^{-1}$ to 43.25 ng $L^{-1}$ for PFAS, 0.00 ng $L^{-1}$ to 21.80 ng $L^{-1}$ for PFCA, and 0.00 ng $L^{-1}$ to 43.25 ng $L^{-1}$ for PFSA | Monitoring of PFAS in key rivers should be done continuously to determine the increase in concentrations and associated health impacts. | Jeong et al. (2025) |
| 16 PFAS | 2020 | The Asan Lake area of South Korea | Air, water, sediments, soils, and fish | Air samples PFCAs (90.5%) sediment samples PFCAs (54.7%), PFOS (30.9%), PFDA (12.7%) soil samples PFCAs (76.6%) PFOS (22.5%), PFOA (24.8%) fish samples PFSAs (62.2%), FOSA (1.1%), | 20.7 pg $m^{-3}$ –98.2 pg $m^{-3}$ in air, 17.7 ng $L^{-1}$–467 ng $L^{-1}$ in water, 0.04 ng $g^{-1}$–15.0 ng $g^{-1}$ dry weight (dw) in sediments, and (n.d.)–12.9 ng $g^{-1}$ dw in soils, and t $\Sigma_{19}$ PFAS in fish r n.d. to 197 ng $g^{-1}$ ww | Fish accumulate long-chain PFAS such as PFOS; monitoring is necessary to decrease risk assessment uncertainty. | Lee et al. (2020) |

| | | | | | | | |
|---|---|---|---|---|---|---|---|
| PFOA, PFOS, PFHxS, PFNA, and PFDeA | 2018 to 2020 | Korea | Serum | | PFOA: 0.050 μg $L^{-1}$, PFOS: -0.056 μg/$L^{-1}$, PFHxS: 0.071: μg $L^{-1}$, PFNA: 0.019 μg $L^{-1}$, PFDeA: 0.017 μg $L^{-1}$ | Lower kidney functioning was associated with PFAS exposure among teenagers in Korea. Long-term studies are needed. | Yun et al. (2023) |
| 17 PFAS, including 10 (PFCAs), 4 PFOS, and 3 perfluoroalkyl sulfonamides | 2016 | Children's Health and Environmental Chemicals in Korea | Breast milk samples | PFOA (88%), PFUnDA (86%), PFNA (63%), PFHxA (40%), PFHS (35%), and PFHpA (26%) | PFOA: 0.072, PFOS: 050, PFPeA: 0.058, PFHxA: 0.047, PFHpA: 0.028 (ng $mL^{-1}$) | Daily intake of PFOS and PFOA through breast milk is low and presents little risk. | Lee et al. (2018) |
| Σ14 PFAS | 2023 | Lactation Counseling Clinic in Korea | Breast milk samples | PFOS, PFOA, and PFDA (100%), PFHxS (87.4%), PFNA (87.0%), and PFHxA (72.9%) | PFOS: 0.05, PFOA: 0.10, PFDA: 0.031, PFHxS: 0.007, PFNA: 0.033 (ng $mL^{-1}$) | High fish consumption is associated with PFOS. PFOA in human milk increased by 278% within 12 years, prompting greater emphasis on PFAS studies and policies. | Kim et al. (2023) |
| 16 PFASs | 2016 | Soonchunhyang University Hospital, Seoul, South Korea | Infants blood samples | PFBA, PFOA, PFNA, PFDA, PFUnDA, PFHxS, PFHpS, and PFOS (>90%) | 2.44–44.7 ng $mL^{-1}$ | The results show that some PFASs can cause thyroid dysfunction, which includes the development of hypothyroidism. | Kim et al. (2016) |

**(c)** Reverse osmosis: This high-pressure membrane filtration technique for PFAS treatment successfully eliminates PFAS because it filters water across a semipermeable membrane that effectively separates PFAS from water (Kang et al., 2025; PFAS Treatment Technologies for Water Utilities and PFAS-Contaminated Sites, Xylem).

## 23.6.6 Legal and policy challenges

The legal framework for PFAS in South Korea is primarily based on several regulations, including the CCA, K-REACH, and the POPs Control Act. Although these frameworks incorporate specific provisions for PFAS, comprehensive regulation is still quite limited. Under the POPs Control Act, South Korea listed PFOS, its salts, and PFOS-F as prohibited substances. The Act prohibits the manufacture, import, export, and use of PFOS, its salts, and PFOS-F, except for specific restrictions and uses permitted by the Stockholm Convention (Risk Reduction Approaches For Pfass-A Cross-Country, Analysis; Guardian et al., 2020).

The K-REACH has been effective since January 1, 2015. K-REACH is expected to modify laws in South Korea by adopting registration, assessment, authorization, and registration requirements analogous to those of the EU REACH legislation and is set to be introduced in 2025. To save businesses money on compliance costs associated with the registration duty for new chemicals created or imported annually, as of January 2025, the annual tonnage of manufactured or imported chemicals between 0.1 and 1 ton will no longer require registration, instead requiring only notification. However, the Ministry of Environment recently added a requirement for risk indications to the reporting process for new compounds under 1 ton, which was previously unnecessary. K-REACH imposes strict obligations on all stakeholders, including chemical producers, importers, consumers, and vendors. Noncompliance can result in fines, including penalty fees of up to 5% of sales and jail terms of up to five years (PFAS, Risk; K-REACH, TheAct).

### *23.6.6.1 Role of industries in resisting regulation*

According to Korea's industrial sector, a complete ban on PFAS may disrupt production and exports without feasible alternatives. Through publications, conferences, and consultations, the Ministry of Trade, Industry, and Energy (MOTIE) has been actively engaging with industry stakeholders to gather feedback. MOTIE's collaborative approach is demonstrated by the comments received thus far from the semiconductor, display, battery, and automotive industries (S. Korea, requests). Concerns were expressed about the need to find substitutes within the 5- or 12-year grace periods, which could cause supply chain disruptions worldwide, postpone the deployment of electric vehicles, and impede attempts to achieve carbon neutrality. A complete prohibition on over 10,000 PFAS types without individual PFAS risk evaluations creates concerns over regulation. The absence of comprehensive PFAS molecular information further complicates the regulatory process.

### *23.6.6.2 Public awareness of per- and polyfluoroalkyl substances in South Korea*

With the assistance of the MOE as part of the national implementation strategy, the Korean government provides information on POPs control policies, environmental POPs monitoring data, and the country's emission inventory. POP data is accessible through the PRTR system. Additionally, it provides information online for the public's right to know. The government encourages private sector forums for academic research and the exchange of information on POPs. Public engagement and awareness are being enhanced by research studies on various PFAS contamination issues, including the presence of PFAS in potable groundwater near US military bases, as well as the HBM program, also known as KoNEHS, a nationwide initiative aimed at reducing the health effects of hazardous chemicals. Additionally, the Ministry of Environment's public consultations regarding regulations and safety limits are playing a key role in raising public awareness and encouraging civic participation in the governance of chemical safety.

### 23.6.7 Comparison with international regulations

South Korea is actively working to reduce, minimize, and eradicate POPs, including particular PFAS compounds such as PFOS, PFOA, and PFHxS, in alignment with the objectives of the Stockholm Convention.

Even with the implementation of PFAS regulations, South Korea's existing PFAS standards are less stringent than those of the United States, the EU, and other significant East Asian nations. The EU restricts 20 PFAS chemicals to 100 ng $L^{-1}$, while the United States limits PFOA and PFOS in drinking water to 4 ng $L^{-1}$ each. Conversely, South Korea limits PFOA and PFOS at 70 ng $L^{-1}$, separately or in combination, and PFHxS at 480 ng $L^{-1}$.

Like EU REACH, South Korea's chemical regulatory system, K-REACH, is governed by the Act on Registration and Evaluation of Chemical Substances. These regulations are designed to ensure the safe management of chemicals and protect the environment and human health from any potential hazards related to chemical substances.

### 23.6.8 Future directions and recommendations

Persistently high PFAS exposure levels underscore the critical need for ongoing periodic surveys and additional monitoring, particularly for emerging compounds like PFASs, which are only recently being studied. A nationwide investigation and suitable corrective measures are urgently needed to manage the identified health risks, as research findings reveal significant observations regarding PFAS contamination in groundwater linked to military-related activities, underscoring urgent environmental and public health concerns in South Korea.

## 23.7 Conclusion

PFAS represent a vast and hazardous class of chemicals known for their extreme persistence, environmental mobility, and potential for bioaccumulation in living organisms. Their widespread use and resistance to degradation pose long-term risk to human health and ecosystems. Nations worldwide follow a variety of PFAS regulations, each with a distinct approach. Some of the essential rules of PFAS include stringent standards, ongoing environmental monitoring, and thorough risk assessment.

In Asian countries such as Japan, the industry faces several obstacles, including nonstandardized tests, inconsistent PFAS definitions, and a shortage of alternative solutions. The need to solve these problems requires international collaboration and adoption of standardized processes, especially definitions, testing procedures, and data exchange. In China, there has been a major improvement in the development of the regulatory framework to address PFAS pollution, which has increased national awareness of environmental and health risks. The incorporation of PFOS/PFOA/PFHxS in control lists at the national level, industry-specific restrictions, and the establishment of drinking water limits are essential examples of a shift towards more sustainable and science-based governance. However, PFAS research in India is still in its early stages; nevertheless, it is gaining momentum. While there is extensive research on PFAS pollution in India, it remains fragmented and lacks proper organization. Thorough studies are required to fully understand the issue and its implications. India has joined the Stockholm Convention, yet it has to establish regulations or oversight mechanisms. Due to regional and global pressures, South Korea imposes strict PFAS laws. Importers and manufacturers in South Korea are required to register, assess, and report, with severe fines imposed for noncompliance. South Korea has mandated chemical accident management plans for hazardous compounds and imposed specific limits on the levels of PFOS, PFOA, and PFHxS in drinking water. South Korea's strategy addresses public and environmental health concerns by striking a balance between flexibility and industry-specific regulations.

Taken together, this indicates that the differences between Asian countries limit the overall effectiveness of PFAS management. To minimize inconsistencies, standardized definitions, testing, and data-sharing platforms are crucial. International collaboration and Asian regional governance, alongside harmonization with global treaties, will play a crucial role in managing the PFAS pollution. Future directions should emphasize the advancement of safer substitutes to PFAS and the incorporation of science into regulatory decisions.

## List of abbreviations

| | |
|---|---|
| **ACN** | Acetonitrile |
| **ADONA** | Ammonium 4,8-dioxa-3H-perfluorononanoate |
| **AFFF** | Aqueous film-forming foams |
| **AGC** | Asahi Glass Company |
| **AICIS** | Australian Industrial Chemicals Introduction Scheme |
| **AUC** | Area under the curve |
| **BDD** | Boron-doped diamond electrodes |
| **BIS** | Bureau Of Indian Standards |
| **BPA** | Bisphenol A |
| **CCA** | Chemical Control Act |
| **COP** | Conference Of The Parties |
| **CPCB** | Central Pollution Control Board |
| **CSCL** | Chemical Substances Control Law |
| **CSIR** | Council Of Scientific And Industrial Research |
| **CTAB** | Cetyltrimethylammonium bromide |
| **DEHP** | Di(2-Ethylhexyl) phthalate |

| | |
|---|---|
| **DHEA** | Dehydroepiandrosterone |
| **diPAP** | Polyfluoroalkyl Phosphate Diester |
| **DNA** | Deoxyribonucleic Acid |
| **DONA** | 4,8-dioxa-3H-perfluorononanoic acid |
| **DWTP** | Drinking water treatment plant |
| **ECHA** | European Chemicals Agency |
| **EDCs** | Endocrine-disrupting chemicals |
| **EPA** | Environmental Protection Agency |
| **EtOH** | Ethanol |
| **EU** | European Union |
| **EXTEND** | Extended Tasks On Endocrine Disruption |
| **FASA/FASAA** | Perfluoroalkane sulfonamide/perfluoroalkane sulfonamido acetic acid |
| **FDIA** | Perfluoroalkanes with Di-Iodine terminal substituents. |
| **FMF** | Fluoropolymer manufacturing facility |
| **FOSA/PFOSA** | Perfluorooctanesulfonamide |
| **FOSAA** | Perfluorooctane sulfonamido acetic acid |
| **FOSE** | Perfluorooctane sulfonamido ethanol |
| **FOSE/FOSE** | Perfluorooctane sulfonamido ethanol (Or sulfonamide ethanol) |
| **FSCJ** | Food Safety Commission Of Japan |
| **FSSAI** | Food Safety And Standards Authority Of India |
| **FTCA** | Fluorotelomer carboxylic acid |
| **FTI** | Fluorotelomer iodide |
| **FTMAC** | Fluorotelomer methacrylate |
| **FTO** | Fluorotelomer Olefin |
| **FTOH** | Fluorotelomer Alcohol |
| **FTSA** | Fluorotelomer sulfonic acid |
| **FTUA/FTUCA** | Fluorotelomer unsaturated carboxylic acid |
| **GAC** | Granular activated carbon |
| **GB** | Guobiao (National Standards Of The People's Republic Of China) |
| **HBM** | Human biomonitoring |
| **HFPO** | Hexafluoropropylene oxide |
| **HPLC-MS/MS** | High-performance liquid chromatography–tandem mass spectrometry |
| **IIT** | Indian Institute Of Technology |
| **IITR** | Indian Institute Of Toxicology Research |
| **IMS** | Ion mobility spectrometry |
| **IPE** | Ion-pairing extraction |
| **ISO** | International Organization for Standardization |
| **JECS** | Japan Environment And Children's Study |
| **JIS** | Japanese Industrial Standards |
| **KECO** | Korea Environment Corporation |
| **KoNEHS** | Korean National Environmental Health Survey |
| **K-REACH** | Korean Registration, Evaluation, Authorization And Restriction Of Chemicals |
| **LC-IMS-QTOF-MS** | Liquid chromatography–ion mobility spectrometry–quadrupole time-of-flight–mass spectrometry |
| **LC-MS/MS** | Liquid chromatography–tandem mass spectrometry |
| **LC-QTOF-MS** | Liquid Chromatography–quadrupole time-of-flight mass spectrometry |
| **LOQ** | Limit of quantification |
| **MCL** | Maximum contaminant level |
| **MDA** | Method Detection limit |
| **MEE** | Ministry Of Ecology And Environment |
| **MeOH** | Methanol |
| **METI** | Ministry Of Economy, Trade And Industry |
| **MFDS** | Ministry Of Food And Drug Safety |

| | |
|---|---|
| **MHLW** | Ministry Of Health, Labor And Welfare |
| **MOE** | Ministry Of Environment |
| **MOTIE** | Ministry Of Trade, Industry And Energy |
| **MQL** | Minimum quantifiable limit |
| **NA** | Not available |
| **ND** | Not detected |
| **NEERI** | National Environmental Engineering Research Institute |
| **NGT** | National Green Tribunal |
| **NIER** | National Institute Of Environmental Research |
| **NIES** | National Institute For Environmental Studies |
| **NIP** | National Implementation Plan |
| **OPFR** | Organophosphate Flame Retardant |
| **PAHs** | Polycyclic aromatic hydrocarbons |
| **PCB** | Polychlorinated biphenyl(S) |
| **PFA** | Perfluoroalkyl acid |
| **PFAA** | Perfluoroalkyl acid(S) |
| **PFAS** | Per- and polyfluoroalkyl substances |
| **PFBA** | Perfluorobutanoic acid |
| **PFBS** | Perfluorobutane sulfonic Acid/Perfluorobutanesulfonate |
| **PFC** | Perfluorinated compound(S) |
| **PFCA** | Perfluoroalkyl carboxylate/perfluoroalkyl carboxylic Acid |
| **PFCAs** | Perfluoroalkyl carboxylic acids |
| **PFD/PFDA** | Perfluorodecanoic acid |
| **PFDA** | Perfluorodecanoic acid/perfluorodecanoate |
| **PFDOA/PFDoDA** | Perfluorododecanoic Acid |
| **PFDS** | Perfluorodecanesulfonic acid |
| **PFH** | Perfluorohexane sulfonate |
| **PFHPA** | Perfluoroheptanoic acid |
| **PFHS** | Perfluorohexanesulfonic acid |
| **PFHxA** | Perfluorohexanoic acid |
| **PFHxS** | Perfluorohexane sulfonic acid |
| **PFNA** | Perfluorononanoic acid |
| **PFOA** | Perfluorooctanoic acid |
| **PFOB** | Perfluoro-N-octyl bromide |
| **PFOS** | Perfluorooctane sulfonate |
| **PFOSF** | Perfluorooctane sulfonyl fluoride |
| **PFPeA** | Perfluoropentanoic acid |
| **PFSA** | Perfluoroalkane sulfonate/perfluoroalkane sulfonic acid |
| **PFTrDA** | Perfluorotridecanoic acid |
| **PFUnDA** | Perfluoroundecanoic acid |
| **POP** | Persistent organic Pollutant(S) |
| **POPRC** | Persistent Organic Pollutants Review Committee |
| **POPs** | Persistent organic pollutants |
| **PRTR** | Pollutant release and transfer register |
| **REACH** | Registration, Evaluation, Authorization And Restriction Of Chemicals (EU Regulation) |
| **SCC** | Solid Waste And Chemicals Management Center |
| **SPEED** | Strategic Programs On Environmental Endocrine Disruptors |
| **US** | United States |
| **USA** | United States Of America |
| **VOCs** | Volatile organic compounds |
| **WPPA** | Water Pollution Prevention Act |
| **WWTP** | Wastewater treatment plant |

# References


Acceptable Purposes: PFOS and PFOSF. Available from: https://chm.pops.int/Implementation/Exemptions/AcceptablePurposesPFOSandPFOSF/tabid/794/Default.aspx. (accessed 14.07.25.).

Act on the Regulation of Manufacture and Evaluation of Chemical Substances - English - Japanese Law Translation. Available from: https://www.japaneselawtranslation.go.jp/en/laws/view/3350/en. (accessed 14.07.25.).

Action Taken by Three Ministries against Violations of Chemical Substances Control Law. https://www.env.go.jp/en/headline/133.html. (accessed 15.07.25.).

Air quality in South Korea. Available from: https://www.iqair.com/in-en/south-korea. (accessed 15.07.25.).

Alazaiza, M.Y.D., Alzghoul, T.M., Ramu, M.B., Abu Amr, S.S., Abushammala, M.F.M., 2025. PFAS contamination and mitigation: a comprehensive analysis of research trends and global contributions. Case Studies in Chemical and Environmental Engineering 11, 101127. Available at https://doi.org/10.1016/j.cscee.2025.101127.

All POPs listed in the Stockholm Convention. Available from: https://www.pops.int/TheConvention/ThePOPs/AllPOPs/tabid/2509/Default.aspx#:~:text=Annex%20A%20(Elimination)Annex%20B%20(Restriction). (accessed 14.07.25).

Ambient (Outdoor) Air Pollution. https://www.who.int/news-room/fact-sheets/detail/ambient-(outdoor)-air-quality-and-health. (accessed 14.07.25.).

Arima, Y., Okayasu, Y., Yoshioka, D., Nagai, Y., Kobayashi, Y., 2024. Multiphoton-driven photocatalytic defluorination of persistent perfluoroalkyl substances and polymers by visible light. Angewandte Chemie (International Ed. in English) 63 (42), e202408687. Available at https://doi.org/10.1002/anie.202408687.

Arimura, T.H., 2024. In: Arimura, T.H., Hibiki, A. (Eds.), Air Pollution in Japan and the World BT - Introduction to Environmental Economics and Policy in Japan. Springer Nature Singapore, Singapore, pp. 107–122. Available at https://doi.org/10.1007/978-981-97-2187-0_8.

Asakawa, D., Lin, H., Ruan, Y., Taniyasu, S., Yeung, L.W.Y., Tojo, T., et al., 2025. Characterizing seasonal patterns, gas-particle partitioning, and potential sources of airborne per- and polyfluoroalkyl substances in Japan. Environmental Science & Technology 59 (10), 5145–5154. Available at https://doi.org/10.1021/acs.est.4c14109.

Atsuko, I., Chihiro, M., Sachiko, I., Ait, B.Y., Keiko, Y., Reiko, K., 2023. The Hokkaido study on environment and children's health: overview of chemical exposures in utero and children's health until school age. ISEE Conference Abstracts 2025 (1). Available at https://doi.org/10.1289/isee.2023.SA-084.

Baluyot, J.C., Reyes, E.M., Velarde, M.C., 2021. Per- and polyfluoroalkyl substances (PFAS) as contaminants of emerging concern in Asia's freshwater resources. Environmental Research 197, 111122. Available at https://doi.org/10.1016/j.envres.2021.111122.

Bao, J., Liu, L., Wang, X., Jin, Y.-H., Dong, G.-H., 2017. Human exposure to perfluoroalkyl substances near a fluorochemical industrial park in China. Environmental Science and Pollution Research 24 (10), 9194–9201. Available at https://doi.org/10.1007/s11356-017-8620-6.

BfS - Fukushima - Environmental Impact of the Fukushima Accident: Radiological Situation in Japan. Available from: https://www.bfs.de/EN/topics/ion/accident-management/emergency/fukushima/environmental-consequences. (accessed 14.07.25).

Binu, K.R., Yamashita, N., Prabhasankar, V.P., Praveenkumarreddy, Y., Shenoy, J.K., Balakrishna, K., 2022. In: Kumar, M., Mohapatra, S. (Eds.), A First Report of Perfluoroalkyl Substances (PFAS) in a Large West-Flowing River in Southern India BT - Impact of COVID-19 on Emerging Contaminants: One Health Framework for Risk Assessment and Remediation. Springer Nature Singapore, Singapore, pp. 3–16. Available at https://doi.org/10.1007/978-981-19-1847-6_1.

Brennan, N.M., Evans, A.T., Fritz, M.K., Peak, S.A., von Holst, H.E., 2021. Trends in the regulation of per- and polyfluoroalkyl substances (PFAS): a scoping review. International Journal of Environmental Research and Public Health 18 (20). Available at https://doi.org/10.3390/ijerph182010900.

Cabinet Decision on the Cabinet Order for the Partial Revision of the Order for Enforcement of the Act on the Evaluation of Chemical Substances and Regulation of Their Manufacture, etc. Available from: https://www.meti.go.jp/english/press/2024/0705_002.html. (accessed 16.07.25.).

Chapter 3- Results of the Environmental Monitoring in FY, 2016. Available from: https://www.env.go.jp/chemi/kurohon/en/http2017e/pdf/05_Chapter3.pdf. (accessed 13.07.25.).

Chaturvedi, M., Joy, S., Gupta, R.D., Pandey, S., Sharma, S., 2023. Endocrine disrupting chemicals (EDCs): chemical fate, distribution, analytical methods and promising remediation strategies – a critical review. Environmental Technology Reviews 12 (1), 286–315. Available at https://doi.org/10.1080/21622515.2023.2205026.

Chemical Management, Act on Confirmation, etc. of Releases Amounts of Specific Chemical Substances in the Environment and Promotion of Improvements to the Management Thereof. Available from: https://www.nite.go.jp/en/chem/aboutcmc.html. (accessed 14.07.25).

Chemicals and Hazardous Waste, Stockholm Convention. Available from: https://www.mofa.go.jp/policy/environment/convention/chemical.html. (accessed 14.07.25.).

Cheng, X., Wang, Y., Zhang, J., Guo, H., Liu, L., Liu, L., et al., 2025. Trends in serum levels of emerging and legacy per- and polyfluoroalkyl substances from 2008 to 2018: a longitudinal study in China. Environmental Health Perspectives. Available at https://doi.org/10.1289/EHP15340.

China Implements First Chemical Park PFOA and PFOS Control Standards. Available from: https://www.cirs-group.com/en/chemicals/china-implements-first-chemical-park-pfoa-and-pfos-control-standards. (accessed 24.07.25.).

China Releases "Reference List of PFOA, Its Salts and PFOA-Related Compounds" | Enviliance ASIA. Available from: https://enviliance.com/regions/east-asia/cn/report_11146. (accessed 16.07.25.).

Civil Code - Japanese/English - Japanese Law Translation. Available from: https://www.japaneselawtranslation.go.jp/en/laws/view/3494/en#je_pt3ch5at1. (accessed 14.07.25).

Climate change increases in importance to citizens around the world. Available from: https://www.ipsos.com/ko-kr/climate-change-increases-importance-citizens-around-world. (accessed 15.07.25.).

Coady, K.K., Biever, R.C., Denslow, N.D., Gross, M., Guiney, P.D., Holbech, H., et al., 2017. Current limitations and recommendations to improve testing for the environmental assessment of endocrine active substances. Integrated Environmental Assessment and Management 13 (2), 302–316. Available at https://doi.org/10.1002/ieam.1862.

Corsolini, S., Sarkar, S.K., Guerranti, C., Bhattacharya, B.D., Rakshit, D., Jonathan, M.P., et al., 2012. Perfluorinated compounds in surficial sediments of the ganges river and adjacent Sundarban Mangrove wetland, India. Marine Pollution Bulletin 64 (12), 2829–2833. Available at https://doi.org/10.1016/j.marpolbul.2012.09.019.

Cousins, I.T., DeWitt, J.C., Glüge, J., Goldenman, G., Herzke, D., Lohmann, R., et al., 2020. The high persistence of PFAS is sufficient for their management as a chemical class. Environmental Science: Processes & Impacts 22 (12), 2307–2312. Available at https://doi.org/10.1039/D0EM00355G.

Crane, R.A., Scott, T.B., 2012. Nanoscale zero-valent iron: future prospects for an emerging water treatment technology. Journal of Hazardous Materials 211–212, 112–125. Available at https://doi.org/10.1016/j.jhazmat.2011.11.073.

CSIR News. Available from: https://csirnews.niscpr.res.in/home/article/832. (accessed 15.07.25.).

CSIR-IITR Leads India's National PFAS Monitoring Push. Available from: https://www.selectscience.net/article/csir-iitr-leads-india-s-national-pfas-monitoring-push. (accessed 15.07.25.).

CSIR-IITR leads India's national PFAS monitoring push. Available from: https://www.csir.res.in/sites/default/files/2025-06/16_to_30_june_2025_final.pdf. (accessed 15.07.25.).

Daikin's Approach to PFAS. Fluorochemicals, Daikin Global. Available from: https://www.daikinchemicals.com/sustainability/pfas.html. (accessed 13.07.25.).

Dasmahapatra, A.K., Williams, C.B., Myla, A., Tiwary, S.K., Tchounwou, P.B., 2023. A systematic review of the evaluation of endocrine-disrupting chemicals in the Japanese Medaka (Oryzias Latipes) Fish. Frontiers in Toxicology 5, 1272368. Available at https://doi.org/10.3389/ftox.2023.1272368.

Economic and Social Commission for Asia and the Pacific Committee on Environment and Development Fifth Session, 2018. Available from: https://www.unescap.org/events/committee-environment-and-development-fifth-session. (accessed 16.07.25.).

Effectiveness of Shimizu PFAS Soil Remediation Technology Confirmed in the U.S. | Company Information, Shimizu Corporation. Available from: https://www.shimz.co.jp/en/company/about/news-release/2024/2024019.html. (accessed 14.07.25.).

Endocrine Disrupting Chemicals. Available from: https://toxicslink.org/endocrine-disrupting-chemicals. (accessed 14.07.25.).

Endocrine Disruptors Regulations and Lists in Japan. Available from: https://www.chemsafetypro.com/Topics/Japan/Endocrine_Disruptors_Regulations_and_Lists_in_Japan.html. (accessed 14.07.25.).

Environmental Issues in Japan and Solutions. Available from: https://energytracker.asia/environmental-issues-in-japan-and-solutions/. (accessed 15.07.25.).

Environmental Pollution in Japan - Statistics & Facts. Available from: https://www.statista.com/topics/7305/pollution-in-japan/#topicOverview. (accessed 15.07.25.).

Exemption From Registration of Chemical Substances. Korea Environment Corporation. Available from: https://www.keco.or.kr/en/index.do. (accessed 15.07.25.).
Fiedler, H., 2008. In: Mehmetli, E., Koumanova, B. (Eds.), Stockholm Convention on Pops: Obligations and Implementation BT - The Fate of Persistent Organic Pollutants in the Environment. Springer, Netherlands: Dordrecht, pp. 3–12.
Food Sanitation Act - English - Japanese Law Translation. Available from: https://www.japaneselawtranslation.go.jp/en/laws/view/3687/en. (accessed 14.07.25.).
Forever Chemicals - An Introduction. Available from: https://www.taxtmi.com/article/detailed?id=13697. (accessed 13.07.25.).
Forever Chemicals - An Introduction. Available from: https://www.taxtmi.com/article/detailed?id=13697. (accessed 15.07.25.).
Forever Chemicals in India: An Emerging Environmental Challenge. Available from: https://www.manufacturingtodayindia.com/forever-chemicals-in-india. (accessed 15.07.25.).
Fujii, Y., Kato, Y., Sakamoto, K., Matsuishi, T., Harada, K.H., Koizumi, A., et al., 2018. Tissue-specific bioaccumulation of long-chain perfluorinated carboxylic acids and halogenated methylbipyrroles in dall's porpoises (phocoenoides dalli) and harbor porpoises (phocoena phocoena) stranded in Northern Japan. The Science of the Total Environment 616–617, 554–563. Available at https://doi.org/10.1016/j.scitotenv.2017.10.033.
Fukushima, T., 2021. CSCL and industrial chemical management framework in Japan. ACS Chemical Health & Safety 28 (6), 410–416. Available at https://doi.org/10.1021/acs.chas.1c00037.
Further Research on PFAS Pollution in China – PFAS Concentrations in the Environment, PFAS Policies, and Their Impacts – Sino-German Environmental Partnership. Available from: https://environmental-partnership.org/news/further-research-on-pfas-pollution-in-china-pfas-concentrations-in-the-environment-pfas-policies-and-their-impacts/. (accessed 19.07.25.).
G, H., G, S., RS, R., R, D., R, P., R, R., 2023. Early detection of emerging persistent perfluorinated alkyl substances (PFAS) along the East Coast of India. The Science of the Total Environment 902, 166155. Available at https://doi.org/10.1016/j.scitotenv.2023.166155.
Gao, C., Quan, F., Qiu, W., Zheng, Y., 2025. Assessing the impact of serum per- and polyfluoroalkyl substance concentrations on immune function in an industrialized region of China. Environ. Heal. (Washington, D.C.) 3 (4), 352–362. Available at https://doi.org/10.1021/envhealth.4c00224.
Gautam, R.K., Mottaghipisheh, J., Verma, S., Singh, R.P., Muthukumaran, S., Navaratna, D., et al., 2025. PFAS contamination in key Indian States: a critical review of environmental impacts, regulatory challenges and predictive exposure. Journal of Hazardous Materials Advances 18, 100748. Available at https://doi.org/10.1016/j.hazadv.2025.100748.
Gomez-Ruiz, B., Gómez-Lavín, S., Diban, N., Boiteux, V., Colin, A., Dauchy, X., et al., 2017. Efficient Electrochemical Degradation Of Poly- And Perfluoroalkyl Substances (PFASs) from the effluents of an industrial wastewater treatment plant. Chemical Engineering Journal 322, 196–204. Available at https://doi.org/10.1016/j.cej.2017.04.040.
Goudarzi, H., Miyashita, C., Okada, E., Kashino, I., Chen, C.-J., Ito, S., et al., 2017a. Prenatal exposure to perfluoroalkyl acids and prevalence of infectious diseases up to 4 years of age. Environment International 104, 132–138. Available at https://doi.org/10.1016/j.envint.2017.01.024.
Goudarzi, H., Araki, A., Itoh, S., Sasaki, S., Miyashita, C., Mitsui, T., et al., 2017b. The association of prenatal exposure to perfluorinated chemicals with glucocorticoid and androgenic hormones in cord blood samples: the Hokkaido study. Environmental Health Perspectives 125 (1), 111–118. Available at https://doi.org/10.1289/EHP142.
Guardian, M.G.E., Boongaling, E.G., Bernardo-Boongaling, V.R.R., Gamonchuang, J., Boontongto, T., Burakham, R., et al., 2020. Prevalence of per- and polyfluoroalkyl substances (PFASs) in drinking and source water from two Asian countries. Chemosphere 256, 127115. Available at https://doi.org/10.1016/j.chemosphere.2020.127115.
Guidelines for Water Quality Monitoring Central Pollution Control Board Parivesh Bhawan East Arjun Nagar, Delhi-32. Available from: https://cpcb.nic.in/openpdffile.php?id=UmVwb3J0RmlsZXMvTmV3SXRlbV8xMTZfR3VpZGVsaW5lc29mIHdhdGVycXVhbGl0eW1vbml0b3JpbmdfMzEuMDcuMDgucGRm. (accessed 15.07.25.).
Guo, Y., Hua, Y., Yang, W., Zhou, J., Wang, L., Dai, X., 2025. The stubborn "chemical invader": challenges and solutions for per- and polyfluoroalkyl substances (PFAS) in Chinese sewage sludge. ACS ES&T Water 5 (3), 1067–1070. Available at https://doi.org/10.1021/acsestwater.5c00065.

Guruge, K.S., Manage, P.M., Yamanaka, N., Miyazaki, S., Taniyasu, S., Yamashita, N., 2008. Species-specific concentrations of perfluoroalkyl contaminants in farm and pet animals in Japan. Chemosphere 73 (1 Suppl), S210–S215. Available at https://doi.org/10.1016/j.chemosphere.2006.12.105.

Guruge, K.S., Yeung, L.W.Y., Li, P., Taniyasu, S., Yamashita, N., Nakamura, M., 2011. Fluorinated alkyl compounds including long chain carboxylic acids in wild bird livers from Japan. Chemosphere 83 (3), 379–384. Available at https://doi.org/10.1016/j.chemosphere.2010.12.010.

Hadibarata, T., Kristanti, R.A., Mahmoud, A.H., 2020. Occurrence of endocrine-disrupting chemicals (EDCs) in river water and sediment of the Mahakam River. Journal of Water and Health 18 (1), 38–47. Available at https://doi.org/10.2166/wh.2019.100.

Hasegawa, K., Motoki, N., Inaba, Y., Toubou, H., Shibazaki, T., Nakayama, S.F., et al., 2024. Maternal exposure to per- and polyfluoroalkyl substances and offspring chromosomal abnormalities: the Japan environment and children's study. Environmental Health Perspectives 132 (9), 97004. Available at https://doi.org/10.1289/EHP13617.

Hong, S., Kim, O.-J., Jung, S.K., Jeon, H.L., Kim, S., Kil, J., 2024. The exposure status of environmental chemicals in south Korea: the Korean National Environmental Health Survey 2018–2020. Toxics. Available at https://doi.org/10.3390/toxics12110829.

Howes, S., Wyrwoll, P., 2012. Moving Toward a New Development Model for East Asia-The Role of Domestic Policy and Regional Cooperation, pp. 55–120. Available from: https://www.eria.org/RPR-FY2011-10.pdf. (accessed 15.07.25.).

Hu, Y., Cheng, H., 2013. Water Pollution during China's industrial transition. Environmental Development 8, 57–73. Available at https://doi.org/10.1016/j.envdev.2013.06.001.

Hung, M.D., Jung, H.J., Jeong, H.H., Lam, N.H., Cho, H.S., 2020. Perfluoroalkyl substances (PFASs) in Special Management Sea Areas of Korea: distribution and bioconcentration in edible fish species. Marine Pollution Bulletin 156, 111236. Available at https://doi.org/10.1016/j.marpolbul.2020.111236.

Hung, M.D., Nga, N.T., Jung, H.J., Shin, Y.-J., Lee, B.-C., Cho, H.S., 2025. Perfluoroalkyl substances (PFASs) in the major rivers of Korea: distribution in the environment and bioaccumulation in crucian carp fish. The Science of the Total Environment 979, 179483. Available at https://doi.org/10.1016/j.scitotenv.2025.179483.

Iguchi, T., Sumi, M., Tanabe, S., 2002. Endocrine disruptor issues in Japan. Congenital Anomalies 42 (2), 106–119. Available at https://doi.org/10.1111/j.1741-4520.2002.tb00859.x.

IIT Madras Study. Item No. 13 Court No. 2 on 19 December 2024. Available from: https://www.greentribunal.gov.in/gen_pdf_test.php?filepath=L25ndF9kb2N1bWVudHMvbmd0L2Nhc2Vkb2Mvb3JkZXJzL0RFTEhJLzIwMjQtMTItMTkvY291cnRzLzIvZGFpbHkvMTczNDk0MjQ4NTE5MjAyMjUzMzY3NjkxZjE1OTg4MjgucGRm. (accessed 15.07.25.).

India PFAS Situation Report-2019. Available from: https://ipen.org/sites/default/files/documents/india_pfas_country_situation_report_mar_2019.pdf. (accessed 13.07.25.).

Itoh, S., Araki, A., Mitsui, T., Miyashita, C., Goudarzi, H., Sasaki, S., et al., 2016. Association of perfluoroalkyl substances exposure in utero with reproductive hormone levels in cord blood in the Hokkaido study on environment and children's health. Environment International 94, 51–59. Available at https://doi.org/10.1016/j.envint.2016.05.011.

Itoh, H., Harada, K.H., Kasuga, Y., Yokoyama, S., Onuma, H., Nishimura, H., et al., 2023. Association between serum concentrations of perfluoroalkyl substances and global DNA methylation levels in peripheral blood leukocytes of japanese women: a cross-sectional study. The Science of the Total Environment 859 (Pt 1), 159923. Available at https://doi.org/10.1016/j.scitotenv.2022.159923.

Jala, A., Adye, D.R., Borkar, R.M., 2023. Occurrence and risk assessments of per- and polyfluoroalkyl substances in tea bags from India. Food Control 151, 109812. Available at https://doi.org/10.1016/j.foodcont.2023.109812.

Japan Designates List of PFOA-Related Compounds as Class I Specified Substances | SGS India. Available from: https://www.sgs.com/en-in/news/2024/12/safeguards-17424-japan-designates-list-of-pfoa-related-compounds-as-class-i-specified-substances. (accessed 15.07.25.).

Japan Establishes New Standards for PFOS and PFOA in Mineral Water | RRMA. Available from: https://rrma-global.org/news-details/japan-establishes-new-standards-for-pfos-and-pfoa-in-mineral-water/MTk4OA. (accessed 14.07.25).

Japan PFAS Situation Report, 2019. Available from: https://ipen.org/sites/default/files/documents/japan_pfas_country_situation_report_apr_2019.pdf. (accessed 13.07.25.).

Japan to Tighten Regulations on Dechlorane Plus to Comply with Stockholm Convention - Foresight. Available from: https://www.useforesight.io/news/japan-to-tighten-regulations-on-dechlorane-plus-to-comply-with-stockholm-convention. (accessed 14.07.25).

Japan Will Ban 138 Perfluorinated Substances by 2025. Available from: https://globalpccs.com/2024/10/29/japan-will-ban-138-perfluorinated-substances-by-2025/. (accessed 14.07.25.).

Japan's "PRTR Act", Eviliance ASIA. Available from: https://enviliance.com/regions/east-asia/jp/jp-chemical/jp-prtr. (accessed 14.07.25.).

Japan's 5th Expert Meeting Proposed 50 ng/L Combined Limit for PFOS/PFOA in Water I Enviliance ASIA. Available from: https://enviliance.com/regions/east-asia/jp/report_12989. (accessed 14.07.25.).

Jeong, C., Jang, H., Jeon, J., 2025. Distribution and seasonal variations of perfluoroalkyl substances (PFAS) in streams within the Changwon region. Environmental Engineering Research 30 (2). Available at https://doi.org/10.4491/eer.2024.367.

Ji, C.H., Lee, S.H., Bae, G.S., Kim, H.W., 2024. Balancing environmental safety and economic feasibility: a review of soil fluorine management strategies in South Korea. Sustainability. Available at https://doi.org/10.3390/su16198391.

Jiao, E., Zhu, Z., Yin, D., Qiu, Y., Kärrman, A., Yeung, L.W.Y., 2022. A pilot study on extractable organofluorine and per- and polyfluoroalkyl substances (PFAS) in water from drinking water treatment plants around Taihu Lake, China: what is missed by target PFAS analysis? Environmental Science: Processes & Impacts 24 (7), 1060–1070. Available at https://doi.org/10.1039/D2EM00073C.

Jin, Y., Andersson, H., Zhang, S., 2016. Air pollution control policies in China: a retrospective and prospects. International Journal of Environmental Research and Public Health 13 (12). Available at https://doi.org/10.3390/ijerph13121219.

Kang, J.-K., Kim, M.-G., Oh, J.-E., 2025. Occurrence and removal of 42 legacy and emerging per- and polyfluoroalkyl substances (PFAS) in drinking water treatment plants in South Korea. Water Res. X 29, 100329. Available at https://doi.org/10.1016/j.wroa.2025.100329.

Kannan, K., Corsolini, S., Falandysz, J., Fillmann, G., Kumar, K.S., Loganathan, B.G., et al., 2004. Perfluorooctanesulfonate and related fluorochemicals in human blood from several countries. Environmental Science & Technology 38 (17), 4489–4495. Available at https://doi.org/10.1021/es0493446.

Kim, D.-H., Kim, U.-J., Kim, H.-Y., Choi, S.-D., Oh, J.-E., 2016. Perfluoroalkyl substances in serum from South Korean infants with congenital hypothyroidism and healthy infants – its relationship with thyroid hormones. Environmental Research 147, 399–404. Available at https://doi.org/10.1016/j.envres.2016.02.037.

Kim, J.H., Moon, N., Lee, J.-W., Mehdi, Q., Yun, M.-H., Moon, H.-B., 2023. Time-course trend and influencing factors for per- and polyfluoroalkyl substances in the breast milk of Korean mothers. Chemosphere 310, 136688. Available at https://doi.org/10.1016/j.chemosphere.2022.136688.

Kim, K., Son, S., An, S., Choe, S., Jung, S., 2025a. Toward strengthened regulation of PFAS in South Korea: health risks, global standards, and policy recommendations. Journal of Korean Society of Environmental Engineers 47 (6), 410–431. Available at https://doi.org/10.4491/KSEE.2025.47.6.410.

Kim, Y., Cho, A., Seo, Y.E., Kho, Y., Choi, K., 2025b. Occurrence of per- and polyfluoroalkyl substances (PFAS) in potable groundwater near military bases in South Korea. Environ. Sci. Technol. Lett. 12 (4), 440–446. Available at https://doi.org/10.1021/acs.estlett.5c00105.

Korean REACH: The Act on the Registration and Evaluation of Chemicals (K-REACH). Available from: https://www.intertek.com/assuris/chemicals/regulatory/k-reach/. (accessed 17.07.25.).

Korean REACH: The Act on the Registration and Evaluation of Chemicals (K-REACH). Available from: https://www.intertek.com/assuris/chemicals/regulatory/k-reach/. (accessed 15.07.25.).

Koulini, G.V., Nambi, I.M., 2024. Occurrence of forever chemicals in Chennai waters, India. Environmental Sciences Europe 36 (1), 60. Available at https://doi.org/10.1186/s12302-024-00881-1.

Koulini, G.V., Vinayagam, V., Nambi, I.M., Krishna, R.R., 2024. Per- and polyfluoroalkyl substances (PFAS) in Indian environment: prevalence, impacts and solutions. Journal of Water Process Engineering 66, 105988. Available at https://doi.org/10.1016/j.jwpe.2024.105988.

K-REACH: The Act on the Registration and Evaluation of Chemical Substances. Available from: https://www.yordasgroup.com/global-notifications/south-korea. (accessed 15.07.25.).

Kumar, S., Singh, P., Verma, K., Kumar, P., Yadav, A., 2022. Environmental issues and their possible solutions for sustainable development, India: a review. Current World Environment 17 (3), 531–541. Available at https://doi.org/10.12944/CWE.17.3.3.

Kumawat, M., Sharma, P., Pal, N., James, M.M., Verma, V., Tiwari, R.R., et al., 2022. Occurrence and seasonal disparity of emerging endocrine disrupting chemicals in a drinking water supply system and associated health risk. Scientific Reports 12 (1), 9252. Available at https://doi.org/10.1038/s41598-022-13489-3.

Kuribara, I., Motoki, T., Matsukami, H., Takahashi, Y., Kuramochi, H., 2024. Atmospheric concentrations of per- and polyfluoroalkyl substances and their emissions at a waste recycling facility producing refuse-derived paper and plastics densified fuel. The Science of the Total Environment 954, 176456. Available at https://doi.org/10.1016/j.scitotenv.2024.176456.

Kurwadkar, S., Dane, J., Kanel, S.R., Nadagouda, M.N., Cawdrey, R.W., Ambade, B., et al., 2022. Per- and polyfluoroalkyl substances in water and wastewater: a critical review of their global occurrence and distribution. The Science of the Total Environment 809, 151003. Available at https://doi.org/10.1016/j.scitotenv.2021.151003.

Lee, S., Kim, S., Park, J., Kim, H.-J., Choi, G., Choi, S., et al., 2018. Perfluoroalkyl substances (PFASs) in breast milk from korea: time-course trends, influencing factors, and infant exposure. The Science of the Total Environment 612, 286–292. Available at https://doi.org/10.1016/j.scitotenv.2017.08.094.

Lee, Y.-M., Lee, J.-Y., Kim, M.-K., Yang, H., Lee, J.-E., Son, Y., et al., 2020. Concentration and distribution of per- and polyfluoroalkyl substances (PFAS) in the Asan lake area of South Korea. Journal of Hazardous Materials 381, 120909. Available at https://doi.org/10.1016/j.jhazmat.2019.120909.

Letter Inviting Participation in the PFOA Stewardship Program, US EPA. Available from: https://www.epa.gov/assessing-and-managing-chemicals-under-tsca/letter-inviting-participation-pfoa-stewardship-program. (accessed 13.07.25.).

Li, J., Del Vento, S., Schuster, J., Zhang, G., Chakraborty, P., Kobara, Y., et al., 2011. Perfluorinated compounds in the Asian atmosphere. Environmental Science & Technology 45 (17), 7241–7248. Available at https://doi.org/10.1021/es201739t.

Li, Y., Liu, Y., Shi, G., Liu, C., Hao, Q., Wu, L., 2022. Occurrence and risk assessment of perfluorooctanoate (PFOA) and perfluorooctane sulfonate (PFOS) in surface water, groundwater and sediments of the Jin River Basin, Southeastern China. Bulletin of Environmental Contamination and Toxicology 108 (6), 1026–1032. Available at https://doi.org/10.1007/s00128-021-03435-w.

Li, W., Tanaka, S., Kitaji, Y., Hashikomi, S., Xu, Y., Ikeo, T., 2024a. Remediation of per- and polyfluoroalkyl substances (PFAS) contaminated soil via soil washing with various water-organic solvents. Journal of Hazardous Materials 480, 135943. Available at https://doi.org/10.1016/j.jhazmat.2024.135943.

Li, J., Duan, W., An, Z., Jiang, Z., Li, L., Guo, M., et al., 2024b. Legacy and alternative per- and polyfluoroalkyl substances spatiotemporal distribution in China: human exposure, environmental media, and risk assessment. Journal of Hazardous Materials 480, 135795. Available at https://doi.org/10.1016/j.jhazmat.2024.135795.

Lin, H., Taniyasu, S., Yamazaki, E., Wei, S., Wang, X., Gai, N., et al., 2020. Per- and polyfluoroalkyl substances in the air particles of Asia: levels, seasonality, and size-dependent distribution. Environmental Science & Technology 54 (22), 14182–14191. Available at https://doi.org/10.1021/acs.est.0c03387.

Listing of POPs in the Stockholm Convention. Available from: https://www.pops.int/TheConvention/ThePOPs/AllPOPs/tabid/2509/Default.aspx. (accessed 15.07.25.).

Liu, L., Qu, Y., Huang, J., Weber, R., 2021. Per- and polyfluoroalkyl substances (PFASs) in Chinese drinking water: risk assessment and geographical distribution. Environmental Sciences Europe 33 (1), 6. Available at https://doi.org/10.1186/s12302-020-00425-3.

Mahinroosta, R., Senevirathna, L., 2020. A review of the emerging treatment technologies for PFAS contaminated soils. Journal of Environmental Management 255, 109896. Available at https://doi.org/10.1016/j.jenvman.2019.109896.

Malisch, R., Fürst, P., Šebková, K., 2023. Persistent organic pollutants in human milk. Persistent Org. Pollut. Hum. Milk 1–683. Available at https://doi.org/10.1007/978-3-031-34087-1/COVER.

Megson, D., Bruce-Vanderpuije, P., Idowu, I.G., Ekpe, O.D., Sandau, C.D., 2025. A systematic review for non-targeted analysis of per- and polyfluoroalkyl substances (PFAS). The Science of the Total Environment 960, 178240. Available at https://doi.org/10.1016/j.scitotenv.2024.178240.

Minatoya, M., Itoh, S., Miyashita, C., Araki, A., Sasaki, S., Miura, R., et al., 2017. Association of prenatal exposure to perfluoroalkyl substances with cord blood adipokines and birth size: the Hokkaido study on environment and children's health. Environmental Research 156, 175–182. Available at https://doi.org/10.1016/j.envres.2017.03.033.

Miyake, Y., Kobayashi, T., Kameya, T., Managaki, S., Amagai, T., Masunaga, S., 2014. Comparison study on observed and estimated concentrations of perfluorooctane sulfonate using a fate model in Tokyo Bay of Japan. Journal of Environmental Science and Health, Part A: Toxic/Hazardous Substances and Environmental Engineering 49 (7), 770–776. Available at https://doi.org/10.1080/10934529.2014.882192.

Moon, N., Heo, S.J., Kim, J.H., 2025. Associations of exposure to 24 endocrine-disrupting chemicals with perinatal depression and lifestyle factors: a prospective cohort study in Korea. Environments. Available at https://doi.org/10.3390/environments12010015.

Moriwaki, H., Takatah, Y., Arakawa, R., 2003. Concentrations of perfluorooctane sulfonate (PFOS) and perfluorooctanoic acid (PFOA) in vacuum cleaner dust collected In Japanese Homes. Journal of Environmental Monitoring: JEM 5 (5), 753–757. Available at https://doi.org/10.1039/b307147m.

Mukherjee Das, A., Janardhanan, R., 2022. Accelerating research and policy on PFAS in India. Environ. Epidemiol. (Philadelphia, Pa.) 6 (2), e199. Available at https://doi.org/10.1097/EE9.0000000000000199.

Murakami, M., Kuroda, K., Sato, N., Fukushi, T., Takizawa, S., Takada, H., 2009. Groundwater pollution by perfluorinated surfactants in Tokyo. Environmental Science & Technology 43 (10), 3480–3486. Available at https://doi.org/10.1021/es803556w.

Murakami, M., Adachi, N., Saha, M., Morita, C., Takada, H., 2011. Levels, temporal trends, and tissue distribution of perfluorinated surfactants in freshwater Fish from Asian countries. Archives of Environmental Contamination and Toxicology 61 (4), 631–641. Available at https://doi.org/10.1007/s00244-011-9660-4.

National Implementation Plan for the Stockholm Convention Republic of Korea Updated in 2019 Contents, 2019. Available from: https://chm.pops.int/Portals/0/download.aspx?d=UNEP-POPS-NIP-KoreaRep-COP7.English.pdf. (accessed 14.07.25.).

National Institute of Environmental Health Sciences. Perfluoroalkyl and Polyfluoroalkyl Substances (PFAS). Available from: https://www.niehs.nih.gov/health/topics/agents/pfc. (accessed 24.07.25.).

New Regulations Targeting Forever Chemicals | IDTechEx Research Article. Available from: https://www.idtechex.com/en/research-article/new-regulations-targeting-forever-chemicals/30920. (accessed 15.07.25.).

NIES Annual Report, National Institute for Environmental Studies. Available from: https://www.nies.go.jp/kanko/annual/ae-30.pdf. (accessed 14.07.25.).

Park, H., Choo, G., Kim, H., Oh, J.-E., 2018. Evaluation of the current contamination status of PFASs and OPFRs in South Korean tap water associated with its origin. The Science of the Total Environment 634, 1505–1512. Available at https://doi.org/10.1016/j.scitotenv.2018.04.068.

Park, Y.T., Chung, E.Y., Chae, C.H., Lee, Y.H., 2024. Association between serum perfluoroalkyl substances concentrations and non-alcoholic fatty liver disease among Korean Adults: a cross-sectional study using the national environmental health survey cycle 4. Annals of Occupational and Environmental Medicine 36, e10. Available at https://doi.org/10.35371/AOEM.2024.36.E10.

Park, M.-S., Kim, H.-C., Kim, W.J., Hong, Y.-C., Choi, W.-J., Hwang, S.-Y., et al., 2025a. Environmental disease monitoring by regional environmental health centers in Korea: a narrative review. Ewha Medical Journal 48 (1). Available at https://doi.org/10.12771/EMJ.2025.E3.

Park, J., Baik, J.-H., Adjei-Nimoh, S., Lee, W.H., 2025b. Advancements in artificial intelligence-based technologies for PFAS detection, monitoring, and management. The Science of the Total Environment 980, 179536. Available at https://doi.org/10.1016/j.scitotenv.2025.179536.

Pawar, D.D. Present Scenario Of Environmental Issues In India. Available from: https://ijppr.humanjournals.com/wp-content/uploads/2017/01/30.Present_Scenario_of_Environmental__Issues_in_India.pdf. (accessed 14.07.25.).

Per and Poly-Fluorinated Chemicals (PFAS), OECD. Available from: https://www.oecd.org/en/topics/sub-issues/risk-management-risk-reduction-and-sustainable-chemistry/per-and-poly-fluorinated-chemicals.html. (accessed 14.07.25.).

Per- and Polyfluoroalkyl Substances (PFAS), Australian Industrial Chemicals Introduction Scheme (AICIS). Available from: https://www.industrialchemicals.gov.au/consumers-and-community/and-polyfluoroalkyl-substances-pfas. (accessed 14.07.25.).

Per- and Polyfluoroalkyl Substances (PFAS) - ECHA. Available from: https://echa.europa.eu/hot-topics/perfluoroalkyl-chemicals-pfas. (accessed 14.07.25.).

Perera, D.C., Meegoda, J.N., 2024. PFAS: the journey from wonder chemicals to environmental nightmares and the search for solutions. Applied Sciences. Available at https://doi.org/10.3390/app14198611.

Peritore, A.F., Gugliandolo, E., Cuzzocrea, S., Crupi, R., Britti, D., 2023. Current review of increasing animal health threat of per- and polyfluoroalkyl substances (PFAS): harms, limitations, and alternatives to manage their toxicity. International Journal of Molecular Sciences 24 (14). Available at https://doi.org/10.3390/ijms241411707.

Persistent Pollutants Control Act. Available from: https://elaw.klri.re.kr/eng_service/lawView.do?hseq=60960&lang=ENG. (accessed 15.07.25.).

PFAS National Primary Drinking Water Regulation. Available from: https://www.federalregister.gov/documents/2024/04/26/2024-07773/pfas-national-primary-drinking-water-regulation. (accessed 15.07.25.).

PFAS Risk Management Strategies for Businesses. Available from: https://www.3eco.com/article/article/webinar-recap-simplifying-pfas-risk-management-and-compliance-for-businesses/. (accessed 15.07.25.).

PFAS Treatment Technologies for Water Utilities and PFAS-Contaminated Sites | Xylem US. Available from: https://www.xylem.com/en-us/making-waves/water-utilities-news/pfas-treatment-technologies/. (accessed 15.07.25.).

PRTR Substances: PRTR Information Plaza Japan. Available from: https://www.env.go.jp/en/chemi/prtr/about/substances.html. (accessed 14.07.25.).

REACH Regulation - European Commission. Available from: https://environment.ec.europa.eu/topics/chemicals/reach-regulation_en. (accessed 14.07.25.).

REACH24H Korea Contributes Expertise to Korean Ministry of Environment's PFAS Regulatory Consensus Working Group. Available from: http://en.reach24h.com/news/company-news/reach24h-korea-contributes-expertise-to-korean-ministry-of-environments-pfas-regulatory-consensus-working-group. (accessed 15.07.25.).

Regarding the "Review of Environmental Standards for the Protection of Human Health Related to Water Pollution" (5th Report), Ministry of the Environment. Available from: https://www.env.go.jp/press/108066.html. (accessed 14.07.25.).

Regarding the PFAS concern and the fluorine products of AGC SEIMI CHEMICAL. Available from: https://www.seimichemical.co.jp/eng/product/fluoro/PFOA/. (accessed 14.07.25.).

Reply On Behalf Of Respondent No. 1 Central Pollution Control Board (CPCB). Available from: https://www.greentribunal.gov.in/sites/default/files/news_updates/Reply%20on%20behalf%20of%20Respondent%20No.%201%20Central%20Pollution%20Control%20Board%20(CPCB)%20in%20OA%20No.%20360%20of%202024.pdf. (accessed 15.07.25.).

Report of the Persistent Organic Pollutants Review Committee on the work of its eighteenth meeting. Available from: https://chm.pops.int/Portals/0/download.aspx?d=UNEP-POPS-POPRC.18-11Add.4.English.pdf#:~:text=At%20its%20eighteenth%20meeting%2C%20in%20decision%20POPRC%2D18/5%2C%20the%20Persistent%20Organic%20Pollutants%20Review. (accessed 14.07.25.).

Requests for Extension of an Exemption in the Register. Available from: https://www.pops.int/Implementation/Exemptions/Requestsforextensionofanexemption/tabid/9978/Default.aspx. (accessed 14.07.25.).

Review of Water Treatment Systems for PFAS Removal - Concawe. Available from: https://www.concawe.eu/publication/review-of-water-treatment-systems-for-pfas-removal/. (accessed 15.07.25.).

Risk Reduction Approaches For Pfass-A Cross-Country Analysis. Available from: https://www.oecd.org/content/dam/oecd/en/publications/reports/2015/09/risk-reduction-approaches-for-per-and-polyfluoroalkyl-substances-pfass-a-cross-country-analysis_51521362/ed861d08-en.pdf. (accessed 15.07.25.).

Rokni, L., Rezaei, M., Rafieizonooz, M., Khankhajeh, E., Mohammadi, A.A., Rezania, S., 2023. Effect of persistent organic pollutants on human health in South Korea: a review of the reported diseases. Sustainability. Available at https://doi.org/10.3390/su151410851.

Ross, I., McDonough, J., Miles, J., Storch, P., Kochunarayanan, P., Kalve, E., et al., 2018. J. A review of emerging technologies for remediation of PFASs. Remediat. J. 28, 101–126. Available at https://doi.org/10.1002/rem.21553.

Ruan, Y., Lalwani, D., Kwok, K.Y., Yamazaki, E., Taniyasu, S., Kumar, N.J.I., et al., 2019. Assessing exposure to legacy and emerging per- and polyfluoroalkyl substances via hair - the first nationwide survey in India. Chemosphere 229, 366–373. Available at https://doi.org/10.1016/j.chemosphere.2019.04.195.

S. Korea requests EU review ban on dangerous chemical, non-tariff barriers. Available from: https://en.yna.co.kr/view/AEN20231031006400320. (accessed 15.07.25.).

Schug, T.T., Johnson, A.F., Birnbaum, L.S., Colborn, T., Guillette, L.J.J., Crews, D.P., et al., 2016. Minireview: Endocrine Disruptors: Past Lessons and Future Directions. Molecular Endocrinology (Baltimore, Md.) 30 (8), 833–847. Available at https://doi.org/10.1210/me.2016-1096.

Sharma, B.M., Bharat, G.K., Tayal, S., Larssen, T., Bečanová, J., Karásková, P., et al., 2016. Perfluoroalkyl substances (PFAS) in river and ground/drinking water of the ganges river basin: emissions and implications for human exposure. Environmental Pollution (Barking, Essex: 1987) 208 (Pt B), 704–713. Available at https://doi.org/10.1016/j.envpol.2015.10.050.

Sharma, B.M., Scheringer, M., Chakraborty, P., Bharat, G.K., Steindal, E.H., Trasande, L., et al., 2022. Unlocking India's potential in managing endocrine-disrupting chemicals (EDCs): importance, challenges, and opportunities. Exposure and Health 1–15. Available at https://doi.org/10.1007/s12403-022-00519-8.

Sharma, N., Kumar, V., Sugumar, V., Umesh, M., Sondhi, S., Chakraborty, P., et al., 2024a. A comprehensive review on the need for integrated strategies and process modifications for per- and polyfluoroalkyl substances (PFAS) removal: current insights and future prospects. Case Studies in Chemical and Environmental Engineering 9, 100623. Available at https://doi.org/10.1016/j.cscee.2024.100623.

Sharma, A., Jorvekar, S.B., Bhowmik, S., Mohapatra, P., Borkar, R.M., 2024b. Comprehensive assessment of per and polyfluoroalkyl substances (PFAS) contamination in groundwater of Kamrup, Assam, India: occurrence, health risks, and metabolomic insights. Environmental Science: Processes & Impacts 26 (9), 1601–1617. Available at https://doi.org/10.1039/d4em00159a.

Sharma, P., Pal, N., Singh, S., Ojha, R., Kumawat, M., Shubham, S., et al., 2025. Bio-monitoring of endocrine disrupting chemicals in human serum: insights from a study in central India. International Journal of Environmental Health Research 1–13. Available at https://doi.org/10.1080/09603123.2025.2502636.

Shi, S., Zhang, Z., Wang, R., Xue, S., Fei, W., Peng, Y., 2025. Composition, source appointment, and biomigration of per- and polyfluoroalkyl substances in the Fu river of Wuhan, China. Environmental Toxicology and Chemistry / SETAC 44 (4), 926–934. Available at https://doi.org/10.1093/etojnl/vgaf008.

Shimizu Commences PFAS-Contaminated Soil Remediation Testing in the United States. Available from: https://www.shimz.co.jp/en/company/about/news-release/2023/2023025.html. (accessed 14.07.25.).

Shin, E.-S., Park, M.-K., Kim, G., Barghi, M., Choi, S.-D., Yang, J., et al., 2022. Dietary exposure and potential human health risk of dioxins in South Korea: application of deterministic and probabilistic methods. Chemosphere 291 (Pt 2), 133018. Available at https://doi.org/10.1016/j.chemosphere.2021.133018.

Shin, M.-W., Kang, H., Kim, S.-H., 2025. Concentrations of serum per- and polyfluoroalkyl substances and lipid health in adolescents: a cross-sectional study from the Korean National environmental health survey 2018–2020. Toxics. Available at https://doi.org/10.3390/toxics13020091.

Singh, A., Singh, G., Singh, P., Mishra, V.K., 2023. Overview of sources, fate, and impact of endocrine disrupting compounds in environment and assessment of their regulatory policies across different continents. Total Environment: Research Themes 7, 100071. Available at https://doi.org/10.1016/j.totert.2023.100071.

Soil Contamination Countermeasures Act - English - Japanese Law Translation. Available from: https://www.japaneselawtranslation.go.jp/en/laws/view/2038/en#je_ch1at1. (accessed 14.07.25.).

South Korea MOE Released the Final List of CMR Substances, Priority Control Substances and Substances Exempted from Registration. Available from: https://kreach.cirs-group.com/en/news/articles/South-Korea-MOE-Released-the-Final-List-of-CMR-Substances-Priority-Control-Substances-and-Substances-Exempted-from-Registration.html. (accessed 15.07.25.).

South Korea's chemical regulations are undergoing significant changes. Available from: https://www.ramboll.com/en-apac/insights/our-sustainability-impact/south-korea-s-chemical-regulations-are-undergoing-significant-changes. (accessed 15.10.25.).

South Korea's Ministry of Environment adds new persistent organic pollutant PFHxS (147 substances), banning handling thereof. Available from: https://enviliance.com/regions/east-asia/kr/report_10622. (accessed 15.07.25.).

Special Act on Fine Dust Reduction and Management," Available from: https://www.me.go.kr/eng/web/board/read.do;jsessionid=bHWFJ8oJbgu4Dq_zIgyRb8aEzaRIo-4KuRpg1xtB.mehome1?pagerOffset=720&maxPageItems=10&maxIndexPages=10&searchKey=&searchValue=&menuId=461&orgCd=&boardId=912830&boardMasterId=522&boardCategoryId=&decorator=. (accessed 14.07.25.).

Stockholm Convention National Implementation Plans (NIPs). Available from: https://www.pops.int/Implementation/NationalImplementationPlans/Overview/tabid/565/Default.aspx. (accessed 15.07.25.).

Stockholm Convention on Persistent Organic Pollutants (POPs). Available from: https://www.pops.int/Countries/StatusofRatifications/PartiesandSignatoires/tabid/4500/Default.aspx. (accessed 14.07.25.)

Stockholm Convention on Persistent Organic Pollutants (POPs). Overview. Available from: https://www.pops.int/TheConvention/Overview/tabid/3351/Default.aspx. (accessed 14.07.25.).

Strategic Programs on Environmental Endocrine Disruptors '98 (SPEED '98), Health & Chemicals, Ministry of the Environment, Government of Japan. Available from: https://www.env.go.jp/en/chemi/ed/speed98/sp98.html. (accessed 14.07.25.).

Study on Lead in Paints in India. 2019 Available from: https://toxicslink.org/wp-content/uploads/2022/08/Lead%20in%20Paints%20in%20India%202019.pdf. (accessed 14.07.25.)

Substances under the Chemical Substances Control Law, Class I specified chemical substance. Available from: https://www.nite.go.jp/chem/jcheck/list6.action?category=211&request_locale=en. (accessed 14.07.25.).

Summary of the Guideline on the Treatment of Wastes Containing Perfluorooctane Sulfonic Acid (PFOS), and Its Salts in Japan. 2013. Available from: https://www.env.go.jp/en/focus/docs/files/201304-89.pdf. (accessed 15.07.25.).

Sunantha, G., Vasudevan, N., 2016. Assessment of perfluorooctanoic acid and perfluorooctane sulfonate in surface water - Tamil Nadu, India. Marine Pollution Bulletin 109 (1), 612–618. Available at https://doi.org/10.1016/j.marpolbul.2016.05.023.

Takada, M., Murakami, M., Ohnuma, S., Shibata, Y., Yasutaka, T., 2024. Public attitudes toward the final disposal of radioactively contaminated soil resulting from the fukushima daiichi nuclear power station accident. Environmental Management 73 (5), 962–972. Available at https://doi.org/10.1007/s00267-024-01938-w.

Tao, L., Ma, J., Kunisue, T., Libelo, E.L., Tanabe, S., Kannan, K., 2008. Perfluorinated compounds in human breast milk from several asian countries, and in infant formula and dairy milk from the United States. Environmental Science & Technology 42 (22), 8597–8602. Available at https://doi.org/10.1021/es801875v.

Tenth Meeting of the Conference of the Parties to the Stockholm Convention. Available from: https://chm.pops.int/TheConvention/ConferenceoftheParties/Meetings/COP10/tabid/8397/Default.aspx. (accessed 14.07.25.).

Tenth Meeting of the Conference of the Parties to the Stockholm Convention. Available from: https://www.pops.int/TheConvention/ConferenceoftheParties/Meetings/COP10/tabid/8397/. (accessed 14.07.25.)

The Basic Environment Law and Basic Environment Plan. Ministry of the Environment, Government of Japan. Available from: https://www.env.go.jp/en/laws/policy/basic_lp.html. (accessed 23.07.25.).

The Chemical Accidents (Emergency Planning, Preparedness and Response) Rules, 1996. Available from: https://www.indiacode.nic.in/ViewFileUploaded?path=AC_CEN_16_18_00011_198629_1517807327582/rulesindividualfile/&file=ca%28eppr%29rules%2C_1996.pdf. (accessed 15.07.25.).

The National Implementation Plan of the Republic of Korea. Available from: https://chm.pops.int/Portals/0/download.aspx?d=UNEP-POPS-NIP-KoreaRep-COP7.English.pdf. (accessed 14.07.25.).

Thennakoon, S.; Lalantha, M.; Senevirathna, D. Development of Effective Removal Methods of PFCs (Perfluorinated Compounds) in Water by Adsorption and Coagulation. 2010. Available from: http://hdl.handle.net/2433/126798. (accessed 15.07.25.).

Toshima, H., Suzuki, Y., Imai, K., Yoshinaga, J., Shiraishi, H., Mizumoto, Y., et al., 2012. Endocrine disrupting chemicals in urine of Japanese male partners of subfertile couples: a pilot study on exposure and semen quality. International Journal of Hygiene and Environmental Health 215 (5), 502–506. Available at https://doi.org/10.1016/j.ijheh.2011.09.005.

Toxics Link Report: Lead in Paints in India, 2018. Available from: https://ipen.org/news/toxics-link-report-lead-paints-india-2018. (accessed 14.07.25.).

Tsai, M.-S., Miyashita, C., Araki, A., Itoh, S., Bamai, Y.A., Goudarzi, H., et al., 2018. Determinants and temporal trends of perfluoroalkyl substances in pregnant women: the Hokkaido study on environment and children's health. International Journal of Environmental Research and Public Health 15 (5). Available at https://doi.org/10.3390/ijerph15050989.

UNEP. List Perfluorooctanoic Acid (PFOA), Its Salts and PFOA-Related Compounds in Annex A to the Stockholm Convention on Persistent Organic Pollutants with Specific Exemptions. Unep/Pops/Cop.9/Sc-9/12 2019, No. v, 2–4. (accessed 14.07.25.)

UNEP-POPS-NIP-Korea Rep-COP7. Available from: https://chm.pops.int/Portals/0/download.aspx?d=UNEP-POPS-NIP-KoreaRep-COP7.English.pdf. (accessed 15.07.25.).

Wang, F., Zhuang, Y., Dong, B., Wu, J., 2022. Review on per- and poly-fluoroalkyl substances' (PFASs') pollution characteristics and possible sources in surface water and precipitation of China. Water. Available at https://doi.org/10.3390/w14050812.

Wang, B., Hong, W., Wu, Z., Li, X., 2024. Association between exposure to per- and polyfluoroalkyl substances and missed miscarriage: a hospital-based case-control study in Shanghai, China. Environmental Toxicology and Pharmacology 112, 104589. Available at https://doi.org/10.1016/j.etap.2024.104589.

Wang, M., 2025. PFAS contamination disclosures and declining residential land prices in polluted areas of Japan. SSRN Electronic Journal. Available at https://doi.org/10.2139/SSRN.5285846.

Wang, X., Chen, W., Guo, Q., Peng, Z., Sun, Q., Zhao, C., et al., 2025. Current status and risk assessment of perfluoroalkyl acids in surface water and sediments of the yellow river in Shandong, China. Emerg. Contam. 11 (1), 100391. Available at https://doi.org/10.1016/j.emcon.2024.100391.

Washino, N., Saijo, Y., Sasaki, S., Kato, S., Ban, S., Konishi, K., et al., 2009. Correlations between prenatal exposure to perfluorinated chemicals and reduced fetal growth. Environmental Health Perspectives 117 (4), 660–667. Available at https://doi.org/10.1289/ehp.11681.

Waste Management And Public Cleansing Law. Available from: https://www.env.go.jp/en/recycle/basel_conv/files/Waste_Management_and_Public_Cleansing.pdf. (accessed 13.07.25.).

Water Pollution Prevention Act - English - Japanese Law Translation. Available from: https://www.japaneselawtranslation.go.jp/en/laws/view/2815/en#je_ch2at15 (accessed 14.07.25.).

Welcome to the Ministry of Food and Drug Safety | Ministry of Food and Drug Safety. Available from: https://www.mfds.go.kr/eng/index.do (accessed 15.07.25.).

Wen, M., Ma, Z., Gingerich, D.B., Zhao, X., Zhao, D., 2022. Heavy metals in agricultural soil in China: a systematic review and meta-analysis. Eco-Environmental Health 1 (4), 219–228. Available at https://doi.org/10.1016/j.eehl.2022.10.004.

Wu, R., Lin, H., Yamazaki, E., Taniyasu, S., Sörengård, M., Ahrens, L., et al., 2021. Simultaneous analysis of neutral and ionizable per- and polyfluoroalkyl substances in air. Chemosphere 280, 130607. Available at https://doi.org/10.1016/j.chemosphere.2021.130607.

Xu, B., Liu, S., Zhou, J.L., Zheng, C., Weifeng, J., Chen, B., et al., 2021. PFAS and their substitutes in groundwater: occurrence, transformation and remediation. Journal of Hazardous Materials 412, 125159. Available at https://doi.org/10.1016/j.jhazmat.2021.125159.

Xu, C., Yang, Y., Ling, H., Yi, C., Zhang, X., Zhang, R., 2025. Occurrence and human exposure assessment of PFAS in river and groundwater around a closed fluorochemical plant in China. Scientific Reports 15 (1), 16241. Available at https://doi.org/10.1038/s41598-025-01128-6.

Yamashita, N., Yamazaki, E., Taniyasu, S., Hanari, N., Yeung, L.W.Y., 2024. Biochar from paddy field - a solution to reduce PFAS pollution in the environment. Chemosphere 364, 143073. Available at https://doi.org/10.1016/j.chemosphere.2024.143073.

Yamazaki, E., Lalwani, D., Ruan, Y., Taniyasu, S., Hanari, N., Kumar, N.J.I., et al., 2023. Nationwide distribution of per- and polyfluoroalkyl substances (PFAS) in road dust from India. The Science of the Total Environment 892, 164538. Available at https://doi.org/10.1016/j.scitotenv.2023.164538.

Yang, S., Gwak, J., Cha, J., Park, K., Kim, Y., Kim, S.-Y., et al., 2025. Distributions and compositional characteristics of per- and polyfluoroalkyl substances (PFASs) in sediments of the regional seas of South Korea. Marine Pollution Bulletin 217, 118140. Available at https://doi.org/10.1016/j.marpolbul.2025.118140.

Ye, F., Zushi, Y., Masunaga, S., 2015. Survey of perfluoroalkyl acids (PFAAs) and their precursors present in Japanese consumer products. Chemosphere 127, 262–268. Available at https://doi.org/10.1016/j.chemosphere.2015.02.026.

Yeung, L.W.Y., Yamashita, N., Taniyasu, S., Lam, P.K.S., Sinha, R.K., Borole, D.V., et al., 2009. A survey of perfluorinated compounds in surface water and biota including dolphins from the ganges river and in other waterbodies in India. Chemosphere 76 (1), 55–62. Available at https://doi.org/10.1016/j.chemosphere.2009.02.055.

Yu, R.-S., Yu, H.-C., Yang, Y.-F., Singh, S., 2025. A global overview of per- and polyfluoroalkyl substance regulatory strategies and their environmental impact. Toxics. Available at https://doi.org/10.3390/toxics13040251.

Yukioka, S., Tanaka, S., Suzuki, Y., Echigo, S., Kärrman, A., Fujii, S., 2020. A profile analysis with suspect screening of per- and polyfluoroalkyl substances (PFASs) in firefighting foam impacted waters in Okinawa, Japan. Water Research 184, 116207. Available at https://doi.org/10.1016/j.watres.2020.116207.

Yun, J., Jang, E.-C., Kwon, S.-C., Min, Y.-S., Lee, Y.-J., 2023. The Association of Perfluoroalkyl Substances (PFAS) exposure and kidney function in Korean adolescents using data from Korean National Environmental Health Survey (KoNEHS) Cycle 4 (2018-2020): a cross-sectional study. Annals of Occupational and Environmental Medicine 35, e5. Available at https://doi.org/10.35371/aoem.2023.35.e5.
Yun, H., Park, J., Zoh, K.-D., 2024. Target, suspect, and non-target screening of per- and poly-fluoroalkyl substances in wastewater treatment plant effluents in South Korea using ion mobility spectrometry–mass spectrometry. The Science of the Total Environment 957, 177387. Available at https://doi.org/10.1016/j.scitotenv.2024.177387.

# Index

*Note*: Page numbers followed by "*f*" and "*t*" refer to figures and tables, respectively.

## B

## N

## Q



## R

## T

Printed and bound by CPI Group (UK) Ltd, Croydon, CR0 4YY
14/09/2026
02216396-0008